Riparian Zone Hydrology and Biogeochemistry

Benchmark Papers in Hydrology, 5

Riparian Zone Hydrology and Biogeochemistry

Selection, Introduction and Commentary by

T. P. Burt

Department of Geography, University of Durham, UK

G. Pinay

School of Geography, Earth & Environmental Sciences, University of Birmingham, UK

S. Sabater

University of Girona and Catalan Institute for Water Research, Spain

Series sponsored by SAHRA, the Center for Sustainability of semi-Arid Hydrology and Riparian Areas, University of Arizona, USA

International Association of Hydrological Sciences / Association Internationale des Sciences Hydrologiques

Published by the International Association of Hydrological Sciences 2010
IAHS Press, Centre for Ecology and Hydrology, Wallingford, Oxfordshire OX10 8BB, UK

IAHS Benchmark Papers in Hydrology Series, no. 5
ISBN 978-1-907161-09-4

British Library Cataloguing-in-Publication Data.
A catalogue record for this book is available from the British Library.

Production of this book was in part supported by the NSF Center for Sustainability of semi-Arid Hydrology and Riparian Areas (SAHRA) under the STC Program of the National Science Foundation, Agreement no. EAR-9876800 and NSF award DEB–0415977.

Cover photograph A narrow, poorly-drained flood plain on the lower Start stream within the Slapton catchments, South Devon, UK (Photograph: Tim Burt)

IAHS Publications are available from:
IAHS Press, Centre for Ecology and Hydrology, Wallingford, Oxfordshire OX10 8BB, UK
tel: +44 1491 692442; fax: +44 1491 692448; email: jilly@iahs.demon.co.uk

Printed in the UK by 4Word, Bristol.

Foreword to the Benchmark Papers in Hydrology Series

Hydrology is a young science. Following the seminal work by Henry Darcy in the mid-19th century and Robert Horton's influential studies in the first half of the 20th century, knowledge generation in the hydrological sciences exploded through the second half of the 20th century with new discoveries in each of the components of the hydrological cycle. Despite scientific and technological advances, the field of hydrology has been highly fragmented across the engineering–science interface, between basic and applied studies, mathematical and descriptive work, field and laboratory investigations. Hydrological science is now taught in most of the world's major universities, but the teachers and students may be found just as commonly in civil engineering, as in physical geography, forestry, environmental engineering, geology, soil science, or agricultural engineering. Our key journals, textbooks, conference proceedings and monographs are likewise dispersed and scattered across myriad outlets. While Internet searching and .pdf versions of more recent published material make "keeping up" easier, there is a body of important early work that is often ignored by young researchers. As a result, we stand at the beginning of the 21st century in danger of memory loss of our young science.

The objective of the *Benchmark Papers in Hydrology* series is to define and create a world repository of the benchmark papers in hydrology for education, research and practice. In so doing, it is hoped that this series will counter memory loss in our young and highly interdisciplinary science by chronicling the development of ideas leading to current theory in hydrology. Indeed, there are many examples where experiments, ideas, etc. are being "re-invented" because the key papers are not known to new generations of students, scientists and practitioners. This hydrological foundation is what the *Benchmark Papers in Hydrology* series aims to define, publish and distil: the key papers that contributed to the current thinking of the sub-fields of the hydrological sciences. As such, the benchmark papers are hydrological exemplars. They are the best of the contributions that our field has to offer.

The *Benchmark Papers in Hydrology* series interprets the hydrological sciences in the broadest sense. Each volume tackles a specific topic and is edited by a recognized scientific authority. The volume editor makes the initial selection of papers and prepares the Introduction and Commentary to the papers that are finally included in that volume. The series Editorial Board approves the published selection of papers. Articles are reproduced in whole (but sometimes in part) from the key papers of (mostly) the 20th century. Those associated with the *Benchmark Papers in Hydrology* appreciate the International Association of Hydrological Science's lead role in publishing the series and their support of this major effort to disseminate the most important ideas in hydrology to libraries, institutes and individuals worldwide.

Jeffrey J. McDonnell
Series Editor

Other volumes in the IAHS Benchmark Papers in Hydrology Series

1 – ***Streamflow Generation Processes*** by Keith Beven 978-1-901502-53-4

2 – ***Evaporation*** by John H. C. Gash & W. James Shuttleworth 978-1-901502-98-5

3 – ***Groundwater*** by Mary P. Anderson 978-1-901502-74-9

4 – ***Rainfall–Runoff Modelling*** by Keith Loague 978-1-907161-06-3

Acknowledgements

IAHS is grateful to all the publishers and authors who granted permission to reproduce copyright material in this IAHS Benchmark Papers volume.

Special gratitude is due to the American Geophysical Union, the American Society of Agronomy, the American Society of Agricultural Engineers, the American Society of Limnology and Oceanography, the International Society of Limnology and the North American Benthological Society for giving permission to reprint papers from their publications, and also to Professor Alan Hill.

Stephen LeDuc at Oregon State University, USA, Cate Gardner and Penny Perrins at IAHS Press, are thanked for their contributions to preparing the volume.

Production of this book was in part supported by the NSF Center for Sustainability of semi-Arid Hydrology and Riparian Areas (SAHRA) under the STC Program of the National Science Foundation, Agreement no. EAR-9876800 and NSF award DEB-0415977.

SAHRA
www.sahra.arizona.edu

SAHRA promotes sustainable management of water resources not only through research, but by improving dissemination of important science results and products, and by encouraging dialogue about important water issues, research needs and findings among scientists and water managers, policy makers, academics, students, and the general public. The sponsorship and distribution of the IAHS Benchmark Papers in Hydrology furthers the SAHRA mission of transferring knowledge to promote scientifically informed choices by present and future generations of policy makers, water managers, and the general public.

SAHRA is headquartered at the Department of Hydrology and Water Resources at the University of Arizona, USA, but draws on a broad range of academic expertise, disciplines, and institutions to address water science and management questions that are critical in semi-arid and arid regions. Partner institutions include universities in the southwest USA and northern Mexico, US federal agencies and national laboratories.

Contents

Series Editor's Foreword *by Jeffrey J. McDonnell* v

Riparian Zone Hydrology and Biogeochemistry: A Review

1 Introduction 1
2 Landscape ecology, riparian ecotones and emergence of the Riparian Buffer Zone concept 2
3 Hydrology of the riparian zone 3
3.1 Water balance within the riparian zone
3.2 Flow paths within the riparian zone
4 Solute dynamics in the riparian zone 5
4.1 Its role as a conduit
4.2 Its role as a buffer
5 Hyporheic and in-stream processes 7
6 What do we still need to learn about the hydrology and biogeochemistry of riparian zones? 7
6.1 Flow paths and residence time
6.2 Hot spots and connectivity at the landscape scale
6.3 Climate change and riparian zone processes
6.4 New biogeochemical pathways
7 Introduction to the commentaries 10

The Benchmark Papers

A Landscape Ecology 14

A1 ***Cummins, K. W.*** (1974) Structure and function of stream ecosystems. *Bioscience* 24, 631–641. 14

A2 ***Hynes, H. B. N.*** (1975) The stream and its valley. *Verhandlungen Internationale Vereinigung für theoretische und angewandte Limnologie* 19, 1–15 26

A3 ***Jones, J. R., Borofka, B. P. & Bachmann, R. W.*** (1976) Factors affecting nutrient loads in some Iowa streams. *Water Research* 10, 117–122. 42

A4 ***Vannote, R. L., Minshall, G. W., Cummins, K. W., Sedell, J. R. & Cushing, C. E.*** (1980) The river continuum concept. *Canadian Journal of Fisheries and Aquatic Sciences* 37, 130–137. 49

A5 ***Triska, F.J., Sedell, J. R., Cromack, K. Jr, Gregory, S. V. & McCorison, F. M.*** (1984) Nitrogen budget for a small coniferous forest stream. *Ecological Monographs* 54, 119–140. 58

A6 ***Pinay, G. & Décamps, H.*** (1988) The role of riparian woods in regulating nitrogen fluxes between the alluvial aquifer and surface water: a conceptual model. *Regulated Rivers: Research and Management* 2, 507–516. 81

B Hydrology of the Riparian Zone 92

B1 ***Troxell, H. C.*** (1936) The diurnal fluctuation in the ground-water and flow of the Santa Ana River and its meaning. 92
Transactions of the American Geophysical Union 17, 496–504.

B2 ***Hursh, C. R. & Brater, E. F.*** (1941) Separating storm-hydrographs from small drainage-areas into surface- and subsurface-flow. 103
Transactions of the American Geophysical Union 22, 863–871.

B3 ***Dunford, E. G. & Fletcher, P. W.*** (1947) Effect of removal of stream-bank vegetation upon water yield. 114
Transactions of the American Geophysical Union 28, 105–110.

B4 ***Todd, D. K.*** (1955) Ground-water flow in relation to a flooding stream. 121
Proceedings of the American Society of Civil Engineers 81(2), 628-1 to 628-20.

B5 ***Dunne, T. & Black, R. D.*** (1970) An experimental investigation of runoff production in permeable soils. 142
Water Resources Research 6, 478–490.

B6 ***Abdul, A. S. & Gillham, R. W.*** (1984) Laboratory studies of the effects of the capillary fringe on streamflow generation. 157
Water Resources Research 20, 691–698.

C Linking Riparian Zone Hydrology to Solute Transport 167

C1 ***Jackson, W. A., Asmussen, L. E., Hauser, E. W. & White, A. W.*** (1973) Nitrate in surface and subsurface flow from a small agricultural watershed. 167
Journal of Environmental Quality 2, 480–482.

C2 ***Foster, I. D. L. & Walling, D. E.*** (1978) The effects of the 1976 drought and autumn rainfall on stream solute levels. 171
Earth Surface Processes 3, 393–406.

C3 ***Pilgrim, D. H., Huff, D. D. & Steele, T. D.*** (1979) Use of specific conductance and contact time relations for separating flow components of storm runoff. 187
Water Resources Research 15, 329–339.

C4 ***Trudgill, S. T., Pickles, A. M., Smettem, K. R. J. & Crabtree, R. W.*** (1982) Soil-water residence time and solute uptake. 1. Dye tracing and rainfall events. 199
Journal of Hydrology 60, 257–279.

C5 ***McDonnell, J. J.*** (1990) A rationale for old water discharge through macropores in a steep, humid catchment. 223
Water Resources Research 26, 2821–2832.

C6 ***Waddington, J. M., Roulet, N. T. & Hill, A. R.*** (1993) Runoff mechanisms in a forested groundwater discharge wetland. 237
Journal of Hydrology 147, 37–60.

D Biogeochemical Processes and Methods 263

D1 ***Reddy, K. R. & Patrick, W. H. Jr*** (1975) Effect of alternate aerobic and anaerobic conditions on redox potential, organic matter decomposition and nitrogen loss in a flooded soil. 263
Soil Biology and Biochemistry 7, 87–94.

D2 ***Yoshinari, T., Hynes, R. & Knowles, R.*** (1977) Acetylene inhibition of nitrous oxide reduction and measurement of denitrification and nitrogen fixation in soil. *Soil Biology and Biochemistry* 9, 177–183. 272

D3 ***Smith, M. S. & Tiedje, J. M.*** (1979) Phases of denitrification following oxygen depletion in soil. *Soil Biology and Biochemistry* 11, 261–267. 280

D4 ***Mariotti, A., Germon, J. C., Hubert, P., Kaiser, P., Letolle, R., Tardieux, A. & Tardieux, P.*** (1981) Experimental determination of nitrogen kinetic isotope fractionation some principles; illustration for the denitrification and nitrification processes. *Plant and Soil* 62, 413–430. 288

E Riparian Buffering of Surface and Subsurface Flow 307

E1 ***Lowrance, R. R., Todd, R. L. & Asmussen, L. E.*** (1984) Nutrient cycling in an agricultural watershed: I. Phreatic movement. *Journal of Environmental Quality* 13, 22–27. 307

E2 ***Peterjohn, W. T. & Correll, D. L.*** (1984) Nutrient dynamics in an agricultural watershed: observations on the role of a riparian forest. *Ecology* 65, 1466–1475. 314

E3 ***Jacobs, T. C. & Gilliam, J. W.*** (1985) Riparian losses of nitrate from agricultural drainage waters. *Journal of Environmental Quality* 14, 472–478. 325

E4 ***Warwick, J. & Hill, A. R.*** (1988) Nitrate depletion in the riparian zone of a small woodland stream. *Hydrobiologia* 157, 231–240. 333

E5 ***Dillaha, T. A., Reneau, R. B., Mostaghimi, S. & Lee, D.*** (1989) Vegetative filter strips for agricultural nonpoint source pollution control. *Transactions of the American Society of Agricultural Engineers* 32, 513–519. 344

E6 ***McDowell, W. H., Bowden, W. B. & Asbury, C. E.*** (1992) Riparian nitrogen dynamics in two geomorphologically distinct tropical rain forest watersheds: subsurface solute patterns. *Biogeochemistry* 18, 53–75. 353

E7 ***Haycock, N. E. & Burt, T. P.*** (1993) Role of floodplain sediments in reducing the nitrate concentration of subsurface run-off: a case study in the Cotswolds, UK. *Hydrological Processes* 7, 287–295. 377

F In-stream Processes 387

F1 ***Webster, J. R. & Patten, B. C.*** (1979) Effects of watershed perturbation on stream potassium and calcium dynamics. *Ecological Monographs* 49, 1, 51–72. 387

F2 ***Bencala, K. E. & Walters, R. A.*** (1983) Simulation of solute transport in a mountain pool-and-riffle stream: a transient storage model. *Water Resources Research* 19, 718–724. 410

F3 ***Hill, A. R.*** (1983) Nitrate-nitrogen mass balances for two Ontario rivers. In: *Dynamics of Lotic Ecosystems*, edited by T. D. Fontaine and S. M. Bartell, 457–477. Ann Arbor Science Publishers, Ann Arbor, Michigan, USA. 418

F4 ***Duff, J. H. & Triska, F. J.*** (1990) Denitrification in sediments from the hyporheic zone adjacent to a small forested stream. *Canadian Journal of Fisheries and Aquatic Sciences* 47(6), 1140–1147. 440

F5 ***Mulholland, P. J.*** (1992) Regulation of nutrient concentrations in a temperate forest stream: roles of upland, riparian, and instream processes. *Limnology and Oceanography* 37, 1512–1526. 449

F6 ***Holmes, R. M., Fisher, S. G. & Grimm, N. B.*** (1994) Parafluvial nitrogen dynamics in a desert stream ecosystem. *Journal of the North American Benthological Society* 13, 468–478. 465

F7 ***Valett, H. M., Morrice, J. A., Dahm, C. N. & Campana, M. E.*** (1996) Parent lithology, surface–groundwater exchange, and nitrate retention in headwater streams. *Limnology and Oceanography* 41, 333–345. 477

Riparian Zone Hydrology and Biogeochemistry: A Review

1 *Introduction*

We may conclude that in every respect the valley rules the stream
Hynes (1975, Paper A2 in this volume)

Strictly speaking, the riparian zone only includes vegetation along the bed and banks of the river channel (Tansley, 1911). However, in recent decades, the definition of the riparian zone has been extended to include a wider strip of land alongside the channel; very often the flood plain is taken to be coincident with this more broadly defined area (Anderson, 1987). The riparian zone is an ecological boundary or ecotone (Naiman & Décamps, 1990; Fig. 1), a transition area between two adjacent but different ecosystems. Riparian ecotones are important regulators of the movement of energy and material through catchment landscapes and can function either as a conduit or as a barrier (Brinson *et al.*, 1984; Mulholland, 1992, Paper F5; Pinay *et al.*, 1999; Burt & Pinay, 2005). The riparian zone may act as a sink in one case (e.g. eroded soil) but as a source in another (e.g. increased solubility of phosphorus species under anoxic conditions). There is therefore a need to establish the conditions under which these various buffer zone processes work. Fundamental to this is the need to establish the transmission of water through the riparian zone: the precise pathway taken by the water will determine whether water is effectively flowing through the riparian zone, and if its residence time is long enough to allow internal processes to operate effectively (Pinay *et al.*, 2009). Many of the papers reproduced in this book focus on nitrogen buffering in the riparian zone. As Pinay & Décamps (1988, Paper A6) recognised, the spatial distribution of hydrological processes (flow pathways and storage) controls the spatial pattern of nitrogen cycling in the riparian zone (Fig. 1). In other words, the study of nitrogen buffering is an inherently hydrological problem.

The chemistry of river water reflects the biogeochemical processes operating within the river basin. The runoff processes operating on the catchment hillslopes are particularly important since they control the residence time of water within the basin. The riparian zone is perhaps the most important element of the hydrological landscape. Bishop *et al.* (2004) argue that, since the chemistry of water moving downslope is determined at any given point by soil chemistry, the particular role of the riparian soil is to set the stream water chemistry, since this is the last soil in contact with the water before it becomes runoff. Thus, paradoxically, the near-stream zone can act both as a conduit and a barrier, coupling the terrestrial and aquatic environments, but also acting as a barrier between them. Riparian zones are often biogeochemical "hot spots" where hydrological flow paths converge with essential substrates (such as high concentrations of soil carbon and nitrate). These hot spots exhibit disproportionately high reaction rates relative to the surrounding area (McClain *et al.*, 2003) and exert significant controls on water quality.

The riparian landscape is unique among environments because it is a terrestrial habitat, strongly affecting and affected by aquatic environments (Malanson, 1993). Today, there is an intuitive assumption that the condition of the river and the condition of the riparian zone are intimately linked, but it was not always so. A series of papers in the 1970s (Cummins, 1974, Paper A1; Hynes, 1975, Paper A2; Webster & Patten, 1979, Paper F1; Vannote *et al.*, 1980, Paper A4) created a new paradigm for aquatic ecology, borrowing concepts from fluvial geomorphology that stressed the drainage basin as the fundamental unit of analysis (Leopold *et al.*, 1964; Chorley, 1969). The *river continuum concept* (Vannote *et al.*, 1980, Paper A4) is based on the notion of dynamic equilibrium in which the river is viewed as an open (eco)system that is in continuous interaction with its riparian zone and, moving from source to mouth, constantly changing. Yet, this vision of stream functioning was mainly restricted to the river and its banks. Hence, river reaches were classified as heterotrophic in upstream reaches because they were mostly dependent on riparian forest for energy and matter supply; the middle reaches were considered autotrophic as in-stream micro and macrophytes would provide the main food supply; and the lower reaches were characterized as heterotrophic because most of the energy and matter supply originated from upstream, the depth and turbidity within large channels hampering significant primary production. At the same time, another group of researchers developed the spiralling concept (Webster & Patten, 1979, Paper F1; Newbold *et al.*, 1981) which stipulates that, as organic matter and nutrients flow downstream, they are incorporated, released and incorporated again by various groups of living organisms. These two concepts, river continuum and spiralling, were integrated into a river functioning paradigm (Elwood *et al.*, 1983; Minshall *et al.*, 1983). This paradigm was very much focused on the production, cycling and transfer of energy and matter within the stream network, including the riparian zones and the flood plains.

Interestingly, it is at about the same time that another group of researchers started to focus their attention on the riparian zone but from a catchment perspective. While

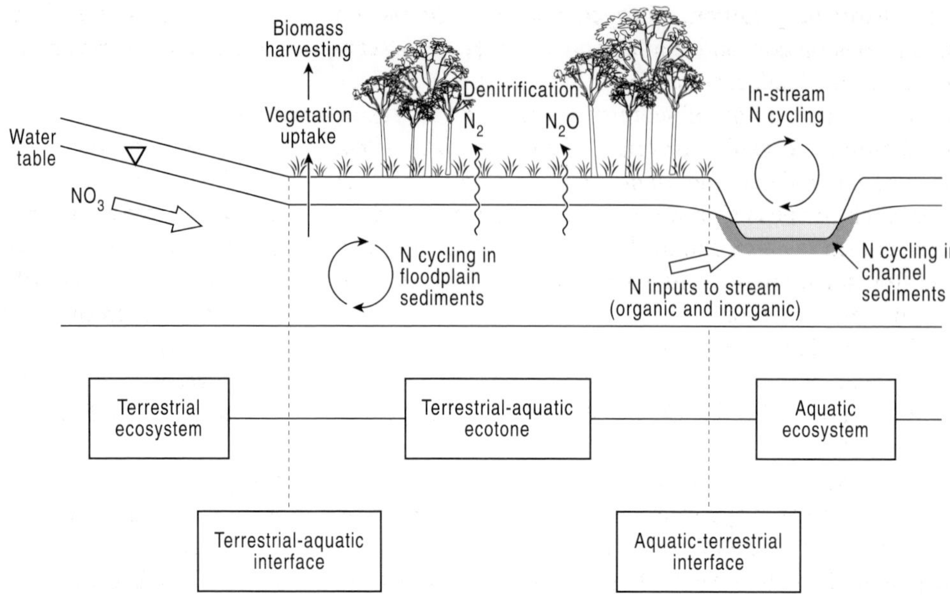

Fig. 1 *Nitrogen cycling within the terrestrial-aquatic ecotone (adapted from Pinay & Décamps, 1988).*

evaluating the transfer of nutrients from upslope into the streams, Peterjohn & Correll (1984, Paper E2) measured a significant decrease in nitrogen concentration in groundwater within the riparian zone. They attributed this decrease to two main processes: plant uptake and denitrification, the latter being the only mechanism that can permanently remove nitrogen from the system. Parallel research in other agricultural contexts confirmed this phenomenon (Lowrance *et al.*, 1984, Paper E1; Jacobs & Gilliam, 1985, Paper E3). Given this intimate connection between terrestrial and aquatic ecosystems, it is hardly surprising that the near-stream zone has become a focus of attention. Indeed, this discovery was enthusiastically taken on board both by ecologists, who realised that they could use this potential nutrient removal as a reason to protect hot spots of biodiversity, and by the agricultural lobby who could use riparian zones to buffer nutrient leaching from intensive agricultural practices, hence protecting water quality. Yet, too much hope was pinned on these rather narrow riparian zones. Soon, they were considered to be the "Swiss knife" of landscape planning, hoping that they could also buffer phosphorus, sediments, pesticides, heavy metals, etc. It turns out, after many hydrological and biogeochemical investigations, that riparian zones are relatively efficient in removing nitrogen, but this is less certain in other cases. Nevertheless, the general assumption is now made that these areas should be maintained in as "natural" a state as possible, so that they are able to buffer the river from its catchment (Bren, 1993).

2 Landscape ecology, riparian ecotones and emergence of the Riparian Buffer Zone concept

Landscapes are an expression of the interaction between climate, topography, network design, and, of course, the land-use history of the river basin. The riparian zone is one unit of this landscape; other units include the river channel, water bodies (small or large), terrestrial ecosystems (farmland, forests, etc.) and human settlements. Amongst these units, the riparian zone has a unique role within the landscape's structure and functioning, derived from its position at the interface between terrestrial and aquatic units connected by means of hydrological flows (Fig. 1). Jones *et al.* (1976, Paper A3) laid the foundation for current landscape ecology research on the relationship between land-use and land cover arrangements and nutrient fluxes, and the role of the riparian zone in particular on reducing nitrogen load.

The occurrence, extent and development of riparian vegetation are related to the geological structure, the geomorphology of the river valley, the soil development, as well as to the hydrological patterns occurring in the catchment. The hierarchical structure of the catchment determines that what is occurring upstream affects the processes downstream (Hynes, 1975, Paper A2; Vannote *et al.*, 1980, Paper A4; Minshall *et al.*, 1983, 1985; Frissell *et al.*, 1986; Alexander *et al.*, 2000) and therefore the structure and function of the riparian zone respond to these changes. As a result of these complex interactions, the riparian zones constitute mosaics in the landscape; their precise role at any given location being determined by their position, extent and development.

This framework justifies that the riparian zone can be viewed as an ecotone across all the landscape patches, where the hydrological flow path is connecting the different units between them. The riparian zone is a key element at the landscape level in the link between hydrology and biogeochemistry. It influences landscape functioning not only within the area itself, but also in the river channel (Pinay & Décamps, 1988, Paper A6; Naiman & Décamps, 1997). These influences are expressed in terms of the material transport from and through the riparian zone, on the biogeochemical cycling of (in particular) carbon, nitrogen and phosphorus, on habitat construction and maintenance within the stream and across the landscape (acting as a biological corridor), and in the interception of sediments, dissolved inorganic and dissolved organic materials. Triska *et al.* (1984, Paper A5) provide an important early example of this sort of research, emphasising both an integrated approach to the catchment system and at the same time showing that relatively small areas of the basin are crucial in terms of nutrient transformation.

The riparian zone provides a buffer zone between the upland and the stream. This effect and function is included under what is often called the riparian buffer zone (RBZ) concept. The vegetation and soil in the riparian zone intercept the materials that flow via surface or subsurface pathways from upland areas through to the water course, and therefore are an important element for the potential improvement of water quality (Gregory *et al.*, 1991). The riparian vegetation also helps regulate the microclimate of riparian and channel ecosystems, by lowering wind speeds and by reducing amounts of solar radiation entering through the canopy, thereby decreasing the water temperature of the stream.

3 Hydrology of the riparian zone

3.1 Water balance within the riparian zone

Given their location (at the foot of valley-side slopes, adjacent to the river channel) and topography (often a wide, flat flood plain), riparian zones are likely to have high water tables, temporarily if not permanently. Even where the substrate is permeable, high water tables are likely to exist, given slope discharge to the riparian zone from one direction and the relatively fixed water level within the river in the other direction. Very low topographic gradients across the flood plain help to sustain waterlogged conditions, especially where the flood plain is wide or the alluvial sediments are fine-grained. Often, the riparian zone is so poorly drained that peat deposits have accumulated, adding to its poorly drained condition still further. Even above the water table, the soil is likely to remain close to saturation given the upward extension of the capillary fringe (Abdul & Gillham, 1984, Paper B6).

Assuming the riparian zone to comprise a distinct storage unit, its water balance may be defined as follows:

Inputs

a. Overland flow from upslope (UOF)
b. Subsurface flow from upslope (USSQ)
c. Precipitation directly on to the riparian zone (RF)
d. Groundwater discharge from local aquifers into the riparian zone (UGW)
e. Seepage from the river channel through the bank (RBS)
f. Overbank flooding from the river to the flood plain surface (ROBI)

Outputs

a. Overland flow from the riparian zone to the river (ROF)
b. Subsurface discharge from the riparian zone to the river (RSSQ)
c. Evaporation from the riparian zone (ET)
d. Percolation from the riparian zone into aquifers below (PERC)

Any imbalance between inputs and outputs must, by definition, involve a change of water storage within the riparian zone (ΔS). The water balance may therefore be expressed as follows (Fig. 2):

$$\mathrm{UOF} + \mathrm{USSQ} + \mathrm{RF} + \mathrm{UGW} + \mathrm{RBS} + \mathrm{ROBI} - \mathrm{ROF} - \mathrm{RSSQ} - \mathrm{ET} - \mathrm{PERC} \pm \Delta\mathrm{S} = 0$$

Of course, not all these processes may be involved, or have the same importance, at any given place or at any particular time. However, the water balance does emphasise how inputs to the riparian zone can originate both from the adjacent slopes and from the river channel. This balance is likely to alter along the river network from headwater tributaries with minimal flood plains to lowland reaches with very wide flood plains.

Few studies have completed water balance computations for riparian zones apart from Wassen *et al.* (1990) and Gilvear *et al.* (1993), who provide them for flood plains. More commonly, the dynamics of water table fluctuation are reported: see, for example, Burt *et al.* (2002a,b). Figure 3 shows water table depths for three elements of the riparian landscape at Trimdon, near

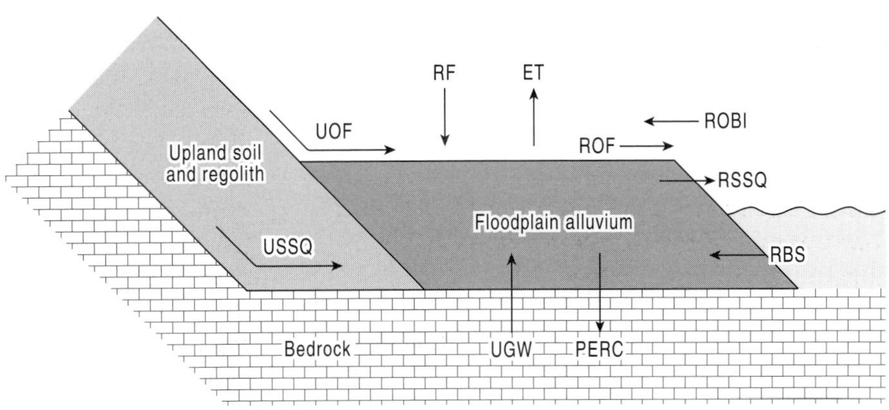

Fig. 2 Components of the water balance of a riparian zone. See text for definitions.

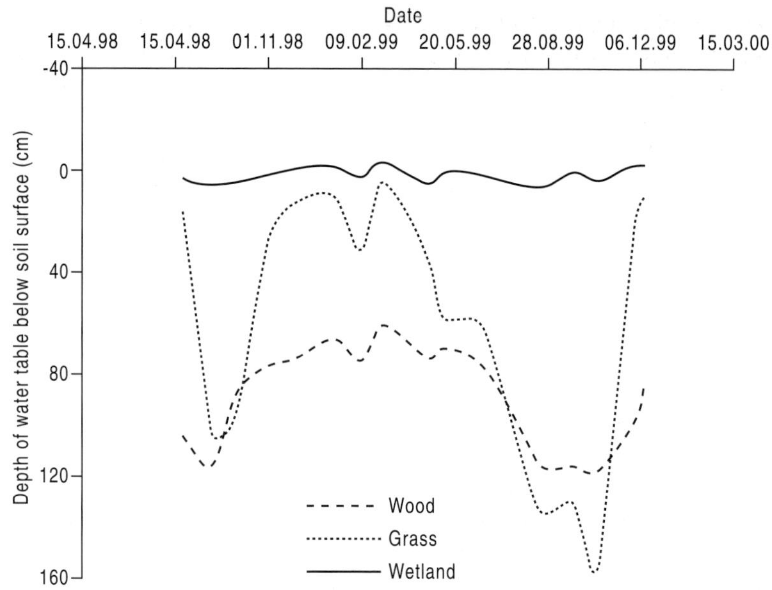

Fig. 3 Water table fluctuations for three elements of the riparian landscape at the NICOLAS field site near Durham, northeast England (source: data from the NICOLAS project – see Matheson, 2001).

Durham in northeast England; this was the location for field experiments conducted as part of the European Union-funded NICOLAS programme (Nitrogen Control by Landscape Structures in Agricultural Catchments – ENV4-CT-97-0395). The seasonal pattern of water table fluctuation is immediately apparent for the wood and grass sites; inflows from adjacent slopes are reduced in summer and continued evaporation lowers the water table even more. At the wetland site, as might be expected, the variation is much more subdued: the water table is only barely below the surface in summer and in winter the site is flooded (i.e. negative values for "depth" to the water table). The wetland is sustained in summer by continued discharge from a nearby spring. The relatively dryness of the woodland is explained by the presence of sandy layers in the soil profile, which aid drainage, as well as by greater losses by evapotranspiration.

3.2 Flow paths within the riparian zone

In headwater catchments, slopes are intimately coupled to streams. There may or may not be a narrow riparian zone with the potential to provide some buffering function between the terrestrial and aquatic ecosystems, but hydrological conditions in the near-stream zone are predominantly controlled by inflows from upslope. Hillslope hydrology has received considerable attention over recent decades: benchmark papers are provided in Beven (2006) and more general reviews in Kirkby (1978) and Anderson & Burt (1990). This body of research is important here for two reasons: first, it provides a detailed understanding of hillslope inputs to the near-stream zone; second, field techniques and experimental design developed for the study of hillslope hydrology have often been replicated in the riparian zone (e.g. Haycock & Burt, 1993, Paper E7). Nor has hillslope hydrological research been limited to the study of flow processes: interest in solute transport on slopes is also directly relevant to the study of buffering processes in the riparian zone, especially in relation to identification of flow paths (e.g. Dunne & Black, 1970, Paper B5; McDonnell, 1990, Paper C5) and for gaining understanding of the evolving chemical signature of water as it moves downslope, through the riparian zone and into the stream (e.g. Kirchner, 2003; Bishop *et al.*, 2004; see also Section 4 below).

In the middle reaches of a river basin, flood plains are wider and there is a balance between inputs to the riparian zone from hillslopes and from the channel (Fig. 2). Bank storage is an important process during flood events, both in-bank and out-of-bank. Hillslope discharge to the riparian zone is held up during flood events but dominates the recession. Concentrated inputs from hillslope hollows can be especially important in maintaining the water table locally (Burt *et al.*, 2002a). One important lesson from hillslope hydrology was the need to map flow paths, rather than assume that flow was straight across the flood plain perpendicular to the channel; given the flood plain gradient, there may be a significant down-valley flow component (Haycock & Burt, 1993, Paper E7). In temperate zones, the main emphasis has been on buffering, as water moves from upslope areas (usually farmland) across the riparian zone to the stream; there is a relatively small buffering capacity for water moving out of the channel during floods. However, in semi-arid areas, water movement from the channel to the parafluvial, hyporheic and riparian zones is a much more important process (Holmes *et al.*, 1994, Paper F6; Harms & Grimm, 2008). Hyporheic zones merge into and form part of the wider riparian buffer zone.

In large river basins, the flood plain becomes an important source of runoff in its own right, and there is little influence from surrounding hillslopes. Drainage of the riparian zone for agricultural "improvement" encourages subsurface flow and may increase nutrient leaching as a result. Slope drainage can by-pass the riparian zone via ditches and drains, further reducing the opportunity for buffering processes to operate. It is thought that rising nitrate concentrations in many rivers in the UK in the 1960s and 1970s was in part caused by extensive land drainage programmes at that time, many of which involved drainage of near-stream land (Green, 1979; Burt *et al.*, 2008).

4 Solute dynamics in the riparian zone

4.1 Its role as a conduit

Towards the end of the 1970s two separate avenues of scientific investigation began to converge. On the one hand, and in light of increasing concern about nitrate losses from soils, research in agricultural science was beginning to undertake experimental studies of leaching in order to understand nutrient losses below the root zone. On the other hand, fluvial geomorphologists were beginning to study links between runoff processes and solute export at the hillslope scale. Together, these themes helped to provide a basis for field studies of riparian zone processes over the next two decades. A review of both approaches is provided in Armstrong & Burt (1993).

Soil scientists tended to work at a smaller scale. They used both intact soil cores (e.g. Bouma *et al.*, 1981; Smettem *et al.*, 1983; Kneale & White, 1984), either in the laboratory or in the field, or specially constructed lysimeters (e.g. Addiscott *et al.*, 1978) to examine leaching losses. Initially, models for solute uptake by water moving through the soil assumed uniform displacement (Horton & Hawkins, 1965), but non-uniform displacement through larger cracks and pores was eventually also taken into account (Germann & Beven, 1981). In fact, macropore flow proved relatively easy to predict, since it will occur when rainfall intensity exceeds the hydraulic conductivity of the soil peds (Trudgill *et al.*, 1982, Paper C4). Not surprisingly, soil scientists soon decided to make field investigations in order to work beyond the idealised conditions of the laboratory. The use of the Rothamsted drain gauges, constructed in 1870, for leaching experiments from 1974 onwards, was a perceptive use of a long-established field installation (Addiscott *et al.*, 1978). In other cases, specially designed lysimeters, such as those at Brimstone Farm (Goss *et al.*, 1993) or North Wyke (Armstrong & Garwood, 1991), in England, allowed one-dimensional observations down soil profiles to be extended to the plot and field scales. However, in most cases these experiments stopped short of the stream and indicated only "plot" losses rather than "sub-catchment" losses. However, there were some exceptions: for example, studies by Bob White and his students at the Oxford University farm at Wytham, England in the early 1980s (White *et al.*, 1983; Haigh & White, 1986), and the early paper on nitrate in surface and subsurface flow by Jackson and colleagues (Jackson *et al.*, 1973, Paper C1). Smettem *et al.* (1983) combined laboratory and field studies; they used intact soil cores to measure soil hydraulic conductivity and to forecast the conditions under which infiltration would by-pass the soil; they then evaluated field response in relation to the thresholds for macropore flow identified in the laboratory. Unfortunately (in one sense), all these studies were conducted on under-drained clay soils; subsurface runoff therefore by-passes the riparian zone. Nevertheless, such studies were important in establishing methodologies for studying solute transport at the hillslope/sub-catchment scale, as well as indicating patterns of nitrate loss in time and space. They also showed what happens when the riparian zone is by-passed and slope drainage moves rapidly from field to stream without moving through any intervening buffer zone.

At much the same time, fluvial geomorphologists were extending studies of hillslope hydrology to incorporate the dynamics of solute erosion. There were two directions to this research: on the one hand, studies of streamflow generation processes were extended to include sediment and solute transport. Keith Beven, in Volume 1 of this *Benchmark Papers in Hydrology Series*, has provided an extensive review of the relevant hydrological literature (Beven, 2006) which formed the basis for this line of enquiry. On the other hand, impetus also came from the study of in-stream solute concentrations at the catchment outlet (e.g. Walling & Foster, 1975; Webb & Walling, 1985); analysis of chemographs suggested profitable lines of enquiry on catchment hillslopes where measurements of soil water (both quantity and quality) could be made to establish links between observed solute variation in-stream and on adjacent slopes. Foster & Walling (1978, Paper C2) provide a good illustration of this catchment-scale approach, with the added interest of capturing post-drought nitrate leaching after a very severe drought. Their inferences about the linkage between subsurface flow and solute transport were shown to be broadly correct within just a few years, as on-slope studies were conducted to complement studies of solute transport in rivers (e.g. Burt, 1979; Burt & Arkell, 1987). Trudgill *et al.* (1982, Paper C4) take a contrasting approach, examining soil water flow in the riparian zone of a woodland slope, but do not link these observations (at least in that publication) to stream solute behaviour. Their work does, however, have the same focus on preferential flow as some other studies already mentioned above. Another study by three of the same authors at a nearby field site (Smettem *et al.*, 1983) does include both slope and stream solute (nitrate) behaviour, but only in the context of an under-drained field rather than a natural hillslope.

McDonnell (1990, Paper C5) and Waddington *et al.* (1993, Paper C6) deserve special mention (and are included in this volume) because of the way in which they provide sufficiently detailed description of hydrological processes operating on a hillslope to allow interpretation of water quality variations in-stream. By so doing, they provide the basis for a hydrologically-based understanding of nutrient cycling in the riparian zone. Both are studies of forested catchments, but with very different topographies. McDonnell's study at Maimai in South Island, New Zealand, was conducted on a very steep slope; although there would be very little opportunity for any buffering of runoff, the paper is important because of its focus on macropore flow and the way in which rapid subsurface flow can still acquire a pre-event signature. Waddington and colleagues studied surface runoff across a wetland; again, there was little opportunity for nitrate depletion (Warwick & Hill, 1988, Paper E4), but mixing was rapid enough for the surface runoff to appear "old". Both McDonnell (1990, Paper C5) and Waddington *et al.* (1993, Paper C6) inform the old–new water debate but from the perspective of this volume, their importance is in linking hydrological processes to flow paths and soil-water residence times.

Irrespective of whether the direction of interest was "bottom up", from stream to slope, or "top down", from slope to stream, the result was a move towards detailed study of soil and slope hydrology and the links between hydrological pathways and solute transport, as exemplified by the McDonnell (1990, Paper C5) and Waddington *et al.* (1993, Paper C6) papers. Inevitably, the riparian zone became the focus for this work since this is the last opportunity for slope drainage water to interact with the soil before it enters the channel (Bishop *et al.*, 2004).

4.2 Its role as a buffer

River ecosystems control the transport of nutrients and organic matter from terrestrial sources (Bennett *et al.*, 2001; Seitzinger *et al.*, 2002; Townsend-Small *et al.*, 2005), produce organic material within aquatic environments, degrade organic matter while transporting it downstream (Hedges *et al.*, 2000) and carry the fingerprint of human activities (e.g. Rosenberg *et al.*, 2000). In that context, flood plain, riparian and in-stream zones are key components of river ecosystems (Hynes, 1975, Paper A2; Naiman & Décamps, 1997; Ren *et al.*, 2000; Fischer *et al.*, 2005; Forshay & Stanley, 2005). In terms of nitrogen, they represent functional nitrogen retention areas which can control and maintain river water quality. Indeed, most of the papers included in this volume (and therefore much of the discussion in this section) are concerned with nitrogen, because this is where the research on solute studies in riparian zones was initially focused. Recent work has broadened coverage to a wider range, particularly phosphorus (e.g. Gburek *et al.*, 2000), but also exchangeable bases (e.g. Burt & Park, 1999), iron, manganese, aluminium (e.g. Park & Burt, 1999a,b), carbon (e.g. Knoepp & Clinton, 2009), rare earth elements (e.g. Dia *et al.*, 2000) and a range of naturally occurring isotopes (e.g. McGlynn *et al.*, 1999; Clément *et al.*, 2003).

In their seminal papers, Lowrance *et al.* (1984, Paper E1), Peterjohn & Correll (1984, Paper E2), and Jacobs & Gilliam (1985, Paper E3) had already considered denitrification, a microbiologically mediated reduction of nitrate into gaseous form, as the main process responsible for the nitrogen buffering capacity of riparian zones. Appropriate denitrification is found at the oxic–anoxic interfaces where oxidized nitrate can be respired/reduced by heterotrophic bacteria under anoxic conditions in the presence of bio-available organic carbon. Riparian zones, due to their position between terrestrial and aquatic systems, act as biogeochemical hot spots, in particular for nitrogen cycling (McClain *et al.*, 2003). Indeed, high biogeochemical processing rates are often sustained by converging hydrological flow paths where each flow carries materials essential to the reaction, i.e. nitrate and carbon, or, more likely in riparian zones, a single hydrological flow path may create a hot spot by transporting the nitrate-limiting reactants to a carbon-rich substrate, i.e. the riparian soil profile.

The occurrence of these biogeochemical hot spots requires consideration of three fundamental elements (Pinay *et al.*, 2002). The first is related to the delivery patterns of carbon and nitrogen inputs along river ecosystems. River systems and their riparian zones can be viewed as open ecosystems dynamically linked longitudinally, laterally and vertically by hydrological and geomorphological processes (Webster & Patten, 1979, Paper F1; Ward, 1989). In riparian zones, the importance of inputs of organic carbon and nitrogen and ammonia via surface flow from upstream increases with the stream order. Indeed, river flood plains are recognized as important storage sites for sediments and associated nutrients mobilized from upstream catchments during floods (He & Walling, 1997; Forshay & Stanley, 2005). Yet, nitrate transfer to riparian zones mainly originates from the adjacent upslope area, whatever the stream order. This implies that active nitrate removal is often located at the interface between upslope and the riparian zone (see Jones *et al.*, 1976, Paper A3 and its commentary in this volume).

The second basic element is that the area of the water–substrate interface (i.e. water–sediment or wetland–upland length of contact) is positively correlated to the efficiency of nitrogen retention and use in river ecosystems. These positive relationships occur both in the main channel itself and in the riparian and flood plain zones (Hill, 1979; Jones & Holmes, 1996; Valett *et al.*, 1996, Paper F7). Hence, the length or the duration of contact between water and these substrates increases the biological use and thereby the total amount of nitrogen processed (Cavanaugh *et al.*, 2006). Empirical evidence shows that the rate of in-stream nitrogen cycling depends on the surface contact of water with sediment as well as on the duration of contact (Pinay *et al.*, 2009).

The third element is related to the role of water levels in riparian zones in nitrogen cycling and, in turn, nitrogen buffering capacity. The frequency, duration, timing and intensity of high water levels, by controlling the duration of soil and sediment oxic and anoxic phases, directly affect nitrogen cycling in riparian soils (Ponnamperuma, 1972; Keeney 1973; Patrick & Tusnem 1982). Indeed, processes involved in nitrogen cycling are very sensitive to the redox status of the soil. For example, organic nitrogen can be transformed to ammonia both under aerobic and anaerobic conditions, but the nitrification process, which requires free oxygen, can only occur in aerated soils or sediments, while denitrification occurs under anoxic conditions. As a consequence, under permanently anaerobic conditions the organic nitrogen mineralization process results in the accumulation of ammonium. On the other hand, Reddy & Patrick (1975, Paper D1) demonstrated that a high frequency of wet and dry conditions enhanced organic matter decomposition and denitrification rates. Therefore, the end products of nitrogen cycling in riparian soils are controlled by the moisture regime (e.g. groundwater table), with important implications for flood plain productivity and management (Cavanaugh *et al.*, 2006).

5 Hyporheic and in-stream processes

The ability of the river system to retain particulate and dissolved material is a direct result of structural complexity within the catchment. For in-channel flow, both meanders and debris dams are crucial elements at the reach scale, combining with shallower areas of riffles and steps. Variations in channel depth generate flow through the hyporheic zones and, during flood events, into the wider riparian zone (Burt *et al.*, 2002a). Even at the habitat scale, materials accumulating in sediments may well be close to other areas where accumulation is minimal. These structural elements contribute a large heterogeneity of water depth, water velocity, material deposition and accumulation, biological communities and biogeochemical processes, both longitudinally and transversally in the river channel. Therefore, the organisms inhabiting the system are as diverse as the processes occurring, and the final outcome is the mixing to a single function. Closely linked to the channel and its water, the *hyporheos* is an area that comprises water-saturated sediments beneath and beside the channel. This area has an active exchange with the surface water, both in terms of dissolved and particulate materials, as well as of biota that travel between the two compartments. Mixing between the surface and hyporheic water masses stimulates biological activity, and is associated with biogeochemical processes different to those occurring in the channel itself. This area is therefore an ecotone between groundwater and surface water (Sabater & Vila 1991).

The recognition that biogeochemical processes occurring within the riparian ecotone could have a parallel beneath and beside the stream channel took a while to be recognized, and was probably related to the technical difficulties involved in its measurement. Triska *et al.* (1989) outlined a conceptual approach to the processes involved in the channel. In these processes, the hyporheic area had a relevant role in terms of nutrient cycling and biotic metabolism. Boulton *et al.* (1992) defined different areas of activity within the hyporheos, differing in physical, chemical and biogeochemical characteristics. The volume immediately beneath the wetted channel surface is known as the *wetted channel hyporheic zone*; the volume beneath dry bars appearing in the active channel is the *parafluvial zone*, whilst the hyporheic area adjacent to the banks and within the flood plain is the *flood plain hyporheic zone*. These distinctions are, in one sense, simply conceptual, and there is a real difficulty in outlining the hyporheic boundaries in practice. However, this can be defined operationally. Conservative tracers (e.g. Bencala, 1984; Triska *et al.*, 1989) and fluorescent dyes (Holmes *et al.*, 1994, Paper F6) have been extensively used for these purposes. The most used are chloride or bromide, which allow the hyporheic water masses to be distinguished. After injection of the tracer, the water mass can be tracked by means of conductivity variations or direct analyses of the solutes, enabling its dynamics (when the hyporheic water mixes with the surface water or moves downstream within the hyporheos) to be identified. Fluorescent dyes can also be helpful in determining water flow paths through the parafluvial area. In all of these cases, the water flow path monitoring is by means of wells similar to those used in the riparian compartment, and therefore is time-consuming and difficult.

The hyporheic zone is mainly a transient storage zone within the stream, where the solutes travelling with the water can be stored or delayed in their flow path beneath the river. This differential water movement and the associated redox differences with respect to the surface waters imply that hyporheic biogeochemical processes are different to those occurring in the surface water compartment (Mulholland, 1992, Paper F5). Overall, these processes have a relevant influence on the surface water quality (Bencala, 1984). Systems with more extensive hyporheic areas can retain and process solutes more efficiently than others with a smaller hyporheic area (Valett *et al.*, 1996, Paper F7). The role of the hyporheic area, in terms of nutrient cycling and biotic metabolism, is by no means negligible in the riparian-stream ecosystem and ongoing research seeks to evaluate its impact and identify its importance within the wider riparian ecotone.

6 What do we still need to learn about the hydrology and biogeochemistry of riparian zones?

Biogeochemists have identified hot spots where hydrological flow paths converge with substrates or other flow paths containing complementary or missing reactants. These hot spots exhibit disproportionately high reaction rates relative to the surrounding matrix (McClain *et al.*,

2003). If riparian zones are indeed hot spots, then there are two aspects of the linkage between hydrological processes and biogeochemical cycling that require better understanding: flow paths and residence times of water within the riparian zone, and identification of hot spots at the landscape scale. In addition to these two points, the influence of climate change on riparian zones and the identification of new biogeochemical pathways are two further areas where we need to develop better understanding.

6.1 Flow paths and residence time

We know that riparian zones are an important interface between terrestrial and aquatic environments and can exert significant controls on water quality and quantity. The extent to which the riparian zone is able to interact with slope runoff depends upon the residence time of water within the riparian zone which, in turn, depends on the dominant flow paths operating in the area. Topography and sedimentary architecture determine which flow paths are operating in the riparian zone. If water has a short residence time, there will be little opportunity for buffer zone processes to operate. For example, Welsch *et al.* (2001) found that although there was more potential for denitrification in areas where nitrate-rich water converged, the throughput of water overwhelmed any denitrification effect; average residence time in the riparian zone was low, preventing anaerobic conditions and significant denitrification. At the other end of the spectrum, where water has a long residence time in the riparian zone, denitrification will be very effective, but the throughput of water is too small to significantly influence river water nitrate levels (e.g. Burt *et al.*, 1999). The optimal condition would seem to be one of intermediate permeability where residence times are sufficient to allow significant amounts of both throughflow <u>and</u> denitrification. To operate effectively, the riparian buffer zone therefore requires a balance between the functions of a conduit and a barrier; this is not just a question of interaction with different soils conditions, but depends on flow-path length and residence time within a distinctive biogeochemical environment (Burt & Pinay, 2005).

All this is well enough known in theory, but difficult to forecast in practice. We require both better three-dimensional (3-D) physically-based modelling of flow paths within the riparian zone and improved field survey techniques to support the modelling. Ground-penetrating radar (Holden *et al.*, 2002) appears to have much to offer, but has been little used in the riparian environment so far. Recent advances in tracer techniques could profitably be applied to the near-stream zone. Tracers (e.g. stable isotopes of water) allow the age or residence time of water within the catchment to be estimated, revealing information about the storage, pathways and source of water in a single measure (McGuire *et al.*, 2005). There is also the opportunity to apply new findings from the study of rare earth elements (REE) to the riparian zone; for example, the cerium (Ce) anomaly in REE series can be used to quantify the percentage of water which has passed through reduced environments rich in organic matter (Duncan & Shaw, 2003; Gruau *et al.*, 2004). Meanwhile, isotopes of carbon and nitrogen can be used to establish the origin of organic matter moving through the riparian zone (Sebilo *et al.*, 2006).

6.2 Hot spots and connectivity at the landscape scale

Experimental catchment studies, generally limited to relatively small basins (usually 1–100 km^2), present a major challenge for wider interpretation due to the necessity for up-scaling to river-basin or larger scales. Environmental management (e.g. for water supply, nature conservation or environmental planning under the EU Water Framework Directive) is typically concerned with impacts at larger basin areas (say, 1000 km^2). It is clear from case studies in small catchments that the degree of functional connectivity between different landscape elements is a key factor in the linkage between hydrology and biogeochemistry and is often crucial in determining the location of hot spots within the landscape (Burt & Pinay, 2005). We contend that the spatial arrangement of land cover (in effect, the way in which the landscape is managed) continues to be significant at an intermediate scale (i.e. somewhere between the small basin and the very large administrative area). It is now feasible to devise modelling platforms using the latest generation of computer simulation models to incorporate a 3-D approach to landscape connectivity in medium-scale drainage basins. For example, Stuart Lane has pioneered the incorporation of hydrologically-intelligent measures of landscape connectivity into multi-scale diffuse pollution models (Lane *et al.*, 2006). The use of agent-based models to track the mobilisation, transport and transformation (and/or deposition) of nutrient species moving through the landscape also seems to offer much potential (Reaney *et al.*, 2007; Reaney, 2008). We envisage that these sorts of models, and others, will become very important in the study of riparian zone hydrology and biogeochemistry in the next decade.

The impact of different hydrological regimes upon both nutrient transformations and delivery pathways is the critical challenge for modelling interactions between landscape components and the "at-a-point" processes of nutrient cycling in both terrestrial and aquatic units. There are three key characteristics that need to be developed for the modelling framework. First, many of the interventions in small agricultural catchments are localised in nature and substantial environmental benefits can accrue from much localised interventions. This points to the need for representation of hydrochemical behaviour to capture localised sink and transformation processes, as well as the dynamics of these processes through time. In effect, some fields generate greater pollution risk than others (Burt & Arkell, 1987), and this spatial variability must be captured in modelling at the catchment scale. Second, although localised processes control the extent to which material is delivered to the drainage network, their role can only be identified by upscaling to larger catchment areas (Lane *et al.*, 2006). Third, the hydrological events that can impact upon the drainage network in terms of the delivery of material have a complex spatio-temporal dynamic. Larger or longer duration events are more likely to connect landscape locations further from the drainage network directly to the drainage network during that event. However, material that does not make it to a stream in a given storm event may become remobilised, either in its original or a transformed form, to be delivered as "old" water in a later event (McDonnell, 1990, Paper C5). Thus, models have to be able to capture this potential time-lag characteristic. In all this, the role of the riparian zone as conduit or barrier remains crucial and it is clear that modelling platforms must develop innovative ways of capturing the sources, transport, sinks and transformation of individual nutrient species if they are to enable the effects of different landscape interventions upon in-stream water quality processes to be tested. We envisage that the development of novel approaches to modelling the role of riparian zones within the wider landscape will be an important aspect of catchment science in the coming years.

6.3 Climate change and riparian zone processes

That the active riparian area expands and contracts with changes in water storage is well known. It is larger during high flows, but it contracts during dry periods (Beven & Kirkby, 1979). Although this is part of a natural process, being climate-driven and therefore a widespread occurrence, extreme examples are seen in the tropical, semi-arid and arid rivers on the one hand, and in large flood plains on the other. The knowledge and capacity to foresee the consequences for these systems in a scenario of global change is even more difficult than in the relatively well-known temperate systems (Pinay *et al.*, 2007).

In arid and semi-arid regions, aerobic and anaerobic conditions in sediments and soils alternate in the shift from dry to saturated (Baldwin & Mitchell, 2000). The return of water to a previously dry stream bed is a "hot biogeochemical moment" (McClain *et al.*, 2003), when many processes re-start; moreover, few of these processes are well quantified. In a well advanced drought, the phreatic layer within the RBZ can descend, amplifying the consequences. Early leaf abscission of riparian woody plants (Sabater *et al.*, 2001) is one of the first to occur. An extended depression of the water table in the riparian zone may subsequently impair the riparian vegetation, damaging trees, and favouring replacement by shrubs and herbs. Soil moisture retention may be also affected, since vegetation controls the permeability near the surface. Implications for soil and hydrological dynamics within the RBZ have not been yet fully described as part of this process. Groundwater extraction and overexploitation may dramatically accelerate it.

In tropical zones, alterations to hydrological connectivity and water flow paths in response to land use change (e.g. shifting from forest to pasture, transformation to agricultural landscapes) remain to be determined. Though effects on fauna (invertebrates and fish) have already been observed (Lorion & Kennedy, 2009a,b), fine-scale analysis of hydrological alterations has not yet been achieved.

In large flood plains the hydrological connections are very complex. Flooding is the most important event along large alluvial rivers and triggers a variety of processes. Floods not only increase the aquatic surface, area but also reshape aquatic and terrestrial habitats and maintain their complexity. Bertoldi *et al.* (2009) identified various hydrological thresholds in a flood plain segment along the Tagliamento River (NE Italy). In that system, an elimination of annual flood events may lead to an increase in vegetated areas, therefore altering the flow-inundation regime of the entire river flood plain complex. Determining the fine-scale hydrology of these large-scale systems remains a challenge from both the technical and modelling points of view.

Identifying the effect of large-scale disturbances (both natural and due to climate and global change) on RBZ hydrology is an important research goal for the future. Fire (Pettit & Naiman, 2007) may be frequent in RBZs in arid and semi-arid zones, but may also occur in tropical regions. Fires may not only affect vegetation but also soils, and therefore may deeply influence the hydrology of the RBZ. Flooding events occurring after fires may be highly destructive of soil and river banks, and may reduce water infiltration and lead to deep erosion (Shakesby & Doerr, 2006). In semi-arid regions, regulation and water extraction may strongly affect the river regime as well as connectivity within the RBZ (and in the wider landscape beyond). Salinisation of aquifers and groundwater can be a complicating factor in the relationships between biogeochemistry and hydrology (Lamontagne *et al.*, 2006).

6.4 New biogeochemical pathways

In a recent review, Burgin & Hamilton (2007) wondered whether we have given too much attention to denitrification as the main nitrogen removal process in aquatic ecosystems. Indeed, the recent use of stable isotope techniques has allowed other microbiological processes which can remove nitrogen to be discovered.

Anaerobic ammonium oxidation (Anammox) is a process by which ammonia, combined with nitrite under anaerobic conditions, produces N_2 (Jetten *et al.*, 1998). This process appears to be very important in deep seas (Engstrom *et al.*, 2009) but also in estuarine environments (Nicholls & Trimmer, 2009). Although it seems that this process is limited by simple organic compounds normally found in large quantities in soils and freshwater sediments, virtually nothing is known about its occurrence and importance in riparian zones.

Dissimilatory nitrate reduction to ammonium (DNRA) is another microbial process reducing nitrate to ammonia (Tiedje *et al.*, 1982). This fermentation process appears to be important in humid tropical soils (Silver *et al.*, 2001), tightening the nitrogen cycle and limiting nitrate leaching. So far, little attention has been given to this process in riparian zones because it does not involve complete removal of nitrogen from the system since it accumulates as ammonia. Yet, recent potential DNRA activity measurements suggest that this can be a significant process in riparian zones (Matheson *et* al., 2002; Davis *et al.*, 2008). Further attention is needed, especially under wet tropical conditions, in order to test if the riparian nitrogen cycling model developed under temperate conditions is valid and could be promoted in fast changing tropical environments.

Another DNRA pathway is linked to sulphur oxidation (Dannenberg *et al.*, 1992). Some sulphur-oxidizing bacteria can use nitrate to oxidize sulphide (H_2S) to sulphate (SO_4^{2-}). Nitrate can then be reduced to N_2 or ammonia. This pathway has been exploited as a water treatment method for removing nitrate. However, its natural occurrence in riparian zones has not yet been confirmed. Indeed, to date, the only two field studies which have measured such nitrate reduction have only measured potentials since they injected nitrate into the sulphur-rich reduced zone via the push-pull methodology (Whitmire & Hamilton, 2005; Burgin & Hamilton, 2008). The challenge now is to determine where nitrate and sulphur can co-exist in order to evaluate the real importance of this dissimilation pathway in riparian zones.

Elemental cycles have often been studied separately. The above-mentioned examples demonstrate that this is not the case in the real world where nitrogen, sulphur, carbon and iron cycles, to mention only the major elements, are intimately linked. The complexity of these coupled cycles has been nicely illustrated by Clement *et al.* (2005) who suggested the coupling of the "ferrous wheel" and anaerobic nitrification along a riparian transect. The possible oxidation of ammonia to nitrite by ferric iron in reduced zones could by-pass the aerobic nitrification pathway, allowing nitrite to then be denitrified. Again, this potential pathway needs further *in situ* studies to determine its real importance for removing ammonia from reduced riparian wetlands.

7 Introduction to the commentaries

In contrast to streamflow generation processes (Beven, 2006), the study of riparian zone hydrology and biogeochemistry is a relatively young area of science. Probably, this makes the precise choice of papers to include a little more difficult than might be the case with a more mature subject. Either way, the final choice is bound to reflect the Editors' background and experience and we are sure there will be some disagreement about what is included in this volume. All three of us discovered new material along the way, papers regularly cited by other people that we had never gone back to ourselves: all three of us now know that Cummins (1974, Paper A1) richly deserves its place here, but it was not on our original list. Some other papers failed to make the final list, usually because, on re-reading, we found that they did not seem as original and influential as we remembered them to be.

It is only right that there is a solid foundation in hydrology, both to reflect the science underpinning this research area, and to acknowledge the provenance of this *Benchmark Series*. Nevertheless, it was equally important to include relevant material from a wide range of publications in what these days is termed biogeochemistry but which, when they were written, came from biology, biochemistry and various branches of aquatic ecology. One important way in which these two main threads are intertwined comes through the initial impetus provided by the study of fluvial geomorphology in the USA in the 1950s and 1960s (Leopold *et al.*, 1964). The recognition of the drainage basin as the fundamental unit in fluvial geomorphology created new paradigms in both hydrology and ecology. These come together in the study of biogeochemical cycling in the riparian zone.

Following other editors in this series, we have tried to place each benchmark paper in context, say why we have included it, and then give some perspective on the debates that followed, either immediately or later on. The papers are arranged in sections, to emphasise that many of the themes developed in parallel, rather than as an integrated effort. At the end of each commentary we have added the editor's initials, to show which one of us wrote it. Choice of commentator reflects our varying interests across the fields of hydrology, geomorphology, biogeochemistry and ecology. Dates of publication range from the 1940s to the 1990s: whether a minimum of 15 years since the last paper was published can provide a reasonable historical perspective, remains to be seen. With hindsight, a future volume might look very different. Meanwhile, enjoy the feast!

References

Abdul, A. S. & Gillham, R. W. (1984) Laboratory studies of the effects of the capillary fringe on streamflow generation. *Water Resour. Res.* **20**, 691–698.

Addiscott, T. M., Rose, D. A. & Bolton, J. (1978) Chloride leaching in the Rothamsted drain gauges: influence of rainfall patterns and soil moisture. *J. Soil Sci.* **29**, 305–314.

Alexander, B. N., Smith, R. A. & Schwarz, G. E. (2000) Effects of stream channel size on the delivery of nitrogen in the Gulf of Mexico. *Nature* **430**, 758–761.

Anderson, E. W. (1987) Riparian area definition – a viewpoint. *Rangeland* **9**, 70.

Anderson, M. G. & Burt, T. P. (1990) *Process Studies in Hillslope Hydrology*. Wiley, Chichester, UK.

Armstrong, A. C. & Garwood, E. A. (1991) Hydrological consequences of artificial drainage of grassland. *Hydrol. Processes* **5**, 157–174.

Armstrong, A. C. & Burt, T. P. (1993) Nitrate delivery from agricultural land. In: *Nitrate: Processes, Patterns and Management* (ed. by T. P. Burt, A. L. Heathwaite & S. T. Trudgill), 239–268. Wiley, UK.

Baldwin, D. S. & Mitchell A. M. (2000) The effects of drying and re-flooding on the sediment and soil nutrient dynamics of lowland river-floodplain systems: a synthesis. *Regulated Rivers: Res. & Manage.* **16**, 457–467.

Bencala, K. E., Kennedy, V. C., Zellweger, G. W., Jackman, A. P. & Avanzino, R. J. (1984) Interactions of solutes and streambed sediment 1. An experimental analysis of cation and anion transport in a mountain stream. *Water Resour. Res.* **20**, 1797–1803.

Bennett, E. M., Carpenter, S. R. & Caraco, N. F. (2001) Human impact on erodable phosphorus and eutrophication: a global perspective. *Bioscience* **51**(3), 227–234.

Bertoldi, W., Gurnell, A., Surian, N., Tockner, K., Zanoni, L., Ziliani, L. & Zolezzi, G. (2009) Understanding reference processes: Linkages between river flows, sediment dynamics and vegetated landforms along the Tagliamento River, Italy. *River Res. & Application* **25**, 501–516, doi: 10.1002/rra.1233.

Beven, K. J. & Kirkby, M. J. (1979) A physically based, variable contributing area model of basin hydrology. *Hydrol. Sci. Bull.* **24** 43–69.

Beven, K. J. (2006). *Streamflow Generation Processes*. Benchmark Papers in Hydrology. IAHS Press, Wallingford, UK.

Bishop, K., Seibert, J., Kohler, S. & Laudon, H. (2004) Resolving the Double Paradox of rapidly mobilised old water with highly variable responses in runoff chemistry. *Hydrol. Processes* **18**, 185–189.

Boulton, A. J., Valett, H. M. & Fisher, S. G. (1992) Spatial distribution and taxonomic composition of the hyporheos of several Sonoran desert (Arizona, USA) streams. *Archiv fur Hydrobiologie* **125**, 37–61.

Bouma, J., Dekker, L. W. & Muilwijk, C. J. (1981) A field method for measuring short-circuiting in clay soils. *J. Hydrol.* **52**, 347–354.

Bren, L. J. (1993) Riparian zone, stream and floodplain issues: a review. *J. Hydrol.* **150**, 277–300.

Brinson, M. M., Bradshaw, H. D. & Kane, E. S. (1984) Nutrient assimilative capacity of an alluvial floodplain swamp. *J. Appl. Ecol.* **21**, 1041–1057.

Burgjn, A. J. & Hamilton, S. K. (2007) Have we overemphasized the role of denitrification in aquatic ecosystems? A review of nitrate removal pathways. *Frontiers in Ecology & Environment* **5**, 89–96.

Burgin, A. J. & Hamilton, S. K. (2008) NO_3- driven SO_4^{2-} production in freshwater ecosystems: implications for N and S cycling. *Ecosystems* **11**, 908–922.

Burt, T. P. (1979) The relationship between throughflow generation and the solute concentration of soil and stream water. *Earth Surf. Processes* **4**, 257–266.

Burt, T. P. & Arkell, B. P. (1987) Temporal and spatial patterns of nitrate losses from an agricultural catchment. *Soil Use Manage.* **3**, 138–143.

Burt, T. P. & Park, S. J. (1999) The distribution of solute processes on an acid hillslope and the delivery of solutes to a stream: I. Exchangeable bases. *Earth Surf. Processes Landf.* **24**, 781–797.

Burt, T. P. & Pinay, G. (2005) Linking hydrology and biogeochemistry in complex landscapes. *Progr. Phys. Geogr.* **29**(3), 297–316.

Burt, T. P, Matchett, L. S., Goulding, K. W. T., Webster, C. P. & Haycock, N. E. (1999) Denitrification in riparian buffer zones: the role of floodplain sediments. *Hydrol. Processes* **13**, 1451–1463.

Burt, T. P., Bates, P. D., Stewart, M. D. Claxton, A. J., Anderson, M. G. & Price. D. A. (2002a) Water table fluctuations within the floodplain of the River Severn, England. *J. Hydrol.* **262**, 1–20.

Burt, T. P., Pinay, G., Matheson, F. E., Haycock, N. E., Butturini, A., Clement, J. C., Danielescu, S., Dowrick, D. J., Hefting, M. M., Hillbricht-Ilkowska, A. & Maitre, V. (2002b) Water table fluctuations in the riparian zone: comparative results from a pan-European experiment. *J. Hydrol.* **265**, 129–148.

Burt, T. P., Howden, N. J. K., Worrall, F. & Whelan, M. J. (2008) Importance of long-term monitoring for detecting environmental change: lessons from a lowland river in south east England. *Biogeosciences* **5**, 1529–1535.

Cavanaugh, J. C., Richardson, W. B., Strauss, E. A., & Bartsch, A. (2006) Nitrogen dynamics in sediment during water level manipulations on the Upper Mississippi River. *River Res. & Application* **22**, 651–666.

Chorley, R. J. (1969) The drainage basin as the fundamental geomorphic unit. In: *Water, Earth and Man* (ed. by R. J. Chorley), 77–99. Methuen, London, UK.

Clément, J. C., Aquilina, L., Bour, O., Plaine, K., Burt, T. P. & Pinay, G. (2003) Hydrological flowpaths and nitrate removal rates within a riparian floodplain along a fourth-order stream in Brittany (France). *Hydrol. Processes* **17**, 1177–1195.

Clément, J. C., Shrestha, J., Ehrenfeld J. G. & Jaffe, P. R. (2005) Ammonium oxidation coupled to dissimilatory reduction of iron under anaerobic conditions in wetland soils. *Soil Biol. & Biochem.* **37**, 2323–2328.

Cummins, K. W. (1974) Structure and function of stream ecosystems, *Bioscience* **22**, 719–722.

Dannenberg, S., Kroder, M., Dilling, W. & Cypionka, H. (1992) Oxidation of H_2, organic compounds and inorganic sulfur compounds coupled to reduction of O_2 or nitrate by sulfate-reducing bacteria. *Archiv fur Microbiologie* **158**, 93–99.

Davis, J. H., Horwath, W. R., Steiner, J. J. & Myrold, D. D. (2008) Denitrification and nitrate consumption in an herbaceous riparian area and perennial ryegrass seed cropping system. *Soil Sci. Soc. Am. J.* **72**(5), 1299–1310.

Dia, A., Gruau, G., Olivier-Lauquet, G., Riou, C., Molénat, J. & Curmi, P. (2000) The distribution of rare earth elements in groundwaters: assessing the role of source-rock composition, redox changes and colloidal particles. *Geochimica et Cosmochimica Acta* **64**(24), 4131–4151.

Duncan, T. & Shaw, T. J. (2003) The mobility of rare earth elements and redox sensitive elements in the groundwater/seawater mixing zone of a shallow coastal aquifer. *Aquatic Geochem.* **9**(3), 1380–6165.

Dunne, T. & Black, R. D. (1970) An experimental investigation of runoff production in permeable soils. *Water Resour. Res.* **6**, 478–490.

Elwood, J. W., Newbold, J. D. O'Neill, R. V. & Van Winkle, W. (1983) Resource spiraling: an operational paradigm for analyzing lotic ecosystems. In: *Dynamics of Lotic Ecosystems* (ed. by T. D. Fontaine & S. M. Bartell), 3–27. Ann Arbor Science, Ann Arbor, Michigan, USA.

Engstrom, P., Penton, C. R. & Devol, A. H. (2009) Anaerobic ammonium oxidation in deep-sea sediments off the Washington margin. *Limnol. Oceanogr.* **54**(5), 1643–1652.

Fischer, H., Kloep, F., Wilzcek, S., & Pusch, M. T (2005) A river's liver – microbial processes within the hyporheic zone of a large lowland river. *Biogeochemistry* **76**(2), 349–371.

Forshay, K. J. & Stanley, E. H. (2005) Rapid nitrate loss and denitrification in a temperate river floodplain. *Biogeochemistry* **75**(1), 43–64.

Foster, I. D. L. & Walling, D. E. (1978) The effects of the 1978 drought and autumn rainfall on stream solute levels. *Earth Surf. Processes* **3**, 396–406.

Frissell, C. A., Liss, W. J., Warren, C. E. & Hurley, M. D. (1986) A hierarchical framework for stream habitat classification: viewing streams in a watershed context. *Environ. Manage.* **10**, 199–214.

Gburek, W. J. Sharpley, A. N. Heathwaite, A. L. & Folmar, G. (2000) Phosphorus management at the watershed scale. *J. Environ. Qual.* **29**, 130–144.

Germann, P. F. & Beven, K. (1981) Water flow in macropores. III. A statistical approach. *J. Soil Sci,* **32**, 31–39.

Gilvear, D. J., Andrews, R., Tellam, J. H., Lloyd, J. W. & Lerner, D. N. (1993) Quantification of the water balance and hydrogeological processes in the vicinity of a small groundwater-fed wetland, East Anglia, UK. *J. Hydrol.* **144**, 311–334.

Goss, M. J., Howse, K. R., Lane, P. W., Christian, D. G. & Harris, G. L. (1993) Losses of nitrate-nitrogen in water draining from under autumn-sown crops established by direct drilling or mouldboard ploughing. *J. Soil Sci.* **44**, 35–48.

Green, F. H. W. (1979) Field under-drainage and the hydrological cycle. In: *Man's Impact on the Hydrological Cycle in the United Kingdom* (ed. by G. E. Hollis), 9–17. GeoBooks, Norwich, UK.

Gregory, S. V, Swanson, F. J, McKee, W. A. & Cummins, K. W. (1991) An ecosystem perspective of riparian zones. *Bioscience* **41**, 540–551.

Gruau, G., A. Dia, G. Olivié-Lauquet, M. Davranche, G. Pinay. (2004) Controls on the distribution of rare earth elements in shallow groundwater. *Water Res.* **38**, 3576–3586.

Haigh, R. A. & White, R. E. (1986) Nitrate leaching from a small, underdrained, grassland, clay catchment. *Soil Use Manage.* **2**, 65–70.

Holden, J., Burt, T. P. & Vilas, M. (2002) Application of ground penetrating radar to the identification of subsurface piping in blanket peat. *Earth Surf. Processes Landf.* **27**, 235–249.

Harms, T. K. & Grimm, N. B. (2008) Hot spots and hot moments of C and N dynamics in a semiarid riparian zone. *J. Geophys. Res. – Biogeosci.* **113**, G01020, doi:01010.01029/02007JG000588.

Haycock, N. E. & Burt, T. P. (1993) The role of floodplain sediments in reducing the nitrate concentration of subsurface runoff: a case study in the Cotswolds, England. *Hydrol. Processes* **7**, 287–295.

He, Q. & Walling D. E. (1997) Spatial variability of the particle size composition of overbank floodplain deposits. *Water, Air and Soil Pollut.* **99**, 71–80.

Hedges, J. I., Mayorga, E. Tsamakis, E. McClain, M. E., Aufdenkampe, A., Quay, P., Richey, J. E., Benner, R., Opshal, S., Black, B., Pimentel, T., Quintanilla, J. & Maurice, L. (2000) Organic matter in Bolivian tributaries of the Amazon River: A comparison to the lower mainstream. *Limnol. Oceanogr.* **45**(7), 1449–1466.

Hill, A. R. (1979) Denitrification in the nitrogen budget of a river ecosystem. *Nature* **281**, 291–292.

Holden, J., Burt, T. P. & Vilas, M. (2002). Application of ground penetrating radar to the identification of subsurface piping in blanket peat. *Earth Surf. Processes Landf.* **27**, 235–249.

Holmes, R. M., Jones, J. B., Fisher, S. G. & Grimm, N. B. (1996) Denitrification in a nitrogen-limited stream ecosystem. *Biogeochemistry* **33**, 125–146.

Holmes, R. M., Fisher, S. G. & Grimm, N. B. (1994) Parafluvial nitrogen dynamics in a desert stream ecosystem. *J. North American Benthological Soc.* **13**, 468–478.

Horton, J. H. & Hawkins, R. H. (1965) Flow path of rain from the soil surface to the water table. *Soil Sci.* **100**, 377–383.

Hynes, H. B. N. (1975) The stream and its valley. *Verhandlungen der Internationalen Vereinigung für Theoretische und Angewandte Limnologie* **19**, 1–15.

Jacobs, T. C. & Gilliam, J. W. (1985) Riparian losses of nitrate from agricultural drainage waters. *J. Environ. Qual.* **14**(4), 472–478.

Jackson, W. A., Asmussen, L. E., Hauser, E. W. & White, A. W. (1973) Nitrate in surface and subsurface flow from a small agricultural watershed. *J. Environ. Qual.* **2**, 480–482.

Jetten, M. S. M., Strous, M., van de Pas-Schoonen, K. T., Schalk, J., van Dongen, U. G. J. M., van de Graaf, A. A., Logemann, S., Muyzer, G., van Loosdrecht, M. C. M. & Kuenen, J. G. (1998) The anaerobic oxidation of ammonium. *FEMS Microbiology Reviews* **22**(5), 421–437.

Jones, J. B. & Holmes, R. M. (1996) Surface-subsurface interactions in stream ecosystems. *Trends in Ecology and Evolution* **11**, 239–242.

Jones, J. R., Borofka, B. P. & Bachmann, R. W. (1976) Factors affecting nutrient loads in some Iowa streams. *Water Res.* **10**, 117–122.

Keeney, D. R. (1973) The nitrogen cycle in sediment-water systems. *J. Environ. Qual.* **2**, 15–29.

Kirchner, J. W. (2003) A double paradox in catchment hydrology and geochemistry. *Hydrol. Processes* **17**, 871–874.

Kirkby, M. J. (ed.) (1978) *Hillslope Hydrology*. Wiley, Chichester, UK.

Kneale, W. R. & White, R. E. (1984) The movement of water through cores of a dry (cracked) clay-loam grassland topsoil. *J. Hydrol.* **67**, 361–365.

Knoepp, J. D. & Clinton, B. D. (2009) Riparian zones in southern Appalachian headwater catchments: carbon and nitrogen responses to forest cutting. *Forest Ecology & Manage.* **258**, 2282–2293.

Lamontagne, S., Leaney, F. W. & Herczeg, A. L. (2006) Patterns in groundwater nitrogen concentration in the riparian zone of a large semi-arid river (R Murray, Australia). *River Research & Application* **22**, 39–54.

Lane, S. N., Brookes, C. J., Heathwaite, A. L. & Reaney, S. M. (2006) Surveillant Science: Challenges for the management of rural environments emerging fropm the new generation diffuse pollution models. *J. Agric. Economics* **57**, 239–257.

Leopold, L. B., Wolman, M. G. & Miller, J. P. (1964) *Fluvial Processes in Geomorphology*. Freeman, San Francisco, USA.

Lorion, C. M. & Kennedy, B. P. (2009a) Relationships between deforestation, riparian forest buffers and benthic macro-invertebrates in neotropical headwater streams. *Freshwater Biol.* **54**, 165–180.

Lorion, C. M. & Kennedy, B. P. (2009b) Riparian forest buffers mitigate the effects of deforestation on fish assemblages in tropical headwater streams. *Ecol. Appl.* **19**, 468–479.

Lowrance, R. R., Todd, R. L. & Asmussen, L. E. (1984) Nutrient cycling in an agricultural watershed: I. Phreatic movement. *J. Environ. Qual.* **13**(1), 22–27.

Malanson, G. P. (1993) *Riparian Landscapes*. Cambridge University Press, Cambridge, UK.

Matheson, F. E. (2001) Nitrogen removal and the fate of nitrate in riparian buffer zones. PhD Thesis, Durham University, UK.

Matheson, F. E., Nyugen, M. L., Cooper, A. B., Burt, T. P. & Bull, D. C. (2002) Fate of ^{15}N-nitrate in unplanted, planted and harvested riparian wetland soil microcosms. *Ecol. Engng* **19**, 249–264.

McClain, M. E., Boyer, E. W., Dent, C. L., Gergel, S. E., Grimm, N. B., Groffman, P. M., Hart, S. C., Harvey, J. W., Johnston, C. A., Mayorga, E., McDowell, W. H. & Pinay, G. (2003) Biogeochemical hot spots and hot moments at the interface of terrestrial and aquatic ecosystems. *Ecosystems* **6**(4), 301–312.

McDonnell, J. J. (1990) A rationale for old water discharge through macropores in a steep, humid catchment. *Water Resour. Res.* **26**, 2821–2832.

McGlynn, B. L., McDonnell, J. J., Shanley, J. B. & Kendall, C. (1999) Riparian zone flowpath dynamics during snowmelt in a small headwater catchment. *J. Hydrol.* **222**, 75–92.

McGuire, K. J., McDonnell, J. J., Weiler, M., Kendall, C., McGlynn, B. L., Welker, J. M. & Seibert, J. (2005) The role of topography on catchment-scale residence time. *Water Resour. Res.* **41**, W05002, doi:10.1029/2004WR003657.

Minshall, G. W., Petersen, R. C., Cummins, K. W., Bott, T. L., Sedell, J. R., Cushing, C. C. & Vanotte, R. L. (1983) Interbiome comparison of stream ecosystem dynamics. *Ecol. Monograph* **53**, 1–25.

Minshall, G. W., Cummins, K. W., Petersen, R. C., Cushing, C. E., Burns, D. A., Sedell, J. R. & Vannote, R. L. (1985). Development in stream ecosystem theory. *Can. J. Fisheries & Aquatic Sci.* **42**, 1045–1055.

Mulholland, P. J. (1992) Regulation of nutrient concentrations in a temperate forest stream: roles of upland, riparian, and instream processes. *Limnol. Oceanogr.* **37**(7), 1512–1526.

Newbold, J. D., Elwood, J. W, O'Neill, R. V. & van Winkle, W. (1981) Measuring nutrient spiralling in streams. *Canadian J. Fisheries & Aquatic Sci.* **38**, 860–863.

Naiman, R. J., & Décamps, H. (1997) The ecology of interfaces: riparian zones. *Annual Review of Ecology & Systematics* **28**, 621–658.

Naiman, R. J. & Décamps, H. (1990) *The Ecology and Management of Aquatic-terrestrial Ecotones.* UNESCO & Parthenon, Paris, France.

Nicholls, J. C. & Trimmer, M. (2009) Widespread occurrence of the anammox reaction in estuarine sediments. *Aquatic Microbial Ecology* **55**(2), 105–113.

Park, S. J. & Burt, T. P. (1999a) Identification of throughflow intensity using the distribution of secondary oxides in soils. *Geoderma* **93**, 61–84.

Park, S. J. & Burt, T. P. (1999b) The distribution of solute processes on an acid hillslope and the delivery of solutes to a stream: II. Exchangeable Al^{3+}. *Earth Surf. Processes Landf.* **24**, 851–865.

Peterjohn, W. T. & Correll, D. L. (1984) Nutrient dynamics in an agricultural watershed: observations on the role of a riparian forest. *Ecology* **65**(5), 1466–1475.

Patrick, W. H. & Tusnem, E. (1982) Nitrogen loss from flooded soils. *Ecology* **53**, 735–737.

Pettit, N. E. & Naiman, R. J. (2007) Fire in the riparian zone: characteristics and ecological consequences. *Ecosystems* **10**, 673–687.

Pinay, G., & Décamps, H. (1988) The role of riparian woods in regulating nitrogen fluxes between the alluvial aquifer and surface water: a conceptual model. *Regulated Rivers, Research & Manage.***2**, 507–516.

Pinay, G., Décamps, H. & Naiman, R. J. (1999) The spiralling concept and nitrogen cycling in large river floodplain soils. *Archiv für Hydrobiology*, Suppl. **11**(3), 281–291.

Pinay, G., Clément, J. C. & Naiman, R. J. (2002) Basic principles and ecological consequences of changing water regime for nitrogen cycling in fluvial systems. *Environ. Manage.* **30**(4), 481–491.

Pinay, G., Gumiero, B., Tabacchi, E., Gimenez, O., Planty-Tabacchi, A. M., Hefting, M. M., Burt, T. P., Black, V., Nilsson, C., Iordache, V., Bureau, F., Vought, L. M., Petts, G. E. & Décamps, H. (2007) Patterns of denitrification rates in European alluvial soils under various hydrological regimes. *Freshwater Biology*, **52**, 252–266.

Pinay, G., O'Keefe, T. C., Edwards, R. T. & Naiman, R. J. (2009) Nitrate removal in the hyporheic zone of a salmon river in Alaska. *River Research & Application* **25**, 367–375.

Ponnamperuma, F. N. (1972) The chemistry of submerged soils. *Advances in Agronomy* **24**, 29–96.

Reaney, S. M. (2008) The use of agent based modelling techniques in hydrology: determining the spatial and temporal origin of channel flow in semi-arid catchments. *Earth Surf. Processes Landf.* **33**, 317–327.

Reaney, S. M., Bracken, L. J. & Kirkby, M. J. (2007) Use of the Connectivity of Runoff Model (CRUM) to investigate the influence of storm characteristics on runoff generation and connectivity in semi-arid areas. *Hydrol. Processes* **21**, 894–906.

Reddy, K. R. & Patrick, W. H. (1975) Effect of alternate aerobic and anaerobic conditions on redox potential, organic matter decomposition and nitrogen loss in a flooded soil. *Soil Biology & Biochemistry* **7**, 87–94.

Ren, T., Roy, R. & Knowles R. (2000) Production and consumption of nitric oxide by three methanotrophic bacteria. *Appl. Environ. Microbiol.* **66**, 3891–3897.

Rosenberg, D. M., McCully, P. & Pringle, C. M. (2000) Global-scale environmental effects of hydrological alterations: introduction. *Bioscience* **50**(9), 746–751.

Sabater, F. & Vila, P. B. (1991) The hyporheic zone considered as an ecotone. In: *Homage to Ramon Margalef, or Why There is Such a Pleasure in Studying Nature*, ed. by J. D. Ros & N. Prat), 35–43. Univ. Barcelona, Barcelona, Spain.

Sabater, S., Bernal, S., Butturini, A., Nin, E. & Sabater, F. (2001) Wood and leaf debris input in a Mediterranean stream: the influence of riparian vegetation. *Archiv fur Hydrobiologie* **153**, 91–102.

Sebilo, M., Billen, G., Mayer, B., Grably, M., Billiou, D. & Mariotti, A. (2006) The use of nitrogen and oxygen natural isotopic composition to characterize and quantify nitrification and denitrification processes in the Seine River and estuary. *Ecosystems* **9**, 1–16.

Seitzinger, S. P., Styles, R. V., Boyer, E. W., Alexander, R. B., Billen, G. & Howarth, R. W. (2002) Nitrogen retention in rivers: model development and application to watersheds in the northeastern USA. *Biogeochemistry* **57**(1), 199–237.

Shakesby, R. A. & Doerr, S. H. (2006) Wildfire as a hydrological and geomorphical agent. *Earth Sci. Rev.* **74**, 269–307.

Silver, W. L., Herman, D. J. & Firestone, M. K. (2001) Dissimilatory nitrate reduction to ammonium in upland tropical forest soils. *Ecology* **82**(9), 2410–2416.

Smettem, K. R. J., Trudgill, S. T. & Pickles, A. M. (1983) Nitrate loss in soil drainage waters in relation to by-passing flow and discharge on an arable site. *J. Soil Sci.* **34**, 499–509.

Tansley, A. G. (1911) *Types of British Vegetation*. Cambridge University Press, Cambridge, UK.

Tiedje, J. M., Sextone, A. J., Myrold, D. D. & Robinson, J. A. (1982) Denitrification – Ecological niches, competition and survival. *Antonie van Leeuwenhoek J. Microbiology* **48**(6), 569–583.

Townsend-Small, A., McClain, M. E. & Brandes, J. A. (2005) Contributions of carbon and nitrogen from the Andes Mountains to the Amazon River: Evidence from an elevational gradient of soils, plants, and river material. *Limnol. & Oceanogr.* **50**(2), 672–685.

Triska, F. J., Sedell, J. R., Cromack, Jr K., Gregory, S. V. & McCorison, F. M. (1984) Nitrogenous budget for a small coniferous stream. *Ecol. Monographs* **54**, 119–140.

Triska, F. J., Kennedy, V. C., Avanzino, R. J., Zellweger, G. W. & Bencala, K. E. (1989) Retention and transport of nutrients in a third-order stream in Northwestern California: hyporheic processes. *Ecology* **70**, 1892–1905.

Trudgill, S. T., Pickles, A. M, Smettem, K. R. J. & Crabtree, R. W. (1982) Soil-water residence time and solute uptake. 1. Dye tracing and rainfall events. *J. Hydrol.* **60**, 257–279.

Valett, H. M., Morrice, J. A., Dahm, C. N. & Campana, M. E. (1996) Parent lithology, surface–groundwater exchange, and nitrate retention in headwater streams. *Limnol. & Oceanogr.* **41**, 333–345

Vannote, R. L., Minshall, G. W., Cummins, K. W. Sedell, J. R. & Cushing, C. E. (1980) The river continuum concept. *Can. J. Fisheries & Aquatic Sci.* **37**(1), 130–137.

Waddington, J. M., Roulet, N. T. & Hill, A. R. (1993) Runoff mechanisms in a forested groundwater discharge swamp. *J. Hydrol.* **147**, 37–60.

Walling, D. E. & Foster, I. D. L. (1975) Variations in the natural chemical concentration of river water during flood flows, and the lag effect: some further comments. *J. Hydrol.* **26**, 237–244.

Ward, J. V. (1989) The four-dimensional nature of lotic ecosystems. *J. North American Benthological Society* **8**, 2–8.

Warwick, J. & Hill, A. R. (1988) Nitrate depletion in the riparian zone of a small woodland stream. *Hydrobiologia* **157**, 231–240.

Wassen, M. J., Barendregt, A., Palczynski, A., De Smidt, J. T. & De Mars, H. (1990) The relationship between fen vegetation, gradients, groundwater flow and flooding in an undrained valley mire at Biebrza, Poland. *J. Ecol.* **78**, 1106–1122.

Webb, B. W. & Walling, D. E. (1985) Nitrate behaviour in streamflow from a grassland catchment in Devon, UK. *Water Res.* **8**, 1005–1016.

Webster, J. R. & Patten B. C. (1979) Effects of watershed perturbation on stream potassium and calcium dynamics. *Ecol. Monographs* **49**(1), 51–72.

Welsch, D. L., Kroll, C. N., McDonnell, J. J. & Burns, D. A. (2001) Topographic controls on the chemistry of subsurface stormflow. *Hydrol. Processes* **15**, 1925–1938.

Whitmire, S. L. & Hamilton, S. K. (2005) Rapid removal of nitrate and sulphate in freshwater wetland sediment. *J. Environ. Qual.* **34**, 2062–2071.

White, R. E., Wellings, S. R. & Bell, J. P. (1983) Seasonal variations in nitrate leaching in structured clay soils under mixed land use. *Agric. Water Manage.* **7**, 391–400.

A Landscape Ecology

Commentary on Paper A1

Cummins, K. W. (1974) Structure and function of stream ecosystems.

This paper set the scene for the American way of understanding river ecology as being functional, and opposed to the (mostly) European view based on the taxonomical identity of the organisms in a river. This controversy was finally settled on the American side by the publication and scientific debate following Vanotte *et al.* (1980; Paper A4), a paper which clearly drew on Cummins' paper and its related work. The concepts summarized in Cummins' paper are now being repeated in all textbooks of Limnology and River Ecology with only minor changes. Cummins (1974) is therefore a real benchmark in river ecology, being cited 631 times up to now – but rarely read by modern ecologists. The paper is also remarkable in how the river and the riparian compartments are described. The compartments are completely separated one from each other, where the riparian zone is mostly a food provider rather than a buffer. From today's perspective, this separation (aquatic-based *vs* terrestrial-based) has been – and still is – effective. It has resulted not only in the specialization of scientists, who usually deal with one of the aspects, but also conceptually and academically – so we ought not to ignore it!

The first part of the paper is an argument in favour of the functional perspective in river ecology, where species recognition as the basis for ecological insight (European view) could be problematic. Instead, using the functional grouping of organisms could facilitate process-oriented questions – the basis for the new stream ecology. The input (from the riparian zone) and processing of organic matter in the stream is linked to the existence of certain functional groups of organisms (shredders, collectors, grazers and predators), which form a trophic web based on the use of particulate organic matter (POM) and dissolved organic matter (DOM). In Cummins' view, the riparian zone is a food source, and leaf litter is the basis of the food web in the river. He recommends that catchment-oriented budgets should consider the active players – the functional arrangement of the biological community. This certainly changed the way ecologists understood within-stream processes.

Several predictions and intuitions stated in the paper have been completed 30 years later. Use of the feeding method to separate functional strategies and the arbitrary separation between the micro- and macro-consumers (the mesofauna and the macrofauna) by the particle size of the ingested food, have been used since (e.g. Anderson & Sedell, 1979). The foreseen role (at that time poorly known) of bacteria and fungi in organic matter decomposition is impressive. All these indications opened minds to a mechanistic understanding of stream ecosystems (e.g. Poff, 1997; Strayer *et al.*, 2003). Among these, the relationships between primary productivity and respiration (the P/R ratio) are already indicated in the paper. Other suggestions have proved not so general. Diatoms are seldom heterotrophs in mountain stream ecosystems. The significant presence of primary producers in streams is not always the result of pollution (nutrient-excess), but is also related to light occurrence. However, the real contribution (the when and how) of primary producers in streams took a long time to be recognized, although suggested by some at the time (e.g. Minshall, 1978).

The paper ends with a remarkable approach to management, by trying to connect scientific concepts with the needs of water quality. Cummins (1974) argued for the maintenance of relationships between organic matter fractions and linked functional groups in the stream, as a way to avoid the nuisance growths of plants. He saw that conserving the stream ecology is essential to maintain good water quality.

SS

References

Anderson, N. H. & Sedell, J. R. (1979) Detritus processing by macroinvertebrates in stream ecosystems. *Ann. Rev. Entomology* **24**, 351–377.

Minshall, G. W. (1978) Autotrophy in stream ecosystems. *Bioscience* **28**, 767–771.

Poff, N. L. (1997) Landscape filters and species traits: Towards mechanistic understanding and prediction in stream ecology. *J. North Am. Benthological Soc.* **16**, 391–409.

Strayer, D. L., Beighley, R. E., Thompson, L. C., Brooks, S., Nilsson, C., Pinay, G. N. & Naiman, R. J. (2003) Effects of land cover on stream ecosystems: Roles of empirical models and scaling issues. *Ecosystems* **6**, 407–423.

Vannote, R. L., Minshall, G. W., Cummins, K. W., Sedell, J. R. & Cushing, C. E. (1980) The river continuum concept. *Canadian J. Fisheries & Aquatic Sci.* **37**, 130–137.

Structure and Function of Stream Ecosystems

Kenneth W. Cummins

It has been stated often that the selection of appropriate methodological strategies for ecological investigations are highly dependent upon the question being addressed. Unfortunately this awareness has not become fully operational—questions often being formulated only after data collection. Frequently, traditional ecological studies have been taxonomic inventories of biological communities—information of limited use in answering certain function- and process-oriented questions. Such inventories seem an inappropriate fabric from which to weave important ecological generalizations. As long as the species is assumed to be the basic ecological unit, ignoring for the moment problems in defining such units, the perpetually incomplete state of our taxonomic knowledge will constitute a major constraint for the development of ecological theory.

This assumption, that species recognition is the fundamental prerequisite for ecological insight, has proven particularly troublesome to the study of flowing water (or lotic) ecological systems. For example, immature stages (nymphs, larvae) of stream insects usually dominate the macroinvertebrate role in energy transformations, but the adults (usually males) constitute the coin of species identification. In addition, diatoms, the dominant lotic algae, present particularly difficult taxonomic problems. However, perhaps only the lack of attention paid the microbial components—bacteria and fungi—of natural communities has allowed the taxonomic focus to endure this long. Discrepancies between morphological, physiological, and genetic species definitions of microorganisms make it apparent how untenable a taxonomic foundation for all ecological questions can be. Thus, there is a need to identify functional groups of organisms, at least partially independent of traditional taxonomic determinations, in order to address important process-oriented ecological questions. It is certainly true that we would not now be in a position to search for other units from which to construct our generalizations without the extensive taxonomic (phylogenetic) information presently available. Nor would it be disputed that continued effort in taxonomic discovery and revision will benefit ecology.

Although it was recognized long ago that functional roles were filled by different taxa in similar habitats separated geographically, the concept of a taxonomic foundation for all of ecology has persisted. For example, Shelford (1914, 1937) discussed such analogous species groups in aquatic systems and Lindeman (1942) and others later assumed biotic categorizations on bases other than taxonomic—particularly nutritional habits (trophic levels).

Unfortunately, investigations are often initiated with an uneven and incomplete taxonomic inventory, followed by assignment of organisms to trophic levels on the basis of scant literature generalizations and concluded with poor approximations of exchange rates. Rarely is this descriptive exercise accompanied by the synthetic insight that the trophic level conceptual framework was intended to facilitate. In many instances trophic categorization has proven as restrictive as taxonomic identification in answering process-oriented questions. This is particularly true since most schemes of trophic partitioning cannot deal adequately with the functional roles concerning community metabolism of particulate (detritus) and dissolved organic matter.

The goal of the discussion that follows is to identify some of the process-oriented questions confronting modern stream ecologists—both theoretical and practical, to examine possible methods of qualifying and quantifying functional groups involved in these processes, to speculate on the impact of process-function investigations on the continual evolution of stream ecosystem theory and, finally, to examine the implications of such an approach to stream management strategies. Some interrelationships between these elements in the development of theory and management programs are conceptualized in Figure 1.

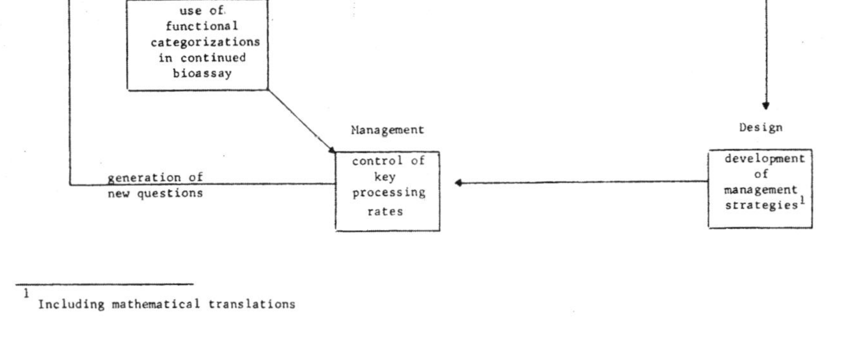

Fig. 1. A schematic representation of the interrelationships between process oriented questions, functional group evaluation, theory generation, and management strategies.

The author is with Kellogg Biological Station, Michigan State University, Hickory Corners, MI 49060.

STREAM ECOSYSTEM STRUCTURE AND FUNCTION

The present status of our knowledge of stream ecosystem structure and function is based on a number of generalizations that have been tested to some degree—primarily in woodland streams of the temperate zone (Teal 1957, Hynes 1961, Nelson and Scott 1962, Egglishaw 1964, 1968, Minshall 1967, Tilly 1968, Hynes 1970, Hall 1971, Cummins et al. 1972, Fisher and Likens 1973, Cummins[1], Petersen and Cummins[2], Boling et al.[3]).

Two intriguing features of such streams are: first, a dependence for the majority of their energy supply on the import of organic matter elaborated in the terrestrial system through which the stream flows (the watershed), and second, the utilization of a great deal of this organic input during the fall-winter period of lowest annual temperatures. That is to say, stream communities are heterotrophic (dependent upon food produced outside the stream) and temperature compensated (having organisms that can process organic matter at reasonable rates below "normal" temperature optima).

A portion of the reduced carbon compounds entering the stream as particles (particulate organic matter or POM) and in solution (dissolved organic matter or DOM), and similar materials produced in the stream itself, are processed to carbon dioxide and nutrients. Certain amounts are bound up for varying periods in resident biological components (standing crops) and the remainder is exported (Fig. 2). As Fisher and Likens (1973) showed for the tiny first order stream, Bear Brook (New Hampshire), 99% of the energy input is imported from the terrestrial surroundings (i.e., it is allochthonous) with only 1% derived from stream photosynthesis by mosses. The inputs were divided between dissolved (47% DOM) and particulate (53% POM) organic matter; 66% of the organic input was estimated to be exported downstream and 34% processed to CO_2.

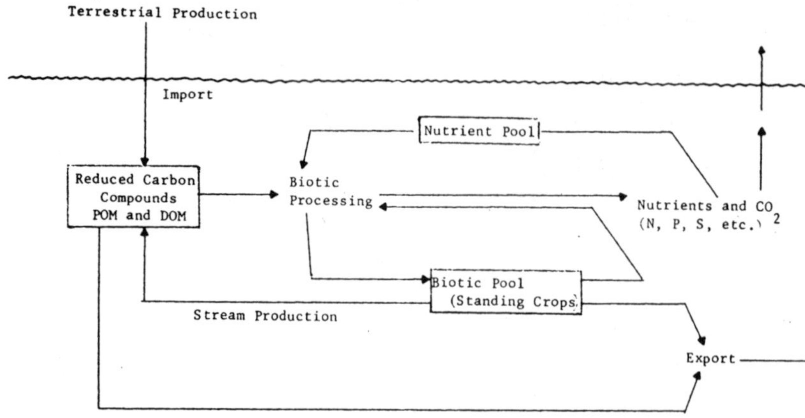

Fig. 2. Diagrammatic representation of the fate of reduced carbon compounds of terrestrial and stream origin: processing to nutrients and CO_2, temporary storage and export.

The organic matter budget produced by Fisher and Likens (1973) is, at present, the most complete annual balance sheet published for any stream system although Sedell et al. (1973) have produced an excellent particulate budget for a stream in the Oregon Cascades. The degree to which the New Hampshire and Oregon streams compare with regard to input, processing, and export of particulate detritus is striking. These investigations entailed periodic assessments of input, standing crop, and export of organic matter, with less attention given the actual processing events.

As suggested in Figure 3, there are a number of functional groups involved in the processing of organic matter, both POM and DOM, in heterotrophic streams. After the introduction of coarse particulate organic matter (CPOM, generally greater than 1 mm in diameter), such as leaves and needles, twigs, branches, bark, nuts, fruits, and flowers, into the stream, two processes occur rapidly.

Soluble organic matter leached from the CPOM enters the DOM pool. Although coarse particulate organic matter releases soluble components during the long residence period while it is being reduced to fine particulate organic matter (FPOM, generally smaller than 1 mm in diameter), the majority of the leaching occurs within 24 hours of initial wetting, even at low temperatures (Petersen and Cummins[2]). This solubilization can be quite appreciable; for example, leaf litter loses 5 to 30% of its dry weight in the first day, depending upon the leaf species in question (Cummins et al. 1972, Petersen and Cummins[2]). The amount of DOM produced also depends on the condition of the CPOM—that is, the degree of leaching completed in the terrestrial environment prior to introduction into the stream system.

The second rapid event is the colonization of CPOM surfaces by microorganisms in transport—cells and spores of bacteria, spores of aquatic hyphomycete fungi, and protozoans (Suberkropp and Klug 1974). A significant portion of the colonization is complete in the first week or two, again depending upon the degree of substrate preconditioning in the terrestrial soil community and the temperature regime. At warm summer stream temperatures (>15°C) the spores and hyphae of terrestrial fungi, present on the CPOM when it enters the water, play a more significant role in the fungal flora.

The leached, microbially colonized CPOM is reduced to FPOM through mechanical disruption wrought by physical abrasion in the turbulent lotic environment and two interdependent community processes—animal feeding and microbial metabolism. The rate of conversion to FPOM is dependent on temperature, extent of terrestrial preconditioning, and qualitative characteristics of the CPOM. For example, a considerable range in processing rates for different leaf species have been reported. Exponential loss rates of less than 0.5% per day are characteristic of "slow" leaf litter such as oaks and hemlock while ash and alder leaves are converted at greater than 1.5% per day (Sedell et al. 1973, Petersen and Cummins[2]).

Animals, identified in Figure 3 as shredders (coarse particle feeders), feed

---

[1] Cummins, K. W. Macroinvertebrates. *In* M. Owens and B. Whitton, eds. River Ecology. Blackwell Sci. Publ., England, in press.

[2] Petersen, R. C., and K. W. Cummins. Leaf processing in a woodland stream ecosystem. *Freshwater Biol.*, in press.

[3] *A.* Boling, R. H., R. C. Petersen, and K. W. Cummins. Ecosystem modeling for small woodland streams. *In* B. C. Patten, Ed. System Analysis and Simulation in Ecology, Vol. 3. Academic Press, N. Y., in press; *and B.* Boling, R. H., E. D. Goodman, J. O. Zimmer, K. W. Cummins, S. R. Reice, R. C. Petersen, and J. A. Van Sickle. Towards a model of detritus processing in a woodland stream. *Ecology*, in press.

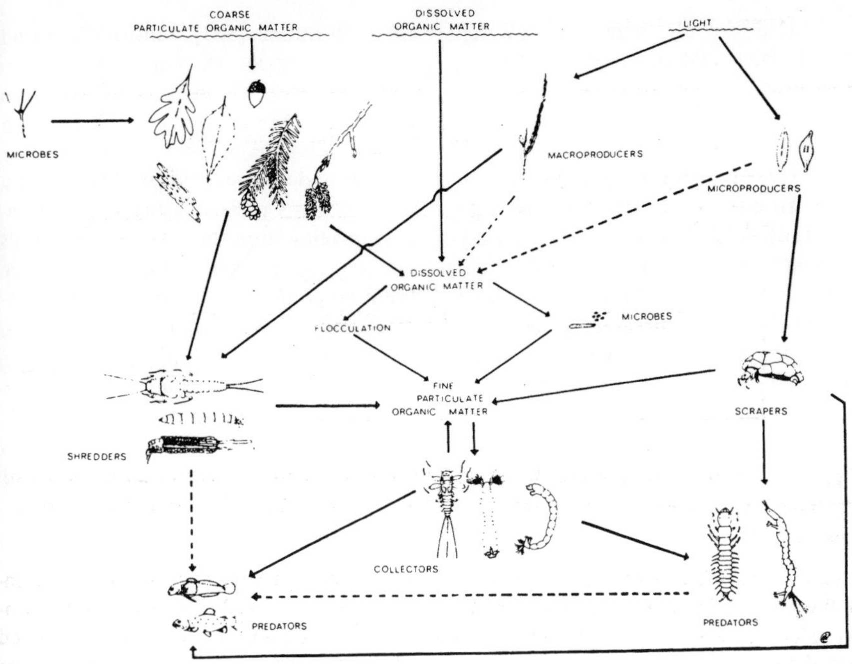

Fig. 3. A conceptual model of stream ecosystem structure and function (modified from Cummins 1973) emphasizing the processing of particulate and dissolved organic matter. CPOM is exemplified as deciduous leaves, coniferous needles, a twig, bark, nut, and flowers. Initial colonization of CPOM is represented by an aquatic hyphomycete fungal spore and the utilization of DOM by rod-shaped and spheroid bacteria. The plant community is represented by diatoms (microproducers) and moss (macroproducers). The animals shown as examples are: shredders—cranefly and caddisfly larvae and a stonefly nymph; collectors—blackfly and midge larvae and a mayfly nymph; scrapers—a caddisfly larva; predators—fishfly and midge larvae and fish (sculpin and trout).

Fig. 4. 1000x scanning electron microscope picture of the surface of a 1-week-old oak leaf incubated in Augusta Creek (Kalamazoo Co., Mich.) from 17 Aug. 73 to 20 Aug. 73. The long fungal spores on the right half of the photograph are the aquatic hyphomycete *Clavatospora tentacula;* the short ones at the upper left are *Lunulospora curvula.* The small ovoid and short rod-shaped bodies are bacteria; microcolonies are particularly visible at the upper left corner. The large diatom frustule shown at the lower left is a *Cymbella* species. (Photograph courtesy of K. F. Suberkropp, Kellogg Biological Station; see also Suberkropp, K. F., and M. J. Klug. Decomposition of deciduous leaf litter in a woodland stream. I. A scanning electron microscopic study. *Microbiol. Ecol.*, in press.)

upon the CPOM converting only about 40% of what they ingest to their own tissues and respiratory CO_2 (Welch 1968). The remainder is egested as feces. This process of reduction in CPOM particle size constitutes a significant contribution to the FPOM pool—approximately 2 to 7 mg of feces per large shredder individual per day (McDiffett 1970, Cummins et al. 1973). During colonization and growth the microbial community also attacks the CPOM which, after leaching, constitutes a high carbon (about 30% cellulose and lignin), low nitrogen substrate. Using the CPOM as a carbon source and dissolved inorganic and organic nitrogen the microflora metabolizes the substrate and grows (Fig. 4), also producing FPOM in the form of fragments and sloughed microbial cells. The microbial biomass associated with CPOM is reflected by increased nitrogen content of stream leaf litter through time (Kaushik and Hynes 1971).

Considerable evidence has accumulated indicating a nutritional dependence by the shredders on the microbial flora of the CPOM rather than the substrate itself (Kaushik 1969, Wallace et al. 1970, Triska 1970, Kostalos 1971, Liston 1972, Mackay and Kalff 1973, Bärlocher and Kendrick 1973a, 1973b). One might liken the microbial tissue to "peanut butter" which only occurs on nutritionally unsuitable "crackers," so that "cracker" ingestion is always a prerequisite to obtaining "peanut butter." Although the microflora can reduce leaf litter coarse particulate organic matter to FPOM in the absence of shredders (Triska 1970, K. F. Suberkropp, Kellogg Biological Station, pers. comm.), the presence of the large particle feeders results in a 20% increase in the conversion of CPOM to FPOM at normal fall-winter stream temperatures (Petersen and Cummins[4], Boling et al.[5]). Also, there can be little doubt that shredder feeding enhances CPOM substrates as microbial colonization sites—for example, in the process of skeletonizing leaves (Fig. 5) by increasing exposed edges and surface areas.

Thus, the FPOM pool is created through physical and biological (shredder and microbial) reduction in CPOM particle size. In addition, FPOM enters directly from the watershed; about 4% of the particulate organic matter input to Bear Brook was estimated to be

[4] See footnote 2, p. 632.
[5] See footnote 3A, p. 632.

Fig. 5. A photocopy of a sugar maple *(Acer saccharum)* leaf skeletonized by the shredder *(Tipula abdominalis* : Diptera : Tipulidae) feeding at 5°C.

FPOM by Fisher and Likens (1973). A significant amount of small particles enters the FPOM pool through activities associated with the DOM pool. Leachate from CPOM as well as DOM from the terrestrial system entering via surface runoff and subsurface groundwater (and aquatic plant and microbial excretions) are converted to FPOM by physical flocculation and microbial assimilation—the microbes themselves being, by definition, fine particles. Physical flocculation of DOM to FPOM can be quite significant and is dependent on such parameters as turbulence, temperature, pH, and various ion concentrations (Lush and Hynes 1973). Cummins et al. (1972) showed the rapid and dramatic effect that bacteria in transport can have on DOM in the form of leaf litter leachate.

Animals feeding on FPOM are referred to in Figure 3 as collectors because of the reaggregation of small particles resulting from their ingestion activities. Various food collecting mechanisms are employed either to filter FPOM from transport in the water or gather fine organic particles from the sediments (Cummins 1973). Collector feeding may increase or decrease particle size within the FPOM pool, but often the feces are similar in size to the ingested particles. The fate of the FPOM is recycling through collector populations, with the associated respiratory loss (25-30% of ingestion [Welch 1968]), microbial metabolism to CO_2, and export.

The carbon fixed in the stream, as indicated previously to be typically less than 1% of the total stream community energy supply, is shown in Figure 3 as photosynthesizing organisms (primary producers) such as algae (usually diatoms) and mosses. Animals specially adapted for removing firmly attached algae from exposed surfaces in running waters are designated scrapers. Available evidence suggests that mosses (and rooted aquatic macrophytes, if they occur) enter the community processing mechanisms as CPOM at times of dieback, being subject to very little animal feeding while alive (Cummins 1973).

The predator category (Fig. 3) functions as a major source of shredder, collector, scraper, and, also, predator mortality. In this view, all nonpredatory mortality is considered physiological death.

FUNCTIONAL GROUPS AND ORGANIC MATTER PROCESSING

Heterotrophy-Autotrophy and Import-Export Relationships

Determination of the rates and efficiencies at which organic matter is processed (converted to CO_2 and nutrients) in running water systems and the factors regulating such rates and efficiencies probably constitutes the primary goal of the present stream research effort. One basic distinction to be made in comparing stream communities is the ratio of heterotrophy to autotrophy; that is, a comparison between organic matter processing (community respiration: O_2 consumption and CO_2 production) and organic matter elaboration (photosynthesis: CO_2 consumption and O_2 production). Of course, the ratio will change with stream order—from first order headwater streams through higher order major rivers, as well as over diurnal and seasonal cycles. The use of a ratio of gross photosynthesis to respiration, placing running water communities on a scale to determine if they are predominantly in a processing (consumption) or producing mode (Fig. 6), was suggested by Odum (1956). Fisher and Likens (1973) expanded the system to include import to and export from the stream system, combining the two ratios in a three dimensional coordinate system.

It appears that organic matter import-export phenomena in streams are subject to considerable interpretation with regard to inclusion in community processing budgets. Obviously, during a spate, organic matter measured in a stream is primarily in an export mode. Quantities transported, particularly CPOM, will depend upon the degree of flooding, or percent bankfull discharge (Leopold et al. 1964). However, the material transported during these short periods, which may constitute the major portion of particulates exported over an entire annual cycle, is often dominated by organic matter from the bank that has been subjected to little or no stream processing and is simply moved through at a rate less than the biological response times.

For example, if a dry leaf that falls on the stream surface and floats on top through a section of stream in a matter of minutes is to be included in an organic matter budget for the section, considerable qualification is necessary with regard to definitions of such terms as annual budget, processing, and efficiency. Methods such as direct micro-

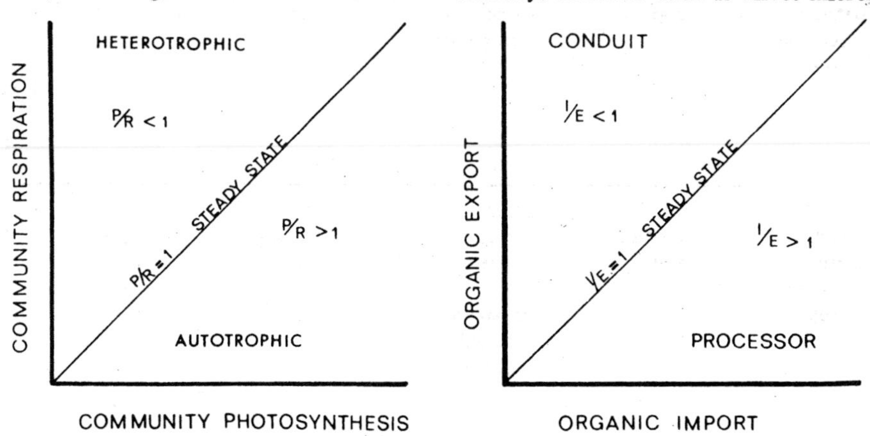

Fig. 6. The relationship between photosynthesis (autotrophy) and respiration (heterotrophy) and import (processor) and export (conduit) of organic matter in stream ecosystems (for a single combination figure see Fisher and Likens 1973).

scopical examination for aquatic hyphomycete fungal spores (Triska 1970), measurements of community respiration per unit organic weight of detritus (Boling et al.[6]), or estimations of microbial biomass using bioluminescent-ATP methods (Chappelle and Levin 1968) could be employed as indices of the percent of such transport material actually subjected to in-stream processing—particularly if such measurements were compared to reference standards incubated in the stream for known periods.

A combination of techniques can be effectively utilized to evaluate the relative importance of heterotrophic vs. autotrophic processes as well as the organic matter storage capacity in streams. Periodic assessment of the average detritus standing crop in a section of stream, primary production (in-stream photosynthesis), and community respiration constitute lotic, process-oriented measurements. When stream community function is to be the focal point of investigations, perhaps such process data should assume priority over watershed-oriented budgets in which the major effort is directed at estimating inputs (import) and outputs (export). In stream process studies, changes in detritus standing crop not accounted for by predicted increases due to primary production, or decreases due to respiration, can be attributed to import or export. The relationships are summarized in Table 1 and some representative data for a small Michigan woodland stream are given in Figure 7. The difference in autotrophy at an open vs. a shaded section of the stream is apparent.

Undoubtedly the ratio of heterotrophy to autotrophy is controlled by light, temperature, organic and inorganic inputs, and flow, with lesser localized effects by invertebrate grazers (scrapers). Shifts from heterotrophy to autotrophy in streams usually involve conversion from the typical diatom-moss (e.g., *Fontinalis*) community and, in some streams, also watercress (*Nasturtium*) beds around the springs, to filamentous green algae (e.g., *Cladophora, Stigeoclonium, Ulothrix,* etc.) and/or beds of rooted aquatic plants. The correlation between maximum light income to streams in the autumn just after leaf-fall and in the spring just prior to leaf-out, and photosynthetic maxima was shown by Hall (1971).

At the Kellogg Biological Station we have demonstrated in paired artificial stream channels that temperature elevation (5°C) can produce a heavy growth of filamentous green algae in one channel when the general fall-winter stream community, light, nutrients, flow, and grazers are the same in both systems. R. L. Vannote and coworkers (pers. comm.) at the Stroud Water Research Center have shown in their superb experimental stream facility that flow-discharge regimes exert important controls on the dominance of primary production in running waters. It appears that spates prevent the establishment of a filamentous algal system, and the attendant reduction in animal diversity, over the annual cycle. Obviously, the "switch" that initiates the trend from heterotrophy to autotrophy is a complex set of interactions and not merely a light controlled change.

Primary Producers

The organisms that contain chlorophyll and are capable of photosynthetic carbon fixation constitute a separable, functional group in stream communities since they collectively represent the internal energy supply for the system. However, within such a category there are two functionally distinguishable components—algae, or microproducers, and vascular plants, or macroproducers (Fig. 3). As stated above, stream vascular hydrophytes (mosses and flowering plants) enter the community processing machinery almost exclusively at the time of dieback, following the same general CPOM pathway as terrestrial plant tissue inputs, probably at faster rates due to their aquatic beginning and more fragile nature. The typical stream microproducers (nonfilamentous algae)

[6] See footnote 3A and B, p. 632.

TABLE 1. Summary of relationships between detritus standing crop, photosynthesis, respiration, and organic matter import and export over a specified time interval (t).

Evaluation system	Measurements			Increase (+) or decrease (-) in g/m² over interval t (t_1 to t_2)	Change in detritus standing crop not accounted for by evaluation system	Conclusions about system not evaluated
	Initial instantaneous (t_1)	Rate estimated for	Final instantaneous (t_2)			
P/R ratio and detritus standing crop	Detritus standing crop	Photosynthetic production Community respiration	Detritus standing crop	+, - or 0	No change	Import = Export (or both = 0)
				+	Positive	Import > Export
				-	Negative	Export > Import
I/E ratio and detritus standing crop	Detritus standing crop	Import of organic matter Export of organic matter	Detritus standing crop	+, - or 0	No change	Photosynthesis = Respiration
				+	Positive	Photosynthesis > Respiration
				-	Negative	Respiration > Photosynthesis

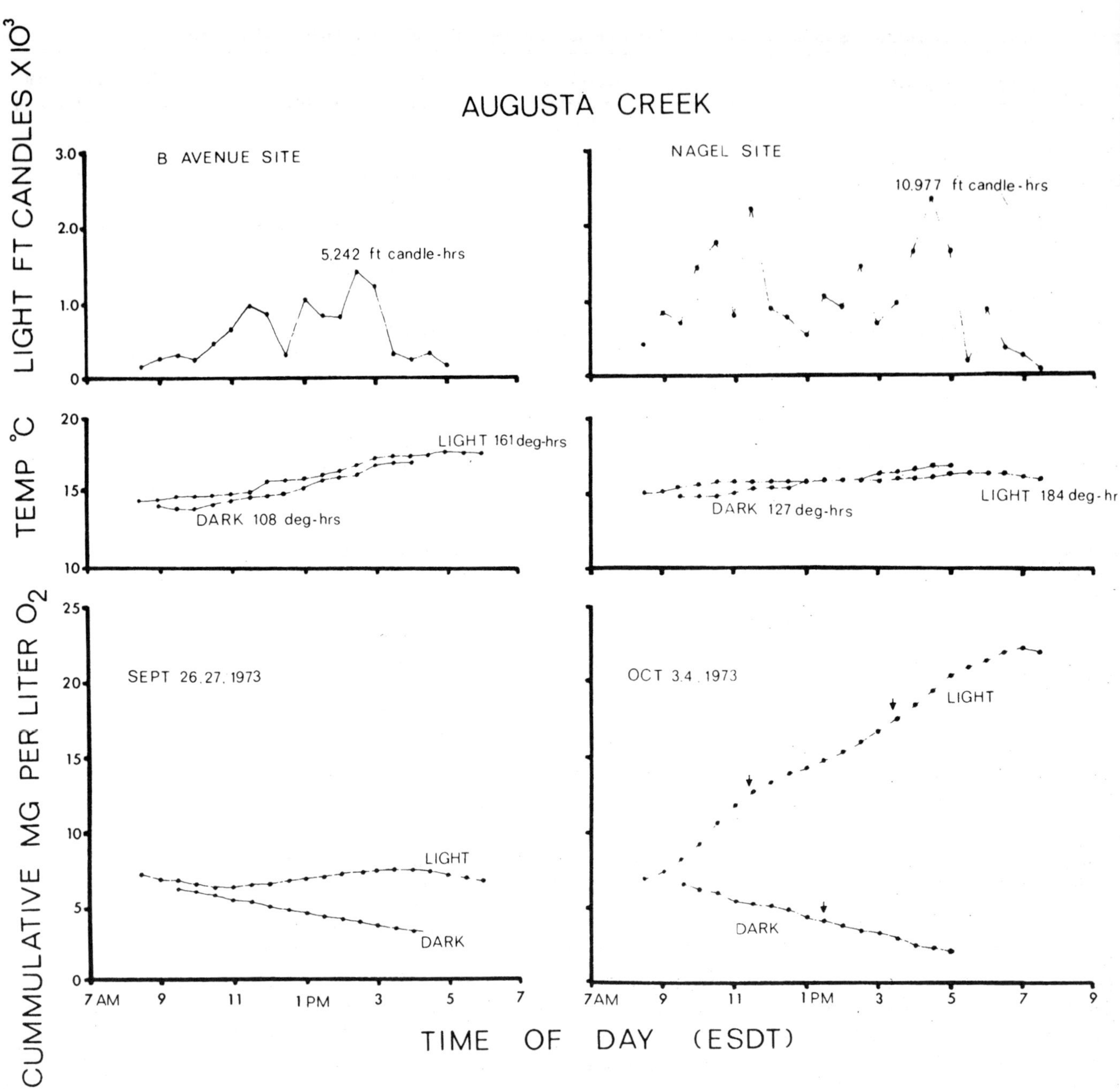

Fig. 7. Comparative data on light, temperature, and oxygen production (light chamber) and oxygen utilization (light and dark chambers) at two sites on Augusta Creek, a small, hardwater brown trout stream in southern Michigan (Kalamazoo and Barry Counties; Manny, B. A., and R. G. Wetzel. Diurnal changes in dissolved organic and inorganic carbon and nitrogen in a hardwater stream. *Freshwat. Biol.*, in press). The B Avenue site is a first order, headwater, heavily shaded section (P/R = 0.5); the Nagel site is a third order, more open stream section about midway along the drainage (P/R = 1.7). Measurements of oxygen concentration were made in closed chambers (holding about 3 liters stream water) using oxygen probes (Yellow Springs Instruments) equipped with a vibrator to maintain water circulation. The chamber bottoms (3.186×10^{-2} m^2) were filled with natural stream sediments ($\bar{x}$ of 1.5 liters) and allowed to incubate in situ for 3 months. The chambers contained 114.1 g (B Ave. Site) and 117.3 g (Nagel Site) organic matter (550°C ash free dry wt). (Data from Donna K. King, Kellogg Biological Station. Autotrophic-heterotrophic relationships in a woodland stream, Ph.D. thesis in preparation.)

are either grazed by scrapers or enter the FPOM pool.

Methods are available for assessing the contribution of primary producers to stream ecosystem metabolism. Changes in dissolved oxygen levels over 24-hour cycles at two points along a stream (upstream-downstream method, e.g., Hall 1971) or in light and dark closed, circulating (Fig. 7) or open flow-through chambers (Hansmann et al. 1971) are frequently measured. Tracer (^{14}C, ^{32}P) techniques are also available either for use in closed systems or for direct applications of small doses to streams (rapidly disintegrating ^{32}P, [Elwood and Nelson[7]]). Standing crop measurements of rooted vascular plants just prior to winter dieback probably provide the best estimate of their contribution to stream organic matter budgets.

Primary production should be measured against a background of light,

[7]Elwood, J. W., and D. J. Nelson. Measurement of periphyton production and grazing rates in streams using a ^{32}P material balance method. *Oikos*, in press.

TABLE 2. A scheme for partitioning stream detritus according to particle size ranges. (Modified from Boling et al. 1974a.)

Detritus Categories: General	Detritus Categories: Specific	Acronym	Approximate size ranges (mm)	Dominant Constituents	Dominant animal detrital feeding group (see Fig. 7)
Coarse particulate organic matter (CPOM)	Large resistant particulate organic matter	RPOM	> 64	Logs, branches, large twigs large sections of bark (processing times > 1 year)	Detritus shredders
	Whole leaf organic matter	LVOM	> 16 < 64	Leaf litter (leaf packs)	Detritus shredders
	Leaf fragment organic matter	LFOM	> 4 < 16	Large leaf, twig, and bark fragments, fruits and nuts, large seeds, buds and flowers, conifer needles	Collector - macrogatherers
	Large particulate organic matter	LPOM	> 1 < 4	Small fragments of plant (and animal) parts	Collector – Macrogatherers Collector – macrofilterers
Fine particulate organic matter (FPOM)	Medium particulate organic matter	MPOM	>0.25 < 1	Plant and animal fragments, feces of large invertebrates	Collector - macrofilterers Collector - microgatherers
	Small particulate organic matter	SPOM	>0.075 <0.25	Plant, animal, and fecal fragments of large invertebrates and feces of small invertebrates	Collector - microfilterers Collector - microgatherers
	Very small particulate organic matter	VPOM	>0.0005 <0.075	Very small detrital fragments and free microorganisms	Collector - microfilterers
Dissolved organic matter		DOM	<0.0005	Organic matter in solution – leachate from plant and animal detritus, microbial and producer excretions	Little or none

temperature, and nutrient (especially P and N) data to permit intersystem comparisons. Cumulative measures of light and temperature, e.g., foot- or meter-candle-hours or days (1m-cand. = 1 lumen/m^2) and degree days, should prove more useful than calendar time intervals. If lower light or thermal limits can be demonstrated, time intervals below these levels should be excluded from the cumulations.

Given that the prototype, first to third order stream is a cool, heavily shaded, heterotrophic system, in-stream photosynthesis normally contributes very little to the total annual energy flux—for example, less than 1% in the New Hampshire stream studied by Fisher and Likens (1973). Larger running waters (fourth order and above) should exhibit a trend toward autotrophy, although nonperturbated prototypes of these larger systems are virtually nonexistent in the temperate zone because of massive man-engendered alterations in watershed vegetation cover, nutrient inputs, and flow (by impoundment).

Microconsumers

Since the nonliving organic matter substrates and the metabolizing microorganisms are never isolated from one another in streams, consideration of these components as functionally separate compartments seems merely academic. At present, compartmentalization of organic substrate and associated microbes by particle size, which is clearly related to processing time, seems to be the most tractable approach.

Inasmuch as a seasonal sequence of reduction in the particle size of organic matter can be demonstrated (Boling et al.[8]), changes in the biochemical nature of particulate organic matter (POM) and in the associated microbial functions associated with particulate degradation would be expected—expectations supported by particle size-dependent respiration measurements (Boling et al.[9] Hargrave 1972). For this reason, it seems fruitful at present to treat detritus particle size classes as functional units, even though organic matter obviously forms a continuum from logs through fine particles, large molecular aggregates and complexes to small molecules. A scheme for which some data have been collected (Boling et al.[8]) is summarized in Table 2.

The DOM size range is primarily in transport while POM size fractions, especially CPOM, are usually concentrated in the sediments. There is evidence that the transport and sediment systems can operate semi-independently (Cummins et al. 1972). Successional changes in biochemistry (Kaushik and Hynes 1968, 1971, Hynes and Kaushik 1969) and microbial flora (Triska 1970,

[8] See footnote 3A and B, p. 632.

[9] See footnote 3A, p. 632.

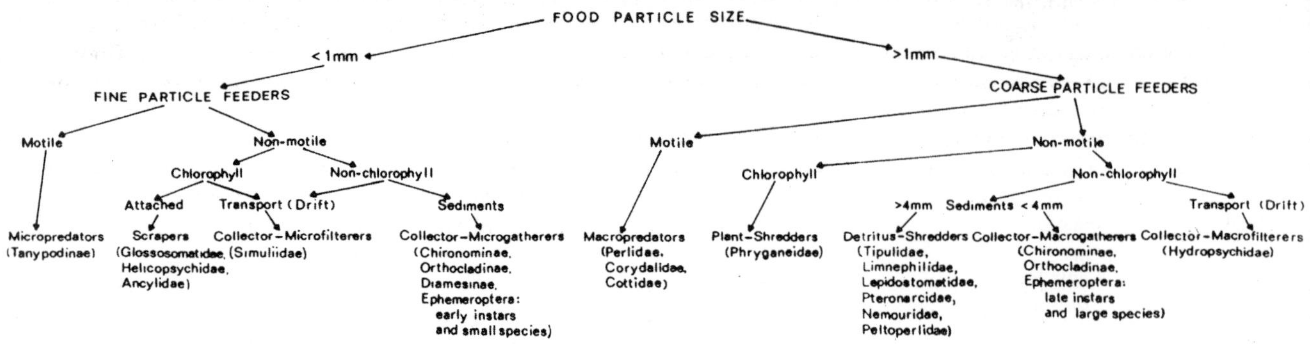

Fig. 8. Functional categorization of stream animals according to particle size and character of food with some selected examples of taxa that, in general, would have representatives included in a given functional category. Trueflies (Diptera)–Midges, Tanypodinae, Chironominae, Orthocladiinae, Diamesinae; blackflies–Simuliidae; craneflies–Tupulidae. Caddisflies (Trichoptera)–Glossosomatidae, Helicopsychidae, Phryganeidae, Limnephilidae, Lepidostomatidae, Hydropsychidae. Mayflies–Ephemeroptera. Stoneflies (Plecoptera)–Perlidae, Pteronarcidae, Nemouridae, Peltoperlidae. Dobson–and fishflies (Megaloptera)–Corydalidae. Limpets (Mollusca, Gastropoda)–Ancylidae. Fish–Cottidae (sculpins).

K. F. Suberkropp and M. J. Klug, Kellogg Biological Station, pers. comm.), termed "conditioning," have been shown for the CPOM category. These initially involve minimal particle size conversion. Similar successional sequences (conditioning) have yet to be detailed for other detritus size fractions.

Although it is true that remarkable amounts of biological activity proceed in streams at low temperatures (late fall through early spring), the processing rates are still under general thermal control. For example, leaf litter processing, taken broadly to include conversion to smaller particles and CO_2, progresses at a rate approximately 20% faster at 10°C than 5°C, as determined in experimental channels populated by the same micro- and macrobiota (Cummins and coworkers, Kellogg Biological Station, unpubl. data). The evolution of much of the stream biota has been largely in response to the major inputs of POM in the autumn, primarily in the form of deciduous leaf litter (Hynes 1963, Ross 1963). The resultant capability to metabolize effectively at temperatures well below normal thermal optima is unparalleled in temperate terrestrial systems.

As new techniques are developed and more data gathered, microbial and biochemical differences which transcend particle size categories will certainly become apparent. In addition to direct observation of microorganisms (especially epifluorescence and scanning electron microscopy, Fig. 4), the isolation and identification of biochemical capabilities of microbial elements will provide critical additional resolution of functional roles. For example, identification and quantification (probably through microcosm and tracer uptake studies) of cellulolytic and/or lignolytic activity of detrital bacteria and fungi in natural streams under *normal* stream temperature regimes should constitute a prime objective for lotic researchers. Fractionation of DOM by molecular weight should permit recognition of microbes associated with certain fractions and concomitant processing rates (Bretthauer 1971, M. J. Klug, Kellogg Biological Station, pers. comm., T. L. Bott and R. A. Larson, Stroud Water Research Center, pers. comm.). Again, if stream studies are process-oriented, new microbial data will be most useful if related to rates at which conversions of coarse particulate to fine particulate to dissolved organics to CO_2 occur.

The presence of microconsumers other than bacteria and fungi in stream detritus is often noted but seldom quantified: for example, protozoans, diatoms, rotifers, microcrustaceans, nematodes, and water mites (the last three overlap in size with early life stages of organisms usually considered macroconsumers [Cummins[10]]). In most instances, only the protozoans and diatoms might constitute significant biomass and metabolic activity relative to the bacteria and fungi. Protozoans serve primarily as particle aggregators functioning both as collectors and, like other microorganisms, as FPOM. Diatoms function as microproducers, although densities on the surfaces exposed to light are usually low, and probably as facultative microconsumers of DOM since they are often found deep in CPOM detrital accumulations in a viable condition. At present, there is little choice but to include protozoans and diatoms along with fungi and bacteria as the processing microbiota of a given detritus fraction.

Macroconsumers

Stream macroconsumers are generally considered to be those animals, dominated by invertebrates, that attain a size of at least 3 to 5 mm at termination of growth (Cummins[10]). The general classification of macroconsumers, primarily according to food particle size shown in Figure 3, has been expanded in Figure 8. As stated previously, the primary role of nonpredator macroconsumers is viewed as particle size conversion—hence the recognition of functional groups on the basis of particle size fractions eaten and method of feeding. Predators are seen as both particle converters, the "particles" being the prey, and as controllers of nonpredator populations. The fortuitous ingestion of animals by shredders, collectors, and scrapers during detrital or algal feeding and detritus and algal intake by predators, primarily as prey gut contents, exemplify inconsistencies resulting when attempts are made to define functional roles beyond the simplistic level of food particle size only (Coffman et al. 1971, Cummins 1973).

Although such inconsistencies may represent little error in estimating the processing rates of organic matter in streams, they suggest that other qualifiers of a particle size classification should be sought. M. J. Klug and coworkers of the Kellogg Biological Station Stream Research Group have been investigating other critera for evaluating macroconsumer functional roles in organic matter processing. A combination

[10] See footnote 1, p. 632.

of data on microbial gut flora, related macroconsumer gut morphology, and food gathering apparatus, together with general food type, constitute promising criteria. Radiotracers can also be used for the rapid, approximate determination of the trophic role of stream animals (e.g. Ball and Hooper 1963, Sedell 1971, Cummins[11]).

Temperature is viewed as controlling the general dimensions of macroconsumer life cycles through influence on metabolism–respiration, feeding, and growth rates. These relationships can be conveniently summarized in terms of cumulative temperature (degree-days) over the growth period (Table 3). Within the overall limits set by temperature, growth rate, and, to some extent, survivorship (excluding predator-engendered losses) are controlled by food quality and quantity. As an example, the effect of in-stream conditioning time (i.e., microbial colonization and growth) and leaf species on feeding by a shredder is shown in Figure 9. There is about a 1 week time lag (at 15°C) before the "slow" leaf species (aspen) is fed upon at the same rate as the "fast" species (ash), but 2 weeks are required before maximum feeding rate is obtained on both types of leaves.

[11] See footnote 1 p. 632.

Management Strategies

The fundamental problem in stream management is clearly "water quality," in the broad sense meaning system quality. Regardless of definition, here lies the challenge–interfacing, in compatible fashion, the self-perpetuating structure and function of running water ecosystems with selfish, "natureless" human goals. Water quality is, in fact, always defined in reference to these goals. For example: Will the system support a particular sport fishery? Will it be a habitat where noxious and pathogenic organisms will flourish? Will it decompose organic wastes or serve merely as an export conduit?

From the data at hand, two points seem clear. First, the maintenance of water quality necessitates the continuance of certain relationships between CPOM, FPOM, and DOM together with the involvement of critical functional ecological groups of both micro- and macroorganisms. Second, unless about one third of the total organic matter input (about one half of the POM) is processed, i.e., converted to CO_2, annually by the stream system and unless in-stream plant growth remains subservient to terrestrial organic matter as the "fuel" to drive the system, the stream in question probably has impaired water quality.

In general, the differences between relatively undisturbed woodland

TABLE 3. Functional categorization of stream animals according to life cycle characteristics with some examples of trophic (by food particle size) and taxonomic groups that would be in each category. For placement of taxonomic groups see Fig. 7.

Generation time (egg to egg)	Designation	Growth period	Approximate degree days	Fine particle feeders (FPF)	Coarse particle feeders (CPF)
Multiple generations per year	Autumnal (Fall-Winter) and Vernal (Spring-Summer)	Sept–March April–August	1300 2500	Baetidae Heptageniidae Leptophlebeidae Glossosomatidae Helicopsychidae Hydropsychidae Chironominae Orthocladiinae Simuliidae	Hydropsychidae Lepidostomatidae
Full year (no "shut-down")	Annual	(Sept–August or April–March	3800	Gastropoda Pelecypoda (Amphipoda)	Perlidae* Tanypodinae* (Amphipoda)
Full Year (with "shut-down")	Annual–Autumnal (Summer "shut-down")	Sept–March	1300	Ephemerellidae Neophylacinae Orthocladiinae	Tipulidae Limnephilidae, e.g. *Pycnopsyche* Lepidostomatidae Nemouridae
	Annual–Vernal (Winter"shut-down")	April–August	2500	Ephemerellidae Brachycentridae Chironominae Orthocladiinae	Limnephilidae, e.g. *Hydatophylax*
More than one year (usually more than 2) (no "shut-down")	Biannual	Autumn–Autumn or Spring–Spring	3800/yr		(Cottidae*)
More than one year (usually 2) (with "shut-down")	Biannual–Autumnal (Summer "shut-down")	Fall–Winter seasons	2600		Pteronarcidae (Peltoperlidae)
	Biannual–Vernal (Winter "shut-downs")	Spring–Summer seasons	5000		Decapoda Corydalidae* Corduligasteridae* Gomphidae*

* Predators

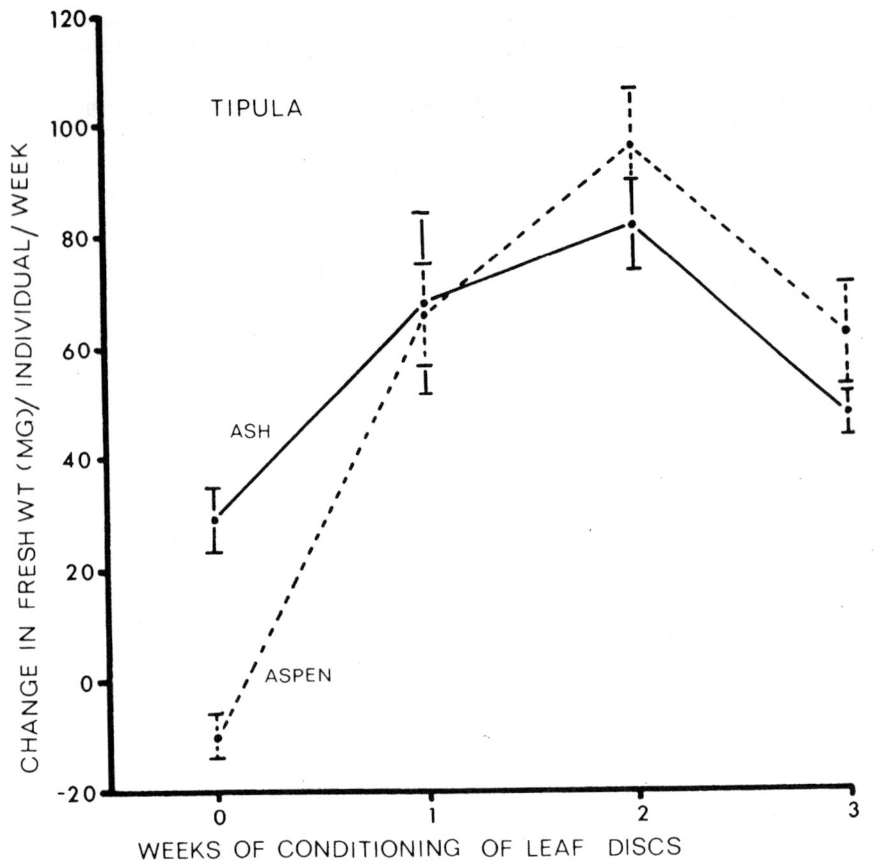

Fig. 9. Effect of in-stream conditioning time and leaf species on feeding by the cranefly larva shredder *Tipula abdominalis* (Diptera : Tipulidae) at 15°C. Each point represents the mean (± 1 SD) of five larvae held individually in tubes (screened at both ends through which water circulated) containing discs of either ash or aspen leaves conditioned 0, 1, 2, or 3 weeks in Augusta Creek. The decline in growth on the 3-week conditioned leaf discs was the result of food limitation. (Data from G. L. Spengler, Kellogg Biological Station.)

streams, characterized by high processing efficiency, and "organically enriched" or "polluted" running water systems of similar dimensions are the size distribution of the organic particles that enter the stream, timing of the inputs, POM retention characteristics of the system, temperature and nutrient regimes, and the presence of key functional groups of organisms. Where appropriate options exist, management strategies should be developed and implemented based on available stream ecosystem theory (Fig. 1). "Water quality" status should be monitored through recognition of the continued appropriate relationships between CPOM, FPOM, DOM, and micro- and macroorganisms.

It is not presently known whether the efficiency with which organic matter is processed in streams can be increased above reported levels (Fisher and Likens 1973, Sedell et al. 1973). Since so few systems have been studied in a fashion permitting comparison, the range of natural efficiencies has yet to be established—clearly, comparison of streams at opposite ends of such a spectrum would be most instructive.

Three general management strategies, singly or in various combinations, seem promising:

(1) changes in the physical nature of the running water system—light (e.g. artificial shading), temperature, aeration, POM retention characteristics, etc.;

(2) changes in organic inputs, particularly particle size distribution; and

(3) changes in the biota, for example shredder population densities.

If, as Fisher and Likens (1973) report for Bear Brook, the general input pattern is for at least half the organic matter to enter as POM, dominated by CPOM, then waste water treatment procedures that produce effluents high in DOM with particulates dominated by FPOM are unlikely to contribute positively to the maintenance of stream "water quality" defined in terms of "natural" organic matter processing rates. The use of retention structures (e.g., rods driven into the stream bottom) to reduce movement and export of CPOM, especially leaf litter, in and through a section of stream should increase total processing, particularly if shredder populations are supplemented by stocking.

Potentially, model development constitutes an important aspect of management strategies, although none, including biological processing, have been utilized in relation to streams. Initial models are currently available which simulate primary production (McIntire 1973) and particulate detritus processing (Boling et al.[12]).

CONCLUSIONS

In summary, the basic features of stream ecosystem structure and function are now established and various functional ecological components and their interrelationships have been defined and initially dimensioned for some representative streams. The search continues for additional or alternative functional criteria to replace partially or totally the previous dependence on classical taxonomic units. Attention is now focused on two apparently generalizable conditions in nonperturbated running waters: The efficient conversion of organic matter, especially particulates, to CO_2 and the maintenance of a minor role played by in-steam plant growth. Clearly the time is at hand to infuse the "new stream ecology" into management strategies directed at our precious running waters.

ACKNOWLEDGMENTS

Contribution No. 256 of the Kellogg Biological Station. The original research and synthesis were supported by National Science Foundation Grant GB-36069X and U.S. Atomic Energy Commission Contract AT(11-1)2002. This paper was possible only through the interchange of ideas with a great many stream ecologists, particularly: Robert C. Petersen, Michael J. Klug, and Keller F. Suberkropp, Kellogg Biological Station; Robin L. Vannote and Thomas L. Bott, Stroud Water Research Center; and James B. Sedell, James D. Hall, Oregon State University, and R. H. Boling, Michigan State University. I am also

[12]See footnote 3A and B, p. 632.

grateful for data and helpful suggestions provided by Robert H. King, Donna K. King, G. Milton Ward, George L. Spengler, and Virginia I. Holt of the Kellogg Biological Station.

REFERENCES

Ball, R. C., and F. F. Hooper. 1963. Translocation of phosphorous in a trout stream ecosystem. Pages 217-228 *in* V. Schultz and A. W. Klement, Jr., ed. Radioecology. Reinhold, New York. 746 pp.

Bärlocher, F., and B. Kendrick. 1973a. Fungi and food preferences of *Gammarus pseudolimnaeus*. *Arch. Hydrobiol.* 72: 501-516.

———. 1973b. Fungi in the diet of *Gammarus pseudolimnaeus* (Amphipoda). *Oikos* 24: 295-300.

Bretthauer, R. 1971. Quantitative estimation of low-molecular ninhydrinpositive matter in waters rich in autumn shed leaves. *Int. Rev. Ges. Hydrobiol.* 56: 123-128.

Chappelle, E. W., and G. V. Levin. 1968. Use of the firefly bioluminescent reaction for rapid detection and counting of bacteria. *Biochem. Med.* 2: 41-52.

Coffman, W. P., K. W. Cummins, and J. C. Wuycheck. 1971. Energy flow in a woodland stream ecosystem: I. Tissue support trophic structure of the autumnal community. *Arch. Hydrobiol.* 68: 232-276.

Cummins, K. W. 1973. Trophic relations of aquatic insects. *Ann. Rev. Entomol.* 18: 183-206.

Cummins, K. W., M. J. Klug, R. G. Wetzel, R. C. Petersen, K. F. Suberkropp, B. A. Manny, J. C. Wuycheck, and F. O. Howard. 1972. Organic enrichment with leaf leachate in experimental lotic ecosystems. *BioScience* 22: 719-722.

Cummins, K. W., R. C. Petersen, F. O. Howard, J. C. Wuycheck, and V. I. Holt. 1973. The utilization of leaf litter by stream detritivores. *Ecology* 54: 336-345.

Egglishaw, H. J. 1964. The distributional relationship between the bottom fauna and plant detritus in streams. *J. Anim. Ecol.* 33: 463-476.

———. 1968. The quantitative relationship between bottom fauna and plant detritus in streams of different calcium concentrations. *J. Appl. Ecol.* 5: 731-740.

Fisher, S. G., and G. E. Likens. 1973. Energy flow in Bear Brook, New Hampshire: an integrative approach to stream ecosystem metabolism. *Ecol. Monogr.* 43: 421-439.

Hall, C.A.S. 1971. Migration and metabolism in a stream ecosystem. *Rept. Univ. North Carolina Inst. Water Res.* 49: 1-243.

Hansmann, E. W., C. B. Land, and J. D. Hall. 1971. A direct method of measuring benthic primary production in streams. *Limnol. Oceanogr.* 16: 822-826.

Hargrave, B. T. 1972. Aerobic decomposition of sediment and detritus as a function of particle surface area and organic content. *Limnol. Oceanogr.* 17: 583-596.

Hynes, H.B.N. 1961. The invertebrate fauna of a Welsh mountain stream. *Arch. Hydrobiol.* 57: 344-388.

———. 1963. Imported organic matter and secondary productivity in streams. *Proc. XVI Int. Congr. Zool.* 3: 324-329.

———. 1970. The ecology of running waters. Univ. Toronto Press, 555 pp.

Hynes, H.B.N., and N. K. Kaushik. 1969. The relationship between dissolved nutrient salts and protein production in submerged autumnal leaves. *Verh. Int. Ver. Limnol.* 17: 95-103.

Kaushik, N.K. 1969. Autumn shed leaves in relation to stream ecology. Ph.D. Thesis, Univ. Waterloo, Ontario, Canada. 219 pp.

Kaushik, N.K., and H.B.N. Hynes. 1968. Experimental study on the role of autumn shed leaves in aquatic environments. *J. Ecol.* 52: 229-243.

———. 1971. The fate of dead leaves that fall into streams. *Arch. Hydrobiol.* 68: 465-515.

Kostalos, M. W. 1971. A study of the detritus pathway: The role of detritus and associated microbiota in the nutrition of *Gammarus minus* (Say) (Amphipoda : Gammaridae). Ph.D. Thesis, Univ. Pittsburgh, Pittsburgh, Pa. 152 pp.

Leopold, L. B., M. G. Gordon, and J. P. Miller. 1964. Fluvial processes in geomorphology. W. H. Freeman and Co., San Francisco. 522 pp.

Lindeman, R. L. 1942. The trophic-dynamic aspects of ecology. *Ecology* 23: 339-418.

Liston, C. R. 1972. Contributions of allochthonous detritus to the energy regime of Doe Run, Meade County, Kentucky. Water Resources Laboratory, University of Louisville, Louisville, Ky. 104 pp.

Lush, D. L., and H.B.N. Hynes. 1973. The formation of particles in freshwater leachate of dead leaves. *Limnol. Oceanogr.* 18: 968-977.

Mackay, R. J., and J. Kalff. 1973. Ecology of two related species of caddisfly larvae in the organic substrates of a woodland stream. *Ecology* 54: 499-511.

McDiffet, W. F. 1970. The transformation of energy by the stream detritivore, *Pteronarcys scotti*. *Ecology* 49: 69-75.

McIntire, D. C. 1973. Periphyton dynamics in laboratory streams: a simulation model and its implications. *Ecol. Monogr.* 43: 399-420.

Minshall, G. W. 1967. Role of allochthonous detritus in the trophic structure of a woodland springbrook community. *Ecology* 48: 139-149.

Nelson, D. J., and D. C. Scott. 1962. Role of detritus in the productivity of rock-out crop community in a piedmont stream. *Limnol. Oceanogr.* 7: 396-413.

Odum, H. T. 1956. Primary production in flowing waters. *Limnol. Oceanogr.* 1: 102-117.

Ross, H. H. 1963. Stream communities and terrestrial biomes. *Arch. Hydrobiol.* 59: 235-242.

Sedell, J. R. The trophic ecology and natural history of *Neophylax concinnus* and *N. oligius* (Trichoptera: Limnephilidae). Ph.D. Thesis, Univ. Pittsburgh, Pittsburgh, Pa. 154 pp.

Sedell, J. R., F. J. Triska, J. D. Hall, N. H. Anderson, and J. H. Lyford. 1973. Sources and fates of organic inputs in coniferous forest streams. Contribution 66, Coniferous Forest Biome, International Biological Program, Oregon State University. 23 pp. (From a Symposium: Synthesis of Coniferous Forest Biome Research, IBP Analysis of Ecosystems, AIBS Meeting, 21 June 1973, Amherst, Mass.)

Shelford, V. E. 1914. An experimental study of the behavior agreement among animals of an animal community. *Biol. Bull.* 26: 294-315.

———. 1937. Animal communities in temperate America. Univ. Chi. Press.

Teal, J. M. 1957. Community metabolism in a temperate cold spring. *Ecol. Monogr.* 27: 283-302.

Tilly, L. J. 1968. The structure and dynamics of cone spring. *Ecol. Monogr.* 38: 169-197.

Triska, F. J. 1970. Seasonal distribution of aquatic hyphomycetes in relation to the disappearance of leaf litter from a woodland stream. Ph.D. Thesis, Univ. Pittsburgh, Pittsburgh, Pa. 189 pp.

Wallace, J. B., W. R. Woodall, and F. F. Sherberger. 1970. Breakdown of leaves by feeding of *Peltoperla maria* nymphs (Plecoptera: Peltoperlidae). *Ann. Amer. Ent. Soc.* 63: 563-567.

Welch, H. E. 1968. Relationships between assimilation efficiencies and growth efficiencies for aquatic consumers. *Ecology* 49: 755-759.

Commentary on Paper A2

Hynes, H. B. N. (1975) The stream and its valley.

The immediate relevance of this paper needs to be placed in the context of the forum and audience of its first presentation. The paper was given by Noel Hynes at the Societas Internationalis Limnologiae (the SIL) during its 1975 meeting at Stuttgart, Germany. Presenting an Edgardo Baldi Memorial Lecture was one of the highest honours that this Society could give to a prominent scientist. The SIL at that time was mostly made up of "pure" limnologists, dealing with the classical approach to the study of lakes and rivers, mostly based on the approach initiated by Naumann and Thienemann. Therefore the paper came as something of a shock: rivers were not to be seen as aquatic ecosystems isolated from their terrestrial environment. Instead, they were considered to be significantly influenced by the catchment area they drained. The paradigm of the paper is clearly advanced by its beautiful title, which focused on the exact meaning of the relationship between the stream and its landscape.

Originally published as part of the proceedings of the SIL, the *Verhandlungen der Internationalen Vereinigung für theoretische und angewandte Limnologie*, this paper is highly cited in spite of the publication policy of the SIL, that is, one publication appearing every four years. A later paper by Hynes in *Hydrobiologia* (Hynes, 1983) placed emphasis on the relevance of groundwater in the ecology of streams.

The 1975 paper was revolutionary in proposing that functional descriptors be used instead of classification factors (e.g. fish zonation), so common at that time. The paper stresses that streams are heterotrophic and derive their energy from terrestrial organic matter. In this sense the paper connected with the early work of Likens and co-workers in the Hubbard Brook experimental watershed (e.g. Likens *et al.*, 1970). Likens himself summarized his findings later on, also in the SIL (Likens, 1984).

Some of the ideas advanced in the paper were at their initial stage, and were later explored and confirmed by stream ecologists. And given that this is mostly the way that stream ecology is approached nowadays, it is obvious that its influence was profound. Some of the hints provided in the paper are outlined here because of their continuing relevance:

- The upslope soil layer, the vegetation and the water flow path show remarkable links regarding nutrients and dissolved organic matter dynamics.
- There occur complex processes between the soil, water and vegetation that are expressed in the ion biogeochemistry, in particular of phosphorus and nitrogen.
- The organic matter quality and quantity is influenced by the biome and even the catchment type, and this affects in turn the consumers relying on this organic matter.

The paper can therefore be seen as a precursor to the paradigm that the surface/groundwater interaction is a biological ecotone, and the place where significant biogeochemical processes occur. In the conclusions of the paper, a warning emerges: the potential changes occurring in the valley will be reflected in the stream. In the original words, "*We may conclude that in every respect the valley rules the stream*".

SS

References

Hynes, H. B. N. (1983) Groundwater and stream ecology. *Hydrobiologia* **100**, 93–99.

Likens, G. E. (1984) Beyond the shoreline: a watershed ecosystem approach. *Verh. Internat. Verein. Limnol.* **22**, 1–22.

Likens, G. E., Bormann, F. H., Johnson, N. M., Fisher, D. W. & Pierce, R. S. (1970) Effects of forest cutting and herbicide treatment on nutrient budgets in the Hubbard Brook watershed-ecosystem. *Ecol. Monographs* **40**, 23–47.

Verh. Internat. Verein. Limnol.	19	1—15	Stuttgart, Oktober 1975

Edgardo Baldi Memorial Lecture

The stream and its valley

H. B. N. Hynes

With 4 figures and 2 tables in the text

These lectures were set up to honour the memory of Dr. Edgardo Baldi, a man who regarded limnology as a science in its own right, one that deals with interrelationships and processes, and with transformations of energy and matter. In that approach he was somewhat ahead of many of us, as the ecosystem concept of ecology was fairly new at the time of his death. The lectures were also intended, as was noted by the late Dr. Umberto D'Ancona who delivered the first of them, to discuss the most vital problems of hydrobiology in the light of their most recent developments. When, therefore, I was invited to address you here, I realized that I ought to attempt to present recent advances in running water ecology in the light of Baldi's concepts of limnology.

Fortunately now is an apposite time to do this, because, within the last few years, there has emerged a growing band of workers who are endeavouring to look at watercourses in a holistic way, rather than as conglomerations of processes, phenomena and taxa; or even, in the current jargon, as victims of "socio-economic problems". Our lacustrine colleagues have gone much further along this way, because in some respects their task was easier. It was possible at first for them to work with the idea of "the lake as a microcosm", and to produce at least a few theories and ideas before they realized that a lake is more than just a basinful of water. In running water there was no possibility of such a head start with a simple theory, so the early steps were more difficult, and we have really only begun to surge forward in this age of interdisciplinary contacts. Indeed, one could say that some of our most important recent discoveries have been of the existence of hydrologists, foresters and soil scientists; which perhaps says something of our innocence.

So my aim during this lecture will be to look at streams not as purely aquatic phenomena, as one can with lakes, but rather to view them as parts of the valleys that they drain. I shall probably fail in my endeavour as my eyes are always drawn to the water, and I like to paddle anyway. I also probably have nothing to say that will be new to my fellow workers in running water, from whom indeed come many of the ideas I shall put forward to-day, especially from my close associates Drs. N. Kaushik, M. Lock and D. Lush. However, perhaps the genuine limnologists among you, by whom I mean the lake folk, with their ancient Swiss lineage, will learn that we are just beginning to catch them up despite the rougher trail that we are following.

Early attempts to understand how streams differ from one another, and hence by implication how they function ecologically, concentrated on local phenomena, such as temperature and flow regime, water chemistry and substratum, to each of which the biota was shown to be clearly related. This led to classifications of watercourses for various purposes, but especially for fisheries. The most comprehensive attempt at such categorization was published by this Society, but it has been accepted in only very general terms. Recently Pennak (1971) has returned to the topic, and suggests that streams should be described by 13 parameters in each of which he has 4 to 6 categories. In this way one could define over 180 million different types of stream, which is really little more useful than listing them by name. It would seem that, just as we have taken so long to appreciate about lakes, classification is not really the road to ecological understanding. Dr. RIGLER has outlined the history of this process at this congress. One does better with processes, particularly those concerned with energetics. But they make a much less tidy classification, and as many of us are biologists by origin we expect the world to be classifiable. For most of us it requires almost a change of creed to admit that it is not — but surely God is no more a taxonomist than he is a mathematician, which is another ecological illusion.

Energy relationships in the water

One recent appreciation — we must not call it a discovery as it has been known since the early days of this Society, and one can find mention of it in the work of one of our founding fathers, Professor AUGUST THIENEMANN — is that streams are basically heterotrophic. They derive most, often nearly all, of their energy from uphill. This topic has been reviewed several times; a few pertinent recent references are EGGLISHAW (1968), FITTKAU (1970), KAUSHIK & HYNES (1971), FISHER & LIKENS (1973), and several papers given at this congress deal with this topic. A recently worked example is shown in Fig. 1, taken from HALL (1972).

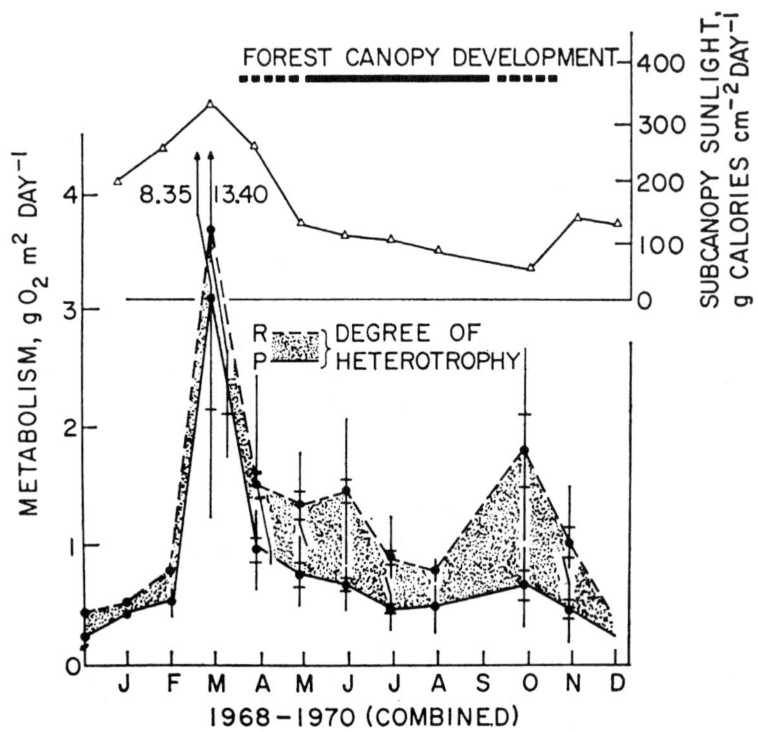

Fig. 1. The oxygen production (P), community respiration (R) and the sunlight recorded month by month at a station on New Hope Creek, North Carolina. The points are mean values, vertical bars are the range of values and horizontal bars are one standard error from the mean. Modified from HALL (1972).

Allochthonous material then undergoes a series of processes in the water that make it available as food for the benthic invertebrates that are so important a feature of all running waters. The significance of heterotrophic processes of this kind has long been appreciated by the soil biologists, and more recently heterotrophy has figured prominently in work on estuaries (ODUM & DE LA CRUZ 1963; DARNELL 1967). It has even become the subject of an article in the popular scientific press (FLOODGATE 1973).

In estuaries, and also apparently in the sea (MANN 1973), much material produced by autochthonous photosynthesis enters the heterotrophic cycle directly, without being eaten in its live state. The same seems to apply to many macrophytes, such as *Rorippa, Potamogeton* and *Ranunculus,* which grow in streams and seem to be rather rarely attacked by herbivores (e. g. KOSLUCHER & MINSHALL 1973). Much then, of the biotic energy that drives stream communities begins in the form of dead material, some from within but, usually, most from outside the stream. In the water it is leached, and it also decays.

Decay of the dead plant material seems to be at least initiated by fungi (KAUSHIK & HYNES 1971), most probably because of their ability to penetrate the tissues and to attack cellulose and lignin. These may be geofungi or the aquatic hyphomycetes that we came to learn about through the work of INGOLD (1966 and earlier papers). It seems from a recent study at Waterloo (BÄRLOCHER & KENDRICK 1973) that the latter are the more important in cold water, and temperature seems to affect their abundance directly (CONWAY 1970). They also seem to be influenced by periods of high flow and to be controlled by the origin of the water, as DUDKA (1971) found their spores to be less abundant in streams arising from swamps on the Kola Peninsula than they were in montane streams.

There is also now good evidence that decaying litter, or more probably the organisms causing the decay, take up a great deal of ionic nitrogen from the water, and that they then convert it to protein. In this way the largely carbohydrate energy of the dead plant material is converted to more suitable food for animals, and the C/N ratio falls. There is now quite a body of literature on this general topic (ODUM & DE LA CRUZ 1963; DARNELL 1967; EHRLICH & SLACK 1969; MATHEWS & KOWALCZEWSKI 1969; KAUSHIK & HYNES 1971; IVERSEN 1973), but there are still many points about the process that need to be resolved. For instance, we still know little about the relative importance of fungi and bacteria, nor if there is a change in their status as decay proceeds. Perhaps a combination of the direct observational approaches of such workers as RODINA (1967), FENCHEL (1970) and MADSEN (1972) with experimental studies such as

Tab. 1. Summary of some comparative data from fast-flowing streams in Scotland. The cotton duck was left in the water for 17 weeks. From EGGLISHAW (1972).

Stream	Nitrate N, mg/l	Loss of tensile strength of cotton, %	Biomass of benthic fauna, g dry wt/m^2	Mean length of 0+ trout at end of season, mm
Buchanty	0.41	>95	1.03	67—76
Fincastle	0.40	77	1.06	—
Shelligan	0.26	95	1.17	59—64
Fender	0.20	71	1.59	69—71
Almond	0.15	20	0.54	52—55
Truim	0.11	15	0.41	48—51

those of KAUSHIK & HYNES (1971) will resolve these questions. But more important from the point of view of the ecology of streams is the finding that rates of decay, which are a crude measure of microbial activity, vary with the species from which the litter arose, the availability of ionic nitrogen, and probably that of phosphate also (NELSON et al. 1969). Other chemical properties of the water, amongst which the level of calcium, or something that varies with it, may also be important (EGGLISHAW 1968; HOFSTEN & EDBERG 1972). All of these characters are derived from the drainage area, and they are readily changed by alteration of the valley. How far-reaching are the implications of such factors in stream ecology is shown by Tab. 1, taken from EGGLISHAW (1972), who measured the loss of tensile strength of strips of cotton duck (in effect the rate of decay of cellulose) placed into several Scottish streams. Possibly, indeed probably, more than ionic nitrogen is involved, but the general message is clear — that the fertility of the valley rules that of the stream in many ways.

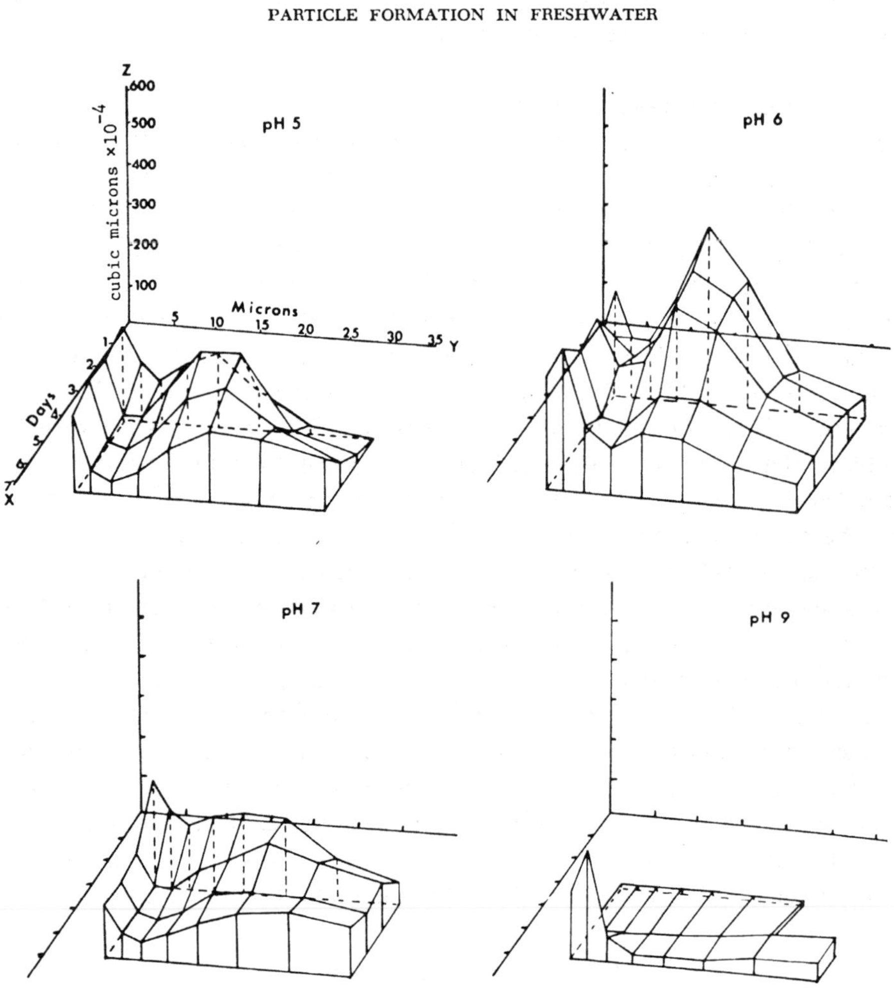

Fig. 2. The formation of particles in leachates of sugar maple leaves, and their changes in sizes and quantity during several days at different acidities. From LUSH & HYNES (1974).

The leaching of dead organic matter has also been subjected to much fairly recent study. It seems that in streams, as in the ocean (RILEY 1970), dissolved organic matter aggregates into small particles which form sites for microbial activity. Work in our laboratory has confirmed that leaf leachates in aerobic fresh water readily form particles under both sterile and nonsterile conditions, that turbulence enhances the process, and that it is much affected by pH (LUSH & HYNES 1974). Fig. 2 shows, as an example of this, the behaviour of leachates of sugar maple leaves at different levels of acidity. The leachate also acts as a buffer, in that its presence tends to bring the reaction of the water towards neutrality, and we have reason to believe that dissolved salts, and perhaps also nutrients, are important in the process of particle formation. This does not occur in distilled water, and addition of nutrients to coloured bog and fenn water is known to remove its colour presumably by precipitation and oxidation. Also WILLIAMS (1968) found that the C/N ratio of organic particles in the Amazon, which carries 20 % of the world's river runoff, was much lower than that of the dissolved organic matter; which perhaps indicates the uptake of nitrogen from the water.

The amount that goes into solution from plant litter may be as high as 25—30 % of the dry weight of leaves from deciduous trees, although it is less for conifers; and extracts from each species tend to behave differently. It has been shown by the Kellogg Laboratory group (CUMMINS et al. 1972; WETZEL & MANNY 1972), that most of a leaf leachate disappeared quite rapidly from the water in an experimental, recirculating, laboratory stream, and that both organic carbon and organic nitrogen were removed. They also noted an increase in fine particles, and in bacteria in suspension, which is in accord with our expectations from our own work. Many bacteria that occur in natural waters are floc formers which entangle small particles, and MADSEN (1972) has suggested that the microbiota that grow on stone surfaces in streams take up dissolved organic matter. Work in progress in our laboratory is showing this to be so. This is in effect, the same phenomenon as the trickling filter at the sewage works, and may indicate the original home of that useful, but empirical, engineering process.

Thus we can picture the dissolved organic matter in a stream as resulting in the first place from leaching, varying in its behaviour according to its specific origin and the chemical characteristics of the water, and perhaps achieving a more or less stable equilibrium between solution and precipitation or adsorption, as suggested by MANNY & WETZEL (1973) for a stream in Michigan. In that study, however, they showed that such features as a small lake and a swamp in the course of the stream caused changes in the water as it passed through them. Moreover, most of the dissolved organic matter goes into solution in or on the soil, which is where most of the litter falls rather than into the stream itself.

The source of the water

Prior to the last few years stream ecologists have been little concerned about the origins of the water. We distinguished between spring-fed and other streams, and between those that respond rapidly to rainfall and those that do not. But we tended to think of basal-flow water as groundwater, and spates as being composed of runoff water. Recently, however, the hydrologists have been producing a whole series of new concepts and these have great implications for stream ecology.

For instance, studies by HEWLETT and his coworkers in the Coweeta Forest, North Carolina, (HEWLETT & NUTTER 1970) show that, in natural watersheds, overland flow is a rare phenomenon. Natural landscapes very rarely produce what hydrologists call "parking-lot hydrographs", and usually considerably less than half the precipitation of heavy storms goes off during the spate that follows. The soil of the valley in fact absorbs much of the water. It then goes on yielding it to the stream for a long period, and the water comes from unsaturated as well as from saturated soil. When rain falls onto a slope of which the soil is already partly drained, it therefore increases the flow from the lower part by displacement. So the yielding portion of the watershed shrinks and expands with drought and rain, and channels and wet spots in the valley get filled very largely from below, with only the rain that falls actually onto them getting into the watercourse without going through the soil. So, in effect, the channel reaches out to tap areas of low storage capacity, but they are mainly filled by water that has been long in the soil and has been displaced by recent rain or meltwater. This general "variable source area" concept has received considerable support among hydrologists, although some maintain that storms also add a fair amount of newly received water from slopes near the channel. Recent papers on this subject are those of DUNNE & BLACK (1970), who note that even when overland flow does occur it is often of old water forced up by displacement, CARSON & SUTTON (1971), who discuss recent ideas on stream flow, and FREEZE (1972), who stresses that the form of the valley slope and the permeability of the soil control the proportion of flood water that is new.

Clearly the hydrologists still have more work to do; but the implications for stream ecology of that which they have already done are, of course very great, because they indicate that most of the water that enters a stream has been for some time in contact with the soil, and that this applies even to at least a large proportion of floodwater. It is also clear that the slope of the valley, the depth and permeability of the soil, and the patterns of precipitation greatly affect the pattern of flow in the stream, which is a prime ecological factor; and more subtly, they effect the dissolved content of the water. As CURRY (1972) stresses, the activity of the water on the slopes, before it enters defined channels, is very important, and the dissolved and suspended load of a river is in large measure a function of the geologic, biologic, and meteorologic conditions within the entire watershed.

Another way in which conditions in a valley can affect stream discharge is, of course, through transpiration. This has been well documented by studies of experimental watersheds before and after felling (e. g. LIKENS et al. 1970). It is not, however, I believe, generally appreciated, at least among stream ecologists that vegetation can control discharge even during the non-growing season, because it dries out the soil during the summer, or dry season, and so makes storage space for winter, or wet-season, rain (HEWLETT & HELVEY 1970).

The source of the organic matter

In forested watersheds the annual amount of litterfall from the trees is well known to be several tons per hectare. Only some of this falls directly into the water, but more blows in from the land. For example, in forest near Hubbard Brook, New Hampshire, 49.1 % of the organic matter that falls from the trees is leaves, but the leafy proportion of the litter that enters the streams is raised to 60 % by blow-in (Gosz et al. 1972; FISHER & LIKENS 1973). Possibly, as this is fairly dense woodland, the amount that blows in is lower than normal. This material also falls seasonally from the trees; for example Gosz et al. (1972) recorded the distribution of litterfall as 49.7 % in autumn, 21.2 % in summer, 22.4 % in winter, and 6.7 % in spring. This is because forest debris is composed of branches, stems, bark, flowers, bud scales and insect frass, as well as leaves. So in the forests of the temperate zone there is a varying but more or less continual supply of coarse particulate matter, and in the tropics even the Amazonian forest has maxima and minima in the rates of litterfall (KLINGE & RODRIGUES 1968). Thus the type of vegetation in the valley very directly affects the supply of material to the stream, and in small streams where long dry seasons and heavy litterfall coincide, as in parts of North America, the litter may cause deoxygenation and be a major factor for living things in the water (SLACK & FELTZ 1968). In arid areas, as in the evergreen oak forest of Arizona, the dry spring-shed leaves may lie almost unleached, until summer rains sweep them into the water (MCCONNELL 1968).

Fine material results from the break-up of litter actually in the water, but some is obviously produced by similar processes on the land. So when rainfall increases the area of the water surface, as described earlier, one would expect some of this matter to enter the stream. In fact, increases in the quantity of fine organic particles in the water at times of flooding have been observed (EGGLISHAW & SHACKLEY 1971), as have increases in bacteria, but one cannot be certain whether this extra material results from wash-in or from a mere increase in turbulence. However, HOBBIE & LIKENS (1973) noticed that the amounts rose after rain in Hubbard Brook tributaries even when there was little increase in flow, so one can assume that at least a considerable part of the increase is caused by the influx of new particles.

Perhaps though some of the extra particles which are found in the water arise from the increased dissolved organic matter that is brought in when the water rises. We have seen that rain water normally remains for some while in contact with the soil. There it clearly picks up leachates from the soil litter, thereby acquiring some dissolved organic matter from almost every dead leaf in the valley. It also acquires organic matter as it drips through the forest canopy, and once in the soil, evaporation and transpiration concentrate it. The result is that water reaching the stream is heavily loaded with dissolved organic matter. FISHER & LIKENS (1973), in their excellent study of a tributary of Hubbard Brook, record mean values of 2.5 to 3.2 mg/l, and state that about 47 % of the total energy input into Bear Brook is in the form of dissolved organic matter. Similar values for concentrations of dissolved organic matter are reported from English chalk streams (WESTLAKE et al. 1970) and somewhat higher ones from a stream in Michigan (MANNY & WETZEL 1973).

Now we know that different species of leaf give rise to different amounts of leachate, so the vegetation influences the total amounts that enter the stream. We can also expect that such compounds as amino-acids and sugars in leachates will be readily attacked in the soil and that tannins and lignins will probably be more persistent. There are also complex interactions between organic matter in soil and various cations, and with clays (SCHNITZER 1971; STEVENSON 1972). So the material that remains in that part of the soil water which is free to move into the stream depends to a great extent upon the nature of the soil, and it may be more or less labile according to the conditions in the valley.

The origin of the ions

It is obvious that the major minerals in the rocks of a valley dominate the inorganic chemistry of the water, but as JOHNSON et al. (1969) have so elegantly demonstrated, the rates at which various ions are delivered to the stream are to a large extent controlled by the terrestrial vegetation. This is illustrated in Tab. 2, taken from their work, which shows that certain elements, e. g. Ca, Mg, K and N, tend to be held up by the forest. Indeed K and N concentrations in the stream water were generally minimal in summer, presumably because of their use by vegetation, and the same has been found for N in English rivers (OWENS et al. 1972), and in North America (FETH 1966).

Tab. 2. The movement of certain nutrient elements through the ecosystem in the watershed at Hubbard Brook, New Hampshire. From JOHNSON et al. (1969).

Element	A Mean transport by stream kg/ha/yr	B Circulated through forest biota kg/ha/yr	% A/B × 100
Na	6.1	1.6	380
S	12.4	4.0	310
Cl	3.6	1.6	220
Al	1.9	1.6	120
Ca	10.6	32.0	33
Mg	2.5	16.0	16
K	1.5	32.0	5
N	1.9	44.0	4

It is clear also that the mere presence of vegetation in the valley gives rise to soil with all its complicated reactions between ions and humic and fulvic acids that were mentioned earlier. These must affect trace elements in the water as they do in soil, but this seems to be still quite unstudied in relation to streams. We may also note, as being very probably significant, that Ca/Na ratios in river water tend to be lower in warmer climates (CURRY 1972), and that groundwater from limestones almost always contains some nitrogen whereas that from granitic rocks often does not (FETH 1966). Perhaps somewhere here lies a key to understanding why calcareous rivers and streams tend to be the most productive.

A further effect of vegetation, already seen in Tab. 2, is that it cycles the biologically important elements from soil to plant to soil. The data of GOSZ et al. (1972), which are those used in the table, show that the forest litter moved 140.4 kg/ha/yr of such elements. This is, of course, to be expected, but of great interest to us is that the elemental composition of the litter of different species varies a great deal, even when the trees are growing in the same area (DAUBENMIRE 1953), and that the rate of loss of these elements from litter is faster in water than it is on the soil (THOMAS 1970). There seem also to be large regional differences. For instance, the orders of the rates of loss of elements from *Eucalyptus* litter from forests in southeast Australia is quite different from that reported from hardwood forest in Russia (ATTIWILL 1968). And, perhaps not surprisingly because of the preponderance of Leguminaceae, tropical forest litter tends to be more nitrogenous than temperate litter (KLINGE & RODRIGUES 1968).

Nitrate is known to be important in heterotrophic processes, and as we have seen it is taken up by decaying leaves, so that the C/N ratio of the material falls. Undoubtedly only a certain amount can be taken up by a given weight of litter. MATHEWS & KOWALCZEWSKI (1968) found that leaves placed in the Thames reached an asymptote when their N content had risen by about 150 %; somewhat similar findings are reported by IVERSEN (1973) for beech leaves in a Danish stream. Similarly, rooted aquatic plants take up and store N (STAKE 1967). We know very little about its release back into the water, which may have considerable implications for running water ecology. Nor is it clear why, despite the increasing use of nitrogenous fertilizers on agricultural land, and its apparently continual loss from the soil, there seems to have been no comparable rise in the nitrate content of river waters in Britain (TOMLINSON 1970; OWENS et al. 1972). Perhaps nitrate is a limiting factor in most situations, but one wonders what becomes of it.

Similarly there are indications that phosphate is important, although EGGLISHAW (1972) did not find it to be so in his experiments with cotton fibres. It is undoubtedly held very tightly within the forest ecosystem where HOBBIE & LIKENS (1973) demonstrated a net gain, from rain, in New Hampshire. However, leaf litter from deciduous trees contains a great deal; so much so that it has been cited as a possible source of eutrophication of lakes (COWAN & LEE 1973). *Eucalyptus*, on the other hand, seems to shed less phosphate with its leaves (ATTIWILL 1968), so here again we find that the type of vegetation is important. Once in the stream, however, phosphate is readily taken up by higher plants (NORMANN 1967; STAKE 1968) and by the organisms growing on dead leaves — but not by the litter itself before it is attacked by microorganisms (NELSON et al. 1969; ELWOOD & NELSON 1972). And, as with nitrate, we know nothing about its release back into the water. Also with this ion, which is so tightly held in the soil and is not renewed there, there is the problem of its continued supply to running water. It has been suggested that fish may be important in carrying it upstream, a topic that is discussed by HALL (1972). It has even been suggested that one ecological function of caimans in Amazonion waters is the release of nutrients at places in the river system where they are scarce (FITTKAU 1970). These are clearly topics that merit much more study.

Undoubtedly further research will show that other nutrients ions are withdrawn from the water by decaying litter. An indication of this is the demonstration by WOODALL (1972) that potassium, which is readily leached from alder leaves, is taken up from the water by such leaves as they decay. One is once more led back to thoughts about the soil and about the chelating properties of fulvic acid and its changing solubilities related to its iron and aluminium contents.

The organic matter in the stream

There is only time to give a very brief sketch of the fate of all this organic matter in the stream itself. It has, however, been clear for some time that where it accumulates on the stream bed it forms an attractive habitat for invertebrates, and one that changes with the season (EGGLISHAW 1968; MACKAY & KALFF 1969). It also forms a major foodstuff that is eaten by a great variety of invertebrates. Most species ingest some, and stream dwellers are opportunistic feeders; a recent paper that stresses this point is KOSLUCHER & MINSHALL (1973). Many fish species, especially in the tropics, are known feed directly on biodetrital matter, and at least one has been shown to select fine particles to eat, thereby probably increasing its take of microorganisms. However, this is a topic into which I must not digress here.

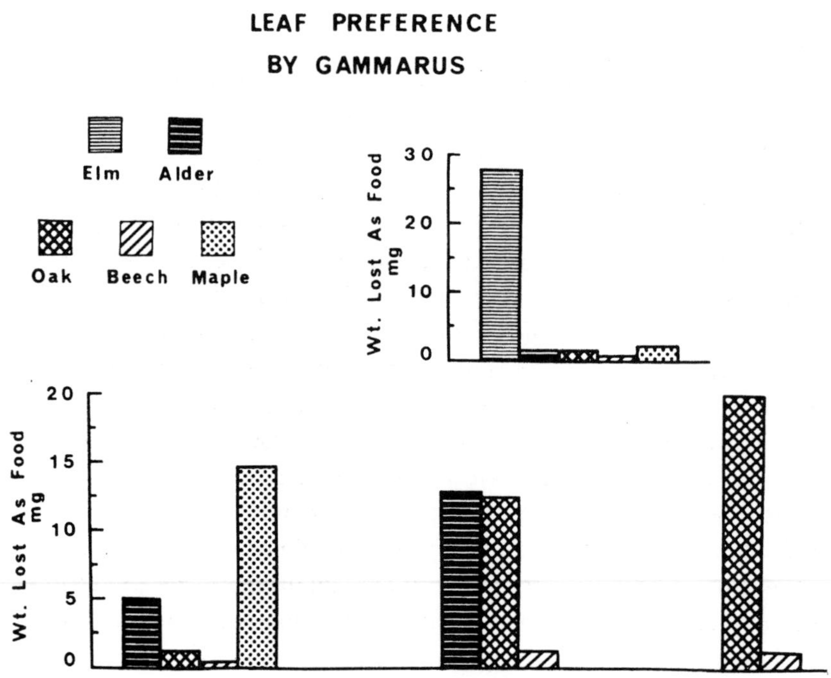

Fig. 3. Preferential feeding by *Gammarus* on autumn shed leaves from different tree species showing the succession of preference when the most favoured species is successively removed. From KAUSHIK & HYNES (1971).

Cummins in particular (1973; Cummins et al. 1973), has stressed that the first creatures to attack litter, which he calls shredders, reduce it to smaller particles that are available to collectors, and other workers have pointed out that shredders comminate much more than they eat (McDiffett 1970). They also produce faeces that are eaten by other organisms or by other members of the same species (Dölling 1962; Egglishaw & Shackley 1971; Woodall 1972; Ladle 1972), and the importance of this type of coprophagy has also recently been stressed with respect to the marine environment (Frankenberg & Smith 1967).

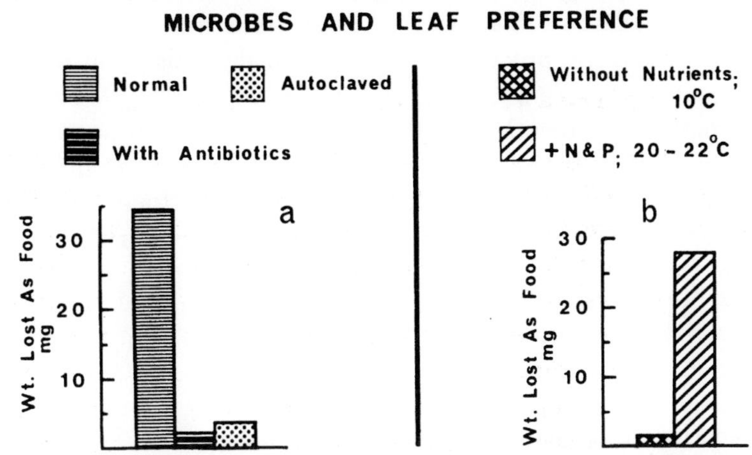

Fig. 4. Preferential feeding by *Gammarus* on elm leaves that had been variously treated to encourage or discourage growth of microorganisms. From Kaushik & Hynes (1971).

However, this whole process which lies at the base of almost all the biotic activity in the water, whether the debris is allochthonous or autochthonous, is subject to two conditions. First the debris must be acceptable as food by at least some shredders, and secondly it must be dietetically suitable for them.

There is now a growing body of literature showing that there are very definite preferences for different types of litter, that the leaves of some species are eaten before those of others, both in the laboratory and in the field (e. g. Dölling 1962; Mathews & Kowalczewski 1969; McDiffett 1970; Wallace et al. 1970; Kaushik & Hynes 1971; Haeckel et al. 1973; Bärlocher & Kendrick 1973 b). An example is shown in Fig. 3, and others were reported at this congress. This results in a successional use of litter from different species, and one can anticipate that litter from mixed woodland produces a more continuously available food supply than does that from a forester's pure stand.

Also, to use Cummins' terminology, the litter must be "processed" before it is shredded, which means it must acquire its complement of microorganisms before it is edible, or even very attractive. Here again there is a new and growing literature showing that faster decaying litter is more attractive than that which decays slowly, and that at any rate a large part of the attraction is the microorganisms, mostly fungi, that grow on it (Kaushik & Hynes 1971; Cummins et al. 1973; Bärlocher & Kendrick 1973 a, 1973 b; Iverson 1973). An example is shown in Fig. 4.

Here also the valley exerts its direct influence through the availability of nitrate, and possibly phosphate, potassium and other nutrients, not to mention such other factors as water temperature, liability to spate, and hence time of residence during which processing can occur, or even size of rock, stone, root or pool, in which the litter may lodge for a while. A smooth channel with few obstructions would allow the litter to leave before it had been processed.

Also, of course, the faeces of the shredders and the debris from their feeding, which are, presumably, still subject to microbial attack, are joined by the fine particles produced from the dissolved organic matter. These also depend fairly directly upon the valley, as leachates of litter from different species of tree behave differently, and the process of precipitation depends upon the chemistry of the water.

Such fine particles, either before they sink, or after resuspension, together with particles from other sources, such as plankton from a lake on the course of the stream, form food for filterers such as blackflies (MACIOLEK & TUNZI 1968; LADLE 1972; LADLE et al. 1972) and net-spinning caddisworms. On, and within, the substratum, they also form food for the collectors, and these also benefit from the faeces of the filterers.

Conclusion

We may conclude then that in every respect the valley rules the stream. Its rock determines the availability of ions, its soil, its clay, even its slope. The soil and climate determine the vegetation, and the vegetation rules the supply of organic matter. The organic matter reacts with the soil to control the release of ions, and the ions, particularly nitrate and phosphate, control the decay of the litter, and hence lie right at the root of the food cycle. One could go on and on, building up an edifice of complexity, all linked and cross-linked in the manner beloved by drawers of food nexes. These relationships are very important, and they are so complex that they will take much unravelling. They do, however, make it clear that every stream is likely to be an individual and thus not really very easily classifiable.

It is also clear that changes in the valley wrought by man may have large effects. Some are obvious and need not be stressed here; but others may be very subtle. For instance, even to replace mixed woodland with one species, or to fell one type of tree selectively, could affect the litter regime and the supply of ions. Even felling trees and leaving them where they lay produced enormous rises in N and P in the water of an experimental watershed (LIKENS et al. 1970; HOBBIE & LIKENS 1973), which would have far reaching effects on litter decay.

WOODALL & WALLACE (1972) have recently shown that there are large biotic differences between streams draining valleys that have been variously altered by man. They attribute these very largely to differing inputs of allochthonous organic matter, and to its specific origin as well as its total amount.

FISHER & LIKENS (1973), in a very thoughtful discussion, suggest that within a stream the accumulated detritus possibly confers stability to the system, in the way that live vegetation may do on land, and they stress that the forest maintains the structural and functional integrity of the stream. MOTTEN & HALL (1972) also conclude that the forest probably dominates any ecological succession a stream might otherwise undergo. It is interesting therefore that English chalk streams, which have been largely deprived of their forests, should still store a great deal of energy as dead organic matter (WESTLAKE et al. 1970), especially as they are well known for their productivity. There the valleys have somehow avoided the consequences of human activity — or were they perhaps even more productive before?

We must, in fact, not divorce the stream from its valley in our thoughts at any time. If we do we lose touch with reality. The real lake is not a basin with two vertical sides as in the textbook. One that is like that, Loch Ness, is so out of line that it harbours monsters. Somewhere, in Australia, there must be a stream with a channel like a gutter, fed by runoff from a landscape paved like a parking lot. There, I predict, will be found the legendary river creature of the aboriginals — the Bunyip.

References

ATTIWILL, P. M. (1968): The loss of elements from decomposing litter. — *Ecology* **49**, 142—145.

BÄRLOCHER, F. & KENDRICK, B. (1973 a): Fungi in the diet of *Gammarus pseudolimnaeus* (Amphipoda). — *Oikos* **24**, 295—300.

— — (1973 b): Fungi and food preferences of *Gammarus pseudolimnaeus*. — *Arch. Hydrobiol.* **72**, 501—506.

CARSON, M. A. & SUTTON, E. A. (1971): The hydrologic response of the Eaton River Basin, Quebec. — *Can. J. Earth Sci.* **8**, 102—115.

CONWAY, K. E. (1970): The aquatic hypomycetes of central New York. — *Mycologia* **62**, 516—530.

COWAN, W. & LEE, G. F. (1973): Leaves as a source of phosphorus. — *Environ. Sci. Technol.* **7**, 853—854.

CUMMINS, K. W. (1973): Trophic relations of aquatic insects. — *Ann. Rev. Ent.* **18**, 183—206.

CUMMINS, K. W. et al. (1972): Organic enrichment with leaf leachate in experimental lotic ecosystems. — *BioScience* **22**, 719—722.

— — (1973): The utilization of leaf litter by stream detritivores. — *Ecology* **54**, 336—345.

CURRY, R. R. (1972): Rivers — a geomorphic and chemical overview. — In: OGLESBY et al. (eds), *River ecology and man.* Academic Press. N. Y. pp. 9—32.

DARNELL, R. M. (1967): Organic detritus in relation to the estuarine system. — *Estuaries*, Amer. Ass. Adv. Sci., Washington D. C. pp. 376—382.

DAUBENMIRE, R. (1953): Nutrient content of leaf litters in the northern Rocky Mountains. — *Ecology* **34**, 786—793.

DÖLLING, L. (1962): Der Anteil der Tierwelt an der Bildung von Unterwasserboden. — *Verh. Zool. Bot. Ges. Wien* **101/102**, 50—85.

DUDKA, I. A. (1971): The abundance of aquatic hypomycetous conidia in rivers and streams throughout the Tershiy coast of the Kola Peninsula. — *Hydrobiol. J.* **7**, 16—20.

DUNNE, T. & BLACK, R. D. (1970): Partial area contributions to storm runoff in a small New England watershed. — *Wat. Resourc. Res.* **6**, 1296—1311.

EGGLISHAW, H. J. (1968): The quantitative relationship between bottom fauna and plant detritus in streams of different calcium concentrations. — J. appl. Ecol. 5, 731—740.

— (1972): An experimental study of the breakdown of cellulose in fast-flowing streams. — *Mem. Ist. ital. Idrobiol.* **29** Suppl., 405—428.

EGGLISHAW, H. J. & SHACKLEY, P. E. (1971): Suspended organic matter in fast-flowing streams in Scotland. I. Downstream variations in microscopic particles. — *Freshwat. Biol.* **1**, 273—286.

EHRLICH, G. C. & SLACK, K. V. (1969): Uptake and assimilation of nitrogen in microecological systems. — *Spec. Techn. Publ. Amer. Soc. Test. Mat.* **448**, 11—23.

ELWOOD, J. W. & NELSON, D. J. (1972): Periphyton production and grazing rates in a stream measured with a ^{32}P material balance method. — *Oikos* **23**, 295—303.

FENCHEL, T. (1970): Studies on the decomposition of organic detritus derived from the turtle grass *Thalassia testudinum.* — *Limnol. Oceanogr.* **15**, 14—20.

FETH, J. H. (1966): Nitrogen compounds in natural water — a review. — *Wat. Resourc. Res.* **2**, 41—58.

FISHER, S. G. & LIKENS, G. E. (1973): Energy flow in Bear Brook, New Hampshire: an integrative approach to stream ecosystem metabolism. — *Ecol. Monogr.* **43**, 421—439.

FITTKAU, E. J. (1970): Role of caimans in the nutrient regime of mouthlakes of Amazon affluents — an hypothesis. — *Biotropica* **2**, 138—142.

FLOODGATE, G. (1973): Where there's muck there's energy. — *New Scientist*, 5 July, 19—20.

FRANKENBERG, D. & SMITH, K. L. (1967): Coprophagy in marine animals. — *Limnol. Oceanogr.* **12**, 443—450.

FREEZE, R. A. (1972): Role of subsurface flow in generating surface runoff. 2. Upstream source areas. — *Wat. Resourc. Res.* **8**, 1272—1283.

GOSZ et al. (1972): Nutrient content of litter fall on the Hubbard Brook Experimental Forest, New Hampshire. — *Ecology* **53**, 769—784.

HAECKEL et al. (1973): *Gammarus fossarum* KOCH als Fallaubzersetzer in Waldbächen. — *Freshwat. Biol.* **3**, 241—249.

HALL, C. A. (1972): Migration and metabolism in a temperate stream ecosystem. — *Ecology* **53**, 585—604.

HEWLETT, J. D. & HELVEY, J. D. (1970): Effects of forest clear-felling on the stream hydrograph. — *Wat. Resourc. Res.* **6**, 768—782.

HEWLETT, J. D. & NUTTER, W. L. (1970): The varying source area of streamflow from upland basins. — *Proc. Sympos. Interdisciplinary Aspects of Watershed Management.* Amer. Soc. Civ. Engrs. New York, pp. 65—83.

HOBBIE, J. E. & LIKENS, G. E. (1973): Output of phosphorus, dissolved organic carbon, and fine particulate carbon from Hubbard Brook watersheds. — *Limnol. Oceanogr.* **18**, 734—742.

HOFSTEN, B. V. & EDBERG, N. (1972): Estimating the rate of degradation of cellulose fibres in water. — *Oikos* **23**, 29—34.

INGOLD, C. T. (1966): The tetraradiate aquatic fungal spore. — *Mycologia* **58**, 43—56.

IVERSON, T. M. (1973): Decomposition of autumn-shed beech leaves in a springbook and its significance for the fauna. — *Arch. Hydrobiol.* **72**, 305—312.

JOHNSON, N. M. et al. (1969): A working model for the variation in stream water chemistry at the Hubbard Brook Experimental Forest, New Hampshire. — *Wat. Resourc. Res.* **5**, 1353—1363.

KAUSHIK, N. K. & HYNES, H. B. N. (1971): The fate of the dead leaves that fall into streams. — *Arch. Hydrobiol.* **68**, 465—515.

KLINGE, H. & RODRIGUES, W. A. (1968): Litter production in an area of Amazonian Terra Firme forest. Part. I. Litter-fall, organic carbon and total nitrogen contents of litter. — *Amazoniana* **1**, 287—302.

KOSLUCHER, D. G. & MINSHALL, G. W. (1973): Food habits of some benthic invertebrates in a northern cool-desert stream (Deep Creek, Curlew Valley, Idaho-Utah). — *Trans. Amer. micr. Soc.* **92**, 441—452.

LADLE, M. (1972): Larval Simuliidae as detritus feeders in chalk streams. — *Mem. Ist. ital. Idrobiol.* **29** Suppl., 429—439.

LADLE, M. et al. (1972): Studies on production and food consumption by the larval Simuliidae of a chalk stream. — *Hydrobiologia* **39**, 429—448.

LIKENS, G. E. et al. (1970): Effects of forest cutting and herbicide treatment on nutrient budgets in the Hubbard Brook watershed-ecosystem. — *Ecol. Monogr.* **40**, 23—47.

LUSH, D. L. & HYNES, H. B. N. (1974): The formation of particles in freshwater leachates of dead leaves. — *Limnol. Oceanogr.* **18**, 968—977.

MACIOLEK, J. A. & TUNZI, M. G. (1968): Microseston dynamics in a simple Sierra Nevada lake-stream system. — *Ecology* **49**, 60—75.

MACKAY, R. & KALFF, J. (1969): Seasonal variation in standing crop and species diversity of insect communities in a small Quebec stream. — *Ecology* **50**, 101—108.

MADSEN, B. L. (1972): Detritus on stones in small streams. — *Mem. Ist. ital. Idrobiol.* **29**, Suppl., 385—403.

MANN, K. H. (1973): Seaweeds: their productivity and strategy for growth. — *Science N. Y.* **182**, 975—981.

MANNY, B. A. & WETZEL, R. G. (1973): Diurnal changes in dissolved organic and inorganic carbon and nitrogen in a hardwater stream. — *Freshwat. Biol.* **3**, 31—43.

MATHEWS, C. P. & KOWALCZEWSKI, A. (1969): The disappearance of leaf litter and its contribution to production in the River Thames. — *J. Ecol.* **57**, 543—552.

MCCONNELL, W. J. (1968): Limnological effects of organic extracts of litter in a southwestern impoundment. — *Limnol. Oceanogr.* **13**, 343—349.

MCDIFFETT, W. F. (1970): The transformation of energy by a stream detritivore, *Pteronarcys scotti* (Plecoptera). — *Ecology* **51**, 975—988.

MOTTEN, A. F. & HALL, C. A. S. (1972): Edaphic factors override a possible gradient of ecological maturity indices in a small stream. — *Limnol. Oceanogr.* **17**, 922—926.

NELSON, D. J. et al. (1969): Estimates of periphyton mass and stream bottom area using phosphorus-32. — *Water Res.* **3**, 367—373.

NORMANN, H. D. (1967): Versuche zur Aufnahme von Phosphat durch *Ranunculus fluitans* LAM. — *Arch. Hydrobiol. Suppl.* **33**, 243—254.

ODUM, E. P. & DE LA CRUZ, A. (1963): Detritus as a major component of ecosystems. — *Bull. Amer. Inst. biol. Sci.* **13**, 39—40.

OWENS, M. et al. (1972): Nutrient budgets in rivers. — *Symp. Zool. Soc. Lond.* **29**, 21—40.

PENNAK, R. W. (1971): Toward a classification of lotic habitats. — *Hydrobiologia* **38**, 321—334.

RILEY, G. A. (1970): Particulate and organic matter in sea water. — *Adv. Mar. Biol.* **8**, 1—118.

RODINA, A. G. (1967): On the forms of existence of bacteria in water bodies. — *Arch. Hydrobiol.* **63**, 238—242.

SCHNITZER, M. (1971): Metal-organic matter interactions in soils and waters. — In: FAUST, S. D. & HUNTER, J. V. (eds), *Organic compounds in aquatic environments.* Marcel Dekker, New York, pp. 297—316.

SLACK, K. V. & FELTZ, H. R. (1968): Tree leaf control on low flow water quality in a small Virginia stream. — *Environ. Sci. Technol.* **2**, 126—131.

STAKE, E. (1967): Higher vegetation and nitrogen in a rivulet in central Sweden. — *Schweiz. Z. Hydrol.* **29**, 107—124.

— (1968): Higher vegetation and phosphorus in a small stream in central Sweden. — *Schweiz. Z. Hydrol.* **30**, 353—373.

STEVENSON, F. J. (1972): Role and function of humus in soil with emphasis on adsorption of herbicides and chelation of micronutrients. — *BioScience* **22**, 643—650.

THOMAS, W. A. (1970): Weight and calcium losses from decomposing tree leaves on land and in water. — *J. appl. Ecol.* **7**, 237—241.

TOMLINSON, T. E. (1970): Trends in nitrate concentrations in English rivers in relation to fertilizer use. — *Wat. Treatm. Exam.* **19**, 277—293.

WALLACE, J. B. et al. (1970): Breakdown of leaves by feeding of *Peltoperla maria* nymphs (Plecoptera: Perltoperlidae). — *Ann. ent. Soc. Amer.* **63**, 562—567.

WESTLAKE, D. F. et al. (1972): The chalk stream ecosystem. — In: KAJAK, A. & HILLBRICHT, A. (eds), *Productivity Problems in Freshwaters.* Warsaw Krakov, pp. 615—635.

WETZEL, R. G. & MANNY, B. A. (1972): Decomposition of dissolved carbon and nitrogen compounds from leaves in an experimental hard-water stream. — *Limnol. Oceanogr.* **17**, 927—931.

WILLIAMS, P. M. (1968): Organic and inorganic constituents of the Amazon River. — *Nature*, Lond. **218**, 937—938.

WOODALL, W. R. (1972): Nutrient pathways in small mountain streams. — Ph. D. Thesis, Univ. of Georgia, Athens. (*Eastern Deciduous Forest Biome Memo Report* 72—58) 118 pp.

WOODALL, W. R. & WALLACE, J. B. (1972): The benthic fauna in four small Southern Appalachian streams. — *Amer. Midl. Nat.* **88**, 393—407.

Commentary on Paper A3

Jones, J. R., Borofka, B. P. & Bachmann, R. W. (1976) Factors affecting nutrient loads in some Iowa streams.

Jones and his colleagues analysed the relationship between stream nutrient concentration and land cover in 34 agricultural drainage basins in Iowa, USA, ranging from 6 to 4000 ha. To their disappointment, they found only two positive relationships between animal density and orthophosphate and ammonia concentrations. They justify this limited relationship between land cover and ammonia and P concentration of the fact that there should be an effect of the landscape arrangement pattern on the fluxes of these nutrients. They then attempted to analyse the drainage basins by determining near stream areas or tile intakes in a spatially explicit manner. This approach significantly increased the number of positive relationships.

Interestingly, their most striking result was not with phosphate or ammonia, but with nitrate. Indeed, they found a significantly negative relationship between the percentage of drainage basin as marshlands and the stream nitrate concentration. Yet, although most of the discussion was related to phosphate and ammonia issues, the authors provided few hints about the potential role of these marshlands in denitrifying incoming nitrate and thus reducing the stream nitrate concentration. They estimated that a wetland area of 2200 ha within the drainage basin could reduce the annual nitrate-N load by 4000 kg.

This paper laid the foundation for current landscape ecology research on the relationship between land use and land cover arrangements and nutrient fluxes, and the role of the riparian zone in particular on reducing nitrogen load. It is sad that this paper has not been widely quoted, especially when reading their discussion on the fact that "*the nutrient concentrations found in a tributary near its mouth represent an integration of hydrologic and chemical processes after derivation from the initial sources. It is unlikely that losses are equal in all parts of the watershed*". Reading this and trying to test it would have saved us precious decades of blind monitoring of nutrient concentrations at the outlet of large streams.

Indeed, not nearly enough attention has been paid so far to the spatial distribution of land cover and land use. Evidence shows large variability of nutrient fluxes from small tributary basins, but little variability from the entire river basin (Omernik, 1976); in other words, the signal to noise ratio is low in large basins but high in smaller tributaries. It follows that "black box" models cannot be used in large river basins to show where, when and by how much a given policy might affect nutrient fluxes as a result of changes in land management, since a policy is not sensitive to local conditions. Once again, this points towards the need for a spatially explicit approach at the landscape scale (Burt & Pinay, 2005).

Therefore, one of the main challenges in this context is to understand the mechanisms by which the spatial arrangement of land cover, including riparian zones, controls the cycling, retention, removal and transfer of nutrients in small catchments.

GP

References

Burt, T. P. & Pinay, G. (2005) Linking hydrology and biogeochemistry in complex landscapes. *Progr. Phys. Geogr.* **29**(3), 297–316.

Omernik, J. M. (1976) The influence of land use on stream nutrient levels. EPA Report 600/3-76-014, US EPA, Washington, USA.

Water Research Vol. 10, pp. 117 to 122. Pergamon Press 1976. Printed in Great Britain.

FACTORS AFFECTING NUTRIENT LOADS IN SOME IOWA STREAMS*

J. R. JONES, B. P. BOROFKA† and R. W. BACHMANN

Department of Animal Ecology, Iowa State University, Ames, Iowa 50010, U.S.A.

(*Received* 24 *June* 1975)

Abstract—The export and concentration of inorganic nitrogen and total phosphorus from 34 watersheds in a northwestern Iowa lake district were measured during March 1971–August 1973. Annual nutrient losses were approximately 0.35 kg ha^{-1} P, 6.7 kg ha^{-1} NO_3-N, and 1.0 kg ha^{-1} NH_3-N. A statistical analysis of the relationship between land-use and plant nutrients was used to determine differences among streams. Animal units in feedlots were significantly correlated with phosphorus and ammonia nitrogen (mg l^{-1} and kg ha^{-1} yr^{-1}). Nitrate nitrogen was negatively correlated with the percentage of watershed in marshland. Tile drainage and surface runoff from grasslands, feedlots, cornfields, and soybean fields were analyzed for nitrogen and phosphorus in spring 1974; mean values are given.

INTRODUCTION

In a recent study of eutrophication in a group of lakes in northwestern Iowa (Jones, 1974), measurements were made of the plant nutrient loads for a number of tributary streams. Consistent differences were noted in the concentrations of phosphorus and nitrogen among streams. To determine the effect of land use on these differences, an inventory of each watershed was undertaken. This paper presents the results of an analysis of the relationship between land use and nutrient output.

METHODS AND MATERIALS

The study area is in the Iowa lakes region in Dickinson County, Iowa, and Jackson County, Minnesota (Fig. 1). It is in the Cary glacial drift of the Des Moines lobe and is enclosed by morainal topography. Seventy per cent of the land is devoted to row-crop agriculture or to livestock production; nearly 20% of the watershed is covered by lakes, ponds or marshes. Populated areas are served by a modern sewage-collection and treatment system that diverts human wastes from the basin.

Water samples were collected from the streams approximately every 10 days from March 1971 to August 1973. At collection time stream flow was estimated by measuring the average stream depth, width and velocity of flow. Additional samples of surface runoff and tile drainage were collected from selected areas within the watershed during spring 1974.

Chemical analyses were made on unfiltered samples. Orthophosphate phosphorus was determined by the stannous chloride method. Ammonia nitrogen was determined by the direct Nesslerization method and nitrate nitrogen by cadmium reduction. Prepared reagents for these analyses were purchased from Hach Chemical Company, Ames, Iowa. Starting in September 1971, total phosphorus analyses (Murphy and Riley, 1962) were made after persulfate oxidation (Menzel and Corwin, 1965). A high correlation ($r = 0.83$) of ortho and total phosphorus values was found after analysis of both forms on 750 samples. A least-squares regression line was calculated to convert early measurements to total phosphorus.

In 1974, samples of surface runoff were filtered through a type HA, 0.45 μm Millipore filter. Filtrate samples were analyzed for total filterable phosphorus (American Public Health Association, 1965) and total particulate phosphorus was obtained by difference.

Boundaries of individual watersheds were delineated from USGS topographic maps, and a land-use inventory

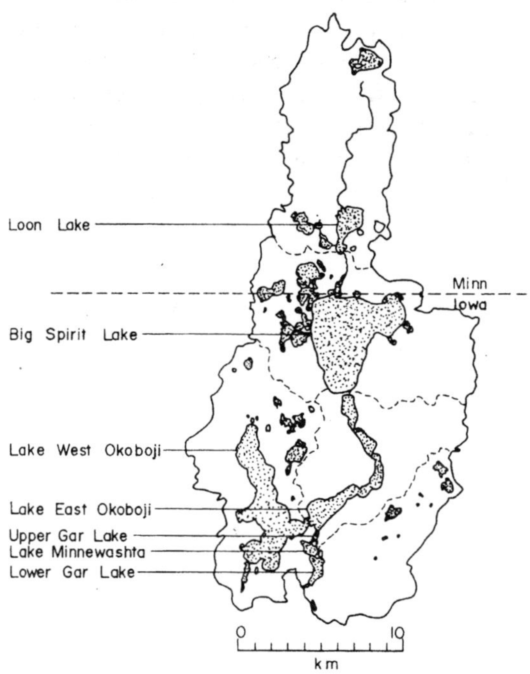

Fig. 1. Map of the watershed. Dashed lines indicate watershed boundaries.

* Journal Paper No. J-8168 of the Iowa Agriculture and Home Economics Experiment Station, Ames, Iowa. Project 1779. Financed in part by the Iowa Great Lakes Water Quality Control Plan and by a grant from the U.S. Department of Interior, Office of Water Resources Research under Public Law 88-379, and made available through the Iowa State Water Resources Research Institute.

† Present Address: Environmental Science and Engineering, Inc., PO Box 13454 University Station, Gainesville, FL 32604, U.S.A.

determined the areas in row crops (corn, soybeans and oats), grasslands (pastured and unpastured), marshland (including permanent water) and urban areas within each watershed. Livestock numbers in each watershed were determined and converted to animal units by using the U.S. Environmental Protection Agency formula (one animal unit being defined as the number of animals required to produce wastes with a biochemical oxygen demand equivalent to one beef steer). One animal unit consists of 1 beef animal, 0.7 dairy animal, 4.5 slaughter hogs, 35 feeder pigs, 12 sheep or lambs, 180 laying hens, or 55 turkeys.

RESULTS

Annual mean concentrations of total P, NO_3-N and NH_3-N from the 34 tributaries samples are given in Table 1. Weighted averages were found by dividing the accumulated total load by the accumulated total discharge. Output of N and P in kg ha^{-1} from 17 watersheds greater than 100 ha in area are given in Table 2. Outputs were determined by integrating flow and concentration measurements over time to yield annual losses from each watershed. Smaller watersheds were omitted from this analysis because most had intermittent flows and, subsequently, nutrient losses may not have been adequately measured by the sampling schedule. The land-use and livestock inventory for the 34 watersheds is presented in Table 3. Totals do not equal 100% because areas in roads, farmsteads and fence rows were not considered.

Table 1. Mean concentrations (mg l^{-1}) of total phosphorus, nitrate nitrogen, and ammonia nitrogen from the 34 tributaries sampled during 1971–1973

Station	P			NO_3-N			NH_3-N		
	1971	1972	1973	1971	1972	1973	1971	1972	1973
1	1.04	0.47	0.35	3.80	6.31	7.10	1.63	1.09	0.64
2	0.07	0.04	0.05	3.94	5.35	10.19	0.15	0.08	0.12
3	0.38	0.25	0.35	1.44	4.19	2.31	0.69	0.25	0.47
5	0.11	0.25	0.10	3.43	2.91	3.51	0.67	0.56	0.73
6	0.12	0.09	1.11	0.15	0.11	0.10	0.84	0.64	0.91
8	0.43	0.45	0.33	7.45	10.52	12.81	0.91	1.52	0.83
9	0.21	0.21	0.16	7.96	12.85	12.38	0.20	0.21	0.41
10	0.22	0.25	0.16	5.14	5.53	9.59	0.44	0.44	0.28
11	0.18	0.50	0.13	4.97	6.54	9.16	0.40	0.77	0.25
12	0.27	0.22	0.28	1.52	0.62	0.93	0.41	0.29	0.69
13	0.20	0.17	0.15	5.12	6.24	7.35	0.24	0.11	0.17
15	0.57	0.18	0.47	3.13	2.82	5.78	0.95	0.29	0.74
16	0.28	0.16	0.14	3.89	3.33	7.93	0.28	0.17	0.21
17	0.62	0.61	0.59	1.04	1.77	0.94	1.09	1.08	1.04
18	0.31	0.22	0.25	2.26	4.88	5.60	0.65	0.21	0.65
19	0.30	0.17	0.14	3.17	4.04	8.00	0.51	0.37	0.27
22	0.17	0.14	0.25	4.49	5.90	7.27	0.52	0.44	0.59
23	0.17	0.10	0.09	2.76	1.95	4.20	0.42	0.21	0.18
24	0.25	0.18	0.16	3.71	5.15	5.69	0.14	0.07	0.15
28	0.92	0.71	0.78	4.54	6.65	4.74	1.46	0.53	2.44
29	0.15	0.13	0.13	0.23	0.14	0.29	0.93	0.51	0.61
30	0.08	0.11	0.27	1.67	1.22	1.00	0.17	0.11	0.53
31	0.24	0.12	0.13	5.46	7.16	4.47	0.32	0.08	0.17
32	0.28	0.32	0.21	0.05	0.04	0.07	0.39	0.24	0.37
33	0.17	0.07	0.11	2.76	4.12	5.11	0.28	0.12	0.26
38	0.55	0.37	0.36	2.74	4.80	7.06	0.77	0.49	0.48
39	2.76	8.18	3.52	3.74	8.00	3.85	2.72	0.90	8.00
40	0.39	0.16	0.19	3.37	4.95	7.49	1.01	0.93	0.55
41	0.23	0.08	0.24	8.24	9.49	9.40	0.75	0.21	0.77
42	2.96	1.33	2.40	3.52	1.91	4.18	4.65	0.89	6.32
44	0.14	0.10	0.07	5.37	7.92	9.97	0.29	0.37	0.26
46	0.23	0.13	0.21	5.11	10.02	10.97	0.30	0.15	0.25
47	0.21	0.12	0.15	5.95	7.91	8.40	0.28	0.06	0.23
48	0.19	0.19	0.14	1.69	2.35	2.78	0.67	0.56	0.60

Table 2. Output (kg ha^{-1}) of total phosphorus, nitrate nitrogen and ammonia nitrogen from 17 watersheds over 100 ha in 1971–1973

Station	P			NO_3-N			NH_3-N		
	1971	1972	1973	1971	1972	1973	1971	1972	1973
1	2.86	0.68	0.89	10.40	9.20	17.96	4.46	1.59	1.61
5	0.47	0.24	0.09	14.31	2.78	3.38	2.81	0.53	0.71
6	0.36	0.04	0.17	0.50	0.04	0.16	2.74	0.29	1.41
8	0.96	0.45	0.64	16.62	10.56	25.31	2.03	1.53	1.64
9	0.47	0.27	0.27	17.71	16.69	20.68	0.45	0.28	0.69
10	0.46	0.41	0.21	10.91	9.03	12.34	0.93	0.72	0.37
11	0.23	0.53	0.18	6.65	6.91	12.32	0.54	0.82	0.34
18	0.50	0.10	0.43	3.65	2.11	9.51	1.04	0.09	1.10
19	0.72	0.21	0.34	7.68	4.88	19.88	1.24	0.45	0.66
22	0.32	0.23	0.48	8.32	9.68	14.09	0.97	0.73	1.14
23	0.38	0.24	0.18	6.33	4.68	8.67	0.95	0.50	0.37
29	0.19	0.06	0.18	0.29	0.07	0.42	1.17	0.24	0.86
33	0.38	0.10	0.39	5.99	5.83	18.12	0.61	0.17	0.94
38	1.29	0.60	0.67	6.42	7.84	13.08	1.80	0.80	0.89
40	0.68	0.23	0.33	5.91	6.97	13.15	1.77	1.32	0.96
41	0.62	0.14	0.55	21.54	16.09	21.81	1.97	0.36	1.78
48	0.15	0.17	0.16	1.30	2.13	3.37	0.51	0.51	0.73

A linear model was used to assess the relationship between land-use practices on the watersheds and the N and P in the streams. A unique contribution, that part of the variability in a particular variable that can be ascribed to a given factor after adjusting for all other measured factors, was judged significant if the probability of the test statistic was equal to or less than 0.01.

Significant correlations between P and N concentrations at 34 stations and the land use variables measured are given in Table 4. Similar significant correlations between P and N losses (kg ha^{-1}) at 17

Table 3. Land use in 34 watersheds

Station	Area (ha)	% Row Crop	% Grasslands	% Marsh	% Urban	Animal Units/ha
1	136	74.3	15.4	0.2	0.0	1.91
2	10	100.0	0.0	0.0	0.0	0.0
3	73	60.3	38.3	0.0	0.0	0.34
5	109	91.7	0.0	3.8	0.0	0.0
6	142	79.6	4.2	15.8	0.0	0.14
8	495	85.0	7.5	0.0	0.0	0.79
9	656	92.0	0.9	0.0	0.0	0.80
10	199	62.3	20.6	0.1	0.0	0.29
11	651	84.8	4.4	0.1	1.1	0.96
12	68	82.3	4.4	0.0	2.9	0.42
13	66	87.9	0.0	0.0	0.0	0.20
15	34	97.1	0.0	0.0	0.0	0.0
16	17	94.1	0.0	0.0	0.0	0.0
17	61	49.2	41.0	4.9	0.0	0.59
18	1175	73.4	19.2	2.9	0.5	0.51
19	277	74.3	10.1	0.4	9.7	0.09
22	732	72.7	22.8	0.5	0.0	0.37
23	125	68.8	21.6	0.0	0.0	0.17
24	36	38.9	41.7	9.7	11.1	0.0
28	12	66.7	0.0	0.0	33.3	4.17
29	1341	56.3	20.2	16.2	4.5	0.29
30	6	0.0	66.7	0.0	33.3	0.0
31	25	76.0	16.0	0.0	8.0	0.0
32	48	8.3	66.7	12.9	10.4	0.09
33	145	55.2	11.0	0.0	29.0	0.05
38	702	71.6	10.4	0.0	13.0	0.39
39	13	61.5	30.7	0.0	0.0	3.77
40	1164	79.0	12.6	0.3	0.0	0.67
41	742	90.4	3.6	0.1	0.0	0.99
42	19	68.4	26.3	0.0	0.0	2.63
44	37	86.5	13.5	0.0	0.0	0.0
46	59	72.9	23.7	0.0	0.0	0.0
47	78	88.5	6.4	0.0	0.0	0.0
48	3917	72.5	12.8	10.1	0.0	0.26

Table 4. Correlation coefficients between phosphorus (mg l^{-1}), nitrogen (mg l^{-1}) and land-use categories from 34 subwatersheds from 1971 to 1973 ($P = 0.01$)

	P	NO_3-N	NH_3-N	Marshlands
Row Crop	-	0.468	-	-0.360
Grasslands	-	-0.395	-	0.278
Marshlands	-	-0.525	-	-
Urban	-	-	-	-
Animal units/ha	0.662	-	0.628	-
P	-	-	0.549	-

Table 5. Correlation coefficients between phosphorus (kg ha^{-1}), nitrogen (kg ha^{-1}) and land-use categories from 17 watersheds from 1971 to 1973 (P = 0.01)

	P	NO_3-N	NH_3-N	Row Crop
Row Crop	-	0.394	-	-
Grasslands	-	-	-	-0.750
Marshlands	-	0.635	-	-
Urban	-	-	-	-0.535
Animal units/ha	0.556	0.422	0.412	-
P	-	-	0.789	-

stations are given in Table 5. A regression analysis of P concentrations was carried out by using animal units per hectare, percentages of each watershed in row crops, grasslands, marshland and urban area, and year classes as independent variables ($R^2 = 0.54$). Only the unique effect of the number of animal units ha^{-1} (AUHA) was significant. The partial regression of phosphorus on AUHA indicated that P increased by 0.67 ± 0.07 mg l^{-1} with each watershed animal unit ha^{-1}.

The same model was used in analysis of NO_3-N (mg l^{-1}) ($R^2 = 0.45$) and NH_3-N (mg l^{-1}) ($R^2 = 0.49$). Percentage of the area in marshland had a significant negative relationship with NO_3-N concentrations, indicating a 0.26 ± 0.07 mg l^{-1} reduction for each percentage point increase in marshland. There were significant differences in NO_3-N concentrations in the streams during the years of study. NO_3-N concentrations were greater in 1973 than in 1971 or 1972. NH_3-N (mg l^{-1}) was significantly related to the number of animal units within the watersheds. The partial regression indicates that NH_3-N was increased by 0.74 ± 0.08 mg l^{-1} for each animal unit per hectare within the watershed. None of the other independent variables made significant unique contributions to the variability in the ammonia or nitrate nitrogen concentrations.

Total P, NO_3-N, and NH_3-N outputs (kg ha^{-1}) from the 17 largest watersheds were used separately in a regression analysis against land-use variables. The R^2 values are 0.50, 0.65 and 0.42, respectively. Year differences were significant in all three analyses. Plant nutrient losses from the watersheds in 1972 were significantly less than in 1971 and 1973. In the analysis, P output was significantly increased by 0.55 ± 0.11 kg ha^{-1}, and NH_3-N output was significantly increased by 0.77 ± 0.23 kg ha^{-1}, for each additional animal unit per hectare in the watershed. NO_3-N (kg ha^{-1}) was significantly decreased by 0.67 ± 0.18 kg ha^{-1} for each percentage point of the watershed in marshland. None of the other independent variables showed significant unique effects in these analyses.

Table 6. Placement of animal units per hectare within the 17 watersheds over 100 ha

Station	Feedlot animal units with surface drainage	Pasture animal units with surface drainage	Feedlot animal units without surface drainage	Pasture animal units without surface drainage
1	1.75	0.16	0.0	0.0
5	0.0	0.0	0.0	0.0
6	0.0	0.0	0.14	0.0
8	0.71	0.0	0.04	0.04
9	0.0	0.0	0.80	0.01
10	0.0	0.24	0.05	0.0
11	0.14	0.01	0.81	0.003
18	0.09	0.07	0.21	0.12
19	0.0	0.06	0.02	0.0
22	0.0	0.05	0.22	0.10
23	0.0	0.10	0.0	0.07
29	0.07	0.11	0.09	0.03
33	0.0	0.02	0.01	0.03
38	0.09	0.06	0.19	0.04
40	0.27	0.15	0.25	0.0
41	0.74	0.06	0.20	0.0
48	0.02	0.12	0.09	0.03

An additional regression analysis was conducted to determine if animal placement within the watersheds influenced P or NH_3-N losses (kg ha^{-1}) in the 17 largest watersheds. This analysis used, as independent variables, year classes and the watershed animal units per hectare divided into four classes: (1) feedlots with drainage to streams or tile intakes, (2) pastures with drainage to streams or tile intakes, (3) feedlots with no surface drainage to streams or tile intakes, and (4) pastures with no surface drainage to streams or tile intakes (Table 6) ($R^2 = 0.54$ for P and $R^2 = 0.59$ for NH_3-N). Differences among years were significant. Total P and NH_3-N outputs in 1972 were less than in 1971 and 1973. The number of feedlot animal units per hectare with drainage to streams or tiles was the only significant variable in either analysis. Partial regression coefficients indicated that 0.61 ± 0.10 kg ha^{-1} P and 0.96 ± 0.18 kg ha^{-1} NH_3-N were associated with each feedlot animal unit per hectare with drainage to a stream or tile intake.

Surface runoff

Each of the streams monitored in 1971–1973 received drainage from a number of different land

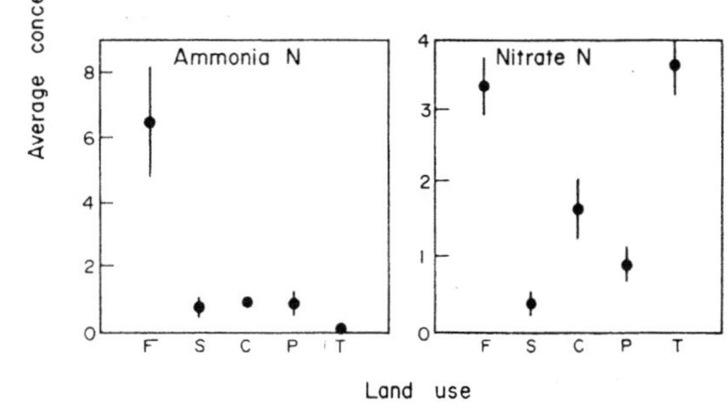

Fig. 2. Mean and standard error of the mean for total phosphorus, particulate phosphorus, ammonia nitrogen and nitrate nitrogen concentrations (mg l^{-1}) in feedlot (F), soybean (S), corn (C), pastures (P) and tile (T) runoff during spring 1974.

Table 7. Results of t-tests of mean plant nutrient concentrations ($mg\,l^{-1}$) in measured runoff from grasslands (G), tiles (T), feedlots (F), pastures (P), corn (C), and soybean (S) fields.[a]

	Comparisons of:			
	Tile vs. F,G,C,S	Feedlot vs. G,S,C	Pasture vs. S,C	Soybean vs. C
Total P	x	x	x	
Filtered P	x	x		
Particulate P	x		x	
NO_3-N	x	x		
NH_3-N	x	x	x	x

[a] x indicates significant difference at P=0.01

uses. To define more closely the nutrient contributions from specific land uses, samples were collected during March–May 1974 (Borofka, 1974) during runoff conditions from 3 tile drains, 13 grasslands, 7 feedlots, 13 cornfields, and 13 soybean fields. Most runoff events occurred during March and April when the ground was frozen or saturated. Figure 2 shows the mean and standard error of the mean for each measured variable. Table 7 summarizes t-tests of mean nutrient concentrations from five types of land runoff. The comparisons made were (1) tile vs the mean of feedlot, soybean, corn, and grassland, (2) feedlot vs the mean of soybean, corn, and grassland, (3) grassland vs the mean of soybean and corn, and (4) soybean vs corn. Analyses were run of logarithms of original data to make the variance more homogeneous across the groups.

Tile drainage had the lowest total P concentration, with only a small fraction as particulate P. NH_3-N values also were lower than in other runoff types. NO_3-N samples were variable, but normally above $2.5\,mg\,l^{-1}$.

Each of the grassland sites had pastured animals during the summer and fall preceding the sampling period. More than half the P was carried by particulate material in grassland runoff. NO_3-N and NH_3-N were normally less than $1\,mg\,l^{-1}$.

Runoff from the feedlots had the highest P values, only one-quarter of which was particulate P. NH_3-N concentrations averaged $6.5\,mg\,l^{-1}$, but were quite variable. Corn and soybean field runoff were similar with about 70% of the total P in the particulate fraction.

DISCUSSION

From analyses of watershed factors associated with nutrient concentrations and loads, livestock are an identifiable source of P and NH_3-N within the studied watersheds. Significant positive relationships were found between the concentration ($mg\,l^{-1}$) ($r = 0.55$) and losses ($kg\,ha^{-1}$) ($r = 0.79$) of P and NH_3-N from the watersheds. Because both are positively correlated with animal units in feedlots with surface or tile drainage, these feedlots are a source of nutrients to the watershed lakes. This is further supported by P and NH_3-N losses being greater in 1971 and 1973, years of high runoff.

Phosphorous and NH_3-N were found in high concentrations in feedlot surface drainages in this and other studies (Miner *et al.*, 1966; McCalla *et al.*, 1969; Gilbertson *et al.*, 1970; Townshend, Black and Janse, 1970; McQuitty, Robertson and Barber, 1971; Taylor *et al.*, 1971; U.S. Environmental Protection Agency, 1971; Edwards, Simpson and Frere, 1972). Others also have found feedlot drainage a high local source of these nutrients in streams (Taylor, 1967; Frink, 1969; Taylor and Kunishi, 1971; Morris, 1973; Muir, Sein and Olson, 1973; Burwell *et al.*, 1974; Loehr, 1974). These nutrients are high in feedlot drainage because NH_3-N results from the hydrolysis of urea (Martin and Goff, 1972) and because P is contained in the solid waste (Taiganides and Hazen, 1966; Vollenweider, 1968).

It is difficult to measure accurately the nutrients lost from feedlots within the watershed because delivery of cattle feedlot wastes often is intermittent and difficult to measure (McQuitty *et al.*, 1971). Most injection to streams occurs in the spring runoff or during periods of rainfall (Miner *et al.*, 1966), which can result in a slug load to the tributary stream (Loehr, 1970). Frequent samples and good measures of the runoff volume of feedlot drainage are necessary to estimate the quantity of nutrients lost.

In this study, the loss of plant nutrients from feedlots with drainage to a stream or tile intake can be estimated from the partial regressions of the number of animal units held in these feedlots on P and NH_3-N. An annual loss of 0.61 kg P per feedlot animal unit was found. Approximately 2630 animal units were held in feedlots having surface drainage within the lake basin studied. The P contribution from these feedlot animal units, therefore, was approximately $1600\,kg\,yr^{-1}$. This was equal to about 16% of the P that annually entered the lakes from the terrestrial watershed (Jones and Bachmann, 1975). Approximately 0.68 kg of NH_3-N is attributed to each feedlot animal, which equaled 6% of the annual terrestrial loss of NH_3-N.

Within the watershed, NO_3-N was negatively related to the percentage of area in marshland. To a lesser extent, grassland areas within the watershed may reduce NO_3-N concentration ($r = 0.39$), but this was not a significant variable in the multiple-regression analysis. Marshlands likely reduce NO_3-N because they are areas of anaerobic decomposition, resulting in denitrification and a gaseous loss of N_2 from these areas. Also, marshlands and many pasture areas do not drain toward streams via subsurface tiles. Tile drainage is high in NO_3-N (Johnson *et al.*, 1965; Erickson and Ellis, 1971; Zwerman *et al.*, 1972; Jackson *et al.*, 1973). Much of the watershed has been tiled to eliminate wetlands and facilitate row-crop agriculture. Watersheds with less tile drainage, such as marshlands, would tend to have lower NO_3-N concentrations and losses. Thomas and Barfield (1974) found NO_3-N in tile effluents much higher than non-tile drainage. In water that did not flow through tiles, the NO · -N was lower originally or was denitrified as water moved through the soil. From the slope of the partial regression of marshland (%) on NO_3-N (kg ha^{-1}) (−0.26 kg ha^{-1} NO_3-N), it was estimated that the 2200 ha of marshland area within the basin reduces the annual NO_3-N loss by 4000 kg.

Statistically, P and N losses have not been identified with the percentage of the watershed in row crops, grassland or urban areas. Yet, our sampling of these specific watersheds indicated high P and NO_3-N and NH_3-N in runoff from these areas. From other studies, pastured watersheds are known to contribute P and NH_3-N to surface waters from cattle wastes (Witzel *et al.*, 1969) and plant leachates (Timmons, Holt and Latterell, 1970; Holt, Timmons and Latterell, 1970; White and Williamson, 1973; Klausner, Zwerman and Ellis, 1974). Phosphorus losses from cropland are the result of soil erosion (Holt *et al.*, 1970; Romkens, Nelson and Mannering, 1973; Schuman, Spomer and Piest, 1973; Taylor, 1967; Timmons, Burwell and Holt, 1968). Erosional losses of P from row-crop watersheds are indicated in this study because a major fraction of the phosphorus lost from cultivated fields was associated with particulate material.

Runoff from urban areas within the watershed contained high concentrations of plant nutrients. Total P concentrations averaged 0.56 mg l^{-1}, NH_3-N averaged 1.21 mg l^{-1}, and NO_3-N averaged 1.83 mg l^{-1} in storm water (Jones, 1974). But urban areas constitute only 4% of the terrestrial watershed, making the total amount of P and N resulting from this source less than 5% of the terrestrial loss.

The nutrient concentrations found in a tributary near its mouth represent an integration of hydrologic and chemical processes after derivation from the initial sources. It is unlikely that losses are equal from all parts of the watershed (Gburek and Heald, 1974). Forces within the watershed tend to decrease P concentrations during movement from the initial nutrient source to the outlet. Soils and sediments scavenge P even within the tributary (Taylor and Kunishi, 1971; Kunishi *et al.*, 1972; and Gburek and Heald, 1974). Much of the cropland within the watershed has less than a 5% slope, and some areas drain to depressions, thus reducing erosion losses of nutrients. These forces tend to mask the input of nutrients from pastures and cultivated fields so that they could not be distinguished statistically.

Annual nutrient losses from the terrestrial watershed were approximately 0.35 kg ha^{-1} P, 6.7 kg ha^{-1} NO_3-N, and 1.0 kg ha^{-1} NH_3-N during 1971–1973. This is within the range of 0.18–1.0 kg ha^{-1} P and 3–90 kg ha^{-1} inorganic N annually lost in runoff from agricultural land as determined by other researchers (Johnson *et al.*, 1965; Vollenweider, 1968; Biggar and Corey, 1969). This annual P loss also compares with P outputs from the Des Moines River basin in Iowa. From July 1971 to August 1973, the average total P loss from this basin was 0.38 kg ha^{-1} yr^{-1} (Baumann, Oulman and Naylor 1973). During the period of study, precipitation contributed 0.32 kg ha^{-1} yr^{-1} of P to the basin (Jones and Bachmann, 1975).

The effect of the major lakes in the basin can be estimated by dividing the annual P loss from the outlet of the lake system by the total area of the basin. This amounts to 0.06 kg ha^{-1} yr^{-1} (Jones, 1974), which is significantly lower than the areal losses of the tributary watersheds. This illustrates the role of lakes in trapping P.

Although livestock were the only factor that could be correlated with P and NH_3-N inputs to the lakes, they were not the sole source of these nutrients. Runoff from grasslands, cultivated land, and urban areas had high nutrient concentrations. The nutrients attributable to feedlot livestock accounted for less than one-sixth of the P and one-tenth of the NH_3-N carried in the inflows from the watersheds. This is a reflection of the high background nutrient levels that could not be attributed to a specific source. Retention of soil material on the cropland by conservation practices would help reduce nutrient losses from the watershed.

Acknowledgements—We wish to thank Messrs. Keith Schardein, Stephen Kilkus, Keith Govro, Randall Maas, and Terry Schwarzenbach for assistance with stream data collection. The Iowa Cooperative Extension Service, U.S. Soil Conservation Service, Agricultural Stabilization and Conservation Service, and U.S. Geological Survey provided information for the land use inventory conducted by William Higgins. Dr. David Cox of the ISU Statistics Laboratory provided assistance with statistical analyses.

REFERENCES

American Public Health Association (1965) *Standard Methods for the Examination of Water and Waste Water*, 12th edition, p. 769. New York, American Public Health Association.

Baumann E. R., Oulman C. & Naylor L. (1973) Water quality studies: Red Rock and Saylorville Reservoirs, Des Moines River, Iowa. *Engineering Research Institute (Publ.)* 73280. Iowa State University, Ames, Iowa.

Biggar J. W. & Corey R. B. (1969) Agricultural drainage and eutrophication, pp. 404–445. *In Eutrophication: causes, consequences, correctives: Proc Symp.* Washington, D.C., National Academy of Sciences.

Borofka B. P. (1974) Characteristics of agricultural runoff to some Iowa lakes. Unpublished M.S. thesis. Library, Iowa State University. Ames. p. 154.

Burwell R. E., Schuman G. E., Piest R. F., Spomer R. G. & McCalla T. M. (1974) Quality of water discharged from two agricultural watersheds in southwestern Iowa. *Water Resour. Res.* **10**, 359–365.

Edwards W. M., Simpson E. C. & Frere M. H. (1972) Nutrient content of barnlot runoff water. *J. Environ. Qual.* **1**, 401–405.

Erickson A. E. & Ellis B. G. (1971) Nutrient content of drainage water from agricultural land. *Mich. Agric. Exp. Sta. Res. Bull.* **31**, 16.

Frink C. R. (1969) Water pollution potential estimated from farm nutrient budgets. *Agron. J.* **61**, 550–553.

Giburek W. J. & Heald W. R. (1974) Soluble phosphate output of an agricultural watershed in Pennsylvania. *Water Resour. Res.* **10**, 113–118.

Gilbertson C. B., McCalla T. M., Ellis J. R., Cross E. O. & Woods W. R. (1970) The effects of animal density and surface slope on characteristics of runoff, solid wastes and nitrate movement on unpaved beef feedlots. *Nebr. Agric. Exp. Sta. Bull.* **SB508.**

Holt R. R., Timmons D. R. & Latterell J. J. (1970) Accumulation of phosphates in water. *J. Agric. Food Chem.* **18**, 781–784.

Jackson W. A., Asmussen L. E., Hauser E. W. & White A. W. (1973) Nitrate in surface and subsurface flow from a small agricultural watershed. *J. Environ. Qual.* **2**, 480–482.

Johnson W. R., Ittihadieh F. I., Daum R. M. & Pillsbury A. F. (1965) Nitrogen and phosphorus in tile drainage effluent. *Soil Sci. Soc. Am. Proc.* **29**, 287–289.

Jones J. R. (1974) Eutrophication of some northwestern Iowa lakes. Unpublished Ph.D. thesis. Library. Iowa State University, Ames, Iowa. 86 pp.

Jones J. R. & Bachmann R. W. (1975) Algal response to nutrient inputs in some Iowa lakes. *Verh. Intl. Ver. Limnol.* **19**, 000–000.

Klausner S. D., Zwerman P. J. & Ellis D. F. (1974) Surface runoff of soluble nitrogen and phosphorus under two systems of soil management. *J. Environ. Qual.* **3**, 42–46.

Kunishi H. M., Taylor A. W., Heald W. R., Gburek W. J. & Weaver R. N. (1972) Phosphate movement from an agricultural watershed during two rainfall periods. *J. Agric. Food Chem.* **20**, 900–905.

Loehr R. C. (1970) Drainage and pollution from beef cattle feedlots. *J. Sanit. Engng Div. Am. Soc. civ. Engr.* **SA6. 96**, 1295–1309.

Loehr R. C. (1974) Characteristics and comparative magnitude of non-point sources. *J. Wat. Pollut. Control Fed.* **46**, 1849–1872.

Martin D. M. & Goff D. R. (1972) The role of nitrogen in the aquatic environment. *Control. Dep. Limnol. Acad. Nat. Sci. Philadelphia* **2**, 46.

McCalla T. M., Ellis J. R., Gilbertson C. B. & Woods W. R. (1969) Chemical studies of runoff following rainfall and snowmelt from beef cattle feedlots. *Agron. Abstr.* 84–85.

McQuitty J. B., Robertson J. A. & Barber E. M. (1971) Feedlot pollution: A literature review. *Res. Bull.* 71–2. Dept. of Ag. Eng. Univ. of Alberta, Edmonton, Canada.

Menzel D. W. & Corwin N. (1965) The measurement of total phosphorus in seawater based on the liberation of organically bound fractions by persulfate oxidation. *Limnol. Oceanog.* **10**, 280–282.

Miner J. R., Lipper R. I., Fina L. R. & Funk J. W. (1966) Cattle feedlot runoff: Its nature and variation. *J. Wat. Pollut. Control Fed.* **38**, 1582–1591.

Morris R. L. (1973) Stream water quality as affected by cattle feedlot runoff: A preliminary report. p. 22. State Hygienic Laboratory, Iowa City, Iowa.

Muir J., Sein E. C. & Olson R. A. (1973) A study of factors influencing the nitrogen and phosphorus contents of Nebraska waters. *J. Environ. Qual.* **2**, 466–470.

Murphy J. & Riley J. P. (1962) A modified single solution method for the determination of phosphate in natural waters. *Analyt. Chim. Acta* **27**, 31–36.

Romkens M. J. M., Nelson D. W. & Mannering J. V. (1973) Nitrogen and phosphorus composition of surface runoff as affected by tillage method. *J. Environ. Qual.* **2**, 292–295.

Schuman G. E., Spomer R. G. & Piest R. F. (1973) Phosphorus losses from four agricultural watersheds on Missouri Valley loess. *Soil Sci. Soc. Am. Proc.* **37**. 424–427.

Taiganides E. P. & Hazen T. E. (1966) Properties of farm animal excreta. *Trans Am. Soc. Agric. Engng* **9**, 374–376.

Taylor A. W. (1967) Phosphorus and water pollution. *J. Soil Wat. Conserv.* **22**, 228–231.

Taylor A. W. & Kunishi H. M. (1971) Phosphate equilibra on stream sediment and soil in a watershed draining an agricultural region. *J. Agric. Food Chem.* **19**, 827–831.

Taylor A. W., Edwards W. M. & Simpson E. C. (1971) Nutrients in streams draining woodland and farmland near Coshocton, Ohio. *Wat. Resour. Res.* **7**, 81–89.

Thomas G. W. & Barfield B. J. (1974) The unreliability of tile effluent for monitoring subsurface nitrate–nitrogen losses from soils. *J. Environ. Qual.* **3**, 183–185.

Timmons D. R., Burwell R. E. & Holt R. F. (1968) Loss of crop nutrients through runoff. *Minnesota Sci.* **24**, 4. 16–18.

Timmons D. R., Holt R. F. & Latterell J. J. (1970) Leaching of crop residues as a source of nutrients in surface runoff water. *Wat. Resour. Res.* **6**, 1367–1375.

Townshend, A. R., Black S. A. & Janse J. F. (1970) Beef feedlot operations in Ontario. *J. Wat. Pollut. Control Fed.* **42**, 195–208.

U.S. Environmental Protection Agency (1971) Role of animal wastes in agricultural land runoff. *Wat. Pollut. Control Res. Ser.* 13020 DGX 08/71.

Vollenweider R. A. (1968) Scientific fundamentals of the eutrophication of lakes and flowing waters, with particular reference to phosphorus and nitrogen as factors in eutrophication. *OECD Technical Report* DAS/CSI/68.27. p. 159.

White E. M. & Williamson E. J. (1973) Plant nutrient concentrations in runoff from fertilized cultivated erosion plots and prairie in eastern South Dakota. *J. Environ. Qual.* **2**, 453–455.

Witzel S. A., Minshall N., Nichols M. S. & Wilke J. (1969) Surface runoff and nutrient losses from Fennimore watersheds. *Agric. Engng* **12**, 338–341.

Zwerman P. J., Greweling T., Klausner S. D. & Lathwell D. L. (1972) Nitrogen and phosphorus content of water from tile drains at two levels of management and fertilization. *Soil Sci. Soc. Am. Proc.* **36**, 134–137.

Commentary on Paper A4

Vannote, R. L., Minshall, G. W., Cummins, K. W., Sedell, J. R. & Cushing, C. E. (1980) The river continuum concept.

This is the most highly cited paper in stream ecology (2232 times according to Google Scholar by May 2009). Because of the influence of all the co-authors and the impetus of American river ecologists who engaged in studies to prove the concept (e.g. Cushing *et al.*; 1983; Minshall *et al.*, 1983; Bott *et al.*, 1985), the impact of this paper was immediate.

In fact, the paper had all of the components necessary to be controversial, in particular for all river ecologists dealing with non-temperate or non-pristine systems. It was innovative in the way that river systems were contemplated, well-written, and intriguing in most of its assertions. The debate was also as immediate as its impact. Papers by Ward & Stanford (1983), Statzner & Higler (1985) and Minshall *et al.* (1985) joined the debate about the concepts. An uncountable number of papers have since checked on the relevance of the River Continuum Concept (RCC) postulates for their particular river systems.

The paper provided a new paradigm on stream and river ecology. Its foundations were in the work of American geomorphologists, Luna Leopold being the most notable (Leopold *et al.*, 1964). These claimed that there is a dynamic equilibrium between the hydrology and the geomorphology of river systems. Essentially, the RCC stated that the functional characteristics of the system were related to its structure, and that this relationship had a spatial framework that justifies the progression, the so-called continuum. The concept therefore develops a "biological analog of the physical system". The RCC proposed that the energy inputs and its forms progressively changed with river size (downstream), and that the biological communities changed accordingly. The community structure and trophic strategies were therefore changing with river size and material inputs. The river metabolism was a result of these relationships and changed from heterotrophy to autotrophy, depending on the availability of organic matter and light. The river habitat and the riparian environment changed accordingly with the river continuum.

The concept also made some predictions on community diversity, equilibrium dynamics and stability. The dynamic equilibrium in the river would result from the maximization of the energy utilization and minimum variation in its use. The balance could produce either storage or leakage of energy. Leakage would result in benefits for downstream communities, which therefore tend to exploit upstream communities. At the time when the RCC was formulated, classical postulates about community succession (Margalef, 1960) were used in the definition of temporal dynamics in rivers. Although rejecting this framework, Vanotte *et al.* did not provide clear arguments on the temporal dimension of river ecosystems (e.g. the expected responses of the ecosystem after disturbances), essential for understanding the structure and function of rivers. Moreover, interactions between the surface water and the groundwater (and their temporal component) as sources for variability were not initially incorporated nor was the hierarchical structure of the river system, with its influence on process and function, clearly outlined. However, other papers – inspired by the RCC postulates – soon filled in some of these gaps. Minshall *et al.* (1983) showed that retention of benthic and transported organic matter decreased according to the size and number of tributaries in the river. Stanford & Ward (1993) expanded on this point, indicating that the hydrological conductivity could change according to the varying porosity of the substratum, and that these shifts mostly occurred at the transitions between river types.

This paper opened the way for the functional understanding of rivers. The framework it provided, though not mentioned originally, allows the incorporation of riparian zone hydrology and biogeochemistry as an added concept to the functioning of the river ecosystem.

SS

References

Bott, T. L., Brock, J. T., Dunn, C. S., Naiman, R J., Ovink, R. W. & Petersen, R. C. (1985) Benthic community metabolism in four temperate stream systems: An inter-biome comparison and evaluation of the river continuum concept. *Hydrobiologia* **123**, 3–45.

Cushing, C. E., Cummins, K. W., Minshall, G. W. & Vannote, R. L. (1983) Periphyton, chlorophyll a, and diatoms of the Middle Fork of the Salmon River, Idaho. *Holarctic Ecology* **6**, 221–227.

Leopold, L. B., Wolman, M. G. & Miller, J. P. (1964) *Fluvial Processes in Geomorphology.* Freeman, San Francisco, USA.

Margalef, R. (1960) Ideas for a synthetic approach to the ecology of running waters. *Int. Revue Ges. Hydrobiol.* **45**, 133–153.

Minshall, G. W., Petersen, R. C., Cummins, K. W., Bott, T. L., Sedell, J. R., Cushing, C. E. & Vanotte, R. L. (1983) Interbiome comparison of stream ecosystem dynamics. *Ecol. Monographs* **53**, 1–25.

Minshall, G. W., Cummins, K. W., Petersen, R. C., Cushing, C. E., Bruns, D. A., Sedell, J. R. & Vannote, R. L. (1985) Development in Stream Ecosystem Theory. *Canadian J. Fisheries & Aquatic Sci.* **42**, 1045–1055.

Stanford, J. A. & Ward, J. V. (1993) An Ecosystem perspective of alluvial rivers: Connectivity and the hyporheic corridor. *J. North Am. Benthological Soc.* **12**, 48–60.

Statzner, B. & Higler, B. (1985) Questions and comments on the River Continuum Concept. *Canadian J. Fisheries & Aquatic Sci.* **42**, 1038–1044.

Ward, J. V. & Stanford, J. A. (1983) The serial discontinuity concept of lotic ecosystems. In: *Dynamics of Lotic Ecosystems* (ed. by T. D. Fontaine & S. M. Bartell), 29–42. Ann Arbor, Michigan, USA.

PERSPECTIVES

The River Continuum Concept[1]

ROBIN L. VANNOTE

Stroud Water Research Center, Academy of Natural Sciences of Philadelphia, Avondale, PA 19311, U

G. WAYNE MINSHALL

Department of Biology, Idaho State University, Pocatello, ID 83209, USA

KENNETH W. CUMMINS

Department of Fisheries and Wildlife, Oregon State University, Corvallis, OR 97331, USA

JAMES R. SEDELL

Weyerhauser Corporation, Forestry Research, 505 North Pearl Street, Centralia, WA 98531, USA

AND COLBERT E. CUSHING

Ecosystems Department, Battelle-Pacific Northwest Laboratories, Richland, WA 99352, USA

VANNOTE, R. L., G. W. MINSHALL, K. W. CUMMINS, J. R. SEDELL, AND C. E. CUSHING. 1980. The river continuum concept. Can. J. Fish. Aquat. Sci. 37: 130–137.

From headwaters to mouth, the physical variables within a river system present a continuous gradient of physical conditions. This gradient should elicit a series of responses within the constituent populations resulting in a continuum of biotic adjustments and consistent patterns of loading, transport, utilization, and storage of organic matter along the length of a river. Based on the energy equilibrium theory of fluvial geomorphologists, we hypothesize that the structural and functional characteristics of stream communities are adapted to conform to the most probable position or mean state of the physical system. We reason that producer and consumer communities characteristic of a given river reach become established in harmony with the dynamic physical conditions of the channel. In natural stream systems, biological communities can be characterized as forming a temporal continuum of synchronized species replacements. This continuous replacement functions to distribute the utilization of energy inputs over time. Thus, the biological system moves towards a balance between a tendency for efficient use of energy inputs through resource partitioning (food, substrate, etc.) and an opposing tendency for a uniform rate of energy processing throughout the year. We theorize that biological communities developed in natural streams assume processing strategies involving minimum energy loss. Downstream communities are fashioned to capitalize on upstream processing inefficiencies. Both the upstream inefficiency (leakage) and the downstream adjustments seem predictable. We propose that this River Continuum Concept provides a framework for integrating predictable and observable biological features of lotic systems. Implications of the concept in the areas of structure, function, and stability of riverine ecosystems are discussed.

Key words: river continuum; stream ecosystems; ecosystem structure, function; resource partitioning; ecosystem stability; community succession; river zonation; stream geomorphology

VANNOTE, R. L., G. W. MINSHALL, K. W. CUMMINS, J. R. SEDELL, AND C. E. CUSHING. 1980. The river continuum concept. Can. J. Fish. Aquat. Sci. 37: 130–137.

De la tête des eaux à l'embouchure, un réseau fluvial offre un gradient continu de conditions physiques. Ce gradient devrait susciter, chez les populations habitant dans le réseau, une série de réponses aboutissant à un continuum d'ajustements biotiques et à des schémas uniformes de charge, transport, utilisation et emmagasinage de la matière organique sur tout le

[1]Contribution No. 1 from the NSF River Continuum Project.

Printed in Canada (J5632)
Imprimé au Canada (J5632)

parcours d'une rivière. Faisant appel à la théorie de l'équilibre énergétique des spécialistes de la géomorphologie fluviale, nous avançons l'hypothèse que les caractéristiques structurales et fonctionnelles des communautés fluviatiles sont adaptées de façon à se conformer à la position ou condition moyenne la plus probable du système physique. Nous croyons que les communautés de producteurs et de consommateurs caractéristiques d'un segment donné de la rivière se mettent en harmonie avec les conditions physiques dynamiques du chenal. Dans des réseaux fluviaux naturels, on peut dire que les communautés biologiques forment un continuum temporel de remplacements synchronisés d'espèces. Grâce à ce remplacement continu, il y a répartition dans le temps de l'utilisation des apports énergétiques. Ainsi, le système biologique vise à un équilibre entre une tendance vers l'utilisation efficace des apports d'énergie en partageant les ressources (nourriture, substrat, etc.), d'une part, et une tendance opposée vers un taux uniforme de transformation de l'énergie durant l'année, d'autre part. A notre avis, les communautés biologiques habitant dans des cours d'eau naturels adoptent des stratégies de transformation comportant une perte minimale d'énergie. Les communautés d'aval sont organisées de façon à tirer profit de l'inefficacité de transformation des communautés d'amont. On semble pouvoir prédire à la fois l'inefficacité (fuite) d'amont et les ajustements d'aval. Nous suggérons ce concept d'un continuum fluvial comme cadre dans lequel intégrer les caractères biologiques prévisibles et observables des systèmes lotiques. Nous analysons les implications du concept quant à la structure, fonction et stabilité des écosystèmes fluviaux.

Received May 14, 1979
Accepted September 19, 1979

Reçu le 14 mai 1979
Accepté le 19 septembre 1979

Statement of the Concept

Many communities can be thought of as continua consisting of mosaics of integrading population aggregates (McIntosh 1967; Mills 1969). Such a conceptualization is particularly appropriate to streams. Several workers have visualized streams as possessing assemblages of species which respond by their occurrences and relative abundances to the physical gradients present (Shelford 1911; Thompson and Hunt 1930; Ricker 1934; Ide 1935; Burton and Odum 1945; Van Deusen 1954; Huet 1954, 1959; Slack 1955; Minshall 1968; Ziemer 1973; Swanston et al. 1977; Platts 1979). Expansion of this idea to include functional relationships has allowed development of a framework, the "River Continuum Concept," describing the structure and function of communities along a river system. Basically, the concept proposes that understanding of the biological strategies and dynamics of river systems requires consideration of the gradient of physical factors formed by the drainage network. Thus energy input, and organic matter transport, storage, and use by macroinvertebrate functional feeding groups may be regulated largely by fluvial geomorphic processes. The patterns of organic matter use may be analogous to those of physical energy expenditure proposed by geomorphologists (Leopold and Maddock 1953; Leopold and Langbein 1962; Langbein and Leopold 1966; Curry 1972). Further, the physical structure coupled with the hydrologic cycle form a templet (Southwood 1977) for biological responses and result in consistent patterns of community structure and function and organic matter loading, transport, utilization, and storage along the length of a river.

Derivation of the Concept

As the cyclic theory for explaining the evolution of land forms and streams (young, mature, ancient) proved unsatisfactory, the concepts gradually were replaced by a principle of dynamic equilibrium (Curry 1972). The concept of the physical stream network system and the distribution of watersheds as open systems in dynamic ("quasi") equilibrium was first proposed by Leopold and Maddock (1953) to describe consistent patterns, or adjustments, in the relationships of stream width, depth, velocity, and sediment load. These "steady state" systems are only rarely characterized by exact equilibria and generally the river and its channel tend toward a mean form, definable only in terms of statistical means and extremes (Chorley 1962); hence, the idea of a "dynamic" equilibrium. The equilibrium concept was later expanded to include at least nine physical variables and was progressively developed in terms of energy inputs, efficiency in utilization, and rate of entropy gain (Leopold and Langbein 1962; Leopold et al. 1964; Langbein and Leopold 1966). In this view, equilibration of river morphology and hydraulics is achieved by adjustments between the tendency of the river to maximize the efficiency of energy utilization and the opposing tendency toward a uniform rate of energy use.

Based upon these geomorphological considerations, Vannote initially formulated the hypothesis that structural and functional characteristics of stream communities distributed along river gradients are selected to conform to the most probable position or mean state of the physical system. From our collective experience with a number of streams, we felt it was possible to translate the energy equilibrium theory from the physical system of geomorphologists into a biological analog. In this analysis, producer and consumer communities characteristic of a given reach of the river continuum conform to the manner in which the river system utilizes its kinetic energy in achieving a dynamic equilibrium. Therefore, over extended river reaches, biological communities should become established which approach equilibrium with the dynamic physical conditions of the channel.

Implications of the Concept

It is only possible at present to trace the broad outlines of the ways the concept should apply to stream ecosystems and to illustrate these with a few examples for which reasonably good information is available. From headwaters to downstream extent, the physical variables within a stream system present a continuous gradient of conditions including width, depth, velocity, flow volume, temperature, and entropy gain. In developing a biological analog to the physical system, we hypothesize that the biological organization in rivers conforms structurally and functionally to kinetic energy dissipation patterns of the physical system. Biotic communities rapidly adjust to any changes in the redistribution of use of kinetic energy by the physcial system.

Stream Size and Ecosystem Structure and Function

Based on considerations of stream size, we propose some broad characteristics of lotic communities which can be roughly grouped into headwaters (orders 1–3), medium-sized streams (4–6), and large rivers (>6) (Fig. 1). Many headwater streams are influenced strongly by the riparian vegetation which reduces autotrophic production by shading and contributes large amounts of allochthonous detritus. As stream size increases, the reduced importance of terrestrial organic input coincides with enhanced significance of autochthonous primary production and organic transport from upstream. This transition from headwaters, dependent on terrestrial inputs, to medium-sized rivers, relying on algal or rooted vascular plant production, is thought to be generally reflected by a change in the ratio of gross primary productivity to community respiration (P/R) (Fig. 2). The zone through which the stream shifts from heterotrophic to autotrophic is primarily dependent upon the degree of shading (Minshall 1978). In deciduous forests and some coniferous forests, the transition probably is approximately at order 3 (Fig. 1). At higher elevations and latitudes, and in xeric regions where riparian vegetation is restricted, the transition to autotrophy may be in order 1. Deeply incised streams, even with sparse riparian vegetation, may be heterotrophic due to side slope ("canyon") shading.

Large rivers receive quantities of fine particulate organic matter from upstream processing of dead leaves and woody debris. The effect of riparian vegetation is insignificant, but primary production may often be limited by depth and turbidity. Such light attenuated systems would be characterized by $P/R < 1$. Streams of lower order entering midsized or larger rivers (e.g. the 3rd order system shown entering the 6th order river in Fig. 1) have localized effects of varying magnitude depending upon the volume and nature of the inputs.

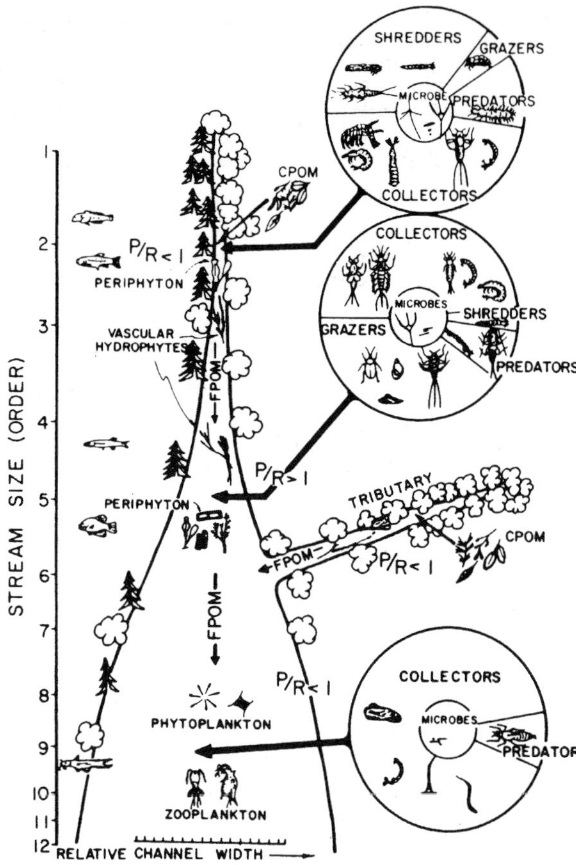

FIG. 1. A proposed relationship between stream size and the progressive shift in structural and functional attributes of lotic communities. See text for fuller explanation.

The morphological–behavioral adaptations of running water invertebrates reflect shifts in types and locations of food resources with stream size (Fig. 1). The relative dominance (as biomass) of the general functional groups — shredders, collectors, scrapers (grazers), and predators are depicted in Fig. 1. Shredders utilize coarse particulate organic matter (CPOM, >1 mm), such as leaf litter, with a significant dependence on the associated microbial biomass. Collectors filter from transport, or gather from the sediments, fine and ultra-fine particulate organic matter (FPOM, 50 μm–1 mm; UPOM 0.5–50 μm). Like shredders, collectors depend on the microbial biomass associated with the particles (primarily on the surface) and products of microbial metabolism for their nutrition. Scrapers are adapted primarily for shearing attached algae from surfaces. The proposed dominance of scrapers follows shifts in primary production, being maximized in midsized rivers

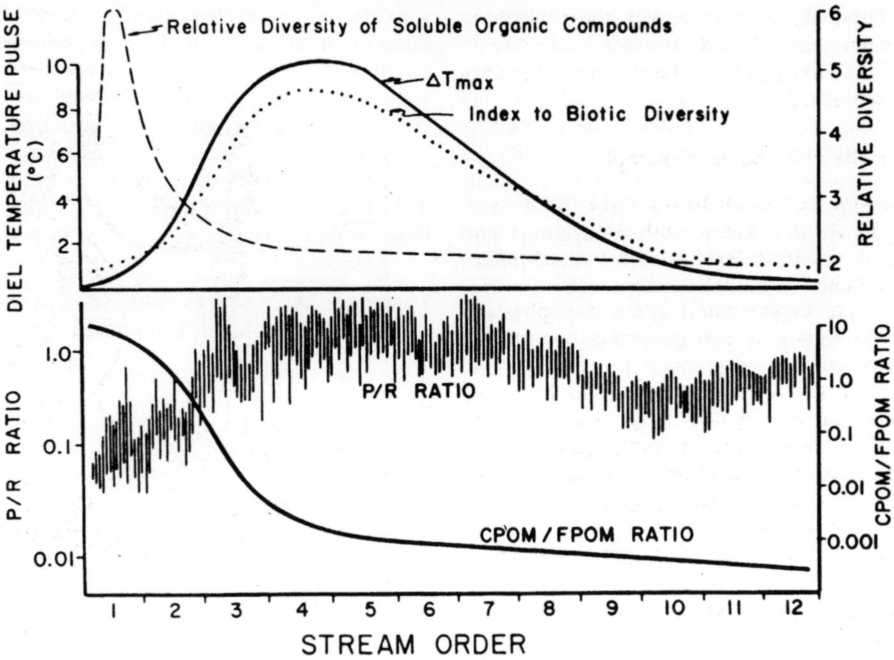

FIG. 2. Hypothetical distribution of selected parameters through the river continuum from headwater seeps to a twelfth order river. Parameters include heterogeneity of soluble organic matter, maximum diel temperature pulse, total biotic diversity within the river channel, coarse to fine particulate organic matter ratio, and the gross photosynthesis/respiration ratio.

with $P/R > 1$. Shredders are hypothesized to be codominant with collectors in the headwaters, reflecting the importance of riparian zone CPOM and FPOM–UPOM derived from it. With increasing stream size and a general reduction in detrital particle size, collectors should increase in importance and dominate the macroinvertebrate assemblages of large rivers (Fig. 1).

The predatory invertebrate component changes little in relative dominance with stream order. Fish populations (Fig. 1) show a shift from cool water species low in diversity to more diverse warm water communities (e.g. Huet 1954). Most headwater species are largely invertivores. Piscivorous and invertivorous species characterize the midsized rivers and in large rivers some planktivorous species are found — reflecting the semi-lentic nature of such waters.

The expected diversity of soluble organic compounds through the continuum is shown in Fig. 2 (dashed line). Headwater streams represent the maximum interface with the landscape and therefore are predominantly accumulators, processors, and transporters of materials from the terrestrial system. Among these inputs are heterogeneous assemblages of labile and refractory dissolved compounds, comprised of short- and long-chain organics. Heterotrophic use and physical absorption of labile organic compounds is rapid, leaving the more refractory and relatively high molecular weight compounds for export downstream. The relative importance of large particle detritus to energy flow in the system is expected to follow a curve similar to that of the diversity of soluble organic compounds; however, its importance may extend further downstream.

Thus the river system, from headwaters to mouth, can be considered as a gradient of conditions from a strongly heterotrophic headwater regime to a seasonal, and in many cases, an annual regime of autotrophy in midreaches, and then a gradual return to heterotrophic processes in downstream waters (Fisher 1977). Major bioenergetic influences along the stream continuum are local inputs (allochthonous litter and light) and transport from upstream reaches and tributaries (Fig. 1). As a consequence of physical and biological processes, the particle size of organic material in transport should become progressively smaller down the continuum (reflected by CPOM:FPOM ratio in Fig. 2, except for localized input of lower order tributaries) and the stream community response reflect progressively more efficient processing of smaller particles.

RIVER ECOSYSTEM STABILITY

Stability of the river ecosystem may be viewed as a tendency for reduced fluctuations in energy flow, while community structure and function are maintained, in the face of environmental variations. This implicitly couples community stability (sensu Ricklefs 1979) to the instability ("noise") of the physical system. In highly stable physical systems, biotic contribution to ecosystem stability may be less critical. However, in widely fluctuating environments (e.g. stream reaches with lage fluctuations in temperature), the biota may assume critical importance in stabilizing the entire system. In this interpretation, ecosystem stability is achieved by a dynamic balance between forces contributing to stabilization (e.g. debris dams, filter feeders, and other retention devices; nutrient cycling) and those contributing to its instability (e.g. floods, temperature fluctuations, microbial epidemics). In systems with a highly stable physical structure, biotic diversity may be low and yet total stability of the stream ecosystem still be maintained. In contrast, systems with a high degree of physical variation may have high species diversity or at least high complexity in species function which acts to maintain stability.

For example, in stream zones experiencing wide diel temperature changes, organisms may be exposed to suboptimum temperatures for significant portions of the day, but over some range in the diel cycle each organism encounters a favorable or optimum temperature range. Under these conditions an optimum temperature will occur for a larger number of species than if the thermal regime displayed minimum variance. Also, in the thermally fluctuating system, many populations have an opportunity to process energy, and as temperatures oscillate around a mean position, various populations may increase or decrease their processing rates. Thus, an important aspect of the predictably fluctuating physical system is that it encompasses optimum conditions for a large number of species. This interplay between physical and biological components can be seen in terms of ecosystem stability by considering the response of total biotic diversity in the river channel as balanced against the maximum diel temperature range (ΔT max) (Fig. 2). Headwater streams in proximity to groundwater supply or infiltration source areas exhibit little variation in ΔT max. With increased distance from subsurface sources and separation of the forest canopy, ΔT max will attain its widest variance because of increased solar input. The ΔT max amplitude is greatly diminished in high order streams due to the buffering effect of the large volume of water in the channel (Ross 1963). In headwater springs and brooks, diversity may be low because biological communities are assembled from those species which can function within a narrow temperature range on a restricted nutritional base; the stability of the system may be maintained by the low amplitude of diel and annual temperature regimes. Total community diversity is greatest in medium-sized (3rd to 5th order in Fig. 2) streams where temperature variations tend to be maximized. The tendency to stabilize energy flow in midsized streams may be aided by high biotic diversity which mitigates the influence of high variance in the physical system as characterized by ΔT max; i.e. variation due to fluctuating thermal regimes should be offset by a high diversity of biota. In large rivers, stability of the system should be correlated with reduction in variance of diel temperature. We wish to emphasize that temperature is not the only factor responsible for the change in community structure; it is simply one of the easiest to visualize. Other factors such as riparian influence, substrate, flow, and food also are important and change in predictable fashion downstream both absolutely and in terms of the relative heterogeneity of each.

Temporal Adjustments in Maintaining an Equilibrium of Energy Flow

Natural stream ecosystems should tend towards uniformity of energy flow on an annual basis. Although the processing rates and efficiencies of energy utilization by consumer organisms are believed to approach equilibrium for the year, the major organic substrates shift seasonally. In natural stream systems, both living and detrital food bases are processed continuously, but there is a seasonal shift in the relative importance of autotrophic production vs. detritus loading and processing. Several studies (Minshall 1967; Coffman et al. 1971; Kaushik and Hynes 1971; MacKay and Kalff 1973; Cummins 1974; Sedell et al. 1974) have shown the importance of detritus in supporting autumn-winter food chains and providing a fine particle base for consumer organisms during other seasons of the year. Autotrophic communities often form the major food base, especially in spring and summer months (Minshall 1978).

Studies on headwater (order 1–3) streams have shown that biological communities in most habitats can be characterized as forming a temporal sequence of synchronized species replacement. As a species completes its growth in a particular microhabitat, it is replaced by other species performing essentially the same function, differing principally by the season of growth (Minshall 1968; Sweeney and Vannote 1978; Vannote 1978; Vannote and Sweeney 1979). It is this continuous species replacement that functions to distribute the utilization of energy inputs over time (e.g. Wallace et al. 1977). Individuals within a species will tend to exploit their environment as efficiently as possible. This results in the biological system (composite species assemblage) tending to maximize energy consumption. Because some species persist through time and because new species become dominant, and these too are exploiting their environment as efficiently as possible, processing of energy by the changing biological system tends to result in uniform energy processing over time. Thus, the biological system moves towards equilibrium by a trade-off between a tendency to make most efficient use of energy inputs through resource partitioning of food, substrate, temperature, etc. and tendency toward a uniform rate of energy processing throughout the year. From strategies observed on small to medium-sized streams (orders 1–5), we propose that biological communities, developed in natural streams in dynamic equilibrium, assume processing strategies involving minimum energy loss (termed maximum "spiraling" by Webster 1975).

Ecosystem Processing Along the Continuum

The dynamic equilibrium resulting from maximization of energy utilization and minimization of variation in its use over the year determines storage or leakage of energy. Storage includes production of new tissue and physical retention of organic material for future processing. In stream ecosystems, unused or partially processed materials will tend to be transported downstream. This energy loss, however, is the energy income, together with local inputs, for communities in downstream reaches. We postulate that downstream communities are structured to capitalize on these inefficiencies of upstream processing. In every reach some material is processed, some stored, and some released. The amount released in this fashion has been used in calculating system efficiency (Fisher 1977). Both the upstream inefficiency (leakage) and the downstream adjustments seem predictable. Communities distributed along the river are structured to process materials (specific detrital sizes, algae, and vascular hydrophytes) thereby minimizing the variance in system structure and function. For example, materials prone to washout, such as flocculant fine-particle detritus, might be most efficiently processed either in transport or after deposition in downstream areas. The resistivity of fine particle detritus to periodic washout is increased by sedimentation in depositional zones or by combination in a matrix with the more cohesive silt and clay sediments. Thus, enhanced retention results in the formation of a distinct community adapted to utilize this material. The minimization of the variance of energy flow is the outcome of seasonal variations of energy input rates (detritus and autotrophic production), coupled with adjustments in species diversity, specialization for food processing, temporal expression of functional groups, and the erosional–depositional transport and storage characteristics of flowing waters.

Time Invariance and the Absence of Succession in Stream Communities

A corollary to the continuum hypothesis, also arising from the geomorphological literature (Langbein and Leopold 1966), is that studies of biological systems established in a dynamically balanced physical setting can be viewed in a time independent fashion. In the context of viewing adaptive strategies and processes as continua along a river system, temporal change becomes the slow process of evolutionary drift (physical and genetic). Incorporation of new functional components into the community over evolutionary time necessitates an efficiency adjustment towards reduced leakage. In natural river systems, community structure gains and loses species in response to low probability cataclysmic events and in response to slow processes of channel development.

The concept of time invariance allows integration of community structure and function along the river without the illusion that successional stages are being observed at a given location in a time-dependent series. The concept of biological succession (Margalef 1960) is of little use for river continua, because the communities in each reach have a continuous heritage rather than an isolated temporal composition within a sequence of discrete successional stages. In fact, the biological subsystems for each reach are in equilibrium with the physical system at that point in the continuum. The concept of heritage implies that in natural river systems total absence of a population is rare, and biological subsystems are simply shifting spatially (visualize a series of overlapping normal species-abundance curves in which all species are present at any point on the spatial axis but their abundance differs from one point to the next) and not in the temporal sense typical of plant succession.

On an evolutionary time scale, the spatial shift has two vectors: a donwstream one involving most of the aquatic insects and an upstream one involving molluscs and crustaceans. The insects are believed to have evolved terrestrially and to be secondarily aquatic. Since the maximum terrestrial–aquatic interface occurs in the headwaters, it is likely that the transition from land to water first occurred here with the aquatic forms then moving progressively downstream. The molluscs and crayfish are thought to have developed in a marine environment and to have moved through estuaries into rivers and thence upstream. The convergence of the two vectors may explain why maximum species diversity occurs in the midreaches.

Conclusion

We propose that the River Continuum Concept provides a framework for integrating predictable and observable biological features of flowing water systems with the physical–geomorphic environment. The model has been developed specifically in reference to natural, unperturbed stream ecosystems as they operate in the context of evolutionary and population time scales. However, the concept should accommodate many unnatural disturbances as well, particularly those which alter the relative degree of autotrophy:heterotrophy (e.g. nutrient enrichment, organic pollution, alteration of riparian vegetation through grazing, clear-cutting, etc.) or affect the quality and quantity of transport (e.g. impoundment, high sediment load). In many cases, these alterations can be thought of as reset mechanisms which cause the overall continuum response to be shifted toward the headwaters or seaward depending on the type of perturbation and its location on the river system.

A concept of dynamic equilibrium for biological communities, despite some difficulties in absolute definition, is useful because it suggests that community structure and function adjust to changes in certain geomorphic, physical, and biotic variables such as stream flow, channel morphology, detritus loading, size of particulate organic material, characteristics of autotrophic production, and thermal responses. In developing a theory of biological strategies along the river continuum, it also should be possible to observe a number of patterns that describe various processing rates, growth strategies, metabolic strategies, and community structures and functions. Collection of extensive data sets over the long profile of rivers are needed to further test and refine these ideas.

Acknowledgments

The ideas presented here have been refined through discussions with our associates in the River Continuum Project, especially T. L. Bott, J. D. Hall, R. C. Petersen, and F. J. Swanson; and other colleagues including S. A. Fisher, C. A. S. Hall, L. B. Leopold, F. J. Triska, and J. B. Webster. We are grateful to J. T. Brock, A. B. Hale, C. P. Hawkins, J. L. Meyers, and J. R. Moeller for their constructive criticism of a draft version of the manuscript. This work was supported by the U.S. National Science Foundation, Ecosystems Studies program under grant numbers BMS-75-07333 and DEB-7811671.

BURTON, G. W., AND E. P. ODUM. 1945. The distribution of stream fish in the vicinity of Mountain Lake, Virginia. Ecology 26: 182–194.

CHORLEY, R. J. 1962. Geomorphology and general systems theory. U.S. Geol. Surv. Prof. Pap. 500-B: 10 p.

COFFMAN, W. P., K. W. CUMMINS, AND J. C. WUYCHECK. 1971. Energy flow in a woodland stream ecosystem. I. Tissue support trophic structure of the autumnal community. Arch. Hydrobiol. 68: 232–276.

CUMMINS, K. W. 1974. Structure and function of stream ecosystems. BioScience 24: 631–641.

CURRY, R. R. 1972. Rivers — a geomorphic and chemical overview, p. 9–31. *In* R. T. Oglesby, C. A. Carlson, and J. A. McCann [ed.] River ecology and man. Academic Press, N.Y. 465 p.

FISHER, S. G. 1977. Organic matter processing by a stream-segment ecosystem: Fort River, Massachusetts, U.S.A. Int. Rev. Hydrobiol. 62: 701–727.

HUET, M. 1954. Biologie, prifils en long et en travers des eaux courantes. Bull. Fr. Piscic. 175: 41–53.

——— 1959. Profiles and biology of Western European streams as related to fish management. Trans. Am. Fish. Soc. 88: 153–163.

IDE, F. P. 1935. The effect of temperature on the distribution of the mayfly fauna of a stream. Publ. Ont. Fish. Res. Lab. 50: 1–76.

KAUSHIK, N. K., AND H. B. N. HYNES. 1971. The fate of dead leaves that fall into streams. Arch. Hydrobiol. 68: 465–515.

LANGBEIN, W. B., AND L. B. LEOPOLD. 1966. River meanders — theory of minimum variance. U.S. Geol. Surv. Prof. Pap. 422-H: 15 p.

LEOPOLD, L. B., AND T. MADDOCK JR. 1953. The hydraulic geometry of stream channels and some physiographic implications. U.S. Geol. Surv. Prof. Pap. 252: 57 p.

LEOPOLD, L. B., AND W. B. LANGBEIN. 1962. The concept of entropy in landscape evolution. U.S. Geol. Surv. Prof. Pap. 500-A: 20 p.

LEOPOLD, L. B., M. G. WOLMAN, AND J. P. MILLER. 1964. Fluvial processes in geomorphology. W. H. Freeman, San Francisco, Calif. 522 p.

MACKAY, R. J., AND J. KALFF. 1973. Ecology of two related species of caddisfly larvae in the organic substrates of a woodland stream. Ecology 54: 499–511.

MCINTOSH, R. P. 1967. The concept of vegetation. Bot. Rev. 33: 130–187.

MARGALEF, R. 1960. Ideas for a synthetic approach to the ecology of running waters. Int. Rev. Gesamten Hydrobiol. 45: 133–153.

MILLS, E. L. 1969. The community concept in marine zoology, with comments on continua and instability in some marine communities: a review. J. Fish. Res. Board Can. 26: 1415–1428.

MINSHALL, G. W. 1967. Role of allochthonous detritus in the trophic structure of a woodland springbrook community. Ecology 48: 139–149.

——— 1968. Community dynamics of the benthic fauna in a woodland springbrook. Hydrobiologia 32: 305–339.

——— 1978. Autotrophy in stream ecosystems. BioScience 28: 767–771.

PLATTS, W. S. 1979. Relationships among stream order, fish populations, and aquatic geomorphology in an Idaho river drainage. Fisheries 4: 5–9.

RICKER, W. E. 1934. An ecological classification of certain Ontario streams. Univ. Toronto Stud. Biol. 37: 1–114.

RICKLEFS, R. E. 1979. Ecology. Chiron Press, Inc. New York, N.Y. 966 p.

ROSS, H. H. 1963. Stream communities and terrestrial biomes. Arch. Hydrobiol. 59: 235–242.

SEDELL, J. R., F. J. TRISKA, J. D. HALL, N. H. ANDERSON, AND J. H. LYFORD JR. 1974. Sources and fate of organic inputs in coniferous forest streams, p. 57–69. *In* R. H. Waring and R. L. Edmonds [ed.] Integrated research in the Coniferous Forest Biome. Bull. Coniferous Forest Biome Ecosystem Analysis Studies. Univ. Washington, Seattle, Wash. 78 p.

SHELFORD, V. E. 1911. Ecological succession. I. Stream fishes and the method of physiographic analysis. Biol. Bull. 21: 9–35.

SLACK, K. V. 1955. A study of the factors affecting stream productivity by the comparative method. Invest. Indiana Lakes Streams 4: 3–47.

SOUTHWOOD, T. R. E. 1977. Habitat, the templet for ecological strategies? J. Anim. Ecol. 46: 337–365.

SWANSTON, D. N., W. R. MEEHAN, AND J. A. MCNUTT. 1977. A quantitative geomorphic approach to predicting productivity of pink and chum salmon in Southeast Alaska. Publ. Pac. N. W. Forest Range Exp. Stn.

SWEENY, B. W., AND R. L. VANNOTE. 1978. Size variation and the distribution of hemimetabolous aquatic insects: two thermal equilibrium hypotheses. Science 200: 444–446.

THOMPSON, D. H., AND F. D. HUNT. 1930. The fishes of Champaign County: a study of the distribution and abundance of fishes in small streams. Bull. Ill. Nat. Hist. Surv. 19: 5–101.

VAN DEUSEN, R. D. 1954. Maryland freshwater stream classification by watersheds. Contr. Chesapeake Biol. Lab. 106: 1–30.

VANNOTE, R. L. 1978. A geometric model describing a quasi-equilibrium of energy flow in populations of stream insects. Proc. Natl. Acad. Sci. U.S.A. 75: 381–384.

VANNOTE, R. L., AND B. W. SWEENEY. 1979. Geographic analysis of thermal equilibria: a conceptual model for evaluating the effect of natural and modified thermal regimes on aquatic insect communities. Am. Nat. 14: (In press)

WALLACE, J. B., J. R. WEBSTER, AND W. R. WOODALL. 1977. The role of filter feeders in flowing waters. Arch. Hydrobiol. 79: 506–532.

WEBSTER, J. R. 1975. Analysis of potassium and calcium dynamics in stream ecosystems on three southern Appalachian watersheds of contrasting vegetation. Ph.D. thesis, Univ. Georgia, Athens, Ga. 232 p.

ZIEMER, G. L. 1973. Quantitative geomorphology of drainage basins related to fish production. Alaska Fish Game Dep. Info. Leafl. 162: 1–26.

Commentary on Paper A5

Triska, F. J., Sedell, J. R., Cromack, K. Jr, Gregory, S. V. & McCorison, F. M. (1984) Nitrogen budget for a small coniferous forest stream.

The H. J. Andrews Experimental Forest has, since it was established in 1948, been a site of intensive experimental research on many ecohydrological processes, including forest–stream interactions. It is located 50 miles east of Eugene, Oregon, in the Cascade Mountains of the Pacific Northwest, USA, and is managed by the US Forest Service in collaboration with Oregon State University and the Willamette National Forest, http://www.fs.fed.us/pnw/exforests/hjandrews/index.shtml. From the 1970s, ecosystem studies expanded with a particular emphasis on hydrology and nutrient cycling within small catchments. This paper summarises some of the work produced there during those active times. Stan Gregory, Frank Triska and Jim Sedell, among others, led an interdisciplinary team that combined approaches in geomorphology, hydrology, ecology and watershed management. This paper is based on the work carried out in Watershed no. 10, although this was not the only area under study at the time. Another stream, the Mack Creek, was involved in river continuum analyses (e.g. Vannote *et al.*, 1980, Paper A4; Minshall *et al.*, 1983).

The research described in this paper was carried out in a first-order basin. In small streams and catchments it is possible to have an effective input–output control. Such a type of work had also started in other low-order streams in the US such as the Hubbard Brook (e.g. Likens, 1970) and Coweeta (e.g. Webster *et al.*, 1990). These approaches had in common the use of whole watersheds, where observational or manipulative experiments could be carried out. Since this approach was rare in Europe, the type and coherence of the data obtained provided a remarkable degree of leadership for American ecologists. The way they approached the study of stream systems was quickly adopted by groups of ecologists elsewhere.

The main aim of the paper was to quantify the organic inputs from the terrestrial environment to the stream, describing the major sources of particulate material. However, a major finding was that the highest input of nitrogen was due to subsurface flow, and this was mostly in the form of organic nitrogen. This result was achieved through a detailed description of all of the relevant compartments involved in the N budget of this forest stream. Hence, the hydrological, meteorological and terrestrial inputs and outputs were evaluated. Considering the biological transformations of nitrogen, the inputs were estimated but denitrification could not yet be evaluated (the use of the acetylene blockage technique was in its infancy). In a forested system such as Watershed no. 10, largely occupied by huge debris of the Douglas fir, the largest pool of nitrogen was in the form of particulate material. The paper showed that the existence of large debris dams was essential in the transformation of the nitrogen from organic to inorganic.

This paper provided a catchment perspective in two respects. The first was the methodological approach that was used to achieve the budget and determine the flows: controlling the different inputs and outputs, as well as the transformations of nitrogen. The second derived from the results obtained. Triska and his colleagues showed that all the processes occurring in the catchment (inputs, outputs, and transformations) occurred in the relatively small area of the stream channel (less than 1% of the watershed area). Further, the stream was not operating in a steady-state annual balance, but mostly episodically. Net retention of nitrogen mostly occurred in summer, a period of intense activity but low flow, and in autumn, when floods reset the system and caused huge changes in transport and accumulation.

The paper highlighted the role of the riparian buffer zone on nitrogen dynamics, by observing the relevance of the area adjacent to the stream channel as the most active in the whole catchment.

SS

References

Likens, G. E. (1970) Effects of forest cutting and herbicide treatment on nutrient budgets in the Hubbard Brook watershed-ecosystem. *Ecol. Monographs* **40**, 23–47.

Minshall, G. W., Petersen, R. C., Cummins, K. W., Bott, T. L., Sedell, J. R., Cushing, C. E. & Vanotte, R. L. (1983) Interbiome comparison of stream ecosystem dynamics. *Ecol. Monographs* **53**, 1–25.

Vannote, R. L., Minshall, G. W., Cummins, K. W., Sedell, J. R., & Cushing, C. E. (1980) The river continuum concept. *Canadian J. Fisheries & Aquatic Sci.* **37**, 130–137.

Webster, J. W., Golladay, S. W., Benfield, E. F., D'Angelo, D. J. & Peters, G. T. (1990) Effects of forest disturbance on particulate organic matter budgets of small streams. *J. North Am. Benthological Soc.* **9**, 120–140.

Ecological Monographs, 54(1), 1984, pp. 119–140

NITROGEN BUDGET FOR A SMALL CONIFEROUS FOREST STREAM[1]

FRANK J. TRISKA
United States Geological Survey, Water Resources Division, 345 Middlefield Road, Menlo Park, California 94025 USA

JAMES R. SEDELL
United States Forest Service, Forest Sciences Laboratory, Oregon State University, Corvallis, Oregon 97331 USA

KERMIT CROMACK, JR.
Forest Research Laboratory, Oregon State University, Corvallis, Oregon 97331 USA

STAN V. GREGORY
Department of Fisheries and Wildlife, Oregon State University, Corvallis, Oregon 97331 USA

AND

F. MICHAEL MCCORISON
United States Forest Service, Tongass National Forest, Box 309, Petersburg, Alaska 99833 USA

Abstract. An annual nitrogen budget is presented for a small stream draining Watershed 10, H. J. Andrews Experimental Forest, Oregon. The role of allochthonous debris in the input, flux, and export of nitrogen is emphasized in the material balance budget. All material entering the stream channel was presumed to enter the water sometime during the year. Material estimates are based on total channel area.

The major annual nitrogen input (1974–1975) was subsurface flow (11.06 g/m^2) as dissolved organic nitrogen (10.56 g/m^2) and nitrate (0.50 g/m^2). Biological inputs of nitrogen amounted to 4.19 g/m^2 as direct terrestrial inputs of: litterfall (1.35 g/m^2), lateral movement (1.78 g/m^2), and throughfall (0.30 g/m^2). Nitrogen fixation on fine wood debris contributed an additional 0.76 g/m^2 based on rates from a nearby watershed. Total nitrogen input was 15.25 g/m^2.

The nitrogen pool was dominated by large amounts of particulate organic matter. Coarse wood constituted 32% of the nitrogen pool (3.80 g/m^2) and fine wood fractions 18% (2.18 g/m^2). The coarse wood fraction greatly influenced stream morphology. Fine organic particulates constituted an additional 40% of the nitrogen pool (4.77 g/m^2).

DON (dissolved organic nitrogen) export (8.38 g/m^2) was less than input, presumably due to biological uptake associated with litter mineralization, sorption, and chemical flocculation. Due to effective retention of particulate inputs by debris dams, biological processing in the particulate nitrogen pool, and uptake and sorption of DON, most particulate organic inputs increased in nitrogen concentration prior to export. Particulate organic nitrogen input (3.13 g/m^2) was greater than export (2.53 g/m^2). Total annual nitrogen output was 11.36 g/m^2, resulting in a gain of 3.89 $g \cdot m^{-2} \cdot yr^{-1}$ to the stream. Thus, the stream was not operating on an annual steady state, but on an input-output regime related to the processing of refractory wood debris and resetting by major storms.

Although particulate and dissolved nitrogen loss per hectare was small for the 10-ha watershed, these losses passed through or were accumulated in a pool encompassing <1% of the watershed area. This concentration of N in the stream allowed establishment of a separate ecosystem whose processing efficiency and capabilities for nutrient cycling were related to the retention capacity of the channel and nutrient quality of inputs within the reach.

Key words: allochthonous; decomposition; detritus; ecosystem; nitrogen; streams; watershed.

INTRODUCTION

Small streams draining heavily forested watersheds depend on organic inputs from the adjacent terrestrial environment (allochthonous) as a source of fixed carbon and nutrients for in situ biological processes. This dependence on terrestrial energy sources has been demonstrated for stream reaches where the forest canopy completely covers the stream, e.g., Teal 1957, Minshall 1967, Fisher and Likens 1973, Sedell et al. 1974, Meyer et al. 1981, Triska et al. 1982. In open streams, however, high in-stream (autochthonous) production can reduce the role of allochthonous inputs, e.g., Minshall 1978, Busch and Fisher 1981, Cushing and Wolf 1982. While the role of allochthonous debris as the energy base of small forested streams is well known, the role of such debris in the stream's nutrient balance has only recently been reported. Meyer et al. (1981) found that falling and blown litter contributed 28% of phosphorus input of Bear Brook,

[1] Manuscript received 7 April 1982; revised and accepted 19 January 1983.

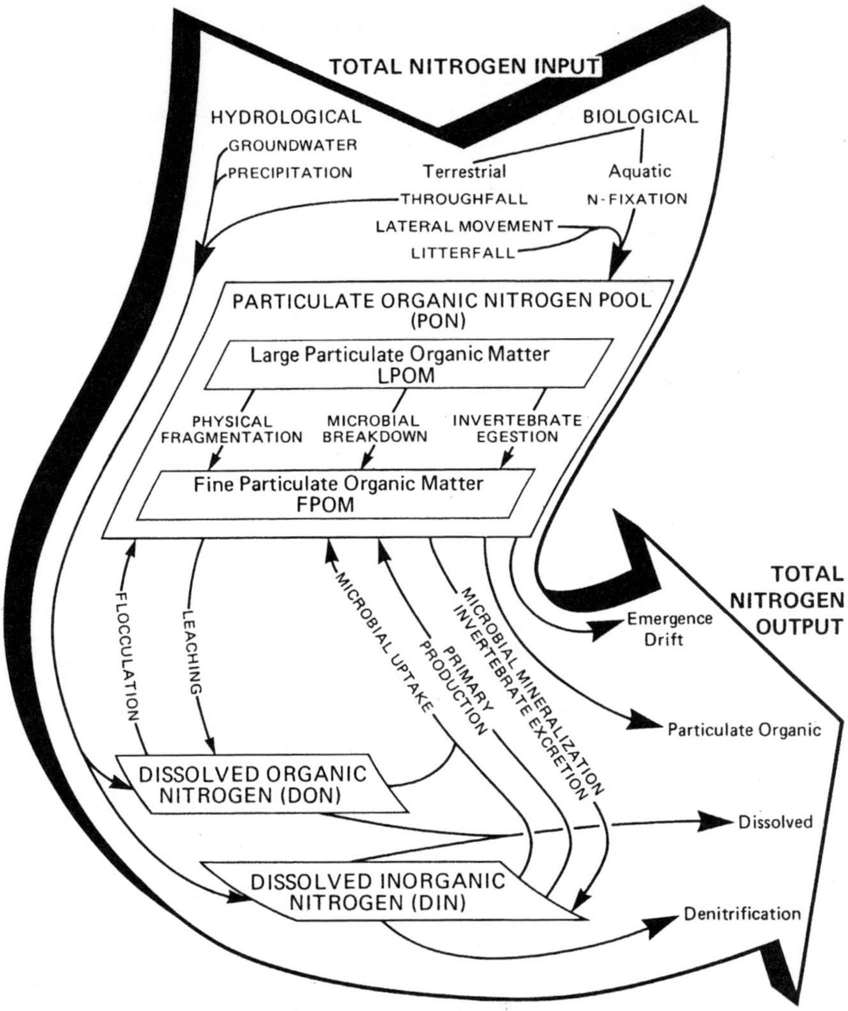

FIG. 1. Schematic drawing of major inputs and pools of nitrogen and nitrogen flux resulting from physical and biological processing of inputs in the watershed stream.

New Hampshire in 1968–1969 and 23% in 1974–1975 (Meyer and Likens 1979). Meyer et al. (1981) also report direct litterfall is ≈11% of the nitrogen budget of the same stream.

In old-growth forests of the Pacific Northwest, nitrogen is an element of interest, since nitrogen concentrations in stream water are often low, and forested watersheds may even exhibit annual net accumulation of nitrogen (Fredriksen 1972, Sollins et al. 1980). Major losses of nitrogen from the forest include leaching and mineralization of organic debris from the litter layer, and direct biological losses such as litterfall. Weathering, an important source of many nutrients, is an insignificant nitrogen source.

Once lost to the stream, nitrogen may be exchanged between the particulate and dissolved components by numerous physical and biological pathways (Fig. 1). For example, dissolved nitrogen can be taken up and immobilized during mineralization of organic carbon by bacteria and fungi. Large particulate organic matter (LPOM) can serve as a nutrient source to detritus-feeding consumers. LPOM can be converted from litter to fine particulate organic matter (FPOM) via invertebrate egestion or physical fragmentation. Particulate organic nitrogen can be converted to dissolved nitrogen via physical leaching or microbial and invertebrate excretion.

The biotic interactions which occur in the particulate organic nitrogen (PON) pool result in an increase in nitrogen concentration and often an increase in absolute levels of nitrogen during decomposition of organic matter (Kaushik and Hynes 1968, 1971, Hynes and Kaushik 1969, Iversen 1973, Hodkinson 1975, Gotto and Taylor 1976, Suberkropp et al. 1976, Triska and Sedell 1976, Davis and Winterbourn 1977, Triska and Buckley 1978, Triska and Cromack 1980, Elwood et al. 1981, J. Rosset and F. Barlocher, *personal communication*). Since nitrogen-enhancing processes are basically microbial and the C:N ratio of litter is high, it has been hypothesized that microbial colonization is instrumental in the passage of fixed carbon and/or nutrients to higher trophic levels in lotic ecosystems (conditioning) (Triska 1970, Barlocher and Kendrick 1973, Petersen and Cummins 1974, Anderson and Grafius 1975, Suberkropp and Klug 1976, Triska and Buckley 1978, Anderson and Sedell 1979, Martin et al. 1980, Triska et al. 1982). Extensive reviews of conditioning and the invertebrate feeding process are provided in Cummins and Klug (1979) and Barlocher and Kendrick (1981).

Despite large inputs of litter debris into small streams, litter and its associated biological processes in streams have not been extensively explored as components of a stream nitrogen budget. Two exceptions include an early study by Bormann et al. (1969), which estimated particulate organic nitrogen export at Hubbard Brook, and Meyer et al. (1981), who measured litterfall and fluvial particulate nitrogen in the nitrogen budget of Bear Brook, New Hampshire. More traditionally, watershed budgets assumed that alterations of the stream's water chemistry or processing by the biota of first-order streams were insignificant, although the potential contribution of exported energy and nutrients to downstream food webs has long been recognized (e.g., Bormann et al. 1969). The objective of this study was to estimate terrestrial nitrogen inputs, both dissolved and particulate, the nitrogen pool associated with organic particulates and instream autotrophs, nitrogen fixation, and other associated within-stream biological processes on the nitrogen budget of a first-order Pacific Northwest stream.

Conceptual framework

We modified the conceptual input scheme of Bormann and Likens (1967, 1970) and Fisher and Likens (1973) to highlight the biological inputs and transformations of nitrogen (Fig. 1). Input to the streams was considered hydrological-meteorological or biological (terrestrial and aquatic) (Fig. 1). Bormann and Likens (1967) defined meteorologic inputs as those which result in direct entrance of organic matter into the stream due to wind, rainfall, and other meteorologic vectors. In addition to precipitation, they included lateral movement, litterfall, and throughfall, which we defined as terrestrial biological inputs. The first of these, lateral movement, is equivalent to the blow-in of Fisher and Likens (1973). Due to the high canopy (60–90 m) and steep slopes (30–60%), wind was a minor input vector. Lateral movement inputs resulted instead from the force of gravity on the steep sidewalls, in conjunction with alternate wetting and drying of litter. Although litterfall was intensified by early autumn rains, abscission occurred over an extended period, particularly for coniferous litter, and was considered a biological rather than a meteorological input. The high canopy minimized direct interception of precipitation. Instead precipitation reached the stream indirectly by canopy drip, which resulted in significant nutrient input as throughfall. The only potential aquatic biological input of nitrogen was dinitrogen fixation. Hydrological-meteorological inputs were composed of particulate and dissolved nitrogen from upstream reaches, and direct subsurface flow to the study reach. Since the entire stream constituted our study reach, upstream import was eliminated as a separate factor. The major output was physical export of both particulate and dissolved nitrogen. Litter storage deep in the sediments was not considered an important factor, as loose mineral sediments were not thick. Nitrogen output due to microbial denitrification was not estimated, since most input-output estimates were made prior to general acceptance of the acetylene inhibition technique. Although quantification of flux within the nitrogen pool was not an objective of the study, estimation of budget parameters contributed significant insight into the impact of biological processes on the stream's nitrogen budget. As most of the stream's energy base was allochthonous and most of those inputs were at least partially processed within the reach (Sedell et al. 1974, Triska et al. 1982), biotic impacts on the nitrogen budget were investigated where possible. The approach was to construct a material balance budget by tallying hydrological-meteorological inputs, and terrestrial and aquatic biological inputs, minus hydrological and biological exports and changes in particulate organic storage. A secondary goal was assessing the contribution of associated biological processes on the stream's nitrogen budget.

Study Area

The study was made at Watershed 10 in the H. J. Andrews Experimental Forest, in the western Cascades of Oregon, USA. At the time of the study the vegetation was dominated with seral 450-yr-old Douglas-fir (*Pseudotsuga menziesii*) with a well-developed understory canopy of western hemlock (*Tsuga heterophylla*). Local environmental variation on the watershed resulted in four major plant community types which are described in detail by Grier and Logan (1977). The watershed covers 10.1 ha and rises from 430 m at the outlet stream gauging station to a maximum watershed elevation of 670 m. Mean annual precipitation is 240 cm (Waring et al. 1978*b*). Overall slope of the stream channel is 45%. Side slopes and headwall, however, range up to 60% because of deep incision of the basin into the main ridge.

Stream discharge typically ranges from ≈0.23 L/s in the summer to peak flows of 140 L/s during the highest winter freshets. Mean discharge over approximately triweekly intervals between 5 March 1974 and 18 March 1975 is presented in Table 9. The uppermost forks are intermittent during summer. Width of the stream chan-

TABLE 1. Mean annual element concentrations in discharge water collected from the weir at Watershed 10. Estimates are from a composite, discharge-proportional water sample analyzed at ≈3-wk intervals.

Source	Nutrient	Mean concentration (mg/L)
R. L. Fredriksen, *personal communication**	NO_3-N	0.006
	Kjeldahl N	0.027
	Total dissolved P	0.101
	Ortho-P	0.042
R. L. Fredriksen (1975)†	Na	1.96
	K	0.339
	Mg	0.834
	Ca	3.20
	Alkalinity (HCO_3^-)	4.17

* 5 March 1974–18 March 1975.
† May 1972–May 1975.

nel ranges from 0.25 m in the upper reaches to 1.0–1.5 m at the watershed's base. Streambed morphology is a "stairstep" series of small pools connected by free-fall zones or riffles running on bedrock. Pools are formed mainly behind accumulations of wood debris. The substrate consists of loose rocks and gravel from weathered tuff and breccia material and bedrock of unweathered tuff and breccia. As the stream was gauged at bedrock, little water was lost due to deep seepage. Water temperature typically ranges between 1° and 15°C (Sedell et al. 1975).

During the year of the study the pH of stream water ranged between 6.4 and 7.4, and bicarbonate alkalinity between 4.1 and 7.34 mg/L (Table 1). Nitrogen concentration was typically low in all forms, and dissolved phosphorus was high relative to nitrogen. Mean dissolved Kjeldahl nitrogen was 0.027 mg/L, and nitrate-nitrogen, 0.006 mg/L. Nitrite and ammonium were mostly undetectable. The nitrate concentration was historically high, due to a large pulse of nitrate of unknown origin, whose maximum concentration was discharged between 14 January and 11 February 1974. Between 28 May 1974 and 26 February 1975, mean nitrate concentration in continuously collected composite samples was 0.001 mg/L (14 samples). Dissolved total phosphate had a mean annual concentration of 0.10 mg/L and orthophosphate a mean annual concentration of 0.04 mg/L. Typical concentrations of other major cations and anions are provided in Table 1.

All inputs which entered the bank-full channel were assumed to enter the water sometime during the year. Total area of stream channel, determined by measuring bank-full width every 10 m, was calculated at 767 m^2. Lateral inputs were a function of total bank length, which was estimated at 1500 m.

Between 1972 and 1975, numerous input-export characteristics were measured: litterfall, lateral movement, standing crop, litter decomposition, seep chemistry, stream chemistry, particulate export, moss cover, insect biomass, and insect emergence. In addition, nitrogen fixation was measured in the stream at Watershed 2, a larger but chemically similar second-order stream in the same forest.

MATERIALS AND METHODS

Terrestrial biological inputs

Nitrogen input by litterfall was estimated with 11 randomly placed 1.0-m^2 litter boxes which sampled 1.5% of channel area. Trap height was 0.5 m above the water. Traps were 15 cm deep and had a removable nylon insert with a mesh opening of 500–800 μm. Litterfall was sampled approximately monthly between March 1972 and March 1975. Data from two full years, May 1972–May 1974, are reported here. Samples were returned to the laboratory, air dried, and sorted into four categories: leaves, needles, CTBW (cones, twigs, bark, wood), and miscellaneous (frass, flower parts, seeds, fruit, etc.). After fractionation, each category was dried at 50°C and weighed. Nitrogen concentration was determined on composite samples by micro-Kjeldahl analysis. Where possible, samples were composited both by year and season.

Lateral movement was sampled by 30 randomly placed rectangular boxes (0.1 m high by 0.3 m deep by 0.5 m wide) with aluminum tray inserts. Traps were placed on the forest floor adjacent to the streambed with the open end (0.5 m) parallel to the stream and facing upslope. Lateral-movement boxes sampled 1% of bankside inputs. Samples were collected monthly and sorted into the five categories described above, dried, and weighed. Seasonally composited samples of each category were analyzed for nitrogen by the micro-Kjeldahl technique. Trays were collected, sorted, and composited for nutrient analysis between March 1973 and May 1975. Two years of data, May 1973–May 1975, are reported.

Throughfall or canopy drip was measured by continuous collection into 20-L polyethylene containers equipped with funnels and strainers of 0.8–1.0 mm mesh (Grier and Logan 1977, L. E. Glenn, *personal communication*), or in precipitation gauges equipped with funnels and glass wool stoppers (Abee and Lavender 1972). Collectors were dispersed both along the stream and in the forest. Samples were taken approximately monthly, filtered (0.8-μm glass fiber filter) and analyzed for total nitrogen (macro-Kjeldahl).

Hydrological-meteorological inputs

The largest nitrogen input to the stream was subsurface runoff of precipitation modified by passing through the litter layer and soil profile. Subsurface runoff was estimated at five seeps, two of which were perennial (Fig. 2). Each seep represented a subbasin based on surface topography, and on the basis of area

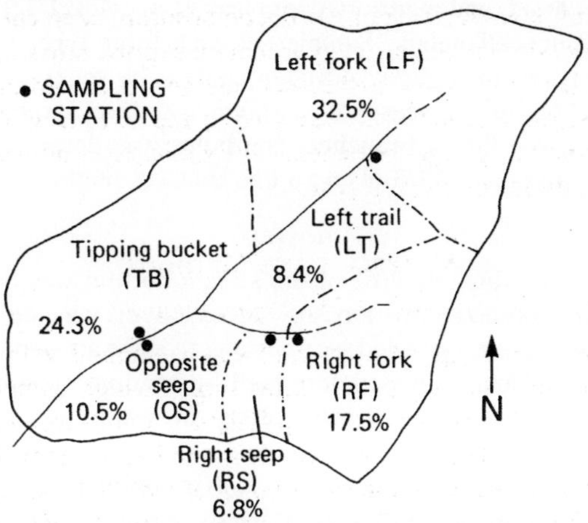

FIG. 2. Map of Watershed 10, indicating the location of major seeps and their estimated contributing areas.

was assigned a proportional contribution of discharge. Subbasin Left Trail (LT) is included with Right Fork (RF) because seepage at LT was partially contaminated by stream water diverted underground by an upstream debris dam, and because the vegetative community (*Rhododendron macrophyllum–Berberis nervosa* type [Grier and Logan 1977]) is similar to RF. Each seep was tapped with a stainless steel pipe to facilitate sampling and minimize biological uptake of nutrients at the seep's surface. Sampling by on-site personnel was approximately biweekly, with intensive sampling during major storms. Most sampling occurred between November and March, the period of greatest precipitation. Seep samples were filtered at a laboratory near the site, then frozen for later analysis of nitrate and total nitrogen (organic nitrogen + ammonium). Total nitrogen was determined by the macro-Kjeldahl method (Jackson 1958), and nitrate by the copper-cadmium reduction method (Noonan and Holcombe 1975, Rand 1976). Extrapolation to N discharge was based on water records from a gauging station at the watershed's base (R. Fredriksen, *personal communication*).

Intermittent seeps were on the stream's right fork. When these seeps ceased flowing, their proportional contribution was determined using the dissolved nitrogen concentrations from seep RF; when all right fork seeps were dry, the most upstream seep (LF, left fork) was used to estimate nitrogen input for the contributing area. These inputs are listed on the data tables as "combined." The most upstream (LF) seep was chosen because it flowed all year, provided the largest contributing area, and had both a north- and south-facing component (Fig. 2). Because nitrate was almost always at or below the limit of detection, fewer samples were analyzed for nitrate than for total nitrogen. To determine inorganic input, samples from two years (1974–1975) were used to enlarge the data base.

Aquatic biological input

The only aquatic biological input of nitrogen was by fixation on various biological substrates, particularly by bacteria associated with wood and moss. Fixation rates were determined on four wood types from Douglas-fir: heartwood blocks, twigs, bark, and wood chips. Wood substrates were dried (50°), weighed, and placed into litter bags (800-μm mesh nylon). Bags were incubated in the stream at Watershed 2, H. J. Andrews Experimental Forest, and removed seasonally for a period of 2 yr, as the study could not be completed at Watershed 10 prior to clearcutting. Upon collection, the sample was divided in half and placed in jars fitted with serum stoppers. Part of the gaseous contents (10 kPa) of the treatment jar was replaced with acetylene generated in the field from calcium carbide. The second jar served as a control. Acetylene gas samples were also checked for potential ethylene contamination. Samples were either incubated in the stream for 24 h or returned to the lab in a cooler filled with stream water and incubated for 24 h at stream temperature. Following incubation, a gas sample was transferred to a nonsterile Vacutainer for analysis by gas chromatography.

To determine nitrogen fixation rates associated with moss, samples were removed from mineral substrates submerged by stream water. Incubation time, temperatures, and gas sampling procedures were identical for both substrates. Gas analysis was undertaken on a gas chromatograph at ambient temperature using a Porpak R 1.8-m (six-foot) column. The theoretical ratio of three moles acetylene reduced to one mole nitrogen fixed was used as a conversion factor (Bergersen 1970). Recent estimates of nitrogen fixation on wood debris from forested environments (Silvester et al. 1982) indicate a ratio of 5:1 may be a more realistic conversion factor for incubations. Thus our results may be an overestimate of actual fixation rates. After incubation, tissue samples were dried at 50°, then weighed. Fixation was calculated per gram dry mass, and the result extrapolated to the known standing crop.

Terrestrial biological exports

Terrestrial biological exports were measured during a dry year, 1972–1973 (167 cm precipitation), a wet year, 1973–1974 (305 cm precipitation), and continued through spring 1975. During the dry year an 80-μm mesh nylon net was placed at the watershed's base (Fig. 2), and organic particles in excess of that size were collected. The net, ≈3 m long, continually sampled the entire discharge. During the wet year (1974) increased water volume necessitated a mesh size of 1000 μm to keep fine organic debris from clogging the net. The net was emptied when filled and samples were refrigerated prior to processing. Samples were separated into coarse particulate organic matter (CPOM) by litter type and fine particulate organic matter

(FPOM) by size class using a Tyler sieve series. Size classes of FPOM were either: large (LFPOM, 1 mm to 250 μm) or medium (MFPOM, 250 μm to 80 μm). As only CPOM was collected during 1974–1975, the year of the budget, FPOM estimates were determined as a proportion of the previous year's export. After sieving, samples were dried (50°) and weighed. Selected storm samples representing a diversity of seasons and discharge volumes were sorted into the four substrate categories used for litterfall and lateral movement. Composited samples were analyzed for nitrogen content using the micro-Kjeldahl determination.

Hydrological-meteorological exports

Hydrological-meteorological output was estimated from grab samples taken at the gauging station each time the seeps were sampled. The samples were filtered and analyzed using identical methods. Additional water samples were collected by a discharge proportional water sampler located at the gauging station (R. Fredriksen, *personal communication*). Continuous samples were composited into a single 20-L polyethylene bottle. Samples were returned to the laboratory, filtered (0.8-μm glass fiber) and analyzed at approximately triweekly intervals. Nitrate and nitrite were determined by the copper-cadmium reduction method, and ammonia and total nitrogen by the macro-Kjeldahl method. Orthophosphate was determined by the molybdate blue method and total phosphorus by the molybdate blue method with ascorbic acid following digestion with persulfate and sulfuric acid (Rand 1976). For both continuous and grab samples, the sum of nitrate, nitrite, and ammonia was subtracted from total nitrogen to estimate dissolved organic nitrogen (DON). Measurements of dissolved export were determined from 5 March 1974 to 18 March 1975, a period of simultaneous intensive sampling, for both grab and continuous samples.

Particulate organic nitrogen pool

Standing crop of particulate organic nitrogen was estimated by taking 15.25 cm diameter (0.018 m^2) cores from the stream bottom. Samples were collected monthly for 18 mo and four times per year for an additional year. Five samples each were removed from riffles and pools on each date. One year's data, May 1973–May 1974, are reported here.

CPOM was removed to a depth of 10 cm by hand and placed in a clean 4-L polyethylene bottle. FPOM was pumped from the bottom of the sampler with a hand-operated pump. Samples were returned to the laboratory, washed through a Tyler sieve series, and sorted to the following size classes: large (LFPOM, 1 mm–250 μm), medium (MFPOM, 250–75 μm), and small (SFPOM, 75–0.45 μm). The smallest size class was determined seasonally. Invertebrates were removed from samples by hand using a dissecting microscope. Coarse particulate organic matter >1 mm was dried, and then sorted into leaves, needles, CTBW, and miscellaneous. Samples of each litter type were analyzed for organic nitrogen content by micro-Kjeldahl technique seasonally. Standing crop of large debris, tree limbs, branches, and boles was determined by Froehlich (1973) using a line transect method.

Mosses

Approximately 20% of the stream bottom was bedrock colonized by mosses. Moss cover was determined using a gridded metre frame to estimate percent cover. Density was determined by random sampling of 1.0-cm squares of moss along the watercourse (J. Lyford, *personal communication*). Moss was analyzed for nitrogen by the micro-Kjeldahl method and for fixation as described previously. Estimates of cover, density, and moss standing crop were normalized to the total channel area.

Other associated biological processes

Change in nitrogen concentration of litter was monitored by a leaf pack assay (Petersen and Cummins 1974). Disappearance rates were determined for needles of Douglas-fir (*Pseudotsuga menziesii*), the most common litter input, and for deciduous litter of big-leaf maple (*Acer macrophyllum*), vine maple (*Acer circinatum*), and alder (*Alnus rubra*).

Leaf litter was collected at abscission, air dried, and strung on monofilament line to produce a 5–15 g leaf pack. Leaf packs were oven dried (50°), weighed, and anchored in the stream using bricks. Packs were oriented upstream with current holding the pack against the face of the brick. Leaf packs were therefore accessible to both biotic and abiotic processes. The result was a large variation in loss of mass due to physical abrasion, decomposition, and shredding by invertebrates.

Three packs were collected monthly from the stream. Leaf packs and bags were washed and insects and ancillary debris were removed by hand. Contents were dried at 50° and reweighed to calculate loss of mass. Dried samples were combined by date and ground through a 0.42-mm (40-mesh) screen for nitrogen and lignin analysis. Nitrogen analysis was undertaken by the micro-Kjeldahl technique. Lignin analysis was determined by the Van Soest (1963, 1965) technique. To obtain some index of humification, nitrogen content complexed to lignin was also determined by Kjeldahl analysis of lignin prepared by the Van Soest technique. To estimate total phosphorus, tissue samples were digested with nitric acid and perchloric acid, then analyzed using essentially the same method as the water samples.

RESULTS

Terrestrial biological inputs

Big-leaf maple and vine maple litter was the major deciduous leaf input to the stream. Although vine maple decomposed faster than big-leaf maple, the two

TABLE 2. Input of nitrogen introduced by litterfall to the stream channel at Watershed 10, H. J. Andrews Experimental Forest. The contribution of coarse wood debris (CWD) and microparticulate nitrogen is excluded.

	Needle		Leaf		CTBW*		Miscellaneous	
	Year 1	Year 2	Year 1	Year 2	Year 1	Year 2	Year 1	Year 2
	Litterfall N input ($mg \cdot m^{-2} \cdot mo^{-1}$)							
May	15.41	37.20	0.22	0.22	18.41	3.13	1.63	4.42
Jun	48.63	64.89	0	3.13	8.10	17.40	5.67	1.08
Jul	73.04	50.39	1.34	1.56	12.97	6.72	21.20	1.67
Aug	84.89	87.48	2.23	6.46	18.73	5.96	3.06	0.56
Sep	89.97	92.73	1.89	5.48	22.22	7.07	3.23	0.59
Oct	116.65	78.40	9.76	46.67	5.18	15.31	3.65	0.91
Nov	122.12	108.71	119.07	84.09	4.56	57.11	4.12	23.52
Dec	39.33	20.59	10.41	1.30	6.32	9.49	2.97	10.13
Jan	50.56	9.71	2.60	0.38	34.89	13.38	15.76	14.88
Feb	14.82	8.92	0.17	0.17	4.00	6.56	1.88	12.90
Mar	21.56	43.13	0.22	3.26	4.69	34.81	1.62	13.95
Apr	15.33	25.35	0.21	0.21	9.93	7.77	2.26	6.51
	Annual input (g/m^2)							
	0.69	0.63	0.15	0.15	0.15	0.19	0.07	0.09

Year 1 (May 1972–Apr 1973)		Year 2 (May 1973–Apr 1974)	
Total annual input	811.16 g	Total annual input	810.65 g
Total annual input	1.06 g/m^2	Total annual input	1.06 g/m^2

* Cones, twigs, bark, wood.

species were combined to estimate nitrogen input. Both were processed faster than coniferous needles (Sedell et al. 1975). Deciduous leaf input was 11% of litterfall nitrogen. Seasonally, most nitrogen input (68%; 80% 1972–1973, 55% 1973–1974) occurred during autumn, particularly November (Table 2). Leaf litter input was intermittent from February through August as leaves entrained by the canopy were slowly released. Partial decomposition in the canopy apparently increased nitrogen concentration, since nitrogen concentration of litter was lower at abscission (Table 3). This factor was considered in calculating nitrogen input. Estimated annual nitrogen input by leaf litter was 0.15 g/m^2 in both years.

Needles were the major nitrogen component (49%) in litterfall (Table 2). Needlefall input began in June with the onset of moisture stress and continued with increasing intensity through November. Input during June and July consisted primarily of hemlock with increasing amounts of Douglas-fir in succeeding months.

Needle litterfall during summer and fall was lowest in nitrogen concentration (Table 3). Presumably most nitrogen was reabsorbed prior to natural abscission (Whittaker et al. 1979). During winter and spring, green needles were a smaller but more concentrated input. Winter input was often storm generated, when wind or wet snow in the canopy removed fine twigs with attached live needles. Annual nitrogen input by needle litterfall was 0.66 g/m^2.

Cones, twigs, bark, and wood (CTBW) accounted for 13% of litterfall nitrogen input (Table 2). Monthly nitrogen input ranged from 3 to 57 $mg \cdot m^{-2} \cdot mo^{-1}$, mostly as a continuous shedding of small twigs. Occasionally more catastrophic events, such as a major limb or bole falling in the vicinity of a litter trap (November 1973), resulted in a large nitrogen input. Nitrogen concentration of twigs, the primary CTBW component, remained nearly constant seasonally (Table 3). Mean annual nitrogen input of CTBW was 0.17 g/m^2.

TABLE 3. Nitrogen concentration (percent) by season, for selected biological substrates from Watershed 10, H. J. Andrews Experimental Forest.

	Needle	Leaf	CTBW*	Miscellaneous
Litterfall				
Winter (Dec–Feb)	0.54	0.60	0.34	0.96
Spring (Mar–May)	0.71	0.70	0.36	0.75
Summer (Jun–Aug)	0.42	0.72	0.31	0.90
Fall (Sep–Nov)	0.46	0.63	0.38	0.98
Lateral movement				
Winter (Jan–Mar)	0.54	0.76	0.30	1.10
Spring (Apr–Jun)	0.77	0.58	0.36	1.05
Summer (Jul–Sep)	0.66	0.65	0.39	1.03
Fall (Oct–Dec)	0.48	0.60	0.29	1.00
Particulate nitrogen pool				
Winter (Dec–Feb)	0.66	0.87	0.29	0.62
Spring (Mar–May)	0.74	0.86	0.24	0.66
Summer (Jun–Aug)	0.53	1.11	0.31	0.80
Fall (Sep–Nov)	0.52	1.16	0.29	1.00
Fine particulate nitrogen pool	Large	Medium		Small
All seasons	0.71	0.75		1.44
Export				
Winter (Dec–Feb)	0.63	0.98	0.48	1.11
Fall (Oct–Dec)	0.56	1.04	0.37	1.47

* Cones, twigs, bark, wood.

TABLE 4. Nitrogen introduced by lateral movement of organic material to the stream channel at Watershed 10, H. J. Andrews Experimental Forest.

	Needle		Leaf		CTBW*		Miscellaneous	
	Year 1	Year 2	Year 1	Year 2	Year 1	Year 2	Year 1	Year 2
	Lateral-movement N input ($mg \cdot m^{-2} \cdot mo^{-1}$)							
May	16.46	22.19	11.33	9.88	19.53	331.55	10.74	35.15
Jun	38.33	20.73	8.88	6.61	91.58	19.44	52.28	17.20
Jul	33.14	18.40	21.36	7.65	38.20	21.76	26.18	17.56
Aug	14.12	22.50	7.05	8.87	31.19	33.25	11.81	23.63
Sep	25.93	23.56	13.70	17.21	47.72	75.23	40.47	9.57
Oct	38.68	25.89	20.45	46.51	50.43	91.24	17.67	25.11
Nov	48.95	28.94	119.66	77.58	101.53	29.66	37.81	20.10
Dec	30.65	30.05	26.59	29.01	29.94	123.61	36.27	34.41
Jan	29.79	23.77	25.92	16.96	41.57	61.75	30.29	20.12
Feb	21.93	21.46	21.43	15.32	26.79	55.78	43.99	18.18
Mar	19.41	15.23	17.43	20.97	176.60	91.14	37.51	22.84
Apr	28.19	11.08	13.74	11.30	43.52	55.62	44.72	21.11
	Annual input (g/m^2)							
	0.35	0.26	0.31	0.27	0.70	0.99	0.39	0.27
	Year 1 (May 1973–Apr 1974)				Year 2 (May 1974–Apr 1975)			
	Total annual input		1335.72 g		Total annual input		1373.54 g	
	Total annual input		1.74 g/m^2		Total annual input		1.79 g/m^2	

* Cones, twigs, bark, wood.

Miscellaneous inputs consisted of all other litter particulates (>800 μm) not easily grouped into the previous categories. The composition was primarily mosses and lichens (mainly *Lobaria* sp.) from the canopy. In season, floral parts, seeds, and insect frass contributed to the miscellaneous category. Nitrogen concentrations were uniformly higher in miscellaneous than in other litter categories (Table 3). Decomposition rate of miscellaneous debris was not determined but was presumably rapid, relative to other litter inputs, due to higher nitrogen concentration. Miscellaneous inputs were 6% of litterfall nitrogen (Table 2), with a mean annual input of 0.08 g/m^2. Microparticulate inputs (MICRO PN [<800 μm]) were not estimated by us but were determined by G. C. Carroll and L. H. Pike of the University of Oregon. Their results are reported in Sollins et al. (1980). Estimates, made from the particulate component of throughfall, were between 50 and 150 $kg \cdot ha^{-1} \cdot yr^{-1}$. We used an intermediate value of 100 $kg \cdot ha^{-1} \cdot yr^{-1}$ dry mass to calculate nitrogen input. Nitrogen concentration, from a composite of 100 of their samples, was 1.75 ± 0.21 (SE) percent of dry mass. This translates to 0.18 $g \cdot m^{-2} \cdot yr^{-1}$, 13% of litterfall nitrogen. A final source of litterfall nitrogen was coarse wood debris (CWD). Sollins et al. (1980) estimate input from logs at 1.1 $kg \cdot ha^{-1} \cdot yr^{-1}$ or 0.11 g/m^2. We used this estimate in our study.

Monthly comparison of inputs between years indicates some variation due in part to differences in collection dates and storm patterns. Seasonally, input amounts of all litter are similar, and on an annual basis are almost identical. Grier and Logan (1977), who estimated litterfall for 2 yr on the terrestrial component of Watershed 10, also found that yearly totals (including green litter) differed from the yearly average by <1%. Thus litterfall provided a remarkably uniform input of biologically fixed carbon and nutrients between years, but with strong seasonal trends.

Lateral nitrogen input

Conifer needles, the primary source of litterfall nitrogen, were the second largest lateral input. Nitrogen from coniferous needles constituted 17% (20% 1973–1974; 15% 1974–1975) of lateral movement (Table 4). Lateral movement of needles had seasonal trends similar to those of litterfall, but was more uniform throughout the year. As with litterfall, lateral needle inputs were highest during autumn. Fall, winter, and spring nitrogen concentrations were similar to those of litterfall, whereas summer concentrations were higher (Table 3). The summer increase may result from partial decomposition on the forest floor, with associated increase in nitrogen concentration. Mean annual nitrogen input by lateral movement of needle litter was 0.31 g/m^2.

Lateral movement of leaf litter added almost as much nitrogen to the stream as lateral needle input and almost twice as much as leaf litterfall. Most lateral leaf input was big-leaf maple and occurred during autumn. Unlike leaf litterfall, significant but decreasing inputs occurred December–May as leaves continued to slide downhill. For example, only 34% of total nitrogen input by leaf litter occurred during November for lateral movement, compared to 68% for litterfall. Nitrogen concentrations from composite samples of lateral movement during autumn were similar to those of litterfall, 0.63% and 0.60% N, respectively (Table 3). Winter concentrations of nitrogen in lateral movement were slightly higher than those in litterfall, but summer concentrations were slightly lower. Leaf litter was 16% of lateral-movement nitrogen and amounted to 0.29 $g \cdot m^{-2} \cdot yr^{-1}$ (Table 4). As with litterfall, miscellaneous organic particulates had the highest nitrogen concentration of any lateral input (1.05%). Nitrogen contribution to the stream was 0.33 $g \cdot m^{-2} \cdot yr^{-1}$, four times the contribution of miscellaneous litterfall. Miscellaneous particulate nitrogen was 19% of lateral movement input and lacked seasonal trends.

The major lateral nitrogen input was CTBW. Various types of wood debris contributed 0.85 $g \cdot m^{-2} \cdot yr^{-1}$, 48% of total lateral movement (Table 4). Seasonal composite samples ranged between 0.29 and 0.39% total nitrogen concentration (Table 3). Concentrations were lowest during late autumn and highest during summer. CTBW was the most variable debris category, due to the heavier mass of individual pieces, and random input. With the exception of CTBW, lateral movement was only 75% as great in year 2 as in year 1. The reason for the apparent annual difference is unknown, and examination of monthly input (Table 4) indicates no special trends. Total annual nitrogen input was nearly identical due to a large input of wood debris in May 1974. Overall, the data indicate that surface erosion provided a steady contribution of organic debris. This more uniform input scheme mitigated the strong seasonal input of leaf litterfall and helped insure a constant source of organic debris for biotic processing.

Dissolved nitrogen input

Throughfall

Dissolved inputs consisted of subsurface flow and throughfall. Throughfall was a potential source of biologically labile nitrogen. Only 5–30% of precipitation reached the stream by direct interception. The remainder was nutrient enriched by dripping through the canopy, which provided 20–32 m^2 of leachable foliar surfaces per square metre of forest floor (Grier and Logan 1977, Waring et al. 1978*a*). This estimate of foliar surface did not include a high density of nitrogen-fixing lichens that inhabited the overstory. Abee and Lavender (1972) estimated mean annual throughfall nitrogen inputs at 0.34 $g \cdot m^{-2} \cdot yr^{-1}$ in a study of six old-growth stands. Highest nitrogen input in throughfall was at 0.40 $g \cdot m^{-2} \cdot yr^{-1}$. Using data from C. C. Grier (*personal communication*) for collections along the stream, throughfall input was 0.40 $g \cdot m^{-2} \cdot yr^{-1}$ during 1974 and 0.35 $g \cdot m^{-2} \cdot yr^{-1}$ during 1973. L. E. Glenn et al. (*personal communication*) estimated throughfall at Watershed 10 to be 0.27 $g \cdot m^{-2} \cdot yr^{-1}$ during 1973 and 0.21 $g \cdot m^{-2} \cdot yr^{-1}$ during 1974 from a set of six sample collectors placed along the stream. The reason for input differences is not known, but may have been related to sample preservation, the number of samplers, or their distribution. Kimmins (1973) indicated that the number of collectors required for adequate estimation of throughfall is larger than the number used in the individual studies cited. All throughfall N input estimates varied within a range of 0.20–0.40 $g \cdot m^{-2} \cdot yr^{-1}$, depending on year and sampler location, with most values at the higher end of that range. An estimate of 0.30 $g \cdot m^{-2} \cdot yr^{-1}$ represents a reasonable estimate of annual throughfall nitrogen input. This estimate is slightly lower than the 0.34 $g \cdot m^{-2} \cdot yr^{-1}$ canopy solution, including stemflow, which was used by Sollins et al. (1980), for their terrestrial nitrogen budget at Watershed 10.

Subsurface inputs

Seep inputs varied seasonally both in volume and nutrient concentration. Differences in seep chemistry were attributed to drainage area and soil depth. Based on seep chemistry, annual dissolved organic nitrogen input from subsurface flow was estimated at 8.49 kg, or 10.56 g/m^2 of channel. Annual nitrate input was an additional 0.38 kg or 0.50 g/m^2 (Table 5). Thus only 5% of dissolved inputs were inorganic, with 95% organic nitrogen. Our calculation of dissolved-nitrogen input may be an underestimate for two reasons. First, seep inputs were estimated by a series of grab samples rather than by continuous monitoring. Because DON concentration varies with streamflow (highest in rising flows, lowest in falling; Fredriksen 1972, Kennedy and Malcolm 1978) grab sampling can miss some initial periods of high nitrogen concentration. Second, nitrate was not present in detectable quantities during many periods of high discharge when even trace quantities would extrapolate to significant input.

Special geological features of many seeps caused nitrogen input to differ from that expected by area alone. For instance, Left Fork (LF) drained 32.5% of the watershed, but contributed 46% of subsurface nitrogen (Fig. 2, Table 5). This was an area of deep soils and permanent discharge. Tipping Bucket (TB), which drained 24% of the watershed area, contributed only 16% of dissolved nitrogen. This area had deep soils and also a subterranean lava dike that slowed runoff and allowed additional time for nitrogen uptake and sorption. Right Seep (RS), draining only 6.8% of watershed area, also contributed 16% of dissolved nitrogen. RS was an area of very shallow soils and rapid runoff. Total nitrogen concentrations at RS during early autumn storms were equivalent to those at LF, but were 2–2.5 times higher during late autumn and early winter. Concentrations at RS were approximately four times higher than TB during equivalent periods. As expected, overall input was related to discharge. Periods when absence of flow resulted in combining of seeps represented only 1% of total subsurface input (Table 5).

TABLE 5. Dissolved-nitrogen input 5 March 1974–18 March 1975 as subsurface runoff estimated at five seeps from Watershed 10, H. J. Andrews Experimental Forest. (···) indicates absence of flow during sample interval.

Discharge interval	Seep: Left fork	Tipping bucket	Right fork	Right seep	Opposite seep	"Combined"*
			Dissolved-nitrogen input (g)			
5–27 Mar	177.23	119.89	53.80	74.16	40.12	
28 Mar–19 Apr	107.45	95.41	69.57	158.79	117.06	
20 Apr–9 May	6.47	2.76	···	···	···	8.55
10–23 May	110.64	36.77	19.60	···	···	12.93
24 May–17 Jun	60.00	34.27	62.94	···	···	41.56
18 Jun–11 Jul	79.15	43.37	···	···	···	32.89
12–31 Jul	22.30	6.78	···	···	···	49.86
1–19 Aug	15.42	1.89	2.01	···	···	1.33
20 Aug–9 Sep	182.04	0.12	3.73	···	···	2.46
10–30 Sep	123.79	4.72	2.45			1.61
1–11 Oct	21.84	0.18	0.40	4.30	0.46	
12–25 Oct	21.84	0.18	0.40	4.30	0.46	
26 Oct–13 Nov	56.83	13.86	33.49	18.74	···	13.32
14 Nov–2 Dec	999.85	145.58	161.45	110.26	···	64.20
3–24 Dec	939.41	492.22	441.11	493.53	194.57	
24 Dec–15 Jan						
16 Jan–3 Feb	707.85	299.91	150.42	468.99	97.20	
4–26 Feb						
27 Feb–18 Mar	62.85	14.45	···	···	···	83.15
NO_3N	62.96	170.65	100.04	14.32	36.17	
Seep total	3757.92	1483.01	1101.41	1347.39	486.04	311.86
Total dissolved nitrogen			8487.43			

* When seeps ceased flowing, their proportional contribution was determined using the dissolved-nitrogen concentrations from seep RF, which was still flowing. (See Methods: Hydrological-meteorological inputs).

To assess the significance of dissolved nitrogen flux seasonally we emphasized discharge patterns and major particulate input events rather than calendar seasons. So defined, the four seasons of our study were approximately: Autumn (1 October–2 December), Winter (3 December–19 April), Spring (20 April–17 June), Summer (18 June–30 September). The autumn season, 1 October–2 December, was the most active period of dissolved nitrogen retention. DON input was 1.73 kg and output 0.82 kg, indicating a net retention of almost 0.9 kg DON by various physical, chemical, and biological mechanisms. Biological retention was facilitated by high particulate organic loading. In litterfall ≈32% of needle litter nitrogen and 87% of leaf litter nitrogen fell during these 2 mo. Autumn also coincided with the beginning life cycle of many invertebrates. Anderson and Cummins (1979) note that the life cycle of many shredder invertebrates is "keyed to the autumnal pulse of leaf fall with major growth during late autumn and winter." Potential biotic interactions were further facilitated by low discharge (<3% annual discharge), which maximized water contact with biologically active surfaces. This combination of factors presumably promoted microbial uptake on litter substrates, resulting in nitrogen retention with subsequent passage to higher trophic levels.

A second period of high nitrogen retention was summer, 18 June–30 September (66%). In summer, subsurface nitrogen input was estimated at 0.58 kg and output 0.19 kg. Only 6% of annual DON input and <2% of discharge occurred during summer. As a result, although summer was a period of intense nitrogen flux, it was nearly insignificant to the annual nitrogen budget.

The winter storm season, 2 December–19 April, was characterized by high streamflow and low nitrogen concentration in stream water. This 4½ mo season had 89% of annual discharge. Some 69% of DON input (5.55 kg) and 78% (5.00 kg) of DON output occurred during winter. Despite high winter flows, net retention of DON (0.55 kg) was observed. This is less than half the 1.30 kg retained during the summer–autumn period, when <5% of discharge and 27% of DON was input to the stream.

Spring, 20 April–17 June, was a 2-mo period of rapidly declining discharge. About 6% of annual discharge occurred during spring. DON input and output nearly balanced, 0.39 kg vs. 0.43 kg, respectively. Leaf litter input was especially low in spring and followed the period of maximum surface scouring of epilithic and organic surfaces.

Annual net DON retention by the stream ecosystem was a high but realistic approximation of potential removal by litter decomposition pathways. Annual DON retention was ≈1.70 kg; litter inputs of all types averaged 2.32 kg. Our leaf pack studies indicate an ≈50%

TABLE 6. Nitrogen introduced seasonally by fixation (as measured by acetylene reduction) for four wood substrates. Data collected from incubated substrates at Watershed 2, H. J. Andrews Experimental Forest, and extrapolated to known standing crop.

Substrate type	Seasonal acetylene reduction rates ($nmol \cdot g^{-1} \cdot d^{-1}$) 1 Apr–31 Oct	1 Nov–31 Mar	Nitrogen fixed ($g \cdot m^{-2} \cdot yr^{-1}$)
Twigs	1129	21	0.09
Bark	269	6	0.17
Chips	993	4	0.10
Wood (3–10 cm diameter)	160	8	0.34

concentration gain in nitrogen following leaching and up to 100% nitrogen concentration increase in other litter inputs (Table 3). While much of the apparent gain resulted from carbon mineralization, a 50% gain by litter detritus would require 1.15 kg of nitrogen. Furthermore, the large amount of residual wood debris and fine organic particulates provided additional uptake sites. Therefore, the observed retention of dissolved nitrogen represents a reasonable nitrogen demand for carbon mineralization.

Aquatic biological input

Maximum nitrogen fixation (as measured by acetylene reduction) occurred during late summer (Table 6). Little or no activity was observed from late autumn through spring. Rates of nitrogen fixation on various wood types were extrapolated to the composition of the CTBW pool for wood debris <10 cm in diameter, and for seasons when fixation was observed. Nitrogen fixation was not attributed to large branches and bole wood, due to absence of reliable estimates. More recently, N-fixation has been detected in large boles in the H. J. Andrews Experimental Forest, but confined to discrete sites of activity that are difficult to extrapolate to an areal basis (Silvester et al. 1982). Rates of fixation were low on fine wood debris and required 24-h incubations, which may overestimate fixation (Silvester et al. 1982). Despite low rates, standing crop of CTBW was high, so the potential contribution of nitrogen fixation to the budget was significant. The greatest contribution to the particulate nitrogen pool from nitrogen fixation was associated with branch wood debris (BWD) at 0.34 g/m^2. Branch wood was 3–10 cm diameter. Contributions to the particulate nitrogen pool for small wood substrates were 0.09 g/m^2 for twigs, 0.17 g/m^2 for bark, and 0.10 g/m^2 for small wood chips. Total nitrogen fixation on various wood substrates was estimated at 0.70 $g \cdot m^{-2} \cdot yr^{-1}$, higher than the 0.008 g/m^2 estimated by Meyer et al. (1981) from terrestrial wood studies by Roskoski (1977).

Nitrogen fixation on moss was estimated in June 1975 and June and August 1976. Fixation was assumed to occur primarily during the warm months as observed for wood. Acetylene reduction per unit dry mass averaged 8769 $nmol \cdot g^{-1} \cdot d^{-1}$. Areal conversion of nitrogen amounted to 0.06 $g \cdot m^{-2} \cdot yr^{-1}$. Examination of moss plants did not reveal the presence of algal symbionts; therefore, fixation was presumably accomplished by bacterial association. Bacterial association between a moss and a nitrogen-fixing *Azotobacter* sp. has been reported by Snyder and Wullstein (1973).

Fixation of nitrogen on leaf material collected from the stream produced negative results. However, Howarth and Fisher (1976) have reported low rates of nitrogen fixation on leaf litter in experimental streams. Further evidence of fixation during the initial stages of decomposition of mangrove leaves (Gotto and Taylor 1976) and in cypress litter (Dierberg and Brezonik 1981) indicates fixation can occur at significant rates at that time in the decomposition process.

Particulate nitrogen pool

Because the stream's morphology was extensively controlled by debris dams, these wooden structures represented a large nitrogen pool. The standing crop of coarse wood debris (CWD) >10 cm diameter, amounted to 12 kg/m^2 particulate organic matter. Analysis of heartwood blocks disclosed a total-nitrogen concentration of 0.04%. This estimate is low compared to P. C. MacMillan and J. E. Means (*personal communication*) who estimated nitrogen content of 0.10% in fallen logs in a midaltitude Douglas-fir stand on the H. J. Andrews Experimental Forest. Even using the lower estimate (assuming N accretion after tree fall), the minimum standing crop of nitrogen in the stream from coarse wood debris was 3.8 g/m^2. As the biological role of this long-term pool is virtually unknown, the amount of nitrogen from large bole wood which enters the food chain was not determined. The most likely biological role of coarse wood debris in small streams is formation of FPOM by physical scouring and surface microbial activity. Maximum processing rate of coarse wood debris is 2–5% per year, based on disappearance rates of heartwood blocks (Triska and Cromack 1980) and is far slower for the larger pieces. A secondary role for wood debris is habitat for a wide variety of tunneling and surface-inhabitating invertebrates (Anderson et al. 1978).

Compared to CWD and BWD, CTBW has a much faster turnover rate. Gilson respirometry (Gilson 1963) conducted monthly at ambient stream temperature showed ≈18% annual decomposition for fine twigs. Fine twigs, bark, and wood chip substrates exhibited mineralization of 14–23% loss of mass within the 1st yr of stream incubation. Nitrogen concentration of composite standing crop samples ranged between 0.25 and 0.31% nitrogen by micro-Kjeldahl analysis. Highest nitrogen concentration was observed during summer (0.31%) when microbial activity measured by both respirometry and plate counts was highest. Nitrogen

TABLE 7. Standing crop of nitrogen associated with various particulate organic materials from the stream channel of Watershed 10. Each monthly estimate is the result of 10 core samples (15.25 cm diameter), 5 each from riffles and pools.

	Needle	Leaf	CTBW*	Miscellaneous	Large FPOM (0.25 mm)	Medium FPOM (0.075 mm)
	Nitrogen standing crop, $\bar{x}$ ± SE (g/m²)					
1973						
Mar	0.31 ± 0.19	0.14 ± 0.08	1.63 ± 0.72	0.34 ± 0.19	0.35 ± 0.16	0.58 ± 0.15
Jun	0.19 ± 0.07	0.06 ± 0.05	2.40 ± 0.80	0.41 ± 0.07	0.30 ± 0.07	0.60 ± 0.19
Jul	0.33 ± 0.15	0.03 ± 0.02	0.61 ± 0.20	0.30 ± 0.10	0.62 ± 0.16	1.59 ± 0.30
Aug	0.24 ± 0.06	0.03 ± 0.02	1.76 ± 0.98	0.92 ± 0.40	0.66 ± 0.18	1.07 ± 0.17
Sep	0.26 ± 0.06	0.04 ± 0.02	1.11 ± 0.41	0.67 ± 0.17	0.31 ± 0.06	0.43 ± 0.07
Oct	0.51 ± 0.19	0.09 ± 0.03	1.58 ± 0.60	0.81 ± 0.21	0.33 ± 0.08	0.55 ± 0.12
Nov	0.59 ± 0.13	0.09 ± 0.03	3.70 ± 1.11	2.44 ± 0.59	0.90 ± 0.31	0.64 ± 0.23
Dec	0.34 ± 0.16	0.16 ± 0.09	3.55 ± 1.06	0.48 ± 0.18	0.88 ± 0.31	0.89 ± 0.26
1974						
Jan	0.15 ± 0.08	0.05 ± 0.02	2.94 ± 1.44	0.73 ± 0.29	0.82 ± 0.30	0.95 ± 0.29
Feb	0.08 ± 0.03	0.05 ± 0.02	3.22 ± 1.35	0.32 ± 0.10	0.61 ± 0.15	0.58 ± 0.12
Mar	0.09 ± 0.02	0.09 ± 0.04	2.84 ± 0.87	0.32 ± 0.12	0.59 ± 0.29	0.78 ± 0.45
Apr	0.17 ± 0.06	0.18 ± 0.08	0.84 ± 0.31	1.15 ± 0.40	0.62 ± 0.18	0.74 ± 0.20
Mean annual	0.27	0.08	2.18	0.74	0.58	0.78

* Cones, twigs, bark, wood.

concentration dropped to 0.29% during fall and winter and to 0.24% by spring (Table 3). The small differences in nitrogen concentration probably reflect changes in our sample composition rather than biological activity, since samples were composites of a wide variety of wood types. In controlled litterbag studies, nitrogen concentration does increase during decomposition of fine wood debris (Triska and Cromack 1980). The nitrogen pool in CTBW samples averaged 2.18 g/m² (Table 7).

Standing crop of both coniferous and deciduous leaf litter constituted a source of biologically available nitrogen. Mean standing crop of nitrogen was 0.27 g/m² for needles and 0.08 g/m² for deciduous leaves (Table 7). Proportionally, the nitrogen pool in leaf litter was 77% in needles and 23% in deciduous leaves. This is comparable to input composition of 68% and 32%, respectively, but with some suggestion that needle litter is more effectively transported. As with inputs, concentrations of nitrogen in standing crop varied seasonally. Needle litter N concentrations were lowest June–November, the time of major needle input (Table 3). Nitrogen concentrations were higher winter through spring, as observed in previous decomposition studies of needle litter (Triska et al. 1975). The highest nitrogen concentration occurred during spring, the time of needle litter consumption by invertebrates (Sedell et al. 1975), and may represent an important nitrogen pool for shredding invertebrates during final instars. The needle litter component of the nitrogen pool was largest in October and November, the months of maximum needle input (Table 7). However, coniferous needles were present throughout the year.

Big-leaf maple and vine maple leaves were the most readily available source of nitrogen for leaching and consumption by higher trophic levels. Mean estimate of nitrogen in leaf standing crop was low (0.08 g/m²), despite large inputs during November. Leaves were readily incorporated into small debris dams not sampled by coring the stream bottom. Debris dam measurements were conducted during summer, after leaf decomposition; hence mean standing crop underestimated this contribution. Estimates of nitrogen concentration from summer and fall composite samples had the highest nitrogen concentration of any litter debris.

LFPOM and MFPOM of unidentifiable origin constituted a major component of the stream's nitrogen pool (totalling 1.36 g/m²). LFPOM was especially prominent during late autumn and early winter (November–January) (Table 7). This size-class possibly originated from physical, microbial, and invertebrate processing of deciduous litter that entered during October and November. MFPOM was prominent at the same time, particularly during December and January. A second period of accumulation occurred during low discharge (July and August). MFPOM deposition during midsummer may have resulted from microbial activity on wood, or chemical flocculation. SFPOM, measured quarterly, is not included on Table 7, but it was the largest component of the FPOM nitrogen pool, at 3.41 g/m².

The mean nitrogen concentration of FPOM (Table 3) is based on few samples due to the difficulty of obtaining FPOM free of fine sediment. Nitrogen estimates from FPOM were obtained from export samples, which contained the least inorganic sediment. Nitrogen concentrations were corrected for ash and extrapolated to the core estimates of standing crop. The FPOM nitrogen pool at 4.77 g/m² was the stream's largest pool of particulate organic nitrogen.

Two minor nitrogen pools were moss and invertebrate consumers. Moss, where present, had an ash-free dry mass (AFDM) standing crop of 40 g/m^2. Mean moss AFDM standing crop amounted to 8 g/m^2 for the entire stream bottom, ≈2.5 times the 3.1 g/m^2 reported by Fisher and Likens (1973). Nitrogen concentration of moss averaged 0.92% for nine samples analyzed, or 0.07 g/m^2.

Annual moss primary production was not determined and few literature estimates are available. Fisher and Likens (1973) estimated gross primary production of moss at 2.14 g/m^2 annually at Bear Brook, New Hampshire. Net primary production could not be determined because moss and its surface microbial flora had a respiration rate that exceeded gross primary production. As moss grows slowly, most gross primary production is probably expended as respiration. Tamm (1953) found that nutrient uptake (N, P, K) by the forest moss *Hylocomium splendens* occurred simultaneously with growth, and that the nitrogen content did not decrease in older segments. Assuming the above is true for aquatic mosses, annual increments of growth would occur primarily by direct uptake from stream water. The lifespan of mosses varies by species (Binkley and Graham 1981), and observation of particulate organic export indicated that winter scour removes substantial biomass annually (included in miscellaneous export). Assuming 40% of standing mass represents that year's net production, and that uptake equals the nutrient content, the annual amount of nitrogen captured for elaboration of moss tissue would be only 22.58 g/channel (0.03 g/m^2) over the entire study reach. The amount of nitrogen flux within the moss community for tissue maintenance is unknown.

Invertebrates were partitioned on the basis of functional groups (Cummins 1973, 1974, 1975). Invertebrate data from core samples indicated shredder plus collector functional groups averaged 70% of total invertebrate biomass. Grazers constituted <1% of invertebrate biomass, which reflected the terrestrial nature of the fixed carbon inputs and shading by the canopy. Measurement of ammonia production conducted by Grafius (1977) on *Lepidostoma quercina*, a common shredder invertebrate, was used to compute invertebrate ammonia production. Ammonia-N released under laboratory conditions (15°C) averaged 3.7 mg·g dry mass·d^{-1}. Shredder invertebrates, including *Lepidostoma* sp., constituted most invertebrate biomass in the stream.

Invertebrate biomass was lowest in January due to the small size of early instars and high waterflow that confined invertebrates to protected areas. Maximum mean biomass (dry mass), determined from core samples, was 1.45 g/m^2 during August, a period of low flow. The annual mean of insect biomass was 0.194 g/m^2 of wetted channel. As a nitrogen pool, mean invertebrate biomass amounted to 0.02 g/m^2 channel. Potential production of ammonia was estimated monthly, assuming a wetted area of 300 m^2 during December–May at base flow and 150 m^2 during June–November at base flow. Annual ammonia-nitrogen contributions to the nitrogen pool by invertebrates was estimated at 0.20 kg for the entire stream, or 0.26 g/m^2 channel. Emergence of 470 mg/m^2 dry mass (Grafius 1974) constituted an additional 0.06 g/m^2 of wetted area annually, or 0.02 g/m^2 of stream channel, an insignificant part of the stream's nitrogen budget.

TABLE 8. Nitrogen loss from Watershed 10, 1 October 1973–30 September 1974 resulting from export of particulate organic matter.

CPOM		FPOM	
Substrate	N loss (g/m^2)	Size	N loss (g/m^2)
First quarter (1 Oct–31 Dec)			
Needle	0.11	Large and medium	0.15
Leaf	0.02	Small	0.70
CTBW	0.19		
Misc.	0.15		
Second quarter (1 Jan–31 Mar)			
Needle	0.04	Large and medium	0.13
Leaf	0.01	Small	0.65
CTBW	0.27		
Misc.	0.08		
Third and fourth quarter			
Combined	0.003	Large and medium	0.005
		Small	0.025

Nitrogen export

Major nitrogen loss from the stream occurred during winter periods of high discharge. Some 95% of the dissolved organic nitrogen left the stream between October and April. Annual dissolved organic nitrogen losses amounted to 6.42 kg for the watershed or 8.38 g/m^2 of stream bottom (Table 9). Nitrate loss was estimated at 0.33 kg for the watershed, or 0.43 g/m^2 of stream channel. This estimate is based on grab samples at the base of the stream, collected simultaneously with seep samples. Our grab sample estimates were comparable to the continuous sampler estimate of 6.07 kg DON (R. Fredriksen, *personal communication*) or 7.92 g/m^2.

Because particulate organic matter was effectively captured and decomposed within the stream, processing of particulate organic inputs was high and output was dependent on total runoff. Most particulate organic output (Table 8) occurred during the first quarter (October–December). In this period, the first major storms flush the stream of debris accumulated during low summer discharge. Virtually all measurable particulate output occurred between October and March.

Output of nitrogen as needle litter amounted to 0.15 g/m^2, or 115 g/yr. Output as intact needles was only 16% of input. Deciduous leaves were incorporated even more readily into debris dams, so only minor amounts

TABLE 9. Export of dissolved organic nitrogen (DON) based on water samples collected at the base of Watershed 10, 5 March 1974–18 March 1975. (· · ·) indicate no DON export samples were taken 25 December 1974–15 January 1975 or 4–26 February 1975. DON concentration 16 January–3 February was used to compute DON export for the total interval.

Discharge interval	Mean discharge (L/s)	DON export (g)
5–27 Mar 1974	13.66	181.77
28 Mar–19 Apr	5.94	702.61
20 Apr–9 May	1.73	11.37
10–23 May	4.47	162.12
24 May–17 Jun	1.61	252.27
18 Jun–11 Jul	0.74	130.06
12–31 Jul	0.35	32.18
1–19 Aug	0.22	1.41
20 Aug–9 Sep	0.14	13.91
10–30 Sep	0.09	11.35
1–11 Oct	0.03	1.70
12–25 Oct	0.03	1.70
26 Oct–13 Nov	0.37	131.15
14 Nov–2 Dec	2.47	684.11
3–24 Dec	11.97	2890.00
25 Dec–15 Jan 1975	17.60	· · ·
16 Jan–3 Feb	8.03	1161.61
4–26 Feb	12.02	· · ·
27 Feb–18 Mar	8.61	59.56
Total		6428.88

were exported as recognizable leaf fragments. Particulate nitrogen loss as leaf litter amounted to only 0.03 $g \cdot m^{-2} \cdot yr^{-1}$ or 7% of annual deciduous leaf input. The amount of nitrogen lost as CTBW varied, depending on discharge. In 1974–1975, with discharge of 173 cm, 0.46 $g \cdot m^{-2} \cdot yr^{-1}$ was lost as woody debris, ≈33% of the annual input. This is the estimate used in the budget. In 1973, a very dry year (80 cm discharge), virtually no wood was exported.

Nitrogen not exported as intact litter was either mineralized, leached, or converted to FPOM. FPOM nitrogen output in the large and medium size-classes alone amounted to 0.28 g/m^2 channel, a larger nitrogen loss than needle and leaf material combined. Estimates of small FPOM nitrogen output were not undertaken at Watershed 10 prior to clearcutting. This fraction constituted >80% of transport in Devil's Club Creek, another first-order stream of similar gradient in the H. J. Andrews Experimental Forest (Naiman and Sedell 1979). Assuming the nitrogen concentration of 1.44% nitrogen and equal susceptibility to export as Devil's Club Creek, estimated nitrogen loss was 1.38 $g \cdot m^{-2} \cdot yr^{-1}$. Total nitrogen export was 11.36 $g \cdot m^{-2} \cdot yr^{-1}$.

DISCUSSION

Early carbon budgets made by stream ecologists (Teal 1957, Minckley 1963, Minshall 1967, Tilly 1968) emphasized the role of fixed carbon from allochthonous sources as the biological energy base of small forested streams. They concluded that carbon fixed within the terrestrial system dominated that fixed within the aquatic system to provide the stream's energy base. Until the work by Fisher and Likens (1973), dissolved organic carbon was not considered a significant input to the carbon budget of streams, and no previous studies considered inorganic carbon as a component of the carbon budget. As a result, carbon budgets have been valuable in understanding the energetics of streams, but are not nutrient budgets for carbon.

Early budgets for nutrients other than carbon often took the opposite approach. These nutrient budgets for watersheds have followed the logic developed by Bormann and Likens (1967): basically, that loss of dissolved nutrients from the terrestrial component of a watershed can be measured at its base. Emphasis was placed on geological outputs, primarily dissolved inorganic nutrients (Bormann et al. 1968, Likens et al. 1970), while particulate and dissolved organic nitrogen and potential fixation and transformation of nitrogen between the soil system and the base of the watershed were deemphasized.

More recently Woodall (1975), in a study of small mountain streams, indicated salamanders and crayfish were major agents of nutrient flux in the detritus pool and may constitute an important nutrient sink. Hynes et al. (1974) also speculated on the role of leaves in both releasing nutrients and serving as a nitrogen sink. Meyer and Likens (1979) and Meyer et al. (1981) report significant input of phosphorous and nitrogen from allochthonous organic debris to Bear Brook, New Hampshire. Thus the flux of allochthonous debris and associated physical, chemical, and biologic processes can constitute or modify significant components of stream nutrient budgets.

Biological input-output

In this study, terrestrial biological inputs (litterfall, lateral movement, and throughfall) contributed ≈20% of the annual nitrogen budget, and once within the stream represented a biologically available source of particulate nitrogen, especially leaf litter inputs. Virtually all species of leaf litter entering the stream had the potential for microbial mineralization within 1 yr (Sedell et al. 1974, 1975, Triska et al. 1975, Triska and Sedell 1976). Although mineralization of litter-associated nitrogen could be accomplished entirely by microbial decomposition, enzymatic reduction of size class (Suberkropp 1980), physical fragmentation, and invertebrate consumption can restrict direct microbial mineralization to <25% of leaf litter inputs (Cummins and Klug 1979).

The proportion of leaf litter nitrogen subjected to physical or chemical transformation in the particulate nitrogen pool depends upon the stream's retention capacity. In this study, bed roughness (generated largely by organic debris dams) captured ≈87% of nitrogen associated with intact leaf litter. This effective retention by the stream's morphology allowed nitrogen associated with litter to enter biological pathways despite the exporting influence of unidirectional flow.

Once litter entered the water, processing occurred by both physical and biological vectors, including leaching, microbial mineralization, and reduction in size class. Four days of leaching at 5°C in three springs of the Washington Cascades resulted in a loss of 42%, 62%, and 50% of the nitrogen capital for vine maple, Douglas-fir, and big-leaf maple litter, respectively (Triska and Sedell 1976). These species provide the overwhelming predominance of leaf litter inputs to the stream at Watershed 10, and represent a potential contribution of 0.16 kg DON from deciduous leaf litter, and 0.48 kg DON from coniferous leaf litter.

Despite nitrogen loss by leaching, a seasonal increase in nitrogen concentration was typically observed on all litter substrates (Table 3). Nitrogen concentration of various litter substrates was lowest at abscission and increased as decomposition proceeded. Nitrogen in needle litter even had an absolute gain from the low concentration which resulted from leaching (Triska and Buckley 1978). Maximum absolute nitrogen content typically occurred at 30–40% loss of mass for leaf packs of Douglas-fir needle litter.

Post-leaching replenishment of nitrogen presumably occurred by uptake of dissolved nitrogen as stream water passed over microbially colonized surfaces. While most DON was undoubtedly refractory, significant uptake, biological utilization, and subsequent output in a different form has been reported (Cummins et al. 1972, Wetzel and Manny 1972). The composition of DON (labile vs. refractory) for the estimated 0.83 g/m^2 leachate and 10.56 g/m^2 subsurface input is unknown; however, only 8.38 g/m^2 was exported. Inorganic nitrogen uptake is also active on microbially colonized litter surfaces (Hynes and Kaushik 1969, Howarth and Fisher 1976). However, low seepwater concentrations minimized its potential impact in this study.

Increase in nitrogen concentration does not guarantee biological availability. Complexation with refractory carbon, particularly lignin and phenolics, results in residues characterized by: (1) an increase in nitrogen content particularly associated with lignin, (2) decrease in particle size, and (3) resistance to decomposition. Such residues have been reported in lotic environments by Suberkropp and Klug (1976), Suberkropp et al. (1976), and Odum et al. (1979). After 33 d of stream incubation at Watershed 10 the percent of litter nitrogen associated with the lignin fraction increased from 13 to 40% in alder, 17 to 67% in Douglas-fir, 9 to 26% in vine maple, and 24 to 74% in big-leaf maple. Nitrogen complexation has been observed to inhibit consumer assimilation despite lowering of C/N ratio (Ward and Cummins 1979), although Martin et al. (1980) report that some invertebrate species may have evolved a digestive strategy of high gut pH to utilize nitrogen-lignin complexes.

Petersen and Cummins (1974) estimate invertebrate processing of 15–20% of leaf litter inputs to a third-order stream in Michigan. Such litter consumption by invertebrates results in both passage of nutrient to higher trophic levels and fragmentation of egested materials. At Watershed 10, low invertebrate biomass in the stream indicates a smaller role for invertebrates. Consumption rates for Douglas-fir litter reported by Grafius (1977) varied from 0.10 $mg \cdot mg^{-1} \cdot d^{-1}$ (*Lepidostoma quercina*) to 0.84 $mg \cdot mg^{-1} \cdot d^{-1}$ (*L. unicolor*) depending on invertebrate instar, temperature, food density, and litter conditioning time. Assuming a consumption rate of 0.3 $mg \cdot mg^{-1} \cdot d^{-1}$, litter nitrogen turnover would be 0.13 g/m^2 of wetted area. Based on a 300 m^2 wetted area, invertebrate turnover was only ≈4% of leaf litter nitrogen. Since only 18% of needle litter nitrogen and 5% of leaf litter nitrogen was exported in the form of intact litter, physical fragmentation and microbial processing were presumably more important proportionally at our site than in more productive higher order streams.

A potentially important loss of nitrogen not determined in this study was microbial denitrification. Hill (1979, 1981) reported denitrification was a major nitrogen sink in Duffin Creek, Ontario, and recent investigations indicate 80–90% of nitrate removed from Duffin Creek sediments is via denitrification (A. R. Hill 1983 and *personal communication*). Kaushik and Robinson (1976), Sain et al. (1977), and Chatarpaul et al. (1979, 1980) also report significant denitrification in laboratory experiments using fine stream sediments. Background NO_3-N in the above studies generally exceeded 1000 μg/L. Swank and Caskey (1982) report denitrification in a second-order mountain stream following disturbance by road building and clearcutting in stream reaches impacted by high sediment loading. Background NO_3-N concentrations in these upland areas were raised to 160–200 μg/L from a predisturbance concentration of ≈20 μg/L. Denitrification was not estimated in our study since most measurements were made prior to general acceptance of the acetylene inhibition technique. Although denitrification is of potentially great importance in streams dominated by fine sediments and high nitrate loads, some subsequent estimates indicate it may be less important in highly oxygenated, gravel-bottom streams with low nitrate concentration. Meyer et al. (1981) indicated denitrification was probably insignificant at Bear Brook, New Hampshire, based on low levels of activity observed in denitrifier cultures and microcosm studies by Sloane (1979). Triska and Oremland (1981) and J. H. Duff (*personal communication*) report significant denitrification associated with algal mats from high-nitrate waters (>800 μg/L) but insignificant denitrification in mats (J. H. Duff, *personal communication*) or sediments (R. S. Oremland, *personal communica-*

tion) from a pristine stream environment (20–40 μg/L). Ventullo and Rowe (1982) also report denitrification associated with epilithic communities in streams; however, exogenous addition of nitrate was required to induce N_2O production. Hill (1981), who compared five streams characterized by high nitrogen depletion during transport, reported that high nitrate concentration appeared to be the main factor accounting for high nitrate loss. Nitrate was nearly undetectable in input seepage water (<10 μg/L) at our site and constituted <5% of nitrogen output. Channel disturbance was also minimal. As a result the role of denitrification is unknown, but probably not a significant component of the stream's nitrogen budget.

Ammonia is produced as the major nitrogen waste of both microbial mineralization and consumer metabolism. Ammonia concentrations in stream water were low on both an instantaneous and annual basis, mostly below the limits of detection. Total annual production of ammonia nitrogen by stream invertebrates (0.26 $g \cdot m^{-2} \cdot yr^{-1}$) and microbial metabolism was probably followed by rapid sorption or uptake on detrital surfaces. Within the stream network lies some potential for nitrification, as nitrate concentration does increase in a downstream direction (Gregory1980).

Hydrological-meteorological input-output

Although our goal was to examine in detail the role of allochthonous organic particulates as a nitrogen source, 77% of nitrogen export was in the form of dissolved nitrogen. The dissolved nitrogen fraction consisted of DON (95%), nitrate (5%), and insignificant amounts of nitrite and ammonia. This dissolved nitrogen composition contrasts with Bear Brook, New Hampshire, where 87% of nitrogen transport was nitrate and ammonia, and only 13% was DON (Meyer et al. 1981). The major difference between sites was the absence of nitrate in our stream. Two potential sources of nitrate at Bear Brook include higher nitrate input in precipitation (nitrate concentration doubled between 1955–1979, Bormann and Likens 1979), and nitrification on the watershed. Vitousek et al. (1982) estimated nitrogen mineralization by a trenching experiment at 17 sites in four states. They reported the highest potential nitrate production in a New Hampshire northern hardwood site, while net nitrate production lagged by months at a poor site Douglas-fir forest in Washington. Meyer et al. (1981) also reported nitrification activity on selected stream substrates from experiments at Bear Brook by Sloane (1979). Vitousek et al. (1982) suggest at least nine processes are potentially responsible for delaying solution losses of nitrate. In the absence of a controlled study, the mechanism for low nitrate input-output at Watershed 10 is unknown.

Element limitation

Low dissolved nitrogen and high dissolved phosphorus (relative to nitrogen) concentrations in stream water (Table 1) support the conclusion that nitrogen limited biological production in the stream at Watershed 10. Elwood et al. (1981) define a nutrient as limiting if an increase in supply of that nutrient affects the parameter in question. A parameter whose value is limited by a given nutrient may also respond to a simultaneously limiting nutrient or to changes in other rate-controlling factors. Using this definition, light was the rate-controlling factor to epilithic algal autotrophs, and nitrogen the limiting nutrient. Gregory (1980), who experimentally added both light and nutrients to the stream at Watershed 10, found no enhancement of primary production or chlorophyll *a* by nitrate addition alone. Standing crop of chlorophyll *a* was 5 times greater when light was added alone, and 20 times greater when both light and nitrate were added. In more open-canopy sections of larger streams in the same forest, Gregory (1980) found significant diel fluctuations in nitrate concentration, whereas orthophosphorus concentration had no diel pattern at any site. Factors controlling primary production by bryophytes are unknown. Mosses are adapted to low light conditions and had nitrogen fixation associated with epiphytic bacteria. Assuming some fraction of nitrogen fixed by epiphytic bacteria is eventually available, bryophytes are presumably less nitrogen limited than epilithic algae.

Another index of nutrient limitation is determination of element ratios. The atomic element ratio of algal-zooplankton tissue composition calculated by Redfield et al. (1963) is 106 C:16 N:1 P, and that of cultured algal tissue is as high as 30 N:1 P (Rhee 1978). The atomic N:P (total dissolved nitrogen : total dissolved phosphorus) at our site was only 0.74:1 and DIN : ortho P was even lower at 0.32:1. This compares with a N:P ratio of 440:1 for fluvial inputs and 291:1 for subsurface inputs at Bear Brook, New Hampshire (Meyer et al. 1981). Our stream water N:P is also lower than the dissolved inorganic nitrogen : soluble reactive phosphorus atomic ratio of ≈30:1 at Walker Branch, Tennessee (Elwood et al. 1981). Both of the above locations were considered phosphorus limited. The low N:P ratios from stream water indicate biotic production is far more likely to be nitrogen limited at our site than streams in the eastern United States.

Low N:P ratio was also reflected in litter inputs, and in litter tissue undergoing decomposition in the stream at Watershed 10. Litter inputs had atomic N:P (Kjeldahl nitrogen : organic phosphorus) ratios of 7:1 for coniferous litter, 7.5:1 for deciduous litter (big-leaf maple), and 26:1 for twig litter. Greater leaching of phosphorus than nitrogen, however, raised N:P to 18:1 for coniferous litter and to 19:1 in deciduous leaf litter (big-leaf maple) as a result of immersion. Low level PO_4-P enrichment experiments (60 μg/L) by Elwood et al. (1981) resulted in N:P ratio of ≈19 after 32 d in red oak leaf packs. They found that mass loss, P content, N content, and respiration were not significantly different than litter exposed to even higher phosphorus enrichment (450 μg/L). Since mean ortho-P concentration in stream water at our site was 42 μg/L, it is not likely that phosphorus was limiting decomposition processes. High atomic C:N ratios in decomposing litter substrates ($\bar{x}$ = 131:1 coniferous litter, 104:1 deciduous litter) instead indicate demand for nitrogen for carbon mineralization from the watershed's nitrogen-dilute waters (DON, 27 μg/L; NO_3-N, 6 μg/L). Previous studies on nitrate addition in controlled experimental channels at 5°C, however, failed to stimulate decomposition (Triska and Sedell 1976). Perhaps as in the case of light for the autotrophic community, some additional factor such as low temperature masked the effect of nitrate addition on litter decomposition in the above study. Based on theoretical element composition nitrogen would be assumed limiting to both autotrophic production and litter decomposition in the stream at Watershed 10.

Nitrogen associated with inorganic particulates

Nitrogen input-output associated with inorganic particulates was not specifically included in our study; however, the significance of erosional processes is currently being examined by F. J. Swanson (*personal communication*). Available data include a mean annual estimate of suspended sediment from Watershed 10 of 90 $kg \cdot ha^{-1} \cdot yr^{-1}$ between 1969–1973 (Fredriksen 1975, Swanson et al. 1982*a*). Suspended sediment export was estimated with a discharge proportional water sampler, which weighted sediment load identically to dissolved constituents. The above estimates include the total suspended component of water samples, and therefore include nitrogen we measured as FPOM. Assuming most sediment export occurs during fall and winter, and in years of greatest discharge, then 0.53% nitrogen concentration ($\bar{x}$ nitrogen concentration fall-winter 1972, Fredriksen 1975) would result in a nitrogen loss of 6.25 $g \cdot m^{-2} \cdot yr^{-1}$, 3.8 times that listed as FPOM. The largest inorganic particulate input to the stream was surface erosion (the inorganic component of lateral movement). Swanson et al. (1982*a*) estimated surface erosion of inorganic particulates at 480 kg/yr to the stream channel at Watershed 10. Assuming a nitrogen content similar to suspended sediment, nitrogen input associated with inorganic particles would be 3.32 $g \cdot m^{-2} \cdot yr^{-1}$. Atmospheric dust input was an additional 0.06 $g \cdot m^{-2} \cdot yr^{-1}$ (Sollins et al. 1980). Using these preliminary figures, surface erosion plus atmospheric input at 3.38 $g \cdot m^{-2} \cdot yr^{-1}$ is less than export of 4.63 $g \cdot m^{-2} \cdot yr^{-1}$ (suspended sediment − FPOM). If large mass movements such as soil creep, or episodic inputs such as root throw or debris torrents are included, however, significant retention of nitrogen associated with the particulate inorganic load is predicted within the channel (Swanson et al. 1982*a*). Since nitrogen is not a component of most minerals, associated nitrogen presumably represents sorption or flocculation of organic nitrogen with fine soil particles, especially clay fractions, which probably participate minimally in nitrogen cycling.

The nitrogen budget

After all major inputs and outputs were considered, a nitrogen budget for the stream was constructed. Total nitrogen input was estimated at 15.25 $g \cdot m^{-2} \cdot yr^{-1}$ (Fig. 3). Hydrological inputs of nitrogen account for 11.06 $g \cdot m^{-2} \cdot yr^{-1}$ or 73% of the input total, 0.50 g/m^2 of which was dissolved inorganic nitrogen. Dissolved nitrogen on an instantaneous basis was never an important component of the standing crop, but DON was 69% of input nitrogen. Biological uptake and/or sorption to debris captured ≈25% of the DON load. DON utilization was greater than indicated by input-output because leaching and decomposition of particulate organic inputs were an additional source of DON. Throughfall amounted to 2% of nitrogen inputs (0.30 $g \cdot m^{-2} \cdot yr^{-1}$) from leaching in the canopy. Direct interception of precipitation was not a significant nitrogen input to the stream. Internal fixation of molecular nitrogen was ≈5% of the nitrogen budget, or 0.76 $g \cdot m^{-2} \cdot yr^{-1}$. Major terrestrial biological sources of nitrogen were CTBW and leaf litter.

Wood debris was stored annually. Stored wood formed debris dams which captured organic particulates, created habitat, and produced FPOM. Leaf and needle litter also contributed to the particulate organic nitrogen pool but intact leaf litter was only 3% of the standing crop. Particulate biological inputs, primarily litterfall and lateral movement, amounted to 3.13 $g \cdot m^{-2} \cdot yr^{-1}$, or 20% of the nitrogen input. Of these sources, leaf and needle litter, the most easily processed particulate organic inputs, contributed 1.41 g/m^2, only 0.18 g/m^2 (or 13%) of which was exported in a recognizable state. The remainder was either leached, converted to FPOM, or mineralized. Nitrogen in wood debris amounted to 1.13 g/m^2 input and 0.46 g/m^2 export; the remainder passed into storage, less that which was leached or fragmented to FPOM. FPOM, the major product of particulate organic processing, had a nitrogen input of 0.18 g/m^2 but amounted to 4.77 g/m^2 on the stream bottom. Total nitrogen output as dissolved components was 78%. This estimate is identical to the total solution losses reported by Bormann et al. (1969) for 1966 at 78% but lower than the 94% solution loss reported in 1965 from Hubbard Brook, New Hampshire, or 96% for Bear Brook, New Hampshire (Meyer et al. 1981). Total nitrogen input exceeded output by 34% in this study, compared to 8% reported by Meyer et al. (1981) at Bear Brook, New Hampshire. The difference is due to storage of nitrogen as refractory particulate organic matter and to the far higher input of dissolved inorganic nitrogen at Bear Brook. Both budgets indicated significant uptake by the stream ecosystem.

Independent estimation of input-output constituents indicated significant particulate organic storage. The particulate component of the nitrogen pool consisted

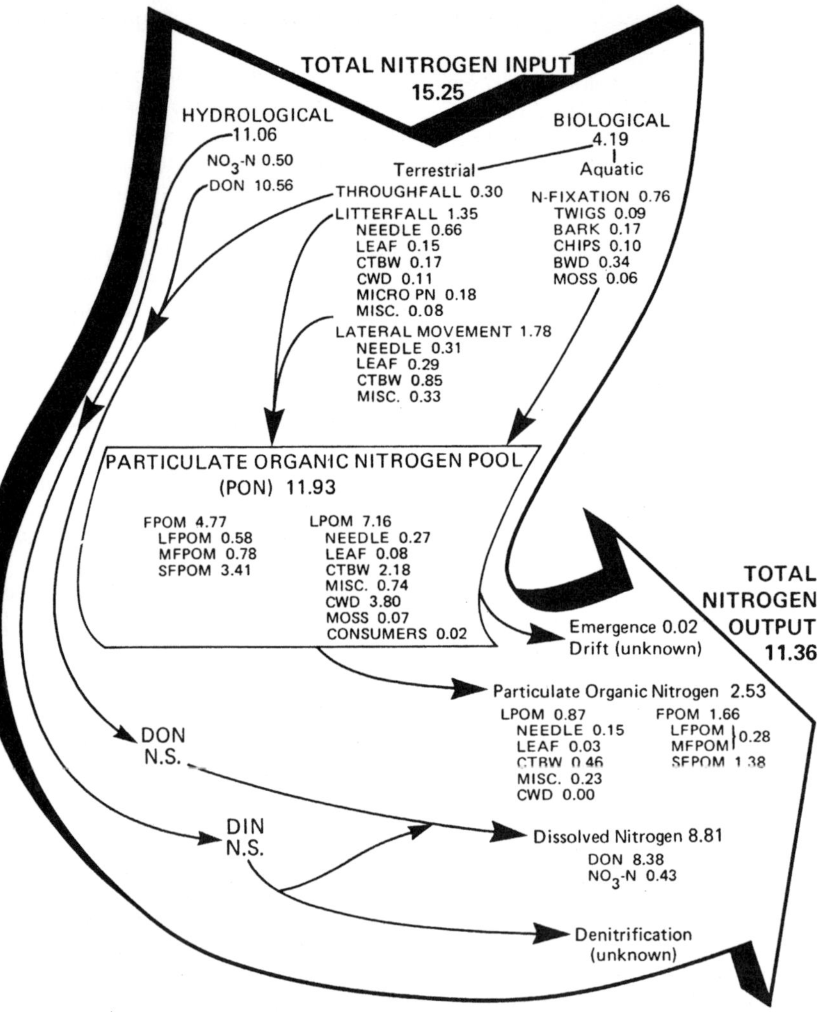

FIG. 3. Nitrogen budget for the stream at Watershed 10 (g/m^2), indicating the source and magnitude of annual nitrogen inputs, mean pool size, and annual exports. Acronyms are defined as follows: DON, dissolved organic nitrogen; DIN, dissolved inorganic nitrogen; CTBW, cones, twigs, bark, and wood (>1.0 mm and <15.0 cm); CWD, coarse wood debris (>15.0 cm diameter); BWD, wood blocks (to estimate N-fixation associated with wood debris); MICRO PN, nitrogen input associated with microparticulate litterfall (<800 μm diameter); LPOM, large particulate organic matter (>1.0 mm diameter); FPOM, fine particulate organic matter (<1.0 mm diameter). DON and DIN as an instantaneous pool are not significant (NS).

of three major components: (1) total wood (51%), 5.98 g/m^2, (2) leaf litter (3%), 0.35 g/m^2, and (3) FPOM (40%), 4.77 g/m^2. Particulate storage of wood debris (CTBW) was an annual event whose magnitude was determined by storm frequency and discharge pattern. In 1974–1975, nitrogen addition was estimated at 0.57 g/m^2 as CTBW, which was 4% of total nitrogen input. CTBW is readily incorporated into the nitrogen pool, has a long turnover time, and is resistant to export (Triska and Cromack 1980, Swanson et al. 1982*b*). Such refractory organic debris has been recognized as an essential stabilizing property of ecosystems (O'Neill et al. 1975, Reichle et al. 1975, Webster et al. 1975).

Nitrogen in coarse wood debris (CWD) was present at a minimum of 3.80 g/m^2, a pool equal to ≈25% of the nitrogen budget. Coarse wood is processed by both biologic and hydrologic means with a time frame in excess of 100 yr. Finer wood components (CTBW) constitute a nitrogen sink of some 2.18 g/m^2, ≈15% of the annual nitrogen budget, and are similarly processed in perhaps 10–20 yr. Due to slow processing times, wood accumulates on the stream bottom, forms habitat, and captures leaf litter (Triska and Cromack 1980, Swanson et al. 1982*b*).

Including the 5% of the nitrogen budget represented by fixation, 34% of the nitrogen input passed into storage during our study year (February 1974–1975). Thus the stream was not operating under annual steady-state conditions as found by Fisher and Likens (1973) for organic carbon. Steady state, if it exists, operates on a time frame related to the decomposition rate of the most significant refractory input and to the cycle of storm events that purge the channel of stored debris. Both Meyer et al. (1981) and Triska et al. (1982) emphasized the importance of year-to-year variation in discharge in examining annual nutrient budgets. Furthermore, Swanson et al. (1982*a*) estimate that massive debris torrents at Watershed 10 (occurrence intervals ≈580 yr) result in export of organic particulate matter equal to almost 30% of the organic budget of all intervening years. Thus events with occurrence intervals of half a millennium help shape the export history of a watershed. Between these long intervals of accumulation and export the large reservoir of refractory detritus aids in the internalization of the nitrogen cycle despite the open nature of the lotic ecosystem.

Summary

Except for potential minor losses to the atmosphere, the stream at Watershed 10, whose channel area was <0.1 ha, transported or accumulated virtually all nitrogen losses of the 10.1-ha watershed. Although induced by physical vectors (gravity, wind, precipitation), almost all nitrogen inputs to the stream were derived from biotic processes in the forest. Major biotically derived inputs (litterfall, throughfall, lateral movement, DON in groundwater, and fixation) constituted >90% of nitrogen input to the stream. Inorganic nitrogen inputs, nitrate, nitrite, and ammonium in subsurface flow constituted <5% of input nitrogen. Of the 92% of organic nitrogen input excluding fixation, 76% was dissolved and 24% was particulate. Although nitrogen loss on a per hectare basis was small for the 10.1-ha watershed, all particulate and dissolved losses passed through or were accumulated in an environment encompassing <1% of the watershed area. This concentration permitted the establishment of a separate ecosystem whose processing efficiency and capacity for nutrient cycling are related to the stream's retention capacity and carbon quality of inputs within the reach.

Although formation of the input-output budget provided only the most minimal glance at internal dynamics, certain potentials for internal nitrogen cycling were observed. Leaf litter, which may lose up to 50% of its initial nitrogen content to the dissolved nitrogen pool, regained lost nitrogen within 40 d by various processes. During biological mineralization, nitrogen was complexed to leaf lignin, a process similar to humification in terrestrial environments. Microbial mineralization, invertebrate consumption, and physical abrasion resulted in the formation of FPOM from wood and litter debris. FPOM generated in this manner was characterized by a nitrogen content of leaf litter and a total fiber content of wood. FPOM constituted a nitrogen sink equal to 32% of the annual nitrogen budget and constituted 66% of the particulate nitrogen export. Biologically preprocessed detritus was, therefore, a major nitrogen contribution to downstream biological communities.

The events described above occurred within a first-order stream of a major drainage. Nitrogen that escapes the first order, while classified as export, constitutes the upstream input to a second-order stream. At each successive stream order the distance between uptake and release of nitrogen, spiraling length (Webster et al. 1975, Webster and Patten 1979, Newbold et al. 1981, Elwood et al. 1983), may vary, and biotic community structure can be modified in response to upstream import (Vannote et al. 1980). Some imported DON becomes available for microbial utilization, and inorganic nitrogen becomes available for uptake by periphyton and riparian vegetation. Nitrogen tied to detrital tissue becomes available to shredding invertebrates and fine particulate organic matter to collector organisms. In addition to leaching and biological processing, nitrogen incorporated in particulate organic matter may be returned to the terrestrial landscape by deposition on flood plains. Physical abrasion in conjunction with biotic activity will result in an overall size reduction of output compared to input, even for wood debris. Through each link in the lotic network, biotic processes will present numerous opportunities for internal cycling before nitrogen lost from the terrestrial landscape finally reaches the river's estuary.

Acknowledgments

The work reported in the paper was supported by a National Science Foundation grant, DEB 74-20744A02, to the Coniferous Forest Biome, Ecosystem Analysis Studies, except for the nitrogen fixation data, which were obtained under National Science Foundation grant DEB 767-21402, Wood Mineralization in the Pacific Northwest. Chemical analyses reported in this paper were performed in the Cooperative Analytical Laboratory under the joint direction of Elly Holcombe and Joanne Kristaponis by agreement supplement number 99 to the Master Memorandum of Understanding between the Forest Service and Oregon State University. We are grateful to Richard Fredriksen for generously furnishing discharge records at Watershed 10, and also unpublished data on total dissolved nitrogen, nitrate, and phosphorus, and to Jack Lyford for unpublished data on the standing crop of moss. We would like to express our gratitude to Barbara Buckley and Linda Roberts for their long hours in both the field and laboratory, and Cliff Dahm for organizing the seep nutrient data. Finally, we would like to thank Ken Cummins and Phil Sollins for reading an early version of the manuscript and Keith Slack and Sam Luoma for reading the revised version and for providing many beneficial suggestions. Products mentioned by brand name are for descriptive purposes only and do not constitute endorsement.

Literature Cited

Abee, A., and D. Lavender. 1972. Nutrient cycling in throughfall and litterfall in 450-year-old Douglas-fir stands. Pages 133–143 *in* J. F. Franklin, L. J. Dempster, and R. H. Waring, editors. Research on coniferous forest ecosystems: a symposium. United States Forest Service Pacific Northwest Range and Experiment Station, Portland, Oregon, USA.

Anderson, N. H., and K. W. Cummins. 1979. Influences

of life histories on the diet of aquatic invertebrates. Journal of the Fisheries Research Board of Canada **36**:335–342.

Anderson, N. H., and E. J. Grafius. 1975. Utilization and processing of allochthonous material by stream Trichoptera. Internationale Vereinigung für Theoretische und Angewandte Limnologie, Verhandlungen **19**:3022–3028.

Anderson, N. H., and J. R. Sedell. 1979. Detritus processing by macroinvertebrates in stream ecosystems. Annual Review of Entomology **24**:351–377.

Anderson, N. H., J. R. Sedell, L. M. Roberts, and F. J. Triska. 1978. The role of aquatic invertebrates in processing of wood debris in coniferous forest streams. American Midland Naturalist **100**:64–82.

Barlocher, F., and B. Kendrick. 1973. Fungi in the diet of *Gammarus pseudolimnaeus* (Amphipoda). Oikos **24**:195–200.

Barlocher, F., and B. Kendrick. 1981. The role of aquatic hyphomycetes in the trophic structure of streams. Pages 743–760 *in* G. C. Carroll and W. T. Wicklow, editors. The fungal community: its organization and role in the ecosystem. Marcel Decker, New York, New York, USA.

Bergersen, F. J. 1970. The quantitative relationship between nitrogen fixation and the acetylene reduction assay. Australian Journal of Biological Science **23**:1015–1025.

Binkley, D., and R. L. Graham. 1981. Biomass production and nutrient cycling of mosses in an old growth Douglas-fir forest. Ecology **62**:1387–1389.

Bormann, F. H., and G. E. Likens. 1967. Nutrient cycling. Science **155**:424–429.

Bormann, F. H., and G. E. Likens. 1970. The nutrient cycles of an ecosystem. Scientific American **223**:92–101.

Bormann, F. H., and G. E. Likens. 1979. Pattern and process in a forested ecosystem. Springer-Verlag, New York, New York, USA.

Bormann, F. H., G. E. Likens, and J. S. Eaton. 1969. Biotic regulation of particulate and solution losses from a forest ecosystem. Bioscience **19**:600–610.

Bormann, F. H., G. E. Likens, D. W. Fisher, and R. S. Pierce. 1968. Nutrient loss accelerated by clear-cutting of a forest ecosystem. Science **159**:882–884.

Busch, D. E., and S. G. Fisher. 1981. Metabolism of a desert stream. Freshwater Biology **11**:301–307.

Chatarpaul, L., J. B. Robinson, and N. K. Kaushik. 1979. Role of tubificid worms on nitrogen transformations in stream sediment. Journal of the Fisheries Research Board of Canada **36**:673–678.

Chatarpaul, L., J. B. Robinson, and N. K. Kaushik. 1980. The effects of tubificid worms on denitrification and nitrification in stream sediment. Canadian Journal of Fisheries and Aquatic Sciences **37**:656–663.

Cummins, K. W. 1973. Trophic relations of aquatic insects. Annual Review of Entomology **18**:183–206.

———. 1974. Structure and function of stream ecosystems. Bioscience **24**:631–641.

———. 1975. Macroinvertebrates. Pages 170–198 *in* B. A. Whitton, editor. River ecology. University of California Press, Berkeley, California, USA.

Cummins, K. W., and M. L. Klug. 1979. Feeding ecology of stream invertebrates. Annual Review of Ecology and Systematics **10**:147–172.

Cummins, K. W., M. L. Klug, R. G. Wetzel, R. C. Petersen, K. F. Suberkropp, B. A. Manny, J. C. Wuycheck, and F. O. Howard. 1972. Organic enrichment with leaf leachate in experimental lotic ecosystems. Bioscience **22**:719–722.

Cushing, C. E., and E. G. Wolf. 1982. Organic energy budget of Rattlesnake Springs, WA. American Midland Naturalist **107**:404–407.

Davis, S. F., and M. J. Winterbourn. 1977. Breakdown and colonization of *Nothofagus* leaves in a New Zealand stream. Oikos **28**:250–255.

Dierberg, F. E., and P. L. Brezonik. 1981. Nitrogen fixation (acetylene reduction) associated with decaying leaves of pond cypress (*Taxodium distichum* var. *nutans*) in a natural and sewage-enriched cypress dome. Applied and Environmental Microbiology **41**:1413–1418.

Elwood, J. W., J. D. Newbold, R. V. O'Neill, and W. Van Winkle. 1983. Resource spiraling: an operational paradigm for analyzing lotic ecosystems. Pages 3–27 *in* W. T. Fontaine III and S. M. Bartell, editors. Dynamics of lotic ecosystems. Ann Arbor Science Publishers, Ann Arbor, Michigan, USA.

Elwood, J. W., J. D. Newbold, A. F. Trimble, and R. W. Stark. 1981. The limiting role of phosphorus in a woodland stream ecosystem: effects of P enrichment on leaf decomposition and primary producers. Ecology **62**:146–158.

Fisher, S. G., and G. E. Likens. 1973. Energy flow in Bear Brook, New Hampshire: an integrative approach to ecosystem metabolism. Ecological Monographs **43**:421–439.

Fredriksen, R. L. 1972. Nutrient budget of a Douglas-fir forest on an experimental watershed in western Oregon. Pages 113–131 *in* J. F. Franklin, L. J. Dempster, and R. H. Waring, editors. Research on coniferous forest ecosystems: a symposium. United States Forest Service Pacific Northwest Range and Experiment Station, Portland, Oregon, USA.

———. 1975. Nitrogen, phosphorus, and particulate matter budgets of five coniferous forest ecosystems in the western Cascades Range, Oregon. Dissertation. Oregon State University, Corvallis, Oregon, USA.

Froehlich, H. 1973. Natural and man caused slash in headwater streams. Loggers Handbook 18, Pacific Logging Congress, Portland, Oregon, USA.

Gilson, W. E. 1963. Differential respirometer of simplified and improved design. Science **141**:531–532.

Gotto, J. W., and B. F. Taylor. 1976. N_2 fixation associated with decaying leaves of the red mangrove *Rhizophora mangle*. Applied and Environmental Microbiology **31**:781–783.

Grafius, E. J. 1974. Conversion of allochthonous detritus by stream detritivores. Thesis. Oregon State University, Corvallis, Oregon, USA.

———. 1977. Bioenergetics and strategies of some Trichoptera in processing and utilizing allochthonous materials. Dissertation. Oregon State University, Corvallis, Oregon, USA.

Gregory, S. V. 1980. Effects of light, nutrients and grazing on periphyton communities in streams. Dissertation. Oregon State University, Corvallis, Oregon, USA.

Grier, C. C., and R. S. Logan. 1977. Old-growth *Pseudotsuga menziesii* of a western Oregon watershed: biomass distribution and production budgets. Ecological Monographs **44**:373–400.

Hill, A. R. 1979. Denitrification in the nitrogen budget of a river ecosystem. Nature **281**:291–292.

———. 1981. Nitrate-nitrogen flux and utilization in a stream ecosystem during low summer flows. Canadian Geographer **3**:225–239.

———. 1983. Nitrate-nitrogen mass balances for two Ontario rivers. Pages 457–477 *in* W. T. Fontaine III, and S. M. Bartell, editors. Dynamics of lotic ecosystems. Ann Arbor Science Publishers, Ann Arbor, Michigan, USA.

Hodkinson, I. D. 1975. Dry weight loss and chemical changes in vascular plant litter of terrestrial origin occurring in a beaver pond ecosystem. Journal of Ecology **63**:131–142.

Howarth, R. W., and S. G. Fisher. 1976. Carbon, nitrogen, and phosphorus dynamics during leaf decay in nutrient-enriched stream microecosystems. Freshwater Biology **6**:1–8.

Hynes, H. B. N., and N. K. Kaushik. 1969. The relationship between dissolved nutrient salts and protein production in submerged autumnal leaves. Internationale Vereinigung für Theoretische und Angewandte Limnologie, Verhandlungen **17**:95–103.

Hynes, H. B. N., N. K. Kaushik, M. A. Lock, D. L. Lush, Z. S. J. Stockner, R. R. Wallace, and D. D. Williams. 1974. Benthos and allochthonous organic matter in streams. Journal of the Fisheries Research Board of Canada **31**:545–553.

Iversen, T. M. 1973. Decomposition of autumn-shed beech leaves in a springbrook and its significance for the fauna. Archiv für Hydrobiologie **72**:305–312.

Jackson, M. L. 1958. Soil chemical analysis. Prentice-Hall, Englewood Cliffs, New Jersey, USA.

Kaushik, N. K., and H. B. N. Hynes. 1968. Experimental study on the role of autumn shed leaves in aquatic environments. Journal of Ecology **52**:229–243.

Kaushik, N. K., and H. B. N. Hynes. 1971. The fate of dead leaves that fall into streams. Archiv für Hydrobiologie **68**:465–515.

Kaushik, N. K., and J. B. Robinson. 1976. Preliminary observations on nitrogen transport during summer in a small spring fed Ontario stream. Hydrobiologia **49**:59–63.

Kennedy, V. C., and R. L. Malcolm. 1978. Geochemistry of the Mattole River of Northern California. United States Geological Survey Open-File Report 78–205, Menlo Park, California, USA.

Kimmins, J. P. 1973. Some statistical aspects of sampling throughfall precipitation in nutrient cycling studies in British Columbian coastal forests. Ecology **54**:1008–1019.

Likens, G. E., F. H. Bormann, N. M. Johnson, D. W. Fisher, and R. S. Pierce. 1970. Effects of forest cutting and herbicide treatment on nutrient budgets in the Hubbard Brook Watershed Ecosystem. Ecological Monographs **40**:23–47.

Martin, M. M., J. S. Martin, J. J. Kukor, and R. W. Merritt. 1980. The digestion of protein and carbohydrate by the stream detritivore *Tipula abdominalis* (Diptera, Tipulidae). Oecologia (Berlin) **461**:360–364.

Meyer, J. L., and G. E. Likens. 1979. Transport and transformations of phosphorus in a forest stream ecosystem. Ecology **60**:1255–1269.

Meyer, J. L., G. E. Likens, and J. Sloane. 1981. Phosphorus, nitrogen, and organic carbon flux in a headwater stream. Archiv für Hydrobiologie **91**:28–44.

Minckley, R. L. 1963. The ecology of a spring, Doe Run, Meade County, Kentucky. Wildlife Monographs **11**.

Minshall, G. W. 1967. Role of allochthonous detritus in the trophic structure of a woodland spring brook community. Ecology **48**:139–149.

———. 1978. Autotrophy in stream ecosystems. Bioscience **28**:767–771.

Naiman, R. J., and J. R. Sedell. 1979. Characterization of particulate organic matter transported by some Cascade Mountain streams. Journal of the Fisheries Research Board of Canada **36**:17–31.

Newbold, J. D., D. W. Elwood, R. V. O'Neill, and W. Van Winkle. 1981. Measuring nutrient spiralling in streams. Canadian Journal of Fisheries and Aquatic Sciences **38**: 860–863.

Noonan, L., and E. Holcombe. 1975. Procedures for chemical analyses of plant and soil samples. Internal Report 160, International Biological Program Coniferous Forest Biome, Oregon State University, Corvallis, Oregon, USA.

Odum, W. E., P. W. Kirk, and J. C. Zieman. 1979. Non-protein nitrogen compounds associated with particles of vascular plant detritus. Oikos **32**:363–367.

O'Neill, R. V., W. F. Harris, B. S. Asmus, and D. E. Reichle. 1975. A theoretical basis for ecosystem analysis with particular reference to element cycling. Pages 28–40 *in* F. G. Howell, J. B. Gentry, and M. H. Smith, editors. Mineral cycling in southeastern ecosystems. Environmental Research and Development Administration Symposium Series (CONF-740513), Washington, D.C., USA.

Petersen, R. C., and K. W. Cummins. 1974. Leaf processing in a woodland stream ecosystem. Freshwater Biology **4**:343–368.

Rand, M. C. 1976. Standard methods for the examination of water and wastewater. 14th edition. American Public Health Association, Washington, D.C., USA.

Redfield, A. C., B. H. Ketchum, and F. A. Richards. 1963. The influence of organisms on the composition of seawater. Pages 26–77 *in* M. N. Hill, editor. The sea. Volume 2. Wiley-Interscience, New York, New York, USA.

Reichle, D. E., R. V. O'Neill, and W. F. Harris. 1975. Principles of energy and material exchange in ecosystems. Pages 27–43 *in* W. H. van Dobber and R. H. Lowe McConnell, editors. Unifying concepts in ecology. Dr. W. Junk, The Hague, The Netherlands.

Rhee, G-Yull. 1978. Effects of N/P ratios and nitrate limitation on algal growth, cell composition, and nitrate uptake. Limnology and Oceanography **23**:10–25.

Roskoski, J. P. 1977. Nitrogen fixation in northern hardwood forests. Dissertation. Yale University, New Haven, Connecticut, USA.

Sain, P., J. B. Robinson, W. N. Stammers, N. K. Kaushik, and H. R. Whitley. 1977. A laboratory study on the role of stream sediment in nitrogen loss from water. Journal of Environmental Quality **6**:274–278.

Sedell, J. R., F. J. Triska, J. D. Hall, N. H. Anderson, and J. H. Lyford. 1974. Sources and fates of organic inputs in coniferous forest streams. Pages 57–69 *in* R. H. Waring, editor. Integrated research in the coniferous forest biome. Proceedings, AIBS Symposium Coniferous Forest Ecosystem. Bulletin Number 5, Coniferous Forest Biome, University of Washington AR-10, Seattle, Washington, USA.

Sedell, J. R., F. J. Triska, and N. S. Triska. 1975. The processing of conifer and hardwood leaves in two coniferous forest streams. Internationale Vereinigung für Theoretische und Angewandte Limnologie, Verhandlungen **19**: 1617–1627.

Silvester, W. B., P. Sollins, T. Verhoeven, and S. P. Cline. 1982. Nitrogen fixation and acetylene reduction in decaying conifer boles: effects of incubation time, aeration and moisture content. Canadian Journal of Forest Research **12**: 646–652.

Sloane, J. 1979. Nitrogen flux in small mountain streams in New Hampshire. Thesis. Cornell University, Ithaca, New York, USA.

Snyder, J. M., and L. H. Wullstein. 1973. Nitrogen fixation on granite outcrop pioneer ecosystems. Bryologist **76**:196–199.

Sollins, P., C. C. Grier, F. M. McCorrison, K. Cromack, Jr., R. Fogel, and R. L. Fredricksen. 1980. The internal element cycles of an old-growth Douglas-fir ecosystem in western Oregon. Ecological Monographs **50**:261–285.

Suberkropp, K. F. 1980. The maceration of deciduous leaf litter by aquatic hyphomycetes. Canadian Journal of Botany **58**:1025–1031.

Suberkropp, K. F., G. L. Godshalk, and M. Klug. 1976. Changes in the chemical composition of leaves during processing in a woodland stream. Ecology **57**:720–727.

Suberkropp, K. F., and M. Klug. 1976. Fungi and bacteria associated with leaves during processing in a woodland stream. Ecology **57**:707–719.

Swank, W. T., and W. H. Caskey. 1982. Nitrate depletion in a second-order mountain stream. Journal of Environmental Quality **11**:581–584.

Swanson, F. J., R. L. Fredriksen, and F. M. McCorrison. 1982*a*. Material transfer in a western Oregon forested watershed. Pages 233–266 *in* R. L. Edmonds, editor. Analysis of coniferous forest ecosystems in the western United States. Hutchinson Ross, Stroudsburg, Pennsylvania, USA.

Swanson, F. J., S. V. Gregory, J. R. Sedell, and A. G. Campbell. 1982*b*. Land-water interactions: the riparian zone. Pages 267–291 *in* R. L. Edmonds, editor. Analysis of coniferous forest ecosystems in the western United States. Hutchinson Ross, Stroudsburg, Pennsylvania, USA.

Tamm, C. O. 1953. Growth, yield and nutrition of a forest moss *Hylocomium splendens*. Middelanden Fran Statens Skogsforskningsinstitut **43**:1–140.

Teal, J. M. 1957. Community metabolism in a temperate cold spring. Ecological Monographs **27**:283–302.

Tilly, L. J. 1968. The structure and dynamics of Cone Spring. Ecological Monographs **38**:167–197.

Triska, F. J. 1970. Seasonal distribution of aquatic hyphomycetes in relation to the disappearance of leaf litter from a small woodland stream. Dissertation. University of Pittsburgh, Pittsburgh, Pennsylvania, USA.

Triska, F. J., and B. M. Buckley. 1978. Patterns of nitrogen uptake and loss in relation to litter disappearance and associated invertebrate biomass in six streams of the Pacific Northwest, USA. Internationale Vereinigung für Theoretische und Angewandte Limnologie, Verhandlungen **20**: 324–332.

Triska, F. J., and K. Cromack. 1980. The role of wood debris in forests and streams. Pages 171–190 *in* R. H. Waring, editor. Forests: fresh perspectives from ecosystem analysis. Oregon State University Press, Corvallis, Oregon, USA.

Triska, F. J., and R. S. Oremland. 1981. Denitrification associated with periphyton communities. Applied and Environmental Microbiology **42**:745–748.

Triska, F. J., and J. R. Sedell. 1976. Decomposition of four species of leaf litter in response to nitrate manipulation. Ecology **57**:783–792.

Triska, F. J., J. R. Sedell, and B. M. Buckley. 1975. The processing of conifer and hardwood leaves in two coniferous forest streams. II. Biochemical and nutrient changes. Internationale Vereinigung für Theoretische und Angewandte Limnologie, Verhandlungen **19**:1628–1639.

Triska, F. J., J. R. Sedell, and S. V. Gregory. 1982. Coniferous forest streams. Pages 292–332 *in* R. L. Edmonds, editor. Analysis of coniferous forest ecosystems in the western United States. Hutchinson Ross, Stroudsburg, Pennsylvania, USA.

Van Soest, P. J. 1963. A rapid method for the determination of cellulose and lignin. Journal of the Association of Official Agricultural Chemists **16**:829–835.

———. 1965. Use of detergents in analysis of fibrous feeds. III. Study of the effect of heating and drying on yield of fiber and lignin in forages. Journal of the Association of Official Agricultural Chemists **48**:785–790.

Vannote, R. L., G. W. Minshall, K. W. Cummins, J. R. Sedell, and C. E. Cushing. 1980. The river continuum concept. Canadian Journal of Fisheries and Aquatic Sciences **37**:130–137.

Ventullo, R. M., and J. J. Rowe. 1982. Denitrification potential of epilithic communities in a lotic environment. Current Microbiology **7**:29–33.

Vitousek, P. M., J. R. Gosz, C. C. Grier, J. M. Melillo, and W. A. Reiners. 1982. A comparative analysis of potential nitrification and nitrate mobility in forest ecosystems. Ecological Monographs **52**:155–177.

Ward, G. W., and K. W. Cummins. 1979. Effects of food quality on growth of a stream detritivore, *Paratendipes albimanus* (Meigen) (Diptera:Chironomidae). Ecology **60**: 57–64.

Waring, R. H., W. H. Emmingham, H. L. Gholz, and C. C. Grier. 1978*a*. Variation in maximum leaf area of coniferous forests in Oregon and its ecological significance. Forest Science **24**:131–140.

Waring, R. H., H. R. Holbo, R. P. Bueb, and R. L. Fredriksen. 1978*b*. Documentation of meteorological data from the Coniferous Forest Biome primary station in Oregon. United States Forest Service General Technical Report **PNW-73**.

Webster, J. R., and B. C. Patten. 1979. Effects of watershed perturbation on stream potassium and calcium dynamics. Ecological Monographs **49**:51–72.

Webster, J. R., J. B. Waide, and B. C. Patten. 1975. Nutrient recycling and the stability of ecosystems. Pages 1–27 *in* F. G. Howell, J. B. Gentry, and M. H. Smith, editors. Mineral cycling in southeastern ecosystems. Energy Research and Development Symposium Series (CONF-740513), Washington, D.C., USA.

Wetzel, R. G., and B. A. Manny. 1972. Decomposition of dissolved carbon and nitrogen compounds from leaves in an experimental hardwater stream. Limnology and Oceanography **17**:927–931.

Whittaker, R. H., G. E. Likens, F. H. Bormann, J. S. Eaton, and T. G. Siccama. 1979. The Hubbard Brook Ecosystem Study: forest nutrient cycling and element behavior. Ecology **60**:203–220.

Woodall, W. 1975. Mineral pathways in small Appalachian streams. Pages 408–422 *in* F. G. Howell, J. B. Gentry, and M. H. Smith, editors. Mineral cycling in southeastern ecosystems. Energy Research and Development Administration Symposium Series (CONF-740513), Washington, D.C., USA.

Commentary on Paper A6

Pinay, G. & Décamps, H. (1988) The role of riparian woods in regulating nitrogen fluxes between the alluvial aquifer and surface water: a conceptual model.

This paper is concerned with nitrogen cycling in riparian woodland in relation to the frequency, intensity and duration of flooding. Given the authors' background as ecologists, it is not surprising that the emphasis is on nitrogen cycling, but it is immediately clear that they adopt a spatially distributed conception of hydrology and that spatial patterns of waterlogging and links between topography and flow paths are crucial controls of the spatial pattern of nitrogen cycling.

The paper builds on a good deal of work in the Garonne Valley and elsewhere on nitrogen cycling in riparian wetlands, woodlands especially, under the supervision of Henri Décamps. This paper is particularly important because it begins to build a bridge between studies of soil microbiology and catchment hydrology, and it is no accident that its appearance in *Regulated Rivers* is significant: yet another illustration that major advances in one field often stem from the introduction of new ideas and approaches from a cognate discipline. One commends the foresight of the journal's editor, Professor Geoff Petts, in introducing this subject matter to an audience of largely hydrologists and river engineers.

Pinay and Décamps present a conceptual model of nitrogen dynamics in the soil of the riparian woodland in relation to the degree of submergence: never submerged, temporarily submerged and always submerged. The focus of the model is microbial denitrification, building on the work of Roger Knowles and others (Yoshinari *et al.*, 1977, Paper D2). The division between aerobic and anaerobic zones determines the balance of nitrification and denitrification in time and space, and the emphasis is not so much on the production of nitrate as its consumption. Indeed, it is clear that there is more interest in allochthonous nitrate, carried in from elsewhere by flowing groundwater, than in autochthonous nitrate, produced *in situ*. It is this emphasis on nitrate sources and flow pathways that is the significant advance, heralding many of the papers that appear later in this volume. Indeed, it can be argued that Fig. 3 is more significant than the generalised model presented in Fig. 1. This shifts the emphasis to the spatial distribution of nitrogen cycling processes across a transect perpendicular to the river within the riparian woodland. Measurements of potential denitrification suggest that nitrate is the limiting factor and that soils could denitrify more nitrate if greater quantities were delivered to the point in question. This has become an important element of our understanding of the functioning of nitrate buffer zones; depletion of nitrate is so immediately effective that the full buffering potential of the riparian zone is not realised (see for example, Burt *et al.*, 1999, Clément *et al.*, 2003). Pinay and Décamps anticipate this interest in buffer zones with their speculations about human impact on the riparian zone. In particular, they note that land drainage will lower water tables and thereby limit the rate of denitrification; river regulation which reduces the waterlogged area along the entire channel length would have a similar effect. They also cite the removal of woodland in reducing carbon supply, but this has proved more contentious (e.g. Haycock & Pinay, 1993).

Only two years after Pinay and Décamps published their paper, an edited volume was published which took this theme forward in a very detailed way (Naiman & Décamps, 1990), emphasising in particular the unique role of the terrestrial-aquatic ecotone as a boundary between terrestrial and aquatic ecosystems. The book, featuring work by many of Pinay and Décamps' colleagues and collaborators, provided a broad perspective of ecotonal communities at the land–water interface, and focused in particular on their management potential. Like Pinay and Décamps, the book includes socio-economic as well as ecological perspectives. At the heart of the book, however, is the link between hydrology and biogeochemistry at the landscape scale and the need to focus on the critical near-stream zone, given its importance in coupling hillslope and channel systems. This is the essence of Pinay & Décamps (1988) too. More recent reviews of links between hydrology, biogeochemistry and ecology at the landscape scale can be found in Tenhunen & Kabat (1999) and Burt & Pinay (2005).

TPB

References

Burt, T. P. & Pinay, G. (2005) Linking hydrology and biogeochemistry in complex landscapes. *Progr. Phys. Geogr.* **29**(3), 297–316.

Burt, T. P., Matchett, L. S., Goulding, K. W. T., Webster, C. P. & Haycock, N. E. (1999) Denitrification in riparian buffer zones: the role of floodplain sediments. *Hydrol. Processes* **13**, 1451–1463.

Clément, J.-C., Aquilina, L., Bour, O., Plaine, K., Burt, T. P. & Pinay, G. (2003) Hydrological flowpaths and nitrate removal rates within a riparian floodplain along a fourth-order stream in Brittany (France). *Hydrol. Processes* **17**, 1177–1195.

Haycock, N. E. & Pinay, G. (1993) Nitrate retention in grass and poplar vegetated buffer strips during the winter. *J. Environ. Qual.* **22**(2), 273–278.

Naiman, R. J. & Décamps, H. (eds) (1990) *The Ecology and Management of Aquatic-Terrestrial Aquatic Ecotones.* UNESCO: Paris and Parthenon Publishing Group, Carnforth, UK.

Tenhunen, J. D. & Kabat, P. (eds) (1999) *Integrating Hydrology, Ecosystem Dynamics and Biogeochemistry in Complex Landscapes.* Wiley: Chichester, UK.

Yoshinari, T., Hynes, R & Knowles, R. (1977) Acetylene inhibition of nitrous oxide reduction and measurement of denitrification and nitrogen fixation in soil. *Soil Biol. & Biochem.* **9**, 177–183.

REGULATED RIVERS: RESEARCH AND MANAGEMENT VOL. 2, 507–516 (1988)

THE ROLE OF RIPARIAN WOODS IN REGULATING NITROGEN FLUXES BETWEEN THE ALLUVIAL AQUIFER AND SURFACE WATER: A CONCEPTUAL MODEL

G. PINAY* and H. DECAMPS

Centre d'Ecologie des Ressources Renouvelables, CNRS, 29, rue Jeanne Marvig, 31055–Toulouse, France

ABSTRACT

Conceptual models are proposed to explain nitrogen fluxes in the soils of riparian zones in relation to the different modalities of water logging and nitrate inputs. Non-submerged, temporarily submerged, and always submerged, sites are considered in both winter and summer. It is shown that capacities for epuration are elevated for nitrate loads from the alluvial aquifer during the lateral transfer of water from agricultural land to the river through the riparian zone. About 30 m of groundwater flow under the riparian wood studied is sufficient to remove all nitrate. Denitrification rates up to 50 mg N_2 per m^2 per day were observed in the field, but a potential for more denitrification exists. The efficiency of the riparian woods as natural filters regulating nutrient transfers to the river deserves more consideration in river management.

KEY WORDS Nitrogen fluxes Alluvial forests Denitrification Conceptual model

INTRODUCTION

For some years wetlands have aroused a new interest in ecology. However, inspite of their large distribution and diversity (mangroves, swamps, riparian zones) they have been ignored as natural ecosystems. Vast floodplain wetlands have been drained to extend agricultural lands (McCormick, 1978; Swift, 1984) even though such utilization has been questioned for a long time (Viosca, 1928).

In Europe (Carbiener, 1970) and in other parts of the world (Petersen *et al.*, 1987) riparian zones along rivers are now considered as part of the natural environment deserving protection or rehabilitation. Their values range from recreational elements of diversification in the landscape (Reimold *et al*, 1980) to natural epuration of diffuse pollution (Dierberg and Brezonik, 1986a). Riparian vegetation may prevent bank erosion (Schlosser and Karr, 1981) and physically control floods (Stearns, 1978; Graf, 1980). However from an ecological point of view, a major role of the riparian zone yet to be clarified refers to land-water interactions and particularly to material transfers (Swanson *et al.*, 1982).

This paper analyses the influence of waterlogged soils within riparian forests on the functioning of the nitrogen cycle on river banks in an agricultural environment (Fustec *et al.*, 1988). Using a general conceptual model of the nitrogen cycle on river banks, several specific models are derived to examine the observed effects of the different modalities of waterlogging and nitrate inputs. These models are in turn utilized to deduce the capacities of epuration of nitrate loads from the alluvial aquifer during the lateral transfer of the water to the river through the riparian zone.

*Present address: I.I.A.S.A., A.2361 Laxenburg, Austria.

0886–9375//88/040507–10$05.00

Received 15 October 1987
Revised 14 December 1987

GENERAL MODEL OF NITROGEN CYCLING IN SOILS WITHIN RIVER CORRIDORS

The loss of nitrogen from an ecosystem results from microbiological denitrification which consists of the reduction of nitrogen oxides to form gazeous products (Knowles, 1981) in the following series:

$$2NO_3^- \rightarrow 2NO_2^- \rightarrow 2NO \rightarrow N_2O \rightarrow N_2 \qquad \text{(Payne, 1973)}.$$

This chain of reactions is closely dependent on environmental conditions, particularly oxygen, temperature, nitrate and organic carbon available, pH, and the presence of microorganisms (Knowles, 1982; Chalamet, 1985). All these parameters allow a partial or complete realization of the nitrate reduction; most of these depend on the nature of the soil, in particular on its drainage characteristics (Labaugh, 1986).

The general cycling of nitrogen in soils submitted to varying degrees of waterlogging is summarized in Figure 1a. Waterlogged soils or sediments show two main zones differing in their functions (Pearsall and

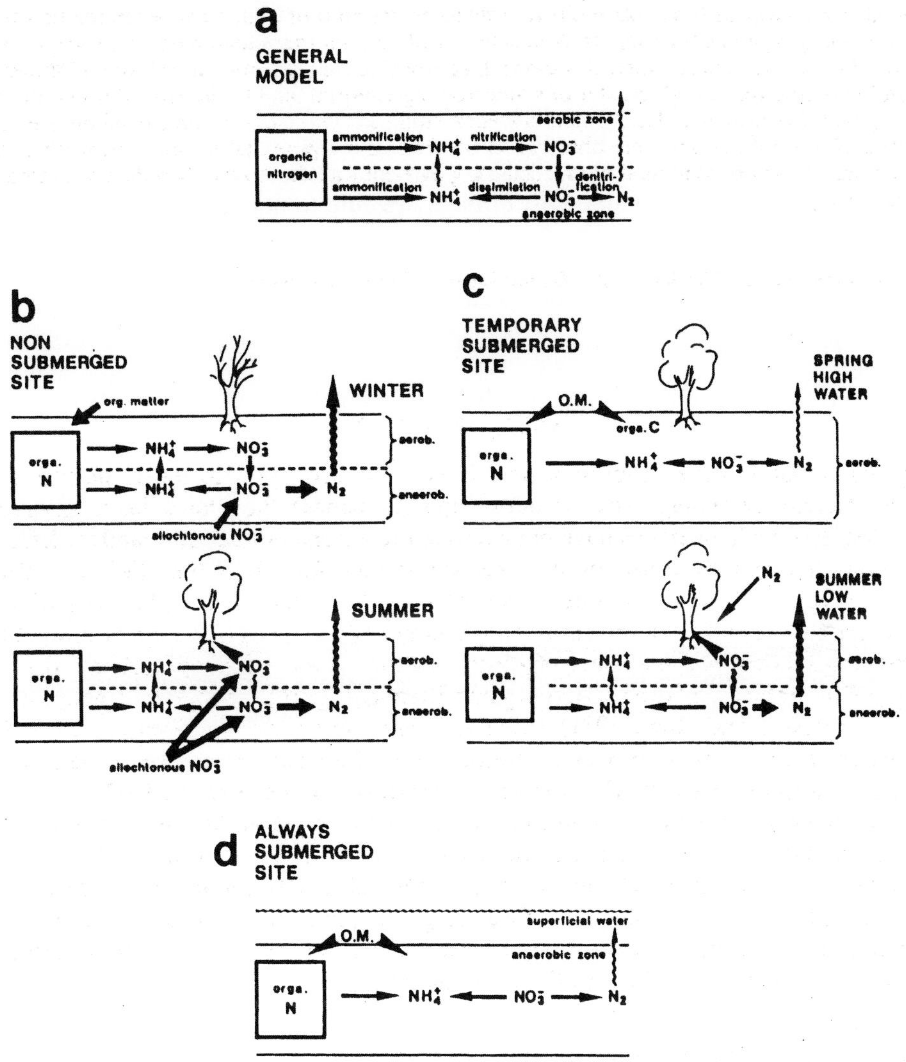

Figure 1. Conceptual models of the nitrogen dynamics in the soil of the riparian wood studied near the river Louge. a: general model; b: non-submerged site (site 7) in winter and in summer; c: temporary submerged site (site 10) in spring high water and in summer low water; d: always submerged site (Site 11)

Mortimer, 1939; Patrick and Delaune, 1972; Reddy and Patrick, 1986). In the aerobic superficial zone, the molecular oxygen may diffuse and more or less supply the biological demand. In the anaerobic saturated deeper zone, the feeble diffusion of the molecular oxygen is not sufficient to supply this biological demand. Therefore, a decrease of the oxidation-reduction potential follows in the deeper zone (Munch and Ottow, 1983). The relative importance of these two zones is a function of the extent of flooding, its duration in time, and the intensity of microbiological processes.

The general model (Figure 1a) allows coexistence of nitrification and denitrification processes in the soil. Most of the nitrate originates from organic matter contained in the soil, as shown by several authors such as Patrick and Tusneem (1972), Keeney (1973), Van Kessel (1977), Smith *et al.*, (1982), Neyra (1982). Ammonification of this organic nitrogen takes place in aerobiosis or in anaerobiosis (Chen *et al.*, 1972; Guthrie and Duxbury, 1978; Baillon *et al.*, 1985). The resulting ammonia diffuses in the aerobic zone and may be mineralized by nitrification only in the superficial zone (Focht and Verstraete, 1977). Such nitrification produces a concentration gradient of nitrate and therefore a diffusion of this nitrate from the aerobic to the anaerobic zone. Most of it is denitrified in this latter zone, part being possibly dissimilated (Tiedje *et al.*, 1982). Dissimilation does not seem to constitute an important loss of nitrogen in the form of gazeous NH_3 as the pH of the soil in riparian woods often ranges from 6·6 and 7·8 and does not allow a significant volatilization of NH_3 (Nelson, 1982). Moreover the diffusion coefficient of ammonia is 6 times less than that of nitrates (Reddy, 1982). It follows that ammonification and diffusion of ammonia control the rate of nitrification and therefore of denitrification in organic soils as indicated by Reddy and Rao (1983). Denitrifying microoganisms are probably located near the interface between aerobic and anaerobic zones (Tiedje *et al.*, 1983). They may coexist in the same sites as strict aerobic organisms (Qi and Hau-Kuei, 1983) and most of them are able to utilize free oxygen when it does exist (Stouthamer, 1967).

The aerobic–anaerobic zonation may be spatial as shown in Figure 1a, but also temperal, the same horizon of the soil being alternatively aerobic and anaerobic depending of its waterlogging and microbial activity.

NITRATE CONCENTRATION WITHIN THE RIPARIAN ZONE

Site area

The area studied is located in the Garonne Valley, about 20 km south of Toulouse (Figure 2). This area is covered by riparian woods extending over 5 ha on the lowest alluvial terrace along the River Louge, a tributary of the River Garonne and above an aquifer which flows about $600\,\mathrm{l\,m^{-2}d^{-1}}$. The upper level of this aquifer is never more than two metres beneath the soil surface (Pinay *et al.*, 1988). The soil is 'brown washed silt and clay' (Hubschman, 1975; Bégon, 1979). The vegetation is similar to that of the Garonne Valley with a willow–poplar community and an ash–elm community depending of the level of the soil (Décamps *et al.*, 1988). Between the cultivated fields and the River Louge, the groundwater passes through a gravel pit lake and then meets the soil of the riparian woods.

Results

The groundwater within the riparian zone was sampled each month for 16 months at 19 sites located on a broad transect from the cultivated fields to the area beyond the woods on the river bank (Figure 2). The nitrate concentrations within the riparian zone diminished clearly from 5 mg l^{-1} $N.NO_3^-$ above the gravel pit lake (sites 1, 2, 3, 4) to 1–5 mg l^{-1} between the gravel pit lake and the riparian woods (site 5); no $N.NO_3^-$ was recorded from samples on the river side of the riparian woods (site 13) (Table I). In fact, about 30 m of groundwater flow under the riparian woods is sufficient for complete epuration. No nitrate is found at the last site sampled after the riparian wood at any time of the year.

The reduction of the nitrate concentration in the groundwater through the gravel pit lake confirms the results already obtained by Bergé (1983). However the epuration capacity of the riparian woods is superior, as a narrow transfer provokes a total disappearance of the nitrate present in the groundwater (Labroue and Pinay, 1986). Microbiological denitrification has been shown to be the main process

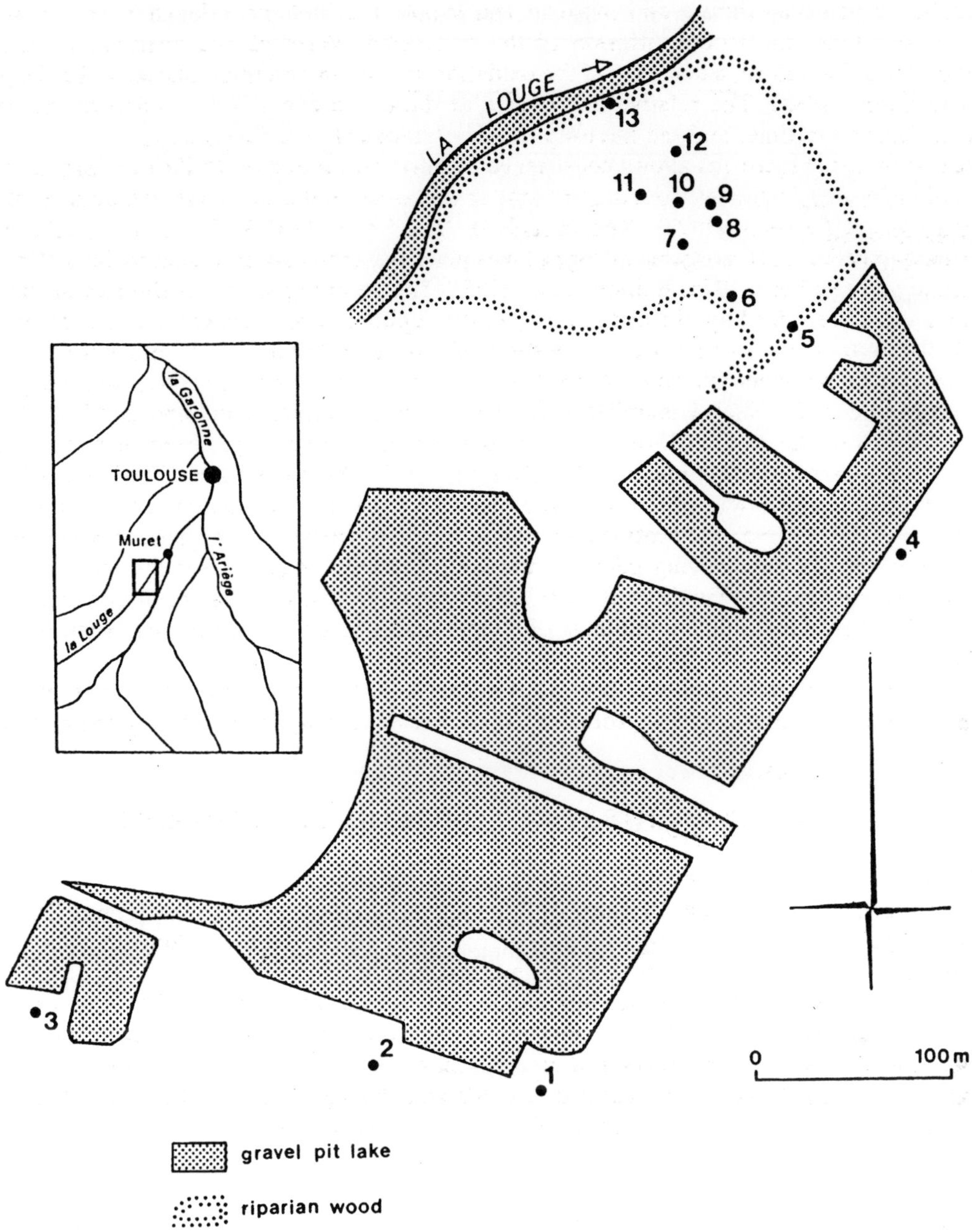

Figure 2. Sites location above the gravel pit lake (1 to 4) and within the riparian wood (5 to 13), utilizied to study the groundwater on its way between the cultivated fields and the River Louge

explaining the reduction of the nitrate load in the groundwater during its flow beneath the riparian wood studied here (Pinay, 1987). Rates of denitrification obtained *in situ* with the enclosure method ranged between 1·5 and 53 mg N_2 emitted per m^2 per day (Pinay, 1987). The differences observed account for the spatial and temporal variability of this process within the riparian wood.

In the riparian zone studied, waterlogging varies in its frequency and intensity due to differences in the topography above a water table near the surface of the soil. In summarizing, the level of the soil diminishes from site 1 which is globally aerobic to site 6, strictly anaerobic, near the River Louge. This

Table I. Nitrate concentration in the groundwater mean and standard deviation of 16 values obtained at a one month interval for the different stations 1–13 shown in Figure 1

	Stations	Mean on 16 months (mg N.NO_3)	Duration of submersion (in months)	Standard deviation
Stations above gravel pit lake	1	5·20	0	0·90
	2	5·00	0	0·42
	3	4·14	0	1·37
	4	2·76	0	1·31
Stations under riparian wood down to the river	5	1·32	0	0·81
	6	0·74	0	0·35
	7	1·00	0	1·05
	8	0·29	0	0·30
	9	0·14	1	0·12
	10	0·00	12	0·00
	11	—	8	—
	12	—	8	—
	13	—	6	—

topographic gradient is parallel to the flow of the water table beneath the riparian woods with its constant input of nitrates.

Application of General Model

The general model may be applied to specific situations according to the degree of waterlogging and nitrate inputs.

Non-submerged sites. This situation was observed in sites such as site 7 upstream from the flow of the watertable (Figure 2). The soil was never totally submerged, and a superficial aerobic zone was present all through the year. Moreover, a constant input of allochthonous nitrate was observed due to groundwater flow. In winter, the allochthonous nitrate from the aquifer is denitrified in the deepest anaerobic zone (Figure 1b). This zone of potential denitrification, between the groundwater and the aerobic zone, occupies an important volume by surface unit. The superficial aerobic zone allows the mineralization of nitrogen to nitrate, the concentration of which increases due to the absence of root absorption during winter. This nitrate is available to denitrifying microorganisms by diffusion to the anaerobic zone, and therefore an increase of the denitrification rate from 6 to 45 mg N_2 emitted per m^2 per day was observed during the winter period (Pinay, 1987).

During the summer period of vegetation growth (Figure 1b), the allochthonous nitrate from the groundwater is denitrified or absorbed by plants and microorganisms. This absorption constitutes a temporary retention of organic nitrogen but not a loss of nitrogen for the ecosystem.

In this model of the nitrate input from the groundwater, the factor limiting denitrification in the deepest zone may be the organic carbon, which is easily assimilated by microorganisms. However, it appears that the diffusion of dissolved carbon is efficient as there is practically no stratification in the aquifer (Pinay, 1986). Therefore, the limiting factor for denitrification appears to be nitrate concentration, in spite of a constant input by the groundwater: it was shown (Pinay and Labroue, 1986) that the potential for denitrification within this deepest zone was seven times larger than that observed (50 mg N_2 emitted per m^2 per day in the field contrasting to 350 mg N_2 experimentally obtained in the laboratory after nitrate input).

Temporary submerged sites. In this situation (e.g. site 10), the varying level of the alluvial aquifer induces waterlogging of the different layers of soil at different times (Figure 1c). The same layer of soil may function as aerobic or anaerobic depending on the hydrological period and its topographical situation. Two separate diagrams explain the nitrogen cycle in these sites according to the hydrological period considered.

During the spring high water (Figure 1c), the organic matter originating from fallen leaves accumulates in the soil. Large concentrations of organic matter, little or not mineralized, are maintained due to a quasi-permanent waterlogging, as indicated by Bruckert *et al.*, (1970) and Kilham and Alexander (1984). Organic nitrogen is mineralized only to ammonia form. Practically no nitrate is available because of the lack of nitrate input from the groundwater downstream to station 9 whatever the time of the year (Pinay, 1986). Moreover, the presence of dissolved reduced manganese and iron in the aquifer under these low sites indicates that, due to the intensity of the processes concerned, nitrate quantity is not sufficient to offset the demand of oxidized molecules. In this case, manganese then iron are utilized as terminal acceptors of electrons. It appears therefore that nitrate constitutes the factor limiting denitrification at the level of these low-lying sites: rates measured during this period were low (2 mg N_2 emitted per m^2 per day).

During the summer low water (Figure 1c), lowering of the alluvial aquifer allows a superficial aerobic zone to develop in which ammonia may be mineralized to nitrate. The latter is absorbed by plants or denitrified. Very high rates of denitrification do not allow nitrate accumulation in the soil (between 30 and 40 mg N_2 emitted per m^2, per day). However, this autochthonous nitrate input is not sufficient to meet the demand of electron acceptors due to the high microbial activity within the soil, and this causes more reduced manganese and iron to be dissolved. Again, available nitrate and not organic carbon constituted the limiting factor for denitrification in these low-lying zones.

Sites always submerged. In this situation a layer of water always persists above the soil (Figure 1d) as in the ancient oxbow of the River Louge. There is no input of allochthonous nitrate by the aquifer to these sites. The development of algae and macrophytes in the superficial water allows the presence of a narrow aerobic zone at the surface of the sediment during summer. But for most of the year, the sediment is quite reduced and, therefore, organic matter and ammonia accumulate. Low rates of denitrification were measured at these sites in each season (1 mg N_2 emitted per m^2, per day). These rates are comparable with the data obtained by other authors (Wijler and Delwiche 1954; Reddy and Patrick, 1975; Brinson *et al.*, 1983). Quasi-permanent reduced conditions in the sediment limit the mineralization of the organic matter, in particular of nitrogen and therefore limit denitrification.

DISCUSSION

The nitrogen cycle changes within the riparian wood of the River Louge according to a gradient which follows the flow of the groundwater. This evolution is related to the allochthonous or autochthonous type of nitrate and organic carbon input and to the variations in the waterlogging of the soil. Figure 3 summarizes the spatial evolution of the different processes in summer, i.e. the most active season. The input of allochthonous nitrate in the aquifer is noticeable until site 9. The denitrified nitrates present in the site downstream to site 10 originate from the mineralization of the *in situ*, accumulated organic matter. Therefore, the nitrate removal capacity of the riparian wood appears far more important than the removal actually realized. This hypothesis is based, on the one hand, on the constant presence of reduced manganese and iron in the aquifer downstream to site 9, and on the other hand on the potential for denitrification rates of up to 50 mg N_2 per m^2 per day observed in the field as compared with the potential for denitrification rates up to 350 mg N_2 per day estimated experimentally. Consequently, about 30 metres of groundwater flow under the riparian wood is enough to remove its nitrate, and the area of the wood downstream to site 9 constitutes a potential for more denitrification. In fact, the low concentration of nitrate in this area causes microorganisms to utilize other oxidized molecules (manganese, iron) as electron acceptors. Thus, it is suggested that a massive input of nitrate to the aquifer under the cultivated

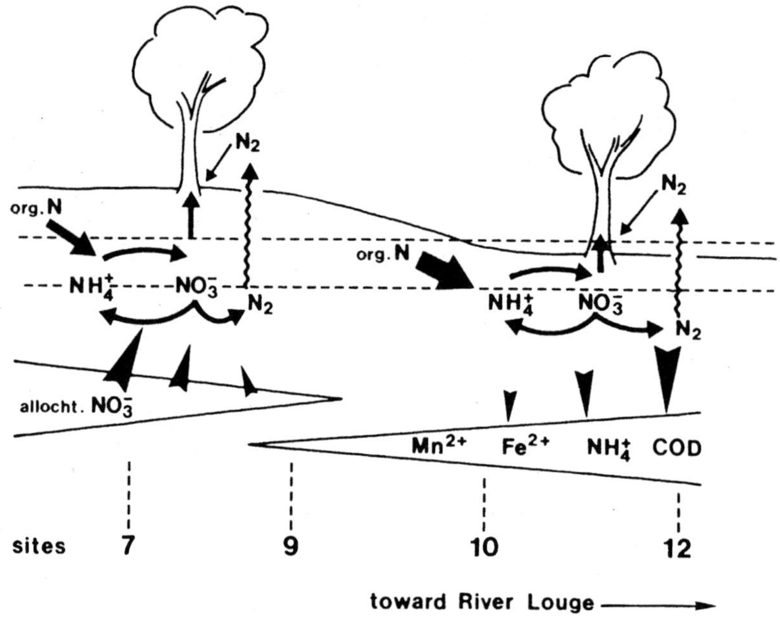

Figure 3. Spatial evolution of the different processes implicated in the nitrogen cycle during summer across a transect perpendicular to the river within the riparian wood.

fields upstream to the riparian wood would increase the denitrification rate in sites 7 to 9, as nitrate constitutes here a limiting factor. Then, the denitrification potential being reached, the excess nitrate would be denitrified in the area downslope from site 9.

The various models proposed to explain the dynamics of nitrogen in the riparian wood of the River Louge indicate the influence of the frequency, intensity, and duration of flooding on the functioning of the riparian woods. Waterlogging due to submersion by flood water from the river will occur more suddenly and its duration will be less than waterlogging due to a levelling up of the water table. The consequences for the functioning of the soil will be different as far as nutrient availability is considered. Moreover, because of the heterogeneity of the riparian zone, the denitrification capacity of the soil will differ considerably at a microscale (Tilton and Kadlec, 1979; Pinay, 1986). In spite of these very local differences, it is clear that, on the whole, riparian woods play a major role in the regulation of nutrient fluxes between subterranean and superficial waters (Jones *et al.*, 1976; Odum, 1978; Yarbro *et al.*, 1984; Lowrance *et al.*, 1984a, b; Pinay, 1987). This role is enhanced by the fact that the filter processes take place in a transitional volume of soil not just in a surface contact (Odum, 1978; Carter-Ewel, 1978; Salo *et al.*, 1986).

Taking into account all these above-mentioned relationships between physical and biological parameters and the nutrient regulation role of the riparian zones, it is possible to underline the major effects of human impacts on the buffering capacity of these alluvial forests. There are three main kinds of impacts: (a) River regulation to prevent flooding, which is based on the isolation of the river bed from its floodplain and a reduction of the waterlogged area all along the stretch of the river; (b) Land drainage in order to lower the watertable reduces the duration of waterlogging and the area of waterlogged soils; (c) Deforestation of riparian forests leads to a reduction in the amount of organic carbon, i.e. the energetic input which allows the process of microbial denitrification. Whatever the impact on the denitrification process, the final effect is a decrease, or even a disappearance, of the buffer capabilities of riparian zones with regard to the transfer of nitrogen between the drainage basin and the river.

CONCLUSION

Other studies have shown the influence of riparian woods, such as the one studied here, on the retention of the nitrate concentrated in the groundwater (Kaushik *et al.*, 1975; Simpson *et al.*, 1983; Yates and Sheridan, 1983; Brinson *et al.*, 1984; Lowrance *et al.*, 1985; Pinay and Labroue, 1986). More precisely, Peterjohn and Correl (1984) have demonstrated that about 19 m of flow under a riparian wood can be sufficient to reduce 90 per cent of nitrate load in the groundwater and at the same time to double the concentrations of ammonia and exchangeable orthophosphate per sediment unit. Similarly, Jacobs and Gilliam (1985a, b) have shown that in about 30 metres of flow beneath a riparian wood, nitrate in the groundwater was eliminated.

Wetlands have been used as a means of tertiary treatment of polluted effluents for several years (Dierberg and Brezonik, 1986b; Dubuc *et al.*, 1986). Fritz and Helle (1986) have shown that this natural epuration is economically more profitable than the conventional chemical epuration, and Deghi *et al.*, (1980) have stressed that it does not adversely affect the stocking capacity of the wetland used. However, the cost of this treatment may be high as it is necessary to convey the effluents to the epurating wetlands (Fritz *et al.*, 1986). In the case of riparian woods along a river, such as the one studied here, the effluent is naturally carried to the wetland by the flow of the groundwater.

It appears that waterlogged conditions are important for the maintenance of riparian systems and the nitrogen cycle dynamics. On the one hand, generally river developments lead to the removal of these ecosystems by filling, dredging, draining, or by managing impoundments in such a way that it cannot allow the maintenance of these alluvial forests. On the other hand the economic value of these semi-wetland areas differs according to the different social interests, and is difficult to estimate. Moreover, owners of riparian lands do not necessarily receive benefit from maintaining them as wetlands (Thibodeau and Ostro, 1981; Kreutzwiser and Pietraszko, 1986; Farber and Costanza, 1987). The efficiency of the riparian woods as natural filters *vis à vis* nutrients has been clearly demonstrated and this adds to the interest in these ecotones as a component of dynamic fluvial landscapes. This fact must be taken into account when considering the protection of riparian habitats in relation to other anthropogenetic uses.

ACKNOWLEDGEMENTS

This work was supported by the Programme Interdisciplinaire de Recherche sur l'Environnement (PIREN) of the Centre National de la Recherche Scientifique. We wish to thank Dr. G. E. Petts for comments on an earlier draft of this paper and Yasmin Taher for correcting the English.

REFERENCES

Baillon, F., Herbauts, J., and Tanghe, M. 1985 'Minéralisation de l'azote dans une séquence de sols hydromorphes de la vallée de la Haute Semois (Belgique)', *Acta Oecologica Oecologia Plantarum*, **6**, 183–192.

Bégon, J. C. 1979. 'Sur la mise en évidence de séquences d'évolution dans les formations limoneuses des terrasses de la Garonne', *C.R. Acad. Sc. Paris*, **288**, série D, 481–484.

Bergé, P. 1983. 'Détection de l'origine des pollutions azotées d'un aquifère alluvial de la vallée de la Garonne par traçage isotopique', *Thèse de 3ème cycle*, Univ. Toulouse, 211 pp.

Brinson, M. M., Bradshaw, H. D., and Holmes, R. N. 1983 'Significance of floodplain sediments in nutrient exchange between a stream and its flood plain', In Fontaine, T. D and Bartell, S. M. (Eds), *Dynamics of Lotic Ecosystems*, Ann Arbor Science, Michigan, U.S.A., 199–220.

Brinson, M. M., Bradshaw, H. B., and Kane, E. S. 1984. 'Nutrient assimilative capacity of an alluvial floodplain swamp', *Journal of Ecology*, **21**, 1041–1057.

Bruckert, S., Dommergues, Y., Weinhard, P., and Boymond, D. 1970. 'Effet litière. III Influence de l'anaérobiose sur la production de composés antimicrobiens hydrosolubles', *Acta Oecologica Oecologia Plantarum*, **5** (2), 137–146.

Carbiener, R. 1970. 'Un exemple de type forestier exceptionnel pour l'Europe Occidentale: la forêt du lit majeur du Rhin au niveau du fossé Rhénan (Fraxino–Ulmetum). Intérêt écologique et biogéographique. Comparaison à d'autres forêts thermophiles', *Vegetatio*, **20**, 97–148.

Carter-Ewel, K. 1978. 'Riparian ecosystems: conservation of their unique characteristics', In *Strategies for Protection and Management of Floodplain Wetlands and Other Riparian Ecosystems. USDA FSGTR*, **WO–12**, 56–62.

Chalamet, A., 1985. 'Effect of environmental factors on denitrification', In Golterman, H. L. (Ed.), *Denitrification in the Nitrogen Cycle*, Plenum Press, New York, 7–29.

Chen, R. L., Keeney, D. R., Graetz, B. A., and Holding, A. J. 1972. 'Denitrification and nitrate reduction in Wisconsin Lake sediment', *Journal of Environment Quality*, **1** (2), 158–162.

Décamps, H., Fortuné, M., Gazelle, F., and Pautou, G. 1988. 'Historical influence of man on the riparian dynamics in a fluvial landscape', *Landscape Ecology*. (submitted).

Deghi, G. S., Carter-Ewel, K., and Mitsch, W. J. 1980. 'Effects of sewage effluent application on litter fall and litter decomposition in cypress swamps', *Journal Applied Ecology*, **17**, 397–408.

Dierberg, F. E. and Frezonik, P. L. 1986a. 'The effect of wastewater on the surface water and groundwater quality of Cypress domes', In Carter-Ewel, K. and Odum H. T. (Eds), University of Florida Press, Gainsville, 83–101.

Dierberg, F. E. and Brezonik, P. L. 1986b. 'Water chemistry of a Florida Cypress dome', In Carter-Ewel, K. and Odum H. T. (Eds), *Cypress Swamps*, University of Florida Press, Gainsville, 34–50.

Dubuc, Y., Janneteau, P., Labonte, R., Roy, C., and Briere, F. 1986. 'Domestic wastewater treatment by peatlands in a northern climate: a water quality study', *Water Resources Bulletin*, **22**(2), 297–303.

Farber, J. and Costanza, R. 1987. 'The economic value of wetlands systems', *Journal of Environmental Management*, **24**, 41–51.

Focht, D. and Verstraete, W. 1977. 'Biochemical ecology of nitrification and denitrification', In Alexander, M. (Ed.), *Advances in Microbial Ecology, vol. 1*, New York, 135–214.

Fritz, W. R. and Helle, S. C. 1986. 'Feasibility of cypress wetland tertiary treatment', In Carter-Ewel, K. and Odum, H. T. (Eds), *Cypress Swamps*, University of Florida Press, Gainsville, 249–258.

Fritz, W. R., Helle, S. C., and Ordway, J. W., 1986. 'The cost of cypress wetland treatment', In Carter-Ewel, K. and Odum, H. T. (Eds), *Cypress Swamps*, University of Florida Press, Gainsville, 239–248.

Fustec, E., Pierre, D., Vallaeys, T., Coetil, F., and Lauga, J. 1988. 'Les transferts de nitrates dans les nappes alluviales de la vallée de la Garonne', (en préparation).

Graf, W. L. 1980. 'Riparian management: a flood control perspective', *Journal of Soil and Water Conservation*, July August, 158–161.

Guthrie, T. F. and Duxbury, J. M. 1978. 'Nitrogen mineralization and denitrification in organic soils', *Soil Sci. Soc. Am. J.*, **42**, 908–912.

Hubschman, J. 1975. 'Les terrasses récentes de la Garonne et leur évolution', *Bull. Ass. Franc. Etude Quaternaire*, **3–4** (44–45), 137–147.

Jacobs, T. C. and Gillaim, J. W. 1985a. 'Headwater stream losses of nitrogen from two coastal plain watersheds', *J. Environ. Qual.*, **14**(4), 467–472.

Jacobs, T. C. and Gillaim, J. W. 1985b. 'Riparian losses of nitrate from agricultural drainage waters', *J. Environ. Qual.*, **14** (4), 472–478.

Jones, J. R., Borofka, B. P., and Bachmann, R. W. 1976. 'Factors affecting nutrient loads in some Iowa streams', *Water Res.*, *10*, 117–122.

Kaushik, N. K., Robinson, J. B., Sain, P., Whiteley, H. R., and Stammers, W. 1975. 'A qualitative study of nitrogen loss from water of a small spring fed stream', *Proc. 10th Canadian Symposium on Water pollution Research, Canada*, 110–117.

Keeney, D. R. 1973. 'The nitrogen cycle in sediment-water systems', *J. Environ. Qual.*, **2** (1), 15–29.

Kilham, O. W. and Alexander, M. 1984. 'A basis for organic matter accumulation in soils under anaerobiosis', *Soil Science*, **137** (6), 419–427.

Knowles, R. 1981. 'Denitrification', *Ecol. Bull.*, **33**, 315–329.

Knowles, R. 1982. 'Denitrification', *Microbiological Reviews*, **46**, 43–70.

Kreutzwiser, D. and Pietraszko, L. J. 1986. 'Wetland values and protection strategies: a study of landowner attitudes in Southern Ontario', *J. Environ. Manage.* **22**, 13–23.

Labaugh, J. W. 1986. 'Wetland ecosystem studies from a hydrologic perspective', *Water Resources Bulletin*, **22** (1), 1–10.

Labroue, L. and Pinay, G. 1986. 'Epuration naturelle des nitrates des eaux souterraines : possibilités d'application au réaménagement des lacs de gravières', *Annls. Limnol.*, **22** (1), 83–88.

Lowrance, R. R., Todd, R. L. and Asmussen, L. E. 1984a. 'Nutrient cycling in an agricultural watershed-I: phreatic movement', *J. Environ. Quality*, **13** (1), 22–27.

Lowrance, R. R., Todd, R. L. and Asmussen, L. E. 1984b. 'Nutrient cycling in an agricultural watershed-II: streamflow and artificial drainage', *J. Environ. Quality*, **213** (1), 27–32.

Lowrance, R. R., Leonard, R. A., Asmussen, L. E., and Todd, R. L. 1985. 'Nutrient budgets for agricultural watersheds in the southeastern coastal plain', *Ecology*, **66** (1), 287–296.

McCormick, J. 1978. 'Ecology and the regulation of freshwater wetlands', In Good, R. E., Whigham, B. F., and Simpson, R. L. (Eds), *Freshwater Wetlands*, Academic Press, New York, 341–355.

Munch, J. C. and Ottow, J. C. G. 1983. 'Réduction bactérienne des oxydes ferriques amorphes et cristallisés', *Science du Sol*, **3–4**, 205–215.

Nelson, D. W., 1982. 'Gaseous losses of nitrogen other than through denitrification', In Stevenson, F. J. (Ed.), *Nitrogen in agricultural soils*. Agronomy monograph, **22**, 327–363.

Neyra, M., 1982. *Nitrification et dénitrification en milieu aquatique. Coexistence dans le sédiment*, Thèse Doctorat 3eme cycle, Lyon I, 101 pp.

Odum, E. P. 1978. 'Ecological importance of the riparian zone', In *Strategies for protection and management of floodplain wetlands and other riparian ecosystems*. USDA FSGTR, **WO-12**, 2–4.

Patrick, W. H. and Delaune, R. D. 1972. 'Characterization of the oxidized and reduced zones in flooded soil', *Soil, Sci. Soc. Ann. Proc.*, **36**, 573–576.

Patrick, W. H. and Tusneem, M. E. 1972. 'Nitrogen loss from flooded soil', *Ecology*, **53** (4), 735–737.

Payne, W. J. 1973. 'Reduction of nitrogenous oxides by microorganisms', *Bacteriol. Rev.*, **37**, 409–452.

Pearsall, W. H. and Mortimer, C. H. 1939. 'Oxidation reduction potentials in waterlogged soils, natural waters and muds', *Journal of Ecology*, *27*, 483–501.

Peterjohn, W. T. and Correll, D. L. 1984. 'Nutrient dynamics in an agricultural watershed: observations on the role of a riparian forest', *Ecology*, **65** (5), 1466–1475.

Petersen, R. C., Madsen, B. L., Wilzbach, M. A., Magadza, C. H. D., Paarlberg, A., Kullberg, A., and Cummins, K. W. 1987. 'Stream management: emerging global similarities', *Ambio*, **16** (4), 166–179

Pinay, G. 1986. *Relations sol-nappe dans les bois riverains de la Garonne. Etude de la dénitrification*, Thèse de Doctorat, Univ. Lyon I, 200 pp.

Pinay, G. 1987. 'Impact of a riparian forest on the nitrogen content of phreatic water in the Garonne basin', *MAB. 5. UNESCO, Proceedings of a workshop*, Toulouse, April 1986.

Pinay, G. and Labroue, L. 1986. 'Une station d'épuration naturelle des nitrates transportés par les nappes alluviales: l'aulnaie glutineuse', *C. R. Acad. Sc. Paris 302, Serie III*, **17**, 629–632.

Pinay, G., Décamps, H., Arles, C., and Lacassin, M. 1988. 'Topographic influence on carbon and nitrogen dynamic in riverine woods', *Arkiv für Hydrobiologie* (in press).

Qi, Z. and Hau-Kuei, C. 1983 'The activity of nitrifying and denitrifying bacteria in paddy soil', *Soil Science*, **135**(1), 31–34.

Reddy, K. R. 1982. 'Nitrogen cycling in a flooded soil ecosystem planted to rice (Oryza sativa L.), *Plant and Soil*, **67**, 203–220.

Reddy, K. R. and Patrick, W. H. Jr. 1975. 'Effect of alternate aerobic and anaerobic conditions on redox potential, organic matter decomposition and nitrogen loss in a flooded soil', *Soil Biol. Biochem.*, **7**, 87–94.

Reddy, K. R. and Patrick, W. H. 1986. 'Denitrification losses in flooded rice fields', *Fertilizer Research*, **9**, 99–116.

Reddy, K. R. and Rao, P. S. C. 1983. 'Nitrogen and phosphorus fluxes from a flooded organic soil', *Soil Science*, **136**(5), 300–307.

Reimold, R. J. Hardisky, M. A., and Philips, J. H. 1980. Wetland values. A non-consumtive perspective', *J. Environ. Manage.*, **11**, 77–85.

Salo, J., Kalliola, R., Hakkinen, I., Makinen, Y., Niemela, P., Puhakka, M., and Coley, P. D. 1986. 'River dynamics and the diversity of Amazon lowland forest', *Nature*, **332**, 254–258.

Schlosser, I. J. and Karr, I. R. 1981. 'Water quality in agricultural watersheds: impact of riparian vegetation during base flow', *Wat. Res. Bull.*, 17, 233–240.

Simpson, R. L., Good, R. E., Walker, R., and Frasco, B. R. 1983. 'The role of Delaware River freshwater tidal wetlands in the retention of nutrients and heavy metals', *J. Environ. Quality*, **12**(1), 41–48.

Smith, C. J., Delaune, R. D., and Patrick, W. H. Jr. 1982. 'Nitrate reduction in Spartina alterniflora marsh soil', *Soil Sci, Soc. Am. J.*, **46**, 748–750.

Stearns, F. 1978. 'Management potential: summary and recommendations', In Good, R. E., Whigham, D. F., and Simpson, R. L. (Eds), Academic Press, New York, 357–363.

Stouthamer, A. H. 1967. 'Nitrate reduction in Aerobacter aerogenes I. Isolation and properties of mutant strains blocked in nitrate assimilation and resistant against chlorate', *Archiv für Mikrobiologie*, **56**, 68–75.

Swanson, F. J., Gregory, S. V., Sedell, J. R., and Campbell, A. G. 1982. 'Land water interactions: the riparian zone', In Edmonds, R. L. (Ed.) *Analysis of coniferous forest ecosystems in the western United States. US/IBP Synthesis Series*, 14, 267–291.

Swift, B. L 1984. 'Status of riparian ecosystems in the United States', *Water Resources Bulletin*, **20**(2) 223–228.

Thibodeau, F. R. and Ostro, B. D. 1981. 'An economic analysis of wetland protection', *J. Environ. Manage.*, **12**, 19–30.

Tiedje, J. M., Sextone, A. J., Myrold, D. D., and Robinson, J. A. 1982. 'Denitrification: ecological niches, competition, and survival, *Antonie van Leeuwenhock*, **48,** 563–583.

Tilton, D. L. and Kadlec, R. H. 1979. 'The utilization of a freshwater wetland for nutrient removal from secondarily treated waste water effluent', *J. Environ. Qual.*, **8**(3), 328–334.

Van Kessel J. F. 1977. 'Factors affecting the denitrification rates in two water–sediment systems', *Water Research*, **11**, 259–267.

Viosca P. Jr. 1928. 'Louisiana wetlands and the value of their wild life and fishery resources', *Ecology*,**9**(2), 216–229.

Wijler, J. and Delwiche, C. C. 1954. 'Investigation on the denitrifying process in soil', *Plant and Soil*, **5**, 155–169.

Yarbro, L. A., Kuenzler, E. J., Mulholland, P. J., and Sniffen, R. P. 1984. 'Effects of stream channelization on exports of nitrogen and phosphorus from North Carolina coastal plain watershed', *Environ. Manage.*, **8**(2), 151–160.

Yates, P. and Sheridan, J. M. 1983. 'Estimating the effectiveness of vegetated floodplains/wetlands as nitrate–nitrite and orthophosphorus filters', *Agriculture–Ecosystems and Environment*, **9**, 303–314.

B Riparian Zone Hydrology

Commentary on Paper B1

Troxell, H. C. (1936) The diurnal fluctuation in the ground-water and flow of the Santa Ana River and its meaning.

The interaction between vegetation and shallow groundwater is an important part of ecosystem dynamics in the riparian zone. A water table is maintained in the alluvial deposits of the riparian zone, either by inflow from the surrounding catchment area or by inflow from the stream itself, or both. Where the catchment area supplies the alluvial groundwater, there can be significant water losses by transpiration so that the amount reaching the stream is greatly depleted, even though the flow direction is still towards the stream. In cases where the stream provides the alluvial groundwater, discharge will decrease downstream and again evaporation losses can be significant.

From the time that hydrometric measurements first began in semi-arid catchments, it was soon noted that there were diurnal variations in stream discharge, with the implication that streamflow losses by transpiration could be significant. Harold Troxell was not the first American hydrologist to tackle this issue: Walter N. White, who like Troxell also worked for the United States Geological Survey, had earlier worked on the problem in the Escalante Valley in southern Utah (White, 1932) but Troxell refined and developed White's approach, working in particular on the Santa Ana River in California. Bond *et al.* (2002) provide a recent example of research on this phenomenon.

Troxell's purpose was to better understand the transfers between streamflow, groundwater and transpiration and thereby demonstrate that the assumptions adopted by most water engineers of the time were wrong, leading to an underestimation of the actual transpiration loss. Three assumptions are stated in the third paragraph of the paper: that the discharge cycle immediately identifies the time of minimum and maximum transpiration loss; that a curve connecting the points of maximum discharge represents the stream discharge if there is no transpiration; and that the difference between this quantity and the actual discharge equals the transpiration loss.

Troxell's approach combines records of stream discharge, field measurements (wells dug specially for the purpose at 9, 130 and 305 m from the bank of the Santa Ana River) and water balance computations. His graphical analysis of diurnal hydrographs, for both water table and river discharge, extends the method developed by White (op. cit.). The key to his approach is to work on increments of discharge per unit length of the river channel (q) and to take into account both absorption (a) into the bank (i.e. "bank storage") and changes in channel storage (s). The method applies irrespective of which way the alluvial groundwater is flowing, as shown by the graphs of absorption and inflow in Fig. 8. Troxell is able to establish the true pattern and volume of transpiration loss (Fig. 7). He was also able to show that the water table would have been higher had there been no transpiration, which is why simple analysis of diurnal cycles underestimates the true loss. He was able to demonstrate that the stream will fluctuate in unison at all points, and that the diurnal cycle does not progress downstream as others had thought.

A later example of streamflow depletion by riparian vegetation is provided by Meyboom (1964) for streams in the Canadian prairies, where very significant losses of water occur downstream. Although Meyboom commits one of the errors that Troxell sought to eliminate (assuming that the maximum discharge would be the flow if there were no transpiration), the analysis is of some interest in that it establishes the order of magnitude of losses and relates these to the different plant species ("phreatophytes") in the riparian zone. For sites where groundwater flows towards the river, Meyboom notes that the stream may still show downstream gains in discharge but the transpiration losses mean that gains are much lower than would otherwise occur; Meyboom concludes rightly that the effect will be particularly significant in wide river valleys where the water table gradient is very low. There are also examples where the river provides the alluvial groundwater – what Meyboom call "thieving". There is also analysis of substantial losses from a reservoir release (over 80% of the total release); initial bank storage losses do not return to the stream because of transpiration. Meyboom concludes that substantial streamflow depletion can occur under both natural and regulated flow conditions, regardless of whether the stream is influent or effluent and, echoing Troxell's observations from California, that depletion may be much more common in prairie rivers than had hitherto been noticed.

Whilst more recent studies of flood-plain hydrology have included situations where flow is both towards and away from the river channel (e.g. Bates *et al.*, 2000; Burt *et al.*, 2002), most studies of riparian buffer zone processes have only included the former condition. Groundwater quality is improved as the water flows through the riparian zone towards the stream. However, it is clear that in some situations, particularly dryland systems (Holmes *et al.*, 1994, Paper F6 – this volume), water flow from the stream to the flood plain will be the normal condition; buffer zone processes may be less effective in such situations but there is still the potential for some improvement if water is able to move into the flood plain and then, further downstream, back into the channel. This flow pattern would form part of what is now regarded as the hyporheic zone.

TPB

References

Bates, P. D., Stewart, M. D., Desitter, A., Anderson, M. G. & Renaud, J.-P. (2000) Numerical simulation of flood plain hydrology. *Water Resour. Res.* **36**, 2517–2529.

Bond, B. J., Jones, J. A., Moore, G., Phillips, N., Post, D. & McDonnell, J. J. (2002) The zone of vegetation influence on baseflow revealed by diel patterns of streamflow and vegetation water use in a headwater basin. *Hydrol. Processes* **16**, 1671–1677.

Burt, T. P., Bates, P. D., Stewart, M. D., Claxton, A. J., Anderson, M. G. & Price, D. A. (2002) Water table fluctuations within the floodplain of the River Severn, England. *J. Hydrol.* **262**, 1–20.

Holmes, R. M., Fisher, S. G. & Grimm, N. B. (1994) Parafluvial nitrogen dynamics in a desert stream ecosystem. *J. North Am. Benthological Soc.* **13**, 468–478.

Meyboom, P. (1964) Three observations on streamflow depletion by phreatophytes. *J. Hydrol.* **2**, 248–261.

White, W. N. (1932) A method of estimating ground-water supplies based on discharge by plants and evaporation from soil: Results of investigations in the Escalante Valley, Utah. *US Geol. Survey Water-Supply Paper 659*.

THE DIURNAL FLUCTUATION IN THE GROUND-WATER AND FLOW OF THE SANTA ANA RIVER AND ITS MEANING

Harold C. Troxell
(Published with the approval of the Director, U. S. Geological Survey)

In the time alloted for this subject it will be impossible to discuss, in its entirety, all phases of the methods used in computing the loss of water by transpiration from native plant-life along the Santa Ana River. The results of this work are published in Bulletin 44 of the Division of Water Resources, State of California. The present paper is confined to a discussion of the diurnal fluctuations that occur in the flow of the Santa Ana River and the adjacent ground-water.

On Figure 1 have been plotted the gage-height records for a short period (July 5-8, 1935) for eight southern California streams. These records have been selected at random from the large number of records filed in the local office of the United States Geological Survey. Some of these streams have a large drainage-area, others are small; some are short and steep, others flat and long. The points of measurement range from 15 to 2950 feet above sea-level. One of the streams (Mojave River) is in the desert-region; the others drain into the Pacific Ocean. The object of Figure 1 is to demonstrate that the diurnal fluctuation exists and is, to a large degree, similar at most points of measurement. As a rule, the maximum discharge occurs about 10 o'clock in the morning, and the minimum late in the afternoon. It is readily recognized that these fluctuations are in the main caused by the evaporation and transpiration-loss in or adjacent to the stream-channel.

It has been the practice of many engineers to assume that the record during the period of greatest discharge, usually about 10 o'clock in the morning, represents the period of minimum transpiration-losses, and the period of least discharge, in the afternoon, represents the maximum transpiration-losses. Others have connected the points of maximum discharge during the diurnal fluctuation and assumed that the curve thus obtained would represent the probable flow of the stream if there were no losses, also that the difference between this quantity and the actual discharge represents the transpiration-loss. I shall attempt to demonstrate that, as a general rule, these assumptions are entirely in error, with the tendency to give results considerably less than the actual transpiration-loss.

On Figure 2 is plotted a copy of the record of water-stage recorder for four of our regular stations on the Santa Ana River for the period July 15-24, 1932. The station at Riverside Narrows is at the Union Pacific Railroad Bridge near Riverside (Arlington). The Hamner Avenue station is 8.2 miles downstream. The Auburndale Bridge station is 11.8 miles below Riverside Narrows, and the Atchison, Topeka, and Santa Fe Railway Bridge station is 16.6 miles below Riverside Narrows. Well E-3 is 13.8 miles below Riverside Narrows, almost half way between the Auburndale Bridge and Santa Fe Railway Bridge stations. The average velocity of the stream-flow in this section of 16.6 miles of channel during the non-flood period is about 1.5 feet per second, or 1.0 mile per hour. Well E-3 was especially dug for this study, on the river-bank about 30 feet from the River.

The large amount of fluctuation noted in these river-records is due to the scouring and filling of the stream-channel and the passage of sand-waves. Notwithstanding this tremendous amount of minor fluctuation, there is a well-defined daily cycle. These records show considerable similarity. They all approach the maximum flow, as well as the minimum, at approximately the same hour.

The record obtained on the ground-water table at Well E-3 is not unlike the river-records, especially that at the Santa Fe Railway Bridge.

In addition to Well E-3, two other wells were dug on the banks of the Santa Ana River in the same vicinity. Well E-1 was 970 feet north of E-3 and 1000 feet from the River. Well E-2 was on

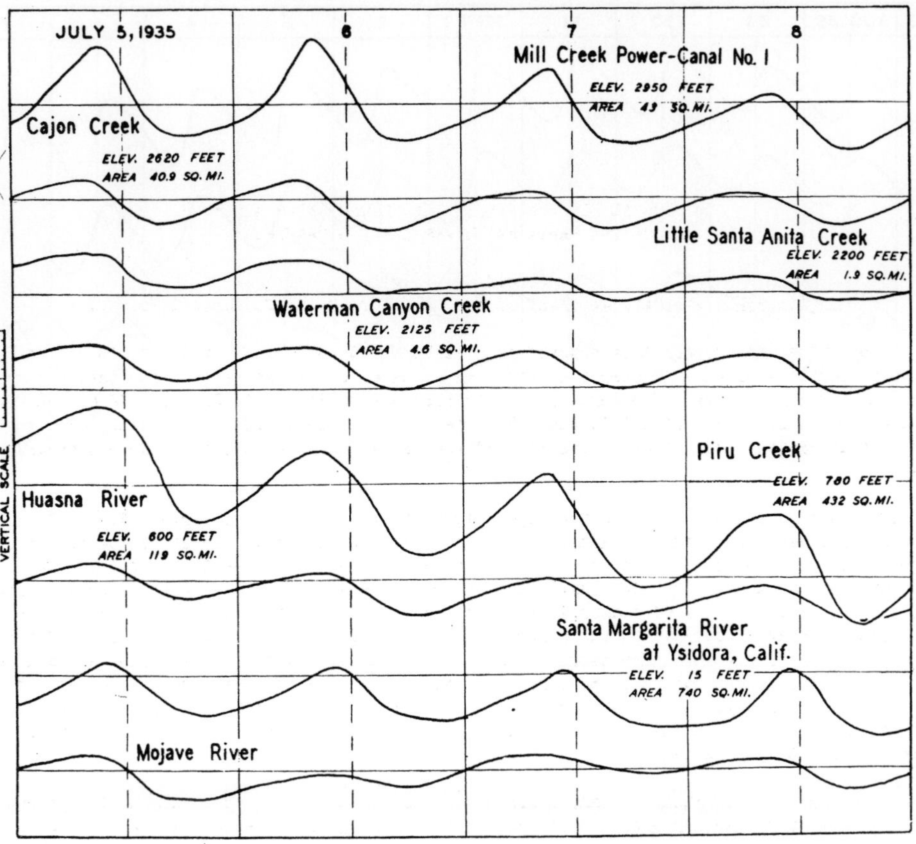

FIG. 1—DIURNAL FLUCTUATIONS IN SOUTHERN CALIFORNIA STREAMS

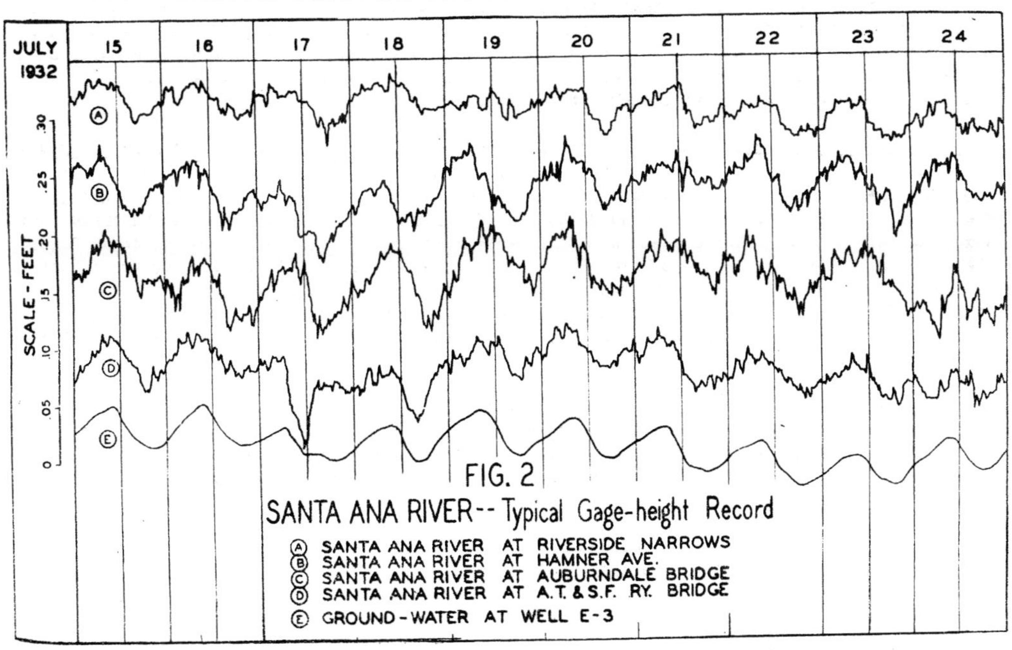

FIG. 2

SANTA ANA RIVER--Typical Gage-height Record

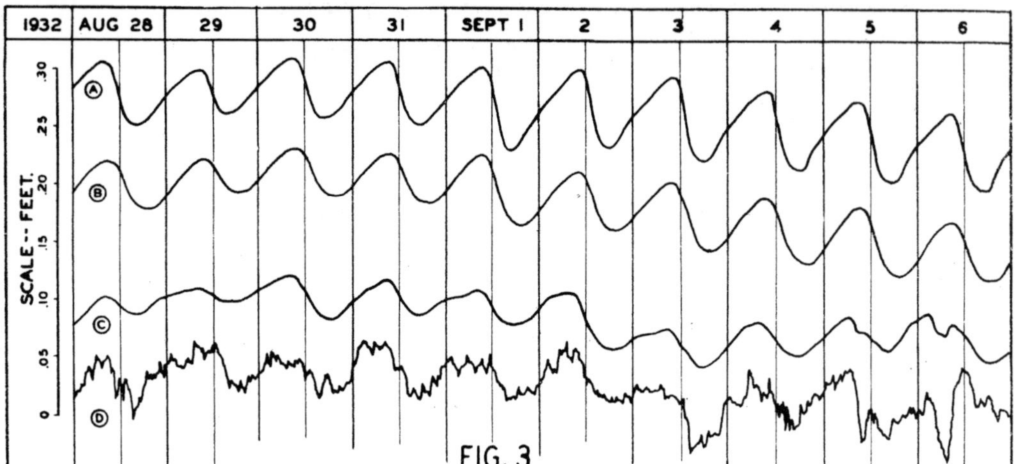

FIG. 3

SANTA ANA RIVER--Typical Ground-Water Record

(A) GROUND-WATER AT WELL E-1
(B) GROUND-WATER AT WELL E-2
(C) GROUND-WATER AT WELL E-3
(D) SANTA ANA RIVER AT A.T. & S.F. RY. BRIDGE

a line between E-1 and E-3, about 430 feet from the River. The record from these wells for the period August 28 to September 6, 1932, is given on Figure 3.

The diurnal fluctuations at this group of wells are all to a certain degree similar to the river-records. The water-table is highest around 10 a.m. and lowest about 6 p.m. At the bottom of Figure 3 is given the record obtained at the Santa Fe Railway Bridge. The records obtained at wells E-2 and E-3 show how the amplitude of this diurnal fluctuation is modified as it approaches the River from Well E-1.

In passing, it should be stated that there is a moderate amount of regulation in the River directly above wells E-1 to E-3. For example, on July 17 the discharge in the River was reduced between 6 a.m. till the afternoon. The record at Well E-3 was modified on that day, notwithstanding the fact that the water-table is usually about 0.2 foot higher than that in the River. Similar effects of regulation are shown on September 5 and 6.

The object of Figures 2 and 3 is merely to show that a diurnal fluctuation occurs both in the River and in the ground-water table adjacent to the River, and that these fluctuations are to a degree similar.

On Figure 4 have been plotted the records of ground-water fluctuation at Well E-1, the gage-height record of the Santa Ana River at the Santa Fe Railway Bridge, and the temperature-records for the 2-day period August 19 and 20, 1932.

It can readily be established that the diurnal fluctuation shown by the record of ground-water at Well E-1 is due, in the main, to transpiration-needs of the plant-life. The water to supply these needs can be drawn from the inflow to the area or from ground-water storage, or from both. Thus, for example, at point c the ground-water table remains at a constant altitude for a short period. This means that during this period the inflow to the ground-water table is sufficient to supply all the transpiration-demands. At point b the transpiration-demands are not only using all the inflow but are also making a heavy drain on the ground-water supply. At point d the transpiration-losses have decreased to such a point that the inflow is again sufficient to supply these needs. At point a the transpiration-needs are much less than the inflow, and as a result the surplus water is placed in storage, causing a rise in the ground-water table. It is thus apparent that the ground-water record is merely an accumulative curve of the rates of ground-water inflow (plus) and the transpiration-use (minus).

The maximum transpiration will occur at the point showing the maximum rate of drop in the ground-water table. This point is indicated by b on the Figure. This point of maximum use oc-

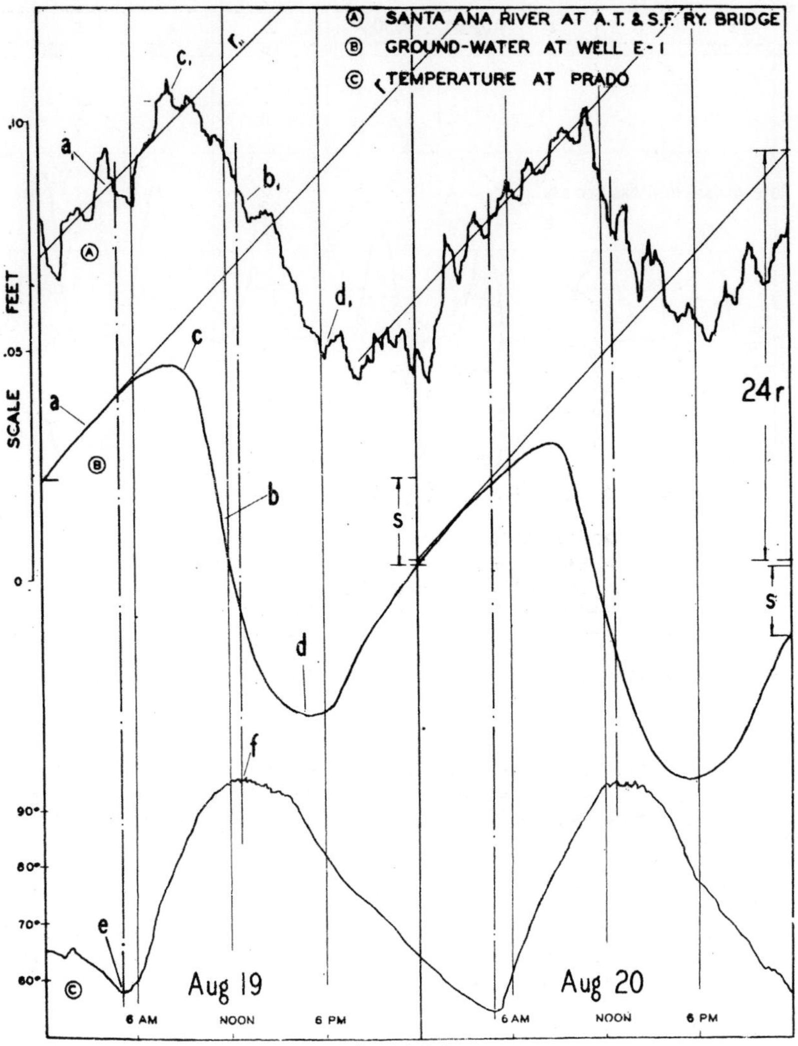

1932

FIG. 4—SANTA ANA RIVER
Effect of Evaporation and Transpiration on River and Ground-water Elevations

curs at or near the time of maximum temperature. Likewise the point of minimum transpiration will occur at the point showing the maximum rate of rise in the ground-water table, such as is indicated by a. This point agrees fairly well with the time of the minimum temperature.

This is in entire agreement with the work of C. A. Taylor (Rainfall-penetration and consumptive use of water, Calif. Div. Water Resources Bull. 33, 1930) at Ontario, California, who demonstrated that transpiration from native water-loving vegetation agreed closely with temperature. Figure 5 gives a section of his record, which shows a fine correlation between the diurnal temperature- and transpiration-cycles.

Walter N. White, in the United States Geological Survey Water-Resources Bulletin of March 10, 1927, suggested that it was possible, by means of records similar to that shown on Figure 4, to determine the daily transpiration. (See also W. N. White, A method of estimating ground-water supplies based on discharge by plants and evaporation from soil: Results of investigations in Escalante Valley, Utah, U. S. Geol. Surv. Water-Supply Paper 659, pp. 1-105, 1932.) If the

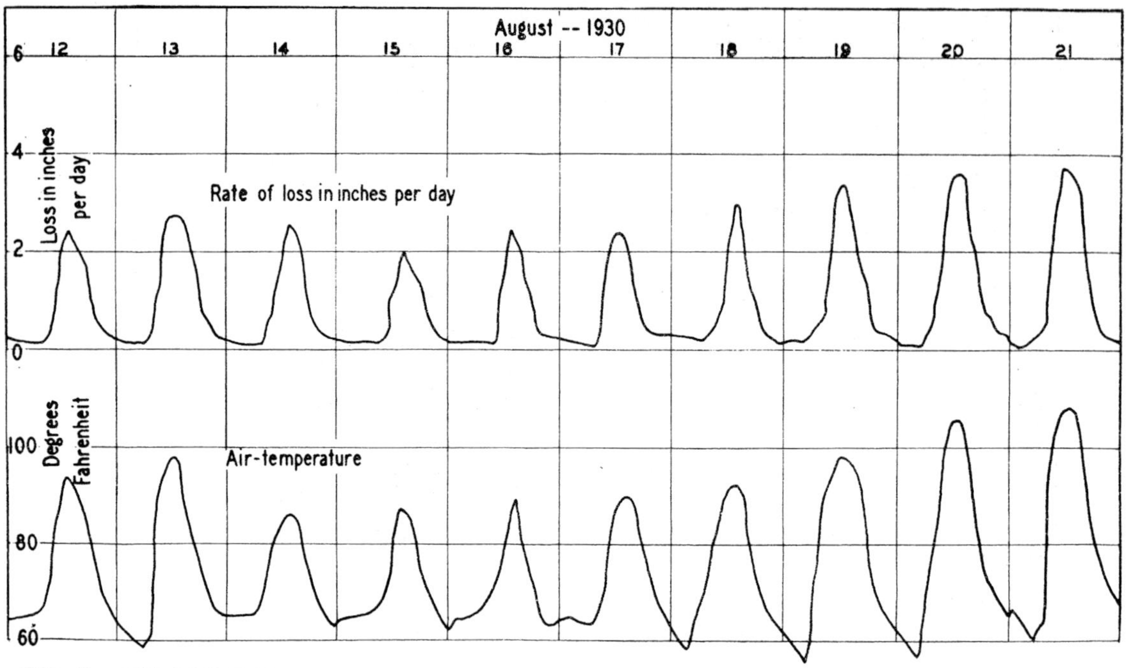

FIG. 5—EVAPORATION-TRANSPIRATION RATE AT ONTARIO WILLOW AND REED TANK AND ITS RELATION TO AIR-TEMPERATURE, BY C. A. TAYLOR BUREAU OF AGRICULTURAL ENGINEERING

transpiration is assumed to be practically zero during the early hours of the morning, say about 2 to 4 o'clock, then the rate of rise in the ground-water table during that period would be the rate of recharge. A tangent was then drawn to the curve at this point and marked r. This represents the rate of recharge of the ground-water table. If this rate were to continue throughout the 24 hours and there were no transpiration or other losses, the water-table would rise an amount equivalent to the distance marked 24r. Transpiration occurred, of course, and instead of rising, the ground-water table dropped the distance s during the day. Consequently, the transpiration-loss would be (24r + s) divided by the specific yield of the soil.

Although these results might provide a satisfactory estimate, they are subject to a certain error. This error is based on the assumption that the rate of recharge r continues as a straight line throughout the 24 hours. On Figure 6 is shown a typical record from a well maintained along the Santa Ana River. To the right of the record is a diagram of a typical well. The presence of the diurnal cycle in the ground-water table indicates that the ground-water is under a certain static head, and if there were no drain on the supply, the ground-water level would approach that head. Early in the season, when transpiration is practically nothing, the altitude of the ground-water table and the static head are the same or nearly the same. As indicated by the diagram, the increase of the transpiration-drain on the ground-water supply causes a cone of depression to form about the trees. As the water-table drops, the distance between the static head and the ground-water increases. This increase in difference of altitude causes an increase in the rate of recharge r, so that at the middle of the transpiration-season the rate of recharge will be a maximum. The record on this Figure clearly shows how the relatively small rate of recharge of June 17-19 increased to a maximum for July 30 to August 1 and then returned to a minimum between November 4 and 6.

Thus it is evident that the rate of recharge r is not a straight line but a curve which ranges from zero at the altitude of the static head to a maximum some distance below that point. By developing a relation between the ground-water altitude and the rate of recharge during the early morning hours, it was possible to develop the shape of this recharge-curve for this particular well. The maximum rate of recharge during August indicated that the static head was about 0.55 foot above the ground-water table for that month. Thus, it is possible that, if there had been no transpiration, the ground-water table would have been 0.55 foot higher.

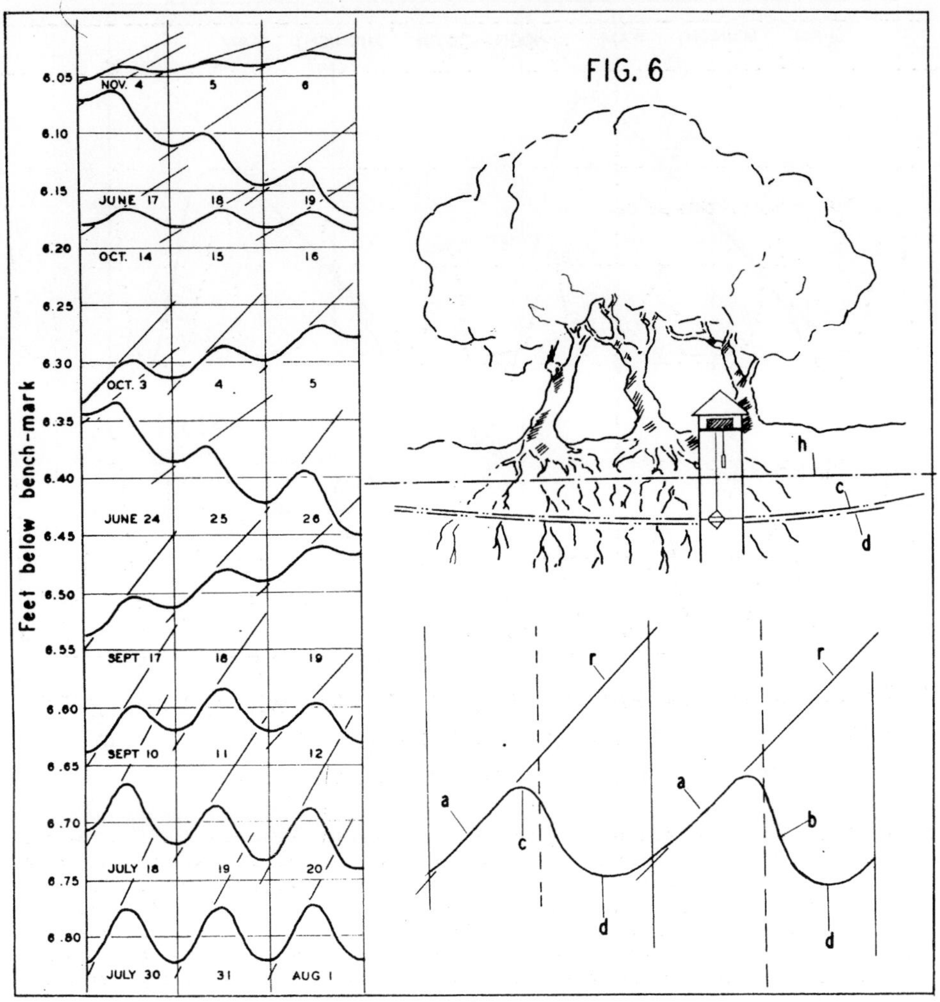

The diurnal cycle as indicated by the ground-water table represents an accumulative curve showing the result of rates of inflow minus rates of transpiration. On the bottom of Figure 7 has been plotted the first derivative of this diurnal ground-water cycle. These rates have been expressed as the average for each hour and have been determined graphically. This curve represents the rate of change in the ground-water table, or the rate of inflow minus rate of transpiration.

It is reasonable to expect that between the hours of midnight and sunrise, the transpiration-losses would be at or near zero. During this period, then, the rate of change in the ground-water curve would equal the rate of inflow. In view of Figure 6 it is not unreasonable to expect that at the point c the rate of recharge will be a minimum and at point d the rate will be a maximum. It is impossible by this method to determine the actual rates of recharge during the transpiration-period. However, the rates for this period can be sketched in with sufficient accuracy to give hourly distribution of the transpiration-losses, which are indicated by the cross-hatched area on Figure 7. As indicated by this Figure, the measurable transpiration-losses start between 5 and 6 a.m., reach a maximum just before noon, and drop to zero shortly after 8 p.m. This agrees fairly closely with the work of Taylor at Ontario and the record from the Taylor evaporimeter.

Most of our streams flow through channels partly filled with recent alluvial deposits. A ground-water table is maintained in these deposits either by inflow from the tributary area or by inflow from the stream itself, or by both. In the first case the movement of the ground-water will be toward the stream, in the second the movement will be away from the stream, and in the

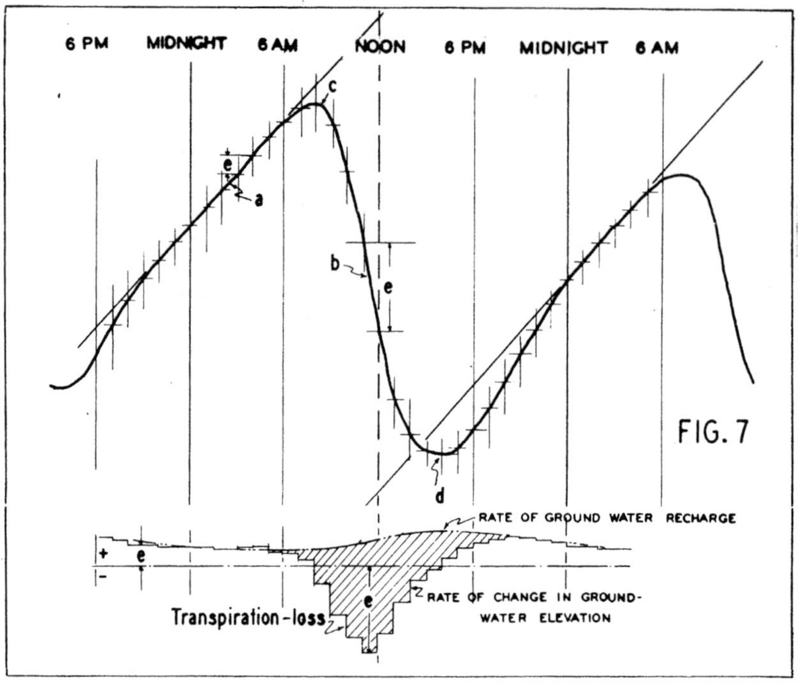
6 PM
MIDNIGHT
6 AM
NOON
6 PM
MIDNIGHT
6 AM
a
b
c
d
e
RATE OF GROUND WATER RECHARGE
RATE OF CHANGE IN GROUND-WATER ELEVATION
Transpiration-loss

FIG. 7

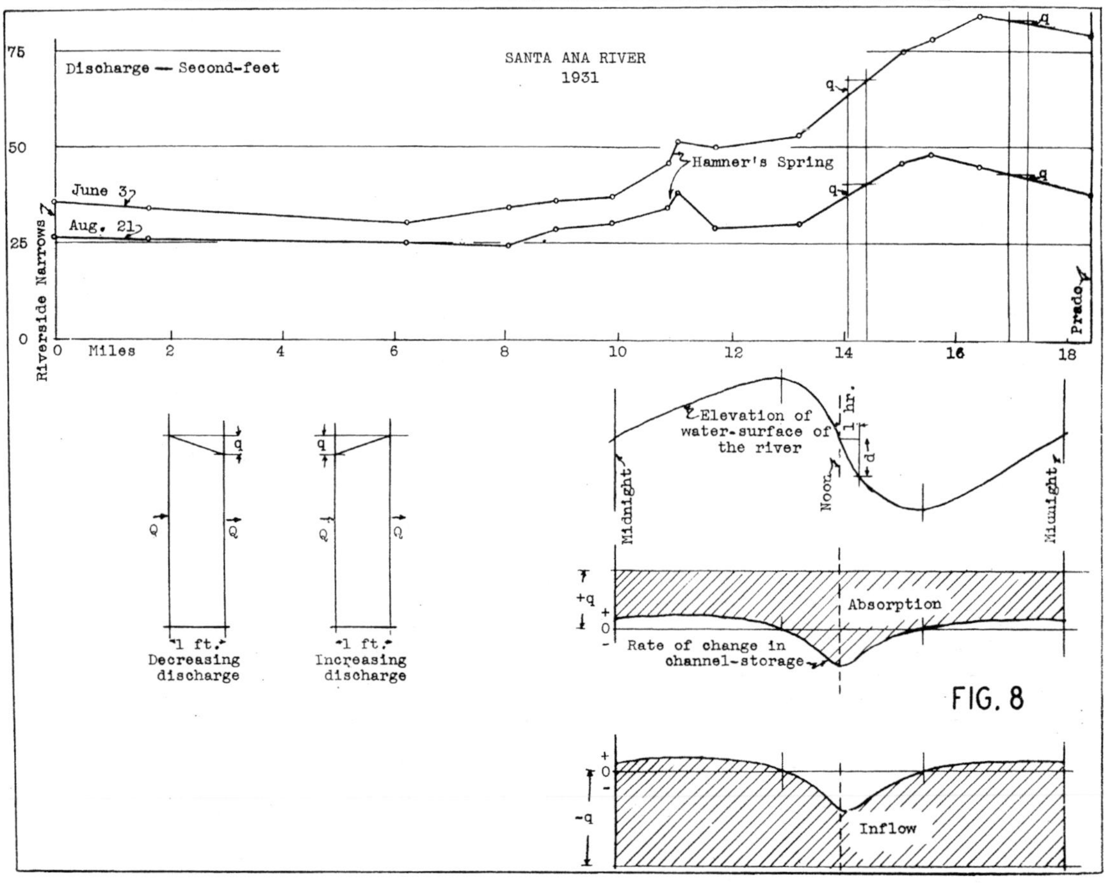
SANTA ANA RIVER
1931
Discharge — Second-feet
June 3
Aug. 21
Hamner's Spring
Riverside Narrows
Prado
Miles
Decreasing discharge
Increasing discharge
1 ft.
Elevation of water-surface of the river
1 hr.
Midnight
Noon
Absorption
Rate of change in channel-storage
Inflow

FIG. 8

third the movement may be in both directions. A riparian plant-life soon establishes itself and draws its water from this ground-water supply.

On the upper part of Figure 8 is a graph showing the discharge at a group of stations on the Santa Ana River for June 3 and August 31, 1931. These observations are the results of a series of simultaneous or nearly simultaneous current-meter measurements made by a group of field-parties from the United States Geological Survey and the Orange County Flood-Control District. These measurements were made between the hours of 8 and 10 a.m., during that part of the diurnal cycle in which the changes in channel-storage are small. As indicated by this graph certain sections of the River show a gain in discharge, while others show a loss. The gage-height records show a marked similarity in the diurnal cycle, both as to time and as to range of stage, notwithstanding the fact that in some sections the River lost discharge, while in others it gained.

Consider for the moment a reach of the stream one foot in length. Assume that this reach is located in a part of the stream in which the rate of flow diminishes progressively downstream. In this event the quantity of water entering this 1-foot reach in a unit of time (Q) will be greater than the quantity leaving it in the same time (Q_1). The difference between Q and Q_1, here called q, entered the reach but failed to leave it as surface-runoff. The tendency is for this quantity q to increase the channel-storage in the 1-foot reach.

In direct opposition to the effect of the increment q is stream-bed absorption. This absorption produces ground-water replenishment. At or about 10 a.m., the water-surface of the stream remains at a constant altitude for a short period, because at this time the quantity of water absorbed in a unit of time (a) exactly equals q in this 1-foot reach. After this period the rate of absorption increases rapidly, and the quantity of water absorbed in a unit of time (a) is equal to the quantity q plus a quantity withdrawn from channel-storage (s), thus lowering the level of the stream in the 1-foot reach. Late in the afternoon a and q again balance, and the water-surface remains constant for the time being. Then follows a period in which q exceeds a and channel-storage is increased. Thus we have the complete diurnal cycle.

It appears, at least to me, that the gage-height record of the stream is nothing more or less than an accumulative curve showing the relation between the positive factor q and the negative factor a. Under these conditions the following equations can be prepared

$$s = (q - a) \qquad \text{or} \qquad a = (q - s)$$

For any 1-foot reach of the stream the above equation is capable of solution, as follows: Let d = hourly change in altitude of the surface of the stream, in feet; b = width of stream, in feet; then, expressed in cubic feet a second, $a = [q - (db/3600)]$.

By means of this equation it is possible to establish the hourly variation in the rate of absorption. Assuming that q remains constant for the 24-hour period, then we have the rate of absorption as indicated graphically by the cross-hatched areas on Figure 8. This curve distinctly demonstrates that the gage-height record of the stream shows that maximum absorption occurs about noon and decreases to a minimum during the night.

In these reaches of the stream the rates of absorption are governed by the altitude of the ground-water table, which is in turn governed to a large degree by transpiration of the riparian plant-life. Thus maximum absorption should occur during the period of maximum transpiration, and minimum absorption during the period of minimum transpiration.

If the stream shows a progressive increase in rate of flow downstream, the reasoning will be very similar to that given for decreasing rate of flow. In this case the quantity of water (Q) entering the 1-foot reach is less than that leaving the reach (Q_1). The increase in rate of flow (q_2) has been taken from the reach. This q_2 becomes the negative part of the equation, in that it tends to reduce channel-storage (s). In direct opposition, the inflow or ground-water recharge (i) to the River tends to increase the channel-storage. Again the gage-height record of the stream becomes the accumulative curve of i (positive) and q (negative). Thus

$$s = (i - q) \qquad \text{or} \qquad i = (s + q) = (db/3600) + q$$

The second cross-hatched area on Figure 8 shows the inflow graphically. The inflow will be a minimum about noon and a maximum in the night. Here again the influence of transpiration is indicated. The inflow to the stream is governed by the condition of the ground-water supply. The heaviest demands by transpiration occur at noon; consequently the inflow to the River will then be a minimum. During the hours of little transpiration the inflow will be a maximum.

After a study of these two cross-hatched areas on Figure 8 it is only reasonable to expect that a stream will fluctuate in unison during the diurnal cycle at all points, much like a long pulsating reservoir. Take, for example, such a stream as the Santa Ana River. In some sections this stream shows an increase in discharge and in other sections a decrease in discharge. The graphs have indicated that for a losing stream the absorption will be a maximum at about noon. The same graphs show that on a gaining stream the inflow will be a minimum at noon. Thus on both a gaining and a losing stream the channel-storage is being depleted at its most rapid rate at noon. The same analysis can be applied to any other part of the diurnal cycle at all stations along the River. So it is not unreasonable to expect that the stream will fluctuate in unison or very nearly in unison at all points.

Owing to the frequent misuse of this diurnal cycle, it has been the purpose of this paper to demonstrate that the diurnal cycle found in the ground-water table and in the stream represents nothing more than an accumulative curve showing the difference between additions to and extractions from storage.

Furthermore, except for a few minor exceptions, it is the intent of this paper to question the assumption, often made, that this diurnal fluctuation progresses downstream. It is believed that there are now filed in the Los Angeles office of the United States Geological Survey sufficient data from southern California streams to prove that the diurnal cycle does not progress downstream.

U. S. Geological Survey,
Los Angeles, California

Commentary on Paper B2

Hursh, C. R. & Brater, E. F. (1941) Separating storm-hydrographs from small drainage-areas into surface- and subsurface-flow.

The inclusion by Keith Beven of a paper by Charles Hursh in the first volume of this series, Benchmark Papers in Hydrology: *Streamflow Generation Processes* (Beven, 2006), recognises Hursh's achievement in establishing the importance of subsurface flow processes at a time when almost all attention was being given to surface runoff (or "surface-stormflow" as Hursh described it in his 1936 paper), particularly through the publications of Robert Horton (Burt, 2008). As the founding Director of the Coweeta Hydrologic Laboratory in North Carolina, USA, Hursh had pioneered a number of research themes, most notably the role of vegetation in controlling water yield (Swank & Crossley, 1988). He established the use of paired catchment experiments as a standard approach to field experiments in hydrology. His interests clearly extended beyond water budgets, however, and there were a series of publications dealing with subsurface flow procedures. Indeed, it was he who first used the term "subsurface-stormflow" (Hursh, 1936).

Hursh's work laid the foundation for John Hewlett's variable source area model of storm runoff generation (Hewlett, 1961; Burt, 2008). This publication by Hursh & Brater is an important link in the development of those ideas (see below). Hursh recognised the importance of understanding subsurface flow in order to better interpret the streamflow hydrograph. The paper begins with an analysis of a storm hydrograph (Fig. 1) in which almost all the storm discharge is generated by precipitation falling directly into the channel itself. Excepting a small amount of rain falling just after the main storm, the total input was 0.64 inch (16.25 mm), equivalent to an average intensity of 1.53 inches (38.9 mm) per hour; the maximum 5-minute intensity was equivalent to 1.68 inches (42.7 mm) per hour. These intensities are too low to generate infiltration-excess overland flow on the permeable forest soils at Coweeta.

Hursh and Brater were the first to establish the importance of direct channel precipitation, but that is hardly relevant here. It is the analysis of the second hydrograph (Fig. 3) that is centrally important to the development of our understanding of near-stream hydrological processes. In this latter case, their analysis shows that only a small proportion of total storm runoff is channel precipitation; almost all is subsurface stormflow. Their analysis distinguishes two types of response: water "in close enough proximity to the stream to contribute a significant amount of percolated water during the period of the storm" (p. 868) and water draining from further upslope takes longer to reach the stream and so supports the gradual recession in stream discharge after the storm hydrograph. This is, of course, an arbitrary distinction in that there is only a single reservoir of soil providing the slope drainage; what Hursh and Brater achieve is to distinguish between near-stream sources that can contribute to the flood runoff and more distant sources where travel times are sufficiently long that drainage is too slow to contribute to "stormflow" and instead sustains "baseflow". Hursh and Brater conclude with an insightful discussion of runoff sources arranged approximately in the order in which they might contribute to the storm hydrograph.

The use of well records to support the graphical analysis is a crucial element of this paper's contribution, providing a bridge between simple hydrograph analysis and process inference. Put another way, Hursh and Brater take us from a "lumped" analysis of whole basin discharge response to a "distributed" perspective of runoff generation, and, crucially from the perspective of this volume, one that emphasises the importance of near-stream sources of subsurface flow to the storm hydrograph. They identify (p. 869) that the rapid increase in slope discharge arises not only from the increasing height of the water table, but also from an increase in the slope of the water table near the stream, resulting in much greater discharge than would be the case if only the former effect happened. This anticipates results from later field studies by Weyman (1973) and Anderson & Burt (1978), and shows the need for careful monitoring of water table fluctuations in the riparian zone.

Hursh's studies at Coweeta laid the foundation for later studies of subsurface flow, including current interest in flow paths and soil water residence times. A clear path can be identified via many of the papers included in Beven (2006) through to those included here (e.g. Haycock & Burt, 1993, Paper E7). Subsurface flow is the driver of many riparian zone processes and in turn the riparian zone controls the stream water quality (Bishop *et al*., 2004). In their analysis of processes and travel times, and in their use of wells to monitor water table fluctuations, Hursh & Brater (1941) must be regarded as a landmark publication in hillslope hydrology, with particular significance to the near-stream zone.

Even more important perhaps than laying the foundation for later studies of subsurface flow, Hursh's work led directly to John Hewlett's formulation of his Variable Source Area (VSA) model, which was also developed at Coweeta. The VSA model was first described in an internal US Forest Service report (Hewlett, 1961), but it was some years before the concept was more widely disseminated (McDonnell, 2009). It is worth quoting Hewlett in some detail, to show the link to Hursh's earlier work:

> *When rain falls on porous forest soil, it enters the ground and either begins to migrate to the nearest stream or is held as "retained" water by the soil particles ... Whether it migrates or is held in place depends chiefly on the character and wetness of the*

soil, which in turn is usually related to its depth and position on slope ... Where it sinks in near a stream and consequently can contribute more to immediate rises in streamflow, it generally will move faster than if it enters the drier slopes and ridges above ... Rainfall influence in producing immediate runoff obviously diminishes with distance from the stream channel ... Under prolonged and heavy rainfall, the stormflow-contributing area contiguous to stream channels may grow wider and wider, depending on the nature and depth of the earth mantle. (Hewlett, 1961).

The concept of saturation-excess overland flow was elaborated by Dunne & Black (1970, Paper B5). The VSA model, developed at Coweeta as a direct result of Charles Hursh's pioneering research, remains one of the central tenets of hillslope hydrology today.

TPB

References

Anderson, M. G. & Burt, T. P. (1978) The role of topography in controlling throughflow generation. *Earth Surface Processes* **29**, 331–334.

Beven, K. J. (2006) *Streamflow Generation Processes*. Benchmark Papers in Hydrology, IAHS Press, Wallingford, UK.

Bishop, K., Seibert, J., Kohler, S. & Laudon, H. (2004) Resolving the Double Paradox of rapidly mobilised old water with highly variable responses in runoff chemistry. *Hydrol. Processes* **18**, 185–189.

Burt, T. P. (2008) Valley-side slopes and drainage basins. I: Runoff and erosion. In: *The History of the Study of Landforms or the Development of Geomorphology*, vol. 4: *Quaternary and Recent Processes and Forms (1890–1965) and the Mid-Century Revolutions*, Chapter 8 (ed. by T. P. Burt, D. Brunsden, R. J. Chorley, N. J. Cox & A. S. Goudie), 325–352. The Geological Society, London, UK.

Dunne, T. & Black, R. D. (1970) An experimental investigation of runoff production in permeable soils. *Water Resour. Res.* **6**, 478–490.

Haycock, N. E. & Burt, T. P. (1993) The role of floodplain sediments in reducing the nitrate concentration of subsurface runoff: a case study in the Cotswolds, England. *Hydrol. Processes* **7**, 287–295.

Hewlett, J. D. (1961) Watershed management. In: *Report for 1961 Southeastern Forest Experiment Station*, 62–66. US Forest Service, Asheville, North Carolina, USA.

Hursh, C. R. (1936) Storm water and absorption. *Trans. Am. Geophys. U.* **17**, 301–302.

McDonnell, J. J. (2009) Hewlett, J. D. & Hibbert, A. R. 1967. Factors affecting the response of small watersheds to precipitation in humid areas. In Sopper, W. E. Sopper and Lull, H. W., editors. Forest Hydrology, New York: Pergamon Press, 275–290. *Progr. Phys. Geogr.* **33**, 288–293.

Swank, W. T. & Crossley, D. A. (1988) *Forest Hydrology and Ecology at Coweeta*. Springer-Verlag, Ecological Studies volume 66.

Weyman, D.R. (1973) Measurements of downslope flow of water in a soil. *J. Hydrol.* **20**, 267–284.

SEPARATING STORM-HYDROGRAPHS FROM SMALL DRAINAGE-AREAS INTO SURFACE- AND SUBSURFACE-FLOW

C. R. Hursh and E. F. Brater
(Presented by C. R. Hursh)

Introduction--To evaluate the effects of different land-management practices on the water economy of small drainage-areas, it is essential that the stream-hydrograph obtained from small drainage-areas be separated into ground-water flow and the several components of storm-flow as accurately as the data permit. The difficulty of drawing a ground-water hydrograph for routine separation of complex storm-hydrographs has been specifically pointed out by numerous workers. For example, W. G. Hoyt [see 1 of "References" at end of paper] states that peaks occurring in such rapid succession that only rarely does surface-runoff have a chance to drain out "make the task of drawing a ground-water hydrograph difficult and the results uncertain." The improvement of present methods and the development of new methods for differentiating ground-water runoff and surface-runoff has been specifically pointed out as one of the deficiencies in hydrologic research [2].

Precipitation and other measurements essential to watershed-studies are being obtained from 30 small drainage-areas supporting continuous stream-flow and ranging in size from three square miles down to 20 acres. These drainages lie within the Coweeta Experimental Forest, an area devoted to forestry-research in the high rainfall-belt of the Southern Appalachian mountains in Macon County, western North Carolina. The mean annual precipitation of 70 inches is well distributed throughout the year. About 50 storm-hydrographs are obtained within a single year for each stream.

Description of experimental drainage-area--For the purpose of discussing some of the results of this study, records from a drainage-area of 40 acres will be used. This drainage, No. 13 on the Coweeta Experimental Forest, is a forested watershed with continuous stream-flow from pervious soils overlying a deeply weathered country rock. Discharge was measured with a right-angled V-notch weir. The lowest elevation of the area is 2,350 feet whereas the highest portion has an elevation of 3,000 feet. The water-yields from this area have been found to be quite similar to the yields of larger drainage-basins in the region.

The mean runoff expressed as cubic feet per second per square mile for this drainage for the period of April, 1936, to January, 1941, is 2.38 csm. The seasonal means are as follows: November, December, January, 2.02 csm; February, March, April, 4.10 csm; May, June, July, 2.09 csm; August, September, October, 1.61 csm.

Methods: General discussion--Analysis of a large number of storm-hydrographs calls for the use of routine practice which may be carried out with a minimum of additional replotting of the original record. It is to determine a satisfactory procedure for routine hydrograph-analysis that a special study has been made of the problems relative to ground-water separation.

As a preliminary to separating ground-water during storms, a normal depletion-curve is prepared for each drainage-area using storm recession-curves. The depletion-curve is then converted to read directly in gage-height over the weir and adjusted to the time-scale of the original field-chart. This adjusted curve is then transferred to transparent cloth and used as an overlay on the original record. Many research-problems appear in the preparation and use of curves of normal stream-depletion which will not be considered in the present discussion. It was found that for practical purposes a composite curve could be used without taking into account seasonal and annual differences. Such differences, although present, were not great enough to be significant within the mechanical errors of drafting the base-curve.

Surface storm-runoff as overland-flow has not been observed on this drainage-area; nevertheless, characteristic flood-hydrographs are produced by heavy rains. A simple hydrograph, such as Figure 1, is produced by an intense summer storm of 0.69 inches falling within a period of 30 minutes and equivalent to an intensity of 0.80 inch per hour for the entire storm. Intensities for the maximum five-minute interval amounted to 1.68 inches per hour and for the maximum ten-minute interval 1.56 inches per hour. The volume of storm-flow discharged as indicated by the area above the base-line 3, Figure 1, is computed to be 1,141 cubic feet. Base-line 3 in this case is merely one of three straight lines drawn from an arbitrary point on the normal recession-curve to points (point a, Fig. 1) on the hydrograph approximately two, three, and four hours after cessation of rainfall.

Measurement of channel-area--The length of the stream-bed for the drainage is 4,006 feet and the average width is 5.2 feet. These dimensions include only the gravelly stream-bed and not the

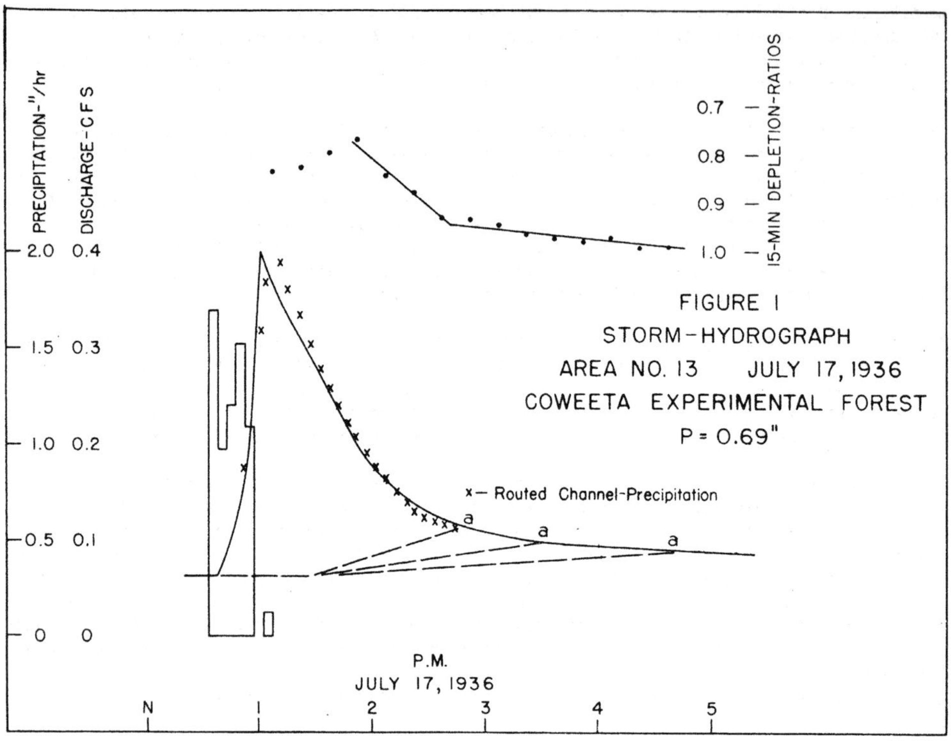

steep heavily vegetated banks of the stream. Channel-precipitation is computed from the area of the stream-bed after making an allowance for vegetative interception, at five per cent. Using this deduction for the 0.69-inch storm producing the hydrograph shown in Figure 1, channel-precipitation is computed to be 1,138 cubic feet. As this amount is practically the same as the total discharge for the storm, it is apparent that the entire discharge above base No. 3, Figure 1, can be accounted for as channel-precipitation. Other types of storm-flow make up only a very small portion of the hydrograph.

It is believed that the importance of channel-precipitation as a component of the flood-hydrograph has not been fully appreciated. Many hydrographs similar to the one above, where practically the entire hydrograph results from channel-precipitation, have been observed on the Coweeta streams. It is possible that this factor should be more fully accounted for even in the case of much larger streams.

Storage-discharge relationship--A storage-discharge relationship may also be employed to demonstrate that the hydrograph shown in Figure 1 comes largely from channel-precipitation. If the correct separation of surface-runoff has been made by base-line 3 in Figure 1, that portion of the hydrograph above this line should be entirely the result of channel-precipitation. Therefore, the depletion of this quantity, after the end of rainfall, is simply the natural draining off of the accumulated channel-storage. Under these conditions there exists a relationship between the quantity of channel-storage (S) present at any instant and the rate of discharge (Q) occurring at that instant. This principle was utilized by Horton [3] to derive hydrographs of overland-flow from stream-flow hydrographs. The storage-discharge relationship is derived from the recession-side of the graph beyond the point of contraflexure. The procedure is as follows: Increments of discharge are added accumulatively, working from right to left, starting at the point where the separation-line meets the recession-curve (point a, Fig. 1). When any particular discharge-increment is added the accumulated quantity represents the amount of channel-storage present at the beginning of that increment. The rate of discharge occurring at that time may be related to the corresponding value of storage. A set of points derived in this manner will represent the relationship between channel-storage and discharge only if the line of separation is drawn so that very nearly all the discharge above the line is a result of channel-storage. If, for instance, the derivation of these points is carried up beyond the point of contraflexure, where the channel-storage is being augmented by channel-inflow or in this case precipitation on the channel, the plotted points immediately fell away from the trend of the lower points. It could be expected, therefore, that a similar distortion of the storage-discharge relationship would occur if the lines of separation were drawn in such a manner that some ground-water discharge is included in the area above the line. This was found to be the case.

Test of trial base-lines of direct runoff--Therefore, the derivation of this relationship was utilized as a test to determine the correctness of a given line of separation. The procedure is illustrated for the hydrograph of Figure 1, where three different base-lines were tried. The storage-discharge relationships derived for all three sets of lines are shown by points plotted on Figure 2. It may be seen that the points determined from base-line No. 3 plot in a consistent curve whereas the others do not. If these points are plotted logarithmically it is found that only the series derived from base-line No. 3 fall on an approximate straight line. It is concluded therefore that the third base-line provides about the correct separation.

Routing channel-precipitation--It is believed that the exactly correct line of separation would be a smooth reverse curve rather than a straight line or a series of straight lines. This point has been more completely discussed in a previous paper [4]. The results of this study serve to strengthen the belief. More recent studies of changes in the water-table of index-wells during heavy storms have indicated that a curved line more nearly represents the actual fact.

The same procedure was applied to four other hydrographs with similar results. The series of points derived from each of the five hydrographs were plotted together logarithmically and were found to produce a consistent relationship. The solid line shown in Figure 2 was derived from a straight line drawn through these five series of points. If this line represents a reasonably correct relationship between channel-storage and discharge it should be possible to route channel-precipitation through the channel and completely reproduce those hydrographs produced entirely by channel-precipitation. The hydrograph shown in Figure 1 was reproduced in this manner, utilizing the diagram of channel-precipitation shown on the Figure. The results of the reproduction are shown by the small x marks. The points are obtained by applying the storage-equation in conjunction with the storage-discharge relationship to successive time-increments.

The storage-equation in its fundamental form is as follows

$$S_1 + P_{ch} - Q = S_2$$

where S_1 = channel-storage at the beginning of a time-interval, S_2 = channel-storage at the end

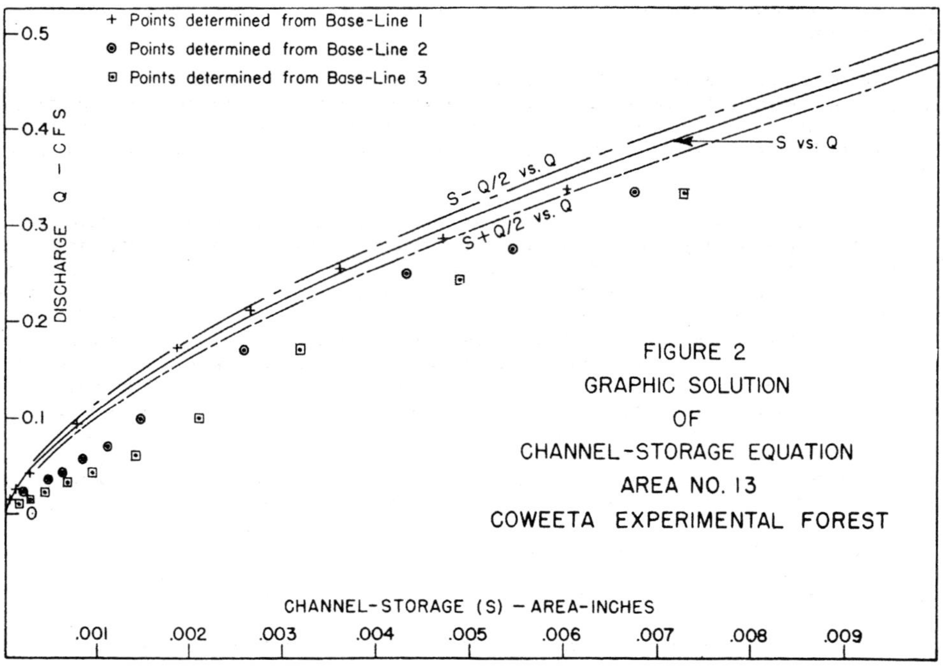

FIGURE 2
GRAPHIC SOLUTION
OF
CHANNEL-STORAGE EQUATION
AREA NO. 13
COWEETA EXPERIMENTAL FOREST

of a time-interval, P_{ch} = increment of channel-precipitation during interval, and Q = average rate of outflow during interval.

If a sufficiently short time-interval is selected it may be assumed that $Q = (Q_1 + Q_2)/2$, where Q_1 and Q_2 are the rates of discharge at the beginning and end, respectively, of a time-interval. The equation may then be rearranged into the following form

$$(S_1 - Q_1/2) + P_{ch} = (S_2 + Q_2/2)$$

The use of the storage-equation in this form to permit a graphical solution was first brought to the writers' attention by C. O. Wisler, Professor of Hydraulic Engineering, University of Michigan.

Graphic solution for routing channel-precipitation--In this form the solution may be accomplished graphically by utilizing the auxiliary curves shown in Figure 2. Knowing the rate of discharge of the beginning of any interval, the amount of P_{ch} occurring during that interval is added to the particular abscissa of the (S - Q/2)-curve which corresponds to that discharge. The abscissa equal to $(S - Q/2 + P_{ch})$ is thus obtained. Vertically above or below this point will be found a point on the (S + Q/2)-curve which indicates the rate of discharge at the end of the interval. This procedure may be quickly carried through for successive intervals. The length of interval was taken as five minutes in this case. It is obvious that care must be taken to keep all the elements of the storage-equation in the same units. It will be noted that the points obtained, Figure 1, give a fairly good reproduction, even for the rising side of the hydrograph. This would seem to be a further indication that the storage-discharge relationship is the correct one and that the graph in this case is chiefly the result of channel-precipitation.

Tests for subsurface storm-flow--It would make little practical difference whether base-line 1, 2, or 3 were selected in the hydrograph of Figure 1, since the amount of discharge above the base-line would not vary by a significant amount considering the accuracy of hydrologic observations. However, the chief purpose of studying such simple graphs is the derivation of information which may aid in the solution of more complex cases. Such a case is the hydrograph shown in Figure 3 resulting from a continuous long rain. The hydrographs shown in both Figures 3 and 1 are midsummer storms. The rainfall-intensities that produced the hydrograph on Figure 3 are not nearly so intense as those producing the hydrograph in Figure 1. Since it has been demon-

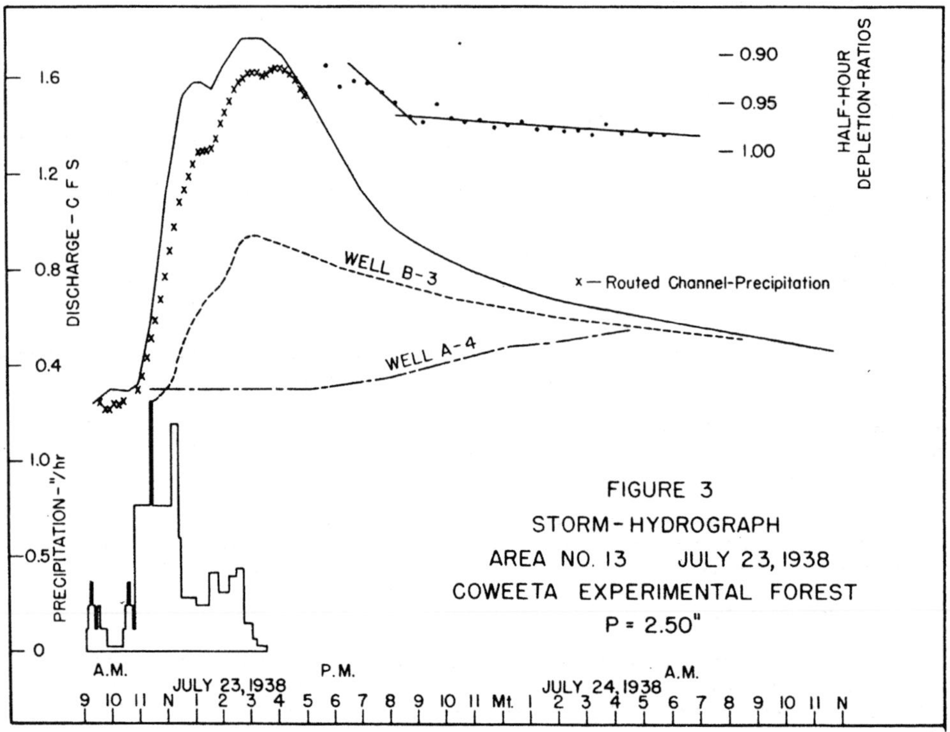

FIGURE 3
STORM-HYDROGRAPH
AREA NO. 13 JULY 23, 1938
COWEETA EXPERIMENTAL FOREST
P = 2.50"

strated that no direct overland storm-flow occurred in the storm of Figure 1, it would follow that precipitation-intensity could not have been great enough to exceed infiltration-rates for the storm of Figure 3. Consequently, on this basis of reasoning, no part of the hydrograph in Figure 3 could have resulted from direct runoff other than channel-precipitation. Evidence to substantiate this belief is obtained by using still other criteria to facilitate the separation of the storm-hydrograph. One of these is the use of the unit-hydrograph or distribution-graph. Another is the use of depletion-ratios. Still another is the observation of changes in the water-table of index-wells.

The hydrograph in Figure 3 is unlike the hydrograph in Figure 1 in that channel-precipitation accounts for only a minor portion of the total flow. This was shown by routing the channel-precipitation through as was done in Figure 1. In this case the points are plotted downward from the actual hydrograph with the intention of establishing the approximate location of the base-line. The time-element involved in locating these points may be somewhat in error. That is, the location of the group of points as a whole could reasonably be shifted to right or left through an increment of a few minutes. However, the quantity of water represented by the area above the points accounts for all the channel-precipitation while all that below the points appears to have resulted from underground storm-flow or ground-water flow.

Use of the unit-hydrograph--The approximate correctness of the line of separation of channel-precipitation in Figure 3 is further substantiated by the use of the unit-hydrograph or distribution-graph. The base-length of the distribution-graph for this drainage-area is about two hours and ten minutes. Effective rainfall ends at about 2:50 P.M. Therefore, according to the distribution-graph, all channel-storage derived from channel-precipitation must have passed the gaging-station by about 5 P.M. This agrees approximately with the results as obtained by routing channel-precipitation.

Use of depletion-ratios--When the discharge depletion-ratios of the recession-side of the hydrograph in Figure 3 are determined by half-hour periods, it is found that at no point is the ratio less than 0.92. (This ratio is obtained by dividing the rate of discharge at any instant by the rate occurring one-half hour earlier. Any other interval could be chosen.) The value for the ordinary dry-weather depletion-curve of discharge for this stream is about 0.9996. For several days following a rain the depletion-ratio varies from 0.92 to 0.99. Depletion-rates for direct runoff known to be of surface-origin are 0.88 or less. In no other part of the several years of records obtained on this stream does the ratio become less than 0.92. The fact that at no point on the recession-curve shown in Figure 3 is the depletion-ratio less than 0.92 may be considered as further evidence that the discharge shown in the hydrograph below the base of channel-precipitation is made up of underground storm-flow.

The ground-water hydrograph--Underground storm-flow has been commented on by numerous writers. It has been referred to in a previous paper [5] as storm-seepage and subsurface storm-flow. Barnes [6] in a discussion of the structure of discharge-recession curves makes use of depletion-ratios and considers that the storm-discharge of streams in the upper Mississippi Valley is made up of the two components, surface-flow and storm-seepage. He has indicated a process by which these two components can be separated by graphical analysis.

Meinzer [7] has repeatedly called attention to the importance of continuous water-level records for interpreting ground-water phenomena. For small drainage-area studies well-records have proven to be invaluable.

In a previous discussion by Hertzler [8] on storm-hydrograph analysis for drainage-areas on the Coweeta Forest, it was assumed that the rise of the ground-water base starts at the beginning of stream-rise. This was based on observations of a limited number of wells. Recent studies of the water-levels of wells actually within the small drainages from which the hydrographs are obtained now indicate that there may be a delay of several hours before a significant rise of ground-water occurs.

So far as storm-water may actually become part of an established water-table, it may also be considered as storm-augmented ground-water. The rapid percolation from this water could then be called augmented ground-water discharge during storm-periods. It is differentiated from recognized ground-water discharge only by its faster percolation-rates or by the time of its reappearance in the stream-channel.

Obviously the underground storm-flow contributing to the lower portion of the recession-side of the storm-hydrograph has more nearly the characteristics of true ground-water, whereas that contributing to the upper portion has more nearly the characteristics of channel-interception.

Use of well-records--Well-records also give an indication as to the nature of the water making up the major portion of the hydrograph shown in Figure 3. Changes in the height of the water-level of two Coweeta Forest wells, but not situated on the drainage-area No. 13, are shown in the same time-scale on the storm-hydrograph. It is apparent that there are different types of rises shown by the two wells. By a study of previous storms Well A-4 is known to rise slowly and to fall slowly and to tend to follow the seasonal depletion-curve of the stream-hydrograph. At the end of the time-period shown in Figure 3, it is still rising and will eventually meet the curve of normal stream-depletion. Such a well appears to be more closely associated with normal stream depletion than with the storm-hydrograph. Well B-3 tends to follow the storm-hydrograph much more closely. The shape of the other well-hydrographs for this storm ranges from low rises similar to Well A-4 to even more rapid rises than Well B-3 but similar to it. Apparently, the hydrograph below the separation-line of channel-precipitation is built up from storm-water temporarily stored along water-tables in close enough proximity to the stream to contribute a significant amount of percolated water during the period of the storm. Whether or not a portion of this water should be called ground-water discharge is apparently a question of definition.

What has been most lacking is an explanation of the nature of underground storm-flow and some proof that it can be related to measurable ground-water phenomena. An explanation has been sought in the changes of water-levels of selected wells. These wells are selected with regard to topography, soil-profile, and proximity to the stream.

Wells to be used for indicators of this separation must have been studied for a sufficient length of time to standardize their performance against the stream. For example, those wells known to fluctuate in keeping with seasonal depletion of the stream are probably the ones that indicate significant water-rises of seasonal ground-water stores. Such a well is indicated by A-4 in Figure 3. Those wells that rise rapidly and produce hydrographs similar to Well B-3 in Figure 3 are the ones that indicate the position of important water-tables contributing to subsurface storm-flow during the storm-period. Consequently, with a previously determined knowledge of the behavior of a particular well in relation to the stream-hydrograph, it is possible to use the well-hydrograph for the separation of storm-augmented ground-water flow.

A possible explanation of the behavior of different wells lies in their relative position with regard to topography and proximity to the surface-stream. The Well J-1, Figure 4, has a water-level ranging from six to nine feet below the ground-surface. It is on drainage-area No.

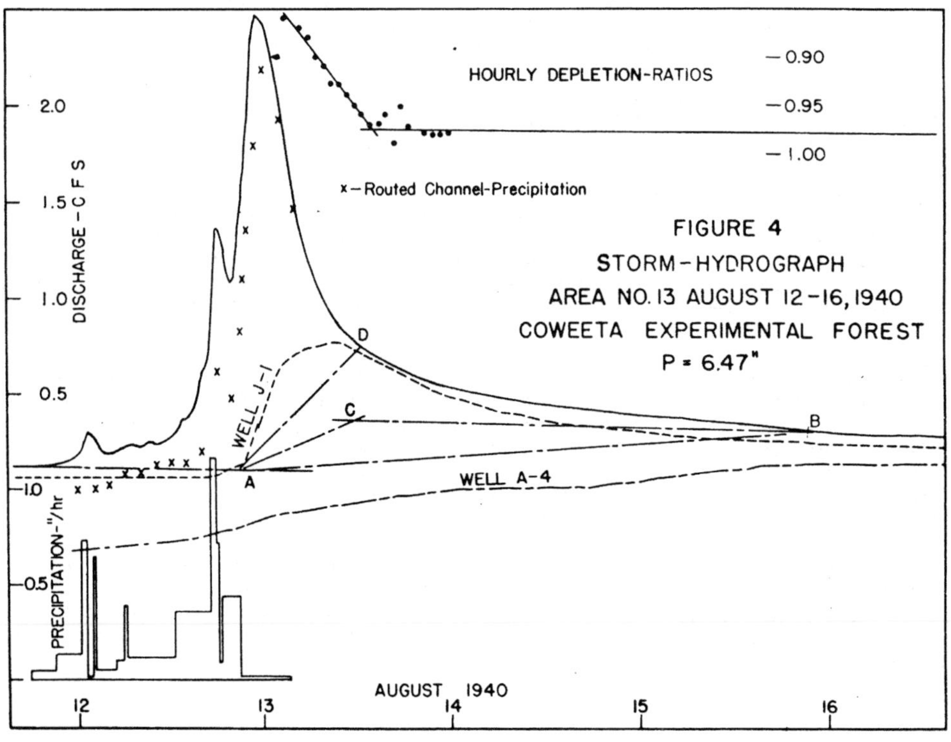

FIGURE 4
STORM-HYDROGRAPH
AREA NO. 13 AUGUST 12-16, 1940
COWEETA EXPERIMENTAL FOREST
P = 6.47"

13 in a narrow but rounded-bottom ravine with moderately steep slopes. The well is about 200 feet from the stream. The soil-profile at the well is porous, colluvial or frost-creep material. Consequently, the soil-profile in which it is located allows the rapid passage of water both vertically and laterally. Ravines filled with such colluvial material may have buried stream-lines that were established before the filling in with the colluvial material and along which water may find an opportunity for rapid movement. The Well J-1, Figure 4, offers ample evidence that there is a rapid rise of an established water-table during moderately heavy storms. Well A-4, not on drainage-area No. 13, has a water-level ranging from nine to 12 feet below the surface and is situated on sloping benchland about 600 feet from the stream.

In the storm shown in Figure 4, and in numerous other storms that have been studied, there appears to be a time coincidence between the maximum height of the water-level in Well J-1 and the sharp break in the depletion-ratios on the recession-side of the hydrograph. This break is believed to be the significant reference-point for drawing in the ground-water hydrograph. At this point, the forms of storm-water most nearly approaching channel-precipitation have passed the gaging-station. The recession-hydrograph shows depletion-ratios in the magnitude of normal ground-water depletion. It is reasonable to believe that the stream is now receiving normal ground-water from higher water-table levels resulting from the storm. The general shape of the hydrograph of Well J-1 suggests that the falling of the water-level of the well tends to follow the shape of the recession of stream-discharges. The fact that the water-level of the Well J-1 falls rapidly indicates that ground-water percolation from certain portions of the drainage-area may be reaching the stream at a relatively rapid rate during the storm-hydrograph.

Hydraulic significance of rise in the water-table--It is recognized that many types of ground-water may contribute to storm-flow. One rough classification of these types might distinguish between that water which actually reaches the water-table and that which does not. The hydrograph shown in Figure 3 is largely of the former type. The cause of this sudden increase in ground-water (gravity-water) discharge may be explained by considering the hydraulics of percolation along with the evidence of changes in the elevation of the water-table as shown by well-records. Two of the more important variables which might affect the discharge are the cross-sectional area through which the water may enter the stream and the slope of the water-table at the stream-edge. Since these steep mountain streams tend to rise very little during a storm, the change in cross-sectional area is not a major factor. However, it is possible that the slope of the water-table at the stream-edge may increase considerably during the storm-period. If, for instance, the slope changed from one in 20 to one in five during the storm, the discharge would be four times as great if flow is laminar or twice as great if flow is turbulent. The evidence indicates that a much greater change than this might occur at the stream-edge where during a long rain the water-table may actually follow the ground-surface for a short distance on each side of the stream. It should be emphasized that it is the slope directly at the water's edge which controls the contribution to the stream. It is quite possible that at some distance from the stream the water-table might temporarily slope away from the stream, but this would have no immediate effect on the rate at which ground-water enters the stream.

Applications and conclusions--For routine practice the sharp break of the recession-hydrograph is used as a reference for a new maximum on the normal depletion-curve extended under the hydrograph from the right. As a rule the location of this break may be quickly found by plotting a few recession-ratios of discharge above normal depletion by half-hour intervals. The actual ratios are not so essential but the trend of the plotted points is an aid to the eye in locating a major break in the recession-curve. In practice this is done quickly on the original chart and kept as a permanent record.

This plotting procedure alone may be used to locate the point of rise of ground-water to a new discharge-level for simple storms without reference to the well-hydrograph. For compound storms where there is a masking of the recession for any precipitation-period by a second rise, the index-well peak can be substituted for the plotted depletion-ratios. In any case the index-well peak may be used as a check on the location of the point in question should it be difficult to locate by other methods. On some storm-hydrographs this point may be located through inspection for practical purposes without further check.

The point A in Figure 4 is located on the normal depletion-curve from the left a time-distance indicated by the beginning of the rise in the water-level of Well J-1, or any other similar well having a storm-peak that coincides with the maximum break on the stream-recession hydrograph. Three lines might then reasonably be drawn for the base of the ground-water hydrograph, A-B, A-C, and A-D.

For the Drainage-Area No. 13 on the Coweeta Forest, it would be logical to draw in the base of ground-water to appear quite similar to the shape of the hydrograph of Well J-1. In practice a straight line is actually used on the head chart because this can be drawn in more easily than a curved line. The difference between the curved line and the straight line in reading the head chart would not be important. This base for ground-water is shown as line A-D, Figure 4. The discharge below this base may be viewed as storm-water that has reached the stream as storm-augmented percolation of gravity-water. The discharge between this base and the base of channel-precipitation may be viewed as rapid seepage through unsaturated soil-profiles, or as coming from naturally wet areas closely associated to the surface-stream.

By applying such procedures as have been discussed it is believed to be possible to obtain a greater degree of uniformity in the routine separation of subsurface-flow in storm-hydrographs than could be obtained by inspection alone.

A complete explanation of storm-seepage on the drainage-area under discussion must await further studies of water-tables under additional conditions of soil-topography. In the light of the present information on the subject, the storm-hydrograph such as is shown in Figure 4 from Drainage-Area No. 13, appears to be made up of storm-water from the following sources, arranged approximately in the order in which they may contribute to the storm hydrograph:

(1) Channel-precipitation.

(2) Contributions from areas of normally shallow water-tables located in close proximity to the stream, and occurring in soil-profiles which are quickly saturated. Where such conditions occur along a stream, it is expected that there will be an actual increase in the width of the channel and subsequent increase in the amount of channel-precipitation. Areas of high water-tables adjacent to spring-heads would be expected to contribute similarly.

(3) Storm-water moving through layers of porous soil-material close to the surface and under considerable gradient and reaching the stream during the period of the storm-hydrograph. Frequently, such layers overlie those that are much less porous, as for example, is the case between A- and B-horizons of many soil-profiles.

(4) Storm-seepage from large bodies of colluvial or other porous soil-material that has filled in along stream-banks. Storm-seepage along natural subterranean drainage-lines that exist in filled ravines and valleys.

(5) Storm-seepage as a result of actually piling up of a high water-table in talus-slopes in close feeding relation to the surface-stream. These talus-slopes are fed by relatively rapid seepage or percolation from steep slopes. The depletion of this temporarily stored storm-water in relatively porous material may account for the recession-side of the hydrograph until its approach to normal depletion-rates.

It is conceivable that all the forms of storm-water mentioned above may contribute to the hydrograph in some orderly fashion related to time. On drainages of uniform topography and soil-mantle this time-factor may become a valuable tool for the study of storm-water. On less uniform drainage-areas any regular appearance of the different types of storm-water into the stream will be more difficult to observe.

References

[1] W. C. Hoyt and others, U. S. Geol. Surv. W.-S. Paper 772, 1936.
[2] National Resources Planning Board, Deficiencies in hydrologic research, 1940.
[3] R. E. Horton, Publication 101, Horton Hydrologic Laboratory, 1935.
[4] E. F. Brater, Trans. Amer. Soc. Civ. Eng., v. 105, pp. 1154-1192, 1940.
[5] C. R. Hursh, Trans. Amer. Geophys. Union, Part II, pp. 301-302, 1936.
[6] B. S. Barnes, Trans. Amer. Geophys. Union, Part IV, pp. 721-725, 1939.
[7] O. E. Meinzer, Trans. Amer. Geophys. Union, pp. 36-37, 1933.
[8] R. A. Hertzler, Civ. Eng., v. 9, pp. 487-489, 1939.

U. S. Forest Service, Appalachian Forest Experiment Station,
Asheville, North Carolina

DISCUSSION

W. G. HOYT (U. S. Geological Survey, Washington, D. C.)--It is with considerable pleasure that I discuss the paper by Messrs. Hursh and Brater entitled "Separating storm-hydrographs from small drainage-areas into surface- and subsurface-flow". The paper marks a distinct advance in our knowledge of what we may call the stream-flow cycle in drainage-basins of significant size in contrast or as a complementary part of the surface- or overland-flow cycle. By this I do not mean that we still do not have a long way to go. The authors have, however, presented some definite factual information concerning the magnitude of subsurface storm-flow. With respect to the surface-runoff phenomena we are deeply indebted to many who, based on the large amount of data collected at experimental areas during recent years, have developed sound, rational concepts of surface runoff as such. Others have developed rational theories and concepts relating more specifically to the movement of water after it has reached river-channels. Still others have struggled with the complexities involved in what is so apparent, namely, the rapid transition of rainfall into flood-flows in areas where apparently there is little, if any, surface- or overland-runoff. It is this phenomenon which Messrs. Hursh and Brater have so clearly described in so far as certain areas in the Coweeta Experimental Forest are concerned. They give quantitative expression to a phenomenon that many have felt for some time has a distinct bearing on flood-flows as such. In general, I believe up to within a relatively few years ago hydrologists looked upon water in the ground mostly with respect to a more or less static water-table, its rise and fall. Hydrograph-analyses such as made by Messrs. Hursh and Brater in conjunction with the results of experimental data indicate that there is a dynamic phase of ground-water flow associated with floods. This dynamic phase is of considerable significance in present problems and is a phenomenon about which we have much to learn. It is in the study of this phenomenon that students of ground-water can be especially helpful.

The facts presented by Messrs. Hursh and Brater seem to indicate, to me at least, the possibility, as Mr. Snyder pointed out in 1939, and also as discussed at the round-table held last year, that it may be necessary when considering stream-flow as such to consider the division of rainfall into surface-runoff and ground-water runoff to take place both on and below the soil-surface instead of entirely above the surface. In other words, in so far as direct flood-runoff in contrast to surface-runoff is concerned, may not many of our phenomena relating to surface-detention, net rainfall-excess, and infiltration relate not only to the surface of the ground but to an indeterminate zone lying at and extending a considerable distance below the ground-surface? In developing such a concept I do not believe we would be in any way retrogressing in our thinking but rather believe we would be taking a step forward with respect to our knowledge of stream-flow from drainage-basins of significant size.

Commentary on Paper B3

Dunford, E. G. & Fletcher, P. W. (1947) Effect of removal of stream-bank vegetation upon water yield.

This paper complements Hursh & Brater (1941, Paper B2), in that the research was also conducted at Coweeta, North Carolina, USA, and at the same time provides a useful counterpart to Troxell (1936, Paper B1), in that both are concerned with transpiration losses in the riparian zone. Additionally, it is of historical interest because, in proposing a management solution to increase stream discharge by cutting down vegetation in the riparian zone, it takes a completely opposite view to what would pertain today, in which the protection of riparian zone is usually of over-riding importance.

Dunford and Fletcher's hypothesis was a simple one: experiments at Coweeta had already demonstrated that clear-cutting an entire forested catchment area could increase the water yield by up to 17 inches (425 mm) per year, roughly one third extra. It was assumed therefore that vegetation having "continuous access to the water table is responsible for losses relatively greater in proportion than the area it occupies". This would be the "stream bank" vegetation because, all things being equal, the water table will be highest at the foot of any slope next to the stream. On Watershed 6 at Coweeta, where the experiment was carried out, this was taken to be the area within 15 feet (4.6 m) elevation of the stream channel. All woody vegetation was cut from an area of 2.62 acres (1.06 ha) or 12% of the catchment area; the length of the cut area was 1600 feet (488 m) and varied in width from 60 to 150 feet (18–46 m); in other words, the entire riparian zone had its woody vegetation cut down.

Although there is some comparison of stream discharge in Watershed 6 before and after the cutting, the main method of assessing the impact of cutting is by means of paired catchments. Although this approach had been pioneered in Switzerland in the 1890s (Engler, 1919, and Burger, 1943, quoted in Pereira, 1973) and at Wagon Wheel Gap in the early part of the 20th century (Bates & Henry, 1928), it was the paired catchment experiments at Coweeta that fully established the method: an initial period of comparison with both catchments in the undisturbed state (the control period) followed by one catchment being "treated" after which comparisons were made between the "control" and the "treated" catchments (Swank & Crossley, 1988). Here Watershed 14 is used as the control catchment. Figure 4 compares the responses of the two streams and shows that not only has the average level of discharge increased for Watershed 6, but that also much of the diurnal fluctuation in flow has been eliminated. It is interesting that Dunford and Fletcher draw on Troxell's (1936, Paper B1) research from a completely different climatic setting in order to interpret their findings: transpiration in the riparian zone was not only responsible for a daily decrease in stream discharge but also for a progressive depletion of the groundwater reservoir in excess of that which would have occurred if there were no transpiration. The conclusion was clear: significant gains in water yield could be achieved with "relatively little effort" by eliminating the transpiration losses from stream-bank vegetation.

Of course, the experiment was performed for good reasons but it is impossible to imagine these findings being applied today. Later research has indicated the multi-functional importance of the riparian zone, and the benefits of increased water yield are unlikely to outweigh the loss of buffer zone functions (although this issuc has bccn controversial in semi-arid countries, where water scarcity is a persistent problem). Nevertheless, the study is an important one, indicating the importance of the near-stream zone for regulating stream discharge response. Methodologically, it also illustrates the advantage of a paired-catchment approach and today most field studies of riparian zone processes involve some element of "control", usually through comparison of plots. Because of the cost, only rarely are field trails extended to whole catchment areas.

TPB

References

Bates, C. G. & Henry, A. J. (1928) Forest and streamflow experiment at Wagon Wheel Gap in Colorado. *US Monthly Weather Review Supp*. no. 36.

Hursh, C. R. & Brater E. F. (1941) Separating storm-hydrographs from small drainage-areas into surface- and subsurface-flow. *Trans. Am. Geophys. U.* **22**, 863–871.

Pereira, H. C. (1973) *Land Use and Water Resources in Temperate and Tropical Climates*. Cambridge University Press, Cambridge, UK.

Swank, W. T. & Crossley, D. A. (1988) *Forest Hydrology and Ecology at Coweeta.* Ecological Studies, volume 66, Springer-Verlag, New York, USA.

Troxell, M. C. (1936) The diurnal fluctuation in the groundwater and flow of the Santa Ana River and its meaning. *Trans. Am. Geophys. U.* **17**, 496–504.

Transactions, American Geophysical Union Volume 28, Number 1 February 1947

EFFECT OF REMOVAL OF STREAM-BANK VEGETATION UPON WATER YIELD

Earl G. Dunford and P. W. Fletcher

Abstract--This is a preliminary report of results of a study wherein vegetation has been cut along the stream bank of a 22-acre watershed on the Coweeta Experimental Forest. Previous measurements on this forest have shown that complete removal of vegetation results in notable gains in water yield because of a reduction in transpiration. The cutting of riparian growth has also resulted in an increase in yield of sufficient magnitude to be significant in water-resource management.

Preliminary examination of the data immediately following cutting shows a virtual elimination of the diurnal fluctuation during the growing season. This indicates that the vegetation immediately adjacent to the stream is making appreciable demands upon groundwater supplies contributing directly to streamflow.

It can be concluded, therefore, that important gains in water yield during the growing season in the southern Appalachian region may be realized with relatively little effort by eliminating the transpiration draft of riparian vegetation. This procedure can be of much practical value during drought years for municipal and industrial watersheds, when even small increases in yield are of unusual importance.

Diurnal variations in runoff have long been associated with the combined draft of evaporation and transpiration upon water in streams and the adjacent groundwater aquifer. This evapotranspiration factor has been termed "consumptive use" by BLANEY, TAYLOR, and YOUNG [see "References" at end of paper, 1930]. Except for water losses through deep seepage, it is almost entirely responsible for the difference between precipitation and runoff within the water year.

The importance of water losses through transpiration and evaporation has been emphasized repeatedly by investigators, particularly in semi-arid regions [BLANEY, TAYLOR, and YOUNG, 1930; BLANEY, TAYLOR, YOUNG, and NICKLE, 1933; TROXELL, 1936; WHITE, 1932; WICHT, 1941; and YOUNG and BLANEY, 1942]. In the southern Appalachian region water resources have been drawn upon heavily to meet the constantly increasing demands of industries and municipalities. In numerous instances supplies have become critically low. This situation has led to an active interest on the part of industrial and municipal watershed managers in the possibilities for augmenting existing supplies during summer drought periods.

On the Coweeta Experimental Forest, studies have already demonstrated that cutting vegetation over the entire surface of an experimental watershed will increase water yield as much as 17 area inches a year [HOOVER, 1944]. This increase, it was concluded, was derived almost entirely from the complete elimination of transpiration. The net effect of evaporation was unchanged because increased soil evaporation was offset by decreased interception. This study [HOOVER, 1944] has afforded an estimate of the maximum effect achieved by reducing transpiration in a superhumid climate.

Following this line of investigation, it was assumed that vegetation having continuous access to the water table is responsible for losses relatively greater in proportion than the area it occupies. Use of water in such locations continues at a high rate throughout the growing season and reaches a maximum during rainless periods when transpiration by plants in drier areas is limited by lowered groundwater tables. Therefore, it was concluded that an important measure of the transpiration draft by vegetation growing in zones of high water tables can be gained by studying the effect of its removal on streamflow in a small experimental watershed.

Transpiration and evaporation of water by stream-bank vegetation have been estimated with some success in the past by measuring the flow of water entering and leaving a given section of stream channel [TROXELL, 1936, and YOUNG and BLANEY, 1942]. It has been pointed out that determinations made in this manner may be subject to non-measurable movements of inflow and outflow. Studies of this nature have been conducted in Temescal Creek, Coldwater Canyon, and the Santa Ana River in Southern California. Daily losses in flow attributed to transpiration and evaporation in Coldwater Canyon varied in amounts from 0.12 to 0.50 acre inches per 1000 feet of canyon bottom. In Temescal Creek losses per day from 12.8 acres increased gradually from 0.02 to 0.64 acre inches from April 16 to May 27, 1929.

The present paper describes the methods of arriving at evapotranspirational losses by a comparison of streamflow before and after cutting of the stream-bank vegetation. It is believed that such a comparison will yield more direct information than the methods employed heretofore. One advantage of the direct treatment method is the fact that non-measurable inflow and outflow seepage do not affect the relative differences in flow due to treatment. On the basis of observations previously made at Coweeta Experimental Forest, it was concluded that measured effects of cutting stream-bank vegetation do not reflect any substantial change in evaporation rates [HOOVER, 1944].

Because of the limited scope of this investigation, the results obtained must be regarded as only indicative of the water volumes utilized by vegetation growing in zones of high water tables. Cutting was confined to a strip of vegetation adjacent to an active stream in a small watershed. No attempt was made to include other possible zones of high water tables existing within the area. This preliminary test will be followed by a more comprehensive study which will include the measure of cutting effects upon water temperatures and aquatic biology.

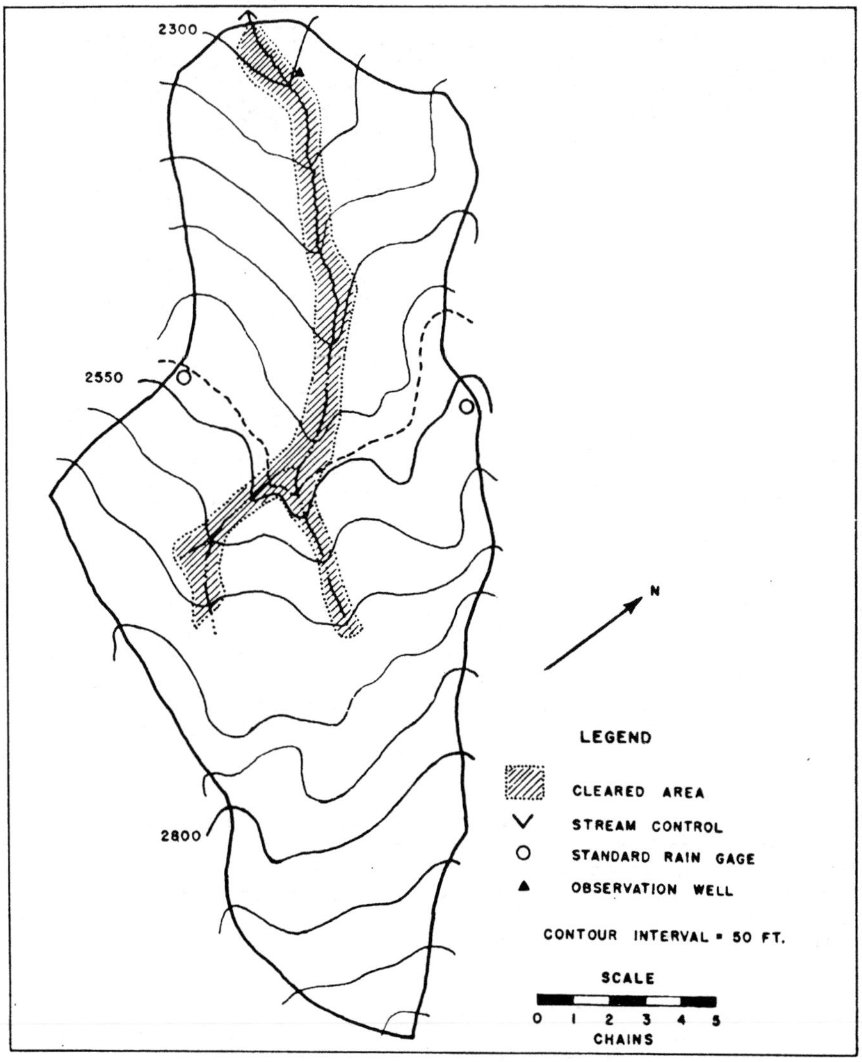

Fig. 1--Topographic map of Area Number 6, Coweeta Experimental Forest, showing area cleared of stream-bank vegetation

The experiment is being conducted on the Coweeta Experimental Forest in the Appalachian Mountains of western North Carolina. The region is one of high rainfall, averaging about 70 inches a year, distributed rather uniformly throughout the year. The growing season extends from April 15 to October 15, with a mean temperature of 65°F. Natural vegetation is abundant and primarily deciduous. For the purposes of investigation a 22-acre watershed designated as Area Number 6 was selected (see Fig. 1). The aspect is northwest and elevations range from 2282 to 2955 feet. Continuous records of streamflow from a sharp crested V-notch weir have been obtained since July 1934.

A treatment area was delineated by assuming that all vegetation at a ground level within 15 feet in elevation above the stream channel had access to the continuous water table. Within those limits was an area of 2.62 acres or 12 per cent of the total watershed area. The length is 1600 feet and the width varies from 60 to 250 feet. During the period of July 21 to 25, 1941, all woody vegetation within this zone was cut, lopped, and scattered (see Fig. 2). The soil surface was disturbed as little as possible and all debris due to cutting was removed from the channel.

Fig. 2--View of Area Number 6 after cutting stream-bank vegetation [photo by U. S. Forest Service]

A complete tally of the vegetation on the cutting area showed a total of 3563 stems larger than 0.5 inch in diameter. The greater portion of this number was represented by an understory of rhododendron, mountain laurel, azalea, witch hazel, black birch, and associated species. The principal tree species in the overstory were chestnut oak, yellow poplar, red maple, and hickory. Sprout growth was quite rapid during the 1942 growing season, one year following treatment. Species such as yellow poplar, red maple, and hickory made an early appearance and grew vigorously. Rhododendron still remains the most abundant species. No further cutting was done following the original treatment.

In the analysis of the effects of cutting, two approaches have been considered. The first of these is based on a comparison of data obtained from the treated watershed alone. The second, introduced as supplementary evidence, employs data from an adjoining watershed serving as an untreated check. Treatment effects derived from either method are subject to variables which cannot be controlled other than by statistical procedures. In view of the preliminary nature of the present study, the possibilities of increasing the accuracy of the estimates by use of statistics were not explored.

A fairly reliable expression of streamflow during rainless periods following cutting can be depicted by a smoothed groundwater depletion curve. Such a curve was developed from data obtained during August and September of 1941, and June to September 1942. A close approximation of the average daily flow is maintained throughout, since it is based on a series of instantaneous midnight readings from the hydrographs. The curve has been synthesized from a group of individual segments of post-treatment hydrographs adjusted horizontally to positions providing the best fit of midnight points. Average readings on a vertical scale at each of these reference points were then obtained, from which the smoothed curve was drawn.

Having an expression of post-treatment stream behavior depicted by the synthesized depletion curve, a measure of treatment effect may then be estimated. In Figure 3 is shown a curve, B, representing the observed flow for a ten-day period just prior to treatment. This period represents the groundwater depletion following a storm producing a peak flow of 7.023 csm. The diurnal fluctuations shown are characteristic of the flow of Area Number 6 prior to treatment. If it can be imagined that the ten-day period shown in Figure 3 occurs immediately following the cutting of stream-bank vegetation, one can assume that the flow from Area Number 6 will be represented by a portion of the synthesized depletion curve, A. After inspection of the hydrograph of the storm preceding curve B, it was estimated that the initial point on the groundwater depletion curve was at a discharge of 1.0225 csm occurring at $08^{h}00^{m}$, June 12, 1942. This can be assumed to represent a point of time immediately following treatment. Post-treatment stream-flow behavior will then follow the pattern shown by a portion of the synthesized depletion curve, A, beginning at a discharge rate of 1.0225 csm.

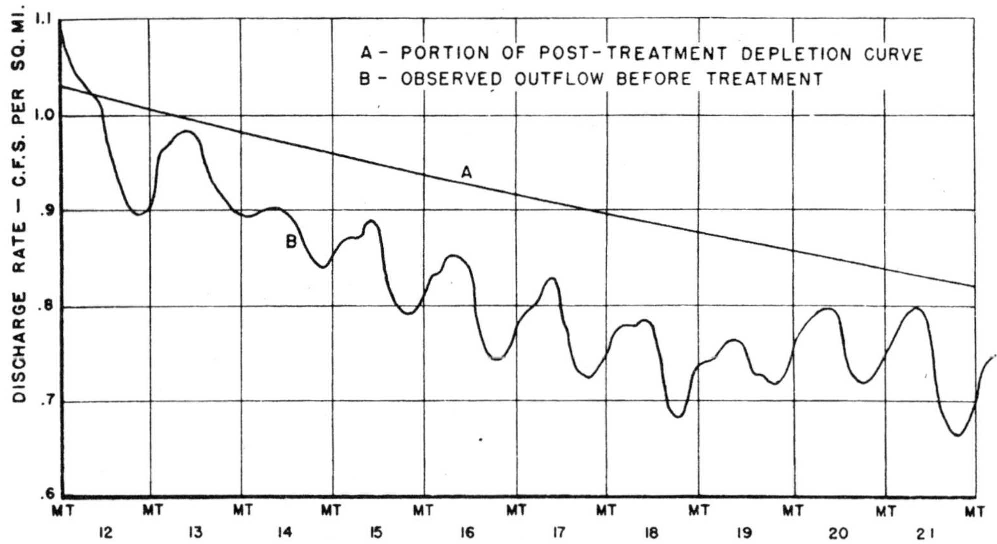

Fig. 3--Estimated effect of cutting stream-bank vegetation, Area Number 6, Coweeta Experimental Forest

Estimated daily increases attributed to cutting vary from 3.8 per cent to 19.0 per cent or a ten-day average of 12 per cent above the measured flow taken from the lower curve. Although some diurnal fluctuation remains after cutting, curve A represents a mean of these daily fluctuations, a fact which does not alter the differences obtained from Figure 3. Indicated treatment differences are shown in Table 1.

An interesting interpretation may be derived from the characteristics of the post-treatment depletion curve A. Its position with respect to pre-treatment depletion curve B in Figure 3 indicates that the transpiration losses are not represented by the daily fluctuations alone. This would tend to bear out some conclusions reached by TROXELL [1936] in his discussion of diurnal fluctuations in the Santa Ana River. He states that the practice of connecting the periods of maximum discharge of diurnal fluctuations and assuming that the difference between the curve thus obtained and the actual flow represents transpiration loss will give results which are too conservative. The interpretation gained from Figure 3 is that transpiration by stream-bank vegetation is not only responsible for a daily depression in the hydrograph but also for a progressive drain on the ground water reservoir in excess of that which would normally occur if there were no transpiration.

A second approach to the problem of estimating the effects of removal of stream-bank vegetation was investigated. This method involves the use of data from a check watershed to aid in the

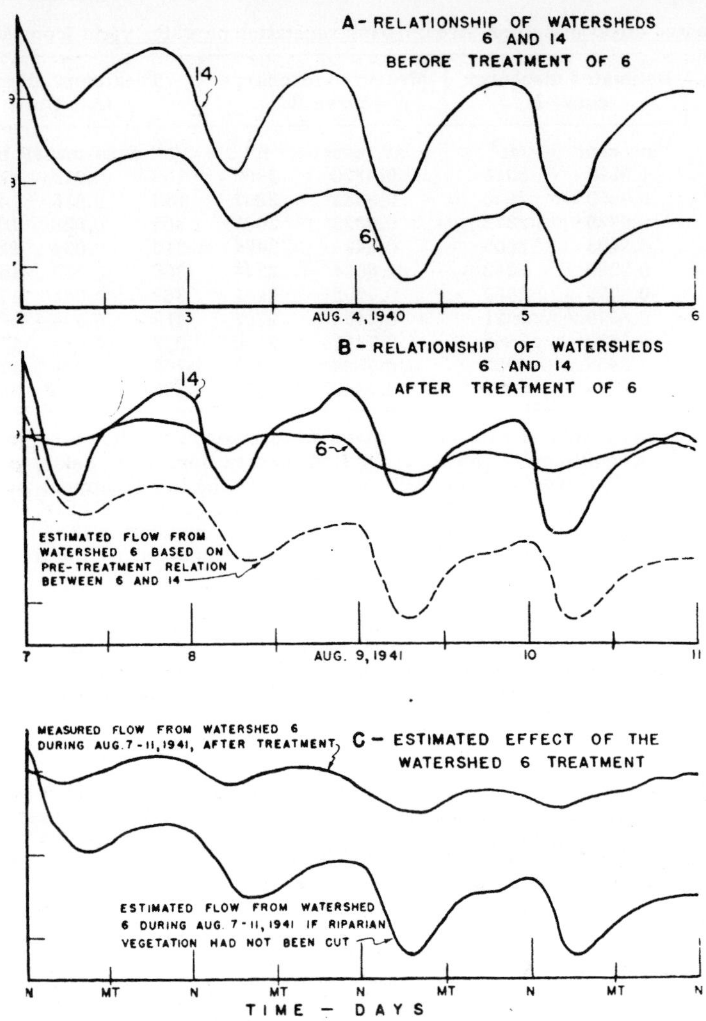

Fig. 4--Estimated increase in water yield due to cutting of stream-bank vegetation, Coweeta Experimental Forest

interpretation of results before and after treatment in Area Number 6. The advantages of this comparison lie in the fact that climatic factors affecting transpiration before and after treatment can be controlled statistically. The following discussion does not fully exploit the advantages of this method, but a more comprehensive investigation will be made of its possibilities in subsequent analyses planned for this study. The graphical representation in Figure 4, however, is indicative of the results which may be found.

The check watershed, Area Number 14, employed in this comparison, is 152 acres in area and lies adjacent to Area Number 6. The general aspect is north and the vegetation along the stream banks is similar to that existing in Area Number 6 prior to cutting. Preliminary investigation indicates that a well-defined relationship exists between the discharge rates of both watersheds. For the purposes of graphical demonstration, two four-day periods were selected, one before and one after treatment. Hourly discharge rates in csm were plotted as shown in Figures 4A and 4B. The discharge rate in Area Number 14 is substantially the same before and after treatment. For Area Number 6, however, treatment appears to have accomplished two things: (1) Raised the average level of discharge rates; and (2) eliminated much of the diurnal fluctuation in flow. The residual fluctuation indicates that there was not a complete removal of the vegetation affecting the groundwater supplies adjacent to the stream.

Table 1--Estimated effect of cutting stream-bank vegetation on water yield from Area Number 6

Day	Estimated discharge curve A		Measured discharge curve B		Treatment effect (A - B)		
	av. csm	ft^3	av. csm	ft^3	ft^3	area inches	pct
1	1.0190	3011	0.9820	2902	109	0.012	3.8
2	0.9950	2940	0.9482	2802	138	0.015	4.9
3	0.9720	2872	0.8822	2607	265	0.028	10.2
4	0.9495	2806	0.8419	2488	318	0.034	12.8
5	0.9280	2742	0.8034	2374	368	0.039	15.5
6	0.9070	2680	0.7753	2291	389	0.041	17.0
7	0.8870	2621	0.7452	2202	419	0.044	19.0
8	0.8670	2562	0.7411	2190	372	0.039	17.0
9	0.8485	2507	0.7597	2245	262	0.028	11.7
10	0.8300	2453	0.7343	2170	283	0.030	13.0

A third step in the graphical comparison gives an estimate of the increase in yields due to treatment. This is accomplished by matching the flow in Area Number 14 before and after treatment as closely as possible and noting the difference between the pre-treatment and post-treatment flows in Area Number 6. This step is represented by the dashed line in Figure 4B and the final comparison in Figure 4C. Increases thus indicated have been computed at 13, 17, and 19 per cent for a three-day period. These compare favorably with those obtained by the depletion-curve method.

It can be concluded, therefore, that important gains in water yield during the growing season in the southern Appalachian region may be realized with relatively little effort by eliminating the transpiration draft of stream-bank vegetation. This procedure can be of much practical value during drought years for municipal and industrial watersheds, when even small increases in yield are of unusual importance.

References

BLANEY, H. F.; TAYLOR, C. A.; and YOUNG, A. A., Rainfall penetration and consumptive use of water in Santa Ana River Valley and Coastal Plain, Calif. Dept. Pub. Works, Div. Water Resources, Bull. 33, 162 pp., 1930.

BLANEY, H. F.; TAYLOR, C. A.; YOUNG, A. A.; and NICKLE, H. G., Water losses under natural conditions from wet areas in Southern California, Calif. Dept. Pub. Works, Div. Water Resources, Bull. 44, 139 pp., 1933.

HOOVER, M. D., Effect of removal of forest vegetation upon water yields, Trans. Amer. Geophys. Union, v. 25, pp. 969-975, 1944.

TROXELL, H. C., The diurnal fluctuations in the ground water and flow of the Santa Ana River and its meaning, Trans. Amer. Geophys. Union, v. 17, pp. 496-504, 1936.

WHITE, W. N., A method of measuring groundwater supplies based on discharge of plants and evaporation from soil, results of investigations in Escalante Valley, Utah, U. S. Geol. Surv. W. S. P. 659-A, 105 pp., 1932.

WICHT, C. L., Diurnal fluctuations in Jonkershoek streams due to evaporation and transpiration, J. So. African For. Assoc., no. 7, pp. 34-49, 1941.

YOUNG, A. A., and BLANEY, H. F., Use of water by native vegetation, Calif. Dept. Pub. Works, Div. Water Resources, Bull. 50, 160 pp., 1942.

Appalachian Forest Experiment Station,
Asheville, North Carolina

(Manuscript received June 3, 1946; presented at the Twenty-Seventh Annual Meeting, Washington, D. C., May 28, 1946; open for formal discussion until July 1, 1947.)

Commentary on Paper B4

Todd, D. K. (1955) Ground-water flow in relation to a flooding stream.

One of America's most distinguished groundwater hydrologists, David Keith Todd's research interest in the subject began with his PhD at Berkeley, California, USA (where he remained on the faculty for 31 years), which focused on transient bank storage of a flooding stream (Todd, 2002). This paper, which derives from his doctoral thesis, combines theory, (hardware) modelling and field observation to address the real-world problems of understanding and managing groundwater resources. It is one of the earliest papers to consider the bank storage effect in any detail. Earlier studies (e.g. Hursh & Brater, 1941, Paper B2) had examined slope discharge in steep catchments where subsurface flows from slope to stream were effectively unidirectional. Todd starts by discussing the groundwater contribution of stream flow during flood periods, but the bulk of the paper concerns flows in both directions, from channel to flood plain aquifer during the rising flood stage and *vice versa* during the falling stage.

Given a lack of field data and the evident complexity of flood plain sediments where some data did exist, Todd immediately chose to approach the problem from a theoretical standpoint. One can imagine that only a few decades later, his approach would have been to employ numerical simulation, but here a hardware model is used instead, to derive fundamental relations that can then be applied to the field situation. A Hele-Shaw viscous fluid model was used to model flows from the river channel into and out of a flood plain. Sine waves of varying amplitude and duration were generated in the reservoir (i.e. the river channel) and allowed to propagate into the flood plain (the narrow slot which Todd confusingly calls the "channel": page 628-2). From a number of experiments, Todd is able to derive equations which are then applied to the field situation in the "Model-Prototype Relations" section. Whilst his initial focus is on the groundwater waves moving into and out of the flood plain, an estimation of the volume of bank storage is also derived. As so often, new emphases arise from modelling work; here, from concentrating initially on inflows and outflows, Todd naturally moves on to consider the volume of water temporarily stored in the flood plain. At the end, his concern is still for groundwater contributions to the river, and more recent work on the attenuation of flood hydrographs because of bank storage is not anticipated.

Todd's approach is notable for combining several different approaches. The use of a hardware model is not surprising; this was an era when both hardware and electrical circuitry were employed as analogues of groundwater flow systems. His model analysis uses physically-based equations to define relations between discharge, amplitude and duration, and to compare with the observed data. There is careful analysis of possible errors and infringements of the underlying assumptions. As already noted, the derived equations are then used to estimate groundwater flows in the field. His example from the Sacramento River shows a concern for groundwater management, in this case to solve seepage problems caused by controlled releases from the Shasta Dam. Notwithstanding the field complexities caused by irregular alluvial deposits, Todd's Fig. 9 provides a helpful comparison with the laboratory results and mathematical analysis. It is interesting that he notes the more subdued response of wells further from the river, but not the fact that there is little delay in the passage of the flood wave out into the flood plain, something that has been observed many times since (e.g. Burt *et al.*, 2002).

Todd's work was the first to provide a theoretical analysis of the bank storage effect and focused attention on the groundwater flow processes in flood plain aquifers associated with the passage of a flood wave. His approach provides a foundation for much later computer simulations of flood plain hydrology (e.g. Bates *et al.*, 2000) as well as for more detailed field studies of near-stream hydrology.

TPB

References

Bates, P. D., Stewart, M. D., Desitter, A., Anderson, M. G., Renaud, J.-P & Smith, J. A. (2000) Numerical simulation of floodplain hydrology. *Water Resour. Res.* **36**, 2517–2530.

Burt, T. P., Bates, P. D., Stewart, M. D., Claxton, A. J., Anderson, M. G. & Price, D. A. (2002) Water table fluctuations within the floodplain of the River Severn, England. *J. Hydrol.* **262**, 1–20.

Hursh, C. R. & Brater, E. F. (1941) Separating storm-hydrographs from small drainage-areas into surface- and subsurface-flow. *Trans. Am. Geophys. U.***22**, 863–871.

Todd, D. K. (2002) About David Keith Todd. *Ground Water* **40**, 673–676.

GROUND-WATER FLOW IN RELATION TO A FLOODING STREAM

David K. Todd[1] J.M. ASCE

Abstract

Quantitative data on ground-water flow during storm runoff periods are scarce. This study presents some information on the time distribution and magnitude of such flows based on model measurements. Idealized hydrographs and aquifers were studied in a Hele-Shaw viscous fluid model. Results of the model analogy are interpreted in terms of bank storage volume and distribution, and magnitude and time distribution of ground-water flow. Seepage bordering the Sacramento River is discussed in relation to the model data.

INTRODUCTION

Little is known of the ground-water contribution of stream flow during flood periods. Data are difficult to obtain in the field because of surface runoff and rainfall contributions, lack of sufficient permeability data, and expensive recording well systems. Then, too, long and irregular intervals between isolated high flows discourage detailed studies.

Because engineering investigations require estimates of total stream flow, the usual procedure in practice is simply to assume some reasonable value for base flow during high flow periods. Various methods have been recommended(1,8,13)[2] for this purpose. It has been stated, and rightly so, that as long as one method is used consistently in any study, errors will be kept to a minimum. Such procedures are justifiable since: (a) detailed, accurate information is lacking; (b) the base flow contribution during a flood period is usually only a small fraction of the total flow; and (c) the accuracy of most stream-flow records does not justify any greater precision.

Previous investigations provide some clues to the true ground-water contribution. Hursh and Brater(5) analyzed field data and used the sharp break of the recession-hydrograph to locate the maximum on the normal ground-water depletion curve. Suter(15) presented data on flood hydrographs and nearby well stages of the Illinois River, showing how well levels rose and fell with the river. In an analytical treatment of waves in unconfined aquifers Werner and Noren(18) showed that harmonic oscillations of a stream were transmitted to the adjacent ground-water table where they were rapidly damped. Jones,(7) in a detailed study of the Upper San Joaquin River Valley, showed by profiles of ground-water levels that ground-water levels up to about one mile from the river were influenced by river levels. Hursh and Barnes(4) discussed the need for distinguishing ground-water flow from surface runoff, and the difficulty of separating interflow from base flow. Also,

1. Asst. Prof. of Civ. Eng., Univ. of California, Berkeley, Calif.
2. Numbers in parentheses refer to references at the end of the paper.

628-1

the question of the amount of temporary bank storage and of its return as subsurface flow was brought out. Based upon watershed studies in Switzerland and laboratory experiments, Roessel[12] indicated that base flow may well be a major part of flood flows, and that different runoff rates may occur from the same quantity of ground-water storage, depending upon the distribution within the aquifer of the water. Both Riesbol[11] and Snyder[14] made important contributions to the subject by their detailed analyses of watershed runoff.

Model Analysis

Because of the absence of field data for analyzing ground-water flows near a stream during periods of storm runoff, a model study was undertaken for this purpose. A Hele-Shaw viscous fluid model was utilized. This model consists of a channel formed by two closely-spaced glass plates through which a mineral oil flows. It can be shown that the laminar flow conditions obtained are analogous to ground-water flow conditions. The model design and applications have been described elsewhere,[16,17] while a similar model has been employed for base course drainage studies.[3] Solitary sine waves of varying amplitudes and periods were generated in a regulating reservoir at one end of the channel and allowed to travel into the channel containing a uniform oil depth. The conditions may be thought of as a model of a flooding stream with a rising and falling stage and having a horizontal water table intersecting the stream channel. Figure 1 shows a schematic diagram of the field conditions reproduced by the Hele-Shaw model analogy.

The model arrangement requires the following assumptions for analogous field conditions: (a) the stage hydrograph is of a sinusoidal shape; (b) an unconfined aquifer with homogeneous isotropic permeability exists adjacent to the stream channel; (c) the stream bank is vertical (corresponding to the reservoir-channel boundary); (d) the flood crest elevation is less than the ground surface so that no overflow occurs; (e) an impermeable horizontal bedrock coincides with the elevation of the stream bed; (f) the stream flow is much larger than the ground-water flow so that the stage is independent of any ground-water flows; (g) the ground-water table is initially horizontal; and (h) compressibility effects are negligible. These assumptions present idealized field conditions; nevertheless they enable quantitative ground-water flow measurements to be made on a model analogy basis. In essence, the model presents by analogy a vertical cross-section perpendicular to a stream channel.

During the time when various sinusoidal oscillations within the regulating reservoir were in progress, lapse-time motion picture records were obtained of the movement of the free surface in the channel. It should be noted that an oscillation is here defined as a fluctuation beginning at the initial minimum oil elevation in the channel, rising to a maximum elevation, and returning to the original minimum level. This corresponds to a sine curve extending from -90° to $+270^{\circ}$. The amplitude is the difference in elevation between the minimum level (-90°) and the maximum level (+90). Records of the oscillations showed that in the rising portion of the oscillation, oil flowed continuously into the channel. A reversal of flow in the second half of the oscillation became evident in that portion of the channel closest to the reservoir as the oil level in the reservoir decreased more rapidly than that in the channel. This process formed a wave crest which travelled from the reservoir into the channel and decreased in amplitude with time. Of interest here is the fact that the area, as viewed from the side of the channel, occupied by the oil above the original constant oil level is analogous to the volume of ground-water storage induced by a flooding stream. From a graph showing the variation of this area with time, the derivative, or slope, of this curve represents the rate of change of area with time, or the ground-water flow rate. To study this flow distribution, 14 independent solitary sinusoidal oscillations in the channel were recorded and analyzed. Oscillation periods ranged from 2 to 6 minutes and amplitudes from 1 to 6 inches (2.5 to 15.2 cm). Included in Table 1, which summarizes all of the basic data, are tabulations of amplitude H, in cm, and period T, in min, for the various runs.

Since the model measurements cover different amplitudes and period of the same sinusoidal oscillation, it was desirable to express the results in terms of a general curve which would be applicable to all of the data. Reasoning that the volume of ground-water (or bank) storage is some function of the oscillation amplitude and period, Figure 2 was prepared. Here the maximum recorded volume V_{max} (from Table I) of storage, expressed in two dimensions as an area with units of cm^2, is plotted against the product of amplitude and period. It can be seen that a linear relationship is reasonably well defined by the model data, indicating that the parameter, V_{max}/HT, may be assumed constant for the various runs. Now since the time from the beginning of the oscillation of V/HT must depend upon the period of the oscillation, the dimensionless time parameter t/T may be introduced. Here V is the volume measured at time t after the start of the oscillation. Curves of V/HT versus t/T were prepared for the 14 runs and superposed; close agreement resulted. A mean curve prepared from the group of curves is shown as the "observed" curve in Figure 3. This volume-time curve begins at zero (since the oil volume below the initial constant oil level is neglected) and rises to a maximum at t/T = 0.56. This instant follows shortly after the time of crest stage in the reservoir (or stream, when speaking of field conditions) at t/T = 0.50. The volume then recedes toward zero, displaying a typical hydrograph shape. The curve is cut off after t/T = 5.0 as model data were not carried beyond this time.

The slope, or derivative, of the volume-time curve represents the ground-water flow. Since the volume to any time t may be expressed as

$$V = \int_{t=0}^{t} Q \, dt \tag{1}$$

where Q is the ground-water flow occurring in any time increment dt, then the derivative of the curve in Figure 3 becomes

$$\frac{\frac{1}{HT}\, dV}{\frac{1}{T}\, dt} = \frac{1}{H} \frac{Q\, dt}{dt} = \frac{Q}{H} \tag{2}$$

Hence, a graph of the parameters Q/H versus t/T should enable the time distribution of the ground-water flow for all of the runs to be compared.

An inspection of the slopes of Figure 3 indicates how the ground-water flow curves will appear. The derivative begins at zero, reaches a positive maximum at the steepest point of rise, returns to zero at the instant of maximum, goes to a minimum at the steepest point on the recession portion of the curve, and finally returns toward zero. For purposes of the ground-water analogy, signs of all ground-water flow curves will be reversed. By so doing ground-water recharge, or stream-flow loss, is given a negative sign; ground-water

TABLE I - SUMMARY OF BASIC DATA FOR SOLITARY

SINUSOIDAL FLOWS

Run No.	Amplitude, H, cm.	Period T, min	Max. Vol., V_{max}, cm^2	Q_{min}, cm^2/min	Q_{max} cm^2/min	t_{min} min.	t_o min.	t_{max} min.
1	7.6	3.0	379	-433	180	1.38	1.75	2.62
2	5.1	3.0	229	-389	129	1.25	1.50	2.25
3	7.6	3.0	419	-541	282	1.12	1.88	2.25
4	10.2	6.0	820	-606	184	2.25	3.20	5.25
5	10.2	2.0	445	-841	242	0.88	1.20	1.88
6	10.2	4.0	651	-587	352	1.50	1.70	3.50
7	5.1	3.0	256	-361	159	1.38	1.70	2.50
8	15.2	3.0	872	-1102	421	1.25	1.80	2.50
9	2.5	3.0	148	-230	30.5	1.12	1.50	2.75
10	10.2	3.0	535	-701	253	1.25	1.70	2.75
11	2.5	3.0	101	-120	35.0	1.38	1.50	2.75
12	12.7	3.0	714	-831	355	1.25	1.75	2.62
13	10.2	3.0	548	-741	268	1.25	1.70	2.50
14	10.2	5.0	742	-608	254	1.62	3.25	4.00

628-4

discharge, or stream-flow gain, becomes positive. This convention is in accord with the usual concept of treating ground-water flow into a stream as a positive contribution.

To verify the above indicated comparison, computed maximum and minimum discharges, Q_{max} and Q_{min}, respectively, were plotted against the oscillation amplitude, H. The minimum discharge may be regarded also as the maximum recharge. The results, using data from Table I, appear as Figure 4 and support the contention that the parameter Q/H may be assumed constant for the various runs. Similarly, if the times, t_{min}, t_o, and t_{max} (see Table I), correspond to Q_{min}, Q_o (zero flow between recharge and discharge), and Q_{max}, respectively, then these times should also bear linear relationships to the oscillation periods. The data plotted in Figure 5 substantiate this supposition—Q_{min} takes place at t/T = o.41, Q_o at t/T = 0.56, and Q_{max} at t/T = 0.86. Based upon these comparisons among the different runs, the 14 discharge curves computed from the volume-time curves likewise showed substantial agreement when plotted in terms of Q/H versus t/T. The mean curve, and derivative of Figure 3, appears as Figure 6.

Although the exact magnitude of ground-water flow adjoining a stream always has been difficult to ascertain, the flow reversal shown by Figure 6 was recognized by earlier investigators. Brater(2) in 1940 stated that when a stream rises above the adjacent ground-water table, "---there must be a temporary reversal in the ground-water gradient and the ground-water contribution could conceivably be negative for a short interval." Other engineers (8, p. 399; 13, p. 266) also have correctly interpreted this phenomenon.

Regarding the scatter of data indicated in the previous figures, errors may be attributed largely to instrumental and analytic sources. The flow rate of the oil between the glass plates is a function of density and viscosity of the oil. Temperature modifies both of these properties; hence all data were corrected from the observed temperature to a standard of 70°F. However, if the oil temperature in the course of an oscillation changed, the single observed temperature would not correctly account for these effects. Errors also may have been introduced by small deviations from the true sinusoidal pattern, since the unsteady flows were manually controlled by inflow and outflow valves connected to the reservoir. Areas of oil were scaled from projections of the 35-mm film record by measuring ordinates at 6-inch intervals along the channel. A grid sheet attached to the back of the glass channel aided in scaling off the ordinates. Measurements were accurate to about 0.15 cm.

The far end of the channel, opposite to the reservoir, was left open except for an overflow lip fixed at the same elevation as the initially constant oil level. This end of the channel could not be sealed without violating the assumed field analogy (Figure 1). The fact that the channel was of a finite length and not of a semi-infinite extent as previously assumed, introduced erroneous volume and flow values. Fortunately, errors did not begin until the oil level had begun to rise and outflow ensued at the far end of the channel. This happened near the middle of the recession portion of the discharge curve. The nature of this limitation may be ascertained from a knowledge of the flow pattern. As the oscillation occurs within the reservoir a wave is propagated into the channel which damps out along the horizontal oil surface at a theoretically infinite distance. Practically, however, the damping rate is very rapid and changes of the oil level become negligible in relatively short distances. The effect of the outflow error was to reduce the oil area within the channel at a faster rate than would have resulted if a longer channel had been used, and to give too large a discharge in the recession portion of the discharge curve.

In an effort to evaluate the magnitude of the error, the recession portion of Figure 6 was plotted on logarithmic coordinates (Figure 7). This showed that the data closely fitted an equation

$$\frac{Q}{H} = 16.2(t/T)^{-2.76} \tag{3}$$

except for a sharp break near t/T = 1.8. Examination of the data showed that this time closely corresponded to the initial outflows at the far end of the channel. Hence it is reasonable to assume that if this flow had not occurred, the logarithmic recession curve, labelled "computed" on Figures 3, 6, and 7, would have continued. It is believed that this computed curve more nearly represents the true conditions pertaining to a semi-infinite aquifer. The flows indicated by the difference between the observed rate and projected logarithmic rate form a small hydrograph with rising and falling limbs. Outflow at the far end of the channel amounted to 9.8 percent of the total inflow volume up to t/T = 5.0.

Figure 8 shows typical dimensionless stage hydrographs of the oil (water table) at distances of 0, 0.5, 1.0, 2.0, and 5.0 feet from the reservoir. The height h occurs at the time t and is always less than H away from the stream channel. These curves illustrate the rapid damping of the oscillation as it travels from the source.

Model-Prototype Relations

To interpret properly the model results they must be expressed in terms of similar prototype conditions. This can be done by assuming length and time scale ratios, and solving for prototype velocities, discharge, and permeability.

The mean velocity of flow in the model may be expressed by

$$V_m = \frac{\rho_m b^2 g}{3\mu_m} \frac{dh}{dx} \tag{4}$$

where ρ_m is the oil density, μ_m is the oil viscosity, b is the half-width of the channel spacing, g is gravity, and dh/dx is the surface slope.[17] Similarly, the velocity of flow in an aquifer may be expressed by the Darcy law as

$$V_p = \frac{\rho_p kg}{\mu_p} \frac{dh}{dx} \tag{5}$$

where ρ_p is the density of ground water, μ_p is its viscosity, and k is the aquifer permeability (expressed as an area). The ratio of these two velocities when governed by the same slope becomes

$$\frac{V_m}{V_p} = V_r = \frac{\rho_m b^2 \mu_p}{3\rho_p k \mu_m} \tag{6}$$

where the subscript "r" denotes a model-prototype ratio. For the mineral oil used in the model, ρ_m = 0.877 gm/cc and μ_m = 1.016 poise at 70°F. The plate spacing was 0.2096 cm, so that b = 0.1048 cm. At 70°F for water,

628-6

ρ_p = 0.998 gm/cc and μ_p = 0.00979 poise. To obtain a value for k in equation (6), V_r must be known. This quantity depends upon the length L_r, and time T_r, scale ratios. Values of L_r = 0.01 and T_r = 0.0003 may be assumed for illustration and to give information of the correct order of magnitude of prototype conditions. Thus, the model channel with a length of 10 feet represents a 1000-ft. distance on one side of and perpendicular to a stream channel, while a 3-inch oscillation is equivalent to a 25-ft river stage fluctuation. Similarly, a 3-min oscillation is comparable to a storm runoff duration of 6.9 days. Inserting these values to find V_r gives

$$V_r = \frac{L_r}{T_r} = \frac{0.01}{0.0003} = 33.33 \tag{7}$$

Using this value in equation (6), together with the other known data, gives a permeability of 9.29 x 10^{-7} cm^2, or 94.2 darcys. A 40-mesh sand, packed to a 40 percent porosity, would have roughly the same permeability (9, p. 113).

If field data happened to be available so as to provide an estimate of k, then V_r could be computed directly. Knowing V_r, L_r and T_r could be selected to study any desired flood amplitude and duration.

Bank Storage

Estimation of the volume of bank storage for assumed prototype conditions may be made from Figure 3. Direct conversion by means of the length scale cannot be made without a further assumption of specific yield. In the model channel the specific yield has been implicitly taken as 100 percent. This is true if the small adhesive oil film on the glass surfaces is neglected. A value of 10 percent specific yield may be assumed as a constant for field conditions. It has been pointed out by Jacob (13, p. 384) that the amount of water draining per unit volume of aquifer during the lowering of a ground-water table is a function of time. The maximum amount drained equals the specific yield. With the constant value of 10 percent specific yield, it is here assumed that the water will flow completely into and out of the aquifer with equal ease. The error incurred by this assumption will be small in this illustration because of the long time involved in the prototype ground-water flow. A large storm runoff of minimum duration would provide the least drainage time and hence give the largest error due to a variable storage coefficient.

As shown by Figure 2 the volume of ground-water recharge, or bank storage, is directly proportional to the amplitude-duration product of the storm runoff. For illustration the previously described 25-ft and 6.9-day runoff may be used. Converting the recharge and discharge from Figure 6 by means of the length and time scales, and multiplying by the specific yield, gives for the maximum recharge 0.00278 cfs per lineal foot of stream channel; 0.00109 cfs/ft for the maximum discharge. The total volume of recharge amounts to 421 ft^3 per lineal foot of stream channel.

Expressing the above results in more meaningful units and taking into account that the identical phenomenon is taking place within the opposite stream bank, the maximum ground-water recharge rate equals 29.3 cfs per mile of stream channel; the maximum discharge, 11.5 cfs/mi. It should be noted also that for this example, the peak stream flow occurs 3.45 days after the start, the maximum recharge occurs after 2.8 days, the flow reverses at 3.9 days, and the maximum discharge occurs after 6.0 days. Measurable return flow to the stream persists for several weeks. The total recharge volume amounts to 102 AF per mile of stream channel.

Bank storage occurs in that portion of the aquifer closest to the stream channel. Previous study[17] has indicated a very rapid decrease in the amplitude with distance inland. Rough calculations for the conditions described in the above illustration indicate that 50 percent of the total bank storage is contained within the first 175 feet of the channel, and 75 percent within 350 ft.

Sacramento River Seepage

An interesting field problem related to the topic of the model study is that of the Sacramento River seepage. During seasons of high runoff, Sacramento River stages below Shasta Dam exceed the adjacent land surface. An extensive levee system ordinarily confines the high flows to within the levees. Landowners of the strip of land adjoining the river and outside the levees, however, contend that "when the water surface is carried well above the outside land surface for any substantial number of days, the pressure forces the water to seep underground and saturate and waterlog the land to an extent that it cannot be worked or pastured and damages the orchard crops and kills the trees".[6] The Corps of Engineers has been studying the problem from a flood control standpoint, while engineers of the Bureau of Reclamation have made extensive measurements of the seepage because some parties have expressed the opinion that controlled releases of Shasta Dam have contributed to the problem. A committee of the California Legislature has recently looked into the problem,[6] and at the present time the California Division of Water Resources is analyzing all available data in preparation for a final report.

This particular problem provides a good illustration of the difficulties typically encountered in applying laboratory results to subsurface field investigations. Data collected by the U. S. Bureau of Reclamation provide an insight into the situation.[10] Shifting of the river channel in the past has produced irregular lenticular deposits of gravel, sand, silt, and clay. The surface material is predominantly fine-grained with scattered gravel and/or sand deposits to be found in the upper 100 feet. Because of the concern for seepage on agricultural land bordering the river, lines of wells were established by the Bureau at several points along the river. The lines extend inland from the levee and water table elevations were observed daily for long periods.

These observations showed that the ground-water table adjacent to the river was governed by rainfall, irrigation, river stage, and geology. The variability of these factors in either space or time indicate the complexity of the problem. The water table approximates the river stage. In summer and other periods with the river stage low, the water table is maintained by irrigation and is higher than the river. Drainage is then toward the river. During periods of high stages, the river level exceeds the water table elevation and recharge, or seepage, occurs. Water table elevations have been observed to change up to 10 feet within 5 days due to varying river levels. Amplitudes of fluctuations decrease inland, extending as far as 1/2 to 1-1/2 miles, depending upon locality. Depths to water table vary in extremes from zero to 25 feet in the river influence area.

Figure 9 shows a small portion of the data. River and nearby water table elevations at West Wilkins Slough for the spring of 1945 are shown. Well No. 1 is located only 100 feet from the levee, Well No. 4 is 600 feet, and Well No. 5 is 1100 feet. Precipitation and well logs are included for completeness. It can be seen that three major river rises occurred, affecting the nearest well most and the farthest well least. Until about April 1 a good correspondence between the river stage and the water table is apparent. Well fluctuations follow those of the river, delayed slightly with time and decreasing with amplitude inland. After April 1, Well Nos. 4 and 5 show a rise not indicated by either the river stage or Well No. 1. Presumably early irrigation is responsible for this deviation since rainfall is negligible.

The river stage continuously exceeded the ground water table elevation (Well No. 1) at this point in the 1944-45 season from November 3, 1944 to April 14, 1945. If the correspondence between the river and wells is taken as an indication of direct connection of permeable strata, a large recharge from the river may be expected. The available well logs, however, indicate rather tight formations so that a quantitative estimate of recharge is difficult to make. No direct application can be made of the model results in this instance because of the vertical recharge of the aquifer by rainfall and irrigation. Then, too, the inland water table may be regulated by Wilkins Slough, which parallels the well line and intersects the river about 900 feet southward, or by inland drainage systems.

Although the model results cannot be applied directly, an interesting comparison should be noted. Field studies by the Corps of Engineers have shown that the amount of seepage is roughly proportional to the number of days river stages are above natural bank levels and to the number of feet the river stage exceeds those levels (6, p. 25). This "foot-days" factor is analogous to the amplitude-period product which was shown by Figure 2 to be directly related to the total ground-water recharge.

Discussion

Although the model data lead to idealized results in terms of field conditions, they nevertheless describe for one of the simplest cases the distribution of ground-water flow. Variation of any of the controlling factors will modify the results herein presented. Storm rainfall is implicitly assumed to have fallen over some portion of the drainage basin upstream from the cross-section studied. Vertical percolation and interflow may be regarded as distinct stream-flow contributors and have been eliminated from the present study. Other hydrograph shapes will yield different ground-water storage and flow curves.

The model results showed that recharge and discharge were directly proportional to the amplitude-period product. In the field this may be interpreted by saying the ground-water recharge and discharge are directly proportional to a stage-duration factor. At a given point on a stream isolated storm events occurring over the drainage area above a given point on a stream would result in approximately constant flood durations, as indicated by the unitgraph theory. In such instances the ground-water recharge and discharge would depend only upon the flood stages, so that by means of a rating curve at that point, an approximate relationship between the stream flow and the ground-water flow could be established.

The model experiments required a horizontal ground-water table in equilibrium with the stream in the prototype situation. This would only rarely occur in the field. The more common situation where the ground-water table and stream are in contact would be an influent or an effluent stream. Here the ground-water table slopes upward or downward to the stream surface, respectively. As a first approximation for such cases the effects may be considered as additive, so that the ground-water flow curve would be shifted upward for effluent streams and downward for influent streams. Depending upon the relative magnitude and direction of the initial ground-water flow to the maximum recharge and discharge rates, flow reversals may or may not result from storm runoff.

The initial ground-water depth was taken as constant for all model runs. For completeness this depth also should be investigated because its variation affects flow rates. For these unsteady flows the ratio of amplitude to depth is a complex function of the flow, and no general statements can be made about the magnitude of this effect.

The ground-water contribution to a storm hydrograph at a given point on a stream is the combined effect of all ground-water flows upstream of that point. Thus the recharge-discharge pattern discussed herein is in phase with the crest stage, so that the flow depends upon the location of the flood crest. A complete analysis involves the integrating time distributions, aquifer and permeability variations, tributary effects, and changes in the storm hydrograph with travel. Although the generalized results presented in Figs. 3 and 6 are applicable only with respect to the assumptions stated, they nevertheless should aid toward a better understanding of ground-water flows bordering on a stream channel. Extending or modifying the analyses to suit particular problems may permit estimates of ground-water flow to be prepared.

ACKNOWLEDGMENTS

Robert G. Dean assisted in the computations and preparation of figures. Engineers of the U. S. Bureau of Reclamation, Sacramento, generously made available the Sacramento River seepage data, while engineers of the Sacramento offices of the California Division of Water Resources, U. S. Corps of Engineers, and U. S. Geological Survey were helpful through their discussions of the problem. Laboratory work was performed under the guidance of J. A. Putnam. J. W. Johnson provided valuable suggestions on the analysis and interpretation of the data.

REFERENCES

1. American Society of Civil Engineers, Hydrology Handbook, Manual of Engineering Practice No. 28, 184 pp., 1949.

2. Brater, E. F., The unit hydrograph principle applied to small watersheds, Trans. American Society of Civil Engineers, v. 105, pp. 1154-1192, 1940.

3. Casagrande, A. and Shannon, W. L., Base course drainage for airport pavements, Trans. American Society of Civil Engineers, v. 117, pp. 792-820, 1952.

4. Hursh, C. R. and Barnes, B. S., Appendix B—Report of sub-committee on subsurface-flow, Trans. American Geophysical Union, v. 25, pt. V, pp. 743-746, 1944.

5. Hursh, C. R. and Brater, E. F., Separating storm-hydrographs from small drainage-areas into surface and subsurface-flow, Trans. American Geophysical Union, v. 22, pt. 3, pp. 863-871, 1941.

6. Joint Committee on Water Problems of the California Legislature, Sacramento River seepage and erosion problems, Sixth Partial Report, pp. 22-29, Sacramento, 1953.

7. Jones, G. H., Hydrology of valley areas adjacent to the upper San Joaquin River, Trans. American Geophysical Union, v. 21, pt. I, pp. 58-78, 1940.

8. Linsley, R. K., Jr., Kohler, M. A., and Paulhus, J. L. H., Applied Hydrology, McGraw-Hill, New York, 689 pp., 1949.

9. Muskat, M., "The flow of homogeneous fluids through porous media," McGraw-Hill, New York, 763 pp., 1937.

10. Plummer, A. K., "Summary of basic data collected during Sacramento River seepage investigations," Sacramento Valley District, U. S. Bureau of Reclamation, Chico, California, 1949 (unpublished).

11. Riesbol, H. S., Some aspects of subsurface water in hydrologic research on agricultural watersheds, Proc. of the Hydrology Conference, Pennsylvania State College, State College, pp. 85-112, 1942.

12. Roessel, B. W. P., Hydrologic problems concerning the runoff in headwater regions, Trans. American Geophysical Union, v. 31, pp. 431-442, 1950.

13. Rouse, H. (ed.), Engineering Hydraulics, J. Wiley and Sons, New York, 1039 pp., 1950.

14. Snyder, F. F., A conception of runoff-phenomena, Trans. American Geophysical Union, v. 20, pt. IV, pp. 725-738, 1939.

15. Suter, M., Apparent changes in water storage during floods at Peoria, Illinois, Trans. American Geophysical Union, v. 28, pp. 425-437, 1947.

16. Todd, D. K., "Investigation of unsteady flow in porous media by means of a Hele-Shaw viscous fluid model," Ph. D. thesis, University of California, Berkeley, 85 pp., 1953.

17. Todd, D. K., Unsteady flow in porous media by means of a Hele-Shaw viscous fluid model, Trans. American Geophysical Union (in press).

18. Werner, P. H. and Noren, D., Progressive waves in non-artesian aquifers, Trans. American Geophysical Union, v. 32, pp. 238-244, 1951.

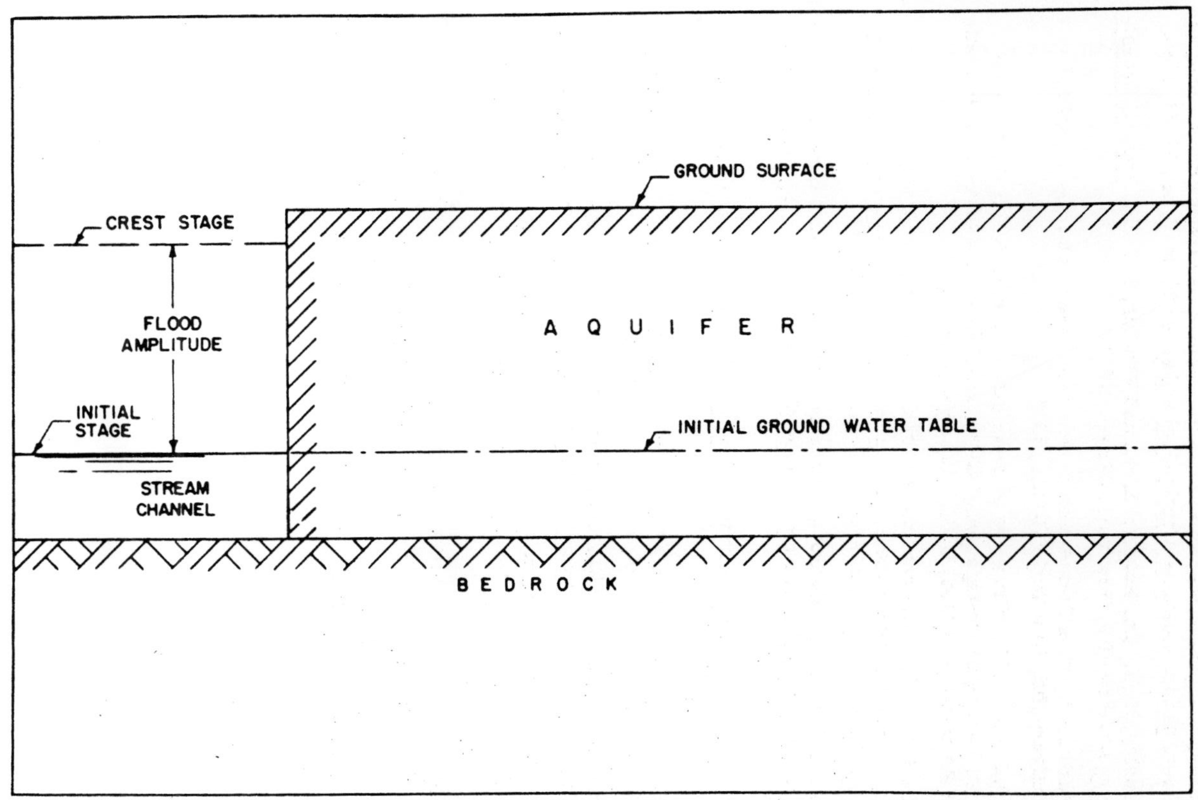

Fig. 1. Schematic diagram of vertical cross-section showing field conditions simulated by the Hele-Shaw model analogy.

628-12

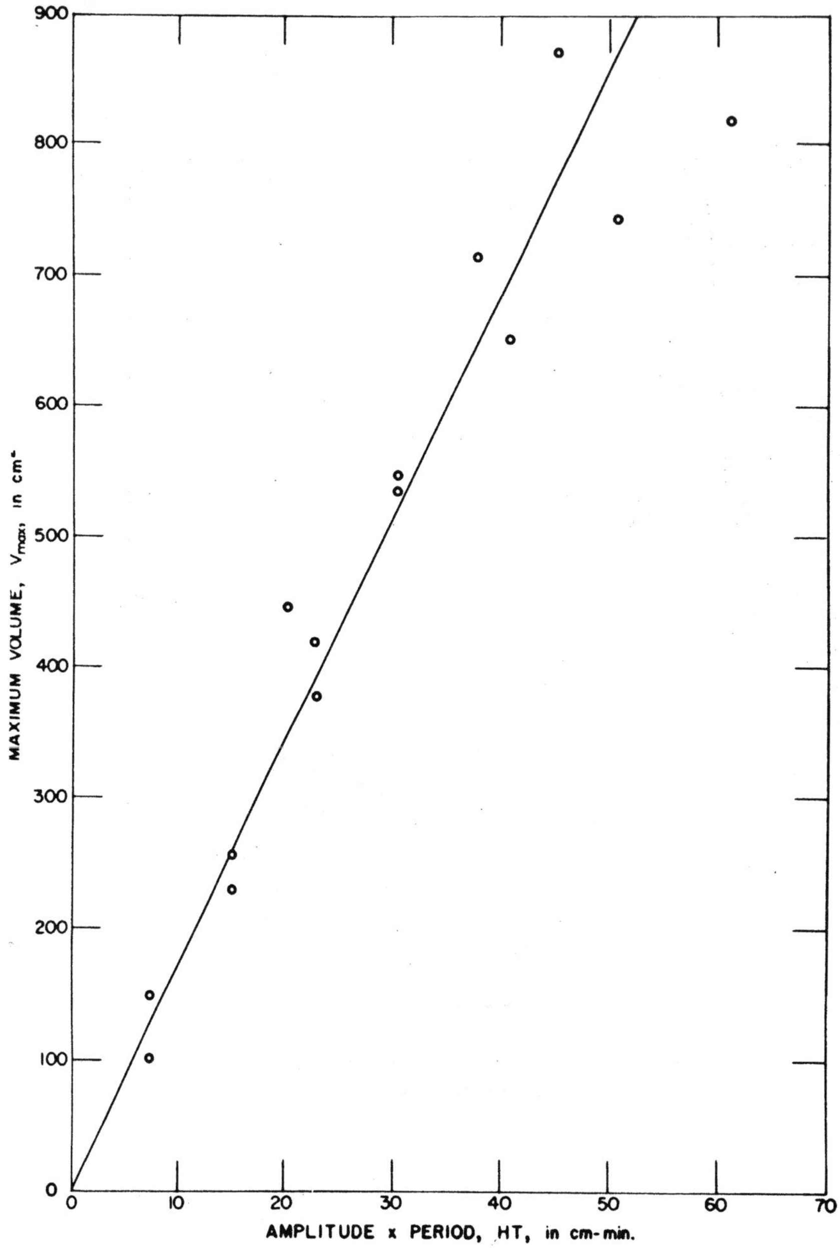

Fig. 2. Relationship of the maximum volume of ground-water recharge to the product of amplitude and period of stream oscillation.

628-13

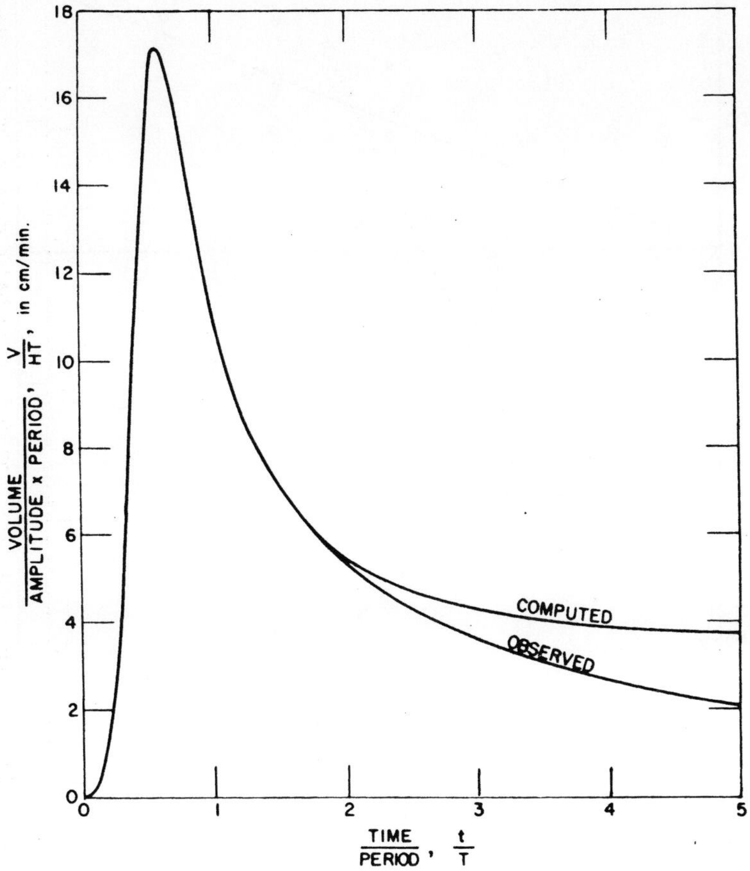

Fig. 3. Mean relationship between ground-water volume and time expressed as V/HT and t/T.

628-14

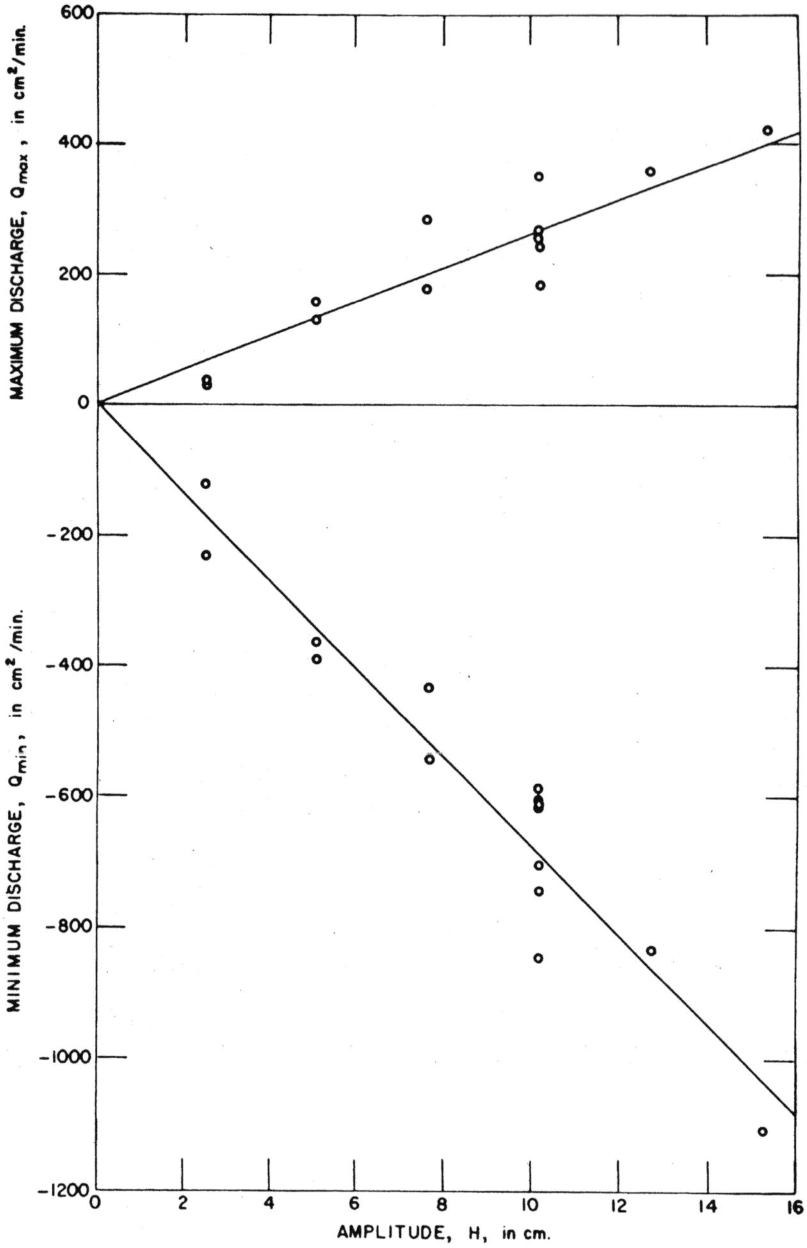

Fig. 4. Relationship of maximum and minimum ground-water discharge to amplitude of stream oscillation.

628-15

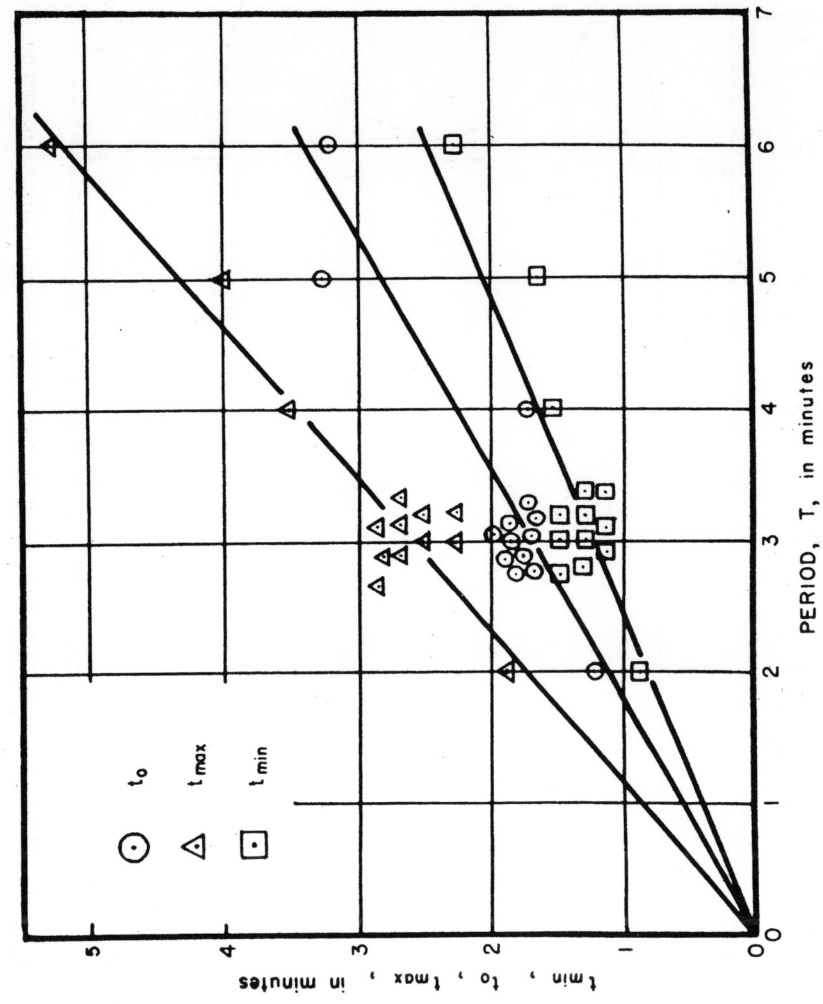

Fig. 5. Relationship of times t_{min}, t_o, and t_{max} to the period of stream oscillation.

628-16

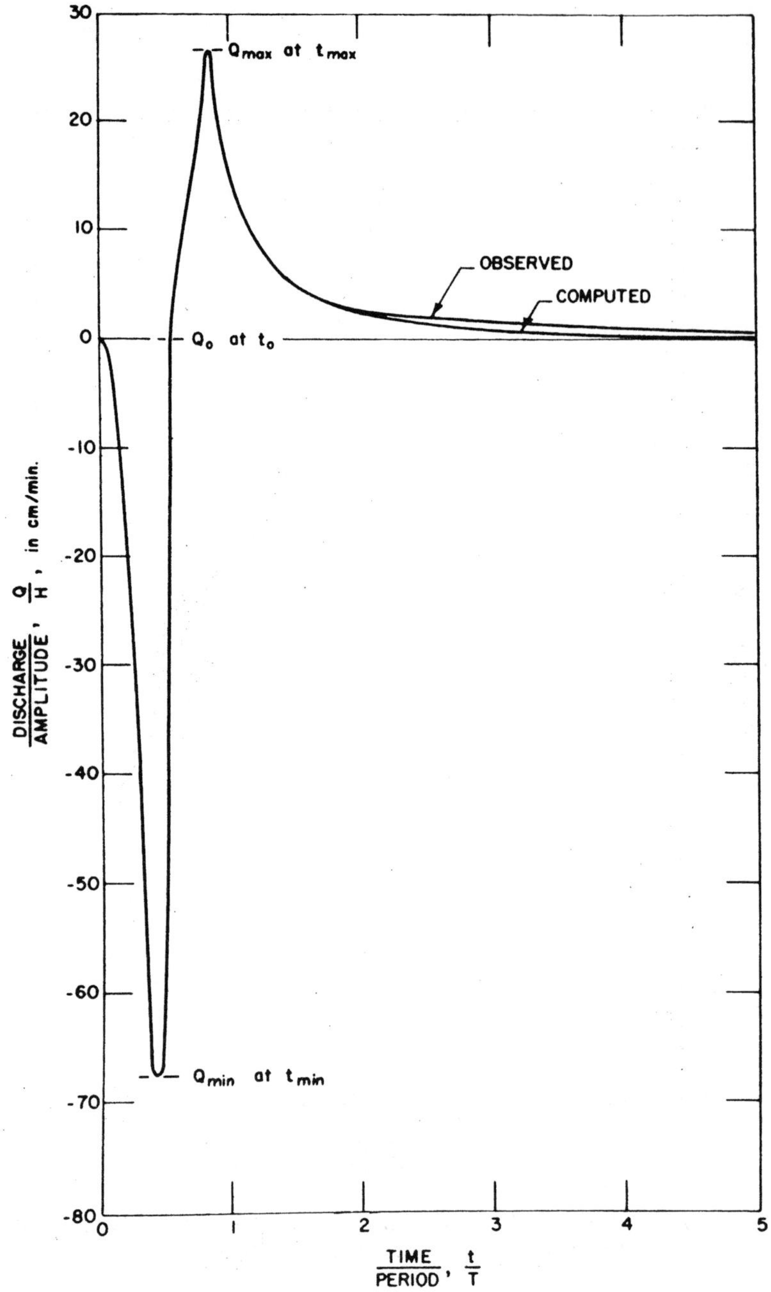

Fig. 6. Mean relationship between ground-water discharge and time, expressed as Q/H and t/T.

628-17

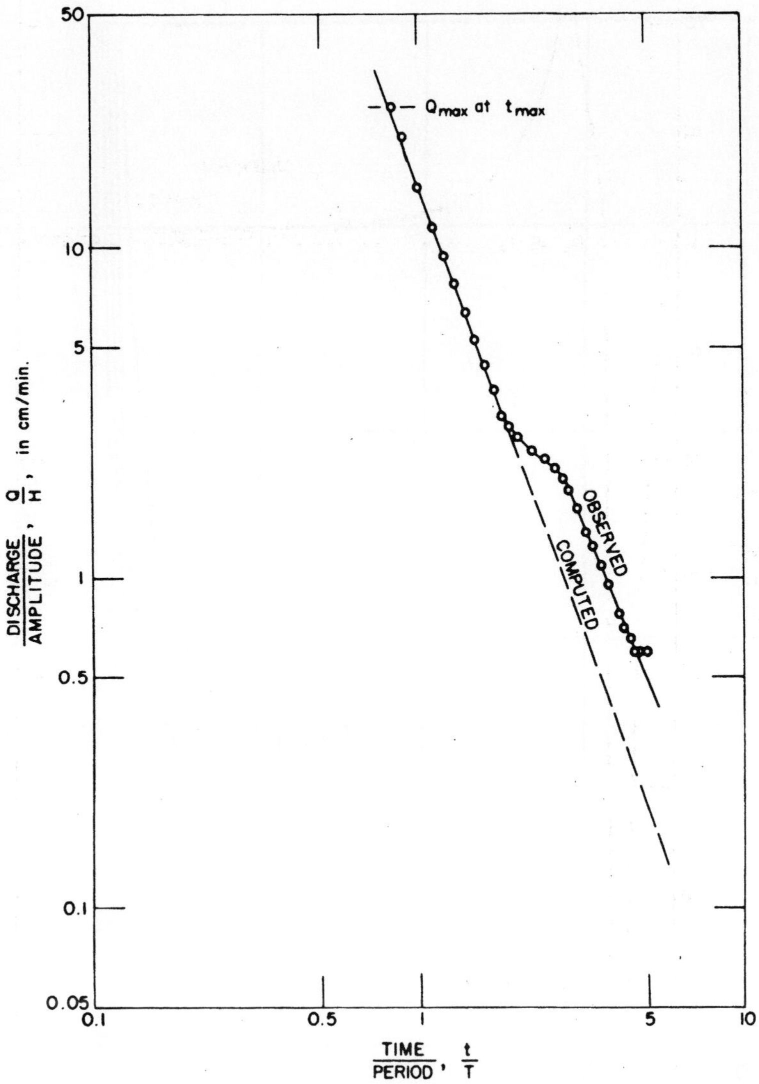

Fig. 7. Logarithmic plot of mean observed and computed ground-water recession curves, expressed as Q/H and t/T.

628-18

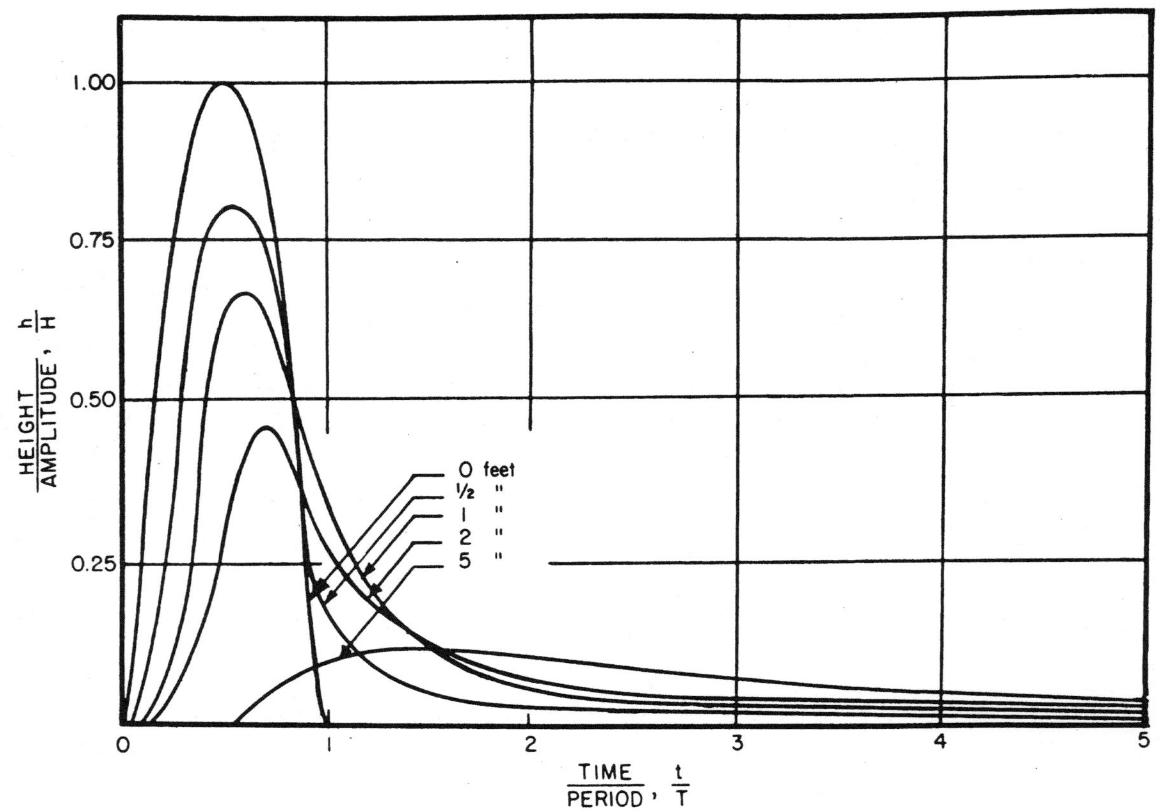

Fig. 8. Mean ground-water table hydrographs at various distances from the stream channel, expressed as h/H and t/T.

628-19

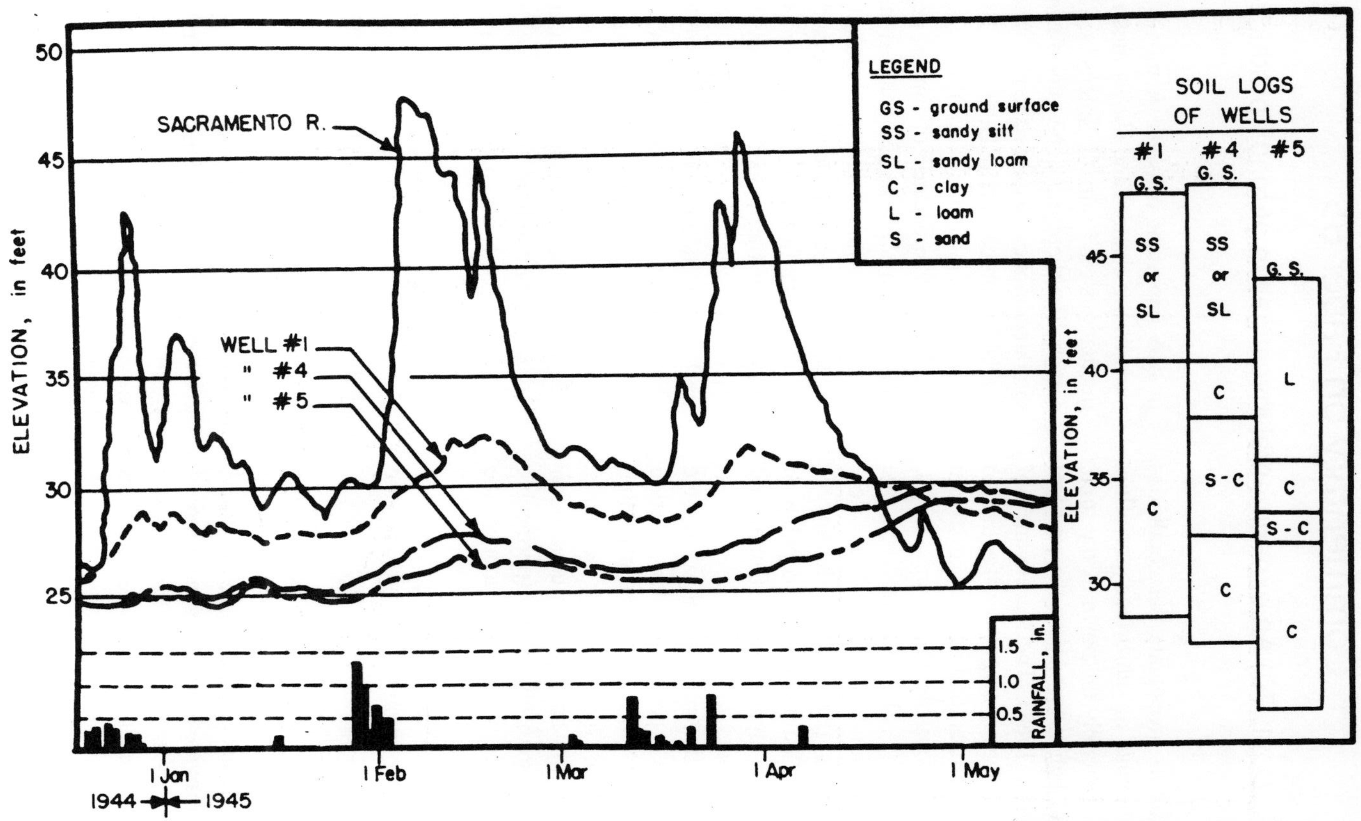

Fig. 9. Stage and ground-water table hydrographs, Sacramento River at West Wilkins Slough, California.

628-20

Commentary on Paper B5

Dunne, T. & Black, R. D. (1970) An experimental investigation of runoff production in permeable soils.

The aim of this study was to investigate the conditions under which significant amounts of subsurface stormflow are produced. These were generally unknown at the time, although field research had begun to address the problem (e.g. Whipkey, 1965; Ragan, 1968). The significance of Dunne and Black's paper is that it was really the first to detail the stormflow processes operating in the near-stream zone. It clearly builds on the variable source area model, first described by John Hewlett (1961), an alternative to the dominant theory of infiltration-excess overland flow propounded by Robert Horton (Burt, 2008). Kirkby & Chorley (1967) claimed that "throughflow is capable of producing runoff peaks in river hydrographs"; Dunne and Black set out to test this assertion.

The first conclusion was that no widespread surface runoff was produced across the experimental slope, during either natural or artificial rainfall, because of the very high infiltration capacity of the soil, over 300 mm per hour. Neither was subsurface stormflow an important contributor to the storm hydrograph, despite conditions (permeable soils, steep slopes) favourable to its production; the throughflow response was subdued and lagged compared to surface runoff production. Instead, overland flow was produced from only a small portion of the hillside: the lower part of the concave plot. This was the only type of storm runoff to make a substantial contribution to the storm hydrograph, and the only form of runoff to exhibit sensitivity to changes of rainfall in a manner similar to streamflow. In a complementary paper (Dunne & Black, 1970b), further detail is provided on the generation of overland flow as a combination of "return flow", water exfiltrating from the soil where the water table intersects the ground surface, and runoff from rain falling directly on to the saturated area and therefore unable to infiltrate ("direct runoff"). Dunne and Black were thus the first to identify what soon became known as "saturation overland flow" (Dunne, 1978). (Various terms have been used in the literature in addition to "saturation overland flow": "saturated" overland flow" (Ward, 1984), "saturation-excess overland flow" (Burt, 1989) and even "Dunne (or Dunnian) overland flow").

It is characteristic of headwater catchments in upland areas that flood plains are narrow or completely absent. In such cases, production of saturation-excess overland flow will be limited to those places where the water table is at or close to the ground surface immediately prior to a storm. As Dunne and Black show, hillslope hollows are particularly important in this regard, because subsurface flow converges downslope, raising the water table nearer the surface than it would be at equivalent slope positions where the contours are straight or convex towards the stream channel. Dunne and Black note that, although return flow has the same origin as subsurface stormflow, it differs in two important respects: it moves much more quickly towards the channel and shows much greater sensitivity to fluctuations in rainfall intensity. During storms and seasonally, there may be significant expansion of the saturated zone, which accords directly to Hewlett's "variable source" area (Dunne & Black, 1970). It is not clear why subsurface stormflow was unimportant at the site, but maybe because of limited depth of the permeable soil horizon restricting soil water storage (and so favouring overland flow rather than throughflow), or perhaps because of the very short slope length involved.

The significance of Dunne and Black's paper for the study of riparian zone hydrology is two-fold. First, it articulated for the first time the detailed process mechanisms underpinning the variable source area concept, indicating the conditions under which overland flow is likely to be produced in the near-stream zone, providing a rapid and large addition to channel discharge. In relation to diffuse pollution transport, overland flow can transport sediment and sediment-bound pollutants from field to stream. Second, the paper drew further attention to the groundwater ridging phenomenon. Although not important at the Sleepers River Experimental Watershed, in other locations there can be a strong relationship between the height of the pre-event water table and the rapid transfer of pre-event water from the riparian zone into the stream (Abdul & Gillham, 1984, Paper B6). This again focuses attention on the spatial variation of storm runoff production and which parts of a riparian zone might be most connected to the channel (Burt & Pinay, 2005). Moreover, Haycock & Burt (1993, Paper E7) suggest that the riparian zone might be short-circuited at times of very high water table, lowering buffer zone efficiency as a result.

Beven (2006) notes that, unfortunately, the Dunne and Black hillslope no longer exists; it was destroyed when Route 2 from Danville to St Johnsburg was driven through the Sleepers River basin in 1975. What a shame that one of the most important sites in hillslope hydrology has been lost for ever!

TPB

References

Abdul, A. S. & Gillham, R. W. (1984) Laboratory studies of the effects of the capillary fringe on streamflow generation. *Water Resour. Res.* **20**, 691–698.

Beven, K. J. (2006) Commentary on Paper 15: Dunne, T. & Black, R. D. (1970b) Partial area contributions to storm runoff in a small New England watershed. In: *Streamflow Generation Processes* (ed. by K. J. Beven), 183–184. Benchmark Papers in Hydrology. IAHS Press, Wallingford, UK.

Burt, T. P. (1989) Storm runoff generation in small catchments in relation to the flood response of large basins. In: *Floods* (ed. by K. J. Beven & P A Carling), 11–36, Wiley, Chichester, UK.

Burt, T. P. (2008) Valley-side slopes and drainage basins. I: Runoff and erosion. In: *The History of the Study of Landforms or the Development of Geomorphology* Chapter 8. (ed. by T. P. Burt,

D. Brunsden, R. J. Chorley, N. J. Cox & A. S. Goudie), volume 4: *Quaternary and Recent Processes and Forms (1890–1965) and the Mid-Century Revolutions*. The Geological Society, London, UK, 325–352.

Burt, T. P. & Pinay, G. (2005) Linking hydrology and biogeochemistry in complex landscapes. *Progr. Phys. Geogr.* **29**(3), 297–316.

Dunne, T. (1978) Field studies of hillslope flow processes. In: *Hillslope Hydrology* (ed. by M. J. Kirkby), 227–293. Wiley, Chichester, UK.

Dunne, T. & Black, R. D. (1970b) Partial area contributions to storm runoff in a small New England watershed. *Water Resour. Res.* **6**, 1296–1311.

Haycock, N. E. & Burt, T. P. (1993) The role of floodplain sediments in reducing the nitrate concentration of subsurface runoff: a case study in the Cotswolds, England. *Hydrol. Processes* **7**, 287–295.

Hewlett, J. D. (1961) Watershed management. In: *Report for 1961 Southeastern Forest Expt. Station*, 62–66. USDA Forest Service, Asheville, NC,.

Kirkby, M. J. & Chorley, R. J. (1967) Throughflow, overland flow and erosion. *IAHS Bulletin* **12**, 5–21.

Ragan, R. M. (1968) An experimental investigation of partial-area contributions. In: *Hydrological Aspects of the Utilization of Water. Reports and Discussions.* IAHS Publ. 76, 241–249. IAHS Press, Wallingford, UK.

Ward, R. C. (1984) On the response to precipitation of headwater streams in humid areas. *J. Hydrol.* **74**, 171–189.

Whipkey, R. Z. (1965) Subsurface stormflow from forested slopes. *IAHS Bulletin* **10**(20), 74–85.

An Experimental Investigation of Runoff Production in Permeable Soils

THOMAS DUNNE

McGill University, Montreal, Canada

RICHARD D. BLACK

Cornell University, Ithaca, New York

Abstract. In an area of low intensity rainfall and permeable soils, three hillside plots were instrumented for a study of runoff-producing mechanisms. Runoff from the plots was measured at the ground surface, the base of the root zone, and in the zone of perennial groundwater seepage. Data on soil moisture, water-table elevation, and piezometric head were also collected during natural and artificial storms. The data showed that, as the infiltration capacity of the soil exceeded the rainfall intensities that occurred and that were applied, overland flow generated by the mechanism described by Horton did not occur. Although soils and topography were those generally thought to be conducive to subsurface stormflow, the runoff produced by this mechanism was too small, too late, and too insensitive to fluctuations of rainfall intensity to add significantly to stormflow in the channel at the base of the hillside. When the water table rose to the surface of the ground, however, overland flow was generated on small areas of the hillside. Only when this overland flow occurred were significant amounts of stormflow contributed to the channel by the hillside. The return periods of storms that would produce such overland flow were found to be very large.

INTRODUCTION

Overland flow on large areas of hillside, described by *Horton* [1945] and others, has not been observed in the 43-square-mile watershed of the Sleepers River in northeastern Vermont. Infiltration capacities of the forested and pastured soils of the area exceed the vast majority of measured rainfall intensities. Total yields of storm runoff are small. The maximum recorded rate of runoff on the Sleepers River Experimental Watershed was 0.10 inches per hour resulting from a rainfall of 3.37 inches in 38 hours on a 16-square-mile catchment. Runoff hydrographs, however, have the same general features as those from other areas where overland flow is observed. They have the same general shape, and respond quickly to rainfall. They show many of the same kinds of relationships between rainfall, runoff, and watershed condition as are recognized in areas known to experience overland flow.

In the absence of evidence of overland flow to supply storm hydrographs, many hydrologists have postulated the mechanism of subsurface stormflow. *Whipkey* [1965, p. 74] defined subsurface stormflow as underground, storm-period flow that reaches the stream channel without entering the general groundwater zone. *Kirkby and Chorley* [1967, p. 7] used the term 'throughflow' to denote 'flow which takes place physically within the soil profile.' *Hewlett and Hibbert* [1965] described their concept of 'translatory flow.' They wrote, 'Above the zone of saturation, we may regard such movement as due to thickening of the water films surrounding soil particles and a resulting pulse in water flux as the saturated zone is approached.'

AIMS OF THE PRESENT STUDY

The magnitude and timing of subsurface contributions to channel runoff and the conditions under which significant amounts of subsurface stormflow are produced are generally unknown. In the absence of universally accepted theoretical tools and detailed information on the physical characteristics of soils and channels, the phenomena listed above can be determined only from intensive field study. To this end, a small catchment in the glaciated upland of the Sleepers River Experimental Watershed near Danville, Vermont, was selected for a detailed study of runoff production in an area of low intensity rainfall and highly permeable soils. The following report will deal only with data on contributions to streamflow from the permeable soils of the steep hillslopes during natural and artificial rainstorms.

THE EXPERIMENTAL AREA

The watershed in which the study was conducted is a 60-acre basin, carved into a glaciolacustrine terrace that lies against a steep, till-covered hillside of well-jointed, calcareous schist. Within this watershed, a hillside was chosen for intensive study (Figure 1). The 0.6-acre hillside has a southerly slope that varies from 30% to 100% and has a relief of 60 feet. The surface outlet for the watershed drains the toe of the slope. The hillside receives no drainage from above as it is separated by a narrow, flat-topped divide from an equally steep slope on its northern side. It has a relatively simple profile and consists of three sections which differ in plan form. To the west is a section whose contours are convex (plot 1); in the center is a concave portion (plot 2); and to the east the contours are straight (plot 3). *Hack and Goodlett* [1960], *Troeh* [1964], and others have suggested that there is a relationship between soil drainage and slope shape. Topography, should, therefore, be reflected in the pattern of antecedent soil moisture which is generally considered to be a major control of runoff production.

Geologically the slope is part of a dissected lake bed (Figure 2) and consists of a 22-foot-thick layer of sand (containing less than 50% silt) overlying 38 feet of varved silt and fine sand (more than 65% silt), which in turn is founded on silty clay till. Dissection and subsequent mass movement of sandy material downslope has resulted in deposition of a 1- to 3-foot-thick layer of highly permeable material over the lower silt loam.

The soils of the slope are well-drained Brown Forest sandy loams on the straight, eastern section, and well-drained Brown Podzolic sandy loams on the concave, central, and convex western portions. The major difference between these soils is that the saturated conductivity, bulk

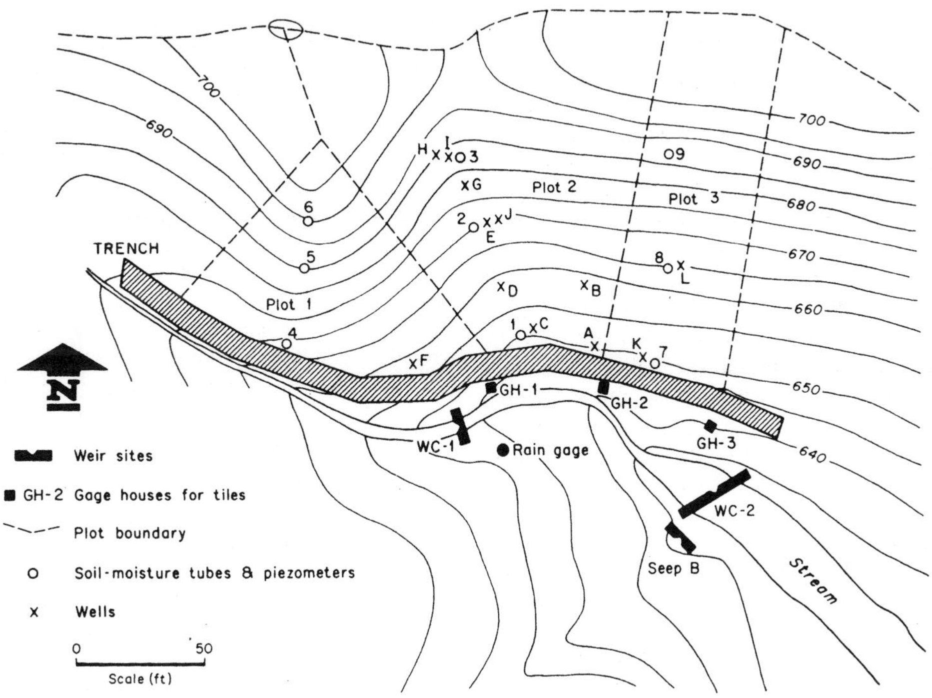

Fig. 1. Map of the hillside studied.

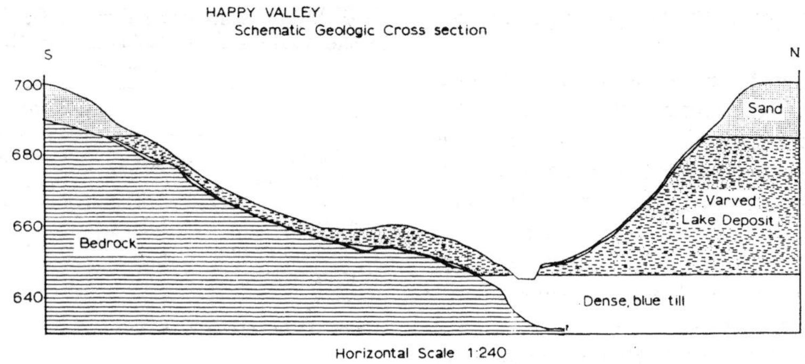

Fig. 2. Schematic geological cross section of the experimental area.

density, structure, and amount of root penetration change abruptly at a depth of 12 to 30 inches in the Brown Podzolic soils. On the straight slope there is a gradual change of soil structure, an increase in density, and a decrease in saturated conductivity as depth increases. Although the slope is now in pasture, pine woodland covered the area until a fire destroyed the woodland approximately 30 years ago. The soil is interlaced with old root holes, worm holes, and structural channels that seem to provide conditions that are conducive to subsurface stormflow.

INSTRUMENTATION

Runoff from the hillside was intercepted by drains installed in a 275-foot-long trench (Figure 1). The depth of the trench varied from 5 to 9 feet depending on the depth of the dense silty-clay till shown in Figure 2. The interceptors were divided into five sections: a 25-foot-long ungaged buffer zone at each end to reduce edge effects, a 125-foot-long section of trench draining the 0.13-acre convex hillside, and two 50-foot-long sections draining the 0.30-acre concave slope and the 0.17-acre straight slope. In each section of trench, runoff was collected at three levels: the soil surface, the base of the root zone, and the zone of perennial groundwater seepage (Figure 3). The flow from each interceptor was measured in gage houses at the eastern end of each section. Discharge was determined from continuous records of stage on calibrated weir slots. The numbering system used to refer to the various flow collectors is shown in Table 1. A roof was built over the trench and surface drain to prevent direct entrance of precipitation.

Channel runoff was gaged continuously at four locations in the 60-acre watershed for comparison with the runoff produced on the hillside plots. One gaging station (WC-1) lay immediately above the outlet from plot 1; another gaging station (WC-2) lay immediately below plot 3 (Figure 1). On the right bank of the stream the slope draining to the reach of channel between the weirs had an area of only 400 square feet.

Natural rainfall was measured with a 6-inch, 24-hour weighing rain gage. During artificial storms a random network of thirty 4-inch-diameter cans was used to measure rainfall. Total amounts of rain were computed as Thiessen-

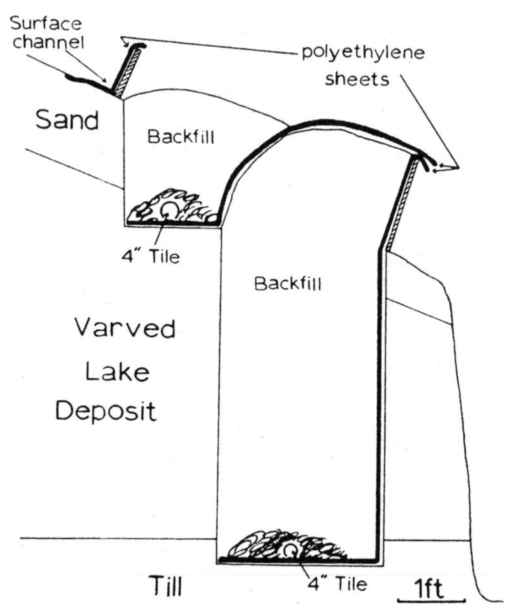

Fig. 3. Cross section of the interceptor trench.

weighted averages of the water caught in these cans.

A nuclear depth probe was used to measure soil moisture. Three access tubes, 7 feet deep, were placed in each plot (Figure 1): one at the base of the slope, one halfway up the slope, and one three-quarters of the way up the slope. Readings were taken weekly throughout the year, daily during snowmelt, before and after natural rainstorms, and several times during and after artificial rainstorms.

Measurements of water-table elevations were made in ¾-inch pipes at the 12 locations shown in Figure 1. Readings were taken weekly throughout the year, several times daily during snowmelt, and every few minutes during natural and artificial rainstorms. Soil water pressure was measured in ¾-inch diameter piezometers at the locations shown in Figure 1. A set of four piezometers at depths of 10, 8, 6, and 4 feet was located at each station. Two-foot-deep piezometers were also installed at locations 1 and 2. Piezometric readings were made at the same time as the water-table measurements.

EVENTS STUDIED

Data were collected in the following types of events:

1. Summer storms.
2. Large, natural autumn storms on unusually wet antecedent conditions.
3. Simulated rainstorms of high return period.
4. Natural rainstorms that occurred so soon after the simulated storms that they could be considered as a part of the same event.

RESULTS

Summer Storms

Thirty-five storms ranging in amount up to 1.09 inches were recorded during the course of the study. The maximum 5-minute and 30-minute intensities of these storms were 3.12 inches and 1.18 inches per hour respectively. The return periods of the storms were less than two years. A larger, more intense storm occurred after one of the artificial rainstorms, but this will be discussed separately.

No significant storm runoff was produced on the hillside in these storms. Neither overland flow nor flow from the base of the root zone were measured. Flow from the lower tiles occurred throughout the summer but did not usually respond to individual rainstorms. Groundwater discharge began to respond slightly after a period of rainy days, but the maximum outflow rate during the summer of 1967 was only 0.0009 cfm per foot of channel, (equivalent to 1.31 cfm for a drainage density of eight miles per square mile, which is an average for the subwatersheds of the Sleepers River Experimental Watershed). As it is a central thesis of this paper that runoff production is spatially nonuniform, to express runoff in inches and inches per hour would therefore be misleading, especially for very small watersheds. In this paper, runoff volumes are expressed in cubic feet and runoff rates are expressed in cubic feet per minute (cfm).

TABLE 1. Numbering System Used in Referring to the Various Flow Collectors in the Interceptor Trench

Level of Collector	Plot: Convex (1)	Concave (2)	Straight (3)
Surface	1-1	2-1	3-1
Base of root zone	1-2	2-2	3-2
Zone of perennial seepage	1-3	2-3	3-3

The reasons for the lack of runoff were the high infiltration capacity and storage capacity of the soils on the hillside. The infiltration capacity of these soils was measured to be greater than 3.15 inches per hour. From June 1, 1967, to October 1, 1967, the storage capacity of these soils exceeded any rainfall event that has been measured at a nearby gage in the last 76 years. Measurement of soil moisture before and after storms showed that rainwater was stored in the upper three feet of soil during the storm and drained slowly to the water table during the next few days. Throughout the summer months, the water table stood 4 to 5 feet below the center of plot 2.

It appears, therefore, that summer storms having a recurrence interval of less than two years do not exceed the soil moisture deficit and replenish the water table during the course of a storm. Nor does rainfall occur with sufficient intensity and duration to form a perched water table within the root zone. Yet streamflow in the watershed responds quickly to every one of these storms.

Large, Natural Autumn Storms

In the early part of October 1967 the soil gradually became wetter, and the water table stood closer to the surface than in the preceding summer months. It was therefore more probable that rainfall would displace soil water in the vadose zone [*Horton and Hawkins,* 1965], causing a response of the water table and consequent subsurface stormflow. Conditions were made even more favorable for this process by adding water to the slope during the testing of an irrigation system and during artificial storms.

In spite of the wet antecedent condition of the hillside, the response of runoff to heavy rains was slight. On October 10, 1967, a 1.35-inch rainstorm with a maximum 30-minute intensity of 0.62 inches per hour (recurrence interval of less than two years) occurred when the water table at well D was only 2.1 feet below the ground surface. Table 2 shows the amounts and lag times of stormflow generated by this rainfall on each of the plots. The total amounts of rain falling on each plot are included for comparison. The data demonstrate that even when the antecedent soil moisture is abnormally high, heavy rains produce a negligible amount of subsurface storm runoff, and that this runoff lags the beginning of rainfall by one or more hours. During the same 10-hour period covered by the data in Table 2, 377 cubic feet of channel runoff were generated between weirs WC-1 and WC-2. Similar results were obtained in other autumn storms.

Another characteristic of the stormflow from this slope was its lack of sensitivity to changes in rainfall intensity, despite the sensitivity of streamflow at the base of the slope. Figure 4 presents a summary of data on rainfall intensity, water-table elevations, and runoff for the 0.70-inch storm of October 18, 1967. This storm occurred on soils that were extremely wet because of irrigations and heavy rains during the preceding two weeks. The water-table depths in wells C, D, and E before the storm were 1.3 feet, 1.35 feet, and 2.7 feet respectively. In spite of the proximity of the water table to the ground surface, only in well D did the water table begin to respond during the first hour of

TABLE 2. Amounts of Rainfall and Stormflow, and the Time Lags between the Beginning of Rainfall and the Initial Response of Flow from each of the Tiles on the Trenched Slope during the 1.35-inch Rainstorm of October 10–11, 1967

	Stormflow,* cubic feet	Stormflow,* inches	Time Lag, minutes
Convex hillside			
Plot 1			
Rainfall	628.	1.35	
Channel 1-1	1.7	0.004	36
Tile 1-2	2.8	0.006	204
Tile 1-3	33.4	0.072	80
Concave hillside			
Plot 2			
Rainfall	1470.	1.35	
Channel 2-1	8.0	0.007	54
Tile 2-2	15.2	0.014	195
Tile 2-3	12.0	0.012	142
Straight hillside			
Plot 3			
Rainfall	679.	1.35	
Channel 3-1	0	0	
Channel 3-2	0	0	
Channel 3-3	1.7	0.003	260

* Stormflow is defined as total flow minus base flow for the 10-hour period from 1900, October 10 to 0500, October 11.

rainfall. Discharge from tile 2–3 followed the same pattern as that of the water-table elevations. It began to increase after channel runoff had declined and did not show the marked sensitivity to changes of rainfall intensity demonstrated by the flow from weir WC-2. Neither overland flow nor flow from the base of the root zone occurred in this storm. The storm produced only 15 cubic feet of stormflow out of the 1516 cubic feet of rain that fell on the three plots.

Large Artificial Rainstorms

Data from natural rainstorms showed that even the concave plot (2) with its steep slope, shallow impeding layer, and unusually wet soil conditions could not produce significant amounts of subsurface stormflow to supply the responsive channel hydrograph in the vast majority of storms that occur on this area. Artificial storms were therefore produced on plots 2 and 3 to study the processes that would occur in rare large storms. The return period of these storms varied from two to greater than 100 years. It is difficult to judge the return period of the antecedent soil moisture conditions, since only one year of soil moisture and groundwater data are available. Weather during that year was normal, however, and conditions before each storm can be roughly compared with those during the rest of the year. The reactions of the plots were different because of the differences in soils and topographic form. Space does not permit presentation of all the data on which the following description is based. For further details the reader is referred to a report by *Dunne* [1969].

On October 25, 1967, a 1.72-inch storm lasting two hours delivered 1902 cubic feet of water to the concave plot (2) at the rate of 15.8 cfm. The return period of this rainfall in the study area is approximately 25 years [*U. S. Weather Bureau,* 1955]. Because of heavy rains and irrigations in the previous month the soil was as wet as it had been during the snowmelt period, which is the wettest time of the year in this area. Before the storm, depths to the water table at wells C, D, and E were 2.1 feet, 1.6 feet, and 2.2 feet respectively.

Measurements of water-table elevations, piezometric head, and soil moisture [*Dunne,* 1969] showed that within half an hour of the onset of rain a perched water table began to develop in plot 2. The water table rose to the surface over the central depression below well E. Runoff data presented in Figure 5 show the marked difference in response of stormflow from the various horizons of the soil. After rain had been falling for 75 minutes, flow from the lower tile (2–3) began to increase slowly from its initial rate of 0.023 cfm to 0.319 cfm two hours after the end of rainfall. Flow from the base of the root zone (tile 2–2) began at the same time as the initial response of tile 2–3. Discharge increased abruptly to a maximum of 0.776 cfm at the end of rainfall and declined more slowly. The flow from the root zone was associated with the rapid rise of the water table into the permeable, sandy horizon at the base of the slope.

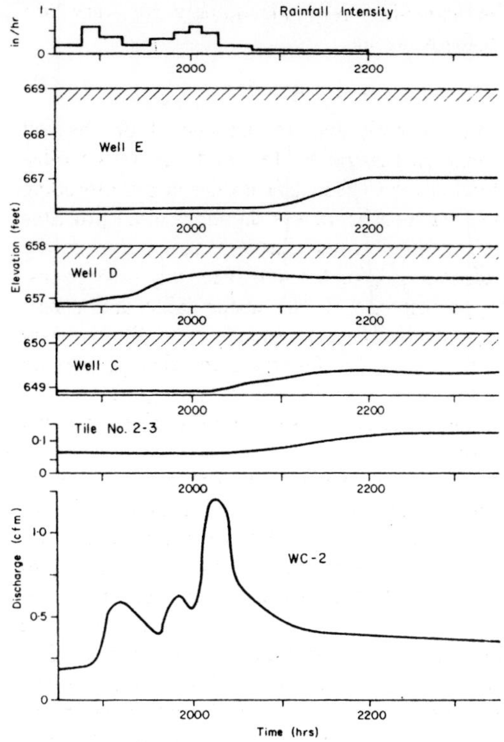

Fig. 4. Rainfall, changes in water-table elevation, and runoff from tile 2–3 and weir WC-2 during and after the 0.70-inch rainstorm of October 18, 1967.

A small trickle of water, deposited directly in the channel by the sprinklers flowed from the surface channel (2–1) during the first 65 minutes of the storm. There was, however, no gene-

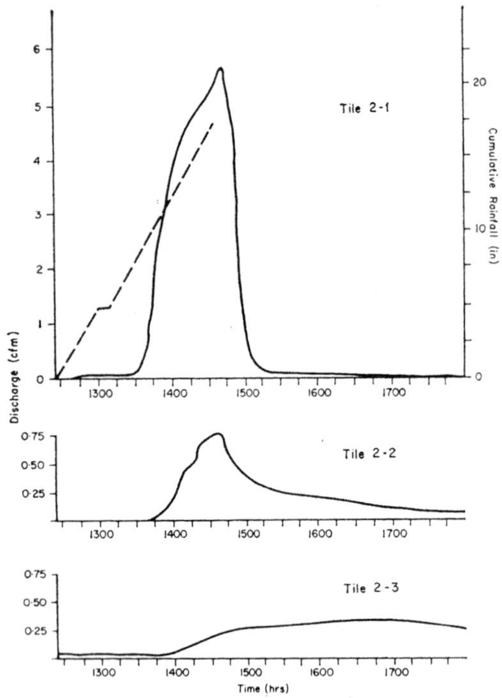

Fig. 5. Discharge from the three tiles draining plot 2 (concave) during and after the artificial storm of October 25, 1967.

ration of Hortonian 'precipitation excess' on the slope as the rainfall intensity did not exceed the infiltration capacity of the soil to produce overland flow. Flow from channel 2–1 suddenly began to increase 65 minutes into the storm. In the 70 minutes from 1330 to 1440 the flow rate increased from 0.027 cfm to 5.615 cfm, and 270 cubic feet of water flowed from the surface channel. Ten minutes after the end of rainfall, the flow began to decline even more rapidly than it had risen. Water emerged from the soil surface over most of the central depression of the plot. The seepage was particularly obvious below slight breaks of slope where the sandy loam thinned.

The water that was measured as overland flow was a combination of 'return flow' from the upper horizon of the soil and of rain that had fallen directly onto the saturated area that developed over a small portion of the hillside in the central concavity of the plot where the water table had reached the ground surface. An attempt was made to estimate the contributions to surface runoff from the two sources mentioned. The area over which water was seen flowing at the ground surface at the end of the storm was mapped. The rate of rainfall onto this area was calculated to yield the contribution of direct precipitation. By the end of the storm 1.364 cubic feet of runoff per minute (24.3% of the peak surface runoff rate) were being produced in this way, while 4.264 cfm were produced by water that had emerged from the soil surface in the central hollow of the concave slope. It is probable that the area covered by overland flow was underestimated, since it was difficult to see some of the thin films of overland flow around the margins of the saturated area. Apart from the possibility of this error, however, the data give values for the separate contributions that are reasonable in the light of similar calculations made for one of the natural storms that produced overland flow (see later).

In the 2-hour period of the artificial storm, plot 3 received 1.94 inches (1096 cubic feet) of water. The return period of a rainfall of this magnitude in the Sleepers River Watershed is approximately 50 years [*U. S. Weather Bureau*, 1955]. In the absence of an impeding layer, a perched water table did not form. The water table varied in depth from 4 to 7 feet in the lower half of the plot and percolation and displacement of water in the unsaturated zone were so slow that the water table did not begin to rise until the end of rainfall. It never reached the ground surface and so no significant stormflow was produced. The only source of runoff was that from the lower tile, which rose to a peak of 0.143 cfm nine hours after the end of the storm.

Overland flow from 8% of the irrigated area completely dominated the pattern of runoff production on the hillside during the 1.72-inch artificial storm, as Figures 6 and 7 show. Although some of the flow had subsurface origins, and may be classified as 'interflow' according to the definition of *Linsley et al.* [1949], the fact that it ran over the surface was of major impor-
and calculation of velocities of subsurface flow
tance. Measurement of overland flow velocities showed that on emerging from the soil surface, the speed of the water increased by a factor of 100 to 500. This allowed a much larger quantity of water to be supplied to the stream channel than could have been removed by subsur-

face stormflow. The establishment of surface runoff on this and other slopes (during other rainfalls), in effect, caused an increase in the drainage density of the watershed which drastically reduced the distance which water had to travel through the soil before it reached the surface. Since velocities of subsurface flow are so slow, even on slopes as steep and permeable as the one described in this report, the distance that water must travel through the soil is an important control of storm runoff production. If the rain had been insufficient to raise the water table to the surface, little storm runoff would have been produced on this slope. Even when the permeable A-horizon was saturated throughout its depth, it could only convey runoff to the channel at the rate of 0.016 cfm per foot of channel. This figure is small by comparison with actual rates of channel runoff production in this 60-acre watershed. Any increase in the efficiency of the evacuation of water would therefore have a large effect on runoff production.

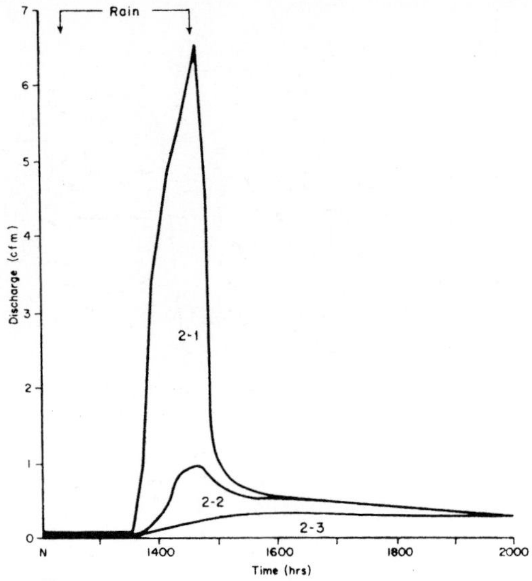

Fig. 6. Combined hydrograph of flows from plot 2 (concave) during the artificial storm of October 25, 1967.

On October 16, 1967, a 1.14-inch artificial storm lasting two hours with a return period of three years yielded similar results. Figure 8 shows that overland flow dominated the storm hydrograph. Soil moisture profiles and water-table elevations showed that the processes were the same as those observed on October 25. The differences in timing and in peak flow rates could be explained in terms of differences of antecedent conditions and rainfall intensity. A 2.41-inch storm lasting two hours was produced on July 17, 1968. Although the return period of this storm is greater than 100 years and antecedent moisture conditions were wetter than normal for early summer, very little storm runoff was produced. A perched water table developed in plot 2, but only intersected the ground surface over a small area near the base of the concavity. By the end of the storm the rate of surface runoff from channel 2–1 was rising rapidly, but the peak rate was only 0.802 cfm. The slope seemed to have just attained the threshold condition at which further rainfall would produce large amounts of runoff. A natural storm occurring immediately after the irrigation confirmed this. On October 5, 1967, an artificial storm occurred in which the important threshold of saturation was not reached. The 1.03-inch storm lasted for one hour on soil conditions that were wetter than normal for early autumn. The amounts of stormflow produced by this and the other artificial storms described are listed in Table 3.

Natural Storms Following Artificial Storms

Two of the artificial storms described previously were followed within a short time by large natural storms. These storms were of shorter duration and higher intensity than the artificial storms, and they occurred on antecedent conditions that were much wetter than those normally encountered by the intense rains of summer. Figure 9 shows the response of runoff at various levels in plot 2 during a 0.88-inch rain that occurred 10 hours after the artificial storm of October 25, 1967. Runoff rates and water-table elevations were receding after the artificial storm, but well readings showed that the water table rose to the surface over the central concavity before the main burst of rain. Overland flow again dominated the hydrograph (Table 4).

One-half hour after the end of the 2.41-inch artificial storm of July 17, 1968, 1.83 inches of rain fell in two bursts; one burst of 3.15 inches per hour lasted for 21 minutes, and the second burst of 2.75 inches per hour lasted for 13 minutes (Figure 10). The return period of this

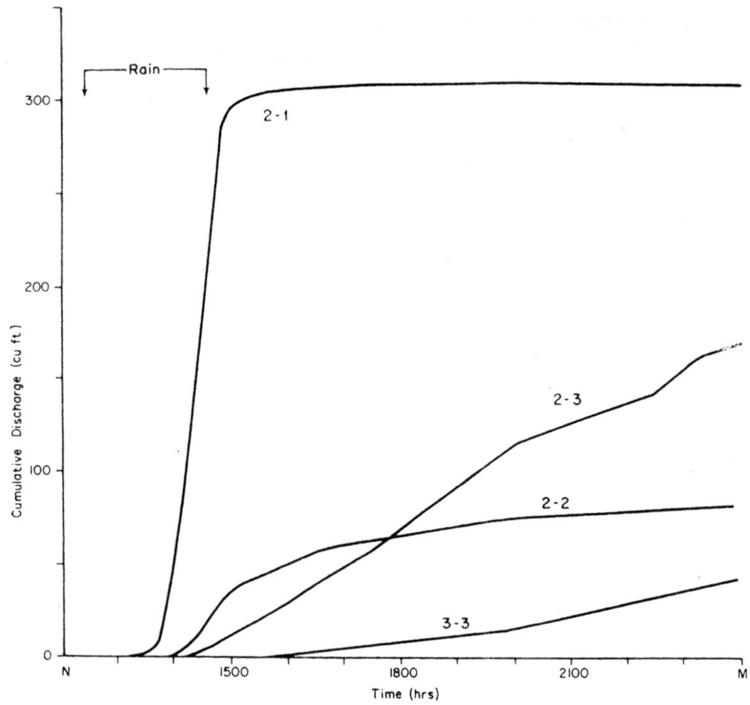

Fig. 7. Cumulative flow for the tiles producing runoff during and after the storm of October 25, 1967.

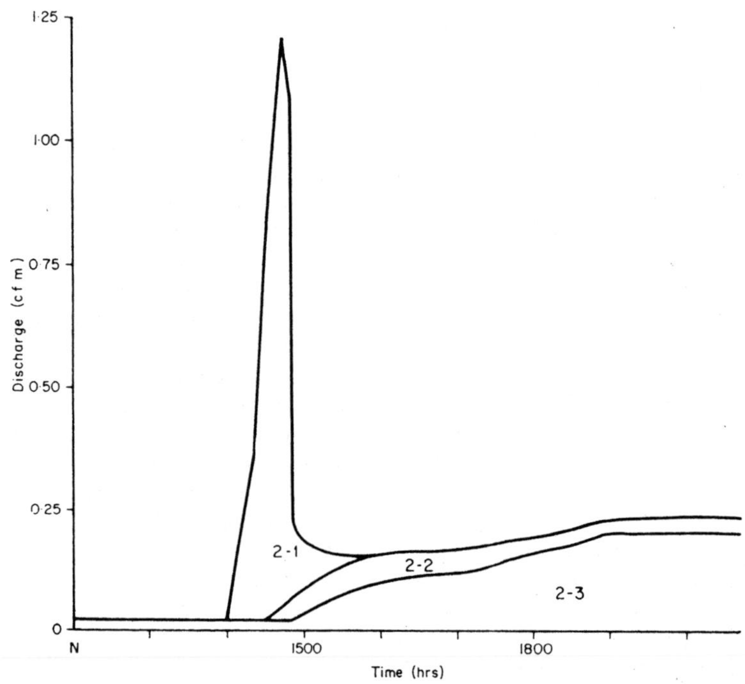

Fig. 8. Combined hydrograph of runoff from plot 2 during the storm of October 16, 1967.

TABLE 3. Amounts of Stormflow in Cubic Feet Produced on Plots 2 and 3 in Artificial Rainstorms*

Source	October 5, 1967	October 16, 1967	October 25, 1967	July 17, 1968
Plot 2				
Rainfall†				
Runoff	1139	1261	1902	1462
Channel 2-1	0	32	309	24
Tile 2-2	0	14	78	?
Tile 2-3	11	46	123	2
Total	11	92	510	26
Plot 3				
Rainfall†				
Runoff	494	536	1096	1109
Channel 3-1	0	0	0	0
Tile 3-2	0	0	0	0
Tile 3-3	1	6	21	0
Total	1	6	21	0

* Stormflow is defined as total flow minus base flow for the period from the beginning of rainfall until six hours after the end of rainfall in the case of the October storms. The stormflow period for the July storm is from the beginning of rainfall until one-half hour after the end of rainfall, when a large natural rainstorm began.

† See text for amounts of rainfall expressed in inches.

storm is approximately 50 years. If the natural and artificial storms are considered as one event, the return period for the combined storm is several hundred years [*U. S. Weather Bureau*, 1955]. Antecedent conditions were abnormally wet for an intense summer storm. The natural rainstorm reinforced the effect of the preceding irrigation on the outflow from tile 2–3. Total stormflow from the lower part of the soil profile, however, was still negligible. Tile 2–2 was leaking during this storm so that no runoff could be collected. Previous irrigations when the root zone was fully saturated, however, had shown that a peak discharge rate of no more than 0.6 cfm to 0.8 cfm could be expected from tile 2–2.

Surface runoff from plot 2 reached the highest level recorded during the study, increasing from 0.060 cfm at the beginning of the storm to 14.305 cfm at the end of the first intense pulse of rain 30 minutes later (Figure 10). After a rapid recession, runoff increased from 0.802 cfm to 12.934 cfm in the second burst of rain. Overland flow was again produced by a combination of return flow and direct precipitation onto the saturated area of the hillside. Mapping of the saturated area allowed a rough calculation of the contributions from both sources. At the end of the first period of intense rain the approximate rate of runoff produced by direct precipitation was 8.289 cfm, while return flow contributed runoff at the rate of 6.016 cfm. By the end of the second burst of rain the contributions were 7.220 cfm from direct precipitation and 5.714 cfm from return flow. The contribution from direct precipitation was controlled by the intensity of the rain and the area of the soil that had become saturated. As this saturated area expanded throughout the storm, the area of the slope over which water could emerge from the ground surface and run quickly downslope increased, resulting in an increase of the contribution to the hydrograph from this source.

In both natural storms, runoff from plot 3 was negligible. Water-table elevations were 5 to 10 feet below the ground surface, and although the water table and piezometric levels rose sharply towards the end of each storm, little subsurface stormflow was measured. The data demonstrated the extreme damping effect of unsaturated percolation and of storage and transmission in the groundwater body. Overland flow did not occur because the water table did not rise to the ground surface. Consequently only a few cubic feet of stormflow were produced.

DISCUSSION

The data collected in this Vermont watershed do not support some of the claims that have

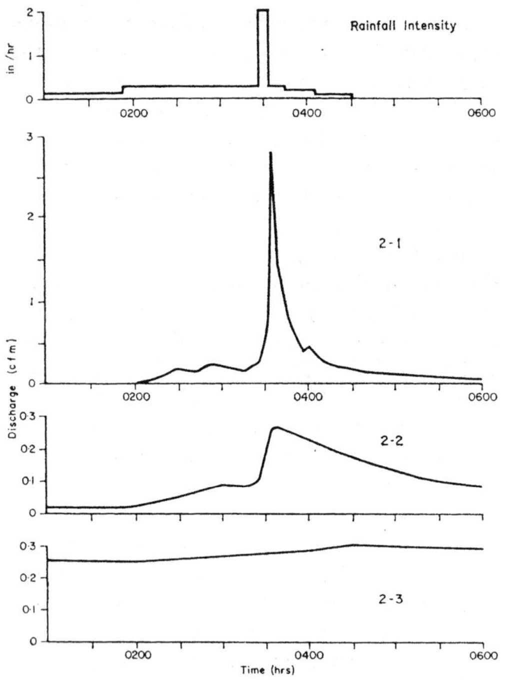

Fig. 9. Rainfall and runoff data from plot 2 for the 0.88-inch natural storm of October 26, 1967.

been made in the recent literature on subsurface runoff production in humid areas. Subsurface stormflow, as defined by the writers quoted in the introduction to this paper, was not an important contributor to the storm hydrograph in this watershed, despite soil conditions that are generally considered ideal for such a mechanism. A similar study carried out on a shallow, poorly drained soil in the same catchment [*Dunne,* 1969] and in another basin in the same area confirmed this conclusion.

On the hillside described in this report, subsurface stormflow produced virtually no runoff during natural storms, in spite of the fact that in the autumn of 1967 the water table was only one or two feet below the ground surface along the central concavity. Even when the water table was so close to the surface, there was a considerable volume of pore space available for storage of rainfall, and very little soil water was displaced into the stream. Storms that generally occur in this region are not of sufficient magnitude to fill this storage capacity and contribute to the water table at rates sufficient to produce large amounts of storm runoff. Even extreme events such as the 3.81-inch rainfall of July 17, 1968, failed to produce stormflow from the straight hillside. On the concave plot, however, the combination of a shallower water table and an impeding layer did allow the generation of subsurface stormflow during irrigations. The runoff produced, however, was remarkable more for its small magnitude and late response than for its importance to total storm runoff. The only runoff-producing mechanism which approached channel runoff in amount and sensitivity was overland flow on a small concave portion of the hillside that comprised 5% to 10% of the total area of plots 2 and 3. This runoff was produced by a combination of return flow and direct precipitation onto the saturated area. Only when water was released from the extreme damping effect of unsaturated and saturated subsurface percolation by breaking through the ground surface could it contribute to channel runoff at a significant rate.

The data on the relative unimportance of subsurface stormflow from the hillside are surprising in view of some recent papers on this subject. *Whipkey* [1965] measured subsurface flow from a 36-inch-deep sandy loam forest soil in Ohio. Although Whipkey did not claim as much, the results he obtained have been accepted as proving that subsurface stormflow is a major contributor to stream hydrographs.

TABLE 4. Amounts of Stormflow in Cubic Feet Produced on Plots 2 and 3 by Natural Rainstorms Following Artificial Rainstorms*

	October 26, 1967	July 17, 1968
Plot 2		
Rainfall†	973	2023
Channel 2-1	64	346
Tile 2-2	46	?
Tile 2-3	45	40
Total	155	386
Plot 3		
Rainfall†	511	1114
Channel 3-1	0	19
Tile 3-2	6	43
Tile 3-3	37	25
Total	43	87

* Stormflow is defined as total flow minus base flow for the period from the beginning of rainfall until six hours after the end of rainfall.

† See text for amounts of rainfall expressed in inches.

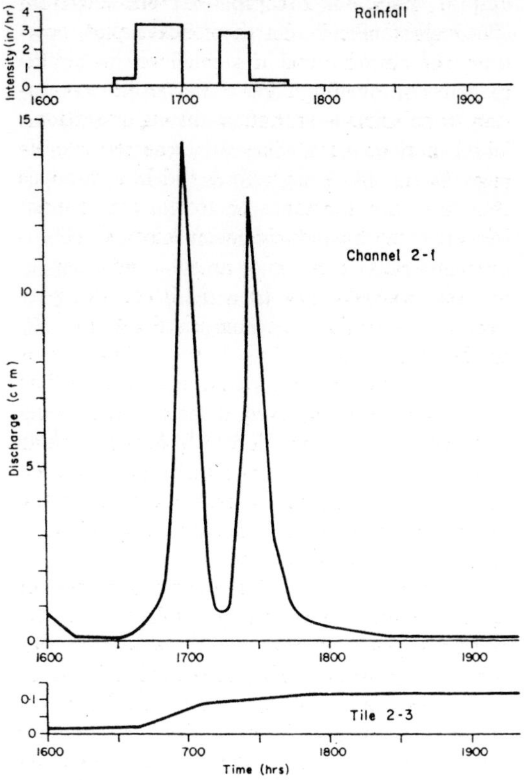

Fig. 10. Rainfall and runoff data for plot 2 (concave) in the natural storm of July 17, 1968.

Kirkby and Chorley [1967, p. 8], for example, claim that Whipkey's data show that '. . . throughflow is capable of producing runoff peaks in river hydrographs.' A closer look at Whipkey's data and comparison with results from the present study throw considerable doubt on the role of subsurface stormflow from hillsides as a major contributor to hydrograph peaks. The available evidence, in fact, suggests that water remaining below the soil surface on its way down hillsides to a stream contributes only a small part of the water supplying the peak of a river hydrograph.

No data are available on the behavior of the stream at the base of the slope Whipkey studied. Nor are any data available on runoff produced on the slope during natural storms. No indication is given in his paper of the extent to which his 'wet' and 'dry' antecedent conditions or the magnitude of his artificial rainfalls represent conditions under which river hydrographs are formed in that region. The area, however, is not too different from the present experimental area to invalidate comparisons. The hillside described in the present paper proved to be more favorable for subsurface stormflow than the one Whipkey studied. The slope used in the present study was much steeper and the impeding layer of the concave plot was generally at a shallower depth. Although under pasture at the time of the study, the slope had been covered by pine woodland 30 years before, and the A-horizon of the soil had an open structure interlaced with root holes and worm holes. The saturated hydraulic conductivity of the A-horizon (estimated from field data by the same method Whipkey used) was greater (13.5 inches per hour) than the one he calculated for the Ohio slope (11.3 inches per hour).

The total amount, peak rate of flow, and responsiveness of subsurface stormflow from the concave slope described in the present study were equal to or greater than those described by Whipkey. On the Ohio slope, a 4-inch storm with an intensity of two inches per hour occurring on a soil profile that had drained from saturation only a few days previously, did not begin to yield runoff until one-half hour after the end of rainfall. The peak flow rate from this storm was 0.016 cfm per foot of hillside. Total runoff in the 24 hours following this storm amounted to only 10% of applied rainfall. On the concave plot used in the present study, a 2-hour long storm of only 1.72 inches also resulted in a peak flow rate of 0.016 cfm per foot of hillside. Ten percent of rainfall left the hillside in the 6-hour period following the end of the storm. The delay before the onset of this flow was smaller than in the example given by Whipkey.

The comparison of results listed above shows that subsurface stormflow on plot 2 in the present study was larger and more responsive than the study described by Whipkey. Comparison of this runoff with channel runoff from the 60-acre watershed, however, showed that subsurface flow from the hillside was a relatively small and rare contributor to the stream hydrograph. The reaction of flow from the trenched hillside during the fall of 1967 when the water table was only 1.3 feet below the soil surface, suggests that even at the base of a hillside where the water table is close to the surface, reaction of the saturated zone was slow by comparison with that of streamflow. Channel runoff data incorporated into another report [*Dunne*, 1969] confirm this.

Significant amounts of stormflow were produced on the hillside by return flow and by direct precipitation onto the saturated area that developed over a small portion of the hillside where the water table reached the ground surface. Although the return flow had the same origin as the flow that remained below the soil surface, its emergence from the ground caused it to have features that were lacking in its subsurface counterpart. The first of these was its greater velocity, which was 100 to 500 times greater than the calculated velocities of subsurface flow. This allowed more water to travel from a larger contributing area in the time available than did the velocity of subsurface flow. The second important feature of the return flow was its greater sensitivity to fluctuations of rainfall intensity, which could be accounted for by the short distance of flow beneath the soil. Only when the water table intersected the soil surface and established conditions suitable for overland flow could large areas of the slope supply water to the channel without the impedance of many feet of subsurface flow. The generation of stormflow by direct precipitation onto the saturated area was also sensitive to fluctuations of rainfall intensity and to the duration of the storm. At the end of rainfall, water drained from the surface and upper few inches of the soil within minutes, and the rate of runoff was drastically reduced. A similar process was observed during artificial storms on an area of shallow, poorly drained soils in the same watershed. In these soils subsurface stormflow contributed between 1% and 5% of the peak runoff rate, while direct precipitation and return flow each contributed almost 50% of the peak rate.

CONCLUSIONS

The findings of this study are in general agreement with the 'partial-area' concept of storm runoff production [*Betson*, 1964; *Ragan*, 1967]. Significant amounts of storm runoff were only produced on small areas of hillside where the water table reached the surface. Subsurface stormflow occurred in large storms but was not an important contributor to total storm runoff despite conditions favorable for its existence. The importance of an area of hillside as a producer of storm runoff depended on its ability to generate overland flow. This latter was the only form of runoff that occurred in sufficient amounts to make a substantial contribution to channel runoff during storms. Furthermore, it was the only form of runoff which exhibited sensitivity to changes of rainfall intensity in a manner similar to channel runoff. A comparison of data from the two irrigated plots also confirms the 'partial-area' concept of storm runoff production.

Acknowledgment. This study was made while the authors were employed at the USDA Sleepers River Experimental Watershed, Danville, Vermont.

REFERENCES

Betson, R. P., What is watershed runoff?, *J. Geophys. Res.*, *69*(8), 1541–1551, 1964.

Dunne, Thomas, Runoff production in a humid area, Ph.D. dissertation, Johns Hopkins University, Baltimore, Maryland, 1969.

Hack, J. T., and J. C. Goodlett, Geomorphology and forest ecology of a mountain region in the Central Appalachians, 66 pp., *U. S. Geol. Survey Prof. Pap. 347*, 1960.

Hewlett, J. D., and A. R. Hibbert, Factors affecting the response of small watersheds to precipitation in humid areas, *Proc. Int. Symp. Forest Hydrol.*, Pennsylvania State University, University Park, 1965.

Horton, J. H., and R. H. Hawkins, Flow path of rain from the soil surface to the water table, *Soil Sci.*, *100*(6), 377–383, 1965.

Horton, R. E., Erosional development of streams and their drainage basins, *Geol. Soc. Amer. Bull.*, *56*, 275–370, 1945.

Kirkby, M. J., and R. J. Chorley, Overland flow, throughflow and erosion, *Int. Assoc. Sci. Hydrol. Bull.*, *12*(2), 5–21, 1967.

Linsley, R. K., M. A. Kohler, and J. H. L. Paulhus, *Applied Hydrology*, 689 pp., McGraw-Hill, New York, 1949.

Ragan, R. M., An experimental investigation of partial area contributions, Berne Symposium, International Association of Science and Hydrology, 1967.

Troeh, F. R., Landform parameters correlated to soil drainage, *Soil Sci. Soc. Amer. Proc.*, *28*, 808–812, 1964.

U. S. Weather Bureau, Rainfall intensity-duration-frequency curves for selected stations in the United States, Alaska, Hawaiian Islands and Puerto Rico, 53 pp., *Tech. Pap. 25*, 1955.

Whipkey, R. Z., Subsurface stormflow from forested slopes, *Int. Ass. Sci. Hydrol. Bull.*, *10*(2), 74–85, 1965.

(Manuscript received August 6, 1969; revised December 29, 1969.)

Commentary on Paper B6

Abdul, A. S. & Gillham, R. W. (1984) Laboratory studies of the effects of the capillary fringe on streamflow generation.

Like Pilgrim *et al.* (1979, Paper C3) and McDonnell (1990, Paper C5), the study by Abdul and Gillham also seeks to resolve the paradox in hillslope hydrology (Kirchner, 2003) whereby the runoff response of hillslopes is rapid but most of the water involved is "old". The paper describes laboratory experiments designed to investigate groundwater–surface water interactions during the process of stormflow generation. It complements the paper by Gillham (1984; see also the review by Beven, 2006) which reports field studies of the same capillary fringe phenomenon.

Ragan (1968) first demonstrated the rapid development of groundwater ridges adjacent to the stream in response to precipitation inputs, generating a rapid increase in groundwater discharge to the stream. Since then, many studies have invoked the capillary-fringe mechanism to account for the large and rapid contribution of pre-event water to the initial stormflow response, starting with Sklash & Farvolven (1979). Interestingly, another benchmark paper in hillslope hydrology, Dunne & Black (1970, Paper B5), also recognised the importance of high water tables in the near-stream zone. In their case, overland flow completely dominated the runoff response; the subsurface stormflow was delayed and subdued compared to the overland flow response. Rapidly rising water tables were a feature of the concave slope section where flow convergence downslope favoured high soil moisture contents and sustained a relatively high water table between rainfall events (Anderson & Burt, 1978).

The height of the capillary fringe above the water table is inversely related to pore size, and so is greater in finer-textured soil. For negative pore water pressures greater (i.e. less negative) than the air intrusion value (also known as the air entry value), the soil remains saturated such that only a very small amount of infiltrating water is required to raise the water table significantly; pore water pressures become positive, the hydraulic gradient is now towards the stream, and pre-event water drains out. Put another way, soil water under tension is rapidly converted to phreatic water (McDonnell & Buttle, 1998). Abdul and Gillham used a laboratory model to investigate the capillary-fringe response to rainfall (from a simulator). The experimental slope, 1.4 m long, was packed with medium-fine sand and set at a 12° slope. Three initial positions of the pre-event water table allowed the influence of the capillary fringe on runoff generation to be explored; in the fourth and fifth experiments, tracers were used to allow the contributions of event and pre-event water to the outflow hydrograph to be estimated. The study showed that the capillary fringe could affect storm runoff generation in two different ways: (1) the rapidly rising water table steepens local hydraulic gradients, causing an immediate increase in groundwater discharge; (2) at the slope foot, surface saturation resulted in overland flow generation, by a mixture of return flow (pre-event water) and direct runoff (rainfall unable to infiltrate). Thus, the mechanism could lead to both event and pre-event water being rapidly transferred to the stream, depending on local conditions (slope, hydraulic conductivity, antecedent conditions) and rainfall intensity.

To further evaluate their interpretation, Abdul and Gillham used a 2-D flow model to simulate the hydraulic response of the two tracer experiments. The flow net for steady state conditions shows clearly the way in which the zone of infiltration across most of the slope is replaced by a seepage face in the lowest part of slope. All the rainfall falling on the seepage face would become overland flow, adding to the exfiltrating water. Upslope, some water would infiltrate and some would become overland flow.

Both the experimental and simulated results show that if the zone of tension saturation extends to the ground surface, or close to it, the application of only a small amount of water can cause an immediate and dramatic rise in the water table. This generates groundwater discharge from the near-stream zone straight away, pre-event water draining into the stream. Later, when the seepage face develops, overland flow will include event water too.

Abdul & Gillham (1984) is regarded as the main "proof of concept" for the capillary fringe groundwater ridging hypothesis (Cloke *et al.*, 2006). The hypothesis has generated a good deal of debate as to whether the mechanism is widely applicable (McDonnell, 1990, Paper C5; McDonnell & Buttle, 1998). Cloke *et al.* (2006) used the ESTEL-2D computer simulation model to simulate the Abdul and Gillham laboratory slope experiment; they then extended the modelling framework to test the applicability of the groundwater ridging mechanism under a wide range of conditions. They concluded that groundwater ridges can indeed develop from the capillary fringe, but this does not necessarily lead to a high proportion of pre-event water discharge, which is mainly controlled by water table height and saturated hydraulic conductivity. Indeed, because field studies have demonstrated much higher proportions of pre-event water than they could model, Cloke *et al.* (2006) conclude that a much wider range of configurations needs to be both modelled and studied in the field, in particular hillslope topography and vegetational effects. As Cloke *et al.* (2006) note, riparian zones are complicated hydrological systems and our knowledge of the riparian spectrum remains incomplete. The paper by Abdul and Gillham may relate to a simplified situation but it has directed our attention to the dynamics of hydrological processes in the near-stream zone and has certainly generated a good deal of subsequent debate!

TPB

References

Anderson, M. G. & Burt, T. P. (1978) The role of topography in controlling throughflow generation. *Earth Surf. Processes* **29**, 331–334.

Beven, K. J. (2006) Commentary on paper 31. Gillham, R. W. (1984) The effect of the capillary fringe on water-table response. In: *Streamflow Generation Processes* (ed. by K. J. Beven), 412–413. Benchmark Papers in Hydrology. IAHS Press, Wallingford, UK.

Cloke, H. L., Anderson, M. G., McDonnell, J. J. & Renaud, J.-P. (2006) Using numerical modelling to evaluate the capillary fringe groundwater ridging hypothesis of streamflow generation. *J. Hydrol.* **316**, 141–162.

Dunne, T. & Black, R. D. (1970) An experimental investigation of runoff production in permeable soils. *Water Resour. Res.* **6**, 478–490.

Gillham, R. W. (1984) The effect of the capillary fringe on water-table response. *J. Hydrol.* **67**, 304–324.

Kirchner, J. W. (2003) A double paradox in catchment hydrology and geochemistry. *Hydrol. Processes* **17**, 871–874.

McDonnell, J. J. (1990) A rationale for old water discharge throug macropores in a steep, humid catchment. *Water Resour. Res.* **26** 2821–2832.

McDonnell, J. J. & Buttle, J. M. (1998) Comment on "A deterministic empirical model of the effect of the capillary-fringe on near stream runoff. 1. Description of the model" by Jayatilaka, C. J. & Gillham, R. W. (*J. Hydrol.* **184**(1996) 299–315). *J. Hydrol.* **207** 280–285.

Pilgrim, D. H, Huff, D. D. & Steele, T. D. (1979) Use of specifi conductance and contact time relations for separating flov components of storm runoff. *Water Resour. Res.* **15**, 329–339.

Ragan, R. M. (1968) An experimental investigation of partial-are contributions. In: *Hydrological Aspects of the Utilization o Water.* IAHS Publ. 76, 241–249. IAHS Press, Wallingford, UK.

Sklash, M. G. & Farvolden, R. N. (1979) The role of groundwater in storm runoff. *J. Hydrol.* **43**, 45–65.

WATER RESOURCES RESEARCH, VOL. 20, NO. 6, PAGES 691–698, JUNE 1984

Laboratory Studies of the Effects of the Capillary Fringe on Streamflow Generation

A. S. ABDUL AND R. W. GILLHAM

Department of Earth Sciences, University of Waterloo

This paper describes laboratory experiments that were conducted to investigate the groundwater–surface water interactions during the process of streamflow generation. In particular, the experiments were designed to examine the role of the capillary fringe in the runoff processes and its influence on the groundwater response to precipitation. The experimental results clearly show that if the zone of tension saturation extends to, or near, ground surface, the application of a small amount of water can cause an immediate rise in the water table, and the magnitude of the rise is much greater than would be expected on the basis of normal specific yield values for sandy materials. Because of the sloping surface of the laboratory model the rising water table caused the rapid generation of hydraulic gradients directed towards the toe of the slope (the stream) resulting in the immediate generation of groundwater discharge to the stream. Two experiments using chloride as a tracer and having rainfall rates of 4.3 and 1.9 cm h^{-1} showed that the discharge of preevent water to the stream preceded event water and that at early times, preevent water was the predominant component of streamflow. In addition, the results of the two tracer experiments show that the significance of preevent water increases with decreasing rainfall rate and that for the 1.9 cm h^{-1} rainfall rate preevent water was the main component of the stream hydrograph. The proportions of preevent and event water in the streamflow after steady conditions were reached were determined largely by the extent of the seepage face. This in turn would be determined by the hydraulic conductivity of the medium, the surface slope, and the rainfall intensity.

INTRODUCTION

The mechanism of streamflow generation has been a controversial topic in hydrology for several decades, with a major source of discussion concerning the relative abundance of surface water and subsurface water in storm hydrographs. In that numerous empirically based models are available for predicting peak flows and total discharge during runoff events, there has been a tendency to view studies of runoff mechanisms to be of greater academic interest than of practical importance. However, as interests shift to the development of models for predicting runoff quality, questions concerning the mechanisms and sources of runoff become increasingly important.

Mechanisms of streamflow generation commonly referred to in the literature include overland flow [*Horton*, 1933], the partial area contribution concept [*Betson*, 1964; *Ragan*, 1968], the variable source area–overland flow concept [*Dunne and Black*, 1970*a*, *b*], and the variable source area–subsurface flow concept [*Hewlett and Hibbert*, 1967]. Hortonian overland flow considers streamflow to be generated as a result of surface runoff from areally extensive portions of a watershed, while the latter mechanisms consider streamflow generation to be controlled by a relatively small proportion (10% or less) of the watershed area. However, the mechanisms listed above, with the exception of the variable source area–subsurface flow concept, have a common characteristic in that they associate the major component of stormflow hydrographs with water of the precipitation event (event water) rather than with water that was in the basin prior to the precipitation event (preevent water). By the variable source area–subsurface flow concept, stormflow is generated through the near stream wetland area by a subsurface translatory flow process which involves the rapid displacement of preevent water by event water. In addition to the references given above, useful reviews of the mechanisms of streamflow generation are given by *Freeze* [1972, 1974].

The use of natural tracers, particularly environmental isotopes such as tritium and oxygen-18, for separating streamflow into "event" and "preevent" waters has raised serious questions concerning the classical concepts of streamflow generation. In particular, in humid regions, tracer studies generally indicate that preevent water constitutes a much larger proportion of storm hydrographs than would be expected on the basis of the generally accepted mechanisms of streamflow generation [*Dincer et al.*, 1970; *Martinec et al.*, 1974; *Martinec*, 1975; *Fritz et al.*, 1976; *Sklash and Farvolden*, 1979]. On the other hand, the results of the tracer studies and the variable source area–subsurface flow concept have been viewed with some degree of skepticism in that a mechanism that could cause the necessarily rapid and large increase in groundwater discharge in response to precipitation events has not been evident.

More recently, *Sklash and Farvolden* [1979] and *Gillham* [1984] have suggested that the capillary fringe could contribute to the rapid groundwater response to precipitation. The capillary fringe is the zone above the water table that remains saturated under negative pressure. Because this zone has little or no storage capacity, if it extends to ground surface, the application of a relatively small amount of water can cause a large and rapid rise in the water table. The water table rise could in turn result in a rapid increase in hydraulic gradient toward the stream and thus a rapid increase in groundwater discharge to the stream. The mechanism is represented schematically in Figure 1. The physical basis of the capillary-fringe effect is described in detail by *Gillham* [1984].

Although the capillary fringe mechanism is consistent with the results of tracer studies of storm flow events and can be argued on the basis of physical principles, there is little direct evidence of the role it plays in the runoff process. This paper describes laboratory experiments that were conducted to investigate the groundwater–surface water interactions during the process of streamflow generation. In particular, the experi-

Paper number 4W0235.
0043-1397/84/004W-0235$05.00

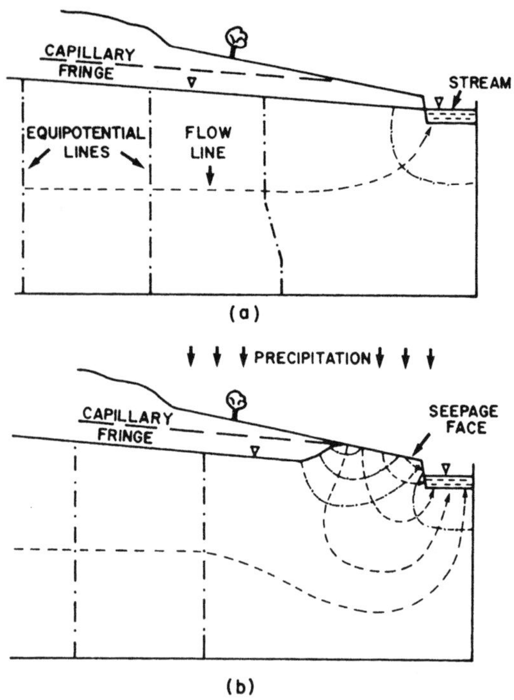

Fig. 1. Schematic showing the proposed effect of the capillary fringe on the discharge of groundwater to surface waters. (*a*) Prior to rainfall. (*b*) Shortly after the onset of rainfall [from *Gillham*, 1984].

ments were designed to examine the role of the capillary fringe in runoff generation processes.

METHODS OF INVESTIGATION

The experimental model consisted of a plexiglass box 140 cm long, 8 cm wide, and 120 cm high (Figure 2). The model was packed with medium-fine sand (U.S. Department of Agriculture classification system) with a uniform 12° slope on the upper surface. A screened tube was inserted at the toe of the slope to drain off free water. This water was considered to represent streamflow.

Six tensiometers, numbered T-1 through T-6 on Figure 2, were positioned in the face of the box for monitoring changes in hydraulic head. The tensiometers were constructed of 1.25-cm OD plexiglass tubing. Gelman Acropor filter membrane with a pore size of 8.0 microns was glued to the end in contact with the soil, and the opposite end was fitted with a plexiglass plug containing two stainless steel ports. One port was used to flush air from the hydraulic system, while the other was connected to a pressure transducer. Each transducer was connected to a stripchart recorder through a voltage amplifier. Thus the monitoring system gave a continuous record of hydraulic head at each of the six monitoring locations.

Rainfall was simulated by a rainulator designed to fit over the top of the model. The bottom of the rainulator was fitted with 0.56-mm ID stainless steel tubes on a 2.6-cm^2 grid. The head of water in the rainulator, and thus the rainfall intensity, was controlled by an adjustable overflow tube.

The sand used in the model was collected from the Perch Lake Basin of Atomic Energy Canada Ltd. at Chalk River, Ontario. The water content–pressure head relation for the sand (Figure 3) was determined by the hanging water column method using the modified Harries apparatus [*Day et al.*, 1967]. The sand sample was placed in a fritted disc buchner (pyrex) funnel that was hydraulically connected to a fixed horizontal microburette. The main drainage and main wetting curves were determined by applying successive increments of negative pressure to an initially saturated sample and successive increments of less negative pressure to the fully drained sample, respectively. The water content corresponding to each value of negative pressure was calculated from the water content at the previous pressure head and the measured volume of outflow. Referring to the main drainage curve of Figure 3, the air entry value for the sand is about −35 cm; however, water content values close to saturation are maintained to a pressure head of about −50 cm. Thus the zone of tension saturation extends from zero to at least 35 cm of negative pressure. In a hydrogeologic context this indicates that the soil would remain saturated for a distance of about 35 cm above the water table. The saturated hydraulic conductivity of the sand was determined from permeameter tests. Initially, the sand in the permeameter was completely saturated by first flushing the permeameter with carbon dioxide, followed by degassed water. After the initial hydraulic conductivity was determined, the sample was drained and then rewet with water that was in equilibrium with the atmosphere, and the hydraulic conductivity was remeasured. In the latter case the water content would not be at complete saturation because of the presence of isolated gas bubbles that would be entrapped during the wetting process. The hydraulic conductivity determined for the fully saturated condition was 9.0×10^{-5} m s^{-1}, while that determined in the presence of entrapped air was 3.5×10^{-5} m s^{-1}.

Initially, the sand in the laboratory model was slowly saturated from the bottom plate (Figure 2) using deaired water. After water was ponded on the entire upper surface, the water table was lowered to the desired position. During the wetting procedure the wetting front remained reasonably uniform, indicating a relatively homogeneous packing, at least in the horizontal direction.

An individual experiment involved the establishment of the

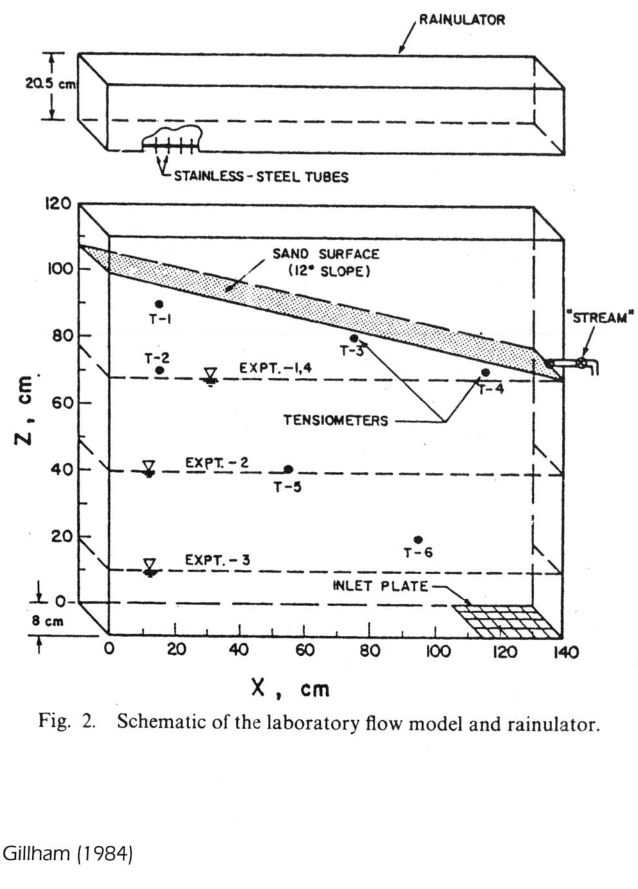

Fig. 2. Schematic of the laboratory flow model and rainulator.

water table at the desired elevation, waiting for hydraulic equilibrium to be established, then initiating rainfall. The hydraulic head and streamflow response were then monitored. Streamflow was determined by measuring the accumulated discharge from the screened tube at the toe of the slope over 30-s intervals. The rainfall rate in experiments 1 through 4 was 4.3 cm h^{-1} or about one third the rewetted saturated hydraulic conductivity.

Five experiments were conducted. The first three were identical except for the initial position of the water table. In experiment 1 the water table was positioned at the toe of the slope (i.e., at the same elevation as the "stream"). In this position, the highest point on the soil surface was about 30 cm above the water table. Thus the triangular section of soil above the water table would be under negative pressure but, based on the drainage curve of Figure 3, would remain saturated. The initial position of the water table in the second experiment was 27 cm beneath the stream (Figure 2). Based on the data from Figure 3, the capillary fringe would extend about 35 cm above the water table and thus the sand would remain saturated in the vicinity of the stream but would exhibit an increasing degree of desaturation to the left of tensiometer T-3. The minimum water content would occur at the soil surface at the left-hand end of the model and would be about 0.3, corresponding to a pressure head of −55 cm. Thus, even with the water table at this depth, the entire medium was at or near saturation.

In experiment 3 the initial water table was positioned at a depth of 57 cm below the stream. The top of the saturated zone (the capillary fringe) would therefore be approximately 22 cm beneath the stream, resulting in the entire soil surface being at a water content less than saturation. Experiments 4 and 5 had the same initial conditions as experiment 1; however, chloride was added as a tracer to the simulated rainfall, and in experiment 5, the rainfall intensity was reduced by a factor of about 2.

RESULTS AND DISCUSSION

Hydraulic Head Response

In experiment 1, with the water table situated at the toe of the slope, the entire flow domain was saturated, although the region above the water table was under negative pressure. Within a few seconds of initiating the rainfall, free water was observed at the toe of the slope and some overland flow was observed over a few centimeters upslope from the stream. As the rainfall continued, the region of overland flow expanded until, by about 3.0 min, there was obvious overland flow for a distance of approximately 80 cm upslope from the stream.

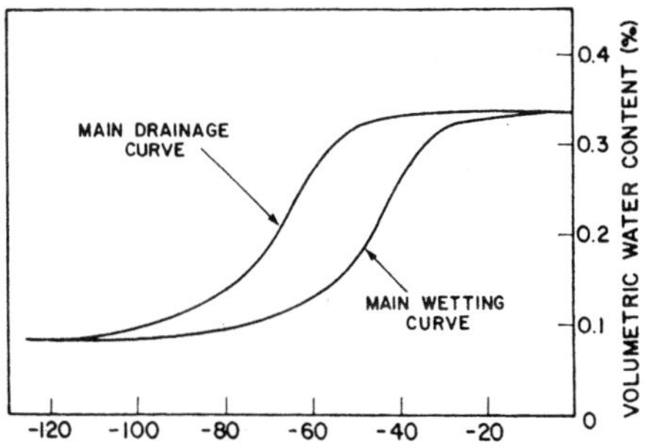

Fig. 3. Water content–pressure head relation measured for the sand used in the laboratory experiments.

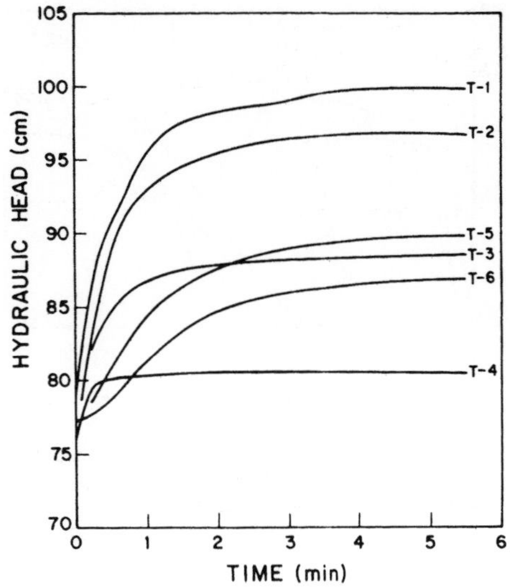

Fig. 4. Hydraulic head response of the six tensiometers for experiment 1.

Drops of rhodamine dye were applied at various points along the surface. For a distance of about 50 cm upslope from the toe, all the dye appeared to be carried away immediately in the surface runoff, while at greater distances an increasing proportion of the dye appeared to be lost to infiltration until, at about 80 to 90 cm upslope, all the dye appeared to infiltrate into the sand. Beyond the region of overland flow, the upper 50 cm of the slope, the sand remained under negative pressure, as evidenced by the small ridges left in the surface by the impinging rain drops. Though admittedly qualitative, the evidence suggests that the seepage face, the zone of zero pressure head at the surface with outwardly directed hydraulic gradients, extends about 50 cm upslope from the toe, and that from 50 to about 85 cm the surface was near zero pressure head but with inwardly directed hydraulic gradients. Within this latter region, a portion of the rainfall would be contributed to overload flow and the remainder would infiltrate.

Figure 4 shows the hydraulic-head response of the six tensiometers. Zero time was the time at which the rainfall was initiated. Zero elevation was chosen as the bottom of the sand model; therefore the point of departure from the hydraulic head axis indicates the initial elevation of the water table. The fact that all curves depart from approximately the same point (78 cm) indicates that the model was under conditions of hydraulic equilibrium prior to the rainfall event. As shown in Figure 4, all tensiometers responded instantaneously to the application of water, and all were approaching a new equilibrium condition about 3.0 minutes after the rainfall was initiated. There were major differences, however, in the rate of response and particularly in the magnitude of response.

The hydraulic head at T-4 increased by about 4 cm during the first 30 seconds of rainfall, then remained relatively constant at a value of about 80.5 cm for the duration of the experiment. Considering its position in the flow system, any further increase in head would indicate standing water on the soil surface; thus the magnitude of the response was limited by the small distance between ground surface and the initial water table position. On the other hand, the hydraulic head at T-1 increased by about 22 cm to a steady state value of 100 cm. Because the elevation of T-1 is 95 cm, the new steady state water table position was above T-1. T-2 responded in a

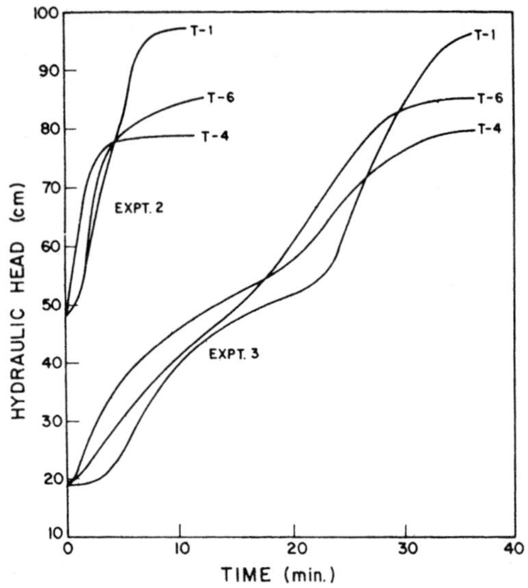

Fig. 5. Hydraulic head response of selected tensiometers for experiments 2 and 3.

manner similar to T-1 but with a slightly reduced magnitude. The difference in the equilibrium hydraulic head values of T-1 and T-2 indicates a vertically downward hydraulic gradient. T-1 and T-2 responded until there was a sufficient hydraulic gradient to transmit the applied water. Thus the magnitude of the response was determined by the surface flux rather than by the surface elevation, as in the case of T-4. The large difference in hydraulic head between T-1 and T-4 indicates groundwater flow from the recharge area (T-1) towards the discharge area (T-4). The steady state hydraulic head values at T-3, T-5, and T-6 are distributed between those at T-1 and T-4 and have magnitudes that are consistent with flow from the recharge area towards the discharge region near the stream.

During the first 3 minutes of rainfall, only 0.2 cm of water was applied, and approximately one half of this was discharged as streamflow. Thus the rapid response of the flow system can be attributed to a negligible storage capacity within the flow system as a result of the sand remaining saturated above the water table. The effect of the capillary fringe on the flow system response is therefore clearly demonstrated.

Although the response was rapid, if the porous medium was totally saturated and if the water and the medium are considered to be incompressible, then the response would theoretically have been instantaneous. Initial tests showed the response time of the monitoring system to be less than 2.0 s, and thus the observed response undoubtedly reflects the actual response of the flow system. Because of the low compressibility of water and the sand matrix, it is unlikely that these could account for the observed lag in the response. A more probable explanation concerns the effect of entrapped air in the porous medium. From data presented by *Gillham et al.* [1976], it can be seen that upon wetting of a sandy geologic material, at zero pressure head, entrapped air may constitute as much as 8% by volume of the porous medium. This air does not occur as a continuous gas phase but rather as isolated and distributed gas bubbles. In the present study, although the medium was saturated with deaired water, water was not flushed through the system to remove any remaining air. Consequently, entrapped air undoubtedly occupied some portion of the pore space. As the pressure became less negative, or more positive, depending upon the position in the flow system, the entrapped air would compress, providing a small increase in the storage capacity of the sand. This change in storage may account for the noninstantaneous hydraulic head response shown in Figure 4.

Figure 5 shows the hydraulic head response for three tensiometers (T-1, T-4, and T-6) for experiments 2 and 3. In experiment 2 the initial water table was at an elevation of 48 cm, or about 27 cm below the stream, resulting in a limited region in the upslope portion of the model that would be at a water content less than saturation. As shown in Figure 5, T-4 responded almost as rapidly as in experiment 1, although the magnitude of the response was much greater as a result of the lower initial position of the water table. The response of T-1 differed somewhat in that there was a slight initial lag due to the presence of the unsaturated zone in the vicinity of T-1. Generally, the rate of response in experiment 2 was not dramatically different from the response in experiment 1. This is not surprising in view of the very limited extent of the unsaturated zone in experiment 2. It should be noted that, as in experiment 1, streamflow commenced within seconds of initiation of rainfall.

The response in experiment 3 (Figure 5) was markedly different from that of the two previous experiments. In this case the initial water table was located about 57 cm below the stream, resulting in a distinct unsaturated zone across the entire surface of the sand. With the initial application of water there was a slight lag in the response of all the tensiometers, with the lag being the greatest at T-1, where the thickness of the unsaturated zone was the greatest. In addition, the rate of response was appreciably less than in experiments 1 and 2, with the steady state values not being approached until about 35 minutes after the rainfall started. The steep rise in T-1, starting at about 24 minutes after the initiation of the rainfall event, almost parallels the rise observed in experiment 2, though displaced in time. This probably corresponds to the time when the capillary fringe approached the soil surface at T-1, thus reflecting the fact that there was little storage capacity remaining in the soil material. In experiment 3, streamflow was not observed to occur until about 30 minutes after initiation of the rainfall event. Because the "stream" was in the same position for all three experiments and because the rainfall intensity was the same, the final steady state values reached by the tensiometers were approximately the same for all three experiments.

A comparison of the results of experiments 3 and 1 demonstrates the dramatic effect that the capillary fringe can have on the hydraulic head response to precipitation and, ultimately, on the generation of streamflow. With the capillary fringe extending to the soil surface, the response of the flow system and streamflow generation were almost instantaneous. With the deeper water table the response was much slower, and streamflow did not commence until the water table rose to the level of the stream.

Water Table Response

The pressure head response at each of the six tensiometers can be calculated by subtracting the elevation head of the respective tensiometers from the corresponding hydraulic head data of Figures 4 and 5. In principle the pressure head values could be interpolated or extrapolated to determine the position of the water table (locus of points having zero pressure head) at any time. In the present study this approach was not feasible because of the rapid and two-dimensional response of the water table and the relatively small number of measurement points. As an alternative, the pressure head values at T-5 were used as a measure of the water-table elevations along a vertical line passing through T-5. This would give an accurate measure of the position of the water table provided there were no vertical hydraulic gradients along the vertical line through T-5. This tensiometer is situated approximately between the recharge and discharge zones, and thus vertical gradients should be small. In addition, the magnitude of the water table response will depend upon the position along the cross section, with the smallest response occurring at the toe of the slope and the largest response occurring in the upslope region. In spite of these uncertainties, the use of T-5 should be sufficient for comparing the water table responses of the three experiments.

Figure 6 shows the elevation of the water table as a function of time for each of the three experiments. As expected, the first experiment had the most rapid response, reaching the new steady state position in about 3 min. In experiment 2 the response was rapid but the rate of rise was slightly lower than in experiment 1. This can be attributed to the slight degree of unsaturation near the surface in experiment 2. As a result of the greater thickness of the unsaturated zone in experiment 3, the rate of response was substantially less than observed in either experiments 1 or 2. The results are consistent with the hydraulic head data, indicating the immediate and rapid response of the water table when the capillary fringe extends to the soil surface and the slower response when there is an unsaturated zone extending to the soil surface.

The amount of water added to the model in order to achieve the maximum water table elevation is the difference between applied rainfall and stream discharge, and these differences are 0.1, 0.7, and 2.8 cm for experiments 1, 2, and 3, respectively. If specific yield is defined as the volume of water added to storage per unit increase in the elevation of the water table, then using the above values of water added, and the average water-table responses of Figure 6, the average effective specific yield values would be 0.007, 0.02, and 0.039 for experiments 1, 2, and 3, respectively. Specific yield for sand of the type used in this experiment would generally be expected to be

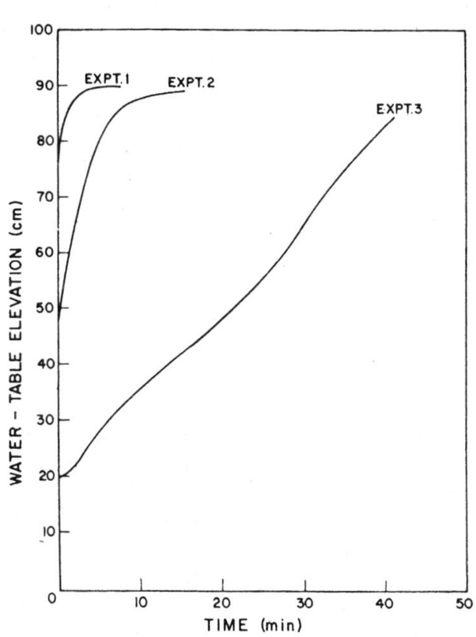

Fig. 6. Water table elevation versus time for experiments 1, 2, and 3.

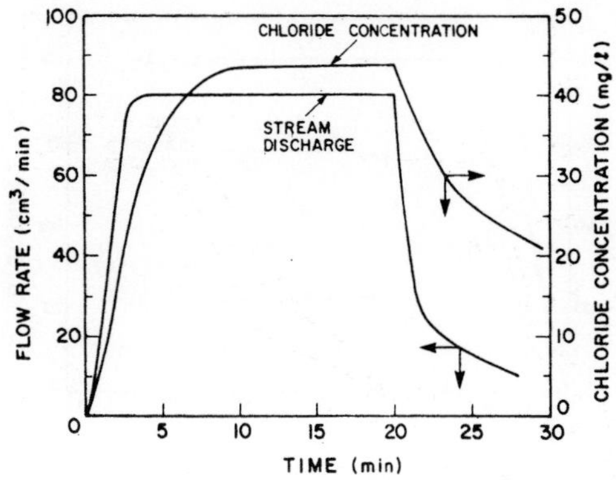

Fig. 7. Stream hydrograph and chlorograph for experiment 4.

within the range of about 0.25 to 0.3. The low values measured in this study and the trend towards increasing values with increasing depth to the water table demonstrate the effects of capillary retention in reducing the storage capacity of the soil material. This effect has been discussed in greater detail by *Duke* [1972]. As a result of the particularly low storage capacity that exists when the zone of tension saturation extends to the soil surface, the application of water results in a rapidly rising water table. Under natural conditions this could cause greatly increased hydraulic gradients towards streams and thus an almost immediate increase in the discharge of preevent water to the stream in response to precipitation.

Stream Flow Response

As a result of sand accumulating in the outflow tube, accurate streamflow data was not obtained for the full duration of experiments 1 through 3. The outflow tube was modified for experiments 4 and 5. Experiment 4 was identical to experiment 1; however, due to the modification, accurate outflow data was obtained, and chloride, at a concentration of 60.6 mg/l, was added to the precipitation to provide a means of separating the outflow into event water and preevent water components.

Figure 7 shows the steamflow hydrograph and the chlorograph for experiment 4. As indicated, streamflow began immediately with the initiation of the rainfall event, rising rapidly to a maximum value at about the same time the hydraulic head values (Figure 4) became constant (about 3 minutes after initiation of rainfall). The maximum outflow rate (80 cm^3/min) is equal to the rainfall rate of 4.3 cm/h. With cessation of the rain at 20 min the streamflow declined, initially at about the same rate as it had risen, then declined more gradually. The slow drainage region may be due to drainage of a small amount of free water on the surface and internal drainage as a result of expansion of air trapped in the pore space.

The trend in the chlorograph is similar to that of the hydrograph, with the concentration of chloride in the effluent increasing to a maximum value of 44.5 mg/l about 12 min after initiation of rainfall. Because the concentration of chloride in the groundwater (preevent water) was initially negligible, the amount of event water in the stream discharge can be calculated by multiplying the streamflow by the ratio of the chloride concentration in the stream to the chloride concentration in the rain (60.6 mg/l). Figure 8 shows the total discharge and the preevent and event water components as determined by the above procedure for experiment 4. At very

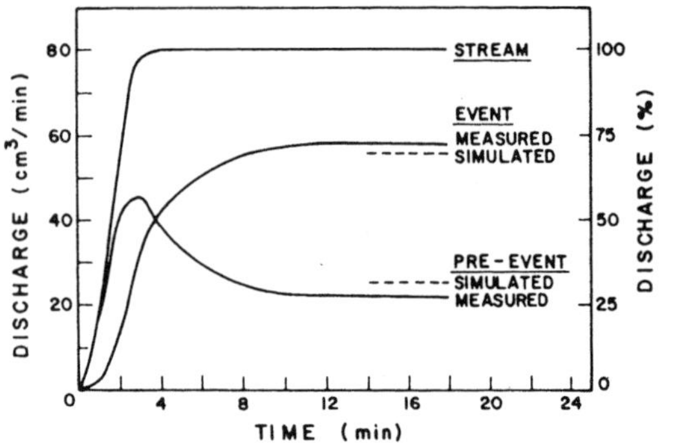

Fig. 8. Measured and simulated event and preevent components of the stream hydrograph for a rainfall rate of 4.3 cm h^{-1}.

early times (less than 2 min), preevent water was the major component of the stream discharge. The preevent discharge reached a maximum of 45.7 cm^3/min at a time of 3 minutes, but its proportion of the total discharge had declined to 57%. During this period the contribution of event water to stream discharge occurs as a result of precipitation falling directly on the seepage face and on the region at zero pressure head extending upslope from the seepage face, then migrating to the stream as surface runoff. At very early times the proportion of the surface that was at zero pressure head was small, and thus there was a very small contribution of event water to the stream. As this region expanded, a progressively larger proportion of the stream water originated as precipitation. After the peak discharge was reached, the contribution to the stream of water that was initially present in the sand declined gradually to a relatively constant value of 22 cm^3/min (28% of the total discharge), while the contribution of event water continued to increase to a relatively constant maximum value of 57.9 cm^3/min (72% of the total discharge). As discussed previously, visual observations indicated that the seepage face occupied about the lower 35% of the sand surface and that a portion of the water falling on an additional 20% of the surface also contributed to surface runoff. Thus the observed runoff of event water (72% of the total discharge) is reasonably consistent with the visual observations.

The total discharge and the preevent and event water contribution determined for experiment 5 are given in Figure 9. The trends are similar to those of Figure 8; however, because of the lower rainfall rate (1.9 cm h^{-1}), the transient period lasted much longer than was observed in experiment 4. When steady state flow conditions were reached, approximately 30% of the surface appeared to be at zero pressure head. This would include both the seepage face and the region beyond the seepage face where there was both infiltration and surface runoff. The hydraulic head values (not shown) remained constant after a time of about 12 minutes, while the preevent and event water contributions remained constant after a time of about 30 min. Comparable times for the high rainfall rate (4.3 cm/h) were 3 minutes and 10 minutes, respectively. The longer transient period is consistent with our previous suggestion that the transient behaviour is the result of the compressibility of entrapped air within the medium.

As in experiment 4, at early time the dominant component of streamflow was preevent water with a gradual increase in the event water contribution to a maximum value of about 13 cm^3/min (34% of the total discharge) at a time of 30 minutes. The steady state proportion of preevent water in the stream (64%) was substantially greater than that observed in experiment 4. It is clear that with the reduced rate of application, the extent of the surface over which runoff occurred was substantially reduced, thus reducing the proportion of event water and increasing the proportion of preevent water in the stream.

Though the trend was somewhat less pronounced in experiment 5, both experiments 4 and 5 show the proportion of preevent water in the stream rising to a maximum value at early time, then gradually declining to the steady state value. As indicated in the previous discussion, and as will be shown further in the following section, the hydraulic head response and the generation of the seepage face and the region of runoff beyond the seepage face appeared to be consistent with the steady state proportions of event and preevent water in the stream. Thus the reason for the early peak in the preevent water contribution is unclear. It is suggested, however, that this effect is the result of mixing of water in the region of surface runoff that extends beyond the seepage face. Within the seepage face, the flux is directed outward, and any event water that penetrates the medium as a result of raindrop impact would be discharged almost immediately to the surface. On the other hand, in the region of zero pressure head that extends beyond the seepage face, a portion of the rainfall migrates as surface runoff and the remainder infiltrates under the influence of inwardly directed hydraulic gradients. In this region, the impact of the raindrops could cause mixing in the upper few millimeters of the sand between the event and preevent waters, causing preevent water to constitute some proportion of the surface runoff. This would result in an erroneously high calculated preevent contribution. This process would operate during the development of the seepage face and would gradually dissipate as the preevent water is depleted from the upper few millimeters of the sand. In experiment 5 the peak preevent discharge did not exceed the steady state preevent discharge by as high a value as was observed in experiment 4, and this is consistent with the reduced extent of the zero pressure head region beyond the seepage face. If this is indeed the explanation, then one must conclude that the calculated values of preevent water discharge during the highly transient early time period of the experiments are somewhat too high. Nevertheless, the steady state values should be quite reliable.

Mathematical Simulations

To further evaluate our interpretations of the physical response and, in particular, to provide further evidence concerning the early peak and the subsequent decline in the pro-

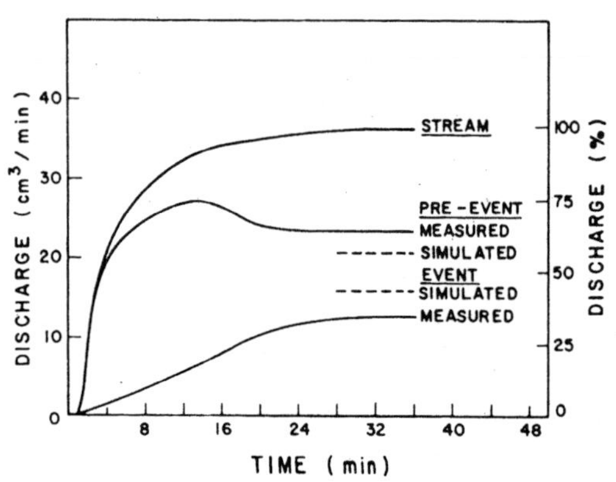

Fig. 9. Measured and simulated event and preevent components of the stream hydrograph for a rainfall rate of 1.9 cm h^{-1}.

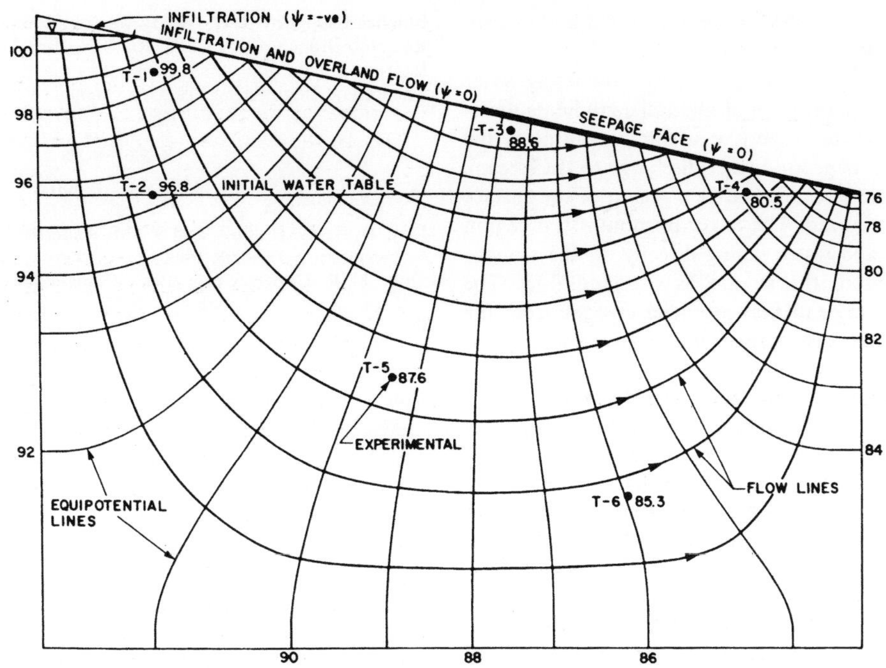

Fig. 10. Flow net for the steady state hydraulic conditions of experiment 1.

portion of preevent water in the streamflow, a two-dimensional saturated-unsaturated flow model was used to simulate the hydraulic response of experiments 4 and 5. A triangular finite element model similar to that described by *Frind et al.* [1976] was used; however, the upper boundary condition was modified to allow for the generation and recession of a seepage face. Though the model is transient, in the present application, because the medium is saturated, it reduces essentially to a steady state saturated flow model that allows for the generation of the seepage face, and the only important parameter required is the saturated hydraulic conductivity. The lateral and lower boundaries were specified as zero flux, while a portion of the upper boundary was a specified flux (equal to the rainfall intensity) and the remaining portion was set at zero pressure head. The proportions of the upper boundary that were under the flux and zero pressure head conditions was adjusted internally in the model and, as expected, was highly sensitive to both the rainfall intensity and the hydraulic conductivity of the medium. Output from the model included the hydraulic head and pressure head distributions and the surface flux and groundwater (preevent) flux to the stream. The model is currently being developed to include hysteresis in the hydraulic parameters and to include the effects of gas compressibility. A description of the complete model will appear in a subsequent paper.

Figure 10 shows the flow net that was generated for experiment 4. This figure shows that the seepage face extends for a distance of about 65 cm upslope from the stream and the region of zero pressure head extends for an additional 45 cm upslope. As indicated previously, all water falling on the seepage face would migrate as overland flow, while a portion falling on the upper zero pressure head region would also migrate as overland flow, and the remainder would infiltrate. The measured hydraulic head values are included in Figure 10 and are in reasonable agreement with the simulated values. The simulated event and preevent contributions to the stream were 64.8 and 25.2 cm^3/min. As indicated on Figure 8, these are very close to the measured values. The hydraulic conductivity value used in the simulation was 5×10^{-5} m s^{-1}. This is slightly larger than the value measured with entrapped air (3.5×10^{-5} m s^{-1}) and less than the value determined under conditions of complete saturation (9.0×10^{-5} m s^{-1}).

Experiment 5 was simulated using the same value of hydraulic conductivity. The results showed that the seepage face extended upslope from the toe for a distance of 50 cm and the region of zero pressure head beyond the seepage face extended an additional 15 cm upslope. The calculated event and preevent contributions were 15.8 and 20.3 cm^3 s^{-1}, respectively and, as shown in Figure 9, are in reasonably good agreement with the measured values.

The hydraulic conductivity, the extent of the seepage face, and the discharge components are all highly interactive. The fact that these characteristics of the numerical simulations are all highly consistent with the measurements and observations of the physical model for both experiments provides strong support for the validity of the experimental results and the previous interpretations of the results.

In that the model gives only steady state predictions, it provides no direct evidence concerning the peak preevent contribution observed at early time. The good agreement between the simulations and the steady state experimental results suggests, however, that the peak is attributable to some early time transient phenomenon. Certainly the mixing of waters in the region upslope from the seepage face remains as a plausible hypothesis.

SUMMARY AND CONCLUSIONS

The experimental and simulated results clearly show that if the zone of tension saturation extends to ground surface, or near ground surface, the application of a small amount of water can cause an immediate rise in the water table. Furthermore, the magnitude of the rise is much greater than would be expected on the basis of normal specific yield values for sandy materials (0.2–0.3). Even when the water table was at a significant depth below ground surface, the storage capacity of the sand was very low, causing the rapid and disproportionate response of the water table.

Because of the sloping surface of the model, the rising water table caused the rapid generation of hydraulic gradients directed towards the toe of the slope (the stream), resulting in the immediate generation of groundwater discharge to the stream. The rising water table also resulted in a region of the surface being at zero pressure head. The lower portion of this region was a seepage face, and water falling directly on the seepage face was contributed to the stream as overland flow. The region of zero pressure head that extended upslope from the seepage face contributed to both overland flow and infiltration.

The experiments using chloride as a tracer showed that preevent discharge to the stream preceded event discharge and that at early times, preevent water was the predominant component of streamflow. At later times, with the development of the seepage face, the event water contribution to the stream increased. At steady state, the preevent contribution to the stream was shown to be strongly dependent upon the rainfall intensity.

The extent to which the laboratory results are applicable to field situations is a question that is difficult to resolve. In particular, the larger scale of field sites and the effects of geologic heterogeneity and vegetative cover are all factors that could not be readily addressed in the laboratory experiments. Acknowledging these uncertainties, the results nevertheless lead one to speculate that under some conditions the capillary fringe can indeed play an important role in the generation of streamflow. These effects would only be realized in situations where the capillary fringe extends to, or near, ground surface in areas in the immedate vicinity of the stream. As a result, the effects of the capillary fringe would generally be most prevalent under humid climatic conditions and during wet periods of the year. Texture, in that it affects the extent of the tension saturated zone, would further determine the occurrence of capillary fringe effects.

As the results of this study show, the capillary fringe may affect the runoff process in two ways. The rapidly rising water table that occurs when the capillary fringe extends to ground surface causes increased hydraulic gradients in the groundwater zone, causing a rapid increase in the discharge of groundwater to the stream. In contrast to this, the rapidly rising water table may result in free surface conditions at the ground surface, resulting in overland flow. (This mechanism was discussed by *Ragan* [1968]). The first process would result in large preevent water contributions to the stream, while the second would result in large event water contributions. The relative importance of these processes would undoubtedly depend upon the rainfall intensity, the surface slope, and the hydraulic conductivity of the geologic material.

Acknowledgments. The authors wish to thank Paul Johnson for his assistance with various aspects of the laboratory experiments. Financial support for this study was provided by the Environmental Research Branch of the Chalk River Nuclear Laboratories of Atomic Energy Canada Ltd. and by an operating grant from the National Science and Engineering Research Council of Canada.

REFERENCES

Betson, R. P., What is watershed runoff?, *J. Geophys. Res.*, *69*(8), 1541–1551, 1964.

Day, P. R., G. H. Bolt, and D. M. Anderson, Nature of soil water, *Agronomy*, *9*, 193–208, 1967.

Duke, H. R., Capillary properties of soils-influence upon specific yield, *Trans. ASAE*, *15*(4), 688–691, 1972.

Dunne, T., and R. D. Black, An experimental investigation of runoff production in permeable soils, *Water Resour. Res.*, *6*, 478–490, 1970*a*.

Dunne, T., and R. D. Black, Partial area contributions to storm runoff in a small New England watershed, *Water Resour. Res.*, *6*, 1296–1311, 1970*b*.

Dincer, T., B. R. Payne, T. Florkowski, J. Martinec, and E. Tongiorgi, Snow-melt runoff from measurements of tritium and oxygen-18, *Water Resour. Res.*, *6*, 110–124, 1970.

Freeze, R. A., Role of subsurface flow in generating surface runoff, 1, Base flow contributions to channel flow, *Water Resour. Res.*, *8*(3), 609–623, 1972.

Freeze, R. A., Streamflow generation, *Rev. Geophys. Space Phys.*, *12*, 627–647, 1974.

Frind, E. O., Gillham, R. W., and J. F. Pickens, Application of unsaturated flow properties in the design of geologic environments for radioactive waste storage facilities, paper presented at the International Conference on Finite Elements in Water Resources, Princeton University, Princeton, N. J., 1976.

Fritz, P., J. A. Cherry, K. V. Weyer, and M. G. Sklash, Runoff analyses using environmental isotopes and major ions, in *Interpretation of Environmental Isotopes and Hydrochemical Data in Groundwater Hydrology*, pp. 111–130, International Atomic Energy Agency, 1976.

Gillham, R. W., The effect of the capillary fringe on water-table response, *J. Hydrol.*, *67*, 307–324, 1984.

Gillham, R. W., A. Klute, and D. F. Heerman, Hydraulic properties of a porous medium: measurement and empirical representation, *Soil Sci. Soc. Am. Proc.*, *40*, 203–207, 1976.

Hewlett, J. D., and A. R. Hibbert, Factors affecting the response of small watersheds to precipitation in humid areas, in *International Symposium on Forest Hydrology*, edited by W. E. Sopper and H. W. Lull, pp. 276–290, Pergamon, New York, 1967.

Horton, R. E., The role of infiltration in the hydrologic cycle, *Eos Trans. AGU*, *14*, 446–460, 1933.

Martinec, J., Subsurface flow from snow-melt traced by tritium, *Water Resour. Res.*, *11*, 496–497, 1975.

Martinec, J., H. Siegenthaler, H. Oescheger, and E. Tongiorgi, New insight into the runoff mechanism by environmental isotopes, *Isot. Hydrol. Proc. Symp.*, *1*, 129–143, 1974.

Ragan, R. M., An experimental investigation of partial-area contributions, *IAHS AISH Publ. 76*, 241–249, 1968.

Sklash, M. G., and R. N. Farvolden, The role of groundwater in storm runoff, *J. Hydrol*, *43*, 45–65, 1979.

A. S. Abdul and R. W. Gillham, Department of Earth Science, University of Waterloo, Waterloo, Ontario, Canada N2L 3G1.

(Received April 20, 1983;
revised February 3, 1984;
accepted February 6, 1984.)

C Linking Riparian Zone Hydrology to Solute Transport

Commentary on Paper C1

Jackson, W. A., Asmussen, L. E., Hauser, E. W. & White, A. W. (1973) Nitrate in surface and subsurface flow from a small agricultural watershed.

The 1960s saw increasing concern about levels of nitrate in surface and groundwater. This was partly in relation to the inefficiency of fertilizer application, but also to worries about drinking water quality and eutrophication. This paper is one of the first to take a catchment-scale approach to measuring nitrate export in response to reports of significant levels of nitrate in drainage waters from N-fertilized fields. The opening section implies that more attention had previously been given to surface runoff than was appropriate, given that nitrate concentrations were usually very low, in contrast to much higher levels in subsurface runoff. Therefore measurements of surface runoff alone could not accurately represent total N losses.

The choice of field site is an interesting one, on the coastal plain at Tifton, Georgia, USA, where the slope angle is very small, only 3.2°. The soil hydrology favours shallow subsurface flow in that the surface soil is a loamy sand with an underlying clay horizon of very low hydraulic conductivity. Since outflow from the experimental sub-catchment was by terracotta tile, we can infer that the whole field was under-drained, presumably more because of the low slope than the low permeability of the surface soil horizon. Discharge was measured using thin-plate weirs and stage recorders whilst water samples were obtained by hand, with the most frequent sampling interval on the rising stage of the storm hydrograph. Subsurface flow is intermittent through the year, with little or no flow in June, and a period of no-flow at the end of the year from September through to December. Of the annual rainfall, 37% is converted into runoff, of which 80% is subsurface. Of the nitrate loss, 99% is in subsurface flow.

It is interesting that the fertilizer application rates are very high; despite this, the level of nitrate export is not excessive, in the range 22–50 kg/ha. If the N content of the 4-12-12 fertilizer is 4% (Tam, 2000), then 36 kg/ha N was applied in March and plus a further 168 kg/ha N in early May, 204 kg/ha N in all. Loss rates are therefore of the order of 10–25%. This degree of N leakage is to be expected in intensively farmed arable areas (cf. Burt & Arkell, 1987). The authors acknowledge that it was not possible to differentiate the amount of nitrate loss associated with fertilizer alone. We now know this is due to several reasons. Mineralisation of organic matter, especially as warm soils re-wet, is likely to be a major source of nitrate, rather than simply nitrate from fertilizer that has not be taken up by plants or soil micro-organisms. If air-dried soils are rewetted, pronounced mineralisation of N and C is frequently observed (Birch, 1960; Haynes & Swift, 1989). It is thought to occur because of a combination of the death of micro-organisms from moisture stress followed by mineralisation on rewetting, and the physical disturbance caused by the drying and wetting process itself. The authors make no comment about the likely source of nitrate from soil that had not been cultivated for four years prior to 1969, but it is now well known that significant long-term stores of organic nitrogen are released by cultivation, resulting in large increases in the nitrate leached (Whitmore *et al.*, 1992). The highest annual N loss is in 1969, the first year of cultivation, rather than in 1970, the wettest year.

This paper by Jackson and colleagues is important for a number of reasons. It establishes the importance of nitrate loss in subsurface flow and by calculating monthly loads, shows how outflow discharge and nitrate export are intimately linked. They also attempt to upscale the results from one small area to the wider drainage basin, by comparing loadings for different source areas. Inputs from the "test watershed", an area intensively farmed for corn, are much greater than for the entire basin, an area of mixed land use. They note the difficulty of extrapolating results from one site to large areas. In a complex landscape, there are many reasons why the average loading for a large basin will be lower than for individual fields, of course, but one reason will be that not all fields drain directly into streams (Burt & Pinay, 2005). Some surface water courses will be buffered by intact riparian zones; other research in the same region of the USA (e.g. Peterjohn & Correll, 1984, Paper E2) soon showed the efficacy of nitrogen transformations in the near-stream zone.

TPB

References

Birch, H. F. (1960) Nitrification in soils after different periods of dryness. *Plant and Soil* **12**, 81–96.

Burt, T. P. & Arkell, B. P. (1987) Temporal and spatial patterns of nitrate losses from an agricultural catchment. *Soil Use and Management* **3**, 138–143.

Burt T. P. & Pinay, G. (2005) Linking hydrology and biogeochemistry in complex landscapes. *Progr. Phys. Geogr.* **29**(3), 297–316.

Haynes, R. J. & Swift, R. S. (1989) Effect of air dried soils on pH and accumulation of mineral N. *J. Soil Sci.* **40**, 341–347.

Peterjohn, W. T. & Correll, D. L. (1984) Nutrient dynamics in an agricultural watershed: observations on the role of a riparian forest. *Ecology* **65**, 1466–1475.

Tam, K. H. (2000) *Environmental Soil Science*, 2nd edition. CRC Press, USA.

Whitmore, A. P., Bradbury, N. J. & Johnson, P. A. (1992) Potential contribution of ploughed grassland to nitrate leaching. *Agriculture, Ecosystems and Environment* **39**, 221–233.

Nitrate in Surface and Subsurface Flow from a Small Agricultural Watershed[1]

W. A. Jackson, L. E. Asmussen, E. W. Hauser, and A. W. White[2]

ABSTRACT

A small agricultural watershed on Cowarts loamy sand in the Georgia Coastal Plains was planted in corn (*Zea mays* L.) each year from 1969 to 1971. Surface and subsurface water samples were collected during each natural rainfall runoff event and NO_3-N was determined.

The initial weighted average NO_3-N concentration was 5 ppm in the subsurface runoff for 3 months before the first planting. The average was 7 ppm in 1969, 10 ppm in 1970, and 9 ppm in 1971. The concentration of NO_3-N in most of the surface runoff for 3 years was less than 1 ppm; the maximum was 3 ppm. Subsurface flow accounted for 80% of the total runoff during the study period. Thus, greater amounts of NO_3-N were discharged in subsurface flow than in surface flow from this watershed.

Additional Index Words: nutrient transport, water quality, environmental pollution.

Nitrogen in plant and animal residues and various forms of nitrogen fertilizers applied to soil are ultimately changed to the nitrate form of nitrogen by natural soil processes in well-drained, aerated soils. Nitrate is highly soluble, and when applied as fertilizer, will move readily into the soil. Except where considerable runoff occurs, nitrate is usually low in surface runoff (4, 8).

Nitrate losses by leaching depend on soil type, management practices, rainfall, and other climatic conditions. Significant levels of NO_3-N have been reported in drainage waters from N-fertilized fields (2, 3).

Coastal Plain soils are generally sandy and have high infiltration rates, but clay often occurs at depths from 91 to 610 cm below the surface. Under this condition, deep surface losses are negligible since clay does not freely transmit water downward (1).

A large percentage of rainfall infiltrates and moves laterally as shallow subsurface flow along the surface of the low permeability zone underlying the sandy topsoil. Since this water returns as streamflow, measurements of surface runoff alone do not accurately represent the total runoff and chemical loss from watersheds. Information regarding the amounts of nitrates being transported from an area by surface and subsurface water is necessary before meaningful conclusions can be made concerning the amount of NO_3-N in drainage water resulting from the use of fertilizer.

The purpose of this research was to determine the amount of nitrate-nitrogen that was transported by surface and subsurface drainage from a small watershed cropped to continuous corn (*Zea mays* L.) for 3 years.

EXPERIMENTAL METHODS

The study area is on the Georgia Coastal Plain Experiment Station, Tifton, Georgia, on 0.34 ha of Cowarts loamy sand with a surface slope of 3.2%. Underlying this loamy sand, at a depth of 92 to 214 cm, is a low permeability zone known as the Hawthorn Formation (1). The subsurface drainage zone has a 3.2% slope.

Subsurface water, which flows along the top of the clay layer, was collected at the lower end of the drainage area by terra cotta tile. The tile depth ranged from 61 to 153 cm, depending on the topography, with 15 cm of gravel above and below the tile. The water from the tile was directed through a 90-degree V-notch weir where a digital recorder monitored the stage of the nearest 0.031 cm at 5-min intervals. Surface runoff was collected in channels around the perimeter of the field and directed through a 30.5-cm H-flume where an analog recorder continuously monitored the stage.

The watershed was planted in corn each year. The land was disked in mid-March, 896 kg/ha of 4-12-12 was broadcast, and the land was redisked. The land was plowed, bedded into modified 81-cm rows, and planted in early April at a rate of 44,478 plants/ha. When the corn was 30 days old, it was sidedressed with liquid ammonium nitrate-urea at a rate of 168 kg N/ha. Herbicides were used to control weeds. After harvest of corn grain, the stalks were chopped and the land was disked and fallowed until spring.

During periods of flow, samples were taken by hand at the V-notch weir and H-flume. Sampling frequency was determined by the stage, with an increasing frequency on the rising side of the hydrograph and decreasing frequency on the recession. The samples were refrigerated at 1.66–4.44C and analysis was complete within a week. Nitrate-N was determined colorimetrically with an Autoanalyzer using a copper-cadmium reduction column (5, 6).

The concentration of NO_3-N was plotted on the discharge hydrograph. The concentration values for 1-hour periods were read from the graph and multiplied by the discharge volume in that period to obtain total load of NO_3-N for each period during the runoff event. The weighted average concentration (WAC) of NO_3-N, reported by month in Table 2, was obtained by dividing the accumulated total load by the accumulated total discharge for the month.

[1]Contribution from Soil, Water, and Air Sciences, Southern Region, ARS, USDA, in cooperation with the Univ. of Georgia Agr. Exp. Sta. Received Jan. 4, 1973.

[2]Chemist, USDA, Watkinsville, Georgia 30677; Geologist and Agronomist, USDA, Tifton, Georgia 31794; and Soil Scientist, USDA, Watkinsville, Georgia, respectively.

RESULTS AND DISCUSSION

The average annual rainfall for a 49-year period at Tifton, Georgia was 120.3 cm (Table 1). Rainfall, runoff, and NO_3-N losses by months for the 3-year study period are shown in Table 2. Annual precipitation was slightly below the 49-year average in 1969 and slightly above in 1970 and 1971. Surface runoff was most numerous and largest in 1970, the year of highest rainfall. Surface runoff results mostly from high-intensity thunderstorms. Subsurface runoff losses were greatest in 1969. For the 3-year period, the combined surface and subsurface runoff averaged 37% of the rainfall; 80% of this runoff was from subsurface flow.

Nitrate-N concentrations were higher in subsurface runoff than in surface runoff for the 3-year period. Weighted average concentrations in drainage water ranged from 5.8 to 12.6 ppm NO_3-N for the study period. The monthly WAC in surface water was markedly lower than the concentration in subsurface flow, ranging from 0.11 to 3.0 ppm with only three values higher than 1.0 ppm.

The land was idle for 4 years prior to 1969, and preliminary drainage water samples collected during the last 3 months of 1968 averaged 4.8 ppm NO_3-N WAC. Following fertilizer application in 1969 and continuing into 1970, NO_3-N concentrations in the drainage water increased. The WAC of NO_3-N was 7.1 ppm during the first crop year and 10.3 ppm in 1970 with a maximum of 12.6 ppm occurring in May. In 1971, the concentration leveled off and the WAC was 9.4 ppm for the year, suggesting that a maximum level had been reached for this cropping system and fertilization practice.

Table 1—Rainfall data—Tifton, Georgia, 49-year average (1923–71)

Month	Rainfall, 1923-1971 avg.
	cm
January	10.13
February	10.16
March	12.01
April	10.39
May	8.92
June	11.73
July	15.39
August	13.34
September	9.27
October	4.90
November	5.00
December	9.04
Total	120.29

Losses of NO_3-N in surface runoff were highest during June to August. Annual NO_3-N losses in surface runoff for 1969, 1970, and 1971 were 0.12, 0.43, and 0.34 kg/ha, respectively.

The two periods with high NO_3-N levels in subsurface discharge coincided with times of high subsurface water losses, the first occurring about March each year and the second in midsummer. Annual NO_3-N losses in subsurface flow for 1969, 1970, and 1971 were 34.1, 49.9, and 22.0 kg/ha, respectively. Thus, 99.1% of the NO_3-N lost in runoff was in subsurface flow.

The NO_3-N losses from the area during the study resulted from a combination of factors. Some of the nitrate losses were related to fertilizer applications. However, samples collected in 1968, before the start of the cropping and fertilization system in 1969, showed appreciable

Table 2—Rainfall, surface and subsurface discharge, NO_3-N concentration, and total load by months

Year	Month	Rainfall	Surface				Subsurface			
			Runoff events	Discharge	Avg. conc. NO_3-N†	Total NO_3-N loss	Runoff events	Discharge	Avg. conc. NO_3-N†	Total NO_3-N loss
		cm		area-cm	ppm	kg/ha		area-cm	ppm	kg/ha
1969	January	4.88	--	--	--	--	1*	1.73	5.80	1.00
	February	7.90	--	--	--	--	1	2.97	6.16	1.83
	March	16.10	--	--	--	--	1	10.54	6.47	6.81
	April	4.50	--	--	--	--	1*	0.71	6.53	0.46
	May	13.61	1	0.25	0.90	0.024	1	3.40	6.69	2.28
	June	3.66	--	--	--	--	--	--	--	--
	July	21.74	9	3.28	0.17	0.056	1	2.34	6.06	1.42
	August	18.42	7	3.43	0.11	0.038	1*	19.08	7.41	14.11
	September	13.41	5	0.30	0.19	0.006	2*	7.29	8.27	6.03
	October	0.71	--	--	--	--	1*	0.23	7.78	0.19
	November	2.16	--	--	--	--	--	--	--	--
	December	10.74	2	0.03	0.21	0.001	--	--	--	--
Total		117.83	24	7.29	0.17	0.125	10	48.29	7.07	34.13
% of total rainfall				6.2				41.3		
% of total runoff				13.0				87.0		
1970	January	7.06	2	trace	0.21	trace	1	0.64	8.09	0.52
	February	12.14	5	0.13	0.21	0.002	2	5.13	9.55	4.89
	March	22.45	8	1.68	0.24	0.040	2	9.60	9.52	9.14
	April	3.43	1	0.15	0.55	0.009	1*	5.49	9.31	5.11
	May	19.10	3	0.53	0.55	0.029	1	2.82	12.57	3.55
	June	4.90	2	0.28	1.11	0.031	1*	5.05	10.67	5.39
	July	16.03	3	0.64	0.95	0.060	--	--	--	--
	August	28.14	11	6.58	0.44	0.226	1	13.36	10.64	14.21
	September	3.78	1	0.10	3.00	0.028	1*	3.38	12.09	4.08
	October	11.33	1	0.03	2.93	0.007	--	--	--	--
	November	1.88	--	--	--	--	--	--	--	--
	December	7.65	--	--	--	--	--	--	--	--
Total		137.90	37	10.11	0.43	0.432	10	45.47	10.31	46.89
% of total rainfall				7.3				33.0		
% of total runoff				18.2				81.8		
1971	January	6.48	--	--	--	--	3	0.20	9.61	0.19
	February	13.26	2	0.18	0.32	0.006	4	2.92	9.42	2.76
	March	13.51	--	--	--	--	3*	5.18	9.20	4.76
	April	9.45	2	0.91	0.31	0.028	3*	0.61	9.32	0.57
	May	12.17	6	1.07	0.35	0.039	2*	3.91	8.86	3.47
	June	16.41	4	2.74	0.40	0.110	2	0.25	6.47	0.01
	July	19.33	7	2.82	0.29	0.082	2	4.07	9.13	3.71
	August	15.65	5	2.21	0.34	0.075	--	--	--	--
	September	1.47	--	--	--	--	--	--	--	--
	October	4.62	--	--	--	--	--	--	--	--
	November	9.78	1	--	0.83	0.001	--	--	--	--
	December	13.28	2	0.25	0.85	0.001	2	6.53	10.04	6.54
Total		135.41	29	9.96	0.34	0.340	21	23.44	9.40	22.01
% of total rainfall				7.4				17.3		
% of total runoff				29.8				70.2		

* Event started in previous month.
† Weighted averages.

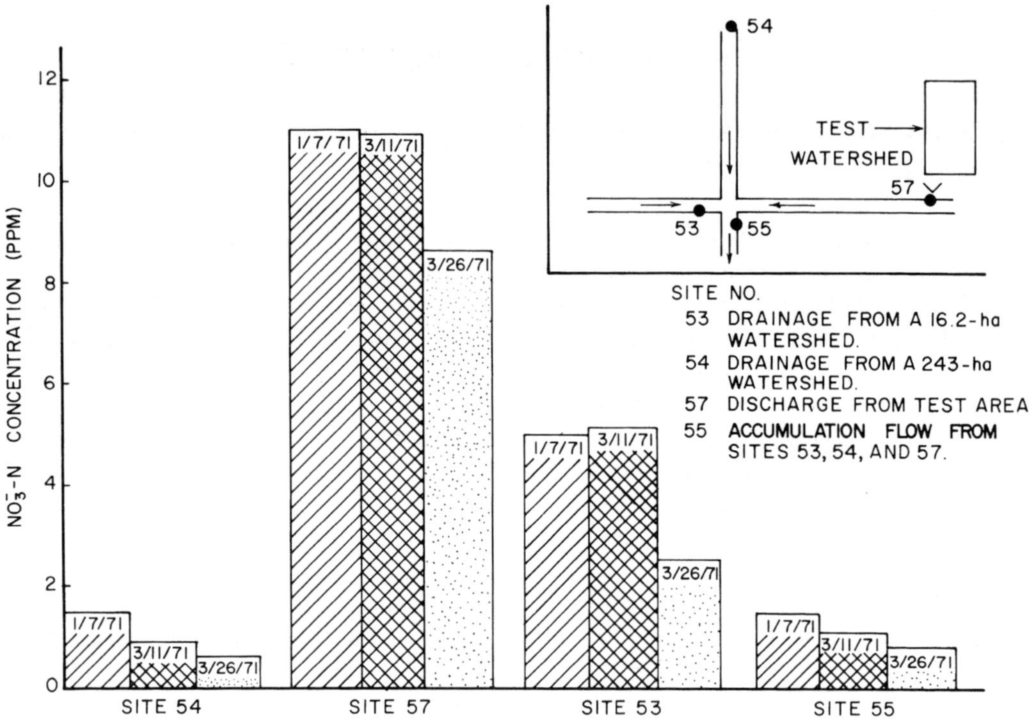

Fig. 1—Simultaneous nitrate-nitrogen concentrations at four sites.

amounts of NO_3-N already present in drainage water from the area. Therefore, under the study conditions, it was not possible to differentiate the amount of NO_3-N losses associated with N-fertilizer alone.

Figure 1 shows the influence of water from a larger and more complex watershed on the NO_3-N levels in water from the test watershed. Site 54 receives both surface and subsurface drainage from a 243-ha watershed consisting of woods, farmland, pasture, and some residential land. Site 53 is a highway ditch beside a 16.2-ha watershed consisting of woods, residential, and agricultural test plots. At Site 57, the discharge from the study area enters the highway ditch. The volume of surface and subsurface flow from Site 54 is about 15 to 20 times that flowing in the ditch at Site 53. Site 55 receives the accumulation of flow from Sites 53, 54, and 57.

Simultaneous samples were taken at these four sites on three different dates and analyzed for NO_3-N (Fig. 1). The NO_3-N concentrations in the water from the 243-ha watershed (Site 54) ranged from 0.6 to 1.5 ppm in the first quarter of 1971. Those from the study area (Site 57) were between 8 and 10 ppm, and those from Site 53 were 2 to 5 ppm. Site 55, the confluent stream of the other three sites, shows the mitigating effect of dilution from the surrounding areas. The concentration at Site 55, ranging from 0.7 to 1.5 ppm, is typical of streams flowing in the Coastal Plain area (7).

Although the study area is representative of the Coastal Plain area, the data are from a point source as opposed to a river basin, which consists of various size watersheds under many cultural practices. Entire watersheds are very seldom subjected to the same overall treatment. The larger the watershed, the less likely it is that the treatments, soil conditions, slopes, and cover are uniform. Thus, observations from specific sites or subwatersheds reflect conditions at that site only and cannot be extrapolated to evaluate environmental effects over large areas.

LITERATURE CITED

1. Asmussen, L. E., and J. C. Ritchie. 1969. Interflow or shallow phreatic flow in the Coastal Plain of Georgia. J. Hydrol. IX(2): 182–193.
2. Johnston, W. R., F. Ittihadieh, R. M. Danm, and A. F. Pillsbury. 1965. Nitrogen and phosphorus in tile drainage effluent. Soil Sci. Soc. Amer. Proc. 29:287–289.
3. Kohl, D. H., G. B. Shearer, and Barry Commoner. 1971. Fertilizer nitrogen: Contribution to nitrate in surface water in a Corn Belt watershed. Science 174:1331–1334.
4. Moe, P. G., J. V. Mannering, and C. B. Johnson. 1968. Loss of fertilizer nitrogen in surface runoff water. Soil Sci. 104:389–394.
5. O'Brien, J. E., and J. Fiore. 1962. Automation in sanitary chemistry—Parts 1 and 2: Determination of nitrates and nitrites. Waste Engr. 33:128–238.
6. Strickland, J. D., C. R. Sterns, and F. A. Armstrong. 1967. The measurement of upwelling and subsequent biological processes by means of the Technicon autoanalyzer and associated equipment. Deep Sea Res. 14:381–389.
7. U. S. Geological Survey. 1965. Quality of surface water. Parts 1 and 2: North Atlantic slope basins, South Atlantic slopes, and East Gulf of Mexico. U. S. Geol. Surv. Water-Supply Pap. No. 1961.
8. White, A. W., A. P. Barnett, W. A. Jackson, and V. J. Kilmer. 1967. Nitrogen fertilizer losses in runoff from crop land tested. Crops Soils Mag. 18(4):28.

Commentary on Paper C2

Foster, I. D. L. & Walling, D. E. (1978) The effects of the 1976 drought and autumn rainfall on stream solute levels.

From the early 1970s onwards, Des Walling of Exeter University (UK) pioneered the study of solute and sediment loads in rivers, working primarily on the River Exe and its tributaries. One of his first PhD students, Ian Foster, studied solute dynamics in a small agricultural catchment of 9.3 km^2 area. His study period coincided with the severe drought of 1975–1976 which affected most of England and Wales (Doornkamp *et al.*, 1978). In order to help explain variations in the concentration of certain dissolved constituents in the main river, measurements were taken on small streams in an attempt to understand the processes involved (Walling & Foster, 1975). The approach, therefore, was to infer processes operating on the catchment hillslopes on the basis of observations taken from small streams. Usually, periods of storm discharge were the focus of attention, analysing the variation of solute concentration over time (the "chemograph") in relation to the stream hydrograph and the behaviour of other solutes. Walling & Foster (1975) comment that "any attempt to account for the nature and timing of the chemograph must consider processes operating over the entire catchment in addition to those in the channel".

Recognising growing concerns about rising trends of nitrate in rivers in southern and eastern England, in the mid-1970s Des Walling and his colleague Bruce Webb had began to make a detailed study of the behaviour of nitrate in the Exe basin. The contrasting behaviour of nitrate compared to many other solutes was soon established, both seasonally and during storm events (Webb & Walling, 1985). Although nitrate generally shows a positive correlation with discharge, concentrations may fall (dilution) or rise (flushing effect) depending on catchment condition; dilution tends to occur in winter storms and *vice versa*. Autumn has the greatest variation in nitrate concentration, with high concentrations characteristic of wet autumns following dry summers.

Foster and Walling note the difficulty of studying extreme events, which by their very nature are hard to anticipate. They also note the difficulty of knowing in advance what aspects of such events should be monitored. It was their great good fortune, therefore, to already be involved in the study of the water quality of an agricultural catchment when the drought occurred. Burt (1994) comments on the ability of long-term monitoring schemes to capture rare events, and while Ian Foster's PhD research was not "long term" in the truest sense, it was certainly long enough to include the whole drought episode, including its aftermath. His field measurement programme was very typical of many at the time, employing flumes or v-notch weirs with stage recorders to measure discharge and a combination of automatic pump samplers and hand sampling to collect water samples for laboratory analysis. The drought itself was one of the most extreme in British history (Doornkamp *et al.*, 1978) and this was reflected in the Exe basin. At Exeter airport, only two months had above-average rainfall in the 16-month period from May 1975 to August 1976 inclusive (Fig. 2(a)). In stark contrast, September and October 1976 both recorded more than twice the monthly average. Thus, when the drought broke in mid-September 1976, there had been two hot, dry summers with a dry winter in between; small wonder therefore that the Yendacott stream dried up for the first time in living memory!

Foster and Walling employ hydrographs and chemographs to compare conditions in the autumns of 1974, 1975 and 1976 and in particular focus on the period from the beginning of September to mid-October 1976. The results presented in Fig. 4 show that solute concentrations in the immediate post-drought period were considerably higher than in previous years, but that because of how wet the 1976 autumn was, concentrations returned to more "normal" levels within six weeks. Nitrate in particular displayed very high concentrations, particularly for the tile drain and woodland stream, when flow resumed (Fig. 4). On the Yendacott stream, nitrate concentrations were initially low when flow resumed in early September but once the heavy rain came in mid-September, both here (Fig. 5) and at Pynes Cottage (Fig. 6), the concentrations increase dramatically, well above today's legal limit for nitrate in surface water (11.3 mg L^{-1} NO_3-N).

Other than being an exemplary case study in hydrograph and chemograph analysis and a rare example of post-drought solute response, Foster and Walling's paper is of particular interest to those interested in riparian buffer zones because of the inferences they draw about the processes involved in the delayed nitrate peaks in the autumn 1976 storm events (Fig. 7). The dominant component of the 22 September nitrate chemograph is the flushing peak that lags some 12 hours behind the peak discharge. As they note, this is reminiscent of the timing of throughflow contributions to storm runoff events, as described for example by Anderson & Burt (1978; included in Beven, 2006). They suggest that the flushing peak is related to a throughflow storm runoff component and that the source of the nitrate is the solution of soluble residues that had accumulated during the drought. They note later in the discussion that the nitrate might also have been produced by mineralisation upon rewetting of dry, warm soils. Later research has confirmed their suppositions, both for solutes in general and nitrate in particular. For example, Burt (1979) showed small increases in stream solute concentration (as indicated by measurements of specific conductance) during times of delayed throughflow generation (Fig. 1), whilst later research in the Slapton Wood catchment (following the lead of Troake & Walling, 1976) confirmed links between throughflow generation and

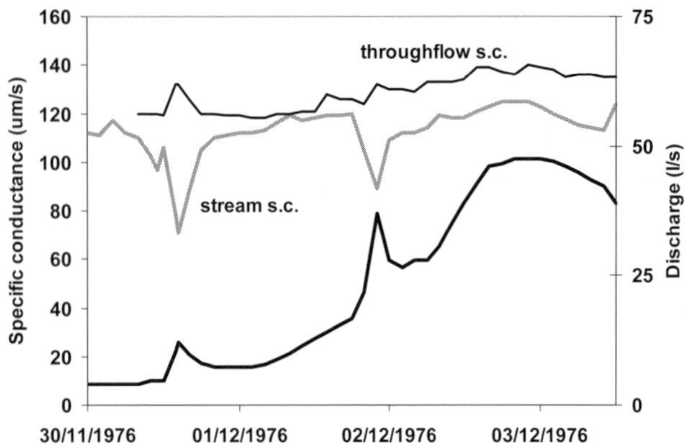

Fig. 1 Changes in stream discharge and specific conductance at Bicknoller Combe. Two storm events are shown: both have an initial discharge peak generated largely by surface runoff with a delayed throughflow peak following about a day later. Here, the second storm occurred at the time of the first throughflow peak. Specific conductance of throughflow increases from 120 on 30/11/1976 to 139 on 3/12/1976, 16% higher. Similar delayed peaks in solute load are seen for the stream water.

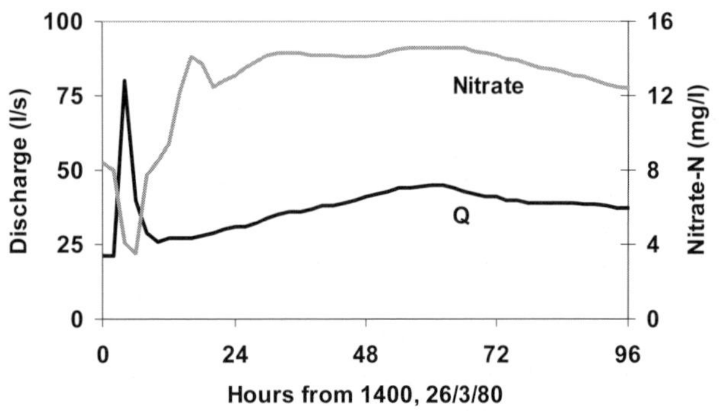

Fig. 2 Stream discharge (L/s) and stream nitrate concentration (mg/L NO_3-N) for the Slapton Wood stream. Note that the delayed nitrate peak precedes the delayed discharge response in this case.

nitrate leaching (Burt & Butcher, 1985; Burt & Arkell, 1987; Fig. 2). Studies such as these lead naturally into more recent work where detailed study of near-stream flow paths and water chemistry has allowed identification of the buffering processes operating in the riparian zone and evaluation of optimal conditions for such processes to function.

TPB

References

Anderson, M. G. & Burt, T. P. (1978) The role of topography in controlling throughflow generation. *Earth Surf. Processes* **29**, 331–334.

Beven, K. J. (2006) *Streamflow Generation Processes. Benchmark Papers in Hydrology*. IAHS Press, Wallingford, UK.

Burt, T. P. (1979) The relationship between throughflow generation and the solute concentration of soil and stream water. *Earth Surf. Processes* **4**, 257–266.

Burt, T. P. (1994) Long-term study of the natural environment: perceptive science or mindless monitoring? *Progr. Phys. Geogr.* **18**, 475–496.

Burt, T. P. & Arkell, B.P. (1987) Temporal and spatial patterns of nitrate losses from an agricultural catchment. *Soil Use & Manage.* **3**, 138–143.

Burt, T. P. & Butcher, D. P. (1985) The role of topography in controlling soil moisture distributions. *J. Soil Sci.* **36**, 469–486.

Doornkamp, J. C., Gregory, K. J. & Burn, A. S. (eds) (1978) *Atlas of Drought in Britain 1975–76*. Institute of British Geographers, London, UK.

Walling, D. E. & Foster, I. D. L. (1975) Variations in the natural chemical concentration of river water during flood flows, and the lag effect: some further comments. *J. Hydrol.* **26**, 237–244.

Webb, B. W. & Walling, D. E. (1985) Nitrate behaviour in streamflow from a grassland catchment in Devon, UK. *Water Res.* **8**, 1005–1016.

EARTH SURFACE PROCESSES, VOL. 3, 393–406 (1978)

THE EFFECTS OF THE 1976 DROUGHT AND AUTUMN RAINFALL ON STREAM SOLUTE LEVELS

I. D. L. FOSTER AND D. E. WALLING

Department of Geography, University of Exeter, Exeter, Devon EX4 4RJ, England

Received 4 August 1977
Revised 19 December 1977

SUMMARY

Operation of a previously existing sampling network during the 1976 drought has provided an opportunity to assess the effects of the drought and subsequent autumn rainfall on the water quality characteristics of a small catchment (9·3 km^2) in East Devon. The availability of records for past years has enabled these data to be compared with longer-term response characteristics.

Analysis indicates that the 1976 drought produced a unique solute response, with levels rising markedly during the autumn flush which resulted from the heavy rainfall following the drought period. Concentrations of several ions increased 3–4 times and NO_3–N levels exhibited even greater increases of up to 50-fold. Some of the factors responsible for these increases and some implications are considered.

KEY WORDS Drought Solute behaviour Nitrates

INTRODUCTION

Although much has been published regarding seasonal variations in solute behaviour and the response of stream solute levels to rapidly changing hydrometeorological conditions during storm runoff events (e.g. Gunnerson (1967), Hendrickson and Krieger (1964), Walling and Foster (1975)), little information is available on the impact of severe drought conditions on the solute dynamics of drainage basins in the humid temperate zone. Such information would be of value to the physical hydrologist or process geomorphologist who seeks to analyse magnitude–frequency parameters, to assess the potential significance of climatic oscillations and changes, and to reconstruct palaeohydrologic environments. Furthermore, the associated changes in water chemistry could be of considerable significance in terms of water use during a time of rapidly dwindling supplies and of pollution from non-point sources. It is to be expected that the behaviour of a drainage basin during the subsequent period of moisture excess would depart markedly from its characteristic response.

The extreme drought conditions experienced throughout the United Kingdom during the summer of 1976 afforded a potential opportunity for assembling information on this topic. However, the design of data collection schemes for extreme conditions of this nature poses many problems. First, the importance of the event may only be appreciated when it is over, secondly, little warning of its onset and duration is available and thirdly, knowledge of the aspects to which attention should be directed may be limited. Doubtless, the experience accumulated during the recent British drought would enable suitable monitoring programme to be designed if the occurrence of another event of similar severity could be predicted. Nevertheless, the information obtained from data collection networks developed for other purposes and functioning at the time of the drought may provide a worthwhile basis for assessing the impact of this rare hydrometeorological event. At the Department of Geography of the University of Exeter, an investigation of the water quality characteristics of an agricultural catchment was in progress at this time. The existence of a detailed monitoring programme and of data from previous more 'normal' conditions affords a valuable opportunity to assess the impact of the drought and the subsequent autumn rainfall on the pattern of solute behaviour.

0360–1269/78/0403–0393$01.00

The study area

The study catchment (Figure 1a) has a drainage area of 9·3 km^2 and the basin ranges in altitude from 21·5 m at the catchment outlet to 235 m on the Raddon Hills which form the northeast divide. It is underlain (Figure 1b) by breccias, conglomerates and sandstones of Permian age which occupy part of the Crediton trough, a depression flanked on both the north and south by Carboniferous Culm Measures.

Two major soil types have been identified within the catchment (Clayden (1971)). Well drained brown earths of the Crediton Series are found in the extreme north of the catchment and similar soils of the Bromsgrove and Bridgenorth Series occur in the southern third of the basin. The central area is occupied by the gleyed and poorly draining brown earths of the Shaldon and Rixdale Series.

With the exception of minor roads and scattered farm buildings, no building development has taken place. Land use is predominantly mixed arable farming with cattle and sheep rearing and rotations of grain and root crops. The majority of the pasture is in leys. Extensive tile drain systems exist in the areas with gleyed and poorly draining soils, especially in the Yendacott subcatchment to the east (Figure 1a). There are some small areas of mixed deciduous woodland, but this occupies less than 4 per cent of the catchment.

The monitoring programme

The monitoring programme was primarily designed to document the major characteristics of the solute response of the streams draining the catchment and some results from the Yendacott subcatchment have been presented by Foster (1977). Water quality and streamflow data have been collected at four measuring sites (Figure 1a) *viz*:

(i) Pynes Cottage at the outlet of the main basin,
(ii) Yendacott at the outlet of a 1·6 km^2 subcatchment,
(iii) A small woodland subcatchment within the Yendacott basin,
(iv) A tile drain outlet within the Yendacott subcatchment.

Automatic pump sampling equipment programmed to operate at half hourly or hourly intervals during storm events and on a regular six-hourly basis has been installed on the two largest streams. At the two remaining sites water samples have been collected manually on a weekly basis and, on occasions, at half hourly intervals during storm events using automatic vacuum samplers. Continuous discharge records are available from all four sites; trapezoidal flumes have been installed at Pynes Cottage and Yendacott and triangular thin-plate weirs are used at the other two locations. Autographic rainfall records are obtained from a Casella natural siphon recorder sited within the catchment.

The water samples collected have been analysed for seven solute parameters, namely, specific electrical conductance ($\mu s. cm^{-1}$), and the concentration (mg/l) of potassium (K), calcium (Ca), sodium (Na), magnesium (Mg), chloride (Cl), and nitrate–nitrogen (NO_3–N). Specific electrical conductance (S.E.C.) was measured using a Portland direct-reading conductivity meter. Chloride and nitrate–nitrogen concentrations were determined using a Technicon Autoanalyser system and the remaining constituents were measured on a Pye-Unicam Atomic Absorption Spectrophotometer.

Records are available from all four sites for the drought period and the ensuing autumn rains although, with the exception of the stream at Pynes Cottage, streamflow ceased during the height of the summer drought. Total record length varies between the measuring locations. Measurements commenced at Yendacott in May 1974, on the woodland subcatchment and at the tile drain outlet in September 1974 and at Pynes Cottage in January 1976. The availability of records for the summers of 1974 and 1975 for the Yendacott site enables useful comparisons to be made with the drought response of 1976.

RESULTS

The hydrometeorological background

In an attempt to isolate some of the factors responsible for the extreme conditions experienced during the summer of 1976, three years of rainfall record from Exeter Airport were compared with the 30-year

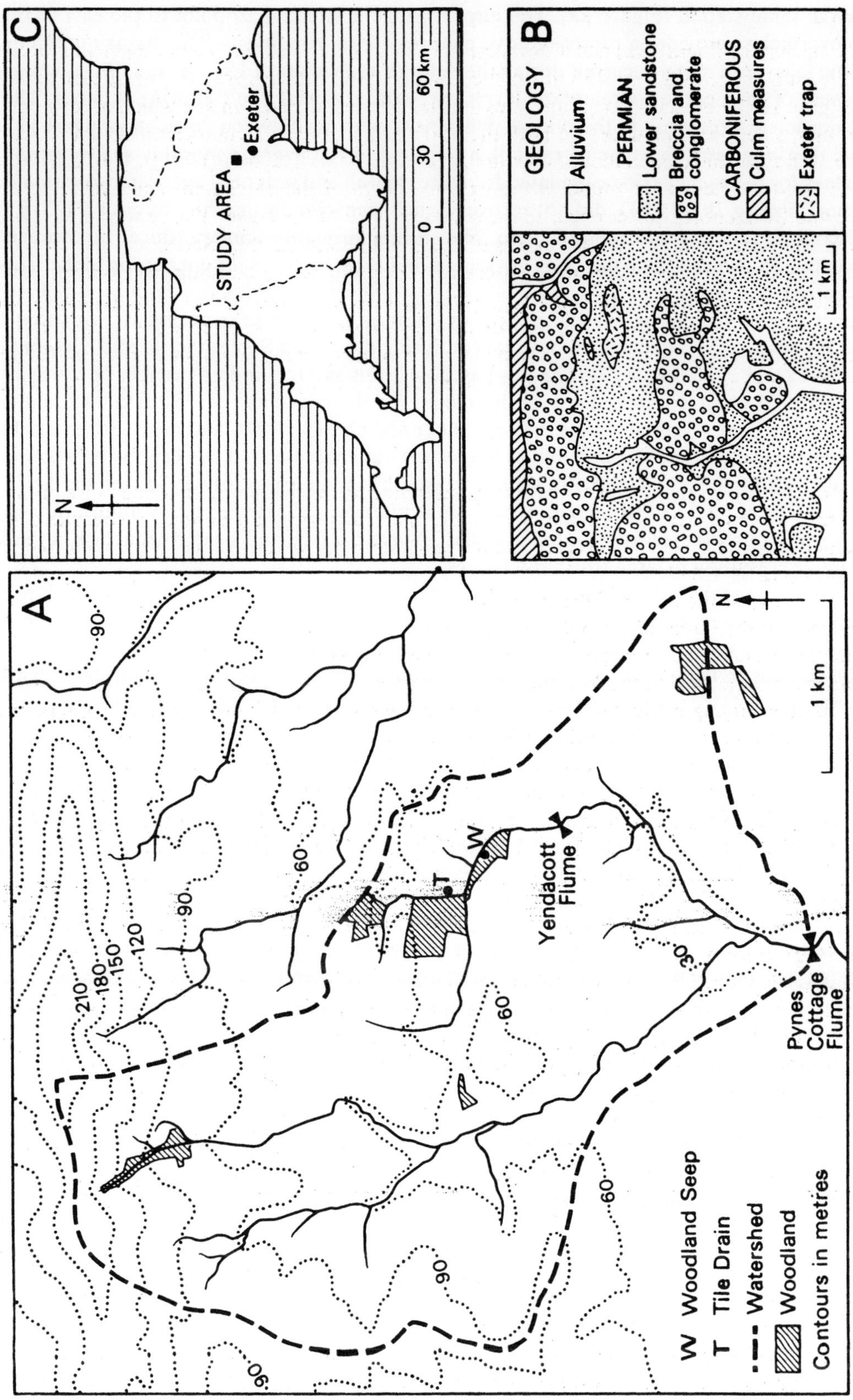

Figure 1. The study catchment, its location and underlying geology

average for the same station (Figure 2a). The airport gauge is situated 10·5 km to the east of the study basin, but nevertheless provides a representative picture of local conditions. The major rainfall deficit in relation to the long-term mean occurred during the autumn and winter of 1975–6 and this is thought to be a primary cause of the very low streamflow levels that existed during the following summer. These low levels were not solely a response to the rainfall deficiency occurring between April and August 1976. The net result of this long-term shortage of rainfall is reflected in the average monthly soil moisture deficit (S.M.D.) values for the same period calculated from the rainfall and potential evapotranspiration data for Exeter Airport (Figure 2b). S.M.D. did not recover to field capacity during the winter of 1975–6 and the deficit for March 1976 was greater than 20 mm. The ensuing drought period produced an average deficit of over 140 mm in August and the S.M.D. record indicates a trend of increasing deficit over the three-year period depicted.

One impact of this increasing rainfall deficiency during the three-year period is evidenced by the streamflow response of the Yendacott subcatchment during the period September to mid-October for the years 1974, 1975 and 1976 (Figure 3). The low baseflow levels and the limited number of storm events in the autumn of 1975 contrast markedly with the record for the previous year. Streamflow data are also presented from the Pynes Cottage gauge for September and October 1976 and the hydrographs for both sites exhibit two major phases. The first half of the record was characterized by very low baseflow levels at Pynes Cottage, whilst the Yendacott stream had ceased flowing during July 1976, for the first time in living memory. The second half of the record at both stations was characterized by intense storm-period activity with a pronounced excess of rainfall over the 30-year average and a rapid reduction in the S.M.D. levels (Figure 2b) resulting in large storm discharges.

It can therefore be seen that the background to the 1976 drought can be traced to the lack of rainfall and the high S.M.D. levels associated with the previous autumn and winter and that the extreme drought conditions of August and early September 1976 were followed by an outstandingly wet autumn. Since many of the streams in the study catchment were dry during the peak of the drought, little can be said about this aspect of solute response. Attention will be focused on the impact on the solute response of the high autumn rainfall falling on a drought-stricken catchment.

Solute levels in the post-drought period: a comparative view

The general patterns of solute response recorded by weekly sampling during the period September to December for the years 1974–6 at Yendacott and for the years 1975 and 1976 at the woodland subcatchment and the tile drain outlet are presented in Figure 4. The availability of three years of data for Yendacott enables the solute levels exhibited during September to December 1976 to be compared with those associated with the more 'normal' conditions of 1974 and the dry autumn of 1975. The 1976 solute data from Yendacott exhibit major increases over previously recorded levels for S.E.C., Ca, Mg, and NO_3–N during the post-drought period, but a less significant differentiation of Na and Cl behaviour. With the exception of NO_3–N, levels of all ions were significantly higher at the end of the 1976 drought than in the previous two years and three measures (S.E.C., Ca, and NO_3–N) subsequently exhibit marked flushing peaks. By the latter part of the period, however, all measures with the exception of NO_3–N and Cl concentrations exhibit levels below those recorded during 1975 and similar to those found in 1974. It would seem that by this point in time the catchment has reverted to a more characteristic wet weather response which in turn contrasts with the solute levels recorded during the dry autumn of 1975.

Results for the period September to December from the tile drain outlet and the woodland subcatchment are also presented in Figure 4, but in these cases the length of record available only permits comparison with the levels recorded during 1975. The woodland stream flowed continuously in 1975, but ceased to flow during August 1976. Solute response throughout the 1975 period evidenced little variation in level for most measures but the onset of flow in October 1976 was marked by a major increase in all concentrations with the exception of Cl (Figure 4b). Concentrations during the onset of runoff at the end of the drought period were frequently more than three times those recorded in previous years. Levels returned to those recorded in the previous year within a period of two to six weeks.

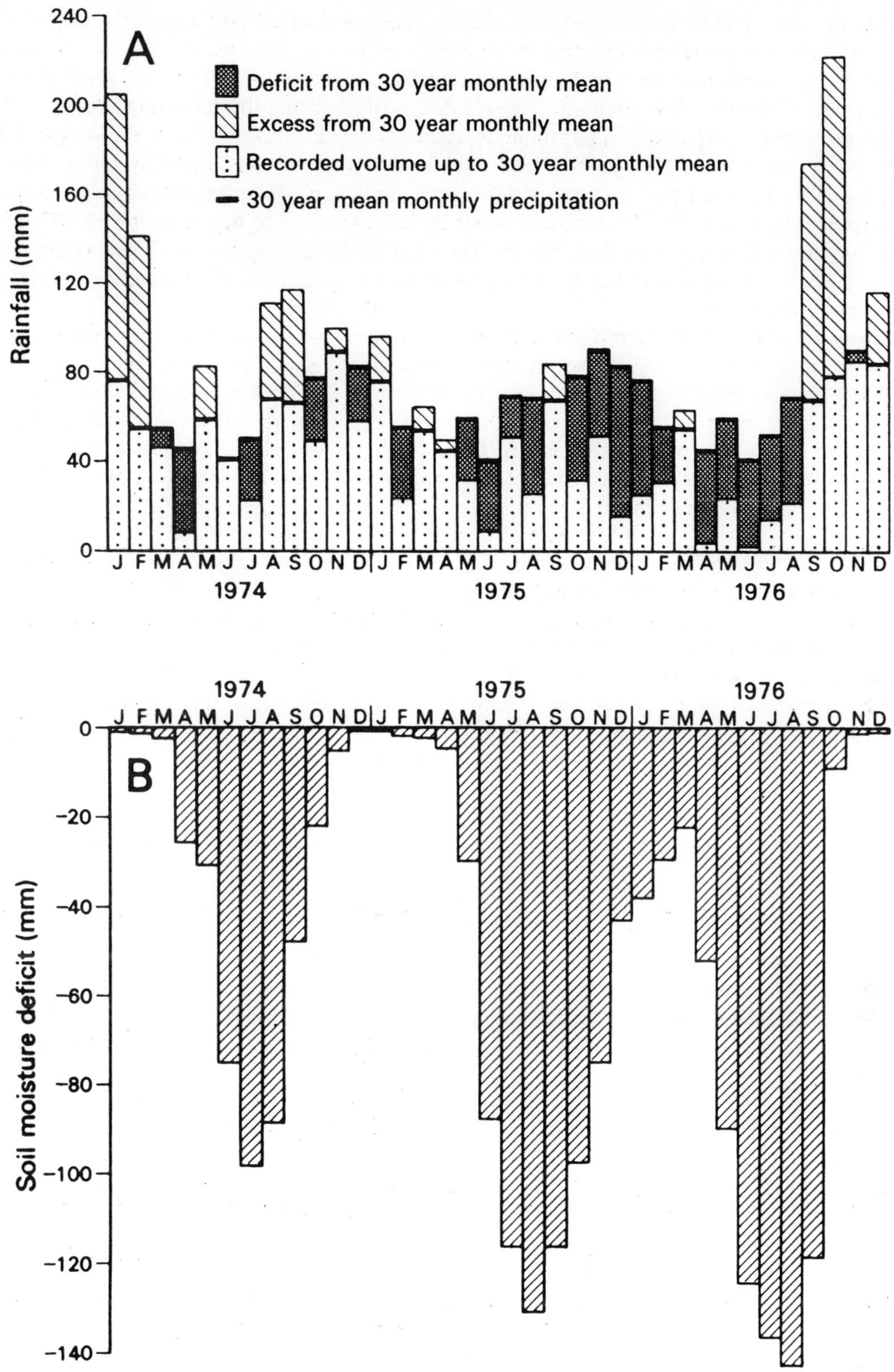

Figure 2. Rainfall and monthly mean soil moisture deficit values at Exeter Airport for the period 1974–6

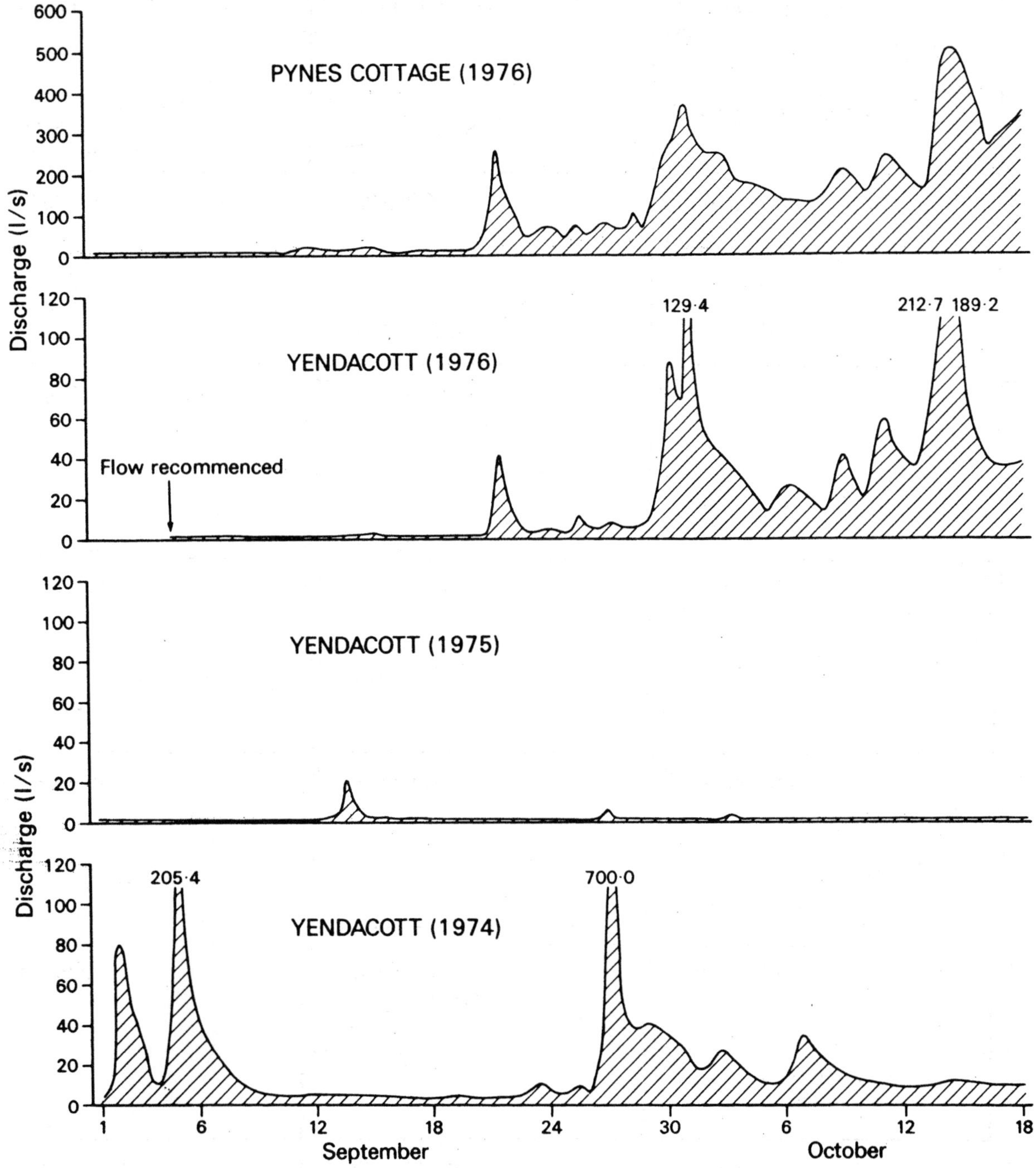

Figure 3. Streamflow at Pynes Cottage during September and early October 1976 and comparison of streamflow at Yendacott, for the same period, 1974–6

The tile drain outlet exhibited an intermittent pattern of flow over the period of record but comparison of solute concentrations during the resumption of flow in the autumn of 1975 and of 1976 suggests that the extreme conditions of 1976 produced a markedly different solute response (Figure 4c). Little variation in solute levels was encountered between October and December 1975, but, like the woodland subcatchment, autumn 1976 produced a considerable increase in concentrations. Most striking was the

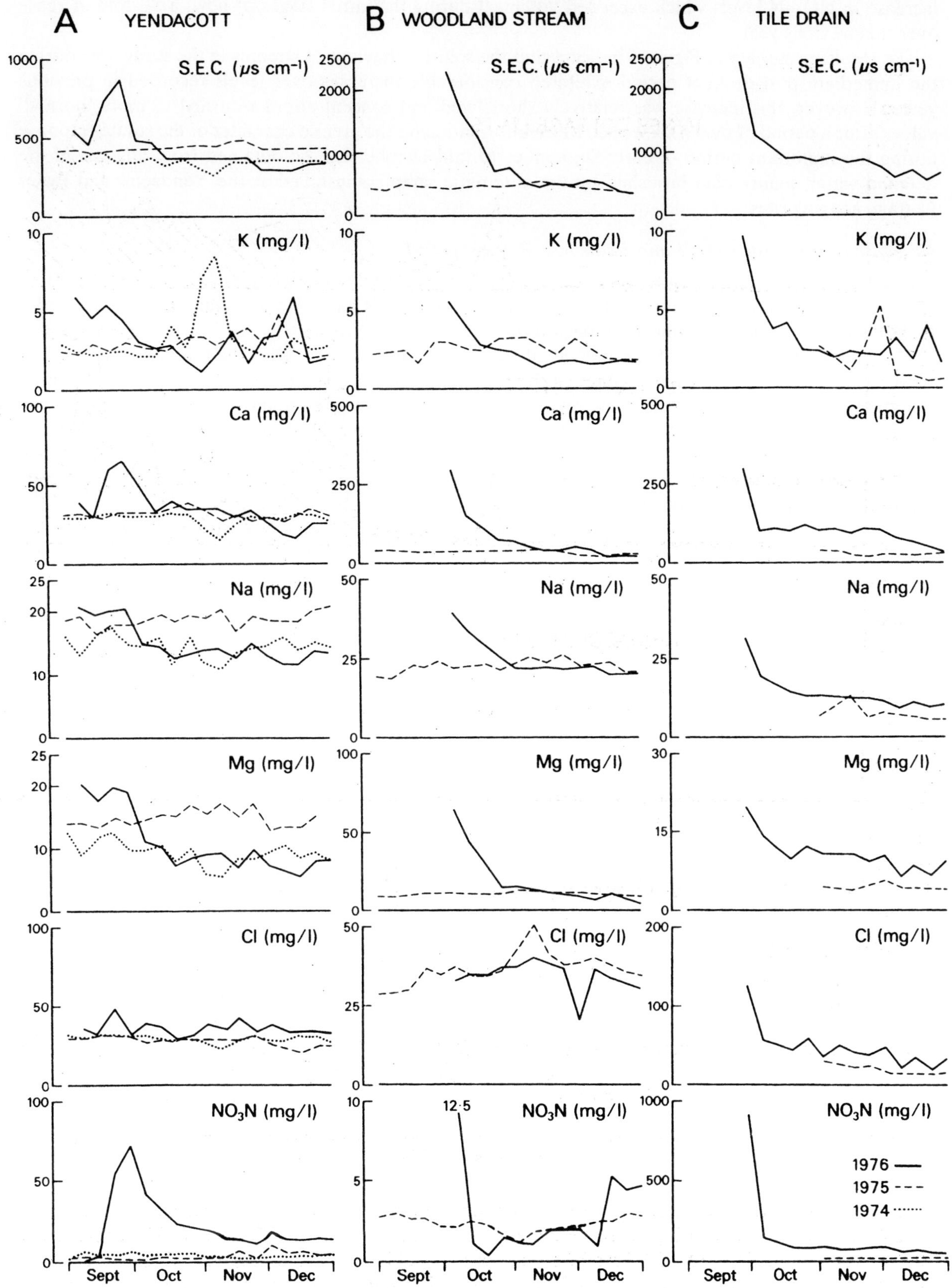

Figure 4. Solute response at three sites during the period September to December 1976, in comparison with previous years levels (based on weekly samples).

increase in NO_3–N levels which exceeded 900 mg/l during the initial stages of flow, a 50-fold increase over the previous year.

The results presented in Figure 4 indicate that the solute behaviour of streams in the study area during the immediate post-drought period exhibited considerable increases over levels recorded in previous years. However, this increase was relatively short-lived and concentrations returned to more 'normal' values after a period of five to six weeks. More information on the precise character of the solute response during the important period of early September to mid-October can be obtained from analysis of the detailed water quality data furnished by the automatic samplers installed at the Yendacott and Pynes Cottage gauging sites.

Response characteristics of the immediate post-drought period

A detailed record of the solute behaviour of the Yendacott basin during September and early October is presented in Figure 5. The records for the same period of 1974 and 1975 are again illustrated in order to indicate the uniqueness of the 1976 response.

The magnitude and range of the individual solute measures during this period demonstrate a close similarity between 1974 and 1975 although the range was slightly greater during 1974 consequent upon the greater diversity of the streamflow regime (Table I). With the exception of K, all solute measures exhibited significantly increased concentrations during 1976, and a greater range of solute levels over the same period. The most significant increases were evidenced by Ca, Cl and NO_3–N which reached maximum levels of 101·0, 75·9 and 99·6 mg/l compared to peak values of 34·2, 36·4 and 8·1 during the same period of the previous two years.

It is clear that the drought had a significant effect on the maximum solute levels recorded in this stream, with NO_3–N levels approaching 13 times those recorded previously. The detailed solute response subsequent to the drought provides further evidence of unique behaviour. With the exception of K

Table I. Minimum, maximum and range of solute concentrations during the period 1 September to 18 October at Yendacott and Pynes Cottage catchments*

Yendacott

Chemical parameter	Minimum			Maximum			Range		
	1974	1975	1976	1974	1975	1976	1974	1975	1976
S.E.C.	260·5	351·4	356·0	403·4	437·3	1039·0	142·9	85·0	683·0
K	2·1	2·3	2·2	7·6	6·9	7·6	5·5	4·6	5·4
Ca	20·3	22·3	31·8	32·0	34·2	101·0	11·7	11·9	69·2
Na	10·4	14·1	12·2	17·6	19·4	25·4	7·2	5·3	13·2
Mg	4·6	9·3	6·7	13·2	15·7	24·0	8·6	6·4	17·3
Cl	25·6	27·8	27·0	32·0	36·4	75·9	6·4	8·6	48·9
NO_3–N	3·1	2·2	0·0	8·1	6·8	99·6	5·0	4·6	99·6

Pynes Cottage

Chemical parameter	1976		
	Minimum	Maximum	Range
S.E.C.	280·0	760·0	480·0
K	3·2	15·2	12·0
Ca	20·0	47·3	27·3
Na	12·5	25·1	12·6
Mg	6·2	20·5	14·3
Cl	20·9	48·1	27·2
NO_3–N	4·9	31·5	26·6

* S.E.C. expressed in $\mu s. cm^{-1}$. Remaining solutes in mg/l.

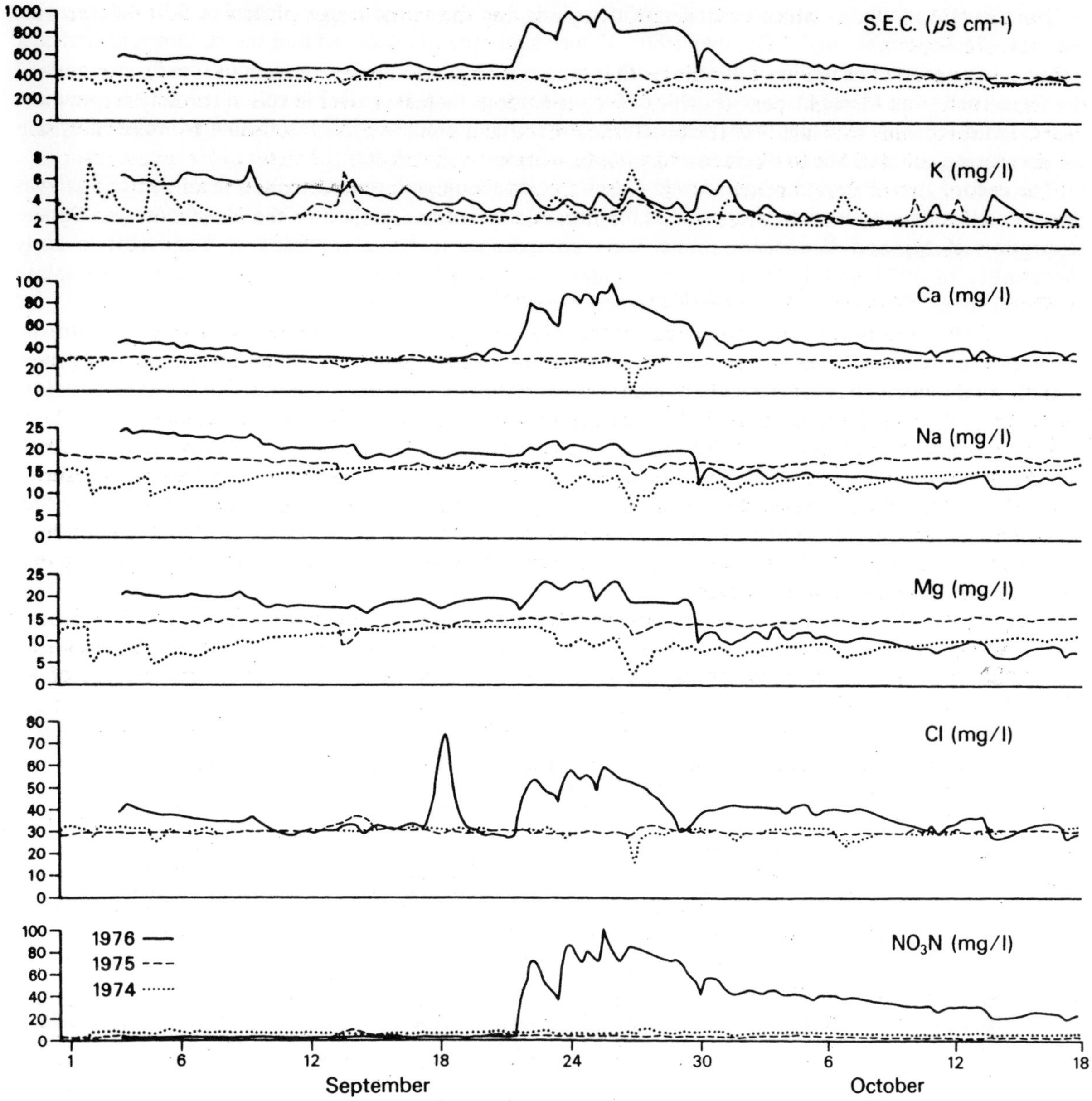

Figure 5. Detailed solute response at Yendacott during September and early October 1976 compared to that of the two previous years

concentrations, the dominant response of this period during 1974 and 1975 was one of simple dilution from the prevailing baseflow concentrations, in response to increasing discharge during storm runoff events. During late September 1976, however, the onset of a period of heavy rainfall and associated storm runoff resulted in an increase in solute levels with increasing discharge. The period of less intense rainfall at the beginning of September 1976 produced no significant storm runoff, but resulted in an increase in background solute concentrations over those of the same period in the previous years for all measures except NO_3–N. During the period 18–19 September 1976, Cl concentrations exhibited a major peak but no other solute measure responded in a similar way.

The increase in solute concentrations with increasing discharge produced a general peak in solute levels between 21 September and 3 October 1976. Values subsequently declined and the stream returned to a more characteristic behaviour of dilution with increasing discharge. A similar trend is clearly emphasized by the record from Pynes Cottage (Figure 6) for the same period. Increases in solute levels in response to increased discharge were highly characteristic of the initial period, but this response was replaced by one of decreasing concentration with increasing flow during early October.

The record from Pynes Cottage affords a useful comparison with that from Yendacott, since flow was sustained throughout the drought period at the former site. During August 1976 all solute levels, with the exception of Mg and K concentrations, were comparable to those recorded at Yendacott in early September of 1974 and 1975, reflecting similar low flow conditions. The general peaking of solute concentrations during late September and early October is well marked by the S.E.C., Ca, Na, Cl, and NO_3–N levels and its timing corresponds closely to that exhibited at Yendacott. However, with the exception of K concentrations, the maximum level and range of the individual solute measures were significantly lower at Pynes Cottage (Table I). This would suggest a dilution of the Yendacott response by water from the western tributaries of the main catchment.

Records obtained for two storm periods during September and October 1976 at Pynes Cottage (Figure 7) can be used to emphasize the different response characteristics associated with the onset of storm rainfall in mid-September and the decline of the post-drought flush in mid-October. For convenience, only NO_3–N, Mg and Cl concentrations have been presented for these events since in combination they demonstrate the major facets of response exhibited by the solute measures under study.

During the two storm hydrographs of the September storm period, there is evidence of a small dilution associated with the rising limb of the hydrograph, but the dominant component of the solute response is the flushing peak which lags the peak discharge by approximately 12 hours. This is very marked in the case of NO_3–N and Cl but is absent from the Mg trace which is dominated by dilution coincident with the discharge peak. Clearly it is not possible to formulate a simple explanatory model which is generally applicable to all solute species. The dilution trend can be accounted for in terms of an input of less concentrated water associated with direct storm runoff. The timing of the flushing peak is reminiscent of the timing of peak throughflow contributions to storm runoff events (e.g. Anderson and Burt (1977), Troake and Walling (1973), Weyman (1970)). It is suggested, therefore that this flushing peak is related to a throughflow storm runoff component (i.e. lateral movement of water through the soil) and that the source of high solute levels is the solution of soluble residues which accumulated in the soil during the drought period, for example as a result of evaporation of soil moisture and capillary rise. That Mg concentrations show no evidence of a flushing peak must remain something of a problem. The rapid decline of solute levels that follows the flushing peak can be ascribed to the gradual reduction and cessation of the throughflow component of storm runoff.

Solute behaviour during the two storm hydrographs of the October storm period (Figure 7) is characterized by a clear dilution response in phase with the discharge peaks. It would appear that most of the readily soluble material had already been removed by this point in time and that the resultant low solute content of direct storm runoff gave rise to the dilution trough. The generally lower levels associated with Mg and Cl concentrations can also be ascribed to the exhaustion of the store of soluble material accumulated during the drought. The relatively high background levels of NO_3–N on which the dilution trend is superimposed can be accounted for in terms of the increased importance of soil water in the baseflow when compared to the drier September conditions. Soil water provides a source of nitrate to sustain these high background levels.

DISCUSSION

The accumulation of readily soluble material during the drought period as a result of the evaporation of soil moisture and associated capillary rise, the drying and scorching of plant growth, the opening of new pathways for moisture movement through the soil by desiccation cracks, and the deposition of dry fallout and leaf exudants uninterrupted by rainfall must largely account for the increase in solute levels

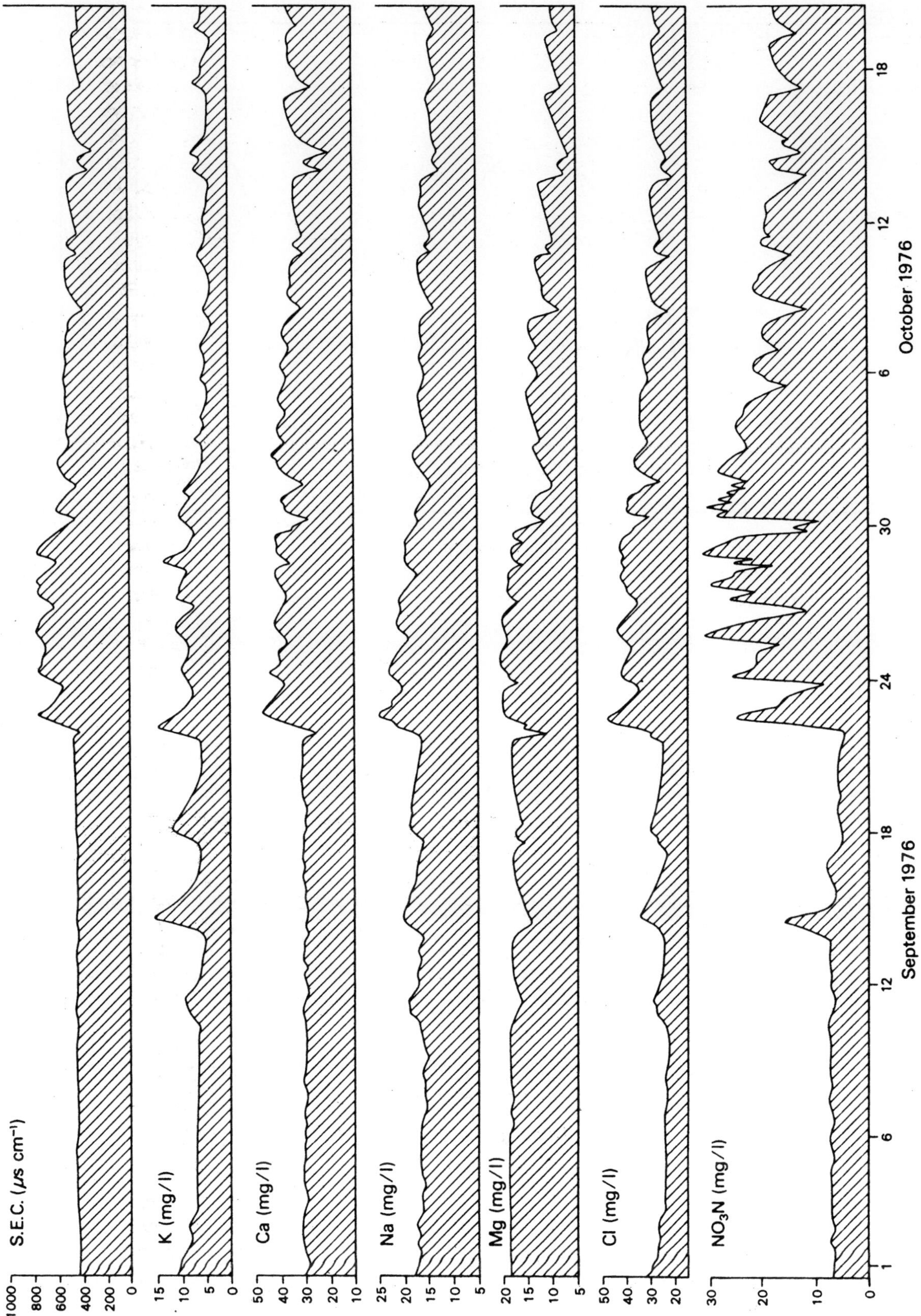

Figure 6. Detailed solute response at Pynes Cottage during September and October 1976

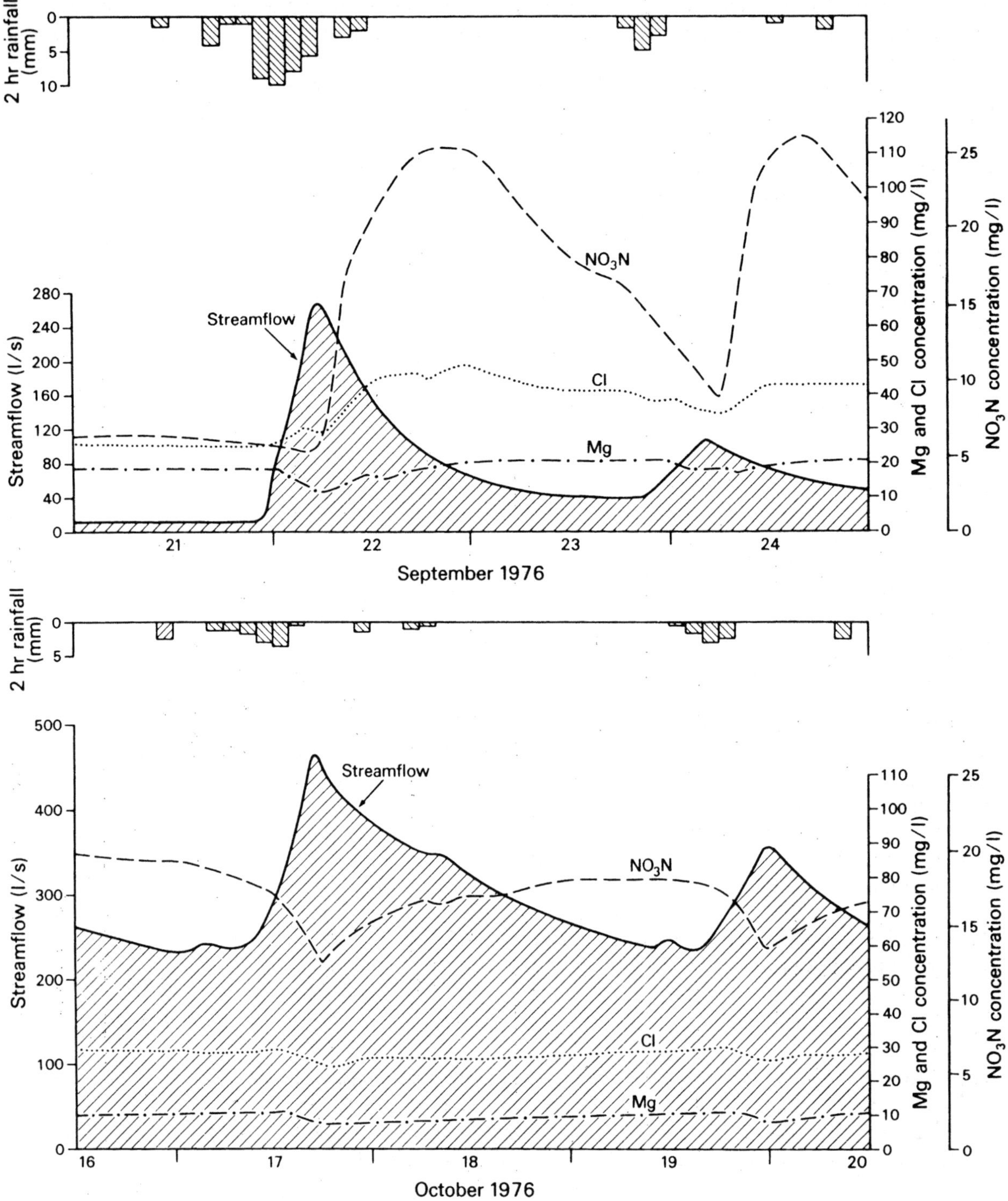

Figure 7. Some characteristics of storm solute response at Pynes Cottage during the beginning and end of the autumn flush 1976

associated with the end of the drought. The dry conditions associated with the autumn, winter and spring of 1975–6 could be important in this context as well as the summer peak drought period.

The variable behaviour of the individual solute measures can be related to the differential impact of these drought factors on their dominant sources. Ionic species supplied largely from atmospheric sources can be expected to respond differently to those originating from weathering of the soil and regolith. Such considerations may in part explain the contrasts between the behaviour of Na and Cl concentrations and those of the other solutes, for atmospheric sources are generally accepted to be important in the case of these two ions (e.g. Holden (1966)). The behaviour of K concentrations during the post-drought period differs markedly from that of other solutes, although increased levels are evidenced. In this case precipitation/vegetation and solute/sediment interactions are important (e.g. Hem (1970)) and consideration must therefore be given to the impact of drought on foliar leaching processes and on the suspended sediment dynamics of the streams. Such considerations may also be important for other solutes. The lack of marked flushing peaks associated with Mg response during the post-drought period presents problems in terms of explanation. Relative solubilities of substances precipitated within the soil during the drought may, however, account for this and other contrasts in solute behaviour.

The effect of the drought on NO_3–N levels deserves special attention, for the associated increases in concentration rise well above those found for other solute parameters. More detailed chemical analyses of the major cations and anions in the streamwaters would enable an ionic balance to be evaluated and the cation concentrations linked to the increase in NO_3–N. Nevertheless, the available data suggest that the high NO_3–N levels are balanced by smaller increases in a number of cations. For these an upper bound of concentration increase of 3- to 4-fold seems applicable, whereas NO_3–N values increased by up to 50-fold. A specific mechanism must be sought to explain the marked rise in NO_3–N concentrations and it is likely that this lies in the interaction between inorganic and organic nitrogen within the soil. There is considerable body of evidence to suggest that severe desiccation and subsequent wetting of the soil can markedly increase rates of mineralization of organic nitrogen. This, coupled with subsequent nitrification (e.g. Russell (1961)) could account for the massive increases in NO_3–N levels found in the streams of the study area, for nitrate is readily leached from the soil. Some of the increases in cation concentrations will therefore be associated with this breakdown of organic material as well as the physical effects of the drought listed previously.

The 1976 drought can thus be seen to have had a very significant impact on solute levels in the streams of the study area. Attention has been focused on concentrations, but equally solute loadings could have been considered. The increased concentration coupled with the high runoff levels of the post-drought period would have resulted in marked increases in solute loads and rates of solute removal from the catchments. These in turn would have important implications for solute budgeting and rates of chemical denudation.

Traditional attempts to model general solute behaviour in terms of bivariate concentration/discharge models (e.g. Hall (1970), Hem (1970), Steele (1968)) face many problems in their failure to incorporate considerations of such factors as seasonal contrasts in basin response, temporal and spatial variability of runoff dynamics and non-stationarity of catchment condition. The impact of drought conditions introduces another feature which cannot be adequately modelled by this approach and serves to demonstrate further the complexity of catchment solute processes.

ACKNOWLEDGEMENT

The authors gratefully acknowledge the support of the Natural Environment Research Council who provided a Postgraduate Studentship (I.D.L.F.) and a Research Grant (D.E.W.) for work on solute dynamics at Exeter. Mr W. J. Bligh provided valuable analytical assistance. S.M.D. and rainfall data for Exeter Airport were kindly made available by the Meteorological Office.

REFERENCES

Anderson, M. G., and Burt, T. P. (1977). 'Automatic monitoring of soil moisture conditions in a hillslope spur and hollow,' *Journal of Hydrology*, **33**, 27–36.

Clayden, B. (1971). *Soils of the Exeter District*, Harpenden, Soil Survey.

Foster, I. D. L. (1977). 'Solute dynamics in a small agricultural catchment,' *unpublished Ph.D. thesis*, University of Exeter.

Gunnerson, C. G. (1967). 'Streamflow and quality in the columbia river basin,' *Proceedings of the American Society of Civil Engineers, Journal of Sanitary Engineering Division*, **93**, 1–16.

Hall, F. R. (1970). 'Dissolved solids—discharge relationships. 1. Mixing models,' *Water Resources Research*, **6**, 845–850.
Hem, J. D. (1970). 'Study and interpretation of the chemical characteristics of natural water,' *United States Geological Survey; Water Supply Paper 1473.*
Hendrickson, G. E., and Krieger, R. A. (1964). 'Geochemistry of natural waters of the Blue Grass Region, Kentucky,' *United States Geological Survey; Water Supply Paper 1700.*
Holden, A. W. (1966). 'A chemical study of rain and stream waters in the Scottish Highlands,' *Freshwater and Salmon Fisheries Research Report 37.*
Russell, E. W. (1961). *Soil Conditions and Plant Growth*, London, Longmans.
Steele, T. D. (1968). 'Digital computer applications of chemical quality studies of surface waters in a small watershed,' *International Association Scientific Hydrology, UNESCO Symposium on Use of Analogue and Digital Computers in Hydrology, Tucson, Arizona*, 203–214.
Troake, R. P., and Willing, D. E. (1973). 'The natural history of Slapton Ley Nature Reserve. VII, the hydrology of Slapton Wood stream: a preliminary report, *Field Studies*, **3**, 719–740.
Walling, D. E., and Foster, I. D. L. (1975). 'Variations in the natural chemical composition of river water during flood flows, and the lag effect: some further comments,' *Journal of Hydrology*, **26**, 237–244.
Weyman, D. R. (1970). 'Throughflow on hillslopes and its relation to the stream hydrograph,' *International Association Scientific Hydrology, Bulletin*, **15**, 25–33.

Commentary on Paper C3

Pilgrim, D. H., Huff, D. D. & Steele, T. D. (1979) Use of specific conductance and contact time relations for separating flow components of storm runoff.

One of the paradoxes of catchment hydrology is the rapid mobilisation of large amounts of "old" (long-residence time) water during times of storm runoff (Kirchner, 2003). The significant contribution of subsurface flow to storm runoff had been well known for decades (Hursh & Brater, 1941, Paper B2). That much of the water involved had resided in the catchment before the storm event was confirmed by publications such as Pinder & Jones (1969) who sought to determine the groundwater contribution to peak discharge on the basis of the runoff chemistry. Since then, a number of varieties of end-member mixing models have been applied in order to achieve hydrograph separation and identify the principal sources of stormflow (see reviews by Bonell, 1993; Buttle, 1994).

The paper by Pilgrim and colleagues is one of the earlier examples of the use of mixing models, but it had significant impact because of the way in which it combined model calculations with laboratory investigations on the residence time of water in the soil. It therefore began to open up the crude "black box" provided by the standard two-component mixing model and moved towards process-based interpretation of chemographs. In so doing, the authors caution against overly simplistic labelling of flow components on the basis of estimates of "old" and "new" water: "stormflow" may include subsurface as well as surface runoff and all subsurface runoff is not "baseflow". Their work had the great advantage that it was underpinned by process observations in the field (Pilgrim & Huff, 1978; Pilgrim *et al.*, 1978), although there is less cross-reference to the field observations than there might have been, perhaps deliberately in order to test out their method. The field plot near Stanford, California, USA, generated all three types of storm runoff – infiltration-excess and saturation-excess overland flow, and subsurface stormflow (which resulted from flow through the macropores rather the soil matrix).

The authors note that the diluting effect of storm flows has been used by many researchers to estimate the different flow components of storm discharge but that there are two complications with this approach: flushing of solutes early in the event, and the increase in solute concentration with increasing time of contact of water in the soil. It is the latter issue that they explore, but they then use their results to analyse the former effect too. Hitherto, the effect of increased residence time on the solute concentration of new water had not been taken into account. By some simple but effective laboratory experiments involving shaking mixtures of soil and deionised water over 7 hours, they are able to calibrate contact time and solute concentration (as indicated by specific conductance). This calibration curve is then used to produce new estimates of flow components for three storms (two natural and one via sprinkler irrigation) and to compare with mixing model estimates of old and new water. Their analysis shows that the subsurface flow comprised mainly new water, as did the surface flow. Only later in the recession of the large fourth may event did the flow become mainly old water; they postulate forcing out of old water from the soil matrix but solute exchange between mobile and immobile water may also have been involved. The contact time–solute concentration curves are also used to give estimates of old water contributions to the early part of the 1 April storm when relatively high solute concentrations were observed on the rising limb of the storm hydrograph ("flushing").

The authors conclude that neglecting the increase in concentration with contact time and using fixed values for the concentration of old and new water (i.e. the "end members" in the mixing model) will probably lead to poor estimates of the relative contribution of flow components, and perhaps more importantly, to incorrect inferences about the sources of storm runoff. As noted earlier, they also caution against overly simplistic labelling of flow components in relation to estimates of old and new water. The need for detailed information or observations about the nature of processes operating is underlined, an advantage that Pilgrim, Huff and Steele had on the basis of their earlier work (Pilgrim & Huff, 1978; Pilgrim *et al.*, 1978) and which is taken forward in more detail in Waddington *et al.* (1993, Paper C6).

TPB

References

Bonell, M. (1993) Progress in the understanding of runoff generation dynamics in forests. *J. Hydrol.* **150**, 217–275.

Buttle, J. M. (1994) Isotope hydrograph separations and rapid delivery of pre-event water from drainage basins. *Progr. Phys. Geogr.* **18**, 16–41.

Hursh, C. R. & Brater, E. F. (1941) Separating storm-hydrographs from small drainage-areas into surface- and subsurface-flow. *Trans. Am. Geophys. U.* **22**, 863–871.

Kirchner, J. W. (2003) A double paradox in catchment hydrology and geochemistry. *Hydrol. Processes* **17**, 871–874.

Pilgrim, D. H. & Huff, D. D. (1978) A field evaluation of subsurface and surface runoff. 1, Tracer studies. *J. Hydrol.* **38**, 299–318.

Pilgrim, D. H., Huff, D. D. & Steele, T. D. (1978) A field evaluation of subsurface and surface runoff. 2, Runoff processes. *J. Hydrol.* **38**, 319–341.

Pinder, G. F. & Jones, J. F. (1969) Determination of the groundwater component of peak discharge from the chemistry of total runoff. *Water Resour. Res.* **5**, 438–445.

Waddington, J. M., Roulet, N. T. & Hill, A. R. (1993) Runoff mechanisms in a forested groundwater discharge swamp. *J. Hydrol.* **147**, 37–60.

Use of Specific Conductance and Contact Time Relations for Separating Flow Components in Storm Runoff

DAVID H. PILGRIM

School of Civil Engineering, The University of New South Wales Kensington, New South Wales 2033, Australia

DALE D. HUFF

Environmental Sciences Division, Oak Ridge National Laboratory, Oak Ridge, Tennessee 37830

TIMOTHY D. STEELE

Water Resources Division, U.S. Geological Survey, Denver, Colorado 80225

The difference between the dissolved-solids concentration in base flow and in storm flow has often been used as the basis for separating components of flow. However, an analysis that explicitly relates the amount of time that runoff water has been in contact with watershed soils to the resulting dissolved-solids concentration shows that simple mass balance chemistry methods for hydrograph separation are misleading. Field studies of surface and subsurface storm flow, when coupled with laboratory determination of the relationship between contact time and dissolved solids content of a soil water mixture, suggest that the residence time of infiltrated water is as short as a few hours in the cases studied. In those cases, hydrograph separation methods based on the simple mass balance equation for the dissolved solids will yield considerable overestimates of the base flow component.

INTRODUCTION

The concentration of dissolved solids and the discharge of a stream are often inversely related: Concentration is low during direct storm runoff periods and high during base flow periods. Relationships of this type are well documented in the literature [e.g., *Hem*, 1970; *Hall*, 1971].

The diluting effect of storm flows has been used by many workers to estimate the relative contributions of different flow components to the total discharge, and to calculate the actual hydrographs of the different components. Separation of the components has been based on the steady state form of the mass balance equation for dissolved solids,

$$C_1Q_1 + C_2Q_2 = C_TQ_T \tag{1}$$

where C is the concentration of dissolved solids, Q is the discharge rate, and 1, 2, and T are the two flow components and the total flow, respectively. Equation (1) is usually solved by assuming that the concentration in each flow component is constant and known. Since both the total flow and its concentration are known and the total flow can be expressed as the sum of the two components, Q_1 and Q_2 can be quantified. However, there are two complications. First, large amounts of solutes are flushed from the soil in subsurface outflow during the early part of storms that occur after a prolonged period with no storm runoff. This happens when newly infiltrated water displaces older water that has resided in the soil for some time and results in high solute concentration. Several studies have documented and discussed this effect [e.g., *Steele*, 1968; *Nakamura*, 1971; *Walling and Foster*, 1975], although the considerable variations in observed results are not well explained. The second complication is the increase in concentration of dissolved solids with increasing time of contact of water with the soil. When precipitation first reaches the watershed soil, solution of soil minerals occurs rapidly. However, the solution rate soon decreases, and days or weeks may pass before effective equilibrium conditions are reached. This effect has not been considered quantitatively in any previous hydrograph separation studies.

The increase of solute concentration or specific conductance with increasing time of contact between water and soil has been recognized qualitatively in a few studies, such as those of *Feth et al.* [1964], *Davis* [1969], *Kennedy* [1971], *Newson* [1971], and *Walling* [1974]. *Shuster and White* [1971] and *Ternan* [1972] also made use of indirect information of this type in studying subsurface flow in karst regions. *Nakamura* [1971] found it necessary to employ a time-varying solute concentration in modeling the specific conductance of runoff from a watershed. However, direct information on the variation of concentration with contact time has not been utilized in the separation of different flow components.

In this paper a quantitative approach that explicitly considers the variation of concentration with contact time is developed and illustrated with data from a field study. The analysis allows valid separation of the components of recorded storm flows based upon the length of time the water constituting each component has been in contact with basin soils. However, as a result of the nature and complexity of flow processes in the field, and also the limitations of the experimental data, the results can be regarded as only semiquantitative. Nonetheless, the inferences that can be drawn from these results are in agreement with our observations of runoff conditions in the field and the findings of others. It also is shown herein that application of the mass balance approach using fixed concentrations for flow components leads to incorrect separation of flow components and invalid inferences regarding sources of flow.

ANALYSIS WITH TIME-DEPENDENT CONCENTRATIONS

If the components of flow in (1) are defined as old and new waters, where the old water has been present for sufficient time

Paper number 8W1179.

to reach an equilibrium concentration C_o, then

$$Q_o + Q_n = Q_T \qquad (2)$$

where subscripts o, n, and T refer to old, new, and total, respectively.

Combining (1) and (2) yields the concentration of the total flow as

$$C_T = C_o - (Q_n/Q_T)(C_o - C_n) \qquad (3)$$

The point of fundamental importance in this expression is that C_n is time dependent, a fact that has not been explicitly incorporated in previous analyses. Thus by developing a time-varying expression for C_n, (3) can be solved for Q_n as a continuous function of time. Then (2) can be used with observed total flow to derive the amounts of old and new water in the hydrograph.

THE FIELD EXPERIMENT

Data on variation of water quality with discharge were collected during a radiotracer study of subsurface storm runoff on a 60 × 160 ft (18.3 × 48.8 m) field plot near Stanford, California. The experimental program has been described by *Pilgrim and Huff* [1978] and *Pilgrim et al.* [1978], so that only the salient features are discussed here. The results described in this paper are based on specific conductance data, with supplemental determinations of inorganic dissolved solids, since concentrations of individual ions were not determined.

A grassed plot was selected with a uniform slope of 30% and other physical characteristics as uniform as could be found in the area. The surface soil was a brown silty loam varying from 10 to 20 inches (250 to 500 mm) deep over the upper half of the plot to 6 to 10 inches (150 to 250 mm) deep at the lower end. This was underlain by silty clay, with claystone and some sandstone at greater depths. Stainless steel troughs were inserted at the soil surface and at various depths in the soil profile to collect surface and subsurface flow at two 6 ft (1.8 m) wide benches excavated into the slope. The water was collected in polyethylene bottles, and flow rates were determined by timing the filling of the bottles. After collection of field samples, laboratory determinations of specific conductance were made. The laboratory temperature approximated 25°C at all times, so that no temperature corrections were made. Selected samples were filtered, and suspended and dissolved solids were determined gravimetrically after drying at 103°C; the inorganic and organic fractions were separated by combustion at 580°C for 30 min [*American Public Health Association*, 1960].

Two storms occurred after instrumentation of the plot was completed in early March 1968. Precipitation totals of 0.35 and 1.00 inches (8.9 and 25.4 mm) were recorded on March 16 and April 1, 1968, respectively, and runoff samples and data were obtained on both occasions. However, neither storm was of sufficient magnitude to carry out the primary tracing experiment [see *Pilgrim and Huff*, 1978]. As the dry season then commenced, the use of irrigation sprinklers was necessary to simulate storm conditions. A total of 7.0 inches (178 mm) of water was applied uniformly at a rate of 0.42 inches/h (10.7 mm/h) between 2000 hours on May 3 and 1252 hours on May 4, 1968.

Summary of Field Results

Surface runoff was recorded at each of the benches in all three events. Outflow of subsurface water from a temporary saturated layer occurred at only one of the troughs for each of the benches. In each case, this was at the trough located at the bottom of the soil A horizon, where the hydraulic conductivity decreased greatly in relation to the surface silty loam. In each event studied, a temporary saturated zone intercepted the surface near the bottom of the plot where the A horizon was shallow, and small seepage springs occurred. The surface flow on the lower end of the plot thus contained some return subsurface flow. In fact, the flows well after cessation of rain were composed entirely of this return subsurface flow.

Pronounced spatial nonuniformity was a feature of the observed flow conditions. The radioisotope tracing results showed large spatial variations in flow patterns over the superficially uniform plot. Small-scale nonuniformity was also of great importance, and most of the subsurface flow occurred in cracks, along roots, and in small holes that apparently resulted from animal or insect activity or decayed roots. As discussed by *Pilgrim et al.* [1978], the elapsed time between bursts of rainfall and the response of subsurface flow was short. The available evidence indicates that this is a common characteristic of subsurface storm flow.

Rainfall, runoff flow rate, specific conductance, and inorganic dissolved solids results for surface and subsurface runoff on April 1 and for subsurface runoff on May 4 are presented in Figure 1 for the bench at the lower end of the plot. These results are similar to those for the bench near the midpoint of the plot and for the other events. In general, results were obtained for only the latter parts of each event when personnel were at the site. The rapid response of subsurface flow to rainfall input is evident in Figures 1*b* and 1*c*.

As was noted earlier, measured values of specific conductance were the primary data for the analysis described here. The inorganic dissolved-solids (DS) concentrations, which were determined for selected samples, were used to check specific conductance observations. Several studies have indicated that the relationships that affect specific conductance of natural waters are complicated. *Steele* [1968], *Hem* [1970], and *McNeal et al.* [1970] have reported different conductance-concentration relationships for individual ionic species, and individual realtionships embodied different types of nonlinearities in some cases. With the variable flow paths of water considered here, and the possible range of rates of solution of individual ions with increasing contact time with the soil, specific conductance can only be considered as an index to water chemistry. Despite these considerations, plots of inorganic DS against specific conductance gave generally consistent but curvilinear relationships, with average deviations of individual results of about 5%. Also, the graphs of specific conductance and inorganic DS in Figures 1*b* and 1*c* are of similar general shapes. Only a few values from samples collected from the ends of recessions, when flow rates were very low, had deviations of the order of 30%. It is thus concluded that specific conductance serves as a satisfactory index for the analysis of variation of solute concentrations described here. This conclusion is supported by the results of *Steele and Matalas* [1974].

The graphs of specific conductance in Figures 1*a*–1*c* show an inverse relationship between conductance and discharge similar to that found in many previous studies. However, the results presented here are different from most previously reported studies because the flows here were collected from either surface or subsurface levels of the soil. The fact that the surface and subsurface flows each demonstrated reduction and then recovery of specific conductance led to the study of the effect of contact time (i.e., the time that water is in contact with soils) reported here.

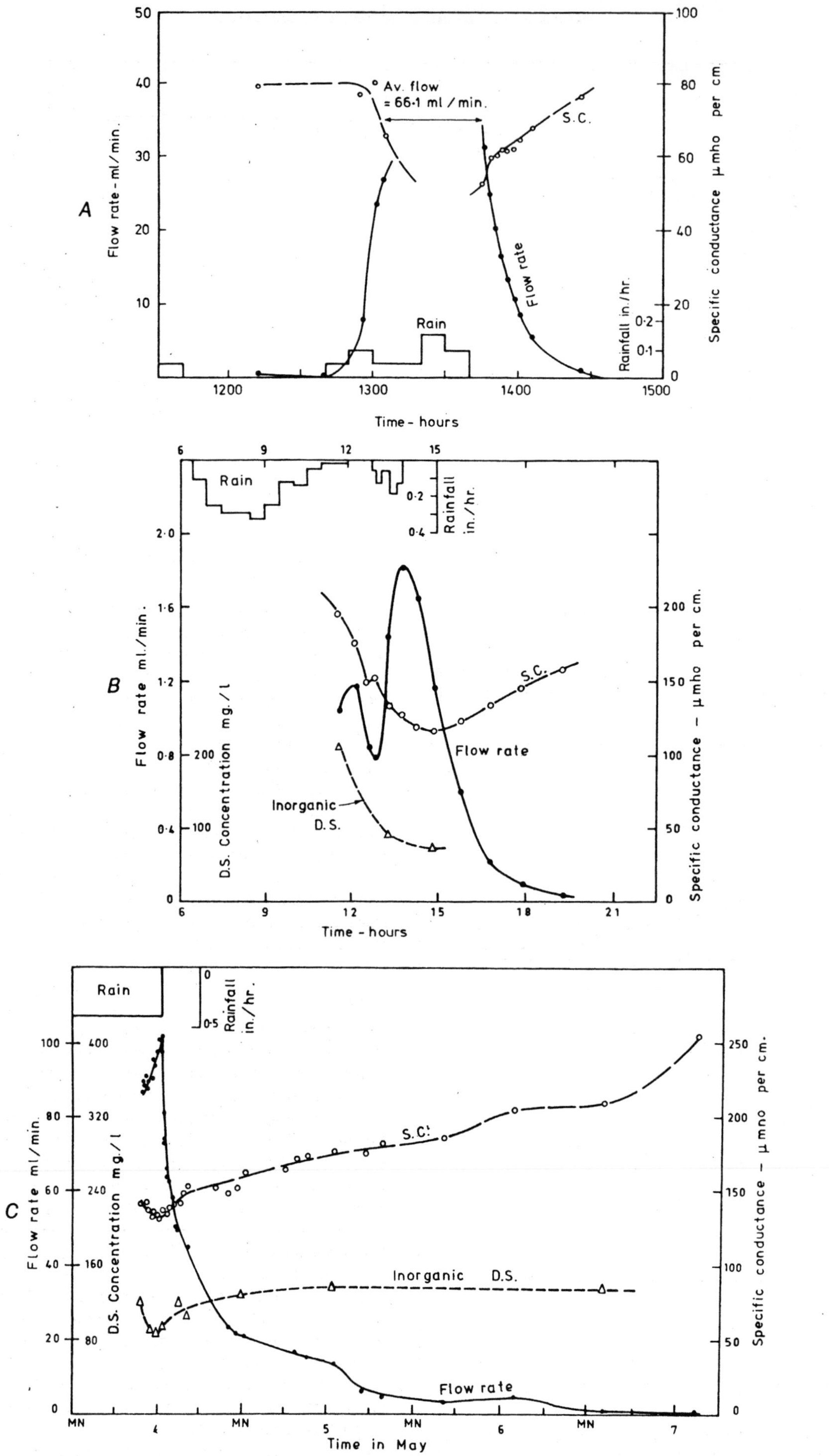

Fig. 1. Rainfall, runoff, specific conductance, and inorganic total dissolved solids for selected runoff events at the bench at the lower end of the plot.

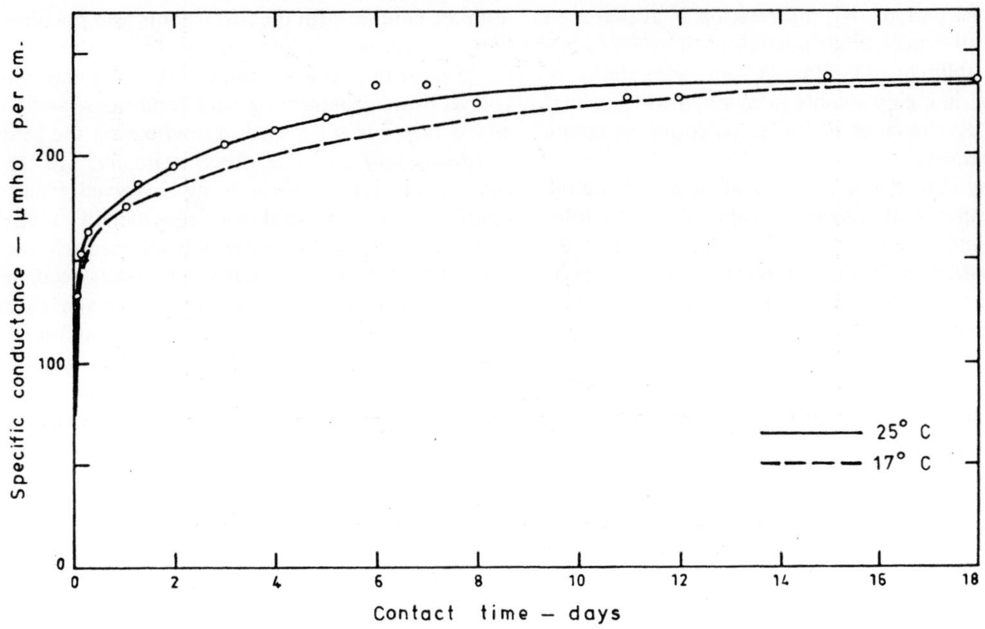

Fig. 2. Relation of average specific conductance and contact time for three soil samples.

CONDUCTANCE–CONTACT TIME DATA

Description of Experiment

A simple laboratory experiment was used to obtain information on the time dependence of specific conductance of water in contact with soil. Three samples of silty loam soil were taken from the top 10 inches (250 mm) of the soil profile and placed in clean 3.5-l polyethylene bottles with tight-fitting screw top lids. Each sample weighed about 2 kg. Approximately 2 l of deionized water was added to each bottle to make a 1:1 soil-to-water mixture by weight. Each bottle was shaken briefly by hand, and about 5 ml of liquid was then drawn off by pipette and tested for specific conductance 1 min after addition of the water. The withdrawn sample was returned, the bottles were again shaken and allowed to stand, and a further specific conductance measurement was made 5 min after addition of the water. This procedure was repeated at increasing intervals for 7 hours after addition of the water. Measurements were then made daily or at 3-day intervals until 18 days had passed since the original addition of the water. The bottles were shaken by hand two or three times each day.

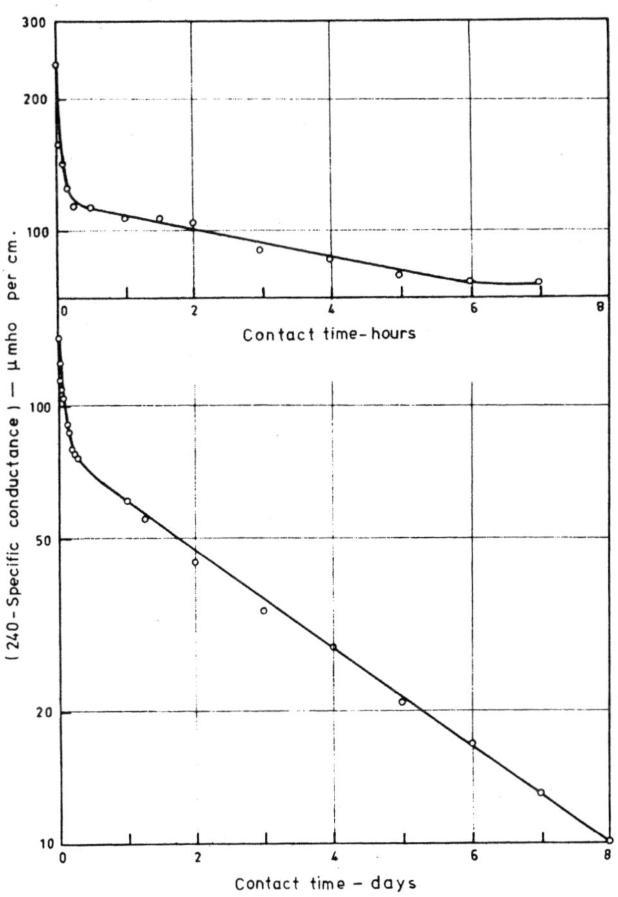

Fig. 3. Semilog plot of the relation of the difference between the equilibrium and actual specific conductance values to contact time at 25°C.

Results

The relationship of the average specific conductance of the three samples with contact time is shown as the solid line in Figure 2. Individual values varied by less than 10% from the average relation. Specific conductance increased very rapidly at first, reaching one third of its final value at 5 min and half of the final value at 15 min. However, it was still increasing slowly at 18 days. Similar results have been obtained in similar leaching tests by *McKeague and Cline* [1963] with silica and by *Kennedy* [1971] with specific conductance, chloride, and silica, although with one soil Kennedy found a decrease in concentrations after about 1 day. In short-term leaching tests, *Davis* [1969] found that at 5 min, concentrations of various ions attained about half their long-term values at 1 hour.

The final specific conductance at equilibrium was assumed to be 240 μmho/cm. The analysis of chemical reaction rate theory led to plotting the logarithm of the difference between the equilibrium value and observed specific conductance values against time. This is shown in Figure 3. Segments of the graph closely approximate straight lines, with sharp breaks at 1 to 10 min after the initial rapid rise and at about 8 hours. Three distinct phenomena seem to describe the rate of solution in the different time periods. No information is available to account for these differences. Some possible reasons are widely differing rates of solubility of different ions, availability of weathered material in highly soluble form when water is first applied, and the development of buffer reactions as solute concentration increases.

Laboratory temperature was constant at or a little below 25°C during the experiment. The relationship shown by solid lines in Figures 2 and 3 thus applies only to this temperature, and allowance must be made for its application at other temperatures. Based on climatic data for Palo Alto, California, soil and water temperatures of 17°C were appropriate to the field studies. The solution rates at this temperature would be about 0.65 of those at 25°C [*Hem*, 1970]. The adjusted specific conductance–contact time relationship for a temperature of 17°C is shown as the dashed line in Figure 2. Little difference is evident in the shapes and positions of the curves for the two temperatures. This results from the fact that for each of the segments of the relation defined by Figure 3 the time to effective equilibrium is still very short in comparison with the time scale of the subsequent segment.

Discussion

The simple experiment described above constituted only a minor part of the overall tracer study. The results can be regarded as only approximating field conditions. In any further work designed specifically to obtain data on the variation of specific conductance or solute concentration with contact time, a more rigorous and representative experimental procedure might be devised, although it would always be difficult to duplicate actual field soil water contact conditions. However, the data obtained from this simple experiment demonstrate the increase in specific conductance and solute concentration with contact time and provide a firm basis for analyzing field variations of conductance, as demonstrated later.

Several factors affecting these experimental results are discussed briefly below.

Shaking of samples. *McKeague and Cline* [1963] investigated dissolved silica in soil solutions and found that shaking the samples produced higher solute concentrations than when the samples were left to stand or when the water flowed with low velocities through the soil. Large differences were only observed for soil containing sand, and this was attributed to the abrasive effect of coarse particles colliding during shaking and providing fine material and fresh surfaces with enhanced solubility. The soil used in the specific conductance experiment described here contained little coarse material. Occasional shaking was adopted as a compromise to ensure mixing of the water and the soil and to simulate field contact conditions, while reducing any abrasive effects. Thus the results are probably not significantly biased by the mixing treatment.

Ratio of water to soil. There is evidence that the concentration of solutes decreases as this ratio increases. *Reitemeier* [1946] and *Eaton et al.* [1960] have shown that concentrations generally are lower in water in saturated soils than in unsaturated soils. *McKeague and Cline* [1963] found that large increases in dilution of soil samples further reduced dissolved silica concentrations and also changed the relation of silica concentration to contact time with the soil. A 1:1 water-to-soil ratio was assumed to approximate field conditions because subsurface flow was observed to result from a saturated layer above the B horizon, with much of the flow occurring in cracks and small holes. The 1:1 ratio also was assumed to be appropriate for surface runoff, where water flowed in a thin film with intimate contact with the surface soil, and possibly some of the flow was actually within the top few millimeters of the soil. Some error in absolute values of the data may result from the 1:1 ratio not representing field conditions, and consideration of this factor is warranted in interpreting the field results.

Adequacy of use of a closed container. In the laboratory experiment, solutes were being dissolved from an isolated sample of soil in what was superficially a closed system, whereas in the field, water moved through the soil in an apparent open system. Thus solution was occurring from different soil grains at different times for a given 'parcel' of water. However, the difference between these conditions and the laboratory experiment was not really one of open versus closed systems, but of spatially distributed versus concentrated systems. In the field the water passing a particular point would have been in contact with other soil over the preceding period, and conversely, other water would have been in contact with the soil at the particular point. If the assumption is made that the soil characteristics were uniform over the plot, the overall conditions for solution in the field would be virtually identical to the short-term location-concentrated conditions of the closed system laboratory experiment. The similarity of the conductance–contact time data for the three samples discussed earlier indicates that the soil characteristics were quite uniform.

Retention of original highly soluble salts. Highly soluble products of weathering and decay of organic matter accumulated since the previous runoff event often result in high initial solute concentrations in runoff. This initial flushing was not simulated in the laboratory experiment where these original highly soluble salts were retained in the closed bottles. This would not be critical here, as the volume of water with relatively high solute concentration retained in the laboratory experiment was very small in comparison with that of the added water. However, it might be desirable in any further experiments to use soil that had been initially washed to remove highly soluble salts, in addition to the unwashed soil used here. Differences in the results would indicate the effects of removal of initial readily soluble material.

Type of water used. Obviously, water added to the soil in the laboratory experiments should be similar to the new water used in the field events. Deionized water was used for the laboratory tests. In terms of specific conductance and solution characteristics, this would have corresponded closely with the rainwater for the storms of March 16 and April 1. However, the tap water used in the sprinkling run of May 4 had a higher initial specific conductance and was less aggressive chemically, which would lead to a less rapid increase in conductance. Arrangements had been made to have the tap water supplied from a surface source, and minimum specific conductance values for surface runoff in the sprinkling event were only slightly higher than those for the runoff events resulting from rainfall. Hence the laboratory data using deionized water should introduce only small errors in the analysis of the sprinkling run of May 4.

In view of the uncertainties resulting from the five factors discussed above, the analysis described here can be considered as only approximate. However, we consider that the simple laboratory experiment simulated field conditions with sufficient accuracy to demonstrate the need for and the validity of the time-dependent analysis of specific conductance. The results of the analysis described in later sections are also in agreement with other observations of flow conditions. In the design of any future work to determine the relation between

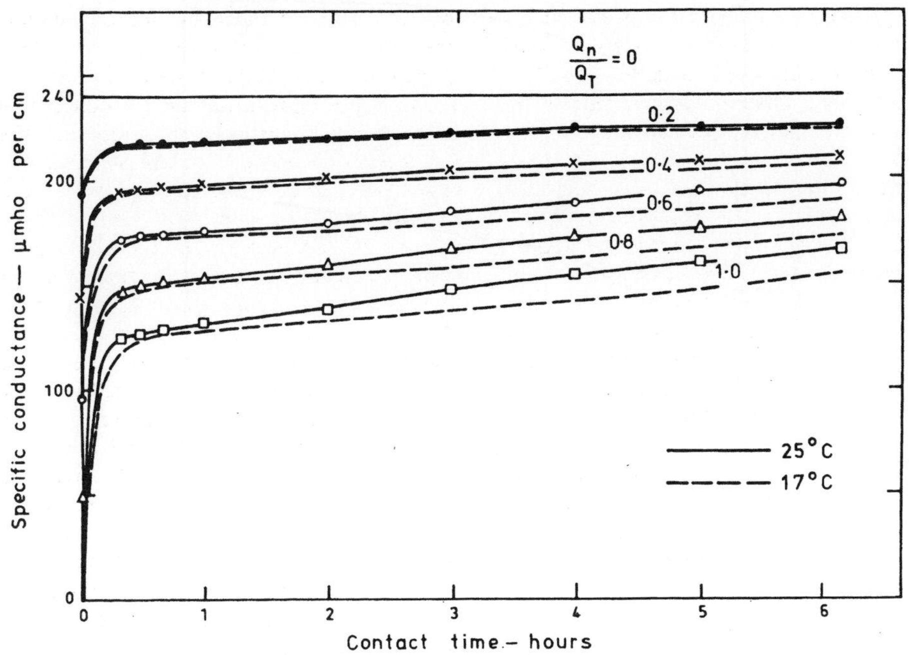

Fig. 4. Relation of specific conductance and contact time for mixtures of new and old water.

specific conductance or solute concentration and contact time, careful consideration should be given to those factors discussed above.

MIXTURES OF NEW AND OLD WATER

The relationship between specific conductance and contact time shown in Figures 2 and 3 can be used with (3) to compute the conductance for flow composed of a mixture of old water in effective equilibrium with the soil and new water that has been recently added to the watershed system. Normally, the old water would be groundwater or soil water that had been held in the unsaturated zone, and the new water would be the outflow from precipitation, either as surface runoff or rapid response subsurface flow. The conductance of mixtures then can be used to estimate the proportions of old and new water, as will be discussed later.

The C values in (1) and (3) refer to concentrations and not to the specific conductance values available here. As was discussed earlier, the relation between specific conductance and DS is not linear throughout the range of data collected in this study. However, because contact times of greater than 5–15 min will generally apply to even new water, Figure 2 shows that only half the total range of specific conductance (i.e., about 120–240 μmho) will be of interest. Over this range the relationship between specific conductance and dissolved solids is approximately linear, and errors of less than 5% are involved in assuming linearity. Thus as a close approximation, specific conductances can be substituted for C in (3).

Assuming a constant value of 240 μmho/cm for C_o in (3), the full lines in Figure 4 show the family of curves of C_T for various values of the ratio Q_n/Q_T of new to total water at a temperature of 25°C. A limited range of contact time of new water (up to 6 hours) is shown in this figure, and extended ranges of time are shown on later diagrams. The relationships shown in Figure 4 are specific to the particular soil composition and apply to a temperature of 25°C. The dashed lines in Figure 4 give the corresponding relations for a temperature of 17°C (based on the dashed curve in Figure 2). This family of curves provides the basis for analysis of the field results.

An alternative mathematical formulation of (3) that can be derived from Figure 3 may be useful and could possibly be used in an algebraic analysis of field data, although the graphics relation provides the most general methodology. The ordinate in Figure 3 represents $(C_o - C_n)$. If C_a represents the intercept of a linear segment of the semilog plot with the vertical axis at time zero, then for the time range where that linear segment applies,

$$C_o - C_n = C_a e^{-kt} \tag{4}$$

Different values of C_a and k would be required for different segments of the relationship shown in Figure 3.

It would also be possible to develop a relationship similar to that in Figure 4 for mixtures of water derived from two storm inputs where insufficient time had elapsed for either component to reach equilibrium. In this case, C_o also would be time dependent.

ANALYSIS OF FLOW COMPONENTS IN RUNOFF RECESSIONS

General Approach

For a runoff event, superposition of the graph of recorded specific conductance on a graphical representation of the relationship expressed by (3) (or (4)), such as that illustrated in Figure 4, provides a means of estimating the relative contributions of flow components with different residence times. The graphical records must be aligned so that the zero time of the conductance–contact time relation coincides with the estimated time of input of the new water contributing to the flow. This analysis is of most general application to the recession parts of runoff events. The earlier portions of the runoff events recorded in the field experiment will be discussed in a subsequent section. It will be assumed that the relationship for a temperature of 17°C applies to the events analyzed.

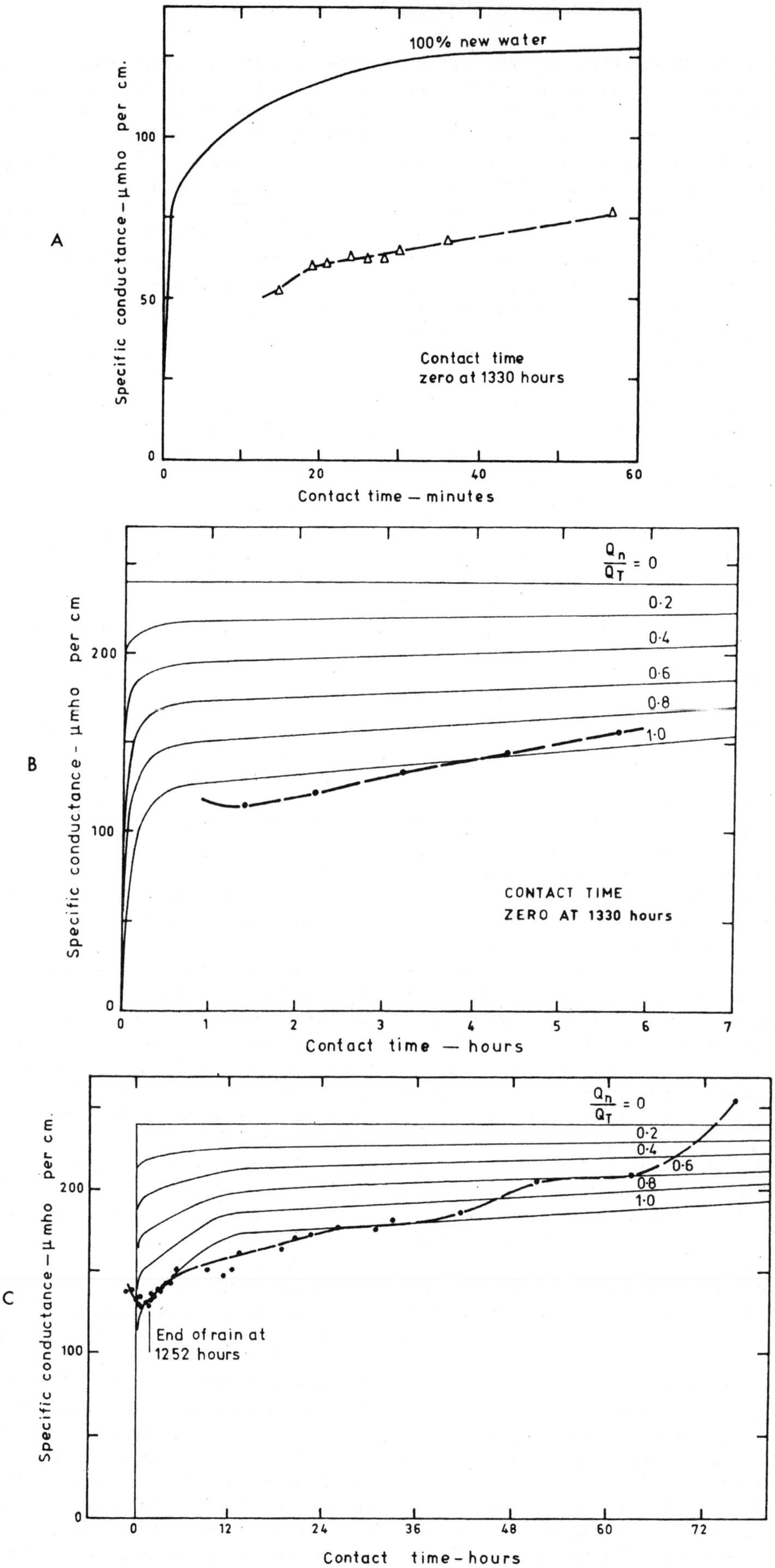

Fig. 5. Specific conductance records from selected runoff events at the bench at the lower end of the plot showing conductance–contact time relations.

A problem in the application of the conductance–contact time relationship to the analysis of the field data arises from the implication that all the new water is added to the soil in the watershed system at a given instant of time. For subsurface water in a real storm event, the water is added at varying rates over the entire duration of the storm, although the water contributing to any subsurface outflow may originate from a much shorter period near the end of the storm after soil water deficiencies are satisfied. For surface runoff, the water would be added only during periods of excess rainfall. Thus it might be more accurate to use a relationship obtained by successively displacing the curves shown in Figure 4 over time increments covering the duration of the effective input of new water and averaging the series of curves so obtained. This would be equivalent to the use of (3) or (4) in a convolution relation. However, this type of approach probably is not justified with practical data, as the effective duration and relative rates of input of new water are not accurately known. In addition, the shape of the curves would not lead to large changes in the form of the relations, in a similar fashion to the small effect of temperature variations. Consequently, the relationships for an instantaneous input of new water have been used here, and in general, zero contact time has been considered to apply to the estimated midpoint of input. If necessary in other applications, time-averaged relationships could be used without changing the principle of the analysis in any way.

Analysis of Runoff Events

The relationships in Figure 4 were used to determine the relative contributions of flow components with different residence times for the three runoff events shown in Figure 1. All of these were recorded at the bench near the lower end of the plot. One represents surface runoff from a natural storm, and the others are subsurface flow from a natural storm and a sprinkling run. Analyses of other recorded runoff events were similar to the examples discussed here.

Figure 5*a* shows the specific conductance record for surface runoff for the storm of April 1 plotted on the same graph as the conductance–contact time relation for 100% new water. The record is plotted so that the zero of the contact time scale coincides with the center of the burst of rain causing the hydrograph rise (at 1330 hours). The recorded specific conductances are much lower than the contact time relations for new water. Thus the surface flow does not appear to have had the same contact opportunity with the soil as was possible in the laboratory experiment. It is also probable that a water-to-soil ratio higher than 1:1 as used here would have been appropriate for determination of the conductance–contact time relation, as the published evidence discussed earlier indicates that higher ratios would decrease the conductance values. However, the increase of recorded specific conductance with time (Figure 5*a*) is of the same form as the derived relation and is consistent with the fact that the runoff does have some contact with the soil.

Results from the other surface runoff events also showed that the recorded specific conductance curve was the same shape as the relation for 100% new water and was initially lower. The data for the other recorded events coincided very closely with the relation for 100% new water. For all surface runoff events, including that illustrated in Figure 1*a*, the analysis indicates that the runoff was composed entirely of new water added to the plot by rainfall during the event. This is consistent with the physical process, even though the surface runoff at the bench at the lower end of the plot included some return subsurface flow.

The variation of specific conductance in subsurface runoff in the storm of April 1 is shown in Figure 5*b*. The curve is located with the contact time zero coincident with the center of the rainfall burst that produced the final hydrograph peak (Figure 1*b*). The curve closely matches the 100% new water relation, indicating that the subsurface flow is composed almost entirely of new water. This conclusion is consistent with the rapid response of the outflow to precipitation inputs. The slight increase with time of observed conductance relative to the contact time curves may represent some small contribution of older water. However, it may be due only to evaporation of the outflow on the collecting trough.

Results for subsurface flow in the sprinkling run of May 4 are shown in Figure 5*c*. The curve is shifted in time to give the best possible fit of the conductance–contact time curves. For an initial period, most of the curve closely matches the 100% new water relation, indicating that the subsurface flow is composed of new water. However, the curve later rises to the 60% new water relation, and the final conductance value is greater than the 100% old water conductance. This may indicate the outflow or gradual forcing from the soil matrix of old water with slow travel times. However, it is more likely that the observed rise in values represents a concentration effect resulting from evaporation of the small runoff volumes on the collecting troughs toward the end of the recession.

The time location of the curve relative to the zero contact time indicates that the average time of infiltration into the soil and hence commencement of contact of the water flowing out in the recession was about $1\frac{1}{2}$ hours before the cessation of rainfall. This seems consistent with the observed lag of about 30 min between a rainfall burst and the peak of the outflow at the bench at the lower end of the plot.

Discussion

The increase in specific conductance during the recessions of both surface and subsurface runoff has been shown to be consistent with the conductance–contact time relationships. These relationships further demonstrated their value in assessing the relative contributions of different components of the flow. Interpretations of the sources of flow and of the flow processes based on the relations are consistent with observations of the flow conditions in the field. These observations and the analysis of specific conductance data provide complementary information; each is useful in interpreting and confirming the other.

Both the surface and subsurface flows were shown to consist mainly of new water, and the increase of specific conductance in the recessions generally corresponds with the expected increase with contact time. Previous studies analyzing data obtained for storm runoff periods in streams generally have considered the increase in specific conductance to result from increasing proportions of base flow.

The data analyzed here were for a 60 × 160 ft (18.3 × 48.8 m) plot and for flows that were collected entirely at the ground surface or at a subsurface horizon, although the surface flows at the bench at the lower end of the plot contained subsurface return flow as well as true surface runoff. However, the data should be typical for small watersheds, and the analysis also would be applicable to small watershed data. In the location studied, the large plot discharged into a small stream, and the total flow in the stream resulting from runoff from the watershed was composed of surface and subsurface flow from areas similar to the study plot. Detailed studies of the runoff process carried out elsewhere and discussed by *Pilgrim et al.* [1978] indicate that subsurface storm flow in layered soils generally occurs in a similar manner. The widespread occurrence of layered soils indicates that the flow conditions investigated here occur frequently. For a small watershed, runoff beginning within an hour or two after the end of a heavy rainfall normally would be classified as base flow or interflow by conventional hydrograph analysis techniques based solely on response times of runoff [e.g., *Linsley et al.*, 1958]. Such analysis takes no account of whether the runoff actually reaches the stream by a surface or subsurface route, and in fact, information of this type normally is not available. The great advantage of the plot data used here was that separate measurements of surface and subsurface runoff were available, and it was for this reason that the plot data were used rather than the data observed for a stream. As was noted earlier, the study was in fact initiated because the surface and subsurface flows individually showed reduction and recovery of specific conductance, rather than a mixture of flows from these sources as generally had been assumed previously.

The upper limit of watershed area to which the type of analysis described here could be applied is difficult to assess. However, application to areas of at least several square kilometers should be realistic. The flows here would consist of water from various surface and subsurface sources. After seeping from the soil or running off the soil surface, the water would flow along the channel system with little contact with the soil and hence little further increase in solute concentration. At any point in the stream channel system, some of the water would have traveled a long distance in the watershed, and some would have originated nearby. In storm periods the average time of channel flow would be 1 or 2 hours at most, and this would cause only a small change of position on the conductance-contact time graphs. Therefore interpretation of flow sources as discussed here still should apply. Interpretation would be more difficult with increasing watershed size, channel flow times, and areal variability. Even approximate interpretation of the analysis described here may not be possible for watersheds over about 100 km² in area.

The analysis does not separate subsurface from surface runoff; rather, it allows separation of old and new water, or waters of two distinct residence times, as long as one component of the water has had close contact with the soil for only a short time (less than about 8 hours for the soil studied here). Examination of other hydrological records and preferably of actual field conditions is necessary to aid in the identification of the flow sources and processes.

FLUSHING EFFECT DURING EARLY FLOW

Solute concentrations and values of specific conductance may be different during rising stages than those at the same discharge in the recession of the same hydrograph. The values during rising stages are usually greater [*Steele*, 1968; *Hendrickson and Krieger*, 1960], although the reverse situation also has been reported [*Toler*, 1965]. Early high values also may be accompanied by a lag between the peak of the hydrograph and the trough of solute concentration. These high early values are attributed to a flushing effect, in which higher concentrations of solutes are washed out by the initial phases of storm runoff. Weathering, leaching, decay of organic matter, and evaporation lead to accumulation of highly soluble material in the soil mass during the period after the previous runoff event. Water held in the soil during this period also is assumed to have reached an equilibrium concentration of solutes. These two sources have been considered to supply the solutes for the flushing by early runoff.

Walling and Foster [1975] were able to correlate the lag between hydrograph peak and solute concentration trough with the magnitude of the flow event, the preceding flow level, the soil moisture deficit, and a seasonal index. Their results quantified the finding of several earlier studies [e.g., *Steele*, 1968; *Kennedy*, 1971] that the flushing effect is highly dependent on the antecedent moisture status. The contact time results of Figure 2 demonstrate further why high soil moisture, reflecting recent appreciable infiltration, leads to low solute concentrations in the water flushed out by subsequent rainfall.

The process depicted in (3) and illustrated in Figure 4 can aid in the interpretation of the flushing effect, as well as in assessing the sources of flow in the hydrograph recession described previously. In the study reported here, few field data

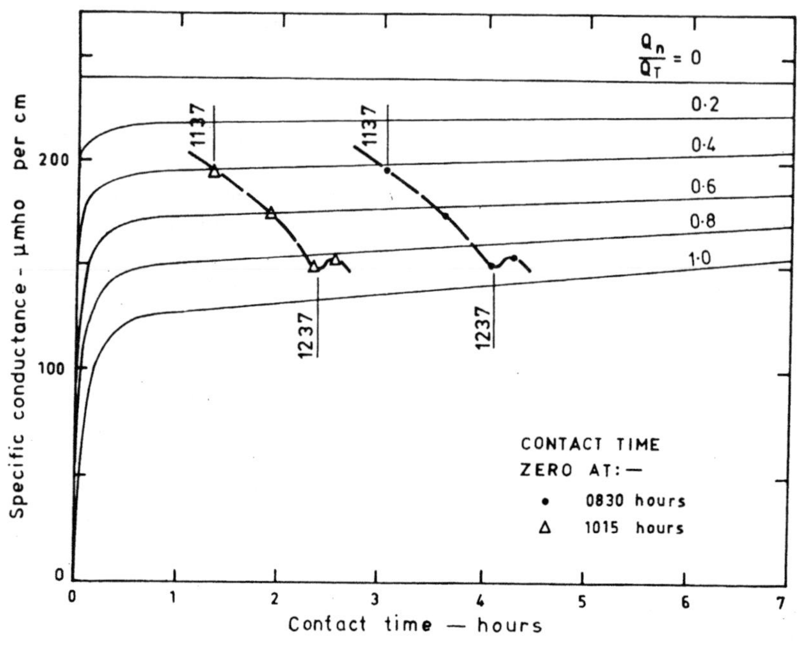

Fig. 6. Specific conductance record for subsurface runoff from 1137 to 1250 hours on April 1, 1968, plotted on conductance-contact time relations.

were collected early during the runoff events, but some evidence of the flushing effect is apparent in the results plotted in Figure 1. Data for the subsurface flow on April 1 (Figure 1*b*) illustrate this point. There is an apparent large flushing effect and an appreciable lag in the trough of specific conductance, which occurs at about 1500 hours. Flow between 1137 and 1250 hours resulted either from the bulk of rainfall with center of mass at about 0830 hours or from the small burst of rainfall centered at 1015 hours. After 1250 hours the bulk of the runoff probably resulted from the subsequent burst of rainfall.

In Figure 6, specific conductance values from 1137 to 1250 hours are plotted on the conductance–contact time graphs in two different positions. These are located so that the zero of the contact time scale coincides with the rainfall center times of 0830 and 1015 hours, discussed above. The different assumed initial time makes little difference to the interpretation. Assuming the water previously in the soil was at effective equilibrium (the previous appreciable rain was on March 16), the total outflow contained about 60% old water at 1137 hours (Figure 6). This old water was rapidly flushed out, so that by 1237 hours the flow was composed of about 90% new water. Figure 5*b* shows that at 1500 hours and at subsequent times the flow was entirely new water. Hence for the entire period between 1250 and 1500 hours, despite an appreciable drop in specific conductance, the flow was composed of new water. The decrease in conductance in this period resulted from a steady decrease in residence time of outflow caused by the rainfall burst centered at 1330 hours [Figure 1*b*]. Interpretation of the flushing effect on the basis of the delay in the trough of specific conductance and solute concentration would thus be suspect. The flushing effect was virtually finished at 1237 hours.

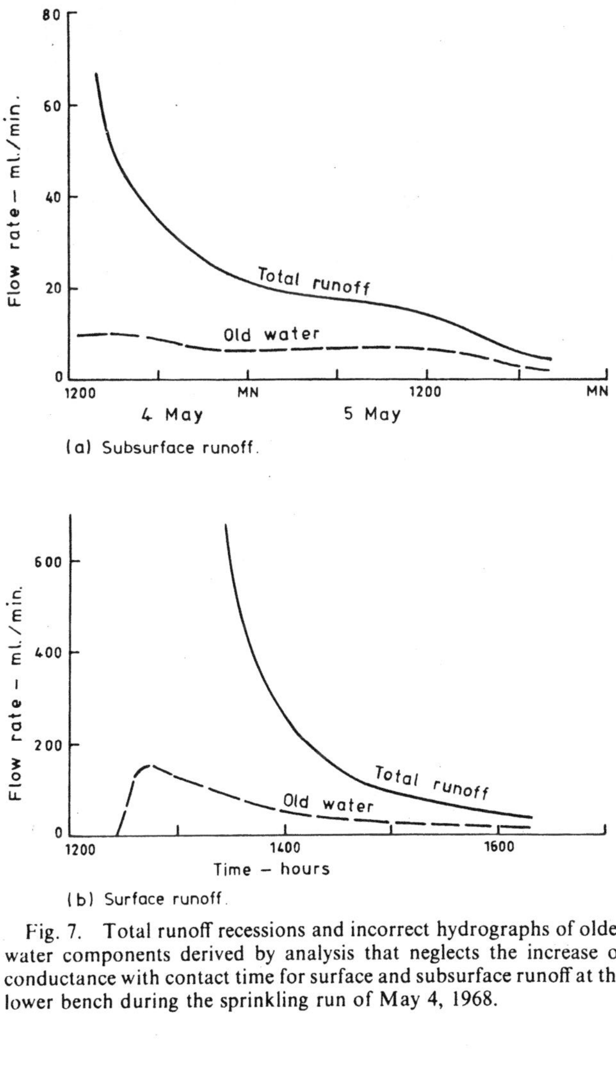

Fig. 7. Total runoff recessions and incorrect hydrographs of older water components derived by analysis that neglects the increase of conductance with contact time for surface and subsurface runoff at the lower bench during the sprinkling run of May 4, 1968.

ANALYSIS BASED ON FIXED VALUES OF SPECIFIC CONDUCTANCE

As was noted earlier, previous separation of flow components using specific conductance data has generally neglected the increase of conductance with contact time. The data from this study for several of the recorded events were analyzed using this assumption to examine the effect of neglect of contact time.

Assuming C_o and C_n have fixed values for older water and new water, respectively, (2) and (3) can be solved to find the flow rate of the older component of the total flow as

$$Q_o = Q_t \frac{(C_t - C_n)}{C_o - C_n} \qquad (5)$$

Normally, the relation between specific conductance and contact time is not known and values of C_o and C_n must be estimated from the observed variations of conductance. From the data in Figure 1 and the data for other events not illustrated here, values of 240 and 120 μmho/cm were assumed for C_o and C_n, respectively. Figure 7 shows the hydrographs of total runoff and the separated component of older water for the subsurface and surface flows for the bench at the lower end of the plot for the sprinkling run of May 4. Surface runoff data for this event have not been shown in previous figures or used to illustrate the other analyses described in this paper. The analysis which considered contact time, described previously, showed that the runoff in both cases actually was composed entirely of new water, even though some of the surface runoff resulted from return subsurface flow. By 1400 hours the surface runoff probably consisted entirely of this return flow.

Without knowledge of the flow processes occurring on the plot it would be tempting to label the hydrographs of older water as 'base flow.' This inference would be invalid here. As was discussed for the analysis involving contact time, the results illustrated here also would apply to flow in the small stream at this location, since the stream flow is the sum of runoff from a watershed composed of areas similar to the study plot.

If the results of the analysis based on the conductance–contact time relations are accepted, it can be concluded that analysis for small watersheds based on specific conductance data without consideration of the increase of values with contact time can lead to poor estimates of the relative contributions of the components of old and new water. This would invalidate inferences regarding the physical sources of flow and their relative contributions, particularly if detailed information or observations regarding the nature of the processes occurring on the watershed were not available. The error of the results may not be obvious, as the shapes of the incorrect separated hydrographs of the older water components in Figure 7 do not look unreasonable. Neglect of the increase of specific conductance with contact time probably will result in overestimates of the contribution of older water.

CONCLUSION

The inverse relation of solute concentration and specific conductance with discharge in storm runoff periods has been widely used to estimate the relative contributions of components of flow with different residence times. Inferences about the physical sources of flow often have been drawn from these estimates. In general, the increase of concentration or specific conductance with increase of contact time of water with soil has not been considered quantitatively in these analyses.

The simple leaching tests described here provide direct information on the variation of specific conductance with contact time, and the derived relation is suited for analysis of field observations from runoff events. The analysis is considered only semiquantitative, as a result of the use of specific conductance rather than concentrations of specific ions, the simple nature of the leaching experiments, and the pilot nature of the field experiment. Despite these limitations, the analysis has been shown to give valuable insights into the runoff processes and to yield interpretations of the field data that are entirely consistent with the other observations and data from the field experiment. The contact time relationships yield information only on the proportions of old and new water and do not give direct information on runoff sources. Thus analyses of this type should be regarded as complementary to other hydrological and field observations.

Specific conductance–contact time relations are of most direct application to the data from recessions of storm runoff events. Also, they assist in the interpretation of the flushing effect in early runoff. The analysis is most useful for small watersheds up to a few square kilometers in area.

Neglect of the increase in conductance with contact time and use of fixed values for the conductance of old and new water probably lead to poor estimates of the relative contributions of the flow components and to incorrect inferences regarding sources of runoff.

Acknowledgments. The study was carried out at Stanford University in California. The authors are indebted to Ray K. Linsley and Paul Kruger for assistance in organization of the project. The Water Resources Planning Management Program, Colorado State University, is thanked for facilitating preparation of this paper. The Environmental Sciences Division of Oak Ridge National Laboratory (operated by Union Carbide Corporation under contract W-7405-eng-26 with the U.S. Department of Energy) also provided assistance in review and final preparation of the manuscript.

REFERENCES

American Public Health Association, *Standard Methods for the Examination of Water and Wastewater Including Bottom Sediments and Sludges*, 11th ed., pp. 213–216, New York, 1960.

Davis, S. N., Silica in streams and ground water of Hawaii, *Tech. Rep. 29*, Water Resour. Res. Center, Univ. Hawaii, Honolulu, 1969.

Eaton, F. N., R. B. Harding, and T. J. Ganje, Soil solution extractions at tenth-bar moisture percentages, *Soil Sci.*, *90*, 253–258, 1960.

Feth, J. H., C. E. Roberson, and W. L. Polzer, Sources of mineral constituents in water from granite rocks, Sierra Nevada, California and Nevada, *Water Supply Pap. 1535-I*, U.S. Geol. Surv., Washington, D. C., 1964.

Hall, F. R., Dissolved solids–discharge relationships, 2, Applications to field data, *Water Resour. Res.*, *7*, 591–601, 1971.

Hem, J. D., Study and interpretation of the chemical characteristics of natural water, 2nd ed., *Water Supply Pap. 1473*, U.S. Geol. Surv., Washington, D. C., 1970.

Hendrickson, G. E., and R. A. Krieger, Relationships of chemical quality of water to stream discharge in Kentucky, 1, Geochemical cycles, *Int. Geol. Congr. Rep. Sess. Norden 21st*, 66–75, 1960.

Kennedy, V. C., Silica variation in stream water with time and discharge, in *Nonequilibrium Systems in Natural Water Chemistry*, *Advan. in Chem. Ser.*, vol. 106, pp. 94–130, 1971.

Linsley, R. K., Jr., M. A. Kohler, and J. L. H. Paulhus, *Hydrology for Engineers*, pp. 156–160, McGraw-Hill, New York, 1958.

McKeague, J. A., and M. G. Cline, Silica in soil solutions, 1, The form and concentration of dissolved silica in aqueous extracts of some soils, *Can. J. Soil Sci.*, *43*, 70–82, 1963.

McNeal, B. L., J. D. Oster, and J. T. Hatcher, Calculation of electrical conductivity from solution composition data as an aid to in-situ estimation of soil salinity, *Soil Sci.*, *110*, 405–414, 1970.

Nakamura, R., Runoff analysis by electrical conductance of water, *J. Hydrol.*, *14*, 197–212, 1971.

Newson, M. D., A model of subterranean limestone erosion in the British Isles based on hydrology, *Trans. Inst. Brit. Geogr. 54*, 55–70, 1971.

Pilgrim, D. H., and D. D. Huff, A field evaluation of subsurface and surface runoff, 1, Tracer studies, *J. Hydrol.*, *38*, 299–318, 1978.

Pilgrim, D. H., D. D. Huff, and T. D. Steele, A field evaluation of subsurface and surface runoff, 2, Runoff processes, *J. Hydrol.*, *38*, 319–341, 1978.

Reitemeier, R. F., Effect of moisture content on the dissolved and exchangeable ions of soils of arid regions, *Soil Sci.*, *61*, 195–214, 1946.

Shuster, E. T., and W. B. White, Seasonal fluctuations in the chemistry of limestone springs: A possible means of characterizing carbonate aquifers, *J. Hydrol.*, *14*, 93–128, 1971.

Steele, T. D., Digital-computer applications in chemical quality studies of surface water in a small watershed, in *The Use of Analog and Digital Computers in Hydrology*, vol. 1, *AIHS Publ. 80*, pp. 203–214, International Association of Scientific Hydrology, Gentbrugge, Belgium, 1968.

Steele, T. D., and N. C. Matalas, Principal component analysis of streamflow chemical data, in *Mathematical Models in Hydrology*, *AIHS Publ. 100*, pp. 355–363, International Association of Scientific Hydrology, Gentbrugge, Belgium, 1974.

Ternan, J. L., Comments on the use of a calcium hardness variability index in the study of carbonate aquifers: With reference to the central Pennines, England, *J. Hydrol.*, *16*, 317–321, 1972.

Toler, L. G., Relation between chemical quality and water discharge in Spring Creek, southwestern Georgia, Geological Survey Research 1965, *Prof. Pap. 525-C*, pp. 209–213, U.S. Geol. Surv., Reston, Va., 1965.

Walling, D. E., Suspended sediment and solute yields from a small catchment prior to urbanization, in *Fluvial Processes in Instrumented Watersheds, Spec. Publ. 6*, edited by K. J. Gregory and D. E. Walling, pp. 169–192, Institute of British Geographers, London, 1974.

Walling, D. E., and I. D. L. Foster, Variations in the natural chemical concentration of river water during flood flows, and the lag effect: Some further comments, *J. Hydrol.*, *26*, 237–244, 1975.

(Received November 10, 1977;
revised September 29, 1978;
accepted October 23, 1978.)

Commentary on Paper C4

Trudgill, S. T., Pickles, A. M., Smettem, K. R. J. & Crabtree, R. W. (1982) Soil-water residence time and solute uptake. 1. Dye tracing and rainfall events.

Although Steve Trudgill's own PhD concerned soil processes and limestone weathering on Aldabra (an island in the Indian Ocean), his contemporaries at Bristol University (in the UK) included people like Darrell Weyman, and he would have been well aware of the rapid developments in hillslope hydrology (reviewed in Beven, 2006), some of which were undertaken at East Twin Brook by Weyman and others from the Bristol group. It was perhaps no surprise therefore that his interests soon developed into the realm of solute transport on hillslopes. At Bristol, he also had close contact with the activities of the "Karst Police" who used dyes to trace water flow through the cavernous limestone aquifers of the Mendip Hills (e.g. Newson, 1973; Smart & Laidlaw, 1977), so an interest in tracing soil water flow paths on slopes was a natural extension of this previous work. At Sheffield University, UK, from the late 1970s onwards, Steve's research group (Keith Smettem, Bob Crabtree and Adrian Pickles) worked both on natural slopes at the Whitwell Wood field site and on under-drained clay soils at nearby Sandbeck (Smettem *et al.*, 1983).

Trudgill *et al.* (1982) use fluorescent dyes to label rainfall infiltrating into a well-structured calcareous woodland soil. Their hillslope plot was some 30 m in length with a slope of 10–15°. They use throughflow troughs inserted in the bank of the ditch at the foot of the slope to intercept slope drainage waters. Stream discharge is measured using a 90° V-notch weir. Water sampling used Rock and Taylor pump water samplers. On the slope an automatic tensiometer system measured soil water potentials whilst soil water was sampled using suction samplers. Indeed, the whole experimental design reflects the accumulated experience of the Bristol Geography Department. Their work on soil-water travel times and solute concentrations builds on a various studies including Pilgrim *et al.* (1979, Paper C3) and includes both the displacement of long-residence-time water and the rapid transfer of soil water via preferential flow paths. The theoretical approach is to predict outflow for both flow types using two simple storage models, one to predict the occurrence of displacement flow through the soil matrix and the other to predict preferential flow. Both models rely on knowledge of the hydraulic conductivity of the soil peds, which was measured using undisturbed soil samples in a small permeameter; if rainfall intensity exceeds the soil ped hydraulic conductivity, then by-pass flow is predicted. An antecedent precipitation index (API) is neatly used to represent soil water storage conditions prior to rainfall.

The central part of the paper concerns the use of fluorescent dyes in soils, to confirm that this is a reliable technique to identify first times of arrival of labelled water bodies during rapid, transient water flows over short distances, notwithstanding the marked adsorption losses suffered by such dyes in soil.

Liquid dye was spread on the soil surface and did not penetrate the soil unless it rained. No long-distance flow was intercepted but the method worked well for studying water flow through soil close to the stream bank. In all cases where outflow happened, this could be predicted using rainfall intensity data: sustained intense rainfall (>3.6 mm per hour for more than two hours) always caused outflow at the throughflow trough. A companion paper (Trudgill *et al.*, 1983) further develops the preferential flow model via column drainage experiments in the laboratory, in order to define the operational soil volume through which preferential flow takes place. In Crabtree & Trudgill (1984), the link between soil solute sources and stream solute response is explored, extending the focus to the hillslope/catchment scale. Storm events produced a rapid input of solute-rich water to the stream. The rapid response seemed to be initiated by stemflow which produced concentrated point-source inputs to the soil, generating by-pass flow. Their model of by-pass flow moving rapidly to the slope foot to displace "old" solute-rich water to the stream very much anticipates the study of McDonnell (1990, Paper C5) at the Maimai catchment, New Zealand. Smettem *et al.* (1983) use a very similar approach to explore links between preferential flow in underdrained clay soils and stream nitrate response.

TPB

References

Beven, K. J. (editor) (2006) *Streamflow Generation Processes*. Benchmark Papers in Hydrology, 183–184. IAHS Press, Wallingford, UK.

Crabtree, R. W. & Trudgill, S. T. (1984) Chemical denudation on a magnesian limestone hillslope, field evidence and implications for modelling. *Earth Surf. Processes Landf.* **10**, 331–341.

McDonnell, J. J. (1990) A rationale for old water discharge through macropores in a steep, humid catchment. *Water Resour. Res.* **26**, 2821–2832.

Newson, M. D. (1973) Rickford and Langford resurgences, Mendip Hills, Somerset, a problem in limestone hydrology. In: *Proc. Univ. Bristol Spelaelogical Society* **13**(1973), 105–112.

Pilgrim, D. H., Huff, D. D. & Steele, T. D. (1979) Use of specific conductance and contact time relations for separating flow components of storm runoff. *Water Resour. Res.* **15**, 329–339.

Smart, P. L. & Laidlaw, I. M. S. (1977) An evaluation of some fluorescent dyes for water tracing. *Water Resour. Res.* **13**(1), 15–33.

Smettem, K. R. J., Trudgill, S. T. & Pickles, A. M. (1983) Nitrate loss in soil drainage waters in relation to by-passing flow and discharge on an arable site. *J. Soil Sci.* **34**, 499–509.

Trudgill, S. T., Pickles, A. M, Smettem, K. R. J. & Crabtree, R. W. (1982) Soil-water residence time and solute uptake. 1. Dye tracing and rainfall events. *J. Hydrol.* **60**, 257–279.

Trudgill, S. T., Pickles, A. M, Smettem, K. R. J. & Crabtree, R. W. (1983) Soil-water residence time and solute uptake. 2. Dye tracing and preferential flow predictions. *J. Hydrol.* **62**, 279–285.

Journal of Hydrology, 60 (1983) 257—279
Elsevier Scientific Publishing Company, Amsterdam — Printed in The Netherlands

[4]

SOIL-WATER RESIDENCE TIME AND SOLUTE UPTAKE
1. Dye Tracing and Rainfall Events

S.T. TRUDGILL, A.M. PICKLES, K.R.J. SMETTEM*[1] and R.W. CRABTREE*[2]
Department of Geography, University of Sheffield, Sheffield S10 2TN (Great Britain)

(Received November 9, 1981; accepted for publication January 20, 1982)

ABSTRACT

Trudgill, S.T., Pickles, A.M., Smettem, K.R.J. and Crabtee, R.W., 1983. Soil-water residence time and solute uptake, 1. Dye tracing and rainfall events. J. Hydrol., 60: 257—279.

Fluorescent dyes have been used to label rainfall infiltrating into a well-structured calcareous woodland soil in situ. Soil-water outflow was monitored using throughflow troughs. The occurrence of dye output from the soil could not be totally predicted by the use of a simple storage model but was related to the occurrence of intense-rainfall events equal to or greater than 3 mm $hr.^{-1}$ lasting for at least 2 hr.

INTRODUCTION

During slow soil-water flow, prolonged solid—solvent contact is allowed and the dissolution of soil chemical constituents is liable to approach chemical equilibrium. During rapid flow, only those chemical elements which are rapidly soluble will be able to maintain high concentrations. Rapid flow will therefore result in the preferential removal of the most rapidly soluble chemical elements and compounds, such as nitrate (Trudgill, 1976, 1977, ch. 5; Trudgill et al., 1981). In structured soils, it is clear that flow rates are not simply related to hydraulic gradient but also to the existence of preferential flow paths which have been identified by numerous workers in both field and laboratory situations (notably J.L. Anderson and Bouma, 1973; Addiscott, 1977; Gaudet et al., 1977; Kanchanasut et al., 1978; Scotter, 1978; Omoti and Wild, 1979a and b; Bouma, 1980; Beven and Germann, 1981a and b; Germann and Beven, 1981; and as reviewed by Trudgill and Briggs, 1979, 1980). Water output from structured soils in response to rainfall events consists of mixtures of rapidly-flowing recent input water moving along these preferential pathways and displaced long residence time water mobilised during rainfall events, both superimposed on any baseflow

Present addresses:
*[1] Department of Soil Science, University of Sydney, N.S.W., Australia.
*[2] School of Geography, University of Leeds, LS2 9JT, Great Britain.

0022-1694/83/0000—0000/$03.00

drainage. Solute uptake into soil water can thus be seen in terms of: (a) near-equilibrium dissolution, ion exchange and diffusion in high-tension water between rainfall events; and (b) dissolution and desorption kinetics, dispersion and miscible displacement during rapid flow in response to rainfall events.

The recognition of the existence of preferential flow paths means that simple displacement models will not be appropriate in the prediction of the occurrence of soil-water outflow, but that outflow may be predicted by the incidence of high-intensity rainfall, as shown for strongly structured soils by Bouma (1980) and Bouma and De Laat (1981). To test the applicability of displacement and rainfall intensity models to the prediction of soil-water outflow and to provide information on soil-water travel time, fluorescent dyes (Smart and Laidlaw, 1977) have been evaluated for use in soils and employed to label infiltrating rain water. Conditions of outflow of dye-labelled water have then been studied. An antecedent precipitation index (API) is considered as a simple predictive tool and physical information models are developed using rainfall-intensity and soil-moisture data.

The complexities of flow paths and travel times of soil water have been emphasised by Haskell and Hawkins (1964) and Horton and Hawkins (1965); Jury (1975a, b) has stressed the dangers of correlating water and solute outputs from tile drains with rainfall inputs without taking a range of travel times into account. The tracing work described in the present paper can be used to distinguish between the arrival of waters of differing residence times at an output point. Such distinctions are fundamental to the interpretation of temporal variations in solute concentrations in soil drainage water.

Soil-water output has been sampled by the use of throughflow troughs and automatic water samplers; temporal variations in solute concentrations have been studied in relation to trough discharge and dye arrival patterns; dye arrival has been studied in relation to rainfall characteristics and soil moisture. Supporting work has also been undertaken on soil-water pathways, on solid—solvent contact time relationships using soil leaching columns and on solute uptake during dissolution experiments, using both rock and soil material. This paper resports on dye response in relation to rainfall conditions; subsequent papers will report on soil-moisture data and on the other observations (Trudgill et al., 1983).

Existing work on soil-water travel times and solute concentrations includes that of Pilgrim et al. (1979) who used experimentally controlled solid—solvent contact times and conductivity measurements as a basis for separating flow components in storm-runoff hydrographs. This parallels other work where chemical hydrographs for soil throughflow water are interpreted in terms of implicit residence-time models, with the occurrence of dilution by quickflow and the increase of solute concentration by the displacement of long-residence-time water (e.g., Burt, 1979). However, the interpretations of such patterns cannot be tested unless water bodies can be distinctively identified in the field. Zimmermann et al. (1967) used hydrogen isotopes to illustrate the discrete movement of soil-water bodies under control conditions and Sklash et al. (1976) used natural variations in ^{18}O-isotope concentrations to interpret storm runoff data. In the present work, fluorescent dyes have been used because of their ease in use and detection. Atkinson (1978) advocates the use of fluorescent dyes in hillslope hydrology to estimate travel times and Mosley (1979) has used fluorescent dyes to demonstrate rapid travel times in forest soils in New Zealand. Tracer dyes were also used by Trudgill et al. (1980) to assist in the interpretation of Ca/Mg ratios in hillslope drainage waters on dolomitic limestone. Adsorption presents a major limiting factor on dye use in soils. Because of adsorption, time to peak concentration and arrival times may show a lag behind other tracers (White, 1977) but in rapid flow systems over short distances first time of arrival may show accord with other tracers, as is shown below for the soils at the present study site. The dyes may be used in the field to identify mobile soil water at an output point, using short trace distances, but not for quantitative recovery analysis.

PREDICTION OF OUTFLOW

If uniform displacement occurs in the soil, successive inputs of rain to the soil will displace a dye front until it emerges at an output point. At a free face (stream bank or throughflow trough) saturation has to occur before outflow occurs. Thus, the simplest model which could predict the occurrence of labelled outflow is a storage model:

$$I = S_{\max} - S_{t_1} \qquad (1)$$

where I = increments of rainfall input for t_1 (dye emplacement time) to t_2 (dye output time) (mm); $S_{\max}$ = soil-moisture storage capacity (mm); and S_{t_1} = soil-moisture storage at t_1 (mm).

Dye output will occur when:

$$I + S_{t_1} = S_{\max} \qquad (2)$$

Rainfall data alone cannot be used as a predictive input to the model since evaporation will decrease the effective input to the soil store. Rainfall data with subtracted measured or calculated evaporation amount or a calculated antecedent precipitation index (API), with an appropriate decay function, can provide suitable estimates of effective input into the soil. Of these the API is simple to apply:

$$\mathrm{API}_n = (\mathrm{API}_{n-1} \cdot K) + P_{n-1} \qquad (3)$$

where API_n = API for day n (mm); API_{n-1} = API for day preceding n (mm); K = decay constant; and P_{n-1} = rainfall in 24 hr. preceding day n (mm) (after Gregory and Walling, 1973, p. 187). Decay constants of between 0.85 and 0.98 are suggested by Gregory and Walling.

In terms of API, the storage model predicts output when:

$$API_I + API_{O-I} + R_0 = S_{max} \quad (4)$$

where API_I = API at dye emplacement (mm); API_{0-I} = net increment in API from input to output (mm); R_0 = rainfall on day of output (mm).

API is used as a surrogate variable for soil-moisture content and is used in this paper because it is relatively easy to obtain for use in a predictive model [Trudgill et al. (1983) consider actual soil-moisture data]. The value of S_{max} can be approximated by measurements of soil-water retention capacity (Hall et al., 1977), that is the amount of water (mm) capable of being held in a soil against a specific force (in this case gravity), though this is not necessarily directly equivalent to the soil-moisture status at which soil-water outflow will occur since retention against gravity was determined under drainage conditions. Nevertheless, it provides a rapidly determinable predictive measurement.

A storage model is derived for uniform flow in homogeneous porous media under ideal conditions and is similar in principle to the model derived by Burns (1975) to predict the depth of nitrate leaching in soils exhibiting no marked preferential flow paths. However, since many workers have shown the significance of preferential flow paths in by-passing much of the water held in the soil, a simple storage model will be invalid if such pathways are present and operational. Outflow will occur in structured soils before S_{max} is reached, especially in relation to localised zones of saturation within the soil. Under field conditions local saturation must occur before macropore flow occurs, either as ped storage capacity is exceeded, leading to outflow (Beven and Germann, 1981a, b) or as ped infiltration capacity is exceeded (Bouma et al., 1978; Edwards et al., 1979). The latter conditions will occur under sustained intense rainfall input (especially under dry cracking clay soils; Bouma, 1980. Bouma and De Laat, 1981). Preferential flow can also occur within peds in capillary sized pores, without the flow in larger macropores envisaged by Beven and Germann (1981a, b). The conditions under which preferential flow may occur are thus varied but they appear to be analogous to those for overland flow (reviewed by Kirkby, 1978). More specifically, it is important to assess not only the infiltration capacity of the bulk soil, but also that of the peds (Bouma, 1980). Preferential flow can be predicted to occur when:

$$RI/C_p > 1 \quad (5)$$

where RI = rainfall intensity (mm hr.$^{-1}$); and C_p = pedal infiltration rate (mm hr.$^{-1}$).

Predictions of the occurrence of dyed outflow from soil will thus be related to the occurrence of intense rainfall, as tempered by the effects of distance of the trace, the continuity of flow pathways and any sorption of the labelled water into the soil surrounding the flow pathway. Thus, in this paper, the occurrence of uniform displacement flow has been predicted from a comparison of API and measured soil-water retention capacity and the occurrence of preferential flow from the comparison of pedal infiltration rate and rainfall intensity.

FIELD SITE, MATERIAL AND METHODS

The field site was at Whitwell Wood, 20 km southeast of Sheffield, U.K., National Grid Reference SK *523788*, at the northern edge of the wood where the hillslope drains to the Bondhay Dyke (Fig. 1). The site supports a 30-yr. old unfertilised stand of beech (*Fagus sylvatica*) planted on Elmton and Aberford soil series (Reeve, 1976) and a narrow riparian strip of semi-natural mixed deciduous trees. The soil series are classifed as well drained. The Elmton series is a rendzina with rubbly bedrock at ~30 cm depth; the Aberford series is a well-structured brown calcareous earth with bedrock at up to 80 cm depth. The former has a loam texture and the latter a fine loam or fine silt texture. Both have variable but small amounts of glacial drift incorporated in the profile. Soil analytical data are given by Reeve (1976) and in Table I. The bedrock is Lower Magnesian Limestone of Permian age, a permeable fractured carbonate aquifer described in other areas by Cairney (1972), Cairney and Hamill (1977), and Aldrick (1978). The area receives a mean annual rainfall of 620 mm and transpiration plus evaporation normally exceeds rainfall from April to September. In the present study, rainfall was monitored using a canopy-level natural siphon recording raingauge and a check gauge, with measurements also being replicated at ground level, with splash-screen mesh, at a nearby non-forested site. Rainfall data quoted below

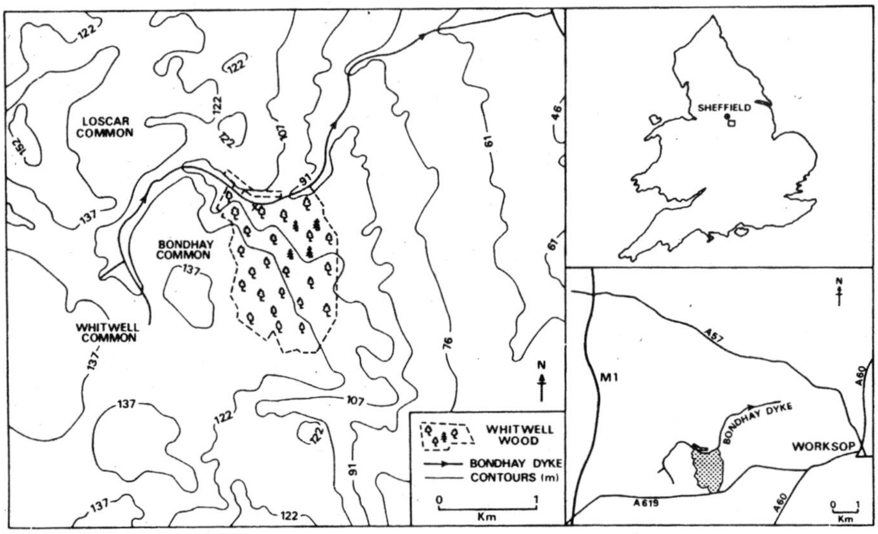

Fig. 1. Location of field study site. Slope plot is at northern edge of woodland (X).

TABLE I

(a) Profile characteristics of Aberford soil series, brown calcareous earth

Depth (cm)	Horizon	
0—20	A	clay loam—silty clay loam, moderate medium subangular blocky structure, common fine fibrous roots
20—30	BW	slightly stony clay loam, moderate medium subangular blocky, rare large pores, highly calcareous
>30	R	Magnesian Limestone, subangular or tabular fragments

(b) Analytical data, soil (1), brown calcareous earth

Horizon	A	Bw
Depth (cm)	0—20	20—30
Sand, 200 μm—2 μm (%)	7	7
60 μm—200 μm (%)	13	15
Silt, 2—60 μm (%)	56	56
Clay, <2 μm (%)	24	22
$CaCO_3$ (%)	4.2	10.8
Organic carbon (%)	2.4	1.8
pH, water, 1:2.5	7.6	7.8
Cation exchange capacity, CEC (meq/100 g)	31.1	23.7

are for the canopy-level gauge. Stemflow and throughfall were also monitored using ten collars and ten troughs installed in a 10-m^2 plot (Wright, 1977).

The hillslope plot monitored is shown in Fig. 2. The slope plot ran for some 30 m at 10—15°, with Elmton soil series above the path and Aberford below, the B horizon of the latter not exceeding 45 cm depth in the plot. Nitrate leaching from the plot has already been described by Trudgill et al. (1981). The stream marked in Fig. 2 is an artificial drainage ditch cut across the slope base, intercepting slope drainage waters; the Bondhay Dyke runs on an alluvial flat to the north of the stream shown, which joins the dyke some 30 m downstream. The source for the stream shown is groundwater seepage, initiated some 10—30 m upstream of the section shown; this flow is augmented by water flow through the soil in response to rainfall events. Soil moisture was monitored on the plot by means of a Scanivalve® automatic recording tensiometer system (M.G. Anderson and Burt, 1977; Burt, 1978a) and a manual tensiometer system (Burt, 1978b). Channel discharge was measured using a thin-plate sharp-crested 90° V-notch weir-and-stage recorder. Soil-moisture suction samplers, with 1-bar ceramic cups (Soil Moisture®, Corp., U.S.A.) were installed on the plot, together with a 20-cm-diameter lysimeter. Two sub-plots, an upper site and a bank site, were used for dye

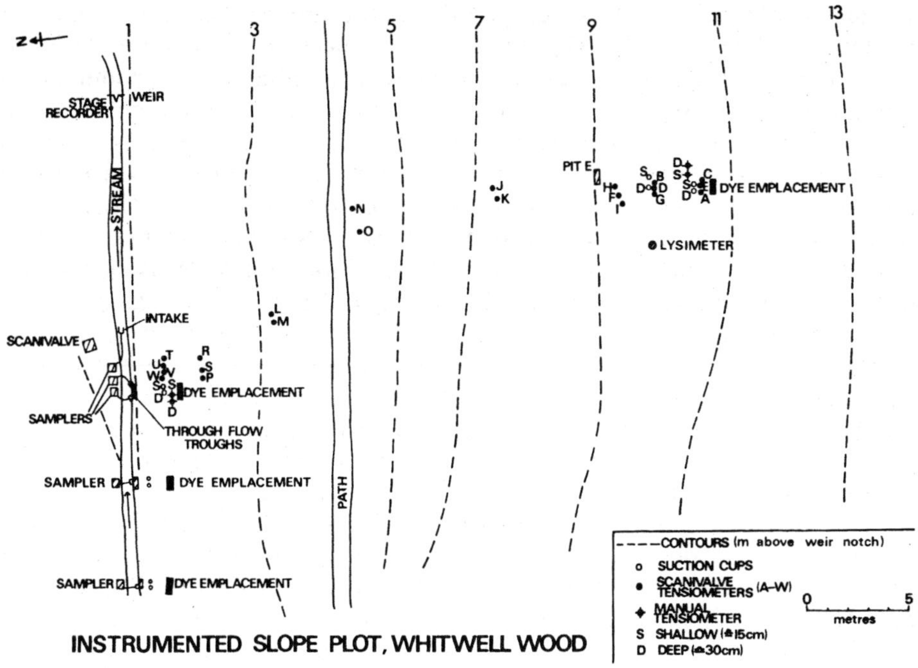

Fig. 2. Layout of instrumented slope plot, with upper site and lower bank site.

tracing, as shown, with throughflow troughs and suction cups for dye detection. The bank site trough installation was in the upslope channel bank where the soil profile is ~45 cm deep. Thirty cm long troughs were inserted directly into the soil without backfilling at 15-, 30- and 45-cm depth and drained under gravity into collection vessels. These were evacuated by the use of automatic water samplers. In the middle trough a Rock and Taylor *48* Interval® peristaltic pump sampler was used, set to remove the water accumulating between sampling events. Throughflow discharge was measured by amount of water present in each bottle in the water sampler. The maximum capacity of the system was 325 ml hr.$^{-1}$, which was only exceeded in one storm during the monitoring period. The sample intervals were 2 h. In the bottom trough and stream vacuum samplers (Automatic Liquid Samplers®, Ltd.) were used. Some contamination of the system is possible with dye residue from the previous sample remaining in troughs and pipes and thus falling dye concentration curves cannot be reliably interpreted. The monitoring period was from November 1978 to February 1980, with the exception of January—February 1979, because of freezing conditions. During the study period the dye emplacement sites and troughs and samplers were moved successively upstream as shown in Fig. 2, in order to provide replicate data. The channel was sampled downstream of the troughs but above the weir pool.

The fluorescent dyes used in this study were Rhodamine WT®, Lissamine

FF® and Amino G Acid®. A comparative evaluation of these tracers is given by Smart and Laidlaw (1977). Fluorescence measurements were made on a Turner III® filter fluorometer, using the filter combinations recommended by Smart and Laidlaw (1977). Alternative names and excitation and emission spectra of the dyes used are given in Table II. All three dyes suffer marked adsorption losses and it follows that the use of dyes to estimate travel times in an adsorptive porous media such as soil should not be considered unless there is adequate observational evidence of their viability. Evidence based on adsorption isotherms derived from batch equilibration tests using the soil under study confirms that adsorption characteristics for disaggregated soil material are similar to those for dilute suspensions of humus and sediments reported by Smart and Laidlaw (1977). However, empirical evidence from intact soil columns shows that adsorption in stuctured soil is considerably less than that predicted by equilibrium adsorption measurements. This is due to the non-equilibrium conditions encountered under rapid flow in structured soils. It can therefore be proposed that fluorescent dyes may be evaluated for the identification of times of arrival of labelled water bodies during rapid, transient water flows over very short distances. During the present work, dyes were placed on the soil surface adjacent to the stream in order to label infiltrating rain water and the dosage was calculated to be high enough to facilitate detection at the output point. Arrival of dyed soil water was monitored at the stream bank 1 m from the emplacement point. This procedure is defended by the evidence presented below.

Adsorption isotherms for disaggregated soil material from the study site are presented in Fig. 3. The data for batch equilibrations following the procedures of Omoti and Wild (1979a). Twenty g disaggregated soil samples (<2 mm mesh, lightly pestled and bulked) were shaken on flask shakers, with three replicates. 1:3 soil to solution ratios were employed. The suspensions were shaken for an equilibrium time period which was calculated from adsorption rate experiments, and then centrifuged at 3000 r.p.m. Corrections for blank fluorescence were made. Rhodamine WT® suffered 90—99% loss as compared to 89% and 33% on humus and kaolinite, respec-

TABLE II

Excitation and emission peaks for fluorescent dyes used (from Smart and Laidlaw, 1977)

Dye	Maximum excitation (nm)	Maximum emission (nm)
Rhodamine WT®	590	580
Lissamine FF®*1	420	515
Amino G Acid®*2	355	445

*1 CI acid Yellow 7®; Lissamine Yellow FF®, Brilliant Sulpho Flavine FF®, Brilliant Acid Yellow 8G®.

*2 7 Amino 1,3-napthalene disulphonic acid.

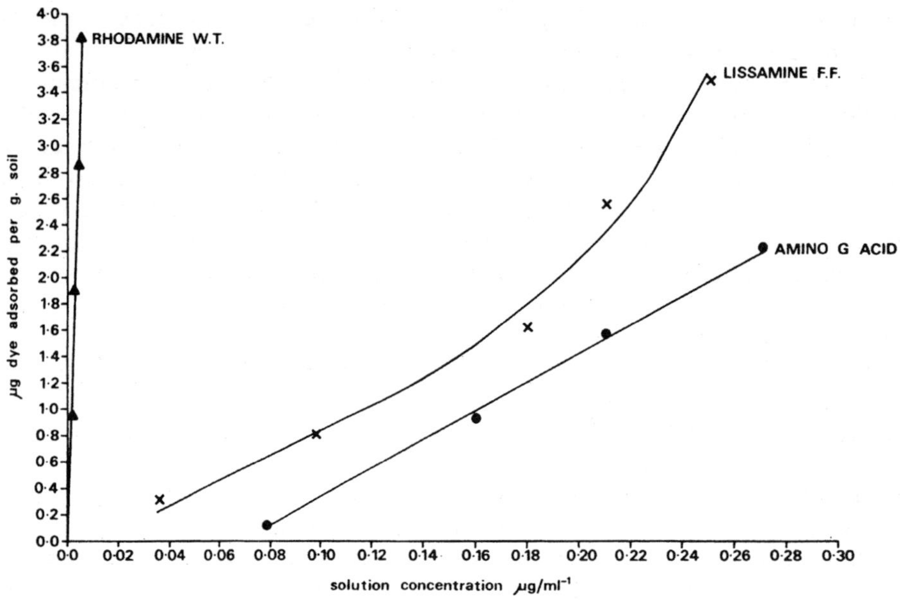

Fig. 3. Equilibrium adsorption isotherms for dyes with disaggregated soil.

tively, reported by Smart and Laidlaw (1977) for dilute suspensions (1:50 substrate to solution ratio). Smart and Laidlaw also report losses of 32% and 4% for Lissamine FF® on humus and kaolinite, respectively, and 61% and 3% for Amino G Acid®, respectively. For the present soils, Lissamine FF® losses were intermediate between Rhodamine WT® and Amino G Acid®.

The isotherms apply to an equilibrium situation, referring specifically to a thermodynamic equilibrium between adsorbate and the adsorbent surfaces. However, when fast flow occurs the adsorption reaction rate may be slow in comparison to flow rate. Reaction kinetics may better describe this situation (Travis and Etnier, 1981) and adsorption rate may be concentration-dependent. In addition, the aggregation of field soils affects the diffusion of large molecules (Barraclough and Nye, 1979) which can become dependent upon the presence of large interaggregate pores. Thus, even under slow flow, in aggregated soils, adsorption data derived from dispersed soil material in batch equilibrations will be an unreliable guide to dye movement in intact aggregated soils and are likely to overestimate the amount of adsorption. This may be additionally complicated by micellisation of dye at high concentrations consequent upon high rates of field application. Micellisation is reported for Rhodamine WT® (Smart, 1972) and at high dye concentrations simple monolayer adsorption models do not apply (Easton et al., 1964; Giles et al., 1964). Micellisation may thus increase apparent adsorption effects. In addition, variations in soil pH may also affect fluorescence (Smart and Laidlaw, 1977) and adsorption mechanisms (Mortland, 1970). It follows that some more realistic tests of dye behaviour in intact stuctured soils are required in order to evaluate the validity of dye use in field soils.

Laboratory soil-column displacement experiments were undertaken to provide breakthrough curves for the dyes under saturated conditions (Fig. 4). Intact soil columns were used, with only the central portion of flow being collected in a fraction collector in order to minimise the effects of any edge flow. The top edges of the soil columns were sealed with puddled clay and the fluorescence data presented are for data with soil-derived background fluorescence subtracted. Saturated conditions were obtained using a constant-head device, with a simultaneous slug injection of all three dyes at concentrations of $50 \mu g\, l^{-1}$ Rhodamine WT®, $100 \mu g\, l^{-1}$ Amino G Acid® and $150 \mu g\, l^{-1}$ Lissamine FF®. One pore volume of the soil was 3252 ml, but breakthrough was almost instantaneous in relation to the presence of preferential flow pathways (Smettem and Trudgill, 1982). There are evident lags between the dyes in time to peak concentrations, precluding the use of dyes as quantitative tracers, even under the rapid-flow conditions specified. However, first times of arrival of all dyes are comparable, despite their differing adsorption characteristics. Compared with losses in batch adsorption tests at equivalent initial concentrations, Rhodamine WT® losses were 90% in the column and 95% in the disaggregated soil, and Lissamine FF®

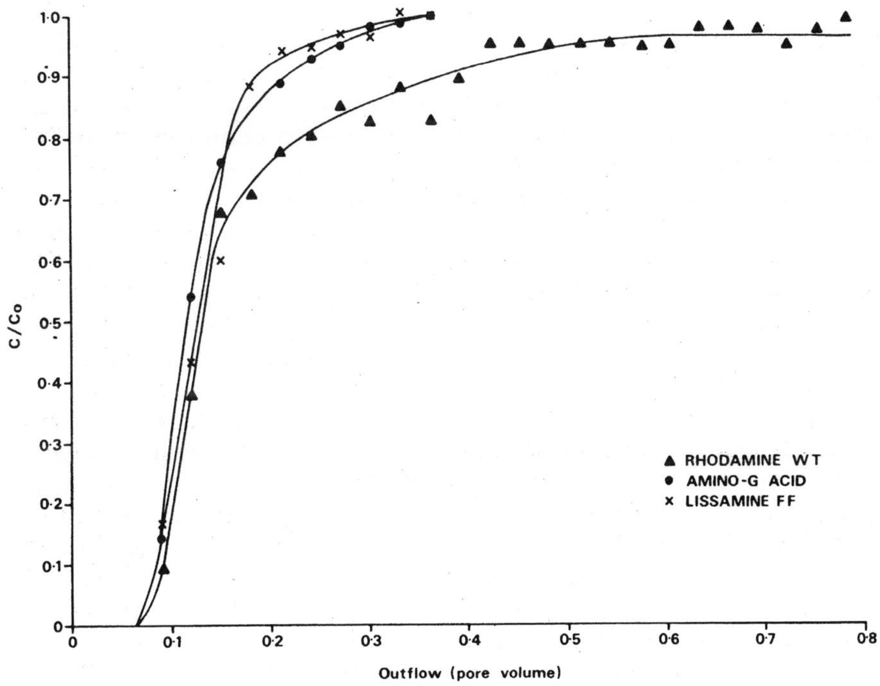

Fig. 4. Breakthrough curves for simultaneously applied dyes in structurally intact soil column from field site. C = concentration of output, C_0 = input concentration.

losses were 40% in the column and 81% in the batch tests. Previous work Smettem and Trudgill (1982), has shown that for study-site soil columns Lissamine FF® had initial breakthrough characteristics similar to chloride, though with a lag to peak concentrations. Field conditions, however, present greater opportunity for adsorption of dyes on soil surface, especially under unsteady flows with pulsed inputs of rainfall spaced over long time intervals. Accordingly, field trials were undertaken using chloride and Rhodamine WT®, the most adsorbed of the three dyes tested. Two hundred and fifty g Analar® $CaCl_2$ was placed on the soil surface on November 16, 1978, 1 m upslope of a stream bank sampling point together with 50 g Rhodamine WT®. The dosage area was 0.3 m long and the trace pathway was 1 m to the bank through soil 0.3 m deep. Minimum dye dosage calculation was made using the formula:

$$A = \mathrm{BD} \cdot V[K(2B) + 2B] \quad (5)$$

where A = application of dye (μg); BD = soil bulk density (g cm^{-3}); V = volume of soil through which trace is to be made (cm^3), assuming a cubic volume under the dye application area and uniform adsorption conditions; B = highest background value recorded prior to trace (μg l^{-1}); and K = distribution coefficient (μg g^{-1} or μg ml^{-1}), K = 850 for Rhodamine WT®, 10.0 for Lissamine FF® and 3.0 for Amino G Acid®, using equilibrium isotherm data (i.e. maximum potential adsorption). In this case, the values for Rhodamine WT® were:

$$A = 1.3 \times 1.2 \cdot 10^5 [850(2 \cdot 0.02) + (2 \cdot 0.02)]$$

with a value of A = 5,310,240 μg or ~5.0—5.5 g. The results are shown in Fig. 5, with samples taken using an automatic water sampler having separate intake hoses for each sample, thus minimising cross-contamination.

If Rhodamine® adsorption affects the arrival time of the dye, then a marked lag can be predicted for the dye arrival, behind the chloride arrival. In fact, the reverse is true, with the samples on December 8, 1978 showing a positive Rhodamine® trace and a lag for chloride. This pattern was not the subject of further investigation but was taken as evidence that Rhodamine WT® adsorption did not delay arrival times relative to chloride.

The data presented above are used as evidence that the fluorescent dyes tested present viable tracers which may be used to identify first time of arrival in the system under study, and thus may be used to test the hypothesis that outflow from the bank should occur under saturated conditions (storage model) and specifically to test whether outflow can occur in response to intense rainfall events, even under unsaturated conditions. Even if adsorption does delay the first time of arrival in long-duration traces, this militates against the acceptance of the intense-rainfall hypotheses that outflow will occur in advance of saturation as adsorption may retard the dye output. The chances of arriving at an erroneous conclusion regarding rapid flow because of tracer behaviour are thus minimised. Since the first times of

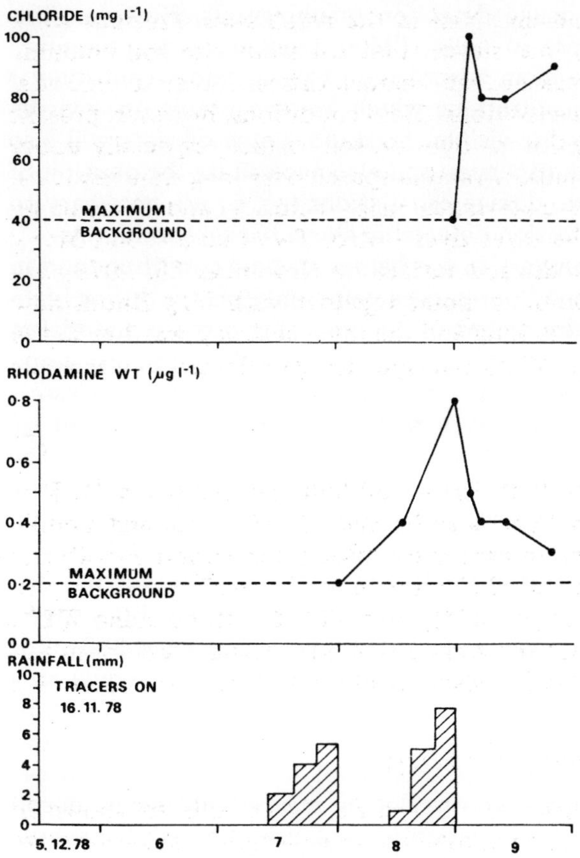

Fig. 5. Field trace using Rhodamine WT® and chloride.

arrival are comparable for all dyes in the soil studied, all three dyes were used at the site with successive traces using Rhodamine WT®, Lissamine FF® and Amino G Acid®. A positive dye trace was taken as twice the maximum background recorded during the storm events prior to the trace.

The occurrence of natural fluorescence in runoff waters has been well documented, largely in relation to the presence of colloidal and dissolved organic matter. Total organic carbon concentration correlates linearly with fluorescence in natural waters between 400 and 600 nm emission wavelengths (Smart et al., 1976). Fulvic acid extracts have a peak emission of 520 nm (Seal et al., 1964) and Smart and Laidlaw (1977) report fluorescence of several organic substances in the range 400—600 nm, with peak emissions at 450—500 nm.

The organic matter emission spectra reported overlap with those of Lissamine FF® (peak 515 nm) and Amino G Acid® (peak 445 nm) but there is little overlap with Rhodamine WT® (peak 580 nm). It is possible to overcome high background fluorescence at dye wavelengths by the use of large enough quantities of dye to make detection unambiguous. However, this may lead to structural collapse in soils (I.M.S. Laidlaw, pers. commun., 1981). In addition, one of the main advantages of using fluorescent dyes is their detectability at low concentrations which are acceptable on environmental grounds and are below the visible threshold. Large dosages will also lead to increased diffusion and dispersion and also necessitate the dilution of samples for analysis. Therefore, dosage calculations (eq. 5) are necessary to estimate the minimum concentration detectable above background levels.

Background fluorescence can vary in the field with flow conditions and in relation to source area. Increased fluorescence commonly occurs during high-flow conditions due to the entrainment of decomposing organic material in surface or near-surface runoff. Periods of high flow are likely to correspond to expected tracer arrival times and it was therefore important to assess background fluorescence variations at a range of flow conditions and especially at high flow. Background fluorescence values for field soil drainage waters, soil suction cup samples and stream water are given in Table III. Soil-water fluorescences were generally higher than stream-water values. Background fluorescence at the Amino G Acid® and Lissamine FF® wavelengths was always higher than that at the Rhodamine WT® wavelengths.

Dye emplacement was on the surface of the soil, within the litter layer. Lissamine FF® and Amino G Acid® were applied as 100 g power. 250 ml of 20% Rhodamine WT® solution were used, being approximately equal to a 50-g dosage. The liquid dye was spread over an 0.3 x 0.15 m area in order to minimise infiltration of the dye solution. Sampling showed that the liquid dye did not penetrate beyond the top 1—2 cm of the soil unless rain occurred. Examination of the soil profile revealed no marked differences in soil properties between the sites over the section of bank used, though the results relate to the sites used and their representativeness of the hillslope as a whole remains untested.

Soil-water retention capacity was measured as the water content remaining after saturation and drainage, using structurally intact soil columns 30 cm long x 20 cm diameter. Hydraulic conductivity of soil was measured on small soil samples of 11.5-cm^2 surface area, using a simple permeameter (Burt, 1978b).

TABLE III

Ranges of values recorded for background fluorescence at dye wavelengths in soil waters (fluorescence converted to $\mu g\ l^{-1}$ equivalent dye concentration)

Site	Rhodamine WT®	Lissamine FF®	Amino G Acid®
Soil drainage waters (throughflow troughs)	0.01– 0.08	10—30	275—300
Suction cups	0.01—0.1	40—100	200—800
Streams	0.01—0.02	1—25	20—100

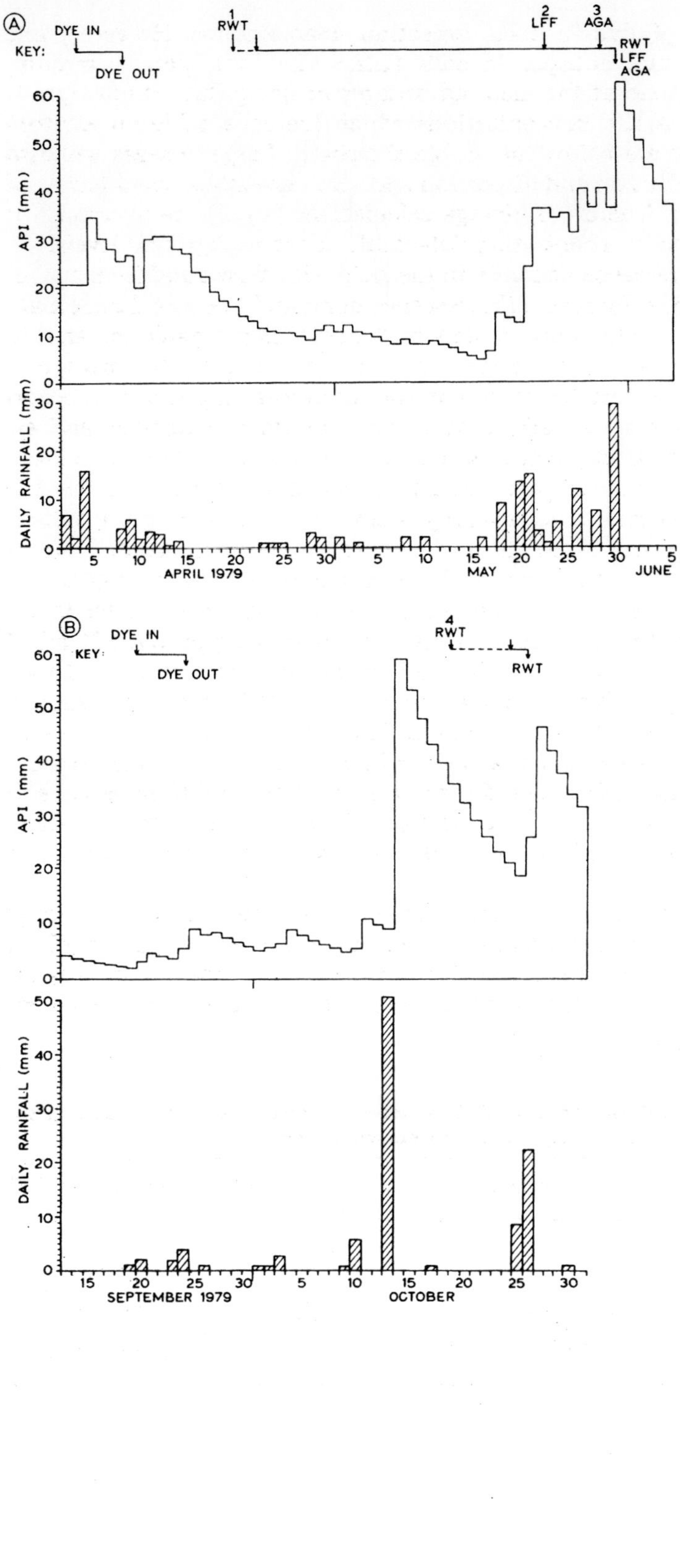
A
DYE IN
KEY:
DYE OUT
1
RWT
2
LFF
3
AGA
RWT
LFF
AGA
API (mm)
DAILY RAINFALL (mm)
APRIL 1979
MAY
JUNE
B
DYE IN
KEY:
DYE OUT
4
RWT
RWT
API (mm)
DAILY RAINFALL (mm)
SEPTEMBER 1979
OCTOBER

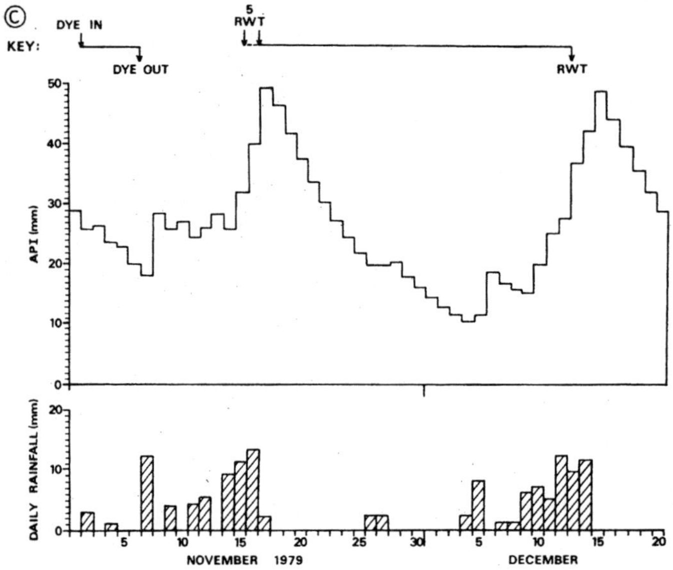

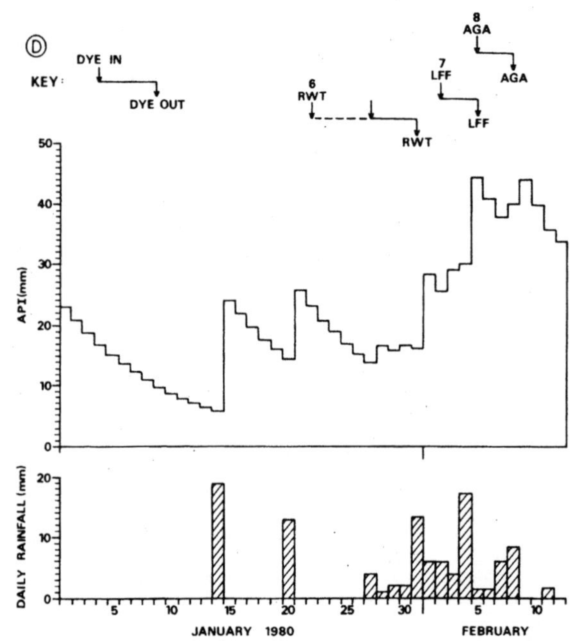

Fig. 6. Dye tracing results for: (A) traces *1—3*, April—June 1979; (B) trace *4*, September—October 1979; (C) trace *5*, November—December 1979; and (D) trace *6—8*, January—February 1980.
RWT = Rhodamine WT®, LFF = Lissamine FF®, AGA = Amino G Acid®.

RESULTS

No flow was ever recorded in the upper site throughflow trough (Fig. 2, pit E), indicating a lack of lateral flow at that position. One dye trace was recorded from the upper-site dye emplacement point to the nearest suction sampler, other samplers were negative. Lysimeter results showed that rapid vertical percolation occurred at this site.

At the bank site, the top trough was permanently dry; the flow in the middle trough was intermittent and in response to rainfall events and the bottom trough flow was sustained. The data reported below are for dye detection in the middle trough. Dye detection ranged from 1 to 37 days from dye emplacement to positive detection in the trough. The individual dye traces are shown in Fig. 6, together with API (using a decay function of 0.9) and daily rainfall. Positive dye traces occur during or just subsequent to high-rainfall events.

Fig. 7 shows the data for the eight dye traces where API_I (mm) is the API for the day of dye emplacement. Assuming that rainfall (minus evaporation) needs to occur to occupy soil-water storage, the API_I data are plotted against the differences in API from input to output (API_{O-I}), plus R_O, the amount of rainfall of the dye output event (assuming no decay of this input). The predicted outflow is at 56 mm, the minimum value in a range of 56—70 mm measured for field capacity in five soil cores. There is no necessary physical reason why the values of API should exactly coincide with soil-water holding

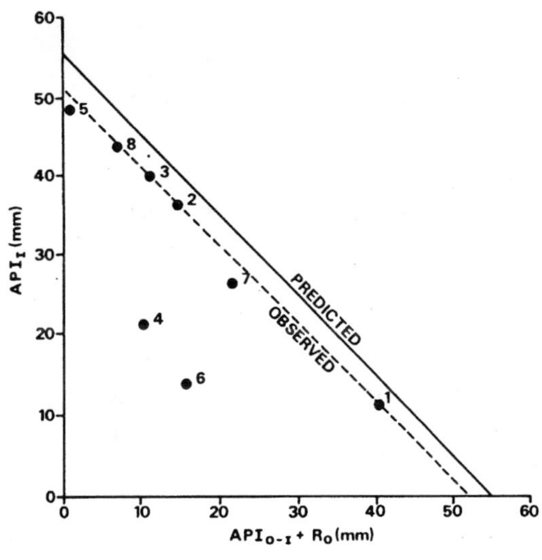

Fig. 7. Storage model for outflow of dyed water. Predicted line is for outflow in relation to measured soil-water retention capacity. API_I = API for dye input (see text for API definition), API_{O-I} = API net increments + R_O = rainfall at output. Data points are for combinations when outflow occurred. Traces *4* and *6* are not predicted by a storage model.

capacity measurements since the choice of the decay function for the API is arbitrary, but it can be seen that the data for six out of eight traces plot close to the predicted line along the relationship:

$$API_{(O-I)} + R_O = -1.025\ API_I + 52.1 \qquad (6)$$

The API model used thus gives a storage value of 52.1 mm, underestimating the minimum measured soil-water retention capacity value by 4 mm. This suggests that a soil-water retention capacity value may be used to predict a maximum value of storage with an error term of ±2%. However, the hypothesis that the storage model can be applied to this soil to predict output is not accepted since two dye trace events lie far outside the scope of the relationship of eq. 6. In addition, it is possible that the relationship is influenced by the fact that all dye output events are coincident with high-rainfall events (Fig. 6). It is pertinent, therefore, to test the applicability of the preferential-flow rainfall intensity model.

If preferential flow occurs in soils when ped infiltration capacity is exceeded, minimum pedal saturated hydraulic conductivities measured for soils from the field site can be used to set the lower limit at which preferential flow occurs. Pedal K_{sat} was measured at $1.0 \cdot 10^{-4}$ cm s^{-1}, equivalent to 3.6 mm hr.$^{-1}$. Thus, it can be predicted that if rainfall intensity exceeds 3.6 mm hr.$^{-1}$, then preferential flow will occur. Whether this will lead to dye

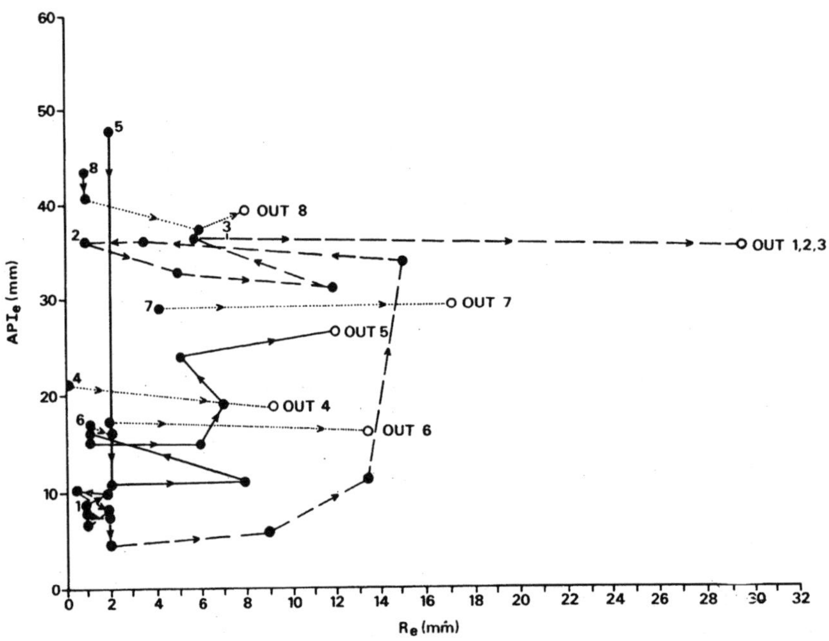

Fig. 8. Plots of history of each trace against API for each rainfall event (API_e) and amount of rainfall event (R_e). *Open circles* indicate outflow of dye.

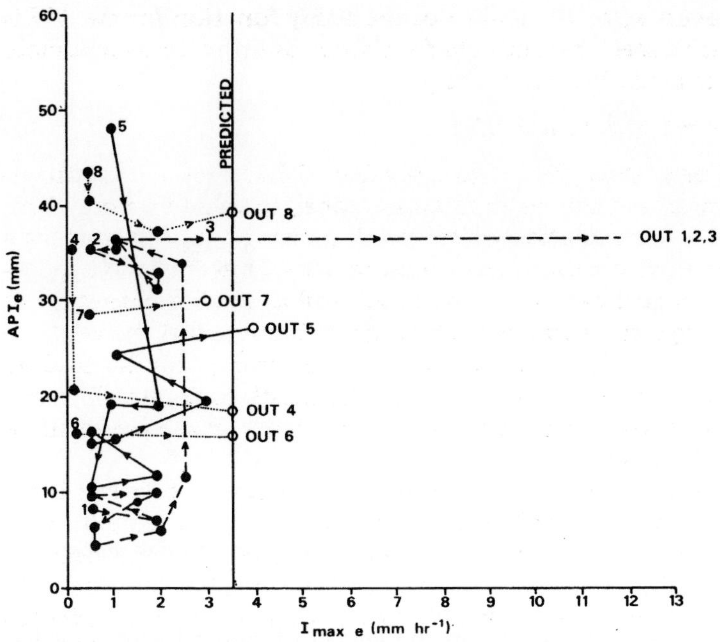

Fig. 9. Trace histories in relation to API for each event (API_e) and maximum intensity for each event ($I_{max\,e}$). *Open circles* indicate outflow of dye.

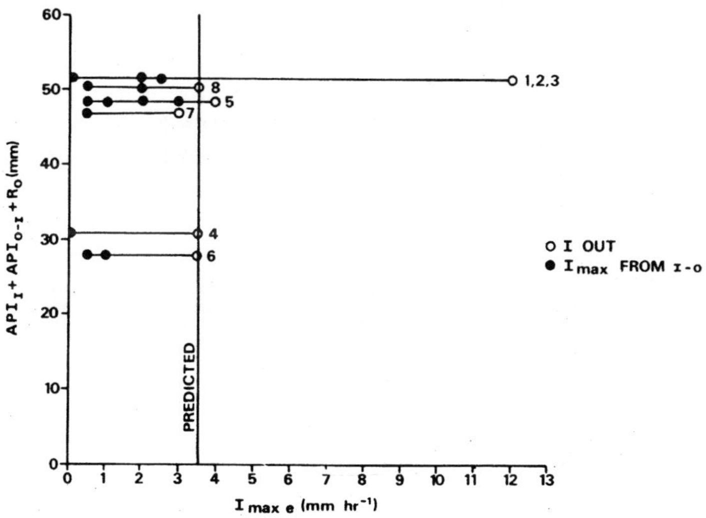

Fig. 10. Dye outflow in relation to calculated soil moisture storage (y-axis) and maximum intensity (x-axis).

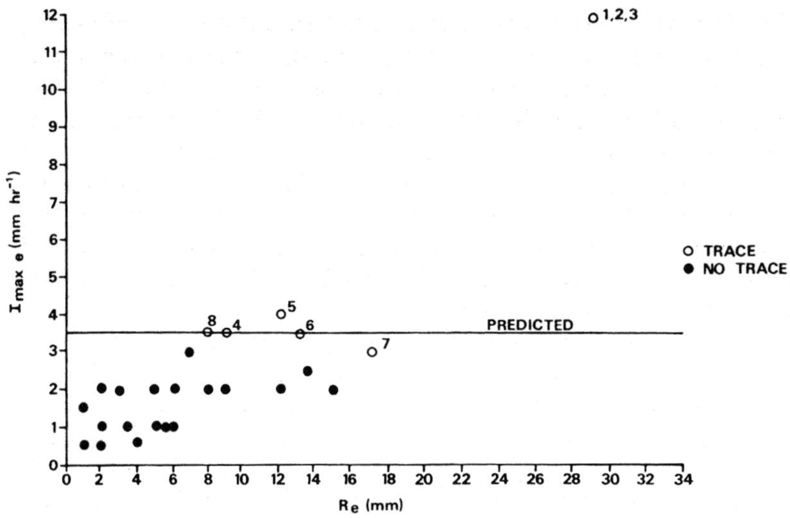

Fig. 11. Dye output in relation to maximum intensity (y-axis) and rainfall amount for each event (x-axis).

output will depend, qualitatively, upon the length of rainfall event needed to sustain flow throughout the trace length and this is difficult to predict quantitatively since the nature of the flow front and pathways are unspecified. The data shown in Fig. 8 assess the role of rainfall amount in leading to dye output; the history of each trace is plotted against API for each rainfall event (API_e) and the amount of each rainfall event (R_e). There is overlap between combinations of antecedent wetness and rainfall which do (open circles) and do not (full dots) give rise to dye output. Although the minimum amounts of rainfall needed to lead to output cannot be specified as rainfall amount is uncontrolled, there is little predictive value in this plot. However, plots showing maximum rainfall intensity occurring between dye emplacement and dye output ($I_{max\ e}$) (Fig. 9) identify a threshold when output occurs which is close to the predicted value of 3.6 mm hr.$^{-1}$. Fig. 10 shows a plot of data for initial API (dye emplacement) API_I, plus rainfall input increment to trace output ($API_{O-I} + R_O$) against maximum rainfall intensity ($I_{max\ e}$) during the trace. An intensity threshold is evident, while there is a wide range of antecedent conditions. An intensity threshold is again seen in Fig. 11 as opposed to an amount threshold.

DISCUSSIONS AND CONCLUSIONS

A storage model could be used to predict six out of eight cases of outflow, but conditions of sustained intense rainfall predict all cases of outflow.

Precise specification of intensity conditions is not simple because of canopy interception and in-storm variations in intensity. Intensities of 3 mm $hr.^{-1}$ occur between dye emplacement and output, without dye output. Under a forest cover, canopy interception has been measured at 1—2 and 2—3 mm for winter and summer rainfall events, respectively. Stemflow occurs, but is not in question in the context of dye entrainment during water entry into the soil as dye emplacement sites were at least 2 m from major stems (but it could be implicated in rapid subsurface flow). If rainfall is sustained, throughfall values were close to canopy rainfall values, that is, during intense rainfall events of 3 mm $hr.^{-1}$ for over 2 hr.

During traces *1—3*, rainfall of close to the output prediction value of 3.6 mm $hr.^{-1}$ occurred for just under 2 hr. without a dye trace on the May 21, 1979, but output occurred during May 30, 1979 when intensities exceeded or equalled 3.6 mm $hr.^{-1}$ for over 2 hr. Output was also seen at values close to or greater than 3.6 mm $hr.^{-1}$ for 2 hr. or over on traces *4*, *5*, *6* and *8*. On one trace *7*, output occurred during sustained rainfall, where $\geqslant$3.6 mm $hr.^{-1}$ was sustained for two separate periods during the event. There is no clear evidence from Fig. 9 that greater API-values mean that lower intensities can lead to output. Strict application of the value of 3.6 mm $hr.^{-1}$ for 1 hr. duration to specify output conditions is a poor predictor of output because although all output conditions fall into this case, there were five cases of this condition where outflow did not occur. However, allowing for interception loss in the first 2 hr., specification of 3.6 mm $hr.^{-1}$ sustained for a minimum of two consecutive hours predicts all outputs, except trace 7 where intensity drops to 1 mm $hr.^{-1}$ between higher intensity values. The cases of May 21, 1979 and October 25, 1980 were close to the limits resolvable on the rain-gauge chart recording system used. It is clear that sustained intense rainfall leads to output and further work should be able to more accurately quantify the threshold conditions. In terms of actual physical conditions, details of canopy interception, soil dye pathways, soil-moisture variations and incremental movement through the soil during successive storms prior to final output are important factors to evaluate and these will be considered in further papers.

ACKNOWLEDGEMENTS

Financial support was provided by the Natural Environmental Research Council (U.K.) (GR3/3459, GT4/78/AAPS/46, GT4/79/AAPS/48), The University of Sheffield Research Fund, The Royal Society and The Carnegie Trust. Assistance was provided by the Forestry Commission, Manchester Polytechnic and Huddersfield Polytechnic. The authors are grateful to Dr. T.P. Burt for comment on the manuscript in draft form.

REFERENCES

Addiscott, T.M., 1977. A simple computer model for leaching in structured soils. J. Soil Sci., 28: 554—563.

Aldrick, R.J., 1978. The hydrogeology of the Magnesian Limestones in Yorkshire between the river Wharfe and the River Aire. Q. J. Eng. Geol., 11: 193—201.

Anderson, J.L. and Bouma, J., 1973. Relationships between saturated hydraulic conductivity and morphometric data of an argillic horizon. Soil Sci. Soc. Am. Proc., 37: 408—13.

Anderson, M.G. and Burt, T.P., 1977. Automatic monitoring of soil moisture conditions in a hillslope spur and hollow. J. Hydrol., 33: 27—36.

Atkinson, T.C., 1978. Techniques for measuring subsurface flow on hillslopes. In: M.J. Kirkby (Editor), Hillslope Hydrology. Wiley, New York, N.Y., pp. 73—120.

Barraclough, D. and Nye, P.A., 1979. The effect of molecular size on diffusion characteristics in soil. J. Soil Sci., 30: 29—42.

Beven, K. and Germann, P., 1981a. Water flow in soil macropores, II. A combined flow model. J. Soil Sci., 32: 15—29.

Beven, K. and Germann, P., 1981b. Water flow in soil macropores, III. A statistical approach. J. Soil Sci., 32: 31—39.

Bouma, J., 1980. Field measurement of soil hydraulic properties characterising water movement through swelling clay soils. J. Hydrol., 45: 149—158.

Bouma, J. and De Laat, P., 1981. Estimation of the moisture supply capacity of some swelling clay soils in The Netherlands. J. Hydrol., 49: 247—259.

Bouma, J., Dekker, L.W. and Wosten, J.H.M., 1978. A case study on infilitration into dry clay soil, II. Physical measurements. Geoderma, 20: 41—51.

Burns, I.G., 1975. An equation to predict the leaching of surface-applied nitrate. J. Agric. Sci., Cambridge, 85: 443—454.

Burt, T.P., 1978a. An automatic fluid-scanning switch tensiometer system. Br. Geomorphol. Res. Group, Tech. Bull. 21. (Geo Abstracts, Norwich).

Burt, T.P., 1978b. Three simple and low-cost instruments for the measurement of soil moisture properties. Huddersfield Polytechnic, Dep. Geogr. Geol., Occ. Pap. 3.

Burt, T.P., 1979. The relationship between throughflow generation and the solute concentration of soil and stream water. Earth Surf. Processes, 4: 257—266.

Cairney, T., 1972. Hydrological investigation of the Magnesian Limestone of south-east Durham, England. J. Hydrol., 16: 323—340.

Cairney, T. and Hamill, L., 1977. Interconnection of surface and underground water resources in southeast Durham. J. Hydrol., 33: 73—86.

Easton, I.A., Giles, C.H. and McKay, R.B., 1964. Association of adsorbed aromatic ions. Chem. Ind., 45: 1863—1864.

Edwards, W.M., Van der Ploeg, R.R. and Ehlers, W., 1979. A numerical study of the effects of noncapillary-sized pores upon infiltration. Soil Sic. Soc. Am. J., 43: 851—856.

Gaudet, J.P., Jegat, H., Vachaud, G. and Wierenga, P.J., 1977. Solute transfer with exchange between mobile and stagnant water, through unsaturated soil. Soil Sci. Soc. Am. J., 41: 665—671.

Germann, P. and Beven, K., 1981. Water flow in soil macropores: 1. An experimental approach. J. Soil Sci., 32: 1—13.

Giles, C.H., Easton, I.A. and McKay, R.B., 1964. Mechanism of adsorption of cationic dyes by alumin and a note on heat exchanges in solution adsorption. J. Chem. Soc., pp. 4495—4503.

Gregory, K. and Walling, D., 1973. Drainage Basin Form and Process. Edward Arnold, London, 458 pp.

Hall, D.G.M., Reeve, M.J., Thomasson, A.J. and Wright, V.F., 1977. Water retention, porosity and density of field soils. Soil Surv., Tech. Monogr., 9, 75 pp.

Haskell, C.C. and Hawkins, R.H.,1964. D_2O ^{24}Na method of tracing soil moisture movement in the field. Soil Sci. Soc. Am., Proc., 28: 724—728.

Horton, J.H. and Hawkins, R.H., 1965. Flow path of rain from the soil surface to the water table. Soil Sci., 100: 377—383.

Jury, W.A., 1975a. Solute travel-time estimates for tile-drained fields, I. Theory. Soil Sci. Soc. Am., Proc., 39: 1020—1024.

Jury, W.A., 1975b. Solute travel-time estimates for tile-drained fields, II. Application to experimental studies. Soil Sci. Soc. Am., Proc., 39: 1024—1028.

Kanchanasut, P., Scotter, D.R. and Tillman, R.W., 1978. Preferential solute movement through larger soil voids, II. Experiments with saturated soils. Aust. J. Soil Res., 16: 269—276.

Kirkby, M.J. (Editor), 1978. Hillslope Hydrology. Wiley, New York, N.Y., 389 pp.

Mortland, M.M., 1970. Clay—organic complexes and interactions. Adv. Agron., 22: 75—117.

Mosley, M.P., 1979. Stremflow generation in a forested watershed, New Zealand. Water Resour. Res., 15: 795—806.

Omoti, V. and Wild, A., 1979a. Use of fluorescent dyes to mark the pathways of solute movement through soils under leaching conditions, 1. Laboratory experiments. Soil Sci., 128: 28—33.

Omoti, V. and Wild, A., 1979b. Use of fluorescent dyes to mark the pathways of solute movement through soils under leaching conditions, 2. Field measurements. Soil Sci., 128: 95—104.

Pilgrim, D.H., Huff, D.D. and Steele, T.D., 1979. Use of specific conductance and contact time relations for separating flow components in storm runoff. Water Resour. Res., 15: 329—339.

Reeve, M., 1976. Soils in Nottinghamshire III, Sheet SK 57 (Worksop). Soil Surv. Rec. No. 33.

Seal, B.K., Roy, R.B. and Mukherjee, S.K., 1964. Fluorescent emission spectra and structure of humic and fulvic acids. J. Ind. Chem. Soc., 41(3): 212—214.

Scotter, D.R., 1978. Preferential solute movement through larger soil voids, I. Some computation using simple theory. Aust. J. Soil Res., 16: 257—267.

Sklash, M.G., Farvolden, R.N. and Fritz, P., 1976. A conceptual model of watershed response to rainfall, developed through the use of oxygen-18 as a natural tracer. Can. J. Earth Sci., 13: 271—283.

Smart, P.L., 1972. A laboratory evaluation of the use of activated carbon for the detection of the tracer dye Rhodamine WT. M.Sc. Thesis, Department of Geography, University of Alberta, Calgary, Alta. (unpublished).

Smart, P.L. and Laidlaw, I.M.S., 1977. An evaluation of some fluorescent dyes for water tracing. Water Resour. Res., 13: 15—33.

Smart, P.L., Finlayson, B.L., Rylands, W.D. and Ball, C.M., 1976. The relation of fluorescence to dissolved organic carbon in surface wates. Water Res., 10: 805—811.

Smettem, K.R.J. and Trudgill, S.T., 1982. An evaluation of some fluorescent and non-fluorescent for use in the identification of water transmission routes in soils. J. Soil Sci., 33(3) (in press).

Travis, C.C. and Etnier, E.L., 1981. A survey of sorption relationships for reactive solutes in soil. J. Environ. Qual., 10(1): 8—17.

Trudgill, S.T., 1976. Rock weathering and climate: quantitative and experimental aspects. In: E. Derbyshire (Editor), Geomorphology and Climate Wiley, New York, N.Y., pp. 59—99.

Trudgill, S.T., 1977. Soil and Vegetation Systems. Oxford University Press, London, 180 pp.

Trudgill, S.T. and Briggs, D.J., 1979. Soil and land potental. Prog. Phys. Geogr., 3(2): 283—299.

Trudgill, S.T. and Briggs, D.J., 1980. Soil and land potential, Prog. Phys. Geogr., 4(2): 262—275.

Trudgill, S.T., Laidlaw, I.M.S. and Smart, P.L., 1980. Soil water residence times and solute uptake on a dolomite bedrock — Preliminary results. Earth Surf. Processes, 5: 91—100.

Trudgill, S.T., Pickles, A.M., Burt, T.P. and Crabtree, R.W., 1981. Nitrate losses in soil drainage waters in relation to water flow rate on a deciduous woodland site. J. Soil Sci., 32: 433—443.

Trudgill, S.T., Pickles, A.M. and Smettem, K.R.J., 1983. Soil-water residence time and solution uptake, 2. Dye tracing and preferential flow predictions. J. Hydrol., 62 (in press).

White, K.E., 1977. Trace methods for the determination of groundwater residence-time distributions. Conf. on Groundwater Quality, Measurement, Prediction and Protection. Water Res. Cent., Stevenage, Sess. 3, Pap. 11.

Wright, I.R., 1977. Net rainfall — below the forest canopy, through fall troughs and stem-flow gauges, Part II. Inst. Hydrol., Rep., 43: 11—17.

Zimmermann, U., Münnich, K.O. and Roether, W., 1967. Downward movement of soil moisture traced by means of hydrogen isotopes. Geophys. Monogr., Am. Geophys. Union, 11: 28—36.

Commentary on Paper C5

McDonnell, J. J. (1990) A rationale for old water discharge through macropores in a steep, humid catchment.

Bishop *et al.* (2004) argue that, since the chemistry of water moving downslope is determined at any given point by soil chemistry, the particular role of the riparian soil is to set the stream water chemistry, since this is the last soil in contact with the water before it becomes runoff. Thus, the riparian zone is perhaps the most important element of the hydrological landscape given that it can decouple the linkage between the major landscape elements, hillslope and channel. In a sense therefore, the riparian zone acts both as a conduit and a barrier, linking the terrestrial and aquatic environments but also acting as a barrier between them (Burt, 2005). It is particularly important therefore to understand solute exchange processes in streamside soils, since this may buffer the impact of hillslope waters on the stream system.

Jeff McDonnell's study of the Maimai catchment in New Zealand was one of the first studies to seek to clarify an important paradox in hillslope hydrology (Kirchner, 2003): how to resolve the fact that the runoff response of hillslopes is rapid but most of the water involved is "old". Put another way, if there is rapid throughput of water through the hillslope to the stream channel, why is the chemical signature of water entering the stream not very similar to that of precipitation? Previous studies of the Maimai catchment had established the importance of macropore flow (Mosley, 1979); that being the case, it might have been expected that storm runoff would be dominated by "new" water, but subsequent studies of stream water chemistry had shown that this was not the case. How then to resolve the issue? McDonnell combined recent advances in hillslope hydrology, both theoretical and practical, to address the problem.

In terms of theory, McDonnell invokes the capillary-fringe effect, promoted by Abdul & Gillham (1984, Paper B6) and others, to account for rapid inputs of soil water to the stream following modest amounts of rainfall. This invokes rapid water table rise in relation to relatively small inputs of water to the soil; slope discharge is proportional to the depth of the saturated zone and so increases proportionately when the water table rises. Most of the water leaving the slope will be "old", since the volume of water stored in the soil prior to the storm event is much greater than the small addition necessary to raise the water table. In larger storms, McDonnell is able to integrate the obvious existence of large volumes of macropore flow on the hillslopes at Maimai with the capillary-fringe effect in the riparian zone. Rising water tables in hillslope hollows (which remain wet because of continuing convergent hillslope drainage during dry periods) feed water into macropores, which in turn drain downslope into the near-stream zone. Whether the macropore water has a "new" or "old" signature is immaterial: the downslope drainage raises the water table close to the stream and predominantly "old" water is shunted into the stream (Anderson & Burt, 1982). In terms of the practicalities, McDonnell's use of tensiometers mirrors a number of studies from the 1970s, and despite the steepness of the slope (steeper than most reviewed in this volume), the evidence of rapid water table rise during storms in the riparian zone is compelling. Thus, McDonnell was able to go a long way in resolving the hydrological paradox: how rapid hydrological response can input "old" water into the stream system (and, by implication, remove the new-water signature, e.g. low pH).

Subsequent studies have refined the McDonnell model. More emphasis is now placed on the storage of old water in bedrock depressions; nevertheless, the basic mechanism of mixing and displacement remains paramount. The latest version of the Maimai hillslope runoff model is shown here in Fig. 1. McGlynn *et al.* (2002) demonstrated that small depressions in the bedrock surface exert significant control on water mobility and mixing. In particular, the bedrock topography appears to determine spatially the pathway of rapid saturated subsurface water flow and tracer breakthrough at the hillslope scale. No wedge-shaped zone of saturation was observed; instead a saturated zone of variable depth was seen to cover the entire slope. Rapid lateral movement within this saturated zone appears to occur along narrow ribbons of highly mobile flow that exploit depressions in the bedrock surface. It is hardly surprising that ideas have developed as more data have been collected at Maimai; nevertheless. McDonnell's 1990 paper remains a classic example of a field study examining the linkage between runoff pathways and slope water chemistry. It has,

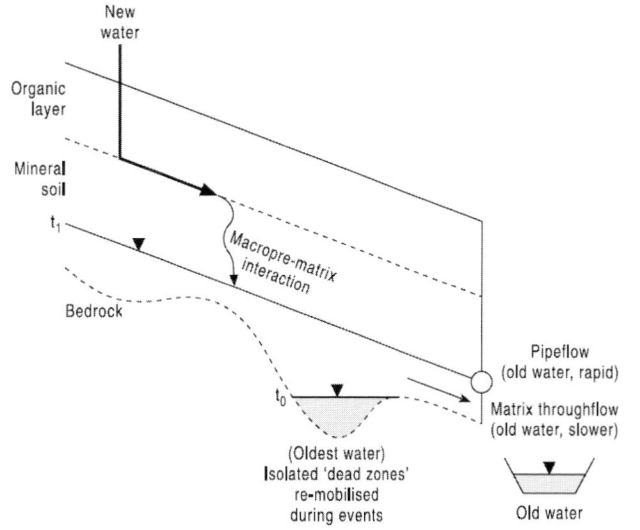

***Fig. 1** The latest version of the Maimai hillslope runoff model (based on Brammer et al., 1995, and McGlynn et al., 2002).*

moreover, been equally important in the development of ideas about how hillslopes function and how riparian zone functions fit into a wider understanding of hillslope hydrology.

TPB

References

Abdul, A. S. & Gillham, R. W. (1984) Laboratory studies of the effects of the capillary fringe on streamflow generation. *Water Resour. Res.* **20**, 691–698.

Anderson, M. G. & Burt, T. P. (1982) The role of throughflow in storm runoff generation: an evaluation of a chemical mixing model. *Earth Surf. Processes Landf.* **7**, 565–574.

Bishop, K., Seibert, J., Kohler, S. & Laudon, H. (2004) Resolving the Double Paradox of rapidly mobilised old water with highly variable responses in runoff chemistry. *Hydrol. Processes* **18**, 185–189.

Brammer, D. D., McDonnell, J. J., Kendall, C. & Rowe, L. K. (1995) Controls on the downslope evolution of water, solutes and isotopes in a steep forested hillslope. *EOS, Trans. Am. Geophys. U.* **76**(46), 268.

Burt, T. P. (2005) A third paradox in catchment hydrology and biogeochemistry – decoupling in the riparian zone. *Hydrol. Processes* **19**, 2087–2089.

Kirchner, J. W. (2003) A double paradox in catchment hydrology and geochemistry. *Hydrol. Processes* **17**, 871–874.

McGlynn, B. L., McDonnell, J. J. & Brammer, D. D. (2002) A review of the evolving perceptual model of hillslope flowpaths at the Maimai catchments, New Zealand. *J. Hydrol.* **257**, 1–26.

Mosley, M. P. (1979) Streamflow generation in a forested watershed. *Water Resour. Res.* **15**, 795–806.

WATER RESOURCES RESEARCH, VOL. 26, NO. 11, PAGES 2821–2832, NOVEMBER 1990

A Rationale for Old Water Discharge Through Macropores in a Steep, Humid Catchment

JEFFREY J. MCDONNELL

Watershed Science Unit, Department of Forest Resources, Utah State University, Logan

Simultaneous observations of rapid preferential flow through macropores and isotopically "old" water displacement remain unresolved in the Maimai (M8) catchment. Continuous, three-dimensional soil moisture energy conditions were monitored in two discrete catchment positions for a series of storm events in 1987. Tensiometric response was related to the soil water characteristic curve, hillslope throughflow, and total catchment runoff. For events yielding ≪2 mm hr^{-1} peak runoff, near-stream valley bottom groundwater systems discharged water volumes sufficient to account for storm period streamflow. This process was assisted by regular low (<−40 cm H_2O) matric potential conditions and rapid filling of available soil water storage. For events yielding >2 mm hr^{-1} peak storm flow, hillslope hollow drainage into steeply sloping first-order channels dominated old water production and most of the catchment storm flow. Highly transient macropore-driven processes of crack infiltration (bypass flow), slope water table development, and lateral pipe flow enabled large volumes of stored water to be delivered to the first-order channel bank at the appropriate time to satisfy catchment storm flow volumes and water isotopic and chemical composition.

1. INTRODUCTION

Progress in understanding processes of storm runoff generation in humid headwater catchments is hampered by discrepancies often found between physical, chemical, and isotopic approaches. Furthermore, the results from chemical and natural isotope separations of stream water into old (pre event) and new (event) water sources often appear to contradict results from hydrometric studies of pathways of water movement on hillslopes.

Recent work in a highly responsive catchment at Maimai (M8) on the west coast of New Zealand has yielded conflicting results regarding the importance of new versus old water in through-flow transmission. Hydrometric studies employing dye tracing and subsurface flow measurement [*Mosley*, 1979, 1982] concluded that rapid flow of new water through macropores was capable of accounting for storm period streamflow. In more recent natural stable isotope and chemical tracing studies, *Pearce et al.* [1986] and *Sklash et al.* [1986] explicitly refuted this earlier interpretation by indicating that old water dominated throughflow in all storm events monitored. On the basis of previous computer simulations [*Sklash and Farvolden*, 1979] and physical principles [*Gillham*, 1984], they suggested that a rapid matrix flow displacement mechanism, occurred through (1) saturated wedges on the lower slopes and (2) groundwater ridges in the valley bottoms. This process involved capillary fringe response [*Ragan*, 1968], which increased local hydraulic gradients and promoted increased gravity drainage of old water to the stream channel.

For this study area it now appears that there may be no need for rapid macropore flow transport of new water to explain the subsurface transmission of water downslope, because stored water is discharged into the stream channel. However, the capillary fringe effect, which limits storage and enables rapid discharge of stored water, has only been clearly demonstrated in laboratory models [*Abdul and Gillham*, 1984; *Stauffer and Dracos*, 1986] and mathematical simulations but has not been well documented in the field. The applicability of these physical and mathematical models to real hillslopes and field soils may be questionable given the latter's heterogeneity.

This paper describes the soil moisture energy conditions at different topographic positions within the M8 catchment and explains the process of rapid exfiltration of isotopically old water into a channel during rainfall events. In companion papers, *McDonnell et al.* [1990], and M. K. Stewart and J. J. McDonnell (Modeling water flow in soils of a steep headwater catchment traced by deuterium, submitted to *Water Resources Research*, 1990 (hereinafter referred to as M. K. Stewart and J. J. McDonnell, 1990)) treat the isotopic behavior of the rainfall, soil water, and channel storm flow.

2. STUDY AREA AND METHODOLOGY

The study was carried out in the Maimai M8 catchment (Figure 1), located in the Tawhai State Forest (45°05′S, 171°48′E), North Westland, New Zealand, described by *Pearce et al.* [1976]. Mean annual gross rainfall in the study area is approximately 2600 mm, producing approximately 1550 mm of runoff from 1950 mm of net rainfall [*Rowe*, 1979]. *Pearce and McKerchar* [1979] note that the catchments are highly responsive to storm rainfall, with 1000 mm (65%) of the mean annual runoff and 39% of the total annual rainfall (*P*) in the form of quickflow (*QF*), as defined by *Hewlett and Hibbert*'s [1967] separation method.

The M8 catchment is underlain by a firmly compacted, moderately weathered, early Pleistocene conglomerate known as the Old Man Gravels. This unit has been described as "effectively impermeable" [*Mosley*, 1979, p. 795], and hydrogeological investigations by D. Bell (personal communication, 1989) indicate that the upper 20 m of this unit form an impermeable capping to the lower conglomerate aquifer systems, with permeabilities and transmissivities of the order of <0.3 m d^{-1} and <15 m^2 d^{-1}, respectively. Slopes are short (<30 m) and steep (mean 34°), with a local relief of 100–150 m. Catchment side slopes consist of regular spurs of

Paper number 90WR01430.
0043-1397/90/90WR-01430$05.00

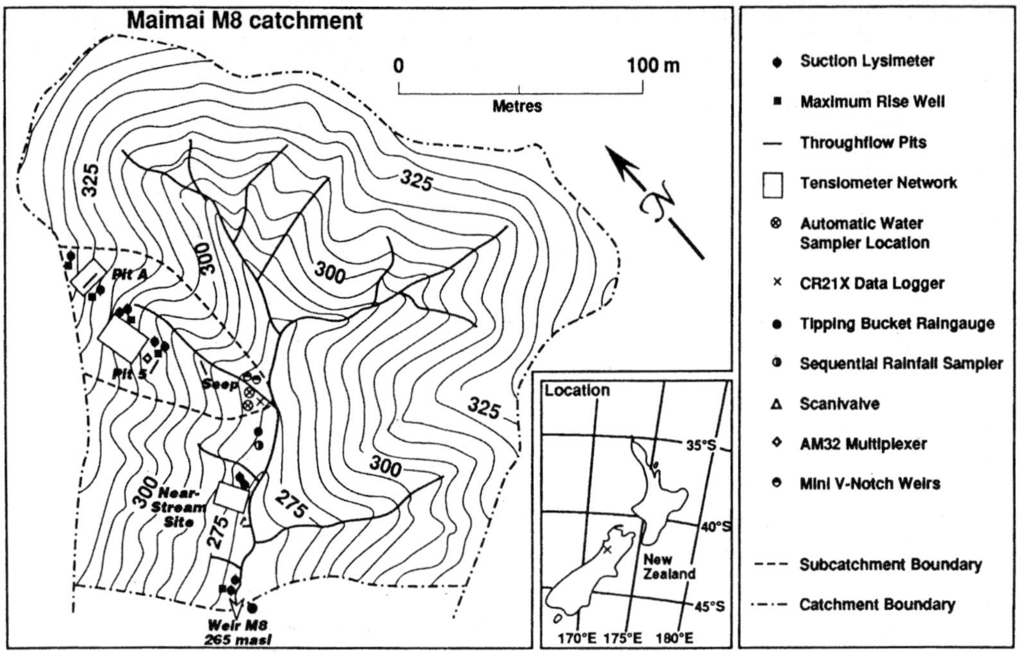

Fig. 1. M8 catchment showing locations of hydrometric instrumentation and sampling sites.

Old Man Gravel bedrock and linear hollows infilled with matrix-rich colluvium. Soils in the catchment have developed from the underlying soft, weathered conglomerate and colluvium and are broadly classified as Blackball Hill soils [*New Zealand Soil Bureau*, 1968; *Mew et al.*, 1975]. They show large spatial variability in depth (mean 0.6 m, range 0.2–1.8 m) and character (from podsolized yellow brown earth (YBE), to mottled YBE, to gley soils) and are dominantly stoney throughout their profiles. Mineral soil horizons are overlain by a thick (mean 170 mm) well-developed upper humic horizon. *Webster* [1977] reported average infiltration capacity of the humic layer and average saturated hydraulic conductivity (K_{sat}) (measured from field cores in the laboratory) of the upper mineral soil of the order of 6100 and 250 mm hr^{-1}, respectively. Limited field measurements of field-saturated hydraulic conductivity (K_{fsat}), using a Guelph permeameter [*McDonnell*, 1989], suggest that conductivity values are highly variable and may range from $\ll$5 mm hr^{-1} in poorly drained hollows to the 250 mm hr^{-1} value reported by Webster on well-drained nose slopes.

The experimental design of M8 (3.8 ha) and an intensively instrumented subwatershed (0.3 ha) have been fully documented [*Pearce et al.*, 1986] and include a complete hydrometric recording system along with eight hillslope throughflow pits located in a variety of soil depth and topographic positions on the slope (Figure 1). Three zones within M8 were intensively sampled and instrumented during September–December 1987 [*McDonnell*, 1989]. The Pit A site (not discussed in this paper) included the deployment of a vertical and horizontal array of 24 recording tensiometers (Scanivalve Model W0602/1P-24T 24 position fluid scanning switch, Sensym SCX15DN pressure transducer) to characterize unsaturated zone response characteristics in an upslope zone known to produce large-event water contributions. The Pit 5 site consisted of an electronically multiplexed and logged array of 32 tensiometers (Campbell 21X logger, Sensym SCX15DN pressure transducers) in a midslope zone to assess the effect of the capillary fringe on saturated wedge development and moisture potentials. The near-stream site was located in a near-stream valley bottom zone and included the same monitoring equipment as Site A, installed along transects both lateral and normal to the channel. The purpose of this site was to monitor the effect of the capillary fringe on near-stream groundwater ridging behavior. Details of tensiometer design and performance are given by J. J. McDonnell (Electronic versus fluid multiplexing in recording tensiometer systems, submitted to *Hydrological Science and Technology,* 1990). Laboratory calibrations of tensiometer tubing pressure response were conducted for a series of tubing materials, diameters, and lengths. Tube material was chosen to maintain negligible pressure attenuation downline.

In addition to hydrometric recording, streamflow, through flow, groundwater, soil water, and precipitation were sampled automatically (Automatic Liquid Sampler 4BSEC) and manually to enable isotopic (deuterium) and chemical (chloride, electrical conductivity) separation of event and pre event water components in storm runoff. Soils were sampled at the near-stream, Pit 5 and Pit A sites. In each case, samples were biased toward stone-free and root-free portions of the horizon because of sampling difficulty. Much of the profile at each site was stoney, and therefore proper undisturbed cores were difficult to extract. Horizon boundaries at each site were distinct, and the Old Man Gravel surface formed an impermeable underlying layer. Table 1 lists the average bulk density and porosity values obtained from each zone, and Figure 2 shows averaged moisture release curves from three replicate cores at each sit, determined using a laboratory tension table apparatus.

3. Mineral Soil Macropores and Preferential Flow

The definition of macropore size classes is quite arbitrary and does not necessarily relate to flow processes [*Beven and*

TABLE 1. M8 Soil Properties for Cores Taken in Three Slope Positions

Site	Total Soil Depth,* m	Sampling Depth, m	BD, g cm^{-3}	θ %	Slope, deg	Topographic Position
Pit A	0.6	0.25	1.5	43	40	upslope hollow
Pit 5	1.5	0.7	0.9	68	34	midslope hollow
Near stream	0.5	0.35	1.3	52	15	valley bottom

*At pit face.

Germann, 1982]. *Luxmoore* [1981] designated three size classes for pores: macro (>1000 μm), meso (10–1000 μm) and micro (<10 μm), where the micropore class corresponds to the soil matrix. In addition to pore size, however, pore structure (including continuity and connectivity) is also of crucial importance to the effective definition of a macropore [*Bouma*, 1981]. As a result, terms such as preferential pathways or macrochannels have been suggested to emphasize the importance of structure on flow dynamics. *Bouma et al.* [1977] have shown that different numbers and sizes of macropores may be effective under different conditions. Therefore a relevant definition of macropores for solute infiltration in a ploughed field may differ from a relevant definition for subsurface flow in a steeply sloping hillslope.

Mosley [1979, 1982] identified preferential flow paths at vertical pit faces in the Maimai catchments for a number of dyed water injection experiments. Assuming that the vertical faces are acceptable samples of the whole soil profile, Mosley's observations show that there are preferred pathways for flow, that is, along cracks and holes in the soil and along live and dead roots and root channels (macropores). In this study, only those pores which are hydrologically effective in terms of channeling flow through the soil and contributing to rapid subsurface flow are of principal concern.

Local M8 catchment conditions are different from many other areas where macropores have been identified because (1) the soil mass is steeply sloping (average 25°–45°), (2) the soil is shallow (average 0.6 m) and underlain by impermeable bedrock, (3) the soil profile is drained at its base by a continuous piping system, and (4) soils remain within 10% of saturation for most of the year. As a result of these conditions, the most hydrologically important macropore types in M8 are considered to be continuous pipes and cracks. The dependence of flow rate on the fourth power of the pore radius means that while the presence of these cracks and pipes may make only a very minor contribution to the total soil porosity, nearly all the rapid flow (at or near saturation) is through these channels [*Bouma and Anderson*, 1973; *Scotter*, 1978].

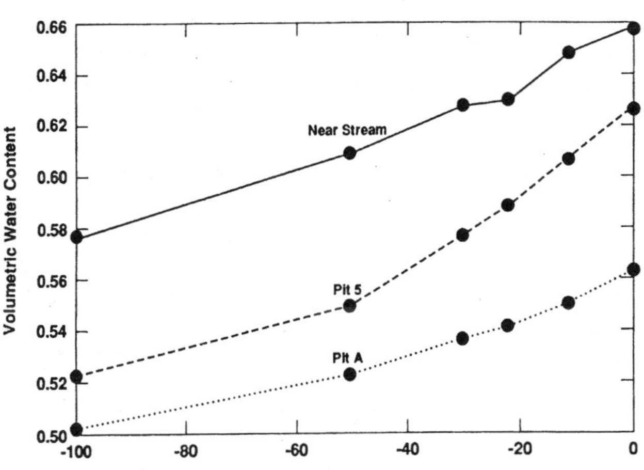

Fig. 2. Moisture release characteristics for soils in the near-stream, Pit 5 and Pit A locations.

3.1. *Pipes and Pipe Flow*

Evidence from a number of recent field monitoring programs, particularly in the United Kingdom, has shown that pipe flow can be a substantial contributor to storm quick flow [*Jones*, 1979; *McCaig*, 1983; *Wilson and Smart*, 1984; *Jones*, 1987]. In the M8 catchment, pipes occur at the mineral soil-Old Man Gravel interface and extend laterally downslope over distances of several tens of meters [*Mosley*, 1979]. This conclusion is supported by the visual observations in this study of pipe outflows at pit faces along the banks of first-order channels and by the fact that hillslope runoff increases rapidly in a downslope direction. Conditions that promote pipe development in the M8 catchment seem to relate to (1) shallow soil depth, (2) underlying impermeable bedrock, and (3) root growth and decay.

Large roots extend vertically through the shallow mineral soil (average 0.6 m depth) but cannot penetrate the underlying conglomerate. Roots then extend laterally over the conglomerate surface (P. Tonkin, personal communication, 1988) for up to several meters. A. Watson (unpublished data, 1988), has shown that the maximum lateral length of 25-year-old *Pinus radiata* root systems in Mangatu Forest, North Island, New Zealand, is 10.4 m. Pipes formed by root growth and subsequent decay may be enlarged by eluviation. Conditions promoting this process in the M8 catchment would include high rainfall, rapid movement of infiltrating water to depth (discussed below), steep slopes and high hydraulic gradients, potentially dispersive soil at the mineral soil base, and the presence of pipe outlets at the first-order channel bank.

As a result of root decay and pore enlargement, a well-connected pipe network has become established in M8, which conducts a large percentage of subsurface storm flow. From limited visual observations at pit faces and along first-order stream banks, pipe outlets range from 3 to 100 mm in diameter.

3.2. *Cracks and Bypass Flow*

In well-structured soils like the clay-rich mottled and gleyed soils in the M8 hollows, flow through continuous cracks, described as channeling [*Beven*, 1981], short circuiting [*Bouma et al.*, 1981], or bypassing [*Smettem et al.*, 1983; *Van Stiphout et al.*, 1987], may result in a deeper penetration of rainfall and solutes than is predicted by uniform displacement [*Thomas et al.*, 1978]. The term "bypass flow" will be used in this paper to describe the vertical movement of free water along continuous cracks (from the mineral soil surface to its base) through an unsaturated or partially saturated soil matrix.

The occurrence and rate of bypass flow will be determined

by (1) total rainfall depth and intensities, (2) prestorm soil moisture content, and (3) soil hydraulic conductivity (K). Bypass flow is a two-domain flow process whereby water movement in vertical cracks is driven by gravity, independent of flow in the soil matrix, which is driven by both gravity and matric (especially capillary) forces [*Germann*, 1986].

At the Pit 5 hollow site, for example, soil was excavated upslope from the pit face to identify any vertical cracking. Regularly spaced vertical cracks, roughly 1–3 mm wide, were observed extending from the mineral soil surface to the Old Man Gravel interface. These cracks showed signs of organic staining along their walls and fine "hairlike" root structures along their entire length. This evidence, together with dye tracer experimental results [*McDonnell*, 1989], suggests that water regularly moves through these channels. Maimai soils, while never dropping significantly below 90% saturation, do occasionally encounter dry periods (of about 3 weeks), when surface cracking can occur (P. Tonkin, personal communication, 1988). In many hillslope hollow zones, upper mineral soils are organic-rich, particularly their midslope sections. Mineral soil surfaces in some situations develop hydrophobicity, which increases the susceptibility to surface cracking.

4. Results of Tensiometric Studies

Tensiometer porous cups were embedded in the soil matrix (away from cracks and pipes) and are assumed to provide information on soil matrix conditions only. By combining the soil water potential data with storm rainfall, streamflow, and hillslope discharge, inferences may be made regarding the nature of subsurface flow. Soil physics considerations are aimed at identifying the rate of development and longevity of the water table, hydraulic gradients, and resulting subsurface water movement. Preferential flow in cracks and pipes, while not directly related to tensiometric response to storm rainfall, can be inferred from soil profile wetting patterns and hillslope discharge.

4.1. *Near-Stream Response*

Tensiometer numbers and porous cup depths for selected tensiometers in the near-stream location are shown in Figure 3. Site numbers are also shown and will be referred to in the following discussion on tensiometric response to storm rainfall. Fifty-eight millimeters of precipitation (P) fell during two short intense bursts on October 29, with peak 10 min intensities of the order of 14 mm hr^{-1}. Streamflow response was rapid, and peak specific discharge was 2.8 mm hr^{-1} (Figure 4). Thirty millimeters of runoff was produced, 29 mm of which was in the form of quick flow (QF). Antecedent precipitation indicies (API_7 and API_{14}, representing 7 and 14 days) were defined by $API_x = \Sigma_{i=1}{}^x Pi/i$, where Pi is the total gross precipitation on the ith day beforehand. API_7 and API_{14} were relatively high (4.7 and 5.0 mm, respectively), and QF/P was 50%. The stream hydrograph had fully recovered from a preceding event on October 24, and base flow conditions were maintained prior to rain input. Pit A through-flow response mirrored that of the main channel and produced a peak discharge of 1250 ml min^{-1}. Seep discharge peaked 3.5 hours after both Pit A and the main channel. Peak flow was only 12.5 ml min^{-1}, and the hydrograph shape showed an extremely steep rise and recession.

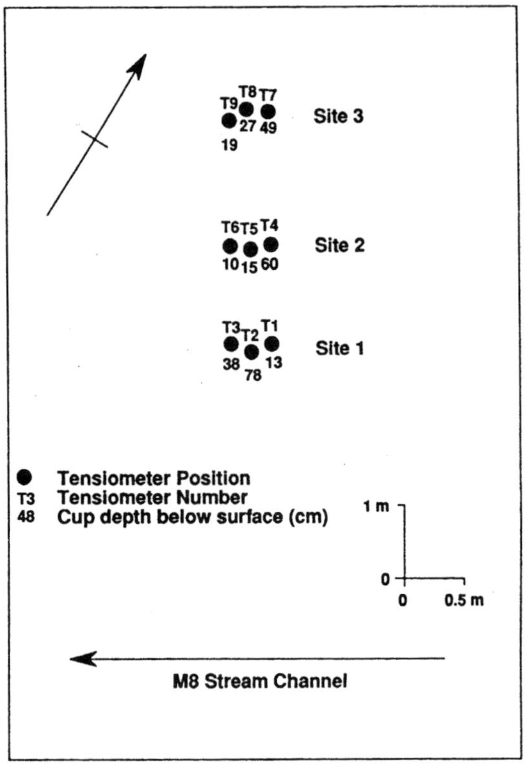

Fig. 3. Tensiometer location and porous cup depths for the near-stream site.

Tensiometric response in the near-stream zone was rapid throughout the soil profile. Soil water potential (ψ) is plotted as a continuous function of time for sites 1–3 (Figure 4). Prestorm ψ ranged from −45 to 10 cm H_2O, with average prestorm suction of about −30 cm H_2O. At sites 1, 2, and 3 all tensiometers responded at the same time (with no appreciable lag with depth) but about 6 hours after the start of the storm (Figures 4*b*-4*d*). In this case, about 90% of the total ψ shift occurred in response to the first rainfall burst at 2400 LT on October 28.

Peak tensiometric response coincided with the stream hydrograph peak. Once peak ψ values were reached, approximately constant values persisted at site 1 and near the surface at site 2 (T6), while the rest gradually declined. Some variation in ψ recession can be seen in Figure 4*b*, where the upslope site 3 tensiometers (T7–T9) showed a more rapid decline from peak positive ψ (0–20 cm H_2O) to negative ψ. Similarly, site 2 ψ recession was more rapid than site 1 ψ recession.

A plot of ψ versus soil depth of T1–T9 provides some indication of the infiltration-groundwater relationship. For t_0, t_1, t_2, and t_3, representing 2330 LT October 28, 0830 LT October 29, 1700 LT October 29, and 0230 LT October 30, a water table was established at each site from previously unsaturated conditions throughout the profile (except site 2). Water table elevation moved from undefined within the zone of measurement to approximately 40 cm from the ground surface at site 1 within 8 hours of the start of the storm. Water table position shifted from 50 to 15 cm from the ground surface at site 2 and from undetected to 28 cm from the ground surface at site 3, each within 8 hours. Total potential (ϕ) characteristics prior to the event showed a strongly lateral component. As the event progressed, ϕ

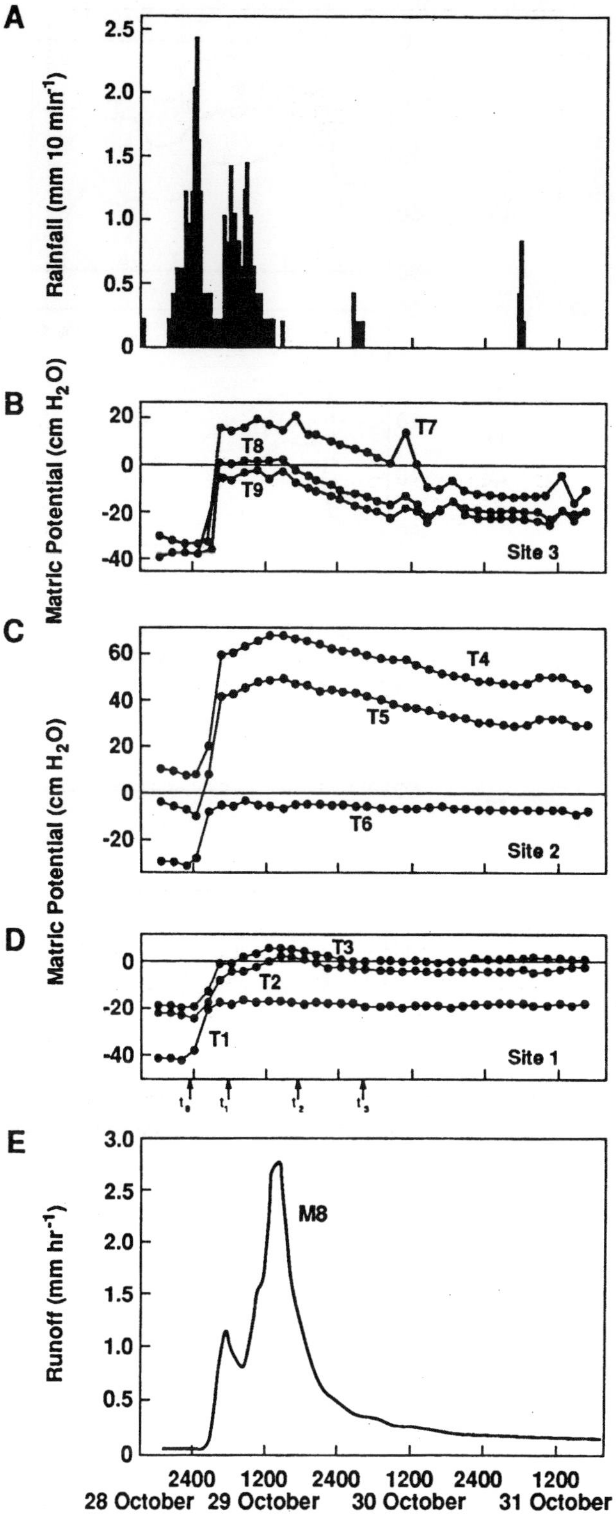

Fig. 4. Hydrometric and tensiometric response for the October 29 event. Precipitation intensity (Figure 4*a*) and M8 runoff (Figure 4*e*) are shown for the catchment, in relation to soil physics response at sites 1, 2, and 3 (Figures 4*d*, 4*c*, and 4*b*, respectively). Note rapidity of response and longevity of positive ψ conditions.

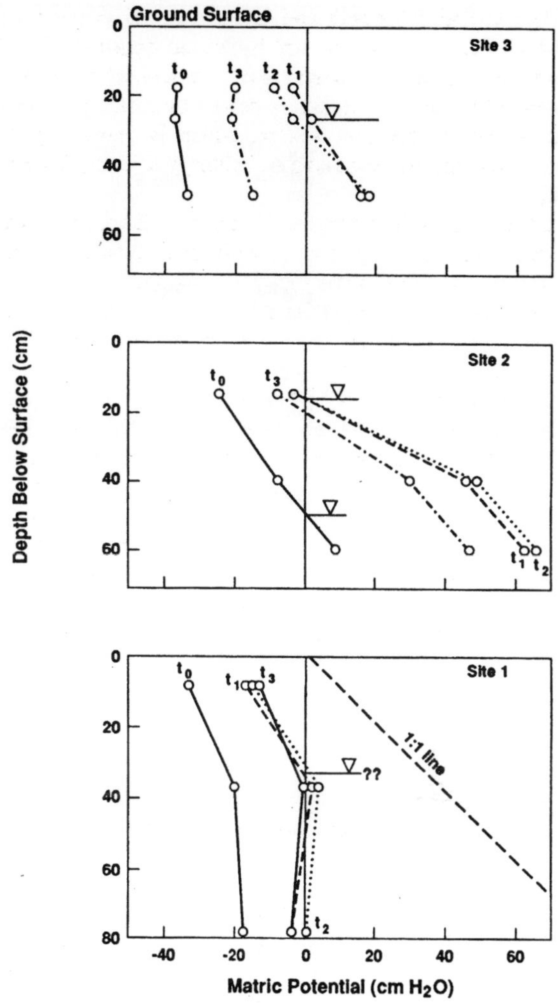

Fig. 5. Near-stream ψ versus depth below surface for the October 29 event.

gradients shifted from lateral to a more downprofile direction (Figure 5). Gradients become more lateral upslope, with downward flow predominating closer to the channel.

In each of the monitored 1987 storms in the near-stream location [*McDonnell*, 1989], ψ response to rainfall input was very rapid through the complete profile for each site location, and water tables were quickly established and maintained at depth. Rapid transition from matric to pressure potential occurred generally within 6 hours of the start of the storm and promoted increased downward ϕ gradients toward the stream channel at site 1. Some lag in ψ response to rainfall input with depth was observed, consistent with the notion of initial downward wetting front propagation through the soil matrix. This pattern rapidly changed to accelerated ψ shift in the lower soil profile, due to perching of groundwater at the mineral soil-Old Man Gravel interface.

Once peak ψ was reached within the profile, values persisted for 1–3 days before gradually returning to prestorm ψ. The maintenance of saturated zones in the near-stream location is a function of the shallow soil cover, underlain by impermeable Old Man Gravels. Additional subsurface flow from upslope zones and low slope angles would also have contributed to persistent low ψ.

4.2. *Midslope Response*

Tensiometer numbers and porous cup depths for selected tensiometers in the Pit 5 location are shown in Figure 6. The

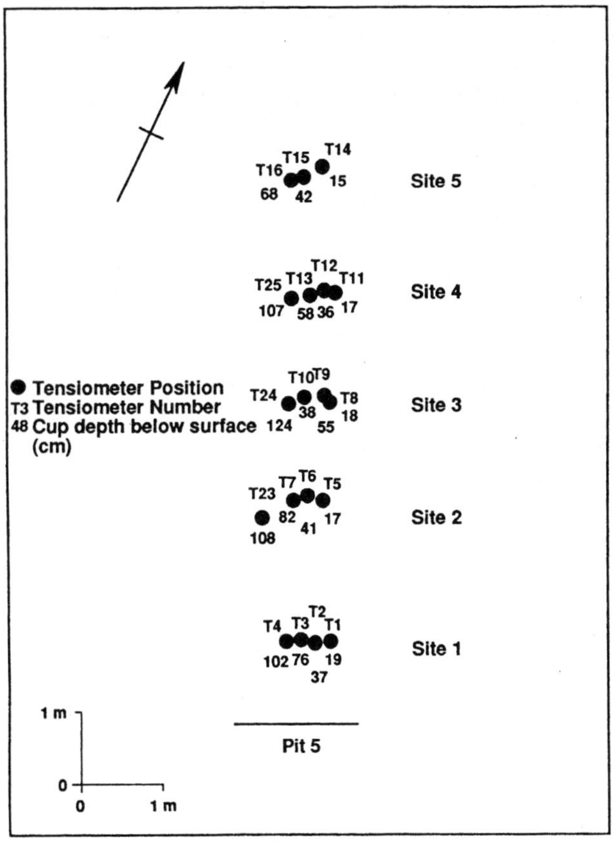

Fig. 6. Tensiometer numbering and porous cup depths for the Pit 5 midslope site.

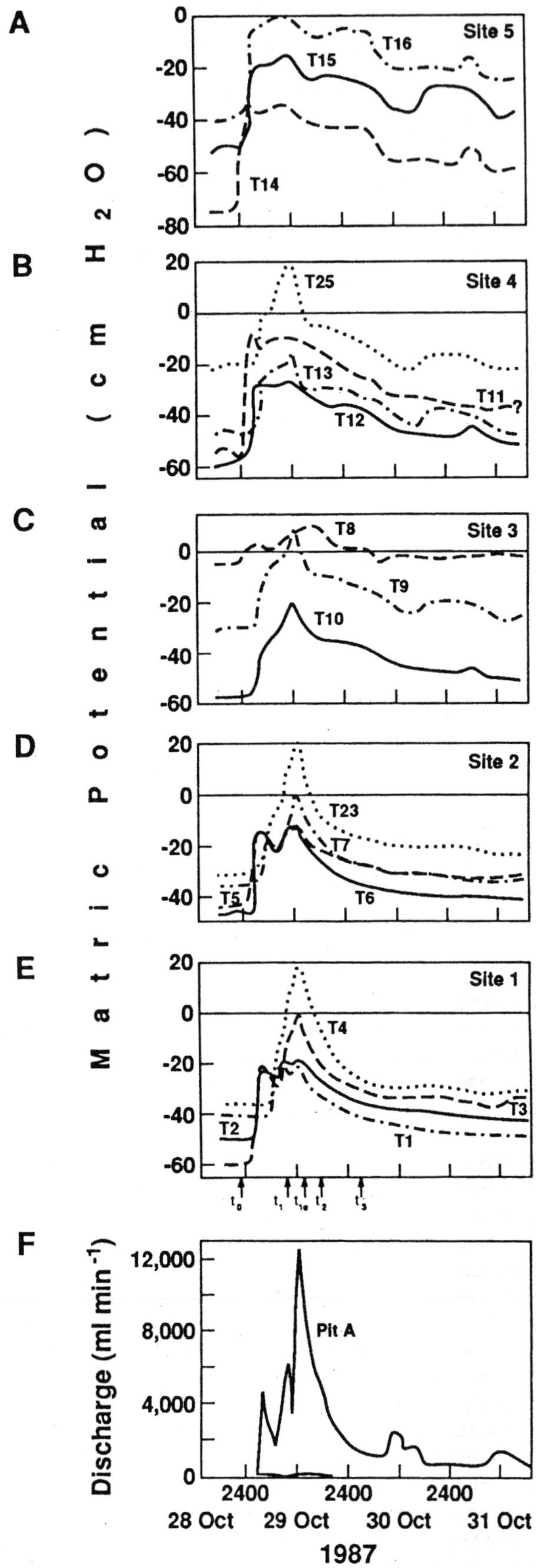

Fig. 7. Hydrometric and tensiometric response for the October 29 event. Pit A throughflow (Figure 7*f*) is shown in relation to soil physics response at sites 1–5 (Figures 7*e*–7*a*, respectively). Note difference in recession of ψ compared to near-stream site.

October 29 storm (already discussed for the near-stream zone) is discussed in detail, and specific time periods t_0, t_1, t_2, etc. (cited above) will also be used for the following discussion of midslope response to storm rainfall. Pit 5 through-flow rates were not available for this event, so $\psi - Q$ comparisons cannot be made. It is assumed, however, that Pit 5 through-flow timing was roughly equivalent to Pit A (Figure 7*f*).

Soil water potential response to storm rainfall at Pit 5 was very rapid throughout the profile, and positive ψ was established at the base of the profile at most sites within 8 hours of the start of the event (Figure 7). The midslope site differed in the timing of response, compared with the near-stream location. In this case, ψ response shapes were closely aligned with the main Pit A hydrograph (Figure 5*f*), exhibiting both a steep rise and recession. In response to the second rainfall burst between 0600 and 1200 LT October 29, sites 1–4 showed an immediate and large shift into positive ψ conditions and with no lag with depth.

At site 1, T1 (19 cm) and T2 (37 cm) reached peak ψ within 3 hours of the first rainfall burst (Figure 7*e*). A second ψ peak of equal magnitude occurred at 1200 LT, corresponding to the second rainfall burst. T3 (76 cm) and T4 (102 cm) showed no ψ peak in response to the first rainfall burst and only began to respond to the second episode. T3 and T4 rose sharply between 2400 LT October 28 and 1200 LT October 29 and then receded at the same rate for another 12 hours before declining more slowly. Low-magnitude ψ was maintained only briefly, as compared to both the near-stream

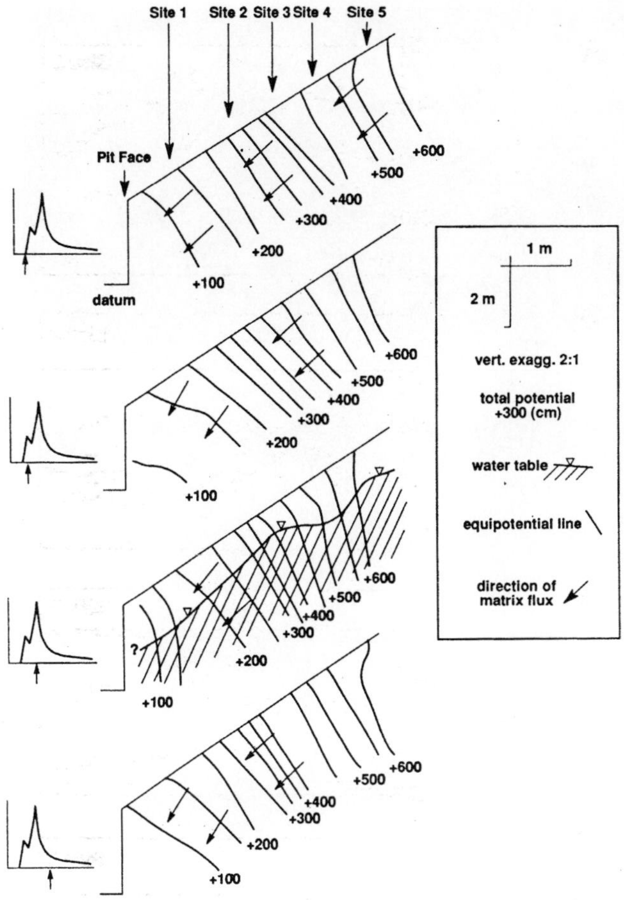

Fig. 8. Cross-sectional diagram (with 2x vertical exaggeration) for the Pit 5 midslope site, indicating ϕ (in centimeters) and corresponding flow lines for time t_0–t_3 during the October 29 event. Note the fixed datum at the bottom of the Pit 5 face.

conditions and the October 24 event conditions at the midslope site.

Tensiometric response to storm rainfall at site 2 was almost identical to site 1 (Figure 7*d*). Moving further upslope, peak ψ response at depth was more subdued and showed a rise to peak, followed by constant low-magnitude ψ. The variation in response, particularly at depth, may be a function of a buildup of saturation near the pit face, as described by *Atkinson* [1978], where a saturated wedge is established upslope of the face. Nevertheless, much of the profile throughout the midslope zone became saturated and maintained positive ψ throughout the catchment's peak runoff response.

Figure 8 shows a very rapid appearance and disappearance of water tables throughout the zone but particularly near the pit face (i.e., sites 1 and 2). Total potential gradients during the October 29 event show a consistent lateral flux component, with some downward flux near the pit face at t_1. The localized downward component shows the effect of the pits' face, but ϕ contours in Figure 8 suggest that for this event at least the influence of the pit may be within about 2 m of the face. Highest ϕ gradients were established at t_{1a} (1200 LT October 29), coinciding with peak ψ and peak pit throughflow. A detectable water table (i.e., somewhere above the deepest tensiometer at each site) was only observed at t_{1a}. This rapid appearance and disappearance of the water table at depth is very different to the near-stream situation, where water tables persist for some time.

Midslope response was highly variable for other monitored 1987 events of different storm magnitudes, intensities, and prestorm soil ψ conditions [*McDonnell*, 1989]. Response seemed to indicate a more erratic infiltration-ψ relationship than the near-stream zone. During monitored low-magnitude events of 5–10 mm rainfall, tensiometric data are consistent with a semiconstant wetting front propagation through the profile and with strong ψ response lags with depth. Although some bypass flow seemed to occur in the upper soil horizon (<50 cm), rainfall depth and soil moisture content was low enough so that the lower soil depths did not receive appreciable moisture from above until streamflow response had subsided. Water tables therefore did not develop at the Pit 5 site, and subsurface storm flow volumes were negligible.

During large magnitude events, matric potential in the lower soil horizons (>75 cm) responded almost instantaneously to infiltrating rain. In each storm, elevational potential dominated total potential computations, and total potential gradients were strongly lateral in the downslope direction. Water table longevity was very short (as compared to the near-stream zone) and showed a close correspondence with Pit 5 through-flow rate. Downslope drainage of perched groundwater was extremely efficient and showed no lag with recorded pit throughflow for storms where these data were available.

5. DISCUSSION

5.1. *Can Near-Stream Valley Bottom Groundwater Sustain Old Water Storm Flow Volumes?*

Tensiometer evidence for the near-stream zone showed that ψ response to rainfall input is rapid, and water tables are quickly established and maintained at depth. Transition from matric to pressure potential occurred generally within 6 hours of the start of the storm and promoted increased downward ϕ gradients toward the stream channel. If the near-stream groundwater ridging mechanism is to explain total catchment runoff, then sufficient volumes of old water must be discharged through this limited near-stream zone to account for measured runoff volumes at the M8 weir. A crude calculation of potential near-stream groundwater flux (q) can be made using measured K_{sat} values from *Mosley* [1979] along with typical near-stream ϕ values (measured from the near-stream tensiometer plot) and computed channel dimensions:

$$q = -K_{sat}\, d\phi/dL \tag{1}$$

where q is the flux rate in millimeters per hour and $d\phi/dL$ is the gradient of ϕ toward the channel. Assuming a K_{sat} of 30–300 mm hr^{-1}, and a typical ϕ gradient of 0.67 (from the October 29, 1987, event, *McDonnell* [1989]), q = 2–20 mm hr^{-1}. If groundwater ridging is important along 125 m of the stream channel (as outlined above) and the saturated depth through which water seeps into the channel is 1 m, then the potential seepage face would be 250 m^2 (assuming water entered from both sides of the channel). Combining the flux rate with the seepage area and then dividing by the catchment area (3.8 ha) produces a near-stream subsurface runoff contribution of 0.13–1.3 mm hr^{-1}. This range of discharge is

Fig. 9. Photograph of pipes venting to the atmosphere along the bank of a steep first-order channel within the M8 catchment. Soils in these positions simply form a face at which water from upslope hollows drains into channel zones.

sufficient to account entirely for the *Pearce et al.* [1986] April 1979 storm but not for the peak of old water discharge for the *Sklash et al.* [1986] September 7–8, 21–22, and 25, 1983, storms or many of the 1987 storms reported by *McDonnell* [1989]. Although a number of simplifying assumptions have been made, the q value suggests that subsurface flux in near-stream valley bottom locations is important in many (but not all) storms.

For events yielding >2 mm hr^{-1} peak storm flow, hillslope hollow drainage into steeply sloping first-order channels dominated old water production and channel storm flow. These processes are seen as distinctly different to valley bottom near-stream behavior. Zones adjacent to first-order channels were characterized by highly incised, steep-sided banks. In many locations the sloping soil mass simply formed a face at which water seeped into the channel (roughly analogous to a pit face). Figure 9 shows an example of this type of channel section, where macropores at the base of the mineral soil layer issue directly into the channel zone. Although this example is from a newly formed stream section (discussed in detail by *McDonnell* [1990]), it characterizes much of the steep first-order channel zones within M8. Hillslope hydrological processes identified at Pit 5 therefore are assumed to be directly comparable to processes controlling water movement to stream areas within these steep first-order channel zones.

5.2. *A Simple Model of Runoff Response*

A model for runoff response in the M8 catchment must satisfy two apparently divergent phenomena: (1) runoff production is extremely rapid, with most flow delivered to stream channels via macropores, and (2) the groundwater (old water) component in channel storm flow is in excess of 85% [*McDonnell*, 1989]. The answer to this problem may simply relate to the large soil water store (350–500 mm) relative to rainfall input during individual storms (typically 25–75 mm). *Sklash et al.* [1986] argue that since less than 30% (in their monitored events) of the rainfall amount appears as quick flow, about 70% of the rainfall either infiltrates or is intercepted, and this is sufficient to supply the subsequent low-flow discharge and transpiration until the next storm. Furthermore, only a small amount of the soil water store must be discharged to obtain the quick flow yield in most storms. Tensiometric evidence from the present study shows that much of the catchment is within 10% of soil saturation for most of the year, and prestorm ψ is generally between -20 and -80 cm H_2O. Therefore all that is required to satisfy the rapidity of catchment runoff response is an efficient transport system to bring stored water quickly from the hillslopes to the surface channels.

Soil moisture content in the M8 catchment increases substantially in the downslope direction because of steep slope angles and relatively permeable soils. The near-stream valley bottom zone, along the main channel, represents the only zone of semipermanent aquifer storage, where $\psi > 0$ cm H_2O in the subsoil. During the initial wetting of the profile, available moisture storage in zones closest to the channel is filled, and resident groundwater begins to discharge into the channel, assisted by groundwater ridging (or mounds) along the channel margins. Soil water overlying the near-stream groundwater ($\psi = -20$ to -40 cm H_2O) is rapidly converted to $+\psi$ and is then displaced into the channel. This process is seen as a near-stream subsurface response and would probably only occur in the valley bottom zone of the main channel section. This subsurface near-stream response is in addition to valley floor saturation overland flow and on-channel precipitation.

As the hydrograph starts to rise, near-stream zones along the steep first-order channels would also start to discharge groundwater into the ephemeral channels. The majority of this input would come from hollow zones (small zero-order basins) upslope from the channel bank. Initially, matrix flow from a growing saturated wedge would contribute to channel inflow, but then as perched water table conditions over the Old Man Gravels developed on the slopes, pipes would conduct the majority of saturated flow downslope. Depending on the total depth of rainfall, hollow saturated zones would expand upslope and increase in local saturated zone thickness, feeding a well-connected system of pipes that would transport the water quickly downslope and into the first-order channels. Even if the pipes were not well connected throughout the slope, kinematic theory suggests that transmission of pressure waves downslope may augment hillslope response if the pipes are locally in zones of saturation and flowing full. At the same time, valley bottom near-stream zones would continue to exfiltrate groundwater into the main channel, with increasing contributions from upslope zones.

At peak flow, water table elevations in the near-stream and hillslope hollow zones would be at their maximum level. Once rainfall inputs ceased, hillslope zones would quickly drain via pipe flow, such that hillslope subsurface flow rates would decrease faster than M8 hydrograph recession. This was observed for both the Pit 5 and Pit A plots. The recession limb of the hillslope-only hydrograph would be like that of a sponge; drainage from the pipes would "shut down" as soon as perched water tables in the hillslope hollows are dissipated. The majority of the subsequent stream recession would be fed by valley bottom near-stream zones, whose deeper aquifer storage (and any additional drainage from upslope) would continue to supply the receding limb. Tensiometric data for the near-stream site showed that once peak ψ was reached within the soil profile, values persisted for 1–3 days before gradually returning to prestorm ψ values.

The simplified model follows the *Hewlett and Hibbert* [1967] notion of an expanding contributing zone through the event and satisfies soil physics data presented earlier. The difference in the two models lies in the mechanism of response, whereby the exceedence of a threshold rainfall depth (and/or intensity) during the hydrograph rising limb converts the system from a matrix-dominated near-stream flux to a pipe flow dominated hillslope hollow system. A reverse switching back to matrix-dominated near-stream flux occurs during the stream recession as soon as perched water tables on hillslopes are dissipated. Although the response described above is almost entirely old water dominated, some new water input to the channel would be derived from direct precipitation onto the limited near-stream saturated area and from headwater responses.

5.3. *A Proposed Mechanism for Macropore Flow of Old Water*

The notion of capillary fringe induced groundwater exfiltration observed in many isotopic investigations has been used to explain large old water storm flow volumes. The basis of these claims has come mainly from laboratory study [e.g., *Abdul and Gillham*, 1984] and numerical simulation [e.g., *Sklash and Farvolden*, 1979]. This has been a dual result of observations made in the laboratory with repacked "sand box" soils, coupled with the algorithmic convenience afforded by the analytic representation of a capillary fringe in the soil's water characteristic. In the soil physics literature, however, the notion of soil having a tension-saturated zone, or capillary fringe, has been questioned. Field retention data have failed to observe significant zones of tension saturation in clays and loams [*Perroux et al.*, 1982], and *Clothier and Wooding* [1983] even question the presence of such zones in laboratory media.

The water characteristics presented in Figure 2 indicate that M8 soils release water with any applied suction and as such do not maintain a discrete tension saturated zone or capillary fringe. Although these are in fact release curves, they provide an estimate of the amount of water still in storage at a given ψ and allow an estimate of how much water is required to saturate the soil. Because M8 soils do not have a capillary fringe, they will actually absorb some water during wetting close to saturation. Thus they have some storage capacity. For example, in the near-stream zone, soils are roughly 0.5 m deep, with antecedent ψ of approximately -40 cm H_2O and porosity of 0.66. Thus some 20 mm of rain could be absorbed by the soil, which is in agreement with processes outlined in Figure 4.

At the Pit 5 midslope hollow site, where soil depths and prestorm suctions are more than double those in the near-stream zone, data in Figure 7 indicate that it may be completely unrealistic to assume that these soils wet up nicely along their $\psi(\theta)$ trajectory, as given by their statically determined, equilibrium representation in Figure 2. So there is a conundrum; why do these soils behave as if they have capillary fringe properties (i.e., limited storage), in terms of measured rapid ψ and water table response? It is likely that the appearance of limited storage arises out of disequilibrium in potential during wetting, which is created by soil inhomogeneities or macropores.

Figure 10 illustrates how in steeply sloping hollow zone, bypass flow leads to the soil truly "releasing" water long before wetting along a measured $\psi(\theta)$ would predict, due to pressure potential disequilibrium within the soil. In this discussion it is important to distinguish between the conductivity of the matrix (K^*) and that of the soil with macropores (K_{sat}). If the flux density of rain, Vo, is greater than K^*, local ponding will eventually occur, leading to vertical bypassing (whether or not Vo is greater than K_{sat}). Therefore it is not unrealistic for 5–10 mm hr^{-1} rainstorms to create localized ponding on a soil purported to have a K_{sat} of 100 mm hr^{-1}. It just means that K^*, the appropriate matrix property, is less

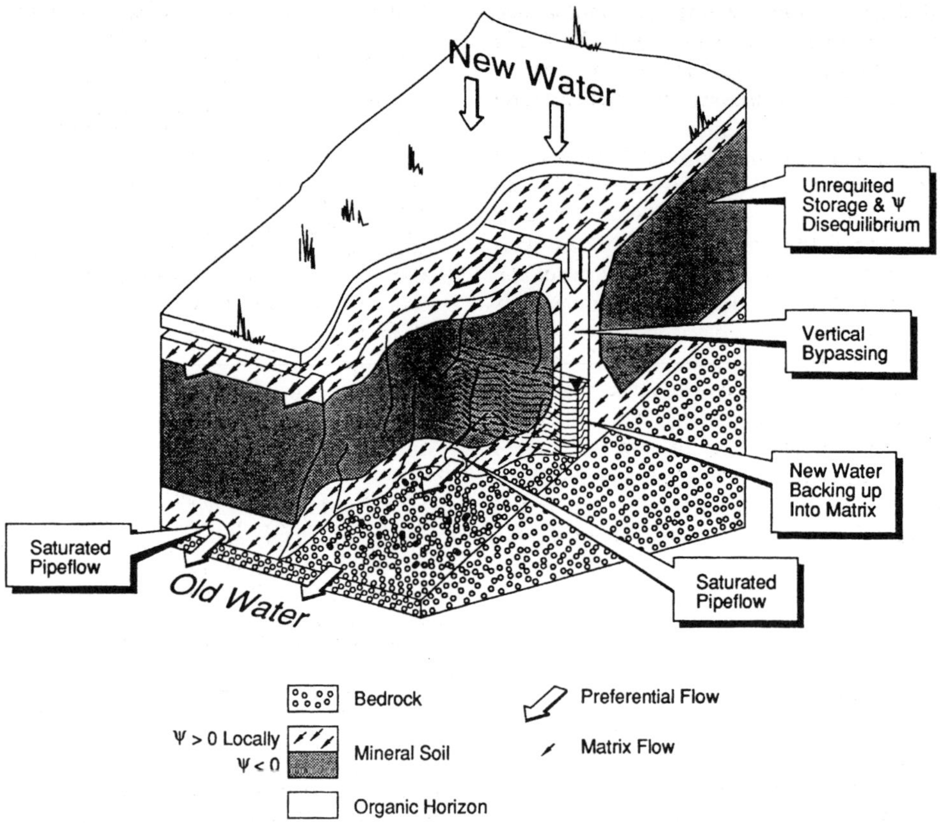

Fig. 10. Conceptual model of runoff production in midslope hollows, combining the idea of unrequited storage and ψ disequilibrium in producing a preferential flow of fully mixed (old) water. In this case, rain flux density exceeds mineral soil surface K^* and moves to depth via vertical cracks. The invading new water perches at the soil-bedrock interface and "backs-up" into the newly saturated matrix. Once free water exists, large pipes in the lower soil zones quickly dissipate transient water tables laterally downslope, producing a rapid through-flow response of well-mixed old water.

than 5–10 mm hr^{-1}. Local bypassing requires minimally that $Vo > K^*$.

During rainfall wetting in the M8 catchment, water infiltrates the organic layer and encounters the mineral soil surface. In some situations [*Mosley*, 1979; *McDonnell*, 1989], $Vo > K^*$, and water flows downslope over the mineral soil surface. In areas with vertical cracks (e.g., Pit 5 midslope hollow), free water on the mineral soil surface flows downslope until it encounters a crack and bypasses to depth. The matrix may still have unrequited storage capacity, but soil water diffusion to unsaturated locales is too slow, relative to its preferential drainage via a crack. Uniform wetting and distribution of infiltrating water may be additionally confounded by the soil not uniformly draining to equilibrium potential beforehand and by hydrophobicity developed along crack walls, as described earlier. The cracks may break capillary continuity, thus limiting matrix drainage, leaving crack walls primed to initiate bypass flow at the beginning of the next rainfall episode. Therefore the pseudo-limited storage response observed at Pit 5, for example, may be a consequence of a dynamic, disequilibrium wetting process, whose interpretation is not possible using static, equilibrium $\psi(\theta)$ information.

As invading new water moves to depth, free water perches at the soil-bedrock interface, and "backs-up" into the matrix, where it mixes with a much larger volume of stored old matrix soil water (Figure 10). Pressure potential evidence from Pit 5 showed that lower-profile tensiometers responded disproportionately, in terms of measured ψ response due to water perching at the mineral soil-bedrock interface. Midslope water tables developed to over 0.5 m of head in the October 29 event. Once established, however, the water table is dissipated by the moderately well-connected system of pipes at the mineral soil-bedrock interface. The relationship between crack infiltration and lateral pipeflow is not linear, because there is a significant time delay between water table perching and subsequent distribution of $+\psi$. This delay is thought to be the key process in shifting new water signatures to that of old water. Isotopic data from Pit 5 through flow [*Sklash et al.*, 1986] showed that old water dominated subsurface flow at these sites by up to 95%. The pipes distribute this mixture of newly bypassed rainfall and mixed stored water downslope to the first-order channel bank. The shift from old water to new water is expected to occur on the slope, as illustrated in Figure 10. M. K. Stewart and J. J. McDonnell (1990) showed that between-storm matrix water varies in age from approximately 1 week at the catchment divide to over 100 days at the main channel margin. *Sklash et al.* [1986] showed that the amount of old water within subsurface storm flow also increased downslope.

Pit 5 through-flow data [*McDonnell*, 1989] showed that throughflow is sensitive to changes in rainfall, in that the preferential flow system ceases to be recharged and drains rapidly once rainfall inputs stop. Even though only a small volume of the soil upslope participates in the runoff episode, both *Sklash et al.* [1986] and *McDonnell* [1989] observed that up to 95% of the water emanating from the Pit 5 face is stored water.

Finally, it should be noted no mention has been made of stemflow effects on bypassing. Rain does not fall directly upon the soil surface where soil physical measurements are made. Rather, it is funneled there by the resident vegetation. These plants may focus the water so that the local flux density of water application to the soil surface may be substantially greater than the recorded rainfall rate. In addition, the focus of this water is at the base of the plant, which is the predominant location of the orifices of plant-engendered macropores. Vegetation may play a critical role in the transmission of rainfall to the surface and its soil transport pathways. Although not specifically studied in this investigation, work is currently underway at the NASA Marshall Space Flight Center to try to quantify these speculations.

6. CONCLUSIONS

Although previous studies in this and other catchments have viewed macropore flow processes to be delivery of new water, this study has shown that macropore (preferential) flow can result in old water displacement. In the M8 catchment, valley bottom groundwater can respond rapidly enough to account for the old water component in small storms and hydrograph rising limbs of larger events. However, discharge through this zone cannot satisfy all hydrographs. At high flow rates, hillslope hollow discharge into steep first-order ephemeral channels dominates runoff and old water storm flow volumes. A crack-pipe model for hillslope runoff is needed to explain the storm flow volumes of the appropriate isotopic composition at the catchment outflow. Although not directly applicable to other hydrologic systems, this work shows the value of combining hillslope and stream-oriented approaches, in an effort to improve our understanding of hydrogeochemical fluxes from forested watersheds. Furthermore, continuous three-dimensional recording of soil water potential conditions greatly illuminates the black box of hillslope flow processes.

Acknowledgments. Support for this project was provided by the American Geophysical Union 1987 Horton Research Grant, New Zealand Commonwealth Scholarship and Fellowship Programme, University Grants Committee, and the University of Canterbury Geography Department. The Institute of Nuclear Sciences and Forest Research Centre are thanked for their analytical services and logistical support throughout the study. B. Clothier provided valuable insight into the discussion, and I. Owens, J. Buttle, C. Taylor, and A. Pearce are thanked for their critical reviews on an earlier draft of the manuscript. I. Owens, M. Bonell, A. Pearce, G. Buchan, and R. Jackson provided very useful input at various stages of the research. NASA Marshall Space Flight Center and Universities Space Research Association provided support during a study leave while this paper was prepared.

REFERENCES

Abdul, A. S., and R. W. Gillham, Laboratory studies of the effects of the capillary fringe on streamflow generation, *Water Resour. Res.*, *20*, 691–698, 1984.

Atkinson, T. C., Techniques for measuring subsurface flow on hillslopes, in *Hillslope Hydrology*, edited by M. J. Kirkby, pp. 73–120, John Wiley, New York, 1978.

Beven, K., Micro-, meso-, macroporosity and channeling flow phenomenoa in soils, *Soil Sci. Soc. Am. J.*, *45*, 1245, 1981.

Beven, K. J., and P. Germann, Macropores and water flow in soils, *Water Resour. Res.*, *18*, 1311–1325, 1982.

Bouma, J., Soil morphology and preferential flow along macropores, *Agric. Water Manage.*, *3*, 235–250, 1981.

Bouma, J., and J. L. Anderson, Relationships between soil structure characteristics and hydraulic conductivity, Field Soil Water Regime, *Spec. Publ. Soil Sci. Soc. Am.*, *5*, 77–105, 1973.

Bouma, J., A. Jongerius, O. Boersma, A. Jager, and D. Schoonderbeek, The function of different types of macropores during saturated flow through four swelling soil horizons, *Soil Sci. Soc. Am. J.*, *41*, 945–950, 1977.

Clothier, B. E., and R. A. Wooding, The soil diffusivity near saturation, *Soil Sci. Soc. Am. J.*, *47*, 636–640, 1983.

Germann, P. F., Rapid drainage response to precipitation, *Hydrol. Processes*, *1*, 3–13, 1986.

Gillham, R. W., The capillary-fringe and its effect on water-table response, *J. Hydrol.*, *67*, 307–324, 1984.

Hewlett, J. D., and A. R. Hibbert, Factors affecting the response of small watersheds to precipitation in humid areas, in *International Symposium on Forest Hydrology*, edited by W. E. Sopper and W. H. Lull, pp. 275–290, Pergamon, New York, 1967.

Jones, J. A. A., Extending the Hewlett model of storm runoff generation, *Area*, *11*, 110–114, 1979.

Jones, J. A. A., The effects of soil piping on contributing areas and erosion patterns, *Earth Surf. Processes Landforms*, *12*, 229–248, 1987.

Luxmoore, R. J., Micro-, meso-, and macroporosity of soil, *Soil Sci. Soc. Am. J.*, *45*, 671, 1981.

McCaig, M., Contributions to storm quickflow in a small headwater catchment—The role of natural pipes and soil macropores, *Earth Surf. Processes Landforms*, *8*, 239–252, 1983.

McDonnell, J. J., The age, origin and pathway of subsurface stormflow in a steep, humid headwater catchment, Ph.D. thesis, 270 pp., Univ. of Canterbury, Christchurch, N. Z., 1989.

McDonnell, J. J., The influence of macropores on debris flow initiation, *Q. J. Eng. Geol.*, in press, 1990.

McDonnell, J. J., M. Bonell, M. K. Stewart, and A. J. Pearce, Deuterium variations in storm rainfall: Implications for stream hydrograph separations, *Water Resour. Res.*, *26*, 455–458, 1990.

Mew, G., T. H. Webb, C. W. Ross, and J. Adams, Soils of Ignangahua Depression, South Island, New Zealand, *Rep. 17*, N. Z. Soil Surv., Christchurch, 1975.

Mosley, M. P., Streamflow generation in a forested watershed, New Zealand, *Water Resour. Res.*, *15*, 795–806, 1979.

Mosley, M. P., Subsurface flow velocities through selected forest soils, South Island, New Zealand, *J. Hydrol.*, *55*, 65–92, 1982.

New Zealand Soil Bureau, General survey of the soils of the South Island, New Zealand, NZ Soil Bur., *Bull. 4*, 404 pp., N. Z. Soil Bur., Wellington, 1968.

Pearce, A. J., and A. I. McKerchar, Upstream generation of storm runoff, in *Physical Hydrology, The New Zealand Experience*, edited by D. L. Murray and P. Ackroyd, pp. 165–192, New Zealand Hydrological Society, Wellington, 1979.

Pearce, A. J., C. L. O'Loughlin, and L. K. Rowe, Hydrologic regime of small, undisturbed beech forest catchments, North Westland, Proceedings of the 3rd Soil and Plant Water Symposium, *Publ. 1-126*, pp. 150–158, N. Z. Dep. of Sci. and Ind. Res., Lower Hutt, 1976.

Pearce, A. J., M. K. Stewart, and M. G. Sklash, Storm runoff generation in humid headwater catchments, 1, Where does the water come from?, *Water Resour. Res.*, *22*, 1263–1272, 1986.

Perroux, K. M., P. A. C. Raats, and D. E. Smiles, Wetting moisture characteristic curves derived from constant-rate infiltration into thin soil samples, *Soil Sci. Soc. Am. J.*, *46*, 231–234, 1982.

Ragan, R. M., An experimental investigation of partial area contribution, *IAHS Publ.*, *76*, 241–249, 1968.

Rowe, L. K., Rainfall interception by a beech-podocarp-hardwood forest near Reefton, North Westland, New Zealand, *J. Hydrol. N. Z.*, *18*, 63–72, 1979.

Scotter, D. R., Preferential solute movement through larger soil voids, 1, some computations using simple theory, *Aust. J. Soil Res.*, *16*, 257–267, 1978.

Sklash, M. G., and R. N. Farvolden, The role of groundwater in storm runoff, *J. Hydrol.*, *43*, 45–65, 1979.

Sklash, M. G., M. K. Stewart, and A. J. Pearce, Storm runoff generation in humid headwater catchments, 2, A case study of hillslope and low-order stream response, *Water Resour. Res.*, *22*, 1273–1282, 1986.

Smettem, K. R. J., S. T. Trudgill, and A. M. Pickles, Nitrate loss in soil drainage waters in relation to bypassing flow and discharge on an arable site, *J. Soil Sci.*, *34*, 499–509, 1983.

Stauffer, F., and T. Dracos, Experimental and numerical study of water and solute infiltration in layered porous media, *J. Hydrol.*, *84*, 9–34, 1986.

Thomas, G. W., R. E. Phillips, and V. L. Quisenberry, Characterization of water displacement in soils using simple chromographic theory, *J. Soil Sci.*, *32*, 32–37, 1978.

Van Stiphout, T. P. J., H. A. J. Van Lanen, O. H. Boersma, and J. Bouma, The effect of bypass flow and internal catchment of rain on the water regime in a clay loam grassland soil, *J. Hydrol.*, *95*, 1–11, 1987.

Webster, J., The hydrologic properties of the forest floor under beech/podocarp hardwood forest, North Westland, M.Sc. thesis, 77 pp., Univ. of Canterbury, Christchurch, N. Z., 1977.

Wilson, C. A., and P. Smart, Pipes and pipeflow in an upland catchment, *Catena*, *11*, 145–158, 1984.

J. J. McDonnell, Department of Forest Resources, Utah State University, Logan, UT 84322.

(Received November 14, 1989;
revised May 31, 1990;
accepted June 27, 1990.)

Commentary on Paper C6

Waddington, J. M., Roulet, N. T. & Hill, A. R. (1993) Runoff mechanisms in a forested groundwater discharge wetland.

The riparian zone is perhaps the most important location along paths of nutrient flux from terrestrial to aquatic ecosystems. Biogeochemical transformations in this zone can affect the degree to which processes in terrestrial ecosystems are translated into changes in the chemistry of surface waters. As noted in the commentary on Pilgrim *et al.* (1979, Paper C3), one of the paradoxes of catchment hydrology is the rapid mobilisation of large amounts of pre-event water during times of storm runoff (Kirchner, 2003). End-member mixing analysis has often been used to estimate the relative contribution of flow components with different residence times and to draw inferences about the physical sources of flow from those estimates. This paper by Waddington and colleagues includes hydrograph separation by use of a standard two-component mixing model, but this is only one part of an analysis that is firmly grounded in field measurements of hillslope hydrology. On that basis, inferences from mixing model results can be corroborated from on-slope observations, something much less commonly achieved than it ought to be.

The study area (approximately 50 km north west of Toronto, Canada) is representative of groundwater-connected wetlands often found in de-glaciated areas of the northern hemisphere. The soils comprise sandy fluvio-glacial deposits overlain by peaty soils. Groundwater discharge maintains high water tables in the low-lying areas, accounting for the development of wetlands in such headwater, valley-floor locations. The research is concerned with the relative contributions to stormflow of groundwater discharge, pipe flow and groundwater ridging in the riparian zone, and it is good to see tightly framed hypotheses guiding the fieldwork. Field instrumentation involved two experimental plots, one streamside and the other in a wetland saturated area. Nests of piezometers were installed across both plots at depths of down to 2.5 m, with tensiometers and suction lysimeters near the surface. Discharge and water chemistry were sampled during 12 natural rain events and five irrigation episodes. A standard two-component end-member mixing model was used to achieve hydrograph separation.

The authors include several cross-sections to show water table position and flow lines. Because there is vertical exaggeration, there is the usual problem that flow lines do not cross lines of equipotential at right angles; nevertheless, a good impression of flow paths is gained, both for the near-stream zone and for the zone of upwelling where groundwater exfiltrates on the edge of the wetland. There is also careful measurement of hydraulic conductivities of wetland soils and sub-soils. Taken together with other measurements, it is possible to gain a clear picture of runoff generation in the saturated areas. Despite the existence of extensive saturated areas, water with a pre-event signature provides almost all the saturated-area discharge.

Groundwater ridging was found to be of marginal importance, even in a large irrigation event. Pipes not connected to areas of surface saturation were also of minor importance in conducting pre-event water to the stream, whereas connected pipes could be an important conduit for saturation-excess overland flow draining to the stream. Saturation-excess overland flow was dominated by pre-event water; since most of this runoff was generated by precipitation falling directly on to the saturated area, this means that event and pre-event water must have mixed rapidly in the saturated zone in order to generate stormflow with a pre-event signature. Without a detailed knowledge of the wetland hydrology and hydrochemistry, it would have been very easy to assume that saturation-excess overland flow was unimportant, on the assumption that saturated areas are thought only to generate surface runoff with an event signature. Calculations of minimum contributing area (Weyman, 1974) would be greatly underestimated as a result. This study shows that the presence of saturation-excess overland flow is not incompatible with a pre-event stormflow signature.

Whilst this study helped take forward research on hydrograph separation and the debate about old and new water, it was equally important in providing a detailed description of processes, hydrological and hydrochemical, operating in the near-stream zone. Warwick & Hill (1988, Paper E4) found an absence of nitrate depletion in the same wetland system. Nitrate removal was limited by two major factors. Firstly, denitrification potentials measured in drainage line substrates were generally low, probably influenced in turn by temperature and the low nitrate concentration in influent groundwater. Secondly, water moved rapidly across the wetland surface to the stream and so there was insufficient residence time for nitrate removal. Hill (1993) also describes nitrate buffer zone processes at the same site. Incoming throughfall had much higher concentrations of nitrate and ammonium than stream baseflow concentrations; a chemical mixing model showed that stream nitrate concentrations were mainly a mixture of throughfall and groundwater in surface stormflow from the wetland. For significant nitrate depletion to occur (even with low influent concentrations), there would need to be a much larger fraction of water moving subsurface. The wetland soils at this site are too impermeable to allow any throughflow to occur, and consequently groundwater exfiltrates to flow rapidly across the wetland surface to the stream. There is clearly opportunity for mixing to occur and for the surface flow to acquire a pre-event signature, but conditions associated with surface runoff do not favour nitrate reduction.

TPB

References

Hill, A. R. (1993) Nitrogen dynamics of storm runoff in the riparian zone of a forested watershed. *Biogeochemistry* **20**, 19–44.

Kirchner, J. W. (2003) A double paradox in catchment hydrology and geochemistry. *Hydrol. Processes* **17**, 871–874.

Pilgrim, D. H, Huff, D. D. & Steele, T. D. (1979) Use of specific conductance and contact time relations for separating flow components of storm runoff. *Water Resour. Res.* **15**, 329–339.

Warwick, J. & Hill, A. R. (1988) Nitrate depletion in the riparian zone of a small woodland stream. *Hydrobiologia* **157**, 231–240.

Weyman, D. R. (1974) Runoff processes, contributing area and streamflow in a small upland catchment. In: *Fluvial Processes in Instrumented Watersheds*. Institute of British Geographers Special Publication no. 6, 33–43.

Journal of Hydrology, 147 (1993) 37–60 37
0022-1694/93/$6.00

[3]

Runoff mechanisms in a forested groundwater discharge wetland

J.M. Waddington, N.T. Roulet*, A.R. Hill

Department of Geography, York University, 4700 Keele Street, North York, Ont. M3J 1P3, Canada

(Received 27 June 1992; revision accepted 6 January 1993)

Abstract

Hydrometric measurements and geochemical hydrograph separation techniques were used in experimental plots to determine the processes that govern stormflow production in a forested swamp situated in a regional groundwater discharge zone. Hydrometric measurements suggested that saturated overland flow from permanently saturated areas created by discharging groundwater was the major storm runoff mechanism. Chemical separation analysis, using a portable irrigation system, indicated that over 70% of the runoff was pre-event water.

Groundwater ridging did not contribute significantly to wetland runoff, whereas preferential flow through pipes accounted for a minor component of the pre-event water at peak streamflow and the majority of the pre-event flow in the recession. Rapid mixing of event water with a significantly larger pool of pre-event water standing in the saturated areas is responsible for the rapid appearance of the pre-event water in surface generated stormflow. The rapid groundwater–surface water mixing in the saturated areas has implications for the interpretation of runoff mechanisms based solely on chemical separation.

Introduction

There have been many studies on the storm response of humid headwater drainage basins (Dunne, 1978). Stable isotope and other chemical separation techniques indicate that stormflow from headwater basins is dominated by water that has resided in the basin before the storm event (Sklash and Farvolden, 1979; Pearce et al., 1986; Sklash et al., 1986; Wels et al., 1991a), yet studies using physical techniques show that overland flow from saturated areas is an important mechanism (e.g. Ragan, 1968; Dunne and Black, 1970; Dunne, 1978). The prominence of pre-event water has been explained by (1) groundwater ridging and the effect of the capillary fringe in the streamside zone (Gillham, 1984; Abdul and Gillham, 1989) and (2) event water flow in

* Corresponding author.

macropores mixing with macropore pre-event water (McDonnell, 1990). This paper describes the production of storm runoff in a forested, headwater swamp where it appears that saturated overland flow is the dominant process, but tracer studies show the presence of a large pre-event component in the stormflow.

Most studies of storm response in humid headwater basins have been conducted in landscapes where the groundwater flux was either very small or absent most of the year. Forested, spring-fed swamps in headwater locations are a significant component of the landscape in regions where large aquifers occur in surficial glacial and fluvio-glacial deposits. Roulet (1990a) found that a large and constant baseflow maintained areas of constant saturation in a headwater swamp. Stormflow response was equal to the direct precipitation on the saturated areas, DPS (Roulet, 1991). O'Brien (1980) examined two New England wetlands located in permanent groundwater discharge areas and found that groundwater dominated stormflow. He suggested that natural pipes directed flow to the stream in one wetland, whereas the rapid transmission of groundwater through a very permeable upper peat layer was the mechanism responsible for the dominance of groundwater in the second wetland. The displacement of pre-event water stored on the surface by snowmelt water has been used to explain the large pre-event signature of snowmelt runoff from a small Canadian Shield basin where wetlands were present (Buttle and Sami, 1992). Bonell et al. (1990) found that pre-event water runoff from shallow groundwater in a New Zealand valley bottom wetland was delivered more rapidly to the basin outlet by groundwater ridging than was event water from saturation overland flow.

The objective of the present study is to determine the dominant runoff mechanisms and the sources of stormflow in a headwater basin wetland situated in a groundwater discharge zone of a regional aquifer. The large groundwater flux in this type of wetland may produce a streamside environment suitable for both groundwater ridging and macropore flow, which could explain the presence of pre-event water in the stormflow. Alternatively, the large amount of groundwater emerging in saturated areas within the wetland could mix with a small volume of rapidly delivered precipitation to yield the pre-event signature of stormflow. A combined hydrometric and geochemical analysis of the response in two experimental plots in the wetland to both natural rainfall and irrigation events is presented to evaluate the following, conflicting, and interdependent research hypotheses: (1) groundwater plays an active role in the generation of stormflow via groundwater ridging in the streamside area and the base of the hillslope; (2) flow in soil pipes and macropores contributes to pre-event stormflow; (3) the processes (1) and (2) are relatively unimportant and rapid mixing of event water with emerging groundwater in the saturated areas is responsible for the large pre-event water signature in the stormflow.

Study area

The study was conducted in a forested, spring-fed, swamp (National Wetlands Working Group, 1988) located approximately 50 km northeast of Toronto, Canada (43°47′N, 79°15′W) on the southern flank of the Oak Ridges moraine. The hydrogeology, hydrogeochemistry, and groundwater of the region were examined by Sibul et al. (1977) and Howard and Beck (1986). The swamp is representative of groundwater-connected wetlands found in areas of glacial deposition in southeastern Canada (Roulet, 1990a), the northeastern USA (Carter and Novitzki, 1988), and the Nordic countries (Calles, 1985).

The swamp (0.031 km^2), located in a first- and second-order stream valley, occupies 2% of the headwater basin (1.6 km^2) (Fig. 1). The swamp soils consist of sandy fluvio-glacial deposits overlain by peaty histosols that increase in thickness from approximately 25 cm at the wetland margin to greater than 150 cm in the near-stream zone. The dominant tree species in the swamp are cedar and hemlock, with dense shrubs and saplings in small clearings of the treed area.

Groundwater entering the wetland through a series of springs at the margin of the wetland and in the wetland itself has created many diffuse saturated zones that join together to form small streamlets. Water is conveyed rapidly across the surface of the swamp to both the first- and second-order streams in 30 min (Roulet, 1990b). At times, the streamlets enter subsurface pipes and re-emerge several meters downslope in other streamlets or as seeps in the stream bank. The areal extent of the saturation is controlled by the microtopography and the constant groundwater input (Roulet, 1990b).

Soil pipes are found along passages adjacent to major tree roots within the swamp, in cavities created by the decay of buried logs or roots, and along the boundary between the organic soils and the underlying fluvio-glacial sands and gravels. Some pipes are 'connected' to the diffuse saturated zones and streamlets whereas others are not linked to the surface hydrologic pathways. The pipes that are 'connected' to surface saturated areas have hydrological responses (Roulet, 1991) and water chemistry (Hill, 1990) similar to the streamlets of the wetland.

Methods

The methods were designed to examine the hydrometric and geochemical stormflow response of the capillary fringe, groundwater mounding and flow in the streamside area, and groundwater and surface water response in the

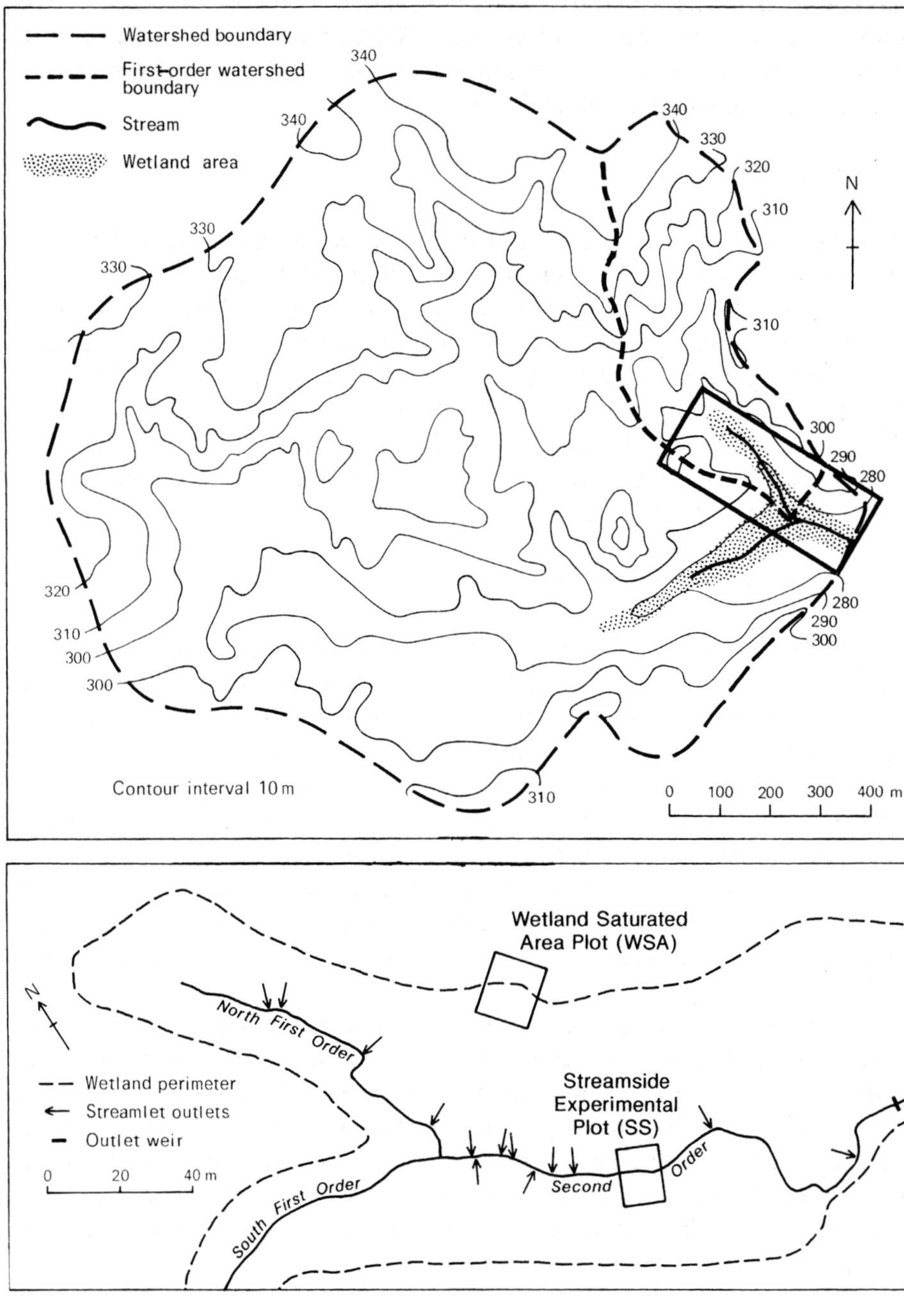

Fig. 1. Study basin and location of experimental plots in the wetland. WSA and SS refer to the wetland saturated area and streamside experimental sites, respectively.

saturated areas of the wetland. Two experimental plots, one streamside (SS) and the other in a wetland saturated area (WSA), were used (Fig. 1). Runoff response from these plots was examined for both natural rain events and simulated rainfall using an irrigation system.

Wetland instrumentation

A thin plate 53° V-notch weir was installed in 1987 at the outlet of the basin. The water level behind the weir was measured continuously by a potentiometric water-level sensor connected to a data-logger. A stage–discharge relationsip was used to calculate the average basin discharge for 15 min intervals during baseflow and 5 min intervals during storm events. Throughfall precipitation was sampled sequentially from two parallel plastic trough collectors of 6.65 m length and 5.0 cm width within the swamp. Samples of precipitation and stream water collected at the outflow throughout storms were analyzed for isotopes (see below).

Streamside area instrumentation

The SS experimental plot was located beside the second-order stream. A 210 m^2 area was monitored during five natural rain events and two irrigation events which occurred between 1 July and 31 October 1990. The site consisted of a single transect of nine piezometer nests installed perpendicular to the stream (Fig. 2). Each piezometer nest consisted of six PVC pipes of 1.25 cm i.d. with a 20 cm slotted point at depths of 30, 50, 75, 100, 150, and 200 cm. On the north bank, two tensiometers were installed at depths of 15 and 30 cm. Groundwater wells (ABS pipe of 3.8 cm i.d.) were installed with each piezometer nest. The perforated portions of all groundwater wells and piezometers were wrapped with 150 μm reinforced filter fabric to prevent siltation.

In the same experimental plot, one unconnected pipe was monitored. The diameter of the pipe was not determined. A thin plate 30° V-notched weir was installed at the discharge point of the pipe (8M) (Fig. 2). A survey of the stream banks, to assess the representativeness of the streamside pipeflow, revealed that there were nine unconnected pipes with baseflow discharge greater than 50 ml s^{-1} as well as numerous very small seepage sites. Many of the pipes contain fluvio-glacial sediments along their walls and appear to bifurcate at several levels.

Instantaneous discharge and water chemistry (Cl^-, Li^+ and $\delta^{18}O$) were measured weekly under baseflow conditions, every 30 min during natural rain events, and every 15 min during irrigation events. Water samples for chemical

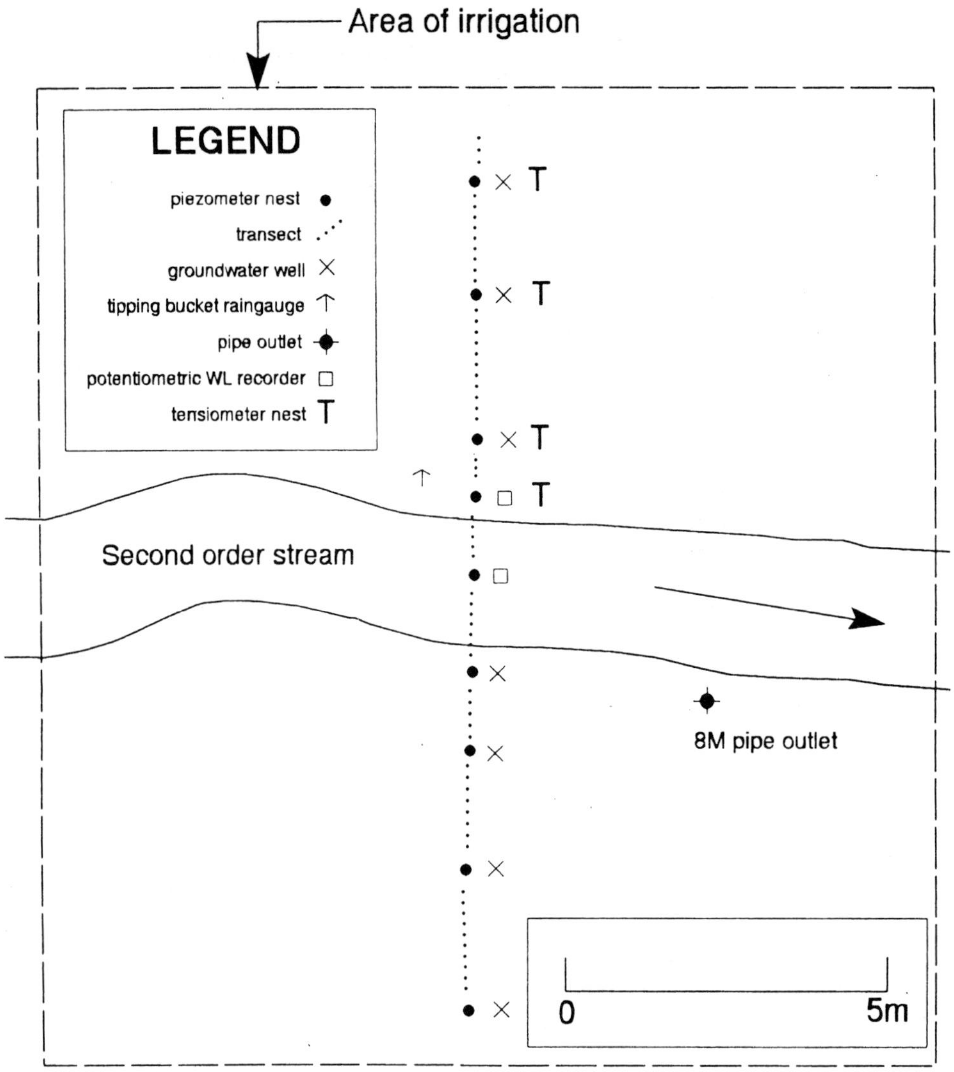

Fig. 2. Location of the instrumentation in the streamside (SS) experimental site. The enclosed rectangle illustrates the location of the irrigation system water application area.

analysis were taken from seven manual precipitation gauges and from the second-order stream downstream of the experimental site every 10 min during storm events.

Wetland saturated area instrumentation

The WSA experimental plot was set up over a saturated groundwater

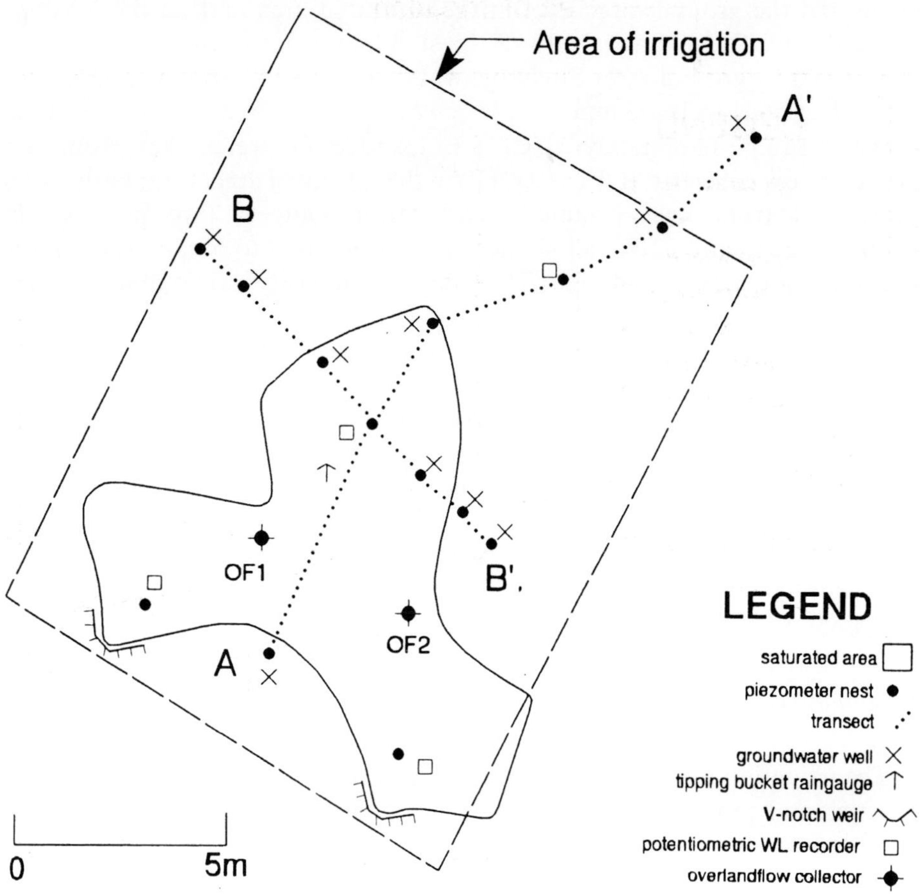

Fig. 3. Location of the instrumentation in the wetland saturated area (WSA) experimental site. The enclosed rectangle illustrates the location of the irrigation system water application area.

discharge region (60 m^2) (Fig. 3). The 190 m^2 experimental site was monitored during 12 natural rain events and five irrigation events between 1 July and 31 October 1990. Two transects (A–A′ and B–B′), comprising 12 piezometer nests with individual piezometers installed to depths of 25, 50, 75, 100, 150, 200, and 250 cm and groundwater wells, were established.

Two 30° V-notch weirs were located at the outflow of the saturated area. The area between the weirs was lined with plastic to ensure that all overland flow discharged via one of the two weirs. Potentiometric water-level sensors monitored the water level continuously behind the weirs and a stage–discharge relationship was established. Two automatic water-level recorders monitored the groundwater level in the saturated area and on the hillslope continuously. The average depth of the standing water in the saturated area maintained by groundwater discharge was determined by spot measurements at 131 locations to be 13 mm.

Water samples for analysis (Cl^-, Li^+ and $\delta^{18}O$) were taken from two overland flow collectors (OFl and OF2) in the saturated area, from both outlet weirs, and from seven manual precipitation gauges. Two porous cup lysimeters at 10 and 20 cm below the surface were used to obtain water in the soil beneath the saturated area. The sampling intensity varied among storm events.

Water levels within the piezometers on the streamside and saturated area plots were monitored weekly under baseflow conditions and every half-hour during storm events using an electronic probe. Groundwater levels were monitored weekly under baseflow conditions and every 5 min during storm events, using a flotation device located in the groundwater well (e.g. Abdul and Gillham, 1989). All water-level sensors and the tipping bucket rain gauge were monitored every 5 min by a data-logger. Hydraulic conductivity was determined by the Hvorslev water-level recovery method using the piezometers (Freeze and Cherry, 1979).

Irrigation system and water chemistry

The irrigation system used for the seven simulated rain events at the two experimental plots has been described in detail by Waddington (1991). It consisted of an in-stream pump, a water distribution system made of lawn sprayers, and a valve that could be used to control intensity. The spatial variability of the irrigation intensity over the plot was 16% and intensities ranged from 4.5 to 14.8 mm h^{-1}, representing rain events with a return period of less than 1–3 years. Lithium bromide was used as a tracer during the irrigation events because lithium has been shown to act as a conservative tracer in this wetland (Waddington, 1991). Lithium concentrations were determined in the laboratory by atomic absorption spectrophotometry, and chloride concentrations were determined by a continuous flow autoanalyzer. $\delta^{18}O$ samples were analyzed by mass spectrometry at the Environmental Isotope Laboratory, University of Waterloo.

Hydrograph separations

The standard two-component hydrograph separation (Sklash and Farvolden, 1979) was used to separate discharge into event and pre-event water fractions by solving the chemical and mass balance equation

$$Q_t C_t = Q_p Q_p + Q_e C_e \quad (1)$$

and as

$$Q_t = Q_p + Q_e \quad (2)$$

the pre-event water fraction, X, is

$$X = (C_t - C_e)/(C_p - C_e) \quad (3)$$

where Q is discharge, C is the tracer concentration, and the subscripts t, p, and e refer to the total streamflow, pre-event water (emerging groundwater and stored water), and event water components, respectively. This approach assumes that a stream receives water from only two reservoirs and that the elements used as tracers are conservative.

The isotopic variations of event water were evaluated by using the incremental intensity mean method (McDonnell et al., 1990). This method considers only the isotope value of rain which has occurred before the time of stream sampling and gives greater weighting to episodes of higher-intensity rain within storms which may produce a large rapid runoff response.

Results

Results of the irrigation and precipitation experiments at the streamside SS and WSA experimental sites are presented below. These results are compared with the response of the entire wetland to natural precipitation events during the study period. In each section below, the general hydrology and general geochemistry are presented, followed by the hydrology and geochemistry of 'typical' events.

Wetland response

During the study period, 36 rain events produced a total of 230 mm of precipitation. The largest event was 23.0 mm, and several 0.2 mm events represented the smallest events. Twelve events which exceeded 3.0 mm of rain were chosen for analysis in the study. These ranged from 3.8 mm to 23.0 mm.

The mean runoff coefficient for the entire wetland was 0.41 ± 0.12. The runoff coefficient in this study is expressed as the ratio of runoff to throughfall depth. As the throughfall is less than open rainfall, the percentage of rainfall converted to runoff does not differ greatly from the 31% observed by Roulet (1991), as he used gross rainfall in his analysis. Maximum discharge during the study period at the second-order stream outlet weir was $83.8\,l\,s^{-1}$ during a 16.8 mm event in which a 15 min interval maximum rainfall intensity exceeded $45\,mm\,h^{-1}$. Stormflow, determined by a baseline to baseline integration of the storm hydrographs, accounted for 85% of the peak discharge in this event, and it ranged from 13 to 85% for the 12 natural events analyzed. Mean baseflow discharge for the second-order stream at the outlet weir was $11.7\,l\,s^{-1}$.

Mean $\delta^{18}O$ values for the outlet stream during baseflow conditions for the months of August to October 1990 were -11.88 ± 0.10‰. Hydrograph separation using $\delta^{18}O$ as a tracer at the second-order stream outlet weir was done for the events of 12 and 18 October. For the 12 October event, $\delta^{18}O$ values for the baseflow before the event, peak flow, and the average incremental intensity mean for precipitation were -11.90‰, -12.31‰, and -15.00‰, respectively. At peak flow ($35.0\,l\,s^{-1}$) the pre-event water fraction of the basin discharge was 0.87.

For the 18 October event, total precipitation was 9.1 mm and $\delta^{18}O$ values for baseflow before the event and the average incremental intensity mean for precipitation were -11.72‰ and -7.26‰, respectively. At peak flow the pre-event water fraction of basin discharge was 0.93. The peak in the event water fraction of stormflow (11%) occurred 15 min after the peak in the stream hydrograph and the basin discharge did not return to 100% pre-event water until 3 h after the end of the rain event (Fig. 4).

Streamside area response

The hydraulic conductivity of the peat in the streamside area ranged from 4×10^{-7} to $2 \times 10^{-6}\,m\,s^{-1}$, and the hydraulic conductivity of the underlying organic silts ranged from 1×10^{-6} to $8 \times 10^{-6}\,m\,s^{-1}$. Table 1 lists the rainfall characteristics and the maximum rise in the water table on the north bank for the two irrigation and five precipitation events. The largest water table rise occurred during the 25 October irrigation event (+26.8 cm), and the minimum water table rise was on 9 October (+1.2 cm). Although a water table rise occurred in all events, only two events produced a water table rise of over 20 cm (25 October and 7 November). The maximum change in soil water potential for the events was 50 cm (Table 1). The most responsive tensiometers were located 4 m from the stream on the north bank. This represents the zone where the small water table formed.

Water chemistry at the outlet of the 8M pipe during baseflow showed a mean chloride concentration of $0.92 \pm 0.05\,mg\,l^{-1}$, similar to concentrations

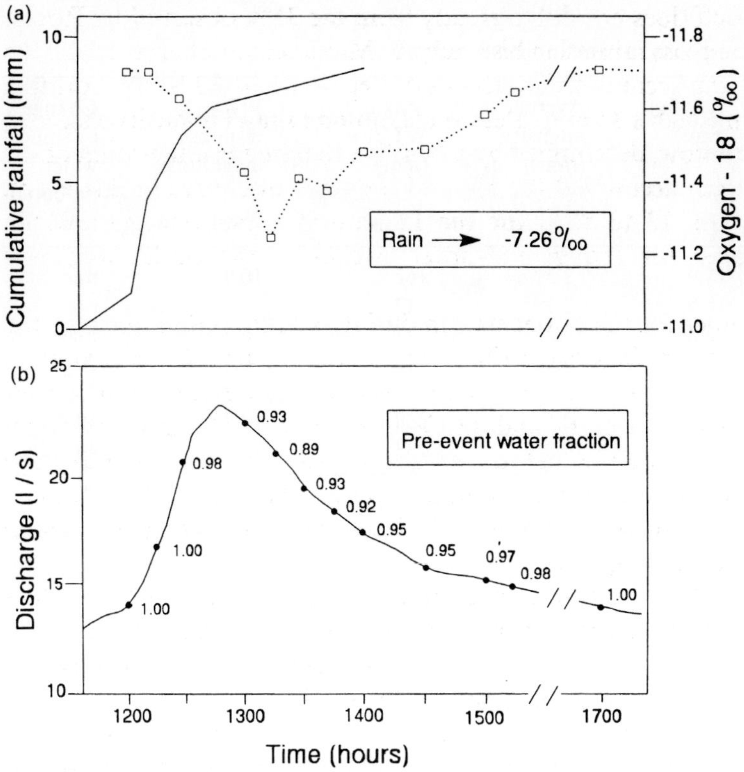

Fig. 4. Hydrograph separation at the wetland outlet weir for 18 October 1990, showing (a) cumulative rainfall (solid line) and the δ ^{18}O of the stream water (open squares), and (b) basin discharge (solid line) and the values of fraction of pre-event water in the basin discharge. Rain δ^{18}O value refers to the average incremental intensity mean ∂^{18}O for precipitation.

typical of deep groundwater entering the wetland (0.6–0.9 mg l^{-1}; Hill, 1990). Mean δ^{18}O for the pipe during baseflow was -11.85‰.

Dye tracing determined that the 8M pipe in the near-stream zone was not connected to a surface streamlet. The mean baseflow discharge from the pipe of 114 $\pm$ 11 ml s^{-1} was one of the largest for an unconnected pipe in the basin.

The event of 25 October was a 2 h irrigation event starting at 11 : 00 h with a constant water intensity of 12.5 mm h^{-1}. The gradients in the near-stream zone remained towards the ground surface for the entire event (Fig. 5(b)), whereas a small recharge area formed behind a groundwater mound on the north bank (Figs. 5(a) and 5(b)). A stagnation point formed in front of the groundwater mound created a hydraulic divide between the near-stream zone and the rest of the wetland; this caused water to be shed into the wetland as

Table 1
Water table (WT) response to rainfall in the streamside area

Date	Rainfall (mm)	Average intensity ($mm\,h^{-1}$)	Maximum WT rise (mm)	WT response magnitude	Change in soil water potential (cm)
14 September	14.9	8.5	162	10.9	n/a
9 October	18.5	2.1	12	0.7	n/a
11 October	3.8	1.3	24	6.3	n/a
12 October	17.5	4.4	110	6.3	20
18 October	9.1	2.3	66	7.3	15
25 October[a]	25.0	12.5	268	10.7	50
7 November[a]	14.8	14.8	233	15.7	40

The water table response magnitude is determined by dividing the maximum water table rise (mm) by the rainfall depth (mm). Change in soil water potential refers to the maximum change in soil water potential at one depth from the beginning of the event to the end of the event.
[a] Irrigation event.

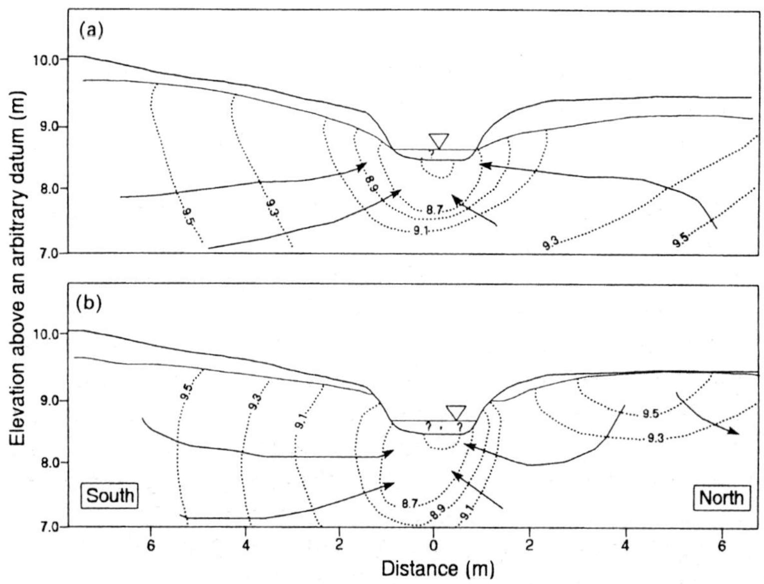

Fig. 5. Near-stream zone response for the 25 October event, showing (a) flow net at 11:00 h and (b) flow net at 13:00 h.

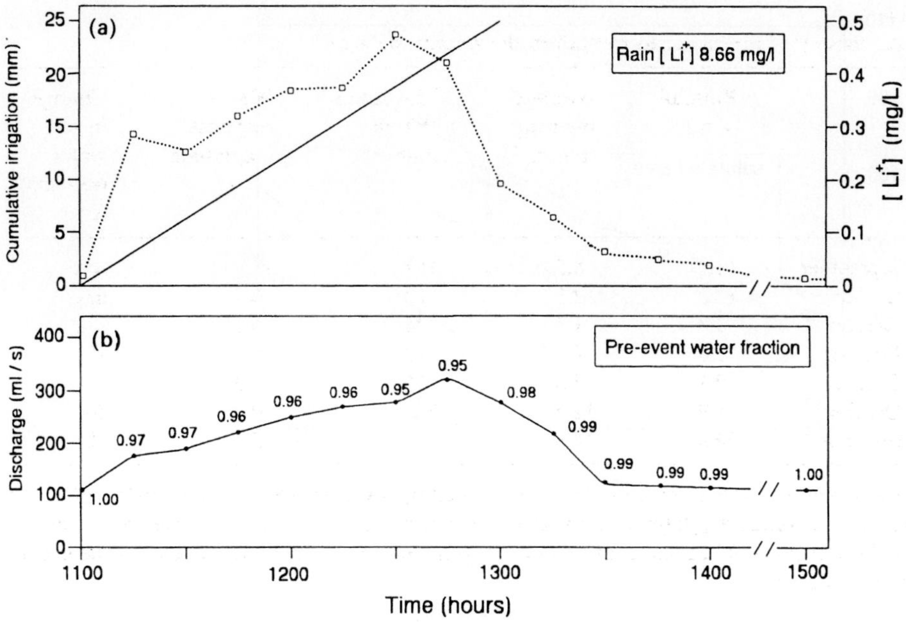

Fig. 6. Near-stream zone response for the 25 October event, showing (a) the cumulative rainfall and pipeflow lithium concentration (dots), and (b) the pipe discharge (line) and values of fraction of pre-event water in the pipeflow.

well as towards the stream. The position of the water table at the end of the event represented the highest water table observed during the study period. The maximum height of increase of the mound in the north bank was 26.8 cm, whereas the water table on the south bank increased by 16.8 cm (Fig. 5(b)). The increase in water level had created a seepage face on both the south and north banks by 13:00 h (Fig. 5(b)).

For the same event, discharge from the 8M pipe rose rapidly from the onset of the event, increasing from 102 ml s^{-1} to a peak of 311 ml s^{-1} (Fig. 6(a)). The peak occurred before the cessation of irrigation, and the recession limb was rapid. Initial lithium concentrations in the pipe were 0.02 mg l^{-1} and the concentrations in the irrigation water ranged from 8.5 to 9.2 mg l^{-1}. Pipe water lithium concentrations rose to 0.47 mg l^{-1} before the peak in discharge and then decreased rapidly to near-zero values at the conclusion of the tracer application (Fig. 6(a)). Hydrograph separation using a mean lithium concentration of 8.66 mg l^{-1} for the irrigation water indicated that the peak flow pre-event water fraction of pipeflow was 0.95 (Fig. 6(b)).

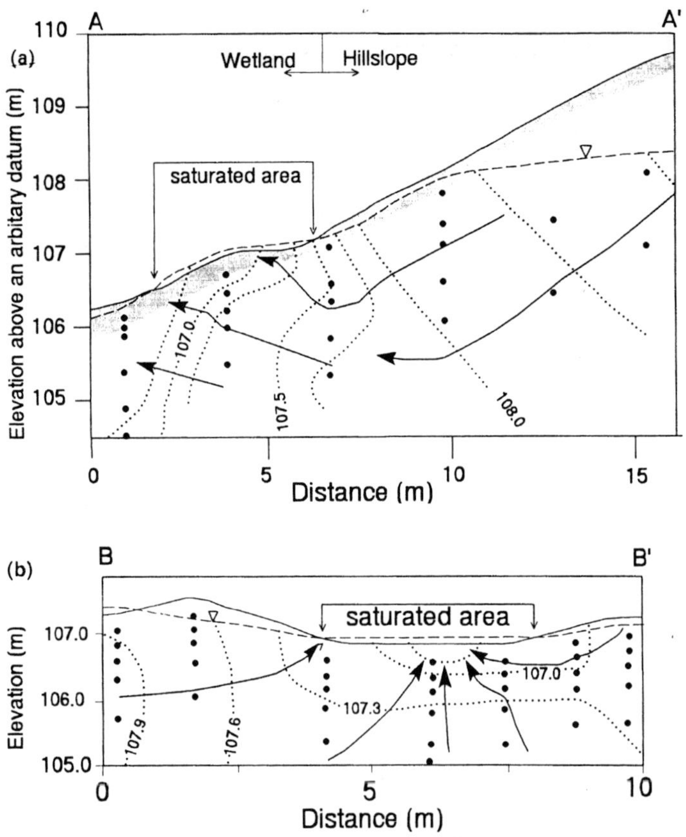

Fig. 7. Saturated area flow nets: (a) transect A–A′ longitudinally across the saturated area; (b) transect B–B′ perpendicular to the saturated area. Dots represent the locations of piezometers, dashed lines represent the position of the water table, dotted lines are equipotential lines, and the arrowed lines designate flow lines.

Saturated area response

Hydraulic conductivity in the WSA experimental plot ranged from 2×10^{-6} on the hillslope to 4×10^{-5} m s^{-1} in the saturated area. Flow nets produced from the 90 piezometers along the two transects in the WSA plot (Fig. 3) did not vary significantly from week to week. The mean flow net is shown in Fig. 7. Both transects A–A′, longitudinal, and B–B′, perpendicular to the saturated area, indicate that the saturated area is located in a localized area of groundwater discharge. The water table rises above the surface at the base of the hillslope and extends downslope into the wetland for a distance of 5 m (Fig. 7(a)). The B–B′ transect flow net (Fig. 7(b)) indicates that the topographical mounds surrounding the saturated area act as local recharge areas.

Table 2
Maximum water table (WT) rise during precipitation and irrigation events at the WSA experimental site

Date	Rainfall (mm)	Saturated area WT rise (mm)	Hillslope WT rise (mm)
8 July	13.7	< 1.8	< 1.8
15 July	23.0	< 1.8	1.8
20 July	3.8	1.8	3.6
22 July	11.9	< 1.8	1.8
24 July	12.6	1.8	1.8
12 August	10.1	1.8	1.8
13 August, A	15.6	1.8	1.8
13 August, B	4.2	1.8	3.6
28 August	12.9	1.0	1.0
13 September	9.0	2.0	1.0
14 September	14.9	3.6	5.4
25 September	8.0	1.0	4.0
27 September	10.3	2.0	7.0
2 October	15.3	4.0	8.0
9 October	18.5	< 1.8	1.8
12 October	17.5	2.0	3.0

Baseflow discharging through the saturated area ranged from 104 to 304 ml s^{-1}, with a mean discharge of 160 $\pm$ 24 ml s^{-1}. Maximum baseflow occurred in September, and minimum discharge occurred during early August. Storm runoff from the WSA experimental plot was examined for 12 natural rain events and five irrigation events. Total storm runoff from the saturated area was determined by a baseflow to baseflow integration of the storm hydrographs. The mean runoff ratio for the saturated area for the 17 events studied was 0.95 $\pm$ 0.10. Maximum discharge observed from the saturated area was 660 ml s^{-1} during the 2 October irrigation event. Stormflow accounted for 54% of the peak discharge for this event and ranged from 44 to 85% for the other 16 events.

The position of the water table was monitored along transect A–A′ during all natural precipitation and irrigation events. Response of the water table was small for both large and small events, with a maximum water table rise of only 8.0 mm (Table 2). The magnitude of water table rise in the hillsope is small because of a lag in the response. Furthermore, the piezometric heads remained virtually unchanged during and after storm events of varying intensity and duration.

With the exception of the piezometers farthest from the base of the slope,

Table 3
Hydrograph separation results for five irrigation events and one natural rain event at the WSA experimental site

Date	Rainfall (mm)	Average intensity (mm h^{-1})	Pre-event water fraction
28 August[a]	12.9	4.8	0.93
13 September[a]	9.0	4.5	0.91
25 September[a]	8.0	8.0	0.88
27 September[a]	10.3	6.2	0.85
2 October[a]	15.3	11.3	0.77
12 October	17.5	4.6	0.93

[a] Irrigation event.

chloride concentrations of the groundwater extracted from the piezometers under baseflow conditions ranged from 1.2 to 3.3 mg l^{-1}. Hill (1990) established that two scales of groundwater contributed water to the basin. Local groundwater had a Cl^- concentration of 1.2–1.8 mg l^{-1}, whereas the regional contribution had a Cl^- concentration of 0.6–0.9 mg l^{-1}; thus the water entering the saturated area is primarily local. Chloride concentrations range from 0.7 to 1.3 mg l^{-1} at the farthest nest, indicating a mixture of regional and local groundwater. Water chemistry at the discharge weirs of the saturated area was similar to that of the groundwater beneath the saturated area ($[Cl^-]$ 1.26 $\pm$ 0.06 mg l^{-1}). $\delta^{18}O$ at the discharge weirs was -11.89 ± 0.05‰ compared with the mean saturated area soil water value of -11.91 ± 0.05‰.

Chemical hydrograph separation was performed for five irrigation events using lithium as a tracer and for one rain event using $\delta^{18}O$. $\delta^{18}O$ values for the 12 October event for baseflow before the event, peak flow, and the average incremental intensity mean for precipitation were −11.87‰, −12.08‰ and −15.00‰, respectively. Pre-event water made up 93% of discharge at peak flow. Separation for the irrigation events was straightforward, as background lithium concentrations of the soil water, groundwater, and saturated area outflow before the events were near zero (0.02 mg l^{-1}), whereas the concentrations of the introduced tracer exceeded 8 mg l^{-1} in all irrigation events. Maximum lithium concentrations coincided with peak flow.

The pre-event water fraction of the saturated area discharge at peak flow ranged from 0.77 to 0.93 (Table 3). There was no statistically significant relationship between total precipitation and the fraction of pre-event water, but an inverse relationship between average precipitation intensity and the

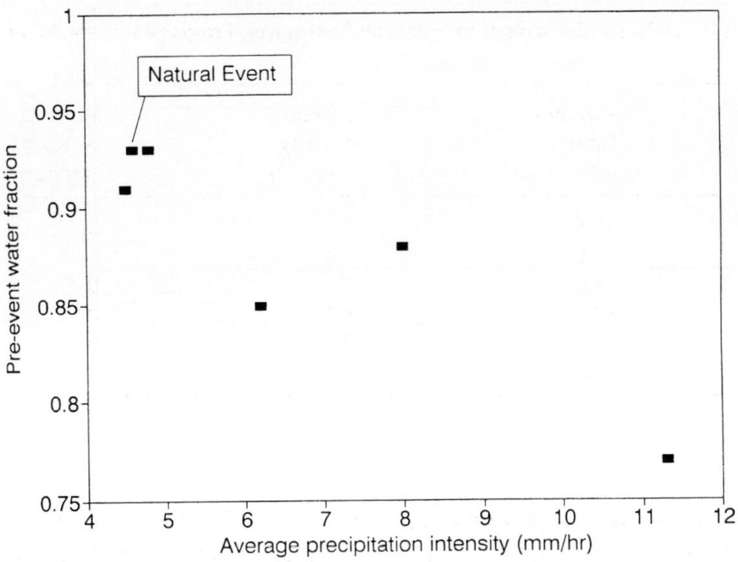

Fig. 8. Average precipitation intensity vs. pre-event water fraction of the saturated area discharge.

magnitude of the pre-event fraction was significant ($r^2 = 0.85$, $P = 0.01$; Fig. 8).

The storm of 25 September, 8.0 mm, is discussed in detail below. Water application to the experimental plot was kept constant for 1 h. Discharge increased rapidly from the onset of precipitation and reached a maximum discharge of 485 ml s^{-1} about 45 min into the event (Fig. 9). Peak flow remained relatively constant until the cessation of the irrigation; this suggests that an equilibrium was reached. The recession of the storm hydrograph was rapid (less than 30 min). The increase in discharge can be attributed to direct precipitation on the saturated areas (DPS), as the total storm runoff from the saturated region during the irrigation event was 8.4 mm, representing a runoff coefficient of 1.05. The maximum water table rise in the center of the saturated area was 1.0 mm at peak flow, whereas the water table increased in the hillslope by only 4.0 mm.

The maximum tracer concentration at the outlet weir was 1.66 mg l^{-1}, compared with a mean lithium concentration for the irrigation water of 13.50 ± 1.1 mg l^{-1}. Hydrograph separation indicates that at peak flow, pre-event water accounted for 88% of the saturated area discharge. The pre-event water fraction remained relatively constant until the cessation of the irrigation event (Fig. 9).

Lithium concentrations at the two overland flow collectors (OF and OF2) were 11–43% higher than those measured at the outlet weirs only 3 m away.

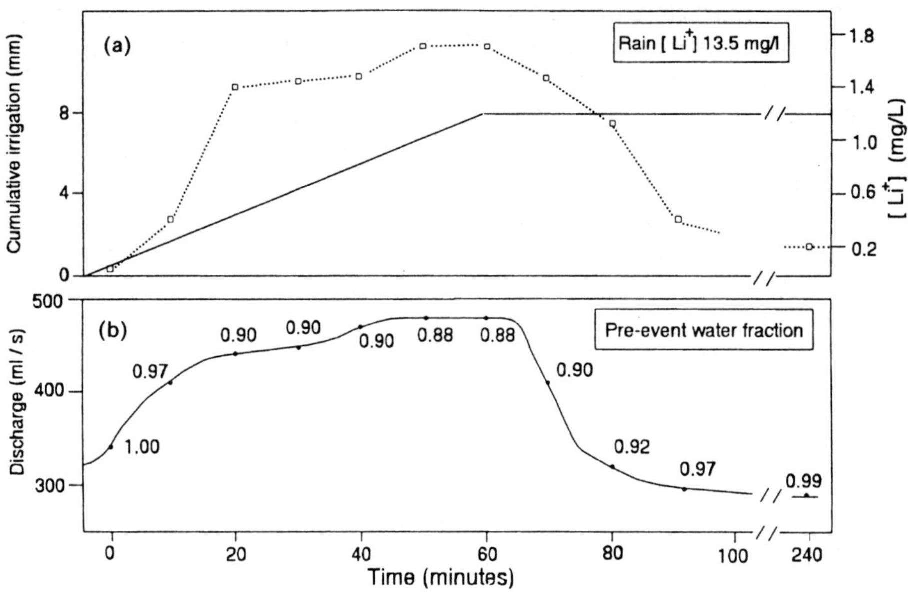

Fig. 9. Hydrograph separation at the WSA weirs for 25 September 1990, showing (a) cumulative irrigation (solid line) and lithium concentration of the saturated area discharge (open squares), and (b) saturated area discharge (line) and values of fraction of pre-event water in the saturated area discharge.

If C_t in the mixing model were replaced with the lithium concentration in the overland flow collectors, then the pre-event fraction would decrease to between 0.68 and 0.75. During the recession limb, lithium concentrations at the outlet weirs became slightly higher than the overland flow values as the remaining event water drained from the saturated area.

Discussion

The results described in the previous section allow an examination of the hypotheses on runoff mechanisms and sources of stormflow. Conventional mechanisms such as groundwater ridging cannot explain the large fraction of pre-event water in the stormflow discharge from this basin. An alternative explanation, the rapid surface water mixing hypothesis, which involves rapid mixing of new rainwater with pre-event water stored on the surface of the wetland, will be proposed.

Evidence of groundwater ridging

Results of the streamside study provide little evidence of groundwater ridging except in the case of the extreme precipitation event of 25 October.

The common characteristics of the hydrological response in the streamside area were: (1) it remained a groundwater discharge zone throughout all events; (2) changes in soil water potential were small; (3) once peaks in hydraulic heads were reached they gradually decreased (several hours) to pre-storm levels. The uncommon feature of the 25 October event was the development of a significant groundwater mound and the creation of a seepage face on the stream banks.

The magnitude of water table response (water table rise/rainfall depth) for all events ranged from 0.7 to 15.7 (Table 1). This in an order of magnitude lower than the 100 and 110 observed in the field by Gillham (1984) and in the laboratory by Heliotis and DeWitt (1987), respectively. The disproportionate water table rise in these studies was attributed to the effect of the capillary fringe. This effect is not evident in most storm events in the present study but, in response to the highest intensity event (greater than $12\,\mathrm{mm\,h^{-1}}$), the water table rose in the streamside zone, although the magnitude was still far less than expected based on the results of the other studies.

To determine if groundwater ridging was capable of supplying a large amount of stormflow to the second-order stream, the potential groundwater flux in the near-stream zone was calculated by combining the streamside flow nets for 25 October (Fig. 5(b)) and measured hydraulic conductivities. The calculated subsurface discharge was between 2.8×10^{-4} and $5.3 \times 10^{-3}\,\mathrm{l\,s^{-1}\,m^{-1}}$ of stream bank in the streamside experimental plot. If this mechanism operates for the total length of the streams in the study basin (approximately 175 m), the subsurface stormflow contribution would vary between 0.1 and $1.2\,\mathrm{l\,s^{-1}}$, representing less than 2% of the basin discharge from a storm of the magnitude of the irrigation (Roulet, 1991). Because the drop in the water table does not occur immediately after the cessation of rainfall and the stream discharge declines very rapidly, subsurface discharge from groundwater ridging becomes a more important contributor of water during the storm recession.

The frequency of groundwater ridging can be determined by looking at the return period of an event with the intensity of the 25 October irrigation event. It represented a once in 3 years event; not uncommon, but certainly too infrequent to explain the presence of pre-event water in all stormflow events.

Evidence of pipe transported 'pre-event' water

Results of the pipeflow experiment show that flow in the unconnected pipe increased significantly during rain events. The pipes, therefore, may represent an important mechanism for drawing water into the near-stream zone from other portions of the wetland. Even if it is assumed that the pipe drains 100%

of the precipitation incident on its source area, then the minimum source area would have to be 165 m^2. If the runoff ratio for the pipe is only 0.5, then the contributing area of the pipe would be twice as large.

The results also indicate that 95% of peak pipeflow was pre-event water for the high-intensity 25 October event. If pipeflow is an important source of storm runoff, then sufficient volumes of pre-event water must be discharged via pipes entering the stream channel during storms. Because baseflow from the 8M pipe is greater than that of the other unconnected pipes entering the stream channel, it can be assumed that the 8M pipe represents a maximum response from unconnected pipes. On this assumption, the peak pipe stormflow for the 25 October event (311 ml s^{-1}) multiplied by the nine large unconnected pipes in the basin yields a total discharge of only 2.8 l s^{-1}, of which most would be pre-event water. As there is a long time lag in the response from the unconnected pipes, they contribute to streamflow during the recession phase of a storm, but the absolute quantity of water delivered to the stream by pipes is sufficient to explain the pre-event component in stormflow.

Evidence of saturated overland flow

The lack of response of the water table and hydraulic heads to storm events in the WSA experimental site (Table 2) indicates that no significant groundwater ridging occurs at the base of the hillslope or adjacent to the saturated area. Given the range of hydraulic conductivities in the saturated area and the maximum water table rise of 8.0 mm, the increase in groundwater discharge per unit meter along the wetland hillslope boundary would be insignificant. The total increase in groundwater discharge of between 1×10^{-5} and 2×10^{-5} l s^{-1} is insufficient to supply the magnitude of pre-event water seen at the outlet weirs of the saturated area.

The results of the irrigation events indicate that the saturated area storm runoff response was rapid, but of short duration. The runoff coefficients of approximately 1.0 confirm that direct precipitation onto the saturated area was the dominant mechanism operating in the groundwater discharge areas of the wetland. These results were not unexpected, as the experimental plot remained saturated throughout the experiment and did not increase or decrease in areal extent. The results of the hydrograph separation, however, indicate that peak flow from the saturated area was dominated by pre-event water (more than 77%). Hydrographic separation of the basin outlet storm discharge showed peak flow pre-event water fractions of over 87–93%, which is consistent with the saturated area response and indicates that the same process is operating at the basin scale.

Examining the results of the irrigation events, a greater portion of event water occurred at the overland flow collectors than at the saturated area outlet weirs. This suggests that either the event water must rapidly mix with the large volume of pre-event water in the saturated area before discharging from the region, or that the event water is being held in storage and displaces pre-event water stored in the saturated area. The large proportions of pre-event water in stormflow and the dependence of the size of the pre-event fraction on precipitation intensity are consistent with rapid mixing of the event water and stored pre-event water in the saturated area.

Rapid surface water mixing hypothesis

If saturated overland flow is the principal runoff mechanism operating in the saturated zone, as deduced above, then an explanation must be provided for the predominantly pre-event signature of the runoff. In the saturated area, the surface reservoir of pre-event water was estimated to be 13 mm in depth. However, this reservoir is also being constantly replenished with pre-event water by discharging groundwater. The discharge from the WSA plot for storms from late August to October averages 300 ml s^{-1}, or 18 mm h^{-1} if based on the area of the saturated zone. This input far exceeds even the most intense input from rainfall or irrigation during the study (i.e. 11.3 mm h^{-1} : 2 October) (see Table 3). This means that even at the most intense rainfall condition studied, rainfall represented only 39% of the input to the saturated area.

With a tracer study in the saturated area it is impossible to know how much of the event water mixes with pre-event water and how much pre-event water is subsequently displaced during the course of a rain or irrigation event. During some irrigation events the water levels in the saturated area reached a steady-state storage level. Under this condition, if it is assumed that the water in the saturated area is well mixed, then the proportion of the change in storage relative to the total stored water can be used as an estimate of the fraction of event water in the stormflow required to explain the large pre-event signature. The pre-event water fractions, employing a steady state, range from 0.76 to 0.93 (Table 4). These values can be compared with the fraction derived from chemical separation. As we are confident in our water level measurements of the saturated area to 1 mm, the minimum ratio of storage change that would indicate that the signatures cannot be explained by mixing alone is 0.07 ($[\Delta WL \pm \text{error}]/[\text{depth of standing water} + \Delta WL \pm \text{error}] = 1\,\text{mm}/(13\,\text{mm} + 1\,\text{mm}) = 0.07$). None of the computed fractions of pre-event water differ by more than 0.06 from the calculated fraction based solely on mixing. We therefore conclude that the rapid mixing of the incoming event

Table 4
Comparison of fraction of event water in saturated area and measured pre-event component in saturated area discharge

Date	Change in storage (Table 2) (mm)	Fraction of total saturated area	Calculated pre-event water fraction	Measured pre-event water fraction	Difference
28 August[a]	1.0	0.07	0.93	0.93	0.00
13 September[a]	2.0	0.13	0.87	0.91	0.04
25 September[a]	1.0	0.07	0.93	0.88	0.05
27 September[a]	2.0	0.13	0.87	0.85	0.02
2 October[a]	4.0	0.24	0.76	0.77	0.01
12 October	2.0	0.13	0.87	0.93	0.06

Calculated pre-event water fraction assumes a steady state. Measured pre-event water fraction values are taken from Table 3.
[a] Irrigation event.

water with an existing surface reservoir of pre-event water explains the large pre-event signature in stormflow from the saturated areas, and the large pre-event signature in basin stormflow.

Conclusions and implications for hydrograph separation

Several mechanisms have been suggested to account for the large and rapid response of pre-event water contribution to stormflow. In this study, groundwater ridging did not contribute to the pre-event stormflow component except during larger, low-return events, whereas preferential flow through pipes responds rapidly enough to account for only a minor amount of the total pre-event stormflow at peak flow. The rapid surface water mixing hypothesis, however, successfully accounts for the large amount of pre-event water in stormflow dominated by saturation overland flow.

The results of this study suggest that there is a problem in interpreting the results of chemical separations when a basin contains saturated areas that are reservoirs of pre-event water. Several studies (e.g. Rodhe, 1987; Wels et al., 1991b) have relied on the event water component from chemical or isotopic hydrograph separations to determine the extent of saturated areas within a drainage basin under the assumption that saturation overland flow is event water. Although these studies found a good relationship between the separation results and their field surveys, calculations of the saturated area of the experimental plot in the present study using the separation technique would yield a saturated area of between 2.1 and 5.6 m^2. If this technique were to be used at the wetland scale, the contributing area derived by chemical separation would be 5–9% of the wetland, in contrast to the surveyed saturated area of 31% (Roulet, 1991). The success of previous studies in relating their independent calculations of saturated area to discharge (e.g. Rodhe, 1987; Wels et al., 1991b) is a function of basin conditions and the size of the input relative to the size of the surface reservoir. The rapid surface water mixing hypothesis may also operate, though to a smaller extent in basins that do not have standing water, but where the soils are porous and saturated. This may in part explain the discrepancies that have appeared in the literature between studies that explain runoff by saturated overland flow and the variable source area concept, but chemical separations indicate a contradictory large subsurface component. Because of rapid surface water mixing, a pre-event stormflow signature can no longer be considered incompatible with the presence of saturated overland flow.

Acknowledgments

Thanks are given to Kirsten Anderson, Angela Travis, Lesley Rigg, Todd Kuiack, and Rosemary Ash for assistance in the field, to K. Sanmugadas for laboratory assistance, and to Craig Allan and Kevin DeVito for advice on the irrigation system and tracer experiments. Thanks are also due to the Metropolitan Toronto and Region Conservation Authority for access to the catchment. The work for this study was completed while J.M.W. was supported by a Natural Science and Engineering Council of Canada (NSERC) Postgraduate Scholarship, and the research was funded by an NSERC operating grant to N.T.R.

References

Abdul, A.S. and Gillham, R.W., 1989. Field studies of the effects of the capillary fringe on streamflow generation. J. Hydrol., 112: 1–18.

Bonell, M., Pearce, A.J. and Stewart, M.K., 1990. The identification of runoff-production mechanisms using environmental isotopes in a tussock grassland catchment, eastern Otago, New Zealand. Hydrol. Processes, 4: 15–34.

Buttle, J.M. and Sami, K., 1992. Testing the groundwater ridging hypothesis of streamflow generation during snowmelt in a forested catchment. J. Hydrol. 135: 53–72.

Calles, U.M., 1985. Deep groundwater contribution to a small stream. Nordic Hydrol., 16(1): 45–54.

Carter, V. and Novitzki, R.P., 1988. Some comments on the relation between ground water and wetlands. In: D.D. Hook et al. (Editors), The Ecology and Management of Wetlands, Vol. 1. Croom Helm, London, pp. 68–86.

Dunne, T., 1978. Field studies of hillslope processes. In M.J. Kirkby (Editor), Hillslope Hydrology. Wiley, Chichester, pp. 227–293.

Dunne, T. and Black, R.D. (1970). Partial area contributions to storm runoff in a small New England watershed. Water Resour. Res., 6(5): 1296–1311.

Freeze, R.A. and Cherry, J.A., 1979. Groundwater. Prentice–Hall, Englewood Cliffs, NJ, 604 pp.
Gillham, R.W., 1984. The capillary fringe and its effect on the water-table response. J. Hydrol., 67: 307–324.
Heliotis, F.D. and DeWitt, C.B., 1987. Rapid water table responses to rainfall in a northern peatland ecosystem. Water Resour. Res., 23(6): 1011–1016.
Hill, A.R., 1990. Groundwater cation concentration in the riparian zone of a forested headwater stream. Hydrol. Processes, 4: 121–130.
Howard, K.W.F. and Beck, P., 1986. Hydrochemical interpretation of groundwater flow systems in Quaternary sediments of southern Ontario. Can. J. Earth Sci., 23: 938–947.
McDonnell, J.J., 1990. A rationale for old water discharge through macropores in a steep, humid catchment. Water Resour. Res., 26(11): 2821–2832.
McDonnell, J.J., Bonell, M., Stewart, M.K. and Pearce, A.J., 1990. Deuterium variations in storm rainfall: implications for stream hydrograph separation. Water Resour. Res., 26(3): 455–458.
National Wetlands Working Group, 1988. Wetlands of Canada, Ecological Classification Series, No. 24. Sustainable Development Branch, Environment Canada, and Polyscience Publications, Montreal, Quebec, 425 pp.
O'Brien, A.L., 1980. The role of groundwater in stream discharges from two small wetland controlled basins in Eastern Massachusetts. Groundwater, 18: 359–365.
Pearce, A.J., Stewart, M.K. and Sklash, M.G., 1986. Storm runoff generation in humid headwater catchments. 1. Where does the water come from? Water Resour. Res., 22(8): 1263–1272.
Ragan, R.M., 1968. An experimental investigation of partial area contributions. IAHS Publ., 76: 241–249.
Rodhe, A., 1987. The origin of streamwater traced by oxygen-18 Uppsala University Department of Physical Geography, Division of Hydrology, Rep. Ser. A 41, 260 pp.
Roulet, N.T., 1990a. The hydrological role of peat-covered wetlands. Can. Geogr., 34(1): 82–83.
Roulet, N.T., 1990b. Hydrology of a headwater basin wetland: groundwater discharge and wetland maintenance. Hydrol. Processes, 4: 387–400.
Roulet, N.T., 1991. Stormflow production in a headwater basin swamp. Nordic Hydrol., 22: 161–174.
Sibul, U., Wang, K.T. and Vallery, D. 1977. Groundwater resources of the Duffin–Rouge River drainage basins. Water Resour. Rep. 8. Ontario Ministry of the Environment, Toronto, Ontario, 77 pp.
Sklash, M.G. and Farvolden, R.N., 1979. The role of groundwater in storm runoff. J. Hydrol., 43: 45–65.
Sklash, M.G., Stewart, M.K. and Pearce, A.J., 1986. Storm runoff generation in humid headwater catchments, 2. A case study of hillslope and low-order stream response. Water Resour. Res., 22(8): 1273–1282.
Waddington, J.M., 1991. Runoff mechanisms in a headwater spring-fed swamp. M.Sc. Thesis, York University, North York, Ont., 99 pp.
Wels, C., Cornett, R.J. and Lazerte, B.D., 1991a. Hydrograph separation: a comparison of geochemical and isotopic tracers. J. Hydrol., 122: 253–274.
Wels, C., Taylor, C.H. and Cornett, R.J., 1991b. Streamflow generation in a headwater basin on the Precambrian Shield. Hydrol. Processes, 5: 185–199.

D Biogeochemical Processes and Methods

Commentary on Paper D1

Reddy, K. R. & Patrick, W. H. Jr (1975) Effect of alternate aerobic and anaerobic conditions on redox potential, organic matter decomposition and nitrogen loss in a flooded soil.

Reddy and Patrick set up a very smart laboratory experiment to test the effect of aerobic–anaerobic conditions on nitrogen cycling using the introduction of ^{15}N-labelled ammonia. Although this is not the first paper written by Patrick on the subject (Patrick & Wyatt, 1964; Patrick & Tusneem, 1972), this is the first experiment which clearly demonstrates the fate of N under alternating aerobic and anaerobic condition. This well-cited paper (approx. 150 citations to date) demonstrates and quantifies that under the regime of 2- and 2-day aerobic–anaerobic periods, two thirds of the labelled ammonium N was lost from the flooded soil during a 128-day incubation. They also determined that shorter aerobic–anaerobic cycles enhanced the rate of organic matter decomposition.

In this paper they did not measure labelled N in the gas but they reasonably assumed that the lost N had been denitrified since it could not be found in organic or inorganic fractions of the incubated sediment. This result provided several exciting implications. It showed that the nitrification process could be effectively activated within a 2-day aerobic period and that two subsequent days under anaerobiosis did not alter its rate for at least 128 days of producing nitrate which were then removed by denitrification. This was confirmed later by Grofmann & Tiedje (1988) when they showed some hysteresis of denitrification during wetting and drying cycles in soil. Moreover, it confirmed that this short-term periodicity reduces the sediment redox values compatible with active denitrification. Moreover, it clearly related the fate of carbon and nitrogen cycles in wetland sediments, revealing the importance of denitrification as a respiration process.

This work generated several new avenues of research, both in terms of fundamental and applied research. For instance, the functioning of most tertiary sewage treatment plants today is based on the application of this short-term aerobic–anaerobic alternation. It is only very recently that the discovery of anammox, a new pathway of anaerobic ammonia oxidation (Kuypers *et al.*, 2003), allows the complete removal of organic nitrogen under anaerobic conditions to be envisaged. This short-term aerobic–anaerobic alternation has been found to support a high rate of organic matter cycling and in turn high primary productivity riparian zones (Brinson *et al.*, 1984). However, short-term alternation of redox conditions was more conducive to nitrous oxide emission, a potent greenhouse gas emitted by both nitrification and denitrification under low oxygen partial pressure (Stevens *et al.*, 1997). A trade-off between maximising denitrification and wetland productivity on the one hand, and limitation of greenhouse gases on the other hand, therefore needs to be found to optimise the biogeochemical potential of wetlands.

GP

References

Brinson, M. M., Bradshaw, H. D. & Kane, E. S. (1984) Nutrient assimilative capacity of an alluvial flood plain swamp. *J. Appl. Ecol.* **21**, 1041–1057.

Groffman, P. M & Tiedje, J. M. (1988) Denitrification hysteresis during wetting and drying cycles in soil. *Soil Sci. Soc. Am. J.* **52**, 1626–1629.

Kuypers, M. M. M, Sliekers, A. O., Lavik, G., Schmid, M., Jorgensen, B. B., Kuenen, J. G., Sinninghe Damsté, J. S., Strous, M. & Jetten, M. S. M. (2003) Anaerobic ammonium oxidation by anammox bacteria in the Black Sea. *Nature* **422**, 608–611.

Patrick, W. H. & Tusneem, M. E. (1972) Nitrogen loss from flooded soil. *Ecology* **53**(4), 735–739.

Patrick, W. H. & Wyatt, R. (1964) Soil nitrogen loss as a result of alternate submergence and drying. *Soil Sci. Soc. Am. Proc.* **28**, 647–653.

Stevens, R. J., Laughlin, R. J., Burns, L. C., Arah, J. R. M. & Hood, R. (1997) Measuring the contributions of nitrification and denitrification to the flux of nitrous oxide from soil. *Soil Biol. & Biochem.* **29**, 139–151.

Soil Biol. Biochem. Vol. 7, pp. 87–94. Pergamon Press 1975. Printed in Great Britain.

EFFECT OF ALTERNATE AEROBIC AND ANAEROBIC CONDITIONS ON REDOX POTENTIAL, ORGANIC MATTER DECOMPOSITION AND NITROGEN LOSS IN A FLOODED SOIL

K. R. REDDY and W. H. PATRICK, JR.

Laboratory of Flooded Soils and Sediments, Department of Agronomy, Louisiana State University, Baton Rouge, Louisiana 70803, U.S.A.

(*Accepted* 28 *June* 1974)

Summary—The effect of several cycles of varying length of alternate aerobic and anaerobic conditions on redox potential, organic matter decomposition and loss of added and native nitrogen was investigated under laboratory conditions in flooded soil incubated for 128 days. Redox potential decreased rapidly when air was replaced with argon for the short-time cycles, but decreased more slowly where the aerobic period was long enough to permit build-up of nitrate. The minimum redox potential reached during the anaerobic period was generally lower for the longer cycles, but in all cases was low enough for denitrification to occur. Rate of decomposition of organic matter was faster in the treatments with a greater number of alternate aerobic and anaerobic periods. Total N (native and applied) losses as high as 24·3 per cent occurred in the treatment with the maximum number of cycles and with alternate aerobic and anaerobic periods of 2 and 2 days. Increasing the durations of the aerobic–anaerobic periods decreased the loss of N. A maximum loss of 63·0 per cent of applied $^{15}NH_4$–N resulted from the shortest (2 and 2 day) aerobic and anaerobic incubation. For soil undergoing frequent changes in aeration status the only labelled N that remained at the end of incubation was found in the organic fraction. Loss of N may have been even greater if labelled inorganic N had not been immobilized by microorganisms decomposing the added rice straw. The greater loss of N resulting from the 2 and 2 day aerobic–anaerobic incubation shows that, in soils where the redox potential falls low enough for denitrification to occur, increasing the frequency of changing from aerobic to anaerobic conditions will increase the loss of N.

INTRODUCTION

It is known that alternate aerobic and anaerobic conditions affect a number of important soil processes. Several studies have shown that the rate and products of decomposition of organic matter are different in flooded and nonflooded soils (Tenny and Waksman, 1930; Alexander, 1961). Decomposition is faster under continuous aerobic conditions than under continuous anaerobic conditions, although inorganic N is usually released earlier to the soil solution under anaerobic conditions. Many workers, namely, Gooding and McCalla, 1945; Stephenson, 1956; Birch, 1960; and others have shown that decomposition of organic matter is stimulated by alternately wetting and drying the soil.

Inorganic N transformations are also influenced by alternate aerobic and anaerobic conditions. Severe N loss has been shown to occur in soils subjected to periods of alternate drained (aerobic) and flooded (anaerobic) conditions (Wijler and Delwiche, 1954; Russell, 1961; Patrick and Wyatt, 1964; MacRae, Ancajas and Salandanan, 1967). Under certain conditions organic N is converted to ammonium N (this process occurs in both aerobic and anaerobic soils), under aerobic conditions the ammonium N is oxidized to nitrate and under anaerobic conditions the resulting nitrate denitrified. Loss of N through sequential nitrification and denitrification is especially high in soils planted to lowland rice, where water management practices sometimes require frequent draining and reflooding.

Russell and Richards (1917) were among the first to recognize the importance of alternate aerobic and anaerobic conditions in determining loss of N. Their studies showed large losses of N from farmyard manure under alternate wetting and drying conditions, but little loss when the material was maintained under either aerobic or anaerobic conditions. Wijler and Delwiche (1954) noted that alternating aerobic and anaerobic conditions should result in greater total N loss from the soil than would be found under continuous anaerobic conditions. Patrick and Wyatt (1964) observed large losses of N (up to 20 per cent of total N or 200 μg/g) as a result of repeated cycles of flooding and drying to field moisture. Tusneem and Patrick (1971) showed that ammonium N was highly unstable under alternate flooded and moist conditions of several weeks' duration, since only trace amount of applied labelled ammonium N was recovered at the end of a 120-day incubation period.

Most of the studies conducted so far show that alternate wetting and drying increases the loss of N from the soil, but these studies do not provide information

Table 1. Showing the changes in total nitrogen as a result of alternate aerobic and anaerobic treatments

Length of aerobic period (CO_2-free air) (Days)	Length of anaerobic period (Ar) (Days)	Number of complete cycles during incubation period	Total nitrogen* (μg/g) At the end of experiment	Net loss	Loss (%)
2	2	32	704	226	24·3
4	4	16	717	213	22·9
8	8	8	730	200	21·5
16	16	4	744	186	20·0
32	32	2	762	168	18·0
64	64	1	810	120	12·9
128	—	Completely aerobic	866	64	6·9
—	128	Completely anaerobic	925	5	0·5

* At the beginning of the experiment soils in all treatments contained 930 μg N/g of soil.

on the redox status of the soil under this changing environment or the minimum periods of aerobic and anaerobic conditions necessary to cause alternate nitrification and denitrification. The present study was designed to show the effect of frequency and length of alternating aerobic and anaerobic periods on loss of both native and added N and on soil organic matter decomposition. Changes in the redox potential of the soil under these alternate aerobic and anaerobic conditions were also measured.

MATERIALS AND METHODS

A Crowley silt loam, sampled from the Rice Experiment Station, Crowley, Louisiana, was used in this study. The soil had a carbon (C) content of 0·702 per cent, a total N content of 805 μg/g and a pH of 5·6. The soil was thoroughly mixed and ground to pass through 0·4 mm mesh. An energy source of 0·5 per cent finely ground rice straw (48 per cent carbon) and 100 μg/g N in the form of $(NH_4)_2SO_4$ containing 10·1458 at % excess ^{15}N were thoroughly mixed with the soil. Aerobic and anaerobic conditions were accomplished by bubbling either CO_2-free air or Ar through the incubation flasks. The various treatments used are shown in Table 1.

Incubation procedure

Duplicate flasks for each treatment were set up as shown in Fig. 1. One hundred and fifty grams of soil was weighed into each of the two incubation flasks along with 300 ml distilled water to give a soil to water ratio of 1:2 and then incubated at 30°C under constant stirring with a magnetic stirrer for a period of 128 days. A platinum electrode was permanently inserted through a rubber stopper to measure the redox potential of the soil suspension. The flasks were connected by means of glass and rubber tubing for continuous

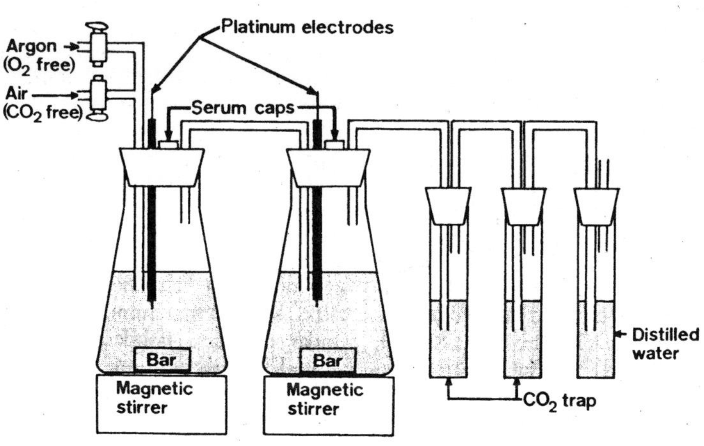

Fig. 1. Diagram of the apparatus used for studying the effect of alternate aerobic and anaerobic conditions on redox potential, organic matter decomposition and nitrogen loss in a flooded soil.

flow of CO_2-free air or Ar for creating aerobic or anaerobic conditions. The CO_2 evolved during microbial respiration was trapped in a 0·5 N KOH solution.

Redox potential measurements

Redox potential measurements were made by connecting the Pt electrode to a pH meter, using a saturated calomel half cell as reference electrode (Patrick and Wyatt, 1964). A complete cell was formed by inserting a salt bridge (glass tube) through a serum cap in the rubber stopper of the flask. Redox potential measurements were made daily during the incubation period of 128 days. The pH of soil suspension at the end of the 128-day incubation period was also measured.

Analytical methods

The CO_2 produced during decomposition of organic matter was measured (Stotzky, 1965). Total C remaining in the soil at the end of incubation period was measured by the dry combustion method (Allison, Bollen and Moodie, 1965). Inorganic and organic N fractions were analyzed by the steam distillation and the modified Kjeldhal methods, respectively, as given by Bremner (1965a, 1965b). The titrated samples containing N at a concentration of approximately 1 mg N per ml were analyzed for ^{15}N using a Dupont Model 21-614 isotope ratio mass spectrometer (Bremner, 1965c).

RESULTS AND DISCUSSION

Redox potential of the soil during alternate aerobic and anaerobic conditions

The redox potential of the soil responded rapidly to changes in the aeration status of the soil (Fig. 2(a–g)). The maximum redox potential under aerobic conditions for all treatments was approximately +600 mV. Displacement of air by Ar caused a rapid decrease in redox potential. The level to which the redox potential fell was approximately +200 mV for the treatment in which the gas was alternated every 2 days and reached lower values when the anaerobic period was extended. For the 4 and 4 day and 8 and 8 day aerobic–anaerobic periods the first decrease in redox potential was greater than subsequent decreases. In the continuously aerobic treatment the potential rose rapidly to approximately +600 mV and remained at this value throughout the 128-day incubation period. For the continuously anaerobic treatment the potential decreased rapidly to approximately −300 mV and remained constant for the entire incubation period. When the length of the aerobic–anaerobic periods was extended beyond 8 and 8 days the potential did not decrease very rapidly when Ar was bubbled through the samples. This slower initiation of reducing conditions was likely due to the presence of nitrate which had accumulated as a result of ammonium oxidation during the relatively long aerobic period. Nitrate has been shown to prevent a rapid decline in redox potential. In the soil that was maintained under a 64 and 64 day aerobic–anaerobic treatment (Fig. 2(f)), the redox potential pattern after introduction of Ar was typical of that observed in a flooded soil containing both O_2 and nitrate (Engler, Patrick and Antie, 1974, in preparation). Rapid depletion of the O_2 as a result of displacement by Ar and consumption by the soil caused an initial decrease in redox potential from +600 mV to +200 mV at which value the potential was stabilized for several weeks. Stabilization of the potential at approximately +200 mV was very likely due to the accumulated nitrate and when this nitrate was denitrified after approximately 100 days the redox potential rapidly fell to approximately −300 mV.

The redox potential values at the end of the incubation period for the various treatments are shown in Fig. 3(a). Since the soil was anaerobic during the last half cycle the potentials are generally reducing, except for the treatment that went through rapid aerobic–anaerobic changes and for the treatment that was maintained under continuously aerobic conditions. There appeared to be a close relationship between the redox potential at the end of incubation and the length of the aerobic–anaerobic periods with a long final anaerobic half cycle being associated with a low redox potential.

The final pH of the soil was influenced by the treatment (Fig. 3(b)). For the soil carried through the various aerobic–anaerobic cycles the final pH was higher for the longer cycles, increasing from pH 6·5 for 2 and 2 day to 7·4 for the 64 and 64 day treatment. The controlling factor determining pH in these treatments was very likely redox potential during the final incubation period since a close inverse relationship between redox potential and pH was observed. For the continuous aerobic and continuous anaerobic treatments the same inverse relationship between redox potential and pH was evident. The soil continuously aerated had a lower final pH than did the original soil, probably resulting from the accumulation of nitrate and sulfate during the long aerobic period. Completely anaerobic conditions resulted in a pH of 7·8, a value somewhat higher than is usually encountered when acid soils are flooded. It should be noted that continuous passage of Ar through the suspension prevented precipitation of carbonate compounds which function to stabilize pH near the neutral point. This suggests that the pH value of flooded soils is sensitive to loss of CO_2. The pH of most reduced soils equilibrated with CO_2 at 1 atm. is 6·1 (Ponnamperuma, Castro and Valencia, 1969).

Decomposition of organic matter as a result of alternate aerobic and anaerobic conditions

The aeration status of the soil had a marked effect on organic matter breakdown. Carbon dioxide evolution for the various treatments is shown in Fig. 4(A–H). There was little difference in CO_2 evolution patterns for aerobic–anaerobic cycles ranging from 2 and 2–32 and 32 days. Carbon dioxide evolution was decreased considerably, however, when the cycle was increased to 64 and 64 days. Aerobic conditions for the

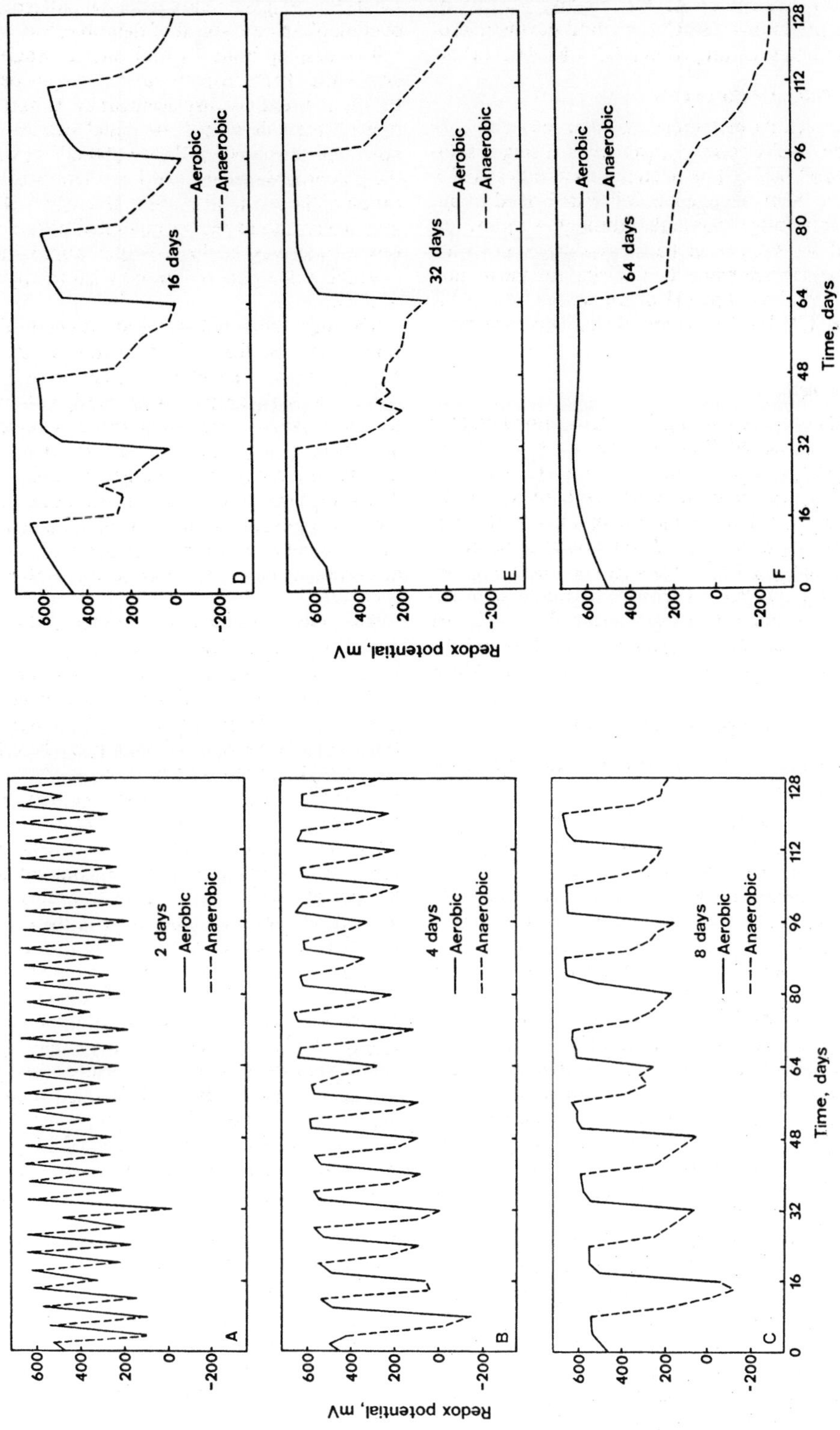
Redox potential, mV
Time, days
2 days
Aerobic
Anaerobic
4 days
8 days
16 days
32 days
64 days
A
B
C
D
E
F

Fig. 2. (A–F).

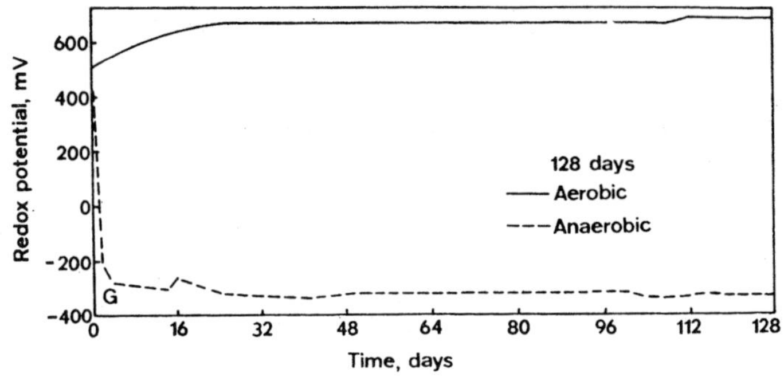

Fig. 2. (A, B, C, D, E, F and G). Changes in redox potential of the soil as affected by alternate aerobic and anaerobic conditions.

entire incubation resulted in approximately the same loss of CO_2 as was observed for most of the aerobic–anaerobic treatments. For the soil kept under continuous anaerobic conditions CO_2 evolution was decreased to approximately half the values obtained for the other treatments. Organic matter breakdown is known to be slower under anaerobic conditions than under aerobic conditions (Acharya, 1935; Alexander, 1961; Tenny and Waksman, 1930). Since C can also be lost from anaerobic soil as CH_4, which was not measured in this experiment, total C loss was calculated from total C content at the beginning and end of the experiment. Carbon loss as determined from dry combustion analysis of organic C at the beginning and end of the experiment showed remarkable agreement with the values obtained by CO_2 evolution (Fig. 5(a)), indicating little C loss as CH_4. The only treatments that showed much discrepancy in C loss between the two methods was the 8 and 8 day aerobic–anaerobic periods and the continuously anaerobic treatment. The slightly greater loss of C for the continuously anaerobic treatment may have been due to CH_4 formation.

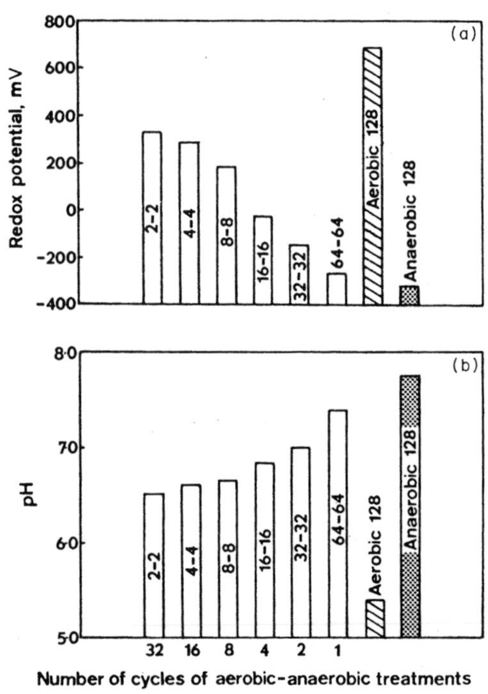

Fig. 3. (a and b). Redox potential and pH of the soil at the end of 128-day incubation period for the various treatments.

Loss of added and native N *as a result of alternate aerobic and anaerobic conditions*

The loss of N as a result of the various aerobic–anaerobic cycles is shown in Table 1 and Fig. 5(b). The greater loss of N took place in the treatments that alternated between aerobic and anaerobic conditions every 2 days. Approximately one-fourth of the total N (24·3 per cent) was lost during 32 of these cycles. Increasing the durations of the aerobic–anaerobic periods decreased the loss of N with approximately 13 per cent loss for the 64 and 64 day treatment. Almost no N was lost in the continuously anaerobic treatment while 7 per cent (64 μg/g) was lost in the continuously aerobic treatment. N is sometimes lost from apparently well aerated soils because of the presence of anaerobic microzones where denitrification can take place, but the continuous stirring and bubbling of air through the soil suspension should have prevented such anaerobic microzones from being established. This small loss of N may have resulted from chemical decomposition of nitrite. Some workers have shown that N losses during the nitrification process are the consequence of chemical decomposition of nitrite under acid soil conditions (Tyler and Broadbent, 1960; Smith and Clark, 1960; Cady and Bartholemew, 1963). No decomposition of nitrite was observed by Greenwood (1962) and Meek and Mackenzie (1965) in alkaline soils. Nitrate formed during the nitrification process might suffer some loss under the low pH conditions of the continuously aerated treatment. In all of the aerobic–anaerobic cycles the redox potential fluctuated between values that were oxidizing enough to support nitrification and reducing enough (below approximately +340 mV) to support denitrification (Pearsall, 1938; Patrick, 1960). The 2 and 2 day aero-

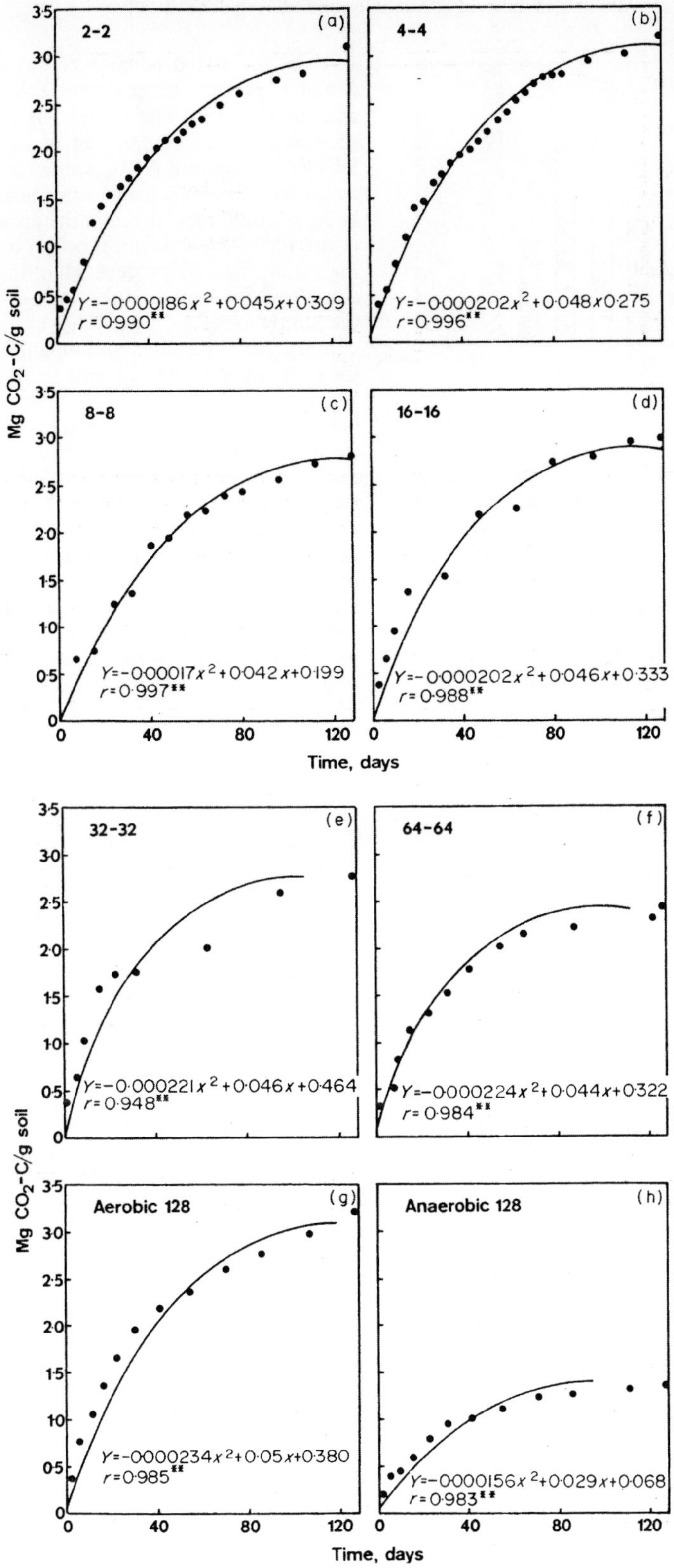

Fig. 4. (a, b, c, d, e, f, g and h). Rate of decomposition of organic matter as influenced by alternate aerobic and anaerobic conditions.

92

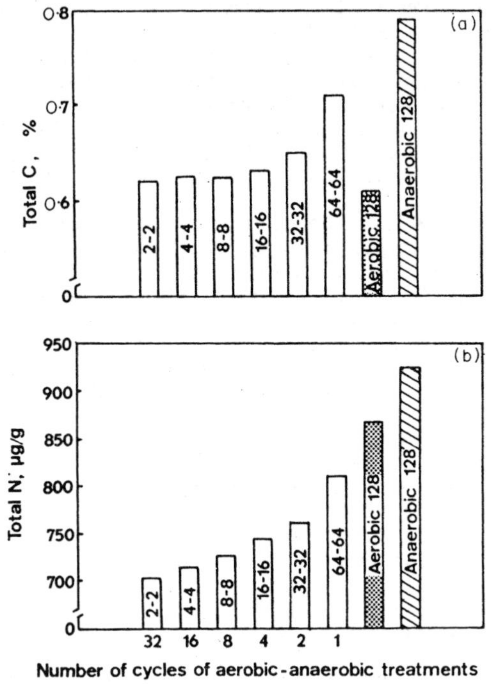

Fig. 5. (a and b). Total carbon and total nitrogen content of the soil at the end of 128-day incubation period as influenced by various treatments.

bic–anaerobic cycle was chosen because it was thought that the duration of the anaerobic half cycle would be too short for the redox potential to fall below the critical value necessary for nitrate reduction. The significant loss of N from the soil subjected to the 2 and 2 day aerobic–anaerobic periods indicates, however, that a 2 day period was long enough for nitrification to occur when air was bubbled through the suspension. A 2 day period was also long enough to support denitrification when Ar was bubbled through. Loss of N during short aerobic and anaerobic periods may be of more significance in nonflooded soils where temporary anaerobic conditions can occur than in flooded soils.

The amount of added labelled N remaining in the soil following incubation for 128 days for the various aerobic–anaerobic treatments is shown in Fig. 6(a). Less than half of the added labelled N (37 per cent in the case of the 2 and 2 day cycle) was recovered in all of the treatments where the soil underwent alternate aerobic and anaerobic incubation. Almost all of the N that was recovered was present in the organic fraction. This was especially true for the treatments with short aerobic–anaerobic periods; no inorganic N was detected in the 2 and 2, 4 and 4 and 8 and 8 day treatments. It is likely that N loss would have been greater if an appreciable part of the added labelled N had not been incorporated into microbial tissue, from which release by decomposition was slow.

All of the added labelled N was recovered in the completely anaerobic treatment, either as organic N or as ammonium N. The completely aerobic treatment showed a loss of 18 per cent of the added 100 μg/g labelled ammonium N. As pointed out above, this loss could have resulted from denitrification in a system being continuously sitrred in the presence of O_2, or as is also likely, from chemical breakdown of nitrite during the nitrification process. All of the inorganic N recovered in the continuously aerobic treatment was in the nitrate form.

Inorganic N (labelled and unlabelled) at the end of the experiment for the various treatments is shown in Fig. 6(b). Little inorganic N, either labelled or unlabelled, was recovered at the end of incubation for treatments with aerobic–anaerobic cycles shorter than 32 and 32 days. For all of the alternating aerobic–anaerobic treatments except the 2 and 2 day treatment, all of the inorganic N was present as NH_4^+. A trace of nitrate was present for the 2 and 2 day treatment. It should be remembered that the last half cycle was under anaerobic conditions. Completely aerobic or completely anaerobic conditions resulted in accumulation of appreciable inorganic N. This N was present as NO_3^- in the aerobic treatment and as NH_4^+ in the anaerobic treatment. For both these treatments and for the alternate aerobic and anaerobic treatments

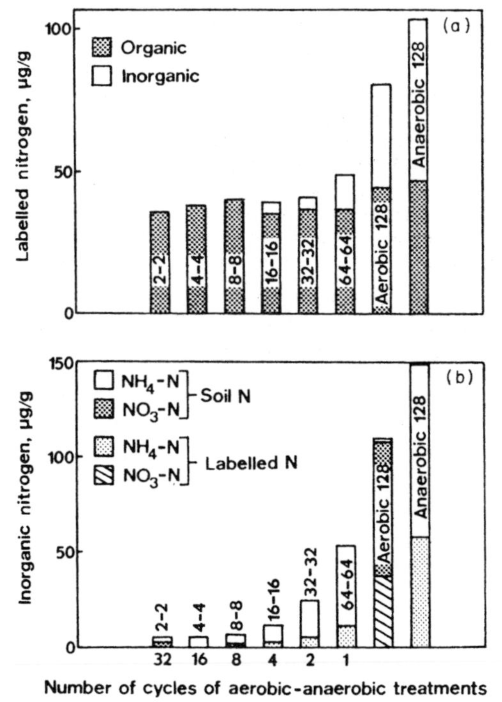

Fig. 6. (a and b). Amount of inorganic nitrogen and labelled nitrogen remaining in the soil following incubation for 128 days under alternate aerobic and anaerobic conditions.

which had inorganic N present at the end of incubation, approximately one-third of the inorganic N was derived from the labelled ammonium N and two-thirds from reactive soil N.

This study shows that increasing the frequency of changing from aerobic to anaerobic soil conditions to 2 days aerobic and 2 days anaerobic increases the loss of N. This loss was especially severe for added ammonium N, with 63 per cent of added labelled ammonium N being lost for the 2 and 2 day treatment during a 128-day incubation. More N would possibly have been lost except for immobilization of part of the added ammonium N by added rice straw.

REFERENCES

ACHARYA C. N. (1935) Studies on the anaerobic decomposition of plant materials. III. Comparison of the course of decomposition under anaerobic, aerobic and partially aerobic conditions. *Biochem. J.* **29,** 1116–1120.

ALEXANDER M. (1961) *Introduction to Soil Microbiology*, Wiley, New York.

ALLISON L. E., BOLLEN W. and MOODIE C. D. (1965) Total Carbon. In *Agronomy 9, Methods of Soil Analysis* (C. A. Black, Ed.), pp. 1346–1366, Am. Soc. Agro., Madison, Wis.

BIRCH H. F. (1960) Nitrification in soils after different periods of dryness. *Pl. Soil.* **12,** 81–96.

BREMNER J. M. (1965a) Inorganic forms of nitrogen. In *Agronomy 9, Methods of Soil Analysis* (C. A. Black, Ed.), pp. 1179–1237, Am. Soc. Agro., Madison, Wis.

BREMNER J. M. (1965b) Total nitrogen. In *Agronomy 9, Methods of Soil Analysis* (C. A. Black, Ed.), pp. 1149–1178, Am. Soc. Agro., Madison, Wis.

BREMNER J. M. (1965c) Isotope-ratio analysis of nitrogen in nitrogen-15 tracer investigations. In *Agronomy 9, Methods of Soil Analysis* (C. A. Black, Ed.), pp. 1256–1286, Am. Soc. Agro., Madison, Wis.

CADY F. B. and BARTHOLEMEW W. V. (1963) Investigations of nitric oxide reactions in soils. *Soil Sci. Soc. Am. Proc.* **27,** 546–549.

GOODING T. H. and MCCALLA T. M. (1945) Loss of CO_2 and NH_3 from crop residues during decomposition. *Soil Sci. Soc. Am. Proc.* **10,** 185–190.

GREENWOOD D. J. (1962) Nitrification and nitrate dissimilation in soil. *Pl. Soil.* **17,** 365–391.

MACRAE I. C., ANCAJAS R. R. and SALANDANAN S. (1967) The fate of nitrate nitrogen in some tropical soils following submergence. *Soil Sci.* **105,** 327–334.

MEEK B. D. and MACKENZIE A. J. (1965) The effect of nitrate and organic matter on aerobic gaseous losses of nitrogen from a calcareous soil. *Soil Sci. Soc. Am. Proc.* **29,** 176–178.

PATRICK W. H. (1960) Nitrate reduction in a submerged soil as affected by redox potential. *Trans. Seventh Inter. Cong. Soil Sci. Comm.* **II,** 494–500.

PATRICK W. H. and WYATT R. (1964) Soil nitrogen loss as a result of alternate submergence and drying. *Soil Sci. Soc. Am. Proc.* **28,** 647–653.

PEARSALL W. H. (1938) The soil complex in relation to plant communities—I. Oxidation-reduction potentials in soils. *J. Ecol.* **26,** 180–193.

PONNAMPERUMA R. N., CASTRO R. V. and VALENCIA C. M. (1969) Experimental study of the influence of the partial pressure of carbon dioxide on the pH values of aqueous carbonate systems. *Soil Sci. Soc. Am. Proc.* **33,** 239–241.

RUSSELL E. J. and RICHARDS E. H. (1917) The changes taking place during the storage of farmyard manure. *J. Agr. Sci.* **8,** 495–563.

RUSSELL E. W. (1961) *Soil Conditions and Plant Growth*, Longmans, Green, London, Ed. 9.

SMITH D. H. and CLARK F. E. (1960) Volatile losses of nitrogen from acid or neutral soils or solutions containing nitrite and ammonium ions. *Soil Sci.* **90,** 86–92.

STEPHENSON I. L. (1956) Some observations on the microbial activity in remoistened air-dried soils. *Pl. Soil.* **8,** 170–182.

STOTZKY G. (1965) Microbial respiration. In *Agronomy 9, Methods of Soil Analysis* (C. A. Black, Ed.), pp. 1562–1565, Am. Soc. Agro., Madison, Wis.

TENNY F. G. and WAKSMAN S. A. (1930) Composition of material organic materials and decomposition in the soil—V. Decomposition of various chemical constituents in plant materials under anaerobic conditions. *Soil Sci.* **30,** 143–160.

TUSNEEM M. E. and PATRICK W. H. JR. (1971) Nitrogen Transformations in Waterlogged Soil, Louisiana State University Agri. Expt. Sta., Bull. 657, pp. 75.

TYLER K. B. and BROADBENT F. E. (1960) Nitrate transformations in California soils. *Soil Sci. Soc. Am. Proc.* **24,** 279–282.

WIJLER J. and DELWICHE C. C. (1954) Investigations on the denitrifying processes in soil. *Pl. Soil.* **5,** 155–169.

Commentary on Paper D2

Yoshinari, T., Hynes, R. & Knowles, R. (1977) Acetylene inhibition of nitrous oxide reduction and measurement of denitrification and nitrogen fixation in soil.

This paper, together with another published a year earlier by the same authors (Yoshinari & Knowles, 1976), provided the foundation for the most popular method of measuring denitrification in the laboratory or *in situ*. It is based on the fortuitous discovery of Fedorova *et al.* (1973), that acetylene (C_2H_2) could be used to measure nitrogen fixation and inhibited nitrous oxide (N_2O) reduction; the later inhibition was confirmed by Yoshinari & Knowles (1976) who tested it on three pure denitrifying cultures. They found that reduction of nitrate or nitrite by pure denitrifying cultures of *Pseudomonas perfectomarinus*, *P. aeruginosa* and *Micrococcus denitrificans* under 0.1 atmosphere of acetylene, resulted in a stochiometric accumulation of nitrous oxide (Yoshinari & Knowles, 1976).

In the present paper (Yoshinari *et al.*, 1977), they confirm that acetylene inhibition of nitrous oxide reduction also occurred in natural denitrifying communities present in soils. They tested the effect of soil moisture on acetylene efficiency to block nitrous oxide reduction. They found that moist soil systems were very sensitive even to 10^{-3} atmospheres of acetylene but the inhibition of nitrous oxide was incomplete at 10^{-5} atmosphere. They attributed this decrease in efficiency to the resistance to gaseous diffusion within the waterlogged soils. A year later, Yeomans & Beauchamp (1978) demonstrated that the inhibition of nitrous oxide reduction in soil in the presence of acetylene decreased after seven days of incubation. Moreover, they determined that the soils which had reduced the nitrous oxide in the presence of acetylene retained this ability for at least three days after the original atmosphere containing acetylene had been removed. Their interpretation was that after a seven-day lag time there had been growth of microbial species which have the ability to reduce nitrous oxide in the presence of acetylene.

The restriction of the use of the acetylene block method after a few days of incubation, because of the limited inhibition of nitrous oxide reduction by acetylene, was further restricted to even shorter periods, i.e. less than 10 hours, by another finding of Hynes & Knowles (1978). They discovered that ammonia oxidation by *Nitrosomonas europea* was inhibited by very low acetylene partial pressure (0.1–10 Pa); this was then confirmed for the soil nitrifying consortium by Berg *et al.* (1982). Therefore, measurement of denitrification activity under an acetylene atmosphere would be underestimated if the nitrate source of the denitrifiers was provided by nitrification, a rather ubiquitous phenomenon in soils where aerobic and anaerobic microsites can co-exist and allow nitrification and denitrification to occur simultaneously. This limitation does not apply though to potential denitrification activity (Smith & Tiedje, 1979, Paper D3) because nitrate is added in excess to the slurry. This later discovery was widely used to decipher the respective contribution of nitrification (low acetylene concentration) and denitrification (high acetylene concentration) in nitrous oxide emission until an innovative method using the determination of N_2O isotopomers (i.e. the intra-molecular distribution of ^{15}N in the linear NNO molecule) provided independent and complementary evidence to elucidate the N_2O production mechanisms (Sutka *et al.*, 2006).

The authors concluded that "*the inhibition of N_2O reduction by C_2H_2 may be a useful tool in permitting the relatively sensitive measurement of both denitrification and N_2 fixation occurring either simultaneously or consecutively in laboratory model systems or in situ in the field*". Indeed, they had great intuition and none of the 800 authors who, to date, have quoted either one or both of these papers initiated by Roger Knowles, or used this method in a wide range of environments, including riparian zones, would deny it!

GP

References

Berg, P., Klemedtsson, L. & Rosswall, T. (1982) Inhibitory effect of low partial pressures of acetylene on nitrification. *Soil Biol. & Biochem.* **14**, 301–303.

Hynes, R. & Knowles, R. (1978) Inhibition by acetylene of ammonia oxidation in *Nitrosomonas europea*. *FEMS Microbiology Letters* **4**, 319–321.

Fedorova, R., Milekhina, E. I. & Il'Yukhina, N. I. (1973) Evaluation of the method of "gas metabolism" for detecting extraterrestrial life. Identification of nitrogen-fixing microorganisms. *Izv. Akad. Nauk. SSSR. Ser. Biol.* **6**, 797–806.

Smith, M. S. & Tiedje, J. M. (1979) Phases of denitrification following oxygen depletion in soil. *Soil Biol. & Biochem.* **11**, 261–267.

Sutka, R. L., Ostrom, N. E., Breznak, J. A., Gandhi, H., Pitt, A. J. & Li, F. (2006) Distinguishing nitrous oxide production from nitrification and denitrification on the basis of isotopomer abundances. *Appl. Environ. Microbiol.* **72**(1), 638–644.

Yeomans, J. C. & Beauchamp, E. G. (1978) Limited inhibition of nitrous oxide reduction in soil in the presence of acetylene. *Soil Biol. & Biochem.* **10**, 517–519.

Yoshinari, T. & Knowles, R. (1976) Acetylene inhibition of nitrous oxide reduction by denitrifying bacteria. *Biochem. Biophys. Res. Commun.* **69**(3), 705–710.

Soil Biol. Biochem. Vol. 9, pp. 177 to 183. Pergamon Press 1977. Printed in Great Britain.

ACETYLENE INHIBITION OF NITROUS OXIDE REDUCTION AND MEASUREMENT OF DENITRIFICATION AND NITROGEN FIXATION IN SOIL

T. YOSHINARI, R. HYNES and R. KNOWLES

Department of Microbiology, Macdonald Campus of McGill University, Ste Anne de Bellevue, Quebec H0A 1C0, Canada

(*Accepted* 22 *October* 1976)

Summary—Reduction of N_2O in moist soil was inhibited completely by 10^{-2} atm C_2H_2 and partially by 10^{-5} atm C_2H_2. The effect of C_2H_4 was 10^4 times less than that of C_2H_2. Denitrification of NO_3^- occurred in anaerobically or aerobically incubated waterlogged soil and in anaerobic but not in aerobic moist soil. In the absence of C_2H_2 there was transient accumulation of N_2O. In the presence of C_2H_2 there was stoichiometric conversion of NO_3^- to N_2O. Some kinetics of the reduction of N_2O and of NO_3^- to N_2O are presented. Denitrification of 1 μg added NO_3^--N.g^{-1} could be measured within 1 h. Stoichiometries of production of N_2O from NO_2^- and NO_3^-, respectively, and production of CO_2 attributable to denitrification were consistent with reported energy yields. Reduction of C_2H_2 to C_2H_4 occurred immediately following complete denitrification of added NO_3^-. The incubation of soil in the presence and in the absence of C_2H_2 thus permits assay of both denitrification and N_2 fixation and provides information on the mole fraction of N_2O in the products of denitrification.

INTRODUCTION

The report that C_2H_2 inhibits the reduction of N_2O during denitrification (Federova *et al.*, 1973) was confirmed by pure culture studies of denitrifying bacteria (Yoshinari and Knowles, 1976; Balderston *et al.*, 1976). Cell suspensions of *Pseudomonas perfectomarinus*, *P. aeruginosa* and *Micrococcus* (*Paracoccus*) *denitrificans* in the presence of 0.1 atm C_2H_2 converted both NO_3^- and NO_2^- stoichiometrically to N_2O with no significant change in the rates of NO_3^- reduction (Yoshinari and Knowles, 1976). In cell suspensions of *P. perfectomarinus*, the inhibition of N_2O reduction by C_2H_2 was reversible and C_2H_4 did not inhibit N_2O reduction (Balderston *et al.*, 1976).

There is no known effect of C_2H_2 on the nitrate- and nitrite-reducing pathways. Azide was shown to inhibit N_2O reduction (e.g., Matsubara and Mori, 1968) but this was equivocal because of the known effect of azide on specific cytochrome systems.

Previous direct assays of denitrification were as follows:

(a) Tracer studies with ^{15}N-labelled compounds (Cady and Bartholemew, 1960; Hauck, *et al.*, 1958; Nömmik, 1956; Starr, *et al.*, 1974). This involves mass spectrometric analysis and is thus relatively insensitive.

(b) Tracer study with ^{13}N-labelled NO_3^- (Gersberg *et al.*, 1976) which involves experimentation adjacent to elaborate cyclotron facilities.

(c) Gas chromatographic analysis (Barbaree and Payne, 1967; Stefanson, 1973). Because of the occlusion of N_2 it is difficult to measure the rate of denitrification in most systems, even when the air N_2 is replaced with an inert gas such as He or Ar.

(d) Gas chromatographic measurement of N_2O reduction (Garcia, 1974). In this case there may be problems with solution of N_2O in the liquid phase and with the measurement of small decrements in N_2O in the presence of relatively large concentrations of N_2O.

Knowles and Denike (1974) studied the effects of NH_4^+, NO_2^- and NO_3^- on C_2H_2 reduction (N_2 fixation) in soil to which glucose was added. However, they failed to detect other than transient accumulation of N_2O since they used successive short periods of exposure to C_2H_2 to assay C_2H_2 reduction rates.

This paper describes the measurement of both denitrification (by C_2H_2 inhibition of N_2O reduction) and N_2 fixation (by reduction of C_2H_2 to C_2H_4) in a soil under various laboratory conditions.

MATERIALS AND METHODS

The characteristics of the St. Bernard sandy loam soil used for the present study were as described previously (Brouzes, *et al.*, 1971; Knowles and Denike, 1974).

Ten-gram portions of moist soil (8 g dry weight) were placed in 50-ml Erlenmeyer flasks. The required amounts of $NaNO_3$, $NaNO_2$, NH_4Cl and glucose dissolved in 1 ml distilled water were added. Final moisture content was 35% of maximum water holding capacity (referred to as moist). The flasks were closed with serum stoppers (Subaseal, England) and evacuated and filled to 1 atm three times with He. Appropriate amounts of C_2H_2, C_2H_4 and N_2O were added by means of a syringe through the stoppers after removing an equivalent volume of He. Flasks were incubated in the dark at 24°C.

The soil contained a small amount of endogenous nitrate (Knowles and Denike, 1974). When it was desired to remove this, the flasks with soil were preincubated with the addition of 0.5 ml distilled water under a He atmosphere for 48 h at 24°C. The flasks were then evacuated and refilled with He, and appropriate amounts of substrates in 0.5 ml distilled water were added by syringe through the stoppers.

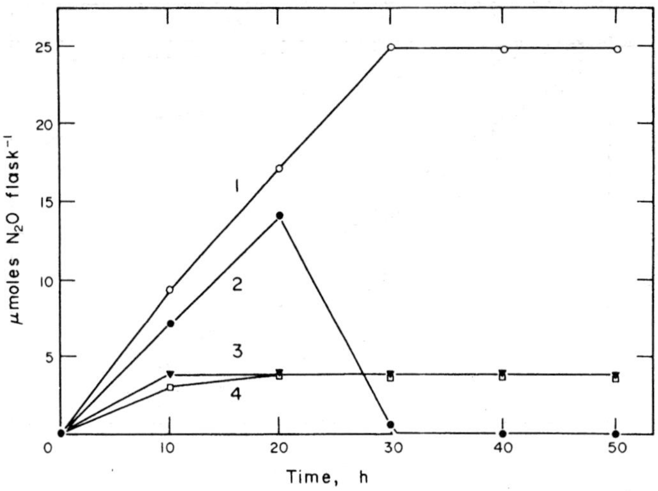

Fig. 1. Production of N_2O by anaerobic soil which was supplemented with glucose (0.1% w/w), NO_3^- (40 μmoles.flask^{-1}) and 0.1 atm C_2H_2 as follows: curve 1, glucose, NO_3^- and C_2H_2; curve 2, glucose and NO_3^-; curve 3, glucose and C_2H_2; curve 4, C_2H_2 alone.

In experiments with waterlogged soil, the total water added to each flask was 5 ml, which was sufficient to cover the soil surface with water. The headspace was ambient air.

At appropriate intervals, 0.2 ml of the gas phase in each flask was removed for gas chromatographic analysis by using a 1 ml syringe with Mininert valve (Precision Sampling Corp., Baton Rouge, Louisiana). Conditions for the measurements of N_2O, CO_2 and C_2H_4 by g.c. were as previously described (Knowles and Denike, 1974).

Calibration for N_2O analyses was carried out by the above procedure except that the flasks with soil were previously sterilized and the water content maintained approximately the same as in the experimental flasks. To these flasks, different amounts of N_2O and 5 ml C_2H_2 were injected. Thus, both the loss due to frequent sample removal and possible N_2O loss by diffusion through the stopper could be corrected.

Calibration for CO_2 and C_2H_4 was carried out by preparing flasks of standard gases to appropriate concentrations. Data are the means of triplicate flasks, except for the kinetic studies in which duplicates were used, and are expressed either as μmoles.flask^{-1} or as μg N.g^{-1} on a dry weight basis (1 μmole.flask^{-1} is equivalent to 1.7 μg N.g^{-1}). The average coefficient of variation was 5%.

RESULTS

In the presence of NO_3^- (40 μmoles flask^{-1}), glucose (0.1%, w/w) and C_2H_2 (0.1 atm)(Fig. 1, curve 1) the formation of N_2O in the moist soil was complete in 30 h with no subsequent reduction of the N_2O,

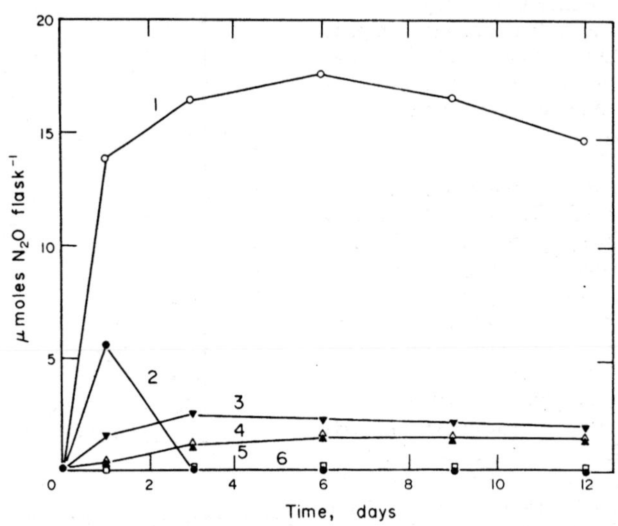

Fig. 2. Production of N_2O by soil supplemented with glucose (0.1% w/w), NO_3^- (40 μmoles.flask^{-1}), NH_4^+ (100 μmoles.flask^{-1}) and 0.1 atm C_2H_2 in air under the following conditions: curve 1, glucose, NH_4^+, NO_3^-, C_2H_2, waterlogged; curve 2, glucose, NH_4^+, NO_3^-, waterlogged; curve 3, glucose, NH_4^+, C_2H_2, waterlogged; curve 4, NH_4^+, C_2H_2, waterlogged; curve 5, C_2H_2, waterlogged; curve 6, NH_4^+, C_2H_2, moist.

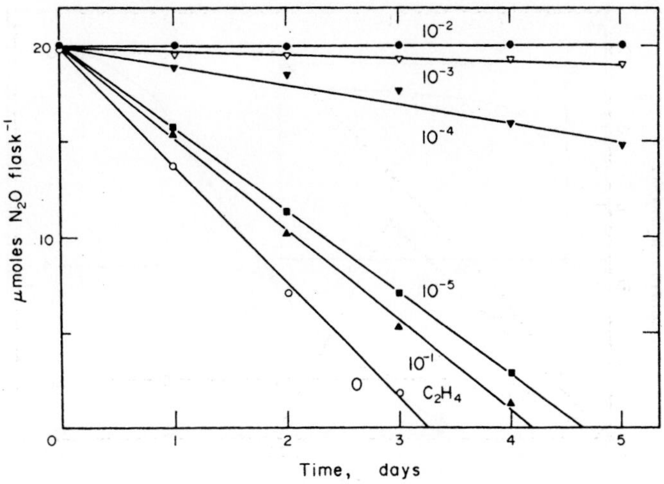

Fig. 3. Reduction of N_2O by anaerobic moist soil in the presence of the indicated concentrations (from 0 to 10^{-2} atm) of C_2H_2 and of 10^{-1} atm C_2H_4. The soil was preincubated for 48 h anaerobically to reduce endogenous NO_3^-.

thus suggesting that 0.1 atm C_2H_2 completely inhibited N_2O reduction. In the absence of C_2H_2 (curve 2) the rate of N_2O production was comparable to that of curve 1 until 20 h but the N_2O was then completely reduced in the next 10 h. This was consistent with N_2O being a transient intermediate in denitrification under these conditions. Curve 4 shows N_2O production (4 μmoles) from endogenous soil NO_3^- in the presence of C_2H_2. The presence of glucose (curve 3) slightly increased the rate of denitrification of this endogenous NO_3^-.

In a similar experiment (shown in Fig. 2) all flasks contained an initially aerobic atmosphere, and all soil samples were waterlogged except for those of curve 6 which were moist. In soil which received NH_4^+, NO_3^-, glucose and C_2H_2 (curve 1) approximately 88% of the NO_3^-–N was converted to N_2O by day 6 and the N_2O subsequently declined, showing that the C_2H_2 inhibition of N_2O reduction was not complete under these conditions. In the absence of C_2H_2 (curve 2) there was transient accumulation of N_2O (only 30% of the added NO_3^-–N) after 1 day and by 3 days the N_2O was completely reduced.

In the presence of C_2H_2 the endogenous NO_3^- was converted to N_2O, the pattern being unaffected by the presence or absence of NH_4^+ (Fig. 2, curves 4 and 5). The presence of NH_4^+, glucose and C_2H_2 (Fig. 2, curve 3) allowed greater production of N_2O. In the moist soil incubated aerobically (curve 6) there was no evidence of denitrification of the endogenous NO_3^-.

The reduction of N_2O (20 μmoles.flask^{-1}) by anaerobic moist soil with no glucose supplement, and in the presence of various concentrations of C_2H_2 is shown in Fig. 3. Reduction of N_2O was inhibited completely by 10^{-2} atm C_2H_2 but only partially by 10^{-5} atm C_2H_2. The inhibitory effect of 10^{-1} atm C_2H_4 was similar to that of 10^{-5} atm C_2H_2. Thus C_2H_4 appears to be 10^4 times less inhibitory than C_2H_2. It may be noted that for each curve the rate of N_2O reduction was roughly constant throughout the experiment.

Various concentrations of N_2O were reduced by anaerobic moist soil supplemented with glucose in the absence of C_2H_2 (Fig. 4). After some initial irregularity, each curve was linear (as also in Fig. 3) suggesting a zero order reaction, but the rates of reduction in separate samples increased somewhat with increasing initial N_2O concentration (see inset). The apparent K_m calculated on the basis of the N_2O concentrations at 5 h and the slopes of the lines after 3 h was of the order of 3–5 μmoles N_2O.flask^{-1} (10–15 μg N_2O–N.g^{-1}). It is to be noted that there was no lag in N_2O reduction in either Figs. 3 or 4.

Various concentrations of NO_3^- were reduced to N_2O by moist anaerobic soil supplemented with 0.1% glucose in the presence of 0.1 atm C_2H_2 (Fig. 5). The

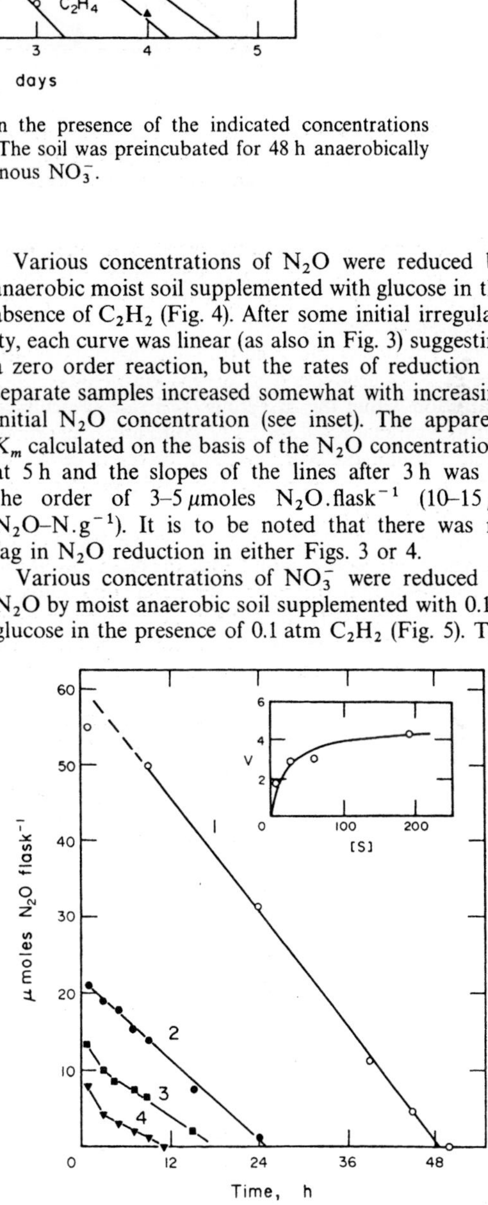

Fig. 4. Reduction of N_2O by anaerobic moist soil which was preincubated 48 h anaerobically to reduce endogenous NO_3^- and then supplemented with 0.1% glucose and the initial N_2O concentrations shown (8 to 60 μmoles.flask^{-1}). The inset shows a plot of v (μg N_2O–N.g^{-1}.h^{-1}) for the period after 3 h and [S] (μg N_2O–N.g^{-1}) at 5 h.

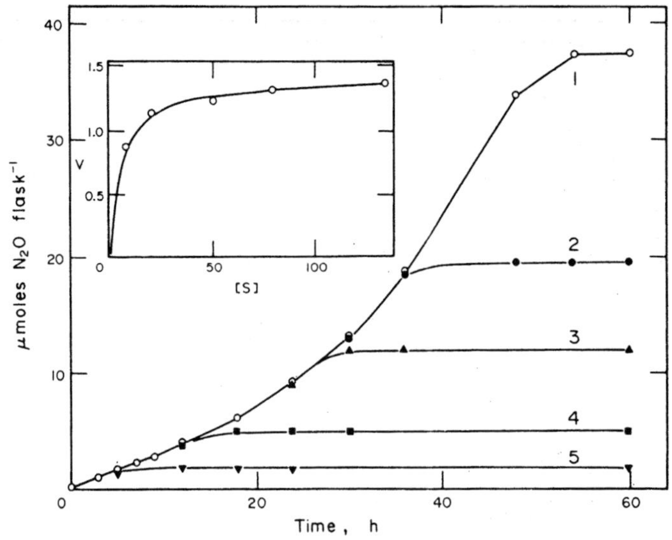

Fig. 5. Production of N_2O by anaerobic moist soil supplemented with 0.1% glucose and 0.1 atm C_2H_2. The initial NO_3^- concentrations in curves 1 to 5 were 71.4, 41.0, 24.0, 7.2 and 0 μmoles.flask^{-1}, respectively. The inset shows a plot of v (μg N_2O–N.g^{-1}.h^{-1}) and [S] (μg NO_3^-–N.g^{-1}) during the first hour of incubation.

production of N_2O ceased at a level corresponding to the amount of NO_3^- added. The inset shows the rate of N_2O production as a function of NO_3^- concentration within the first hour. The rate of denitrification was unaffected by NO_3^- concentrations above about 11 μmoles.flask^{-1} (20 μg N.g^{-1}) and the apparent K_m based on these data was 6 μg NO_3^-–N.g^{-1}.

A calibration curve for N_2O obtained by injecting known amounts of N_2O into flasks containing sterile soil is shown in Fig. 6. Also shown are the amounts of N_2O formed from the different quantities of added NO_3^- during 60 h of incubation in the presence of C_2H_2 (data from Fig. 5). The two lines are parallel and the intercept of the NO_3^- line on the horizontal axis represents the amount of endogenous NO_3^- in the soil. This clearly demonstrates the stoichiometric conversion of NO_3^- to N_2O in the presence of 0.1 atm C_2H_2.

The production of N_2O from various low concentrations of NO_3^- in soil not supplemented with glucose (Fig. 7) showed a pattern similar to that of Fig. 5. The inset shows that the rate of N_2O production during the first hour of incubation was roughly constant at all NO_3^- concentrations above about 1.2 μmoles.flask^{-1} (2 μg N.g^{-1}) and the apparent K_m based on these data was about 0.7 μg NO_3^-–N.g^{-1}. Each amount of NO_3^- was again converted quantitatively to N_2O, though this conversion required about 20 h for the highest amount.

The production of CO_2 from 7 μmoles NO_2^- or NO_3^- by moist anaerobic soil in the presence of C_2H_2

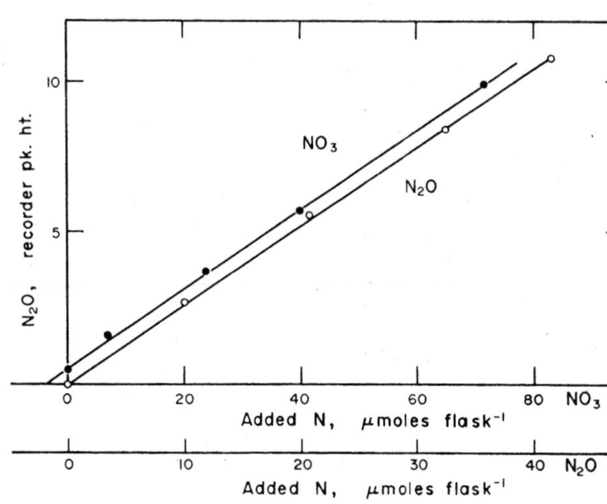

Fig. 6. N_2O detected (in arbitrary recorder peak height units) after the addition of the different amounts of NO_3^- in the experiment of Fig. 5, and after the addition of different amounts of N_2O to flasks containing a similar but sterilized soil (N_2O calibration).

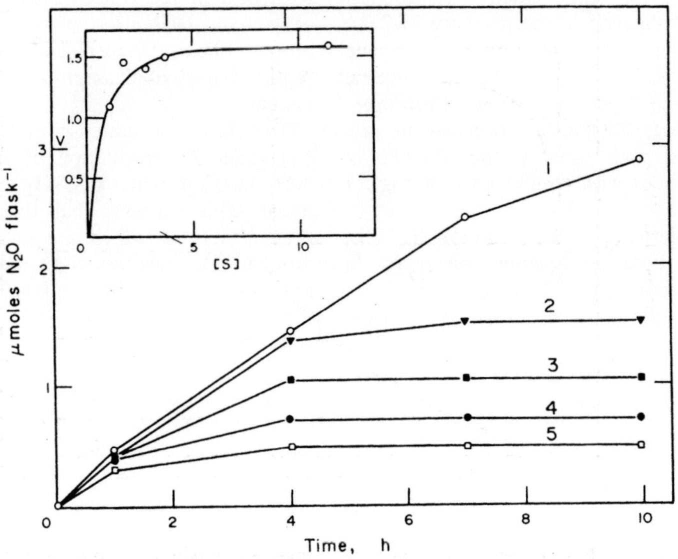

Fig. 7. Production of N_2O by anaerobic moist soil which was pre-incubated 48 hours anaerobically to reduce endogenous NO_3^- and then supplemented with 0.1 atm C_2H_2 and different NO_3^- concentrations, represented by curves 1 to 5, of 7.0, 3.0, 2.0, 1.4, and 1.0 μmole.flask^{-1}, respectively. The inset shows a plot of v (μg N_2O-N.g^{-1}.h^{-1}) and $[S]$ (μg NO_3^--N.g^{-1}) during the first hour of incubation.

and in the presence or absence of 0.1% glucose is shown in Fig. 8. Endogenous production of CO_2 by samples not receiving NO_2^- or NO_3^- was subtracted and the data shown thus represent net CO_2 production (ΔCO_2) attributable to denitrification. When the soil was amended with glucose, the rates of N_2O production with NO_3^- (curve b1) and NO_2^- (curve b3) were approximately the same, whereas the rate of CO_2 production with NO_3^- (a1) was twice that with NO_2^- (a3). In the absence of glucose the rate of N_2O production with NO_3^- (curve b2) was only slightly slower than that with NO_2^- (curve b4), but the rate of CO_2 production with NO_3^- (a2) was again about twice that with NO_2^- (a4).

Both denitrification and N_2 fixation (C_2H_2 reduction) occurred in anaerobic moist soil in the presence of 0.1 atm C_2H_2 and various concentrations of glucose (Fig. 9). Each soil sample was previously incubated under He for 48 h to remove endogenous NO_3^-. Rates of N_2O production (Fig. 9B) were identical in

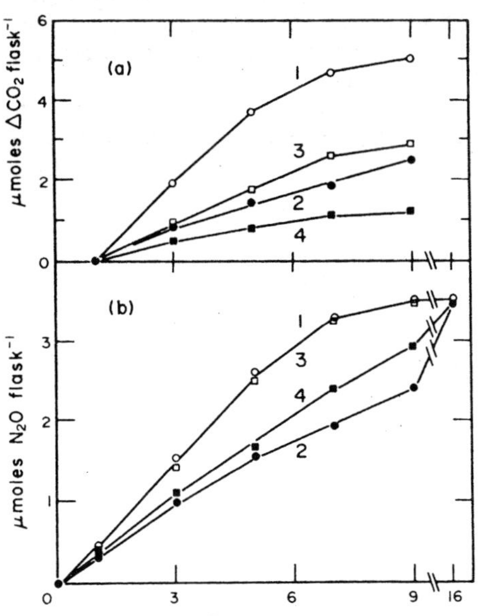

Fig. 8. Production of N_2O and net production of CO_2 attributable to denitrification (i.e. CO_2 production by soil not supplemented with NO_2^- or NO_3^- was subtracted) in anaerobic moist soil in the presence of 0.1 atm C_2H_2, glucose (0.1%) and NO_2^- or NO_3^- (7 μmoles.flask^{-1}) as follows: curve 1, glucose and NO_3^-; curve 2, NO_3^-; curve 3, glucose and NO_2^-; curve 4, NO_2^-.

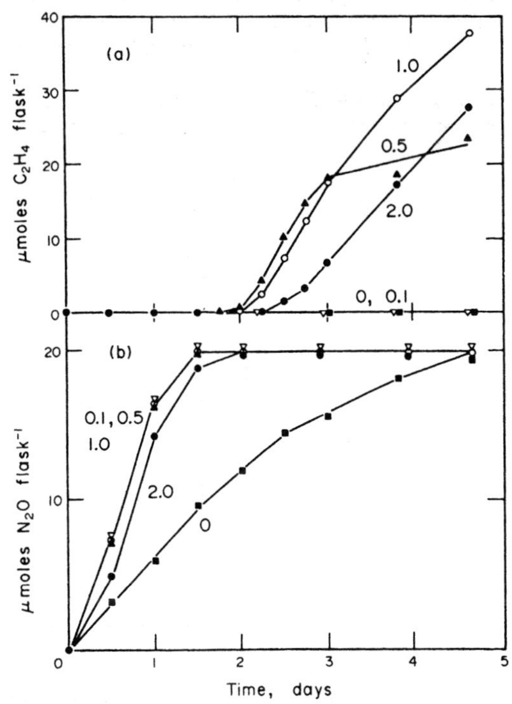

Fig. 9. Production of C_2H_4 and N_2O by anaerobic moist soil in the presence of 0.1 atm C_2H_2, 40 μmoles NO_3^-.flask^{-1} and 0, 0.1, 0.5, 1.0 and 2.0% glucose as indicated.

soil receiving 0.1, 0.5 and 1.0% glucose. Production of N_2O was slightly delayed in the presence of 2.0% glucose and approximately half as rapid in the absence of glucose, under which conditions complete reduction of NO_3^- to N_2O took about three times as long. Furthermore, the gradual decrease in rate of N_2O production with time in the absence of glucose is in agreement with similar data in Figs. 7 and 8.

N_2 fixation, as shown by the reduction of C_2H_2 to C_2H_4 (Fig. 9a), began only after conversion of NO_3^- to N_2O was completed. Production of C_2H_4 was not observed in the absence of glucose or in the presence of 0.1% glucose. Most rapid C_2H_4 production occurred with 0.5 or 1.0% glucose and was relatively delayed with 2.0% glucose.

DISCUSSION

The inhibition of N_2O reduction by C_2H_2 in pure cultures of denitrifying bacteria (Balderston *et al.*, 1976; Yoshinari and Knowles, 1976) was confirmed to occur also in the soil studied. The moist soil system was very sensitive even to 10^{-3} atm C_2H_2, but the inhibition of N_2O reduction in waterlogged soil appeared to be incomplete at 10^{-1} atm C_2H_2. It is possible that the degree of inhibition may depend on the numbers of denitrifiers and also on the resistance to gaseous diffusion.

It was shown that denitrification in soil could be easily and sensitively monitored by measuring the production of N_2O in the presence of C_2H_2. Inhibition by C_2H_4 was 10^4 less than that by C_2H_2 in agreement with observations of Balderston *et al.* (1976).

Soil organic C supported denitrification of NO_3^- to N_2O but the rate somewhat decreased with time. The addition of glucose roughly doubled the rate of denitrification which then remained constant or increased possibly due to proliferation of denitrifying bacteria. Bowman and Focht (1974) reported that 0.5% glucose or more in the liquid phase supported maximum NO_3^- disappearance. In the present study, maximum rate of denitrification was given by 0.1% w/w glucose (0.26% in the liquid phase).

Assimilation of added N was minimal and there was stoichiometric conversion of NO_3^- or NO_2^- to N_2O. The slightly incomplete inhibition of N_2O reduction in the waterlogged soil incubated aerobically made evaluation of the extent of assimilation more difficult. Wijler and Delwiche (1956) also reported negligible assimilation in their denitrifying system.

The curves of reduction of N_2O in the absence of C_2H_2 and of production of N_2O from NO_2^- and NO_3^- in the presence of C_2H_2 appear to be frequently linear. This suggests that these processes are zero order reactions which are saturated with substrate down to relatively low N_2O and NO_3^- concentrations in agreement with Nömmik (1956), Focht (1974) and others. There is no explanation at present for the fact that increasing initial N_2O concentrations were reduced at increasing rates, each of which, however, remained constant during the experiment. The apparent K_m for N_2O reduction calculated from these data was 10–15 μg N_2O–N.g^{-1}. The apparent K_m values for NO_3^- reduction to N_2O in the presence and absence of glucose were 6 and 0.7 μg NO_3^-–N.g^{-1}, respectively.

Comparison with other reports is difficult because of the differing conditions and soil:water ratios used. Table 1 attempts such a comparison of apparent K_m values, all recalculated in terms of dissolved concentrations of N_2O–N or NO_3^-–N. As discussed by Bowman and Focht (1974) and Kohl *et al.* (1976) the availability of organic C makes interpretation of soil denitrification kinetics difficult. The present value for N_2O reduction is lower than that of Garcia (1974) and most of the values for NO_3^- reduction are lower than that of Bowman and Focht (1974). There is some suggestion in the present study that glucose addition raised the apparent K_m value for NO_3^- reduction.

The maximum rates of N_2O reduction observed were in general equal to or slightly greater than the maximum rates of production of N_2O from NO_3^-, in agreement with Nömmik (1956) and Focht (1974).

The data on the production of CO_2 and N_2O in the presence of NO_3^- or NO_2^- permit the derivation of the following equations:

$$2NO_3^- + 2(CH_2O) + 2H^+ \rightarrow 2CO_2 + N_2O + 3H_2O, \quad (1)$$

$$2NO_2^- + (CH_2O) + 2H^+ \rightarrow CO_2 + N_2O + 2H_2O. \quad (2)$$

Furthermore, the data suggest that electrons are

Table 1. Comparison of reported apparent K_m values for denitrification and dissimilatory nitrate reduction. All values were recalculated as dissolved concentrations

Authors	System	Glucose added	Dissolved μg N.ml^{-1} liquid phase
N_2O reduction			
Garcia, 1974	Soil:H_2O (1:1)	–	60
Present study	Soil:H_2O (1:0.38)	+	0.4–0.6
NO_3^- reduction (dissimilatory)			
Bowman and Focht, 1974	Soil:H_2O (1:1)	+	170
Kohl *et al.*, 1976	Soil:H_2O (1:1.7)	–	2.4–29
Fewson and Nicholas, 1961	*Pseudomonas aeruginosa* cell-free extract		7.0
Nason, 1962	*Escherichia coli* cell-free extract		0.22
Present study	Soil:H_2O (1:0.38)	+	17
		–	1.8

transported at equal rates to NO_3^- and to NO_2^-, but that NO_3^- permits the metabolism of twice as much carbohydrate as does NO_2^-. This is consistent with observed and calculated energy yields reported for *Pseudomonas denitrificans* (Koike and Hattori, 1975).

The reduction of C_2H_2 occurred only after denitrification had ceased (c.f. Knowles and Denike, 1974) and was affected by different glucose concentrations as described by O'Toole and Knowles (1973). Although the two processes occurred only successively in this system it is conceivable that they may occur simultaneously in certain natural systems (such as the rhizosphere) where NO_3^- gradients are likely to exist.

It is concluded that the inhibition of N_2O reduction by C_2H_2 may be a useful tool in permitting the relatively sensitive measurement of both denitrification and N_2 fixation occurring either simultaneously or consecutively in laboratory model systems or *in situ* in the field. The production of N_2O in the absence and in the presence of C_2H_2 provides information on the probable mole fraction of N_2O in the products of denitrification.

Acknowledgements—This work was supported by a grant from the National Research Council of Canada. One of us (R.H.) was supported by a scholarship from the National Research Council of Canada.

REFERENCES

BALDERSTON, W. L., SHERR, B. and PAYNE, W. J. (1976) Blockage by acetylene of nitrous oxide reduction in *Pseudomonas perfectomarinus*. *Appl. Environ. Microbiol.* **31,** 504–508.

BARBAREE, J. M. and PAYNE, W. J. (1967) Products of denitrification by a marine bacterium as revealed by gas chromatography. *Mar. Biol. (Berlin)* **1,** 136–139.

BOWMAN, R. A. and FOCHT, D. D. (1974) The influence of glucose and nitrate concentrations upon denitrification rates in a sandy soil. *Soil Biol. Biochem.* **6,** 297–301.

BROUZES, R., MAYFIELD, C. I. and KNOWLES, R. (1971) Effect of oxygen partial pressure on nitrogen fixation and acetylene reduction in a sandy loam soil amended with glucose. In *Biological Nitrogen Fixation in Natural and Agricultural Habitats* (T. A. Lie and E. G. Mulder, Eds.), pp. 481–494. *Pl. Soil, Special Volume.*

CADY, F. B. and BARTHOLEMEW, W. V. (1960) Sequential products of anaerobic denitrification in Norfolk soil material. *Proc. Soil Sci. Soc. Am.* **24.** 477–482.

FEDOROVA, R. I., MILEKHINA, E. I. and IL'YUKHINA, N. I. (1973) Evaluation of the method of "gas metabolism" for detecting extraterrestrial life. Identification of nitrogen-fixing microorganisms. *Izv. Akad. Nauk, SSSR, Ser. Biol.* 1973(6), 797–806.

FEWSON, C. A. and NICHOLAS, D. J. D. (1961) Nitrate reductase from *Pseudomonas aeruginosa*. *Biochim. biophys. Acta* **49,** 335–349.

FOCHT, D. D. (1974) The effect of temperature, pH, and aeration on the production of nitrous oxide and gaseous nitrogen—a zero-order kinetic model. *Soil Sci.* **118,** 173–179.

GARCIA, J.-L. (1974) Réduction de l'oxyde nitreux dans les sols de rizières du Sénégal: mesure de l'activité dénitrifiante. *Soil Biol. Biochem.* **6.** 79–84.

GERSBERG, R. KROHN, K. PEEK, N. and GOLDMAN, C. R. (1976) Denitrification studies with ^{13}N-labeled nitrate. *Science* **192.** 1229–1231.

HARDY, R. W. F., BURNS, R. C. and HOLSTEN, R. D. (1973) Application of the acetylene-ethylene assay for measurement of nitrogen fixation. *Soil Biol. Biochem.* **5,** 47–81.

HAUCK, R. D., MELSTED, S. W. and YANKWICH, P. E. (1958) Use of N-isotope distribution in nitrogen gas in the study of denitrification. *Soil Sci.* **86,** 287–291.

KNOWLES, R. and DENIKE, D. (1974) Effect of ammonium-, nitrite- and nitrate-nitrogen on anaerobic nitrogenase activity in soil. *Soil Biol. Biochem.* **6,** 353–358.

KOHL, D. H., VITHAYATHIL, F., WHITLOW, P., SHEARER, G. and CHIEN, S. H. (1976) Denitrification kinetics in soil systems: the significance of good fits of data to mathematical forms. *J. Soil. Sci. Soc. Am.* **40,** 249–253.

KOIKE, I. and HATTORI, A. (1975) Energy yield of denitrification: an estimate from growth yield in continuous culture of *Pseudomonas denitrificans* under nitrate-, nitrite- and nitrous oxide-limited conditions. *J. Gen. Microbiol.* **88,** 11–19.

MATSUBARA, T. and MORI, T. (1968) Studies on denitrification. IX. Nitrous oxide, its production and reduction to nitrogen. *J. Biochem. (Tokyo)* **64,** 863–871.

NASON, A. (1962) Enzymatic pathways of nitrate, nitrite and hydroxylamine metabolism. *Bact. Rev.* **26,** 16–41.

NÖMMIK, H. (1956) Investigation on denitrification in soil. *Acta Agric. Scand.* **6,** 195–228.

O'TOOLE, P. and KNOWLES, R. (1973) Efficiency of acetylene reduction (nitrogen fixation) in soil: effect of type and concentration of available carbohydrate. *Soil Biol. Biochem.* **5,** 789–797.

STARR, J. L., BROADBENT, F. E. and NIELSEN, D. R. (1974) Nitrogen transformations during continuous leaching. *Proc. Soil Sci. Soc. Am.* **38,** 283–289.

STEFANSON, R. C. (1973) Evolution patterns of nitrous oxide and nitrogen in sealed soil-plant systems. *Soil Biol. Biochem.* **5,** 167–169.

WIJLER, J. and DELWICHE, C. C. (1954) Investigation on the denitrifying process in soil. *Pl. Soil* **5,** 155–169.

YOSHINARI, T. and KNOWLES, R. (1976) Acetylene inhibition of nitrous oxide reduction by denitrifying bacteria. *Biochem. biophys. Res. Commun.* **69,** 705–710.

Commentary on Paper D3

Smith, M. S. & Tiedje, J. M. (1979) Phases of denitrification following oxygen depletion in soil.

This 30-year old study remains as relevant today as it ever was. Indeed, the results presented in this paper highlighted several physiological particularities of the denitrifying community which have led to the widespread use of this method to estimate denitrification activity potential, especially in riparian zones.

The authors conducted a set of laboratory experiments to determine the short-term (up to 10 hours) evolution of soil denitrification rates under anaerobiosis and in the presence of excess nitrate. They used the acetylene method (Yoshinari & Knowles, 1976) to measure the denitrification activity during the soil incubations. Monitoring the rate of denitrification after setting up anaerobic conditions, they measured two distinct phases in the microbial activity. The first (Phase I), which occurred for less than an hour, was related to the de-repression of existing denitrifying enzymes in the bacteria community after the existing free oxygen had been consumed. The second phase (Phase II), which occurred between 4 and 8 hours after anaerobic conditions were set up, was related to the newly synthesised enzyme by the existing denitrifying community.

They tested the hypothesis that Phase I was related to *in situ* denitrification capacity; they conducted another experiment in which they sampled soil before and after irrigation and measured the denitrification rate evolution. They found that denitrification rate measured in Phase I of the sample taken after irrigation was about twice the rate of the sample taken before. They attributed the increase in Phase I rate following irrigation to the partial de-repression of denitrifying enzyme synthesis by reduced soil aeration. It demonstrated that Phase I responds to short-term change in environmental conditions for denitrification activity. On the contrary, they could not measure any significant difference in Phase II rates between samples taken before and after irrigation.

Therefore, Phase I evaluates the cellular denitrifying enzyme activity present at the sampling time and can reflect immediate environmental changes affecting denitrification activity such as soil moisture and aeration. Phase II provides an evaluation of the potential physiological denitrification capacity of the existing denitrifying community. It represents a more robust picture of the current environmental conditions. Combined with new DNA and mRNA based methods (Philippot &Hallin, 2005), the determination of these two phases constitutes a powerful approach to explore the links between denitrifying community structure and function under environmental changes.

In another experiment they demonstrated that denitrification is measurable in aerobic soils. Since denitrifying enzymes are not constitutive, they concluded that denitrification could occur in anaerobic microsites of well-aerated soil or that denitrifying enzymes may persist in functional form in the presence of oxygen. Their experimental design did not allow them to draw any firm conclusion but it is now well known that both hypotheses were correct. This result provided the grounds for further understanding of the hysteresis nature of denitrification (Groffman & Tiedje, 1988) and the rapid reaction of the denitrifying communities to soil waterlogging conditions due to flood or rain events.

This paper had a tremendous impact on ecologists, biogeochemists and landscape ecologists since it provided a robust and cheap tool to assess denitrification potential activity in many different environments, including riparian zones. Although some studies have in the past over-interpreted the results of this method by amalgamating potential and actual activity, it is now routinely used in combination with other field and laboratory measurements to address a difficult problem: measuring denitrification (Groffman *et al.*, 2006).

GP

References

Groffman, P. M. & Tiedje, J. M. (1988) Denitrification hysteresis during wetting and drying cycles in soil. *J. Soil Sci. Soc. Am.* **52**, 1626–1629.

Groffman, P. M., Altabet, M. A., Bohlke, J. K., Butterbach-Bahl, K., David, M. B., Firestone, M. K., Giblin, A. E., Kana, T. M., Nielsen, L. P. & Voytek, M. A. (2006) Methods for measuring denitrification: diverse approaches to a difficult problem. *Ecol. Appl.* **16**(6), 2091–2122.

Philippot, L. & Hallin, S. (2005) Finding the missing link between diversity and activity using denitrifying bacteria as a model of functional community. *Current Opinion in Microbiology* **8**, 234–239.

Yoshinari, T. & Knowles, R. (1976) Acetylene inhibition of nitrous oxide reduction by denitrifying bacteria. *Biochem. Biophys. Res. Commun.* **69**(3), 705–710.

Soil Biol. Biochem. Vol. 11, pp. 261 to 267
Pergamon Press Ltd 1979. Printed in Great Britain

PHASES OF DENITRIFICATION FOLLOWING OXYGEN DEPLETION IN SOIL

M. SCOTT SMITH* and JAMES M. TIEDJE

Departments of Crop and Soil Sciences and of Microbiology and Public Health, Michigan State University, E. Lansing, MI 48824, U.S.A.

(*Accepted* 28 *November* 1978)

Summary—The short-term response of soil denitrification to reduced aeration was studied using the acetylene inhibition method for the assay of denitrification. Two distinct phases of denitrification rate were observed. An initial constant rate, termed phase I, was not decreased by chloramphenicol, was increased slightly or not at all by organic carbon amendment, and lasted for 1–3 h. Phase I was attributed to the activity of pre-existing denitrifying enzymes in the soil microflora. Following phase I the denitrification rate increased; chloramphenicol inhibited this increase. In soils without organic-C amendment a second linear phase, termed phase II, was attained after 4–8 h of anaerobic incubation. The linearity of this phase was attributed to the full derepression of denitrifying enzyme synthesis by the indigenous population and to the lack of significant growth of denitrifiers. Phase I rate was dependent on the initial or *in situ* aeration state of the soil sample; phase II was not. Therefore, phase I may be more directly related to field denitrification rates.

Denitrification rate changes following water saturation of soils in aerobic atmospheres were also examined. Rates were greatly increased by wetting but only after a lag of several hours. Our interpretation is that following wetting of natural soils, anaerobic or partially anaerobic conditions are established by respiration and reduced O_2 diffusion rate; this first eliminates O_2 inhibition then derepresses the synthesis of denitrifying enzymes. Although denitrifying enzymes are apparently present even in relatively dry soils, their activity is low until O_2 inhibition is eliminated. From this evidence we reason that most N is lost from soils during brief periods beginning a few hours after irrigation or a rainfall.

INTRODUCTION

It is generally believed that denitrification occurs in agricultural soils predominantly under saturated or near-saturated conditions (Focht and Verstraete, 1977). Soil denitrification rates have been shown to increase with added water or reduced aeration (Bremner and Shaw, 1958; Ardakani *et al.*, 1977; Pilot and Patrick, 1972). This is consistent with the known physiology of denitrifiers; O_2 both inhibits denitrifying enzyme activity and represses synthesis of new denitrifying enzymes (Payne, 1973). Nonetheless, there are reports suggesting biological production of nitrogen gases in well-aerated soils (Starr *et al.*, 1974; Broadbent and Clark, 1965; Bremner and Blackmer, 1978), though the significance of N loss under these conditions has not been determined. The close relationship between aeration state and denitrification rate indicates that it is important to characterize the short-term soil response to reduced aeration. Until now, methodological limitations have precluded studying the dynamics of the soil response, that is, how rapidly and to what extent denitrification rates change following the onset of anaerobic or partially anaerobic conditions. Nor has the biological component of this response, the short-term reaction of soil denitrifiers to imposed anaerobiosis, been examined. In this paper we address these questions which are important to the basic understanding, and to the potential control and prediction of denitrification.

* Present address, University of Kentucky, Lexington, KY 40506, U.S.A.

MATERIALS AND METHODS

Soils used were Brookston loam, Spinks loamy sand, Miami sandy loam, and Carlisle muck; they have been described by Smith *et al.* (1978). The maximum water holding capacity, as determined by saturating soil in a filter paper funnel, was 43 ml per 100 g for the Brookston and 36 ml per 100 g for the Spinks and Miami. After collection the soils were stored at 2–4°C without drying. However, for many experiments fresh samples were collected from the same sites immediately before experimentation. All soils were at or below field capacity when collected.

The C_2H_2 inhibition method was used to measure denitrification rates; its validity for measuring short-term rates has been established previously (Smith *et al.*, 1978). The methods used here were similar. Briefly: soil to be assayed was made into a slurry and incubated anaerobically in closed flasks. C_2H_2 (0.1 atm) was added to inhibit the reduction of N_2O to N_2. Nitrate was also added to the soil suspensions, 10 μg NO_3^-–$N \cdot g^{-1}$ (soil fresh wt basis) for short-term assays and repeated additions of 200 $\mu g \cdot g^{-1}$ for longer experiments. Rapid attainment of anaerobiosis for phase I assays was ensured by magnetically stirring the suspensions while evacuating and flushing the flasks; incubation was under a He atmosphere. The reported data are means of at least three replicates. Rate data were determined from linear regressions; also reported is the sample standard deviation of the regression, $s_b t$ 0.05, (Snedecor and Cochran, 1967).

To determine denitrification rates in aerobic atmospheres, soils were sieved (<4 mm) to break up large clods and remove debris. This treatment had little effect on smaller, stable aggregates. Soil (35 g) was gently shaken with 0.1 atm C_2H_2 in a stoppered 50 ml centrifuge tube, then compacted by gently tapping the tube.

Chloramphenicol (0.27 mg·g^{-1}) was used to inhibit protein synthesis in soil suspensions. These experiments were done in Brookston soil and with a mixture of 50 g washed silica sand and 5 g Brookston soil. The sand mixture was incubated anaerobically with succinate and NO_3^- several days before the experiment to develop denitrifying populations.

Pure cultures of denitrifiers were grown aerobically to log phase in nutrient broth (Difco, Detroit, MI). They were harvested on 0.45 μm Millipore filters and washed with 3 vol of phosphate buffer (pH 7.0, 50 mM). The filters with the cells were placed in flasks of nitrate broth. The flasks were then made anaerobic and assayed in the presence of C_2H_2 in the same manner as the soil suspensions. Some cultures were also assayed in nitrate broth with 0.5 g·l^{-1} Na thioglycollate added as a reducing agent. The flasks with thioglycollate were evacuated and flushed with He 16 h before inoculation. The denitrifiers had been isolated from soils and were fully characterized during earlier work in this laboratory (T. N. Gamble, unpublished, M.S. Thesis, Michigan State University, 1976). They were: *Pseudomonas fluorescens* biotype II 72, *P. fluorescens* biotype IV 206, *P. stutzeri* 224, *P. aureofaciens* 59, and *Flavobacterium* sp. 175. Numbers refer to our strain designation.

The concentrations of N_2O were measured by g.c. using ^{63}Ni electron capture and microthermistor detectors (Smith *et al.*, 1978). Most probable numbers of denitrifiers were determined with a microtiter system (Rowe *et al.*, 1977). Disappearance of nitrate and nitrite from nitrate broth was determined with diphenylamine.

RESULTS

Two linear phases of denitrification were observed following the imposition of anaerobiosis on aerobic soil (Fig. 1). Phase I lasted from approximately 15 min, the time for N_2O to reach easily measured concentrations, to between 1 and 3 h. Smith *et al.*, (1978) had shown that this initial period is followed by an increase in rate until a second linear phase, phase II, is established. Some phase II data from our earlier work is included in Fig. 1 (inset) to demonstrate the biphasic nature of denitrification rates. The duration of both phases was quite variable among soils. However, the biphasic pattern was observed consistently with both fresh and stored (moist) soil samples when assayed as slurries.

In some experiments attempts were made to eliminate O_2 more rapidly and completely. These included deaerating the solution and flask before adding the soil, varying the duration of evacuating and flushing the flasks, and agitating the suspensions while evacuating. These procedures did not alter the bilinear pattern indicating that residual O_2 or delayed establishment of equilibrium was not responsible for the observed phases. The biological origin of N_2O production was confirmed by finding that autoclaved soils did not evolve N_2O at a significant rate.

The effect of chloramphenicol on denitrification rate is illustrated in Fig. 2. This compound is an inhibitor of protein synthesis but would not be expected to reduce the activity of enzymes already present. Phase I rate was not reduced, but very slightly increased by chloramphenicol. This small increase (which has been consistently observed) could be due to diversion of carbon supply from protein synthesis to respiratory processes. Of greater significance is the effect on the shift from phase I to phase II. Phase II rates were markedly reduced by chloramphenicol. Inactivation of chloramphenicol by soil binding and microbial decomposition would be expected in soil, so this experiment was also conducted with silica sand mixed with a small quantity of soil. In this case the increased rate from phase I to phase II was almost totally eliminated (Fig. 2). Succinate (1%), a non-fermentable substrate, was added to the sand allowing growth and thus accounting for the lack of a distinct linear phase II in the absence of chloramphenicol.

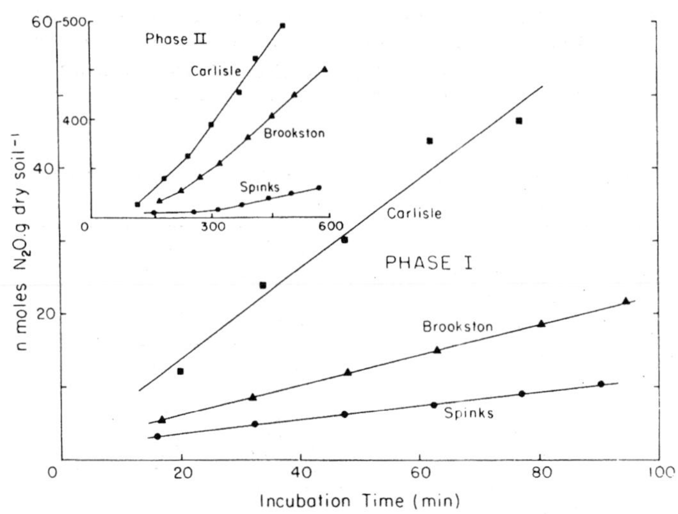

Fig. 1. Evidence for two linear periods of denitrification, shown for three soils, described as phase I and phase II (after 300 min). Points are means of three replicates.

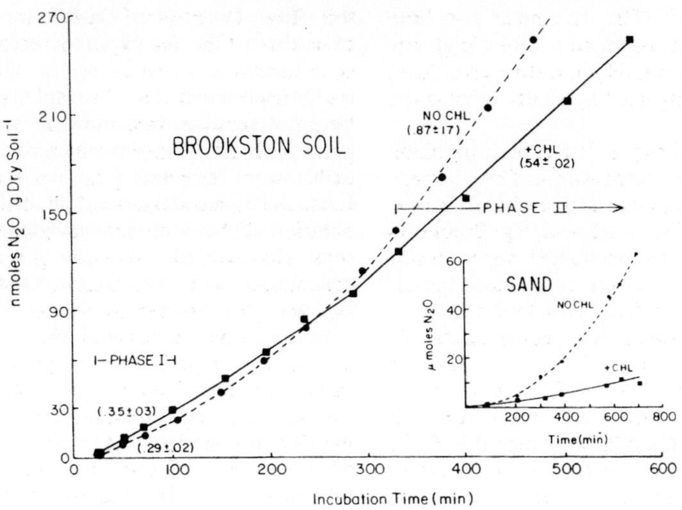

Fig. 2. Effect of chloramphenicol (CHL) on denitrification rates in soil and a sand. Points are the means of three replicates. Numbers in parentheses are rates (nmoles $N_2O \cdot g$ dry $soil^{-1} \cdot min^{-1}$) with 95% confidence limits.

These results indicate that the rate increase following phase I is a result of the synthesis of denitrifying enzymes and suggest that phase I, but not phase II, is a measure of the activity of preexisting denitrifying enzymes.

This conclusion is supported by the effect of glucose additions on denitrification rates. Though the effect was quite dependent on the soil sample, the results in Table 1 are representative of the stimulatory pattern. Phase I rates were not increased by glucose in some soils and only slightly increased in others. Phase I remained linear in glucose-amended soils. Phase II denitrification was often but not always stimulated by glucose. The second linear phase was always of reduced duration and was followed or completely replaced by a logarithmic increase in rate when glucose was added. Without an organic carbon amendment, the denitrification rate remained essentially constant, i.e. phase II persisted, for up to 3 days. It is concluded that during phase I the enzymatic capacity to denitrify is a more important limitation than the supply of electron donor, but that enzyme synthesis and the potential for growth during continued incubation increases the demand for electron donor, making this the more important limiting factor. The data also suggest that a significant increase in number of denitrifiers does not occur in anaerobic soils unless an energy supply is added.

To determine the role of microbial growth in the denitrification rate changes, denitrifiers were enumerated by an m.p.n. procedure after various periods of anaerobic incubation. Increasing numbers could be detected only in soils with added glucose. However, the methods used are not precise enough to detect increases of less than an order of magnitude.

Pure cultures of denitrifiers shifted from aerobic to anaerobic conditions exhibited a distinct lag phase before beginning to denitrify (Fig. 3). Nitrous oxide first appeared between 95 and 205 min after the onset of anaerobiosis; most of the organisms required about 2 h to produce detectable N_2O. *Flavobacterium* sp. 175 grown anaerobically in NO_3^- broth, harvested and assayed in the same manner as the aerobic cultures began producing N_2O immediately. The reducing agent, thioglycollate, did not decrease lag times indicating that neither residual O_2 nor a high initial E_h were delaying denitrification. The lag times correspond quite well with the time of shift from phase

Table 1. Effect of glucose additions on denitrification rate of Brookston soil

Time of anaerobic incubation (h)	Denitrification rate (nmoles $N_2O \cdot g$ dry $soil^{-1} \cdot min^{-1}$)	
	No glucose	Glucose added
0–1½ (Phase I)	0.21 ± 0.03*	0.25 ± 0.03†
6–8	0.56 ± 0.09	0.56 ± 0.08‡
24–27	0.67 ± 0.06	2.82 ± 0.16‡
72–75	0.60 ± 0.07	1.43 ± 0.09‡

* $S_b t$ 0.05.
† 0.15% glucose added.
‡ 0.1% glucose added.

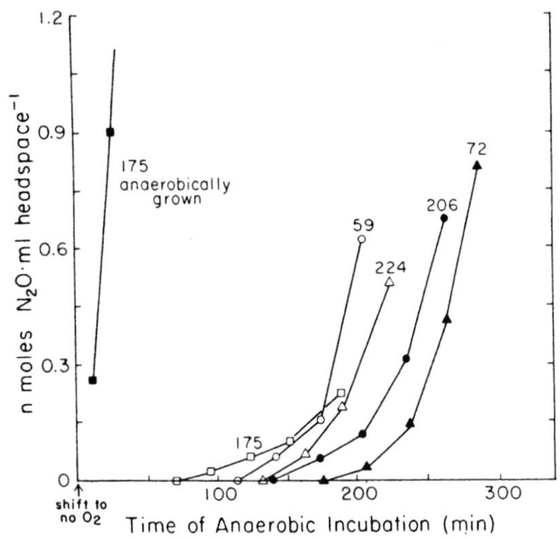

Fig. 3. Derepression of denitrifying enzyme synthesis following shift of five aerobically grown denitrifier strains to anaerobic conditions. Anaerobically grown strain 175 is a control for prior derepression.

I to phase II in anaerobic soils and further support the conclusion that the phase I rate is unaffected by *de novo* enzyme synthesis.

Since phase I is a measure of the activity of *in situ* denitrifying enzymes it should respond to altered soil aeration in the field. This was demonstrated by sampling the Miami soil, planted to corn, immediately before and after 4 h of irrigation. Within 1 h of sampling the soils were made anaerobic and phase I and II rates were determined. The phase I rate of the sample taken after irrigation was approximately twice that of the sample taken before (Fig. 4). However, there was no difference in phase II rates. This comparison serves as a control to show that there were not significant chemical or physical differences between the two samples which could account for the difference in the phase I rates. The increase in phase I rate following irrigation can be attributed to the partial derepression of denitrifying enzyme synthesis caused by reduced soil aeration. It is interesting that a significant amount of phase I denitrification capacity was detected even in the sample taken before irrigation. This soil was at 16% water content and so may be considered reasonably well aerated.

The relationship between phase I rate and soil aeration was further established by adding various amounts of water to 50 g Spinks soil in 125 ml flasks. The flasks were left open for 16 h before phase I rates were determined. Denitrifying activity was present in all the treatments but was greatly increased at water contents above 25% (Table 2).

Nitrous oxide evolution from soils in an aerobic atmosphere, with 0.1 atm C_2H_2 added could also be observed (Fig. 5). Miami soil at 18% and the same soil adjusted to 33% water content (approximately saturation) just before the assay were compared for patterns of N_2O evolution. A very low rate of N_2O evolution, 11 fmoles $N_2O \cdot ml$ headspace$^{-1} \cdot min^{-1}$, was observed from the drier soil and initially from the moistened soil. Approximately 6 h after wetting, the rate began to increase rapidly. Similar results were observed with several other soil samples. The lag period ranged from 4 to 7 h. The O_2/N_2 ratio in the headspace, as determined by g.c. analysis, was reduced by less than 5% during these experiments.

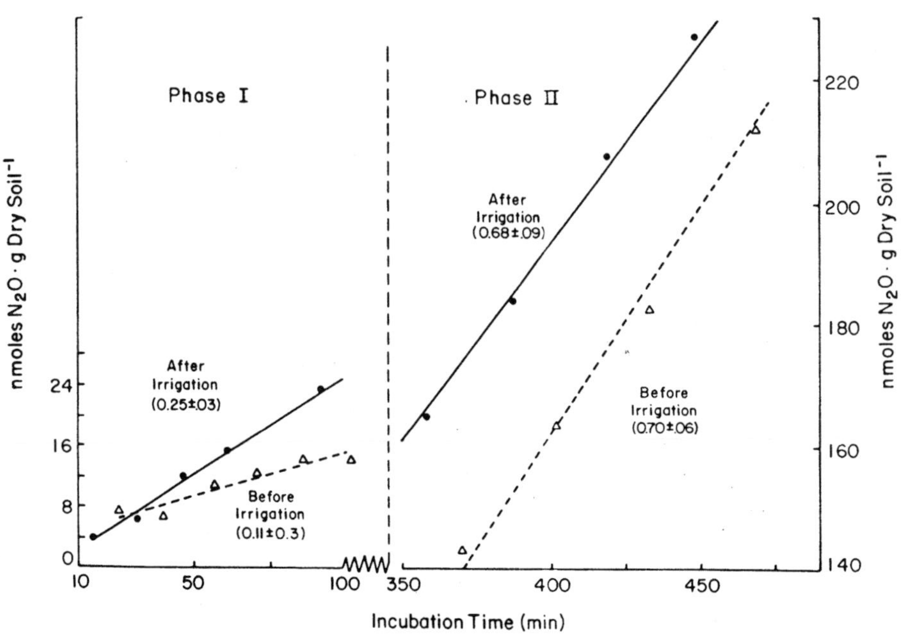

Fig. 4. Phase I and phase II denitrification rates of a Miami soil sampled before and after 4 h of irrigation. Points are the means of three replicates. Numbers in parentheses are rates (nmoles $N_2O \cdot g$ dry soil$^{-1} \cdot min^{-1}$) with 95% confidence limits.

Table 2. Phase I denitrification rate of Spinks loamy sand preincubated aerobically at varying water contents

Gravimetric water content (%)	Denitrification rate (nmoles $N_2O \cdot g$ dry soil$^{-1} \cdot$min^{-1})
6	0.06 ± 0.02*
16	0.06 ± 0.01
27	0.10 ± 0.02
37	0.16 ± 0.06
58	0.38 ± 0.09

* $\pm S_b t$ 0.05.

DISCUSSION

We conclude that the following sequence of events occurs when a soil or soil microsite becomes anaerobic: first, the O_2 inhibition of native denitrifying enzymes is removed, which results in an initial linear phase of denitrification (phase I). After a lag of at least 1 h newly synthesized denitrifying enzymes become functional and the rate increases. At 4–8 h all denitrifiers are fully derepressed so each cell has attained its maximum capacity to denitrify (phase II). At this point denitrification rate can increase only by an increase in number of cells; significant growth will occur only when the supply of available organic carbon is large. These events and their relationship to the phases observed are summarized in Table 3. It should be noted that phases IIa and IIb were referred to as phases II and III in an earlier brief summary (Tiedje *et al.*, 1978). The earlier designation has been changed to avoid the implication that there are always three distinct phases.

The evidence for this explanation of the phases can be summarized as follows: first, various methods of rapidly removing O_2 had no effect, indicating that residual O_2 could not account for the lower initial rate. Chloramphenicol reduced phase II denitrification rates but not phase I, suggesting that the increase was due to enzyme synthesis. The small initial stimulation by glucose relative to subsequent stimulation demonstrates that the capacity to utilize electron donor increases after phase I. Pure cultures of soil denitrifiers exhibited a lag consistent with the duration of phase I before synthesizing denitrifying enzymes. Payne and Riley (1969) observed a slightly shorter lag, 40 min, with the marine denitrifier, *Pseudomonas perfectomarinus*. This could be due to a difference in the physiology of this organism or might be related to a difference in habitat.

Phase I and phase II rates measure completely different factors in soil denitrification. Phase II rate probably corresponds to what is generally called denitrification potential e.g. Balasubramanian and Kanehiro (1976). Our results indicate that this rate is a function of the number of soil denitrifiers and the limitations imposed on them by the energy and electron acceptor supply, pH, and temperature. These factors are often altered by the assay conditions. We have shown that addition of an energy source can have an effect on the rate and the pattern in phase II. It is also probable that moistening of a completely dry soil has effects similar to adding an energy source. Denitrification potential has been shown to correlate very well with mineralizable carbon (Bremner and Shaw, 1958). Phase I, in contrast to phase II or denitrification potential, is sensitive to the aeration state of the native soil. Phase I is essentially an assay of the cellular denitrifying enzyme activity present in the sample and reflects the immediate biological effect of

Table 3. Generalized description of the phases of soil denitrification after the imposition of anaerobic conditions

Phase	Generalized time interval (h) Start	End	Characteristics of rate	Explanation
I	0.25	1–3	Constant	Activity of pre-existing enzymes
Transition	1–3	4–8	Increasing	Active derepression of enzyme synthesis
IIa (low available organic carbon)	4–8	Indefinite	Constant	Indigenous community fully derepressed, no significant growth
IIb (high available organic carbon)	4–8	Indefinite	Logarithmically increasing	Growth of denitrifiers

S.B.B. 11/3—D

changes in soil moisture and aeration—extremely important factors for soil denitrification. Therefore, phase I rate appears to provide more information about *in situ* denitrifying activity than does denitrification potential and should be a more useful approach to the study of soil denitrification.

The importance of derepression of denitrifying enzyme synthesis in nature, as well as in laboratory incubations, is indicated by the increase in phase I rate of soil during field irrigation. Also when water was added to aerobic soils a pattern was observed similar to that in anaerobic incubation, that is, a low initial linear rate was followed by an increase in denitrification rate. It is tempting to attribute this pattern to the same causes in aerobic and anaerobic incubations. Although derepression is undoubtedly involved, other factors must be considered in the aerobic experiments and in field soils. The time required for the generation of anaerobic microsites accounts for the longer initial phase in the aerobic experiments. The difference between anaerobic phase I and phase II was usually less than an order of magnitude, while the rate increase subsequent to wetting of aerobic soils was significantly greater. This appears to be due to removal of O_2 inhibition of existing denitrifying enzymes which would be expected to result in much greater rate increases than would derepression of enzyme synthesis.

Our results demonstrate the value of working with simplified soil systems, in this case stirred anaerobic suspensions. We have been able to isolate and identify the biological effects of reduced soil aeration by controlling physical variables, particularly O_2 and substrate diffusion.

Our data provide some information about when and under what conditions denitrification could be expected to occur. We have demonstrated that denitrifying enzymes are present even in very dry soils. Since denitrifying enzymes are not constitutive (Payne, 1973) this implies that either denitrification can occur in anaerobic microsites of well aerated soils or that denitrifying enzymes may persist in functional form in the presence of O_2. We have also observed evolution of N_2O from the Miami sandy loam at 18% water content in an aerobic atmosphere. If it is assumed that the rates observed in our assays are similar to field rates, then a gross approximation of field N loss can be made. This is done by assuming that the net flux of N_2O from the soil in the tubes, i.e., the rate of accumulation in the headspace divided by the soil surface area, is equal to the flux of N from the soil in the field. Thus, the rate of N loss from the drier Miami sample (Fig. 5) would be low, 69 mg $N \cdot ha^{-1} \cdot day^{-1}$. Some well-aerated soils we assayed aerobically evolved N_2O at significantly greater rates, up to 6 g $N \cdot ha^{-1} \cdot day^{-1}$. These "aerobic" rates are too small to be significant in the N economy of agricultural soils. Yet it is interesting that N_2O evolution was observed from virtually all of the well-aerated soils examined. We have not yet demonstrated what fraction of this N_2O results from biological denitrification; several other mechanisms are possible. Bremner and Blackmer (1978) have suggested that nitrification causes low rates of N_2O production in aerobic soils, presumably by a process observed earlier in *Nitrosomonas* cultures (Yoshida and Alexander, 1970). Non-biological reactions of NO_2^- also might contribute to N_2O production. It is also possible that the soil aggregates were super-saturated with N_2O. Denitrifying conditions in the field well before sampling might have caused accumulations of N_2O within aggregates which slowly equilibrate with the atmosphere. The relative contribution of each of these mechanisms will be difficult to determine. Preliminary experiments were not revealing due in part to variability within and among soil samples, and in part to the difficulty of experimentally isolating the various mechanisms of N_2O production.

In summary, nitrogen gases can be produced more or less continuously by soils. Several mechanisms

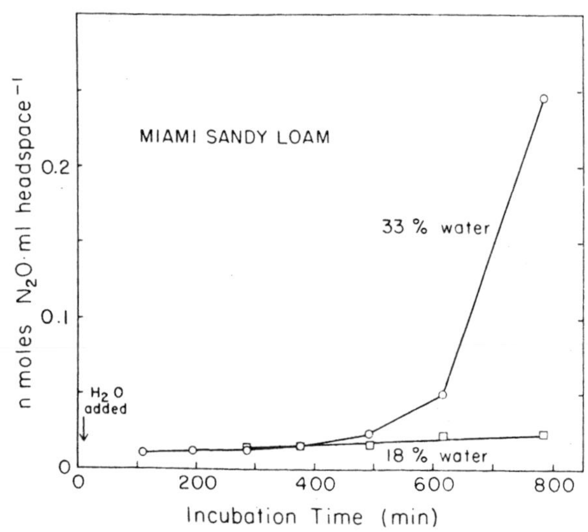

Fig. 5. Lag in denitrification response to adjustment of water content. Points are means of five replicates. Eighteen per cent gravimetric water content is 51% of maximum water holding capacity or approximately field capacity; 33% water is 94% of maximum water holding capacity or approximately saturation.

could contribute to this, but only denitrification in wet soils is likely to result in large N losses. Subsequent to a reduction in soil aeration, denitrification rate is greatly increased, but only after a lag of several hours. Anaerobic or partially anaerobic conditions, established by respiration and reduced O_2 diffusion rate, eliminate O_2 inhibition then derepress the synthesis of denitrifying enzymes. Thus we reason that most nitrogen loss would occur during brief periods beginning a few hours after irrigation or a rainfall and lasting until NO_3^- is depleted or the soil water content decreases.

Acknowledgements—We thank Mary Firestone for a portion of the chloramphenicol work and her ideas on development of the phases concept particularly during the initiation of this project. This work was supported by a National Science Foundation Grant and USDA Regional Research Project NE-39. Published as Journal Article No. 8604 of the Michigan Agricultural Experiment Station.

REFERENCES

Ardakani M. S., Fluehler H. and McLaren A. D. (1977) Rates of nitrate uptake with sudangrass and microbial reduction in a field. *Soil Sci. Soc. Am. J.* **41,** 751–757.

Balasubramanian V. and Kanehiro Y. (1976) Denitrification potential and pattern of gaseous N loss in tropical Hawaiian soils. *Trop. Agric.* **53,** 293–303.

Bremner J. M. and Blackmer A. M. (1978) Nitrous oxide: Emission from soils during nitrification of fertilizer nitrogen. *Science* **199,** 295–296.

Bremner J. M. and Shaw K. (1958) Denitrification in soil. II. Factors affecting denitrification. *J. agric. Sci.* **51,** 40–52.

Broadbent F. E. and Clark F. E. (1965) Denitrification. In *Soil Nitrogen.* (W. V. Bartholomew and F. E. Clark, Eds), pp. 344–359. American Society of Agronomy, Madison.

Focht D. D. and Verstraete W. (1977) Biochemical ecology of nitrification and denitrification. In *Advances in Microbial Ecology.* (M. Alexander, Ed.), pp. 135–214. Plenum Press, New York.

Payne W. J. (1973) Reduction of nitrogenous oxides by microorganisms. *Bact. Rev.* **37,** 409–452.

Payne W. J. and Riley P. S. (1969) Suppression by nitrate of enzymatic reduction of nitric oxide. *Proc. Soc. exp. Biol. Med.* **132,** 258–260.

Pilot L. and Patrick W. H. (1972) Nitrate reduction in soils: Effect of soil moisture tension. *Soil Sci.* **114,** 312–316.

Rowe R., Todd R. and Waid E. J. (1977) Microtechnique for most-probable-number analysis. *Appl. envir. Micro.* **33,** 675–680.

Smith M. S., Firestone M. K. and Tiedje J. M. (1978) acetylene inhibition method for short-term measurement of soil denitrification and its evaluation using ^{13}N. *Soil Sci. Soc. Am. J.* **42,** 611–615.

Snedecor G. W. and Cochran W. G. (1967) *Statistical Methods*, 6th edn, p. 138. Iowa State Univ. Press, Ames.

Starr J. L., Broadbent F. E. and Nielsen D. R. (1974) Nitrogen transformations during continuous leaching. *Soil Sci. Soc. Am. Proc.* **38,** 283–289.

Tiedje J. M., Firestone M. K., Smith M. S., Betlach M. R. and Firestone R. B. (1978) Short-term measurement of denitrification rates in soils using ^{13}N and acetylene inhibition methods. In *Proceeding in Life Sciences-Microbial Ecology* (M. W. Loutit and J. A. R. Miles, Eds), pp. 132–137. Springer-Verlag, Berlin.

Yoshida T. and Alexander M. (1970) Nitrous oxide formation by *Nitrosomonas europaea* and heterotrophic microorganisms. *Soil Sci. Soc. Am. Proc.* **34,** 880–882.

Commentary on Paper D4

Mariotti, A., Germon, J. C., Hubert, P., Kaiser, P., Letolle, R., Tardieux, A. & Tardieux, P. (1981) Experimental determination of nitrogen kinetic isotope fractionation: some principles; illustration for the denitrification and nitrification processes.

Microbial denitrification comprises a series of intermediary steps reducing several nitrogen oxides, including nitrous oxide, a potent greenhouse gas, and ultimately N_2. Wada *et al.* (1975) measured a net fractionation resulting from this stepwise reduction of $\varepsilon p/s = -20‰$. A few studies then used this strong ^{15}N depletion in the residual nitrate substrate to evaluate *in situ* denitrification activity in soil (Blackmer & Bremner, 1977), wetland (Domenach & Chalamet, 1977) and deep aquifers (Mariotti & Letolle, 1977).

Yet, in the paper presented here, Mariotti and colleagues provide the first comprehensive principles for determining the nitrogen kinetic isotope fractionation to be used for nitrification and denitrification. This highly cited paper (over 430 citations to date) set up the foundation for a non-invasive method of evaluating the existence and the rates of these two microbiological processes in different environments.

The method became very popular in the late 1980s due to the combination of several related research outputs. Atmospheric nitrogen was found to be a reliable standard for natural ^{15}N with a $^{15}N/^{14}N = 0.3677\ 10^{-3}$ (Mariotti 1983). During denitrification, oxygen is also fractionated with a constant $\varepsilon p/s = -8.0‰$ (Böttcher *et al.*, 1980). Dual measurement of nitrogen and oxygen isotope composition of nitrate became an excellent tool to prove the existence of denitrification. Simultaneously, the development of mass spectroscopy techniques became financially more accessible to the point that this tool is now easily and cheaply available in most ecological and geochemical laboratories.

Recently, this method has been applied extensively along groundwater flow paths to determine the spatio-temporal pattern of denitrification rates in riparian zones (Clement *et al.*, 2003). Moreover, the low plant uptake fractionation allows determination of whether and when riparian vegetation contributes to groundwater nitrate retention. The main limitation to the method is that the nitrate source should be infinite (constant input) in order to obtain a significant discrimination. Therefore, natural isotope abundance is most useful to track denitrification activity in areas with constant nitrate input to areas conducive to denitrification occurring, i.e. reduced biogeochemical hot spots (McClain *et al.*, 2003). A very recent achievement now allows researchers to decide whether N_2O is produced from nitrification or denitrification (Sutka *et al.*, 2006). It should permit further understanding and quantification of the driving forces regulating the emission of this greenhouse gas in riparian zones, and so optimize riparian zone management to buffer diffuse nitrogen pollution by reducing N_2O emission rates.

GP

References

Blackmer, A. M. & Bremner, J. M. (1977) Nitrogen isotope discrimination in denitrification of nitrate in soils. *Soil Biol. & Biochem.* **9**, 73–77.

Böttcher, J., Strebel, O. & Kölle, W. (1980) Using isotope fractionation of nitrate nitrogen and nitrate oxygen for evaluation of denitrification in a sandy aquifer. *J. Hydrol.* **114**, 413–424.

Clement, J. C., Holmes, R. M., Peterson, B. J. & Pinay, G. (2003) Isotopic investigation of denitrification in a riparian ecosystem in western France. *J. Applied Ecology* **40**, 1035–1048.

Domenach, A. M. & Chalamet, A. (1977) Rapports isotopiques naturels de l'azote. I. Premiers résultats : sols de Dombes. *Revue d'Ecologie & Biologie du Sol* **14**, 279–287.

Mariotti, A. (1983) Atmospheric nitrogen is a reliable standard for natural N-15 abundance measurements. *Nature* **303**(5919), 685–687.

Mariotti, A. & Letolle, R. (1977) Application de l'étude isotopique de l'azote en hydrologie et hydrogéologie – analyse des résultats obtenus sur un exemple précis: le bassin de Mélarchez (Seine et Marne, France). *J. Hydrol.* **33**(1–2), 157–172.

McClain, M. E., Boyer, E. W., Dent, C. L., Gergel, S. E., Grimm, N. B., Groffman, P. M., Hart, S. C., Harvey, J. W., Johnston, C. A., Mayorga E., McDowell, W. H. & Pinay, G. (2003) Biogeochemical hot spots and hot moments at the interface of terrestrial and aquatic ecosystems. *Ecosystems* **6**, 301–312.

Sutka, R. L., Ostrom, N. E., Breznak, J. A., Gandhi, H., Pitt, A. J. & Li, F. (2006) Distinguishing nitrous oxide production from nitrification and denitrification on the basis of isotopomer abundances. *Appl. & Environ. Microbiol.* **72**(1), 638–644.

Wada, E., Kadonaga, T., and Matsuo, S. (1975) 15N abundance in nitrogen of naturally occurring substances and global assessment of denitrification from an Isotopic Viewpoint. *Geochem. J.* **9**, 139–148.

EXPERIMENTAL DETERMINATION OF NITROGEN KINETIC ISOTOPE FRACTIONATION: SOME PRINCIPLES; ILLUSTRATION FOR THE DENITRIFICATION AND NITRIFICATION PROCESSES

by A. MARIOTTI*, J. C. GERMON**, P. HUBERT***, P. KAISER****, R. LETOLLE*, A. TARDIEUX**** and P. TARDIEUX****

KEY WORDS

Denitrification Nitrification Nitrogen isotope fractionation Nitrogen-15 natural abundance

SUMMARY

A few principles relative to the presentation and use of nitrogen stable isotopic data are briefly reviewed. Some classical relationships between the isotope composition of a substrate undergoing a single-step unidirectional reaction, are introduced.

They are illustrated through controlled experiments on denitrification in a soil, and through nitrification by pure cultures of *Nitrosomonas europaea*. In the latter case, the isotope fractionation is calculated from the isotopic composition of the residual substrate, then of the product and the result is shown to be statistically the same for the two procedures.

The isotopic enrichment factor for denitrification is -29.4 ± 2.4‰ at 20°C, and -24.6 ± 0.9‰ at 30°C; for nitrification this factor is -34.7 ± 2.5‰ under the experimental conditions employed.

INTRODUCTION

Many agricultural and biochemical studies use isotopic tracers. Interpretations of variation in stable isotope abundance require a precise knowledge of the isotope fractionation occurring during physical, chemical or biochemical transformations undergone by the compound in question. We present here a method for the expression and calculation of kinetic isotope fractionation associated with transformations in the nitrogen cycle. Such methods have been standard in isotope geochemistry for H, O, C, S for several decades. The need for such a standard method comes from the variety of ways of presenting and discussing data published by agronomists and soil scientists using ^{15}N natural variations for the study of the nitrogen cycle in terrestrial ecosystems.

* Université Pierre et Marie Curie, Laboratoire de Géologie dynamique, T. 26, 4E, 4 place Jussieu, 75230 Paris cedex 05.

** I.N.R.A., Laboratoire de Microbiologie des Sols, 7 rue Sully, 21934 Dijon cedex.

*** E.N.S.M., Centre d'Informatique Géologique, 35 rue St Honoré 77305 Fontainebleau.

**** Institut Pasteur, Chaire de Microbiologie, I.N.A., 9 rue de l'Arbalète, 75005 Paris.

Plant and Soil 62, 413–430 (1981). 0032-079X/81/0623-0413$01.35.

TERMINOLOGY

Expression of isotope data. Standard

^{15}N abundance is expressed in atom per cent by the ratio:

$$A = \frac{^{15}N}{^{15}N + {}^{14}N} \cdot 100$$

The value of A for atmospheric nitrogen is 0.3663 ± 0.0004 atom $^{15}N\%$[10].

If, as a convention, the abundance of the heavy isotope is referred to that of the lighter one, a dimensionless parameter may be defined, the isotope ratio R. For nitrogen: $R = {}^{15}N/{}^{14}N$. Due to the small abundance of ^{15}N in atmospheric nitrogen and the small variations of ^{15}N abundance in the nitrogen cycle, the following approximation holds:

$$R = \frac{^{15}N}{^{14}N} \simeq \frac{^{15}N}{^{15}N + {}^{14}N}$$

Absolute abundance of ^{15}N cannot be measured with very great precision, but small differences in isotope ratio between two compounds are easily measured with high accuracy by using mass spectrometers fitted with double ion-collection, and double inlet systems equipped for rapid switching between standard and samples[17,18]. Technical reasons have led to the use of the so-called 'δ' notation to express results. For nitrogen, this relative unit is defined as:

$$\delta^{15}N = \frac{R_{sample} - R_{standard}}{R_{standard}}$$

As natural variations of ^{15}N are small, a common convention is to express δ in parts per thousand:

$$\delta^{15}N‰ = \delta^{15}N \times 1000$$

A $\delta^{15}N$ value of +10‰, for example, corresponds to a sample with an isotope ratio one per cent higher than that of the standard against which it is expressed. Some authors[11,20] use a slightly different expression:

$$\delta_a{}^{15}N = \left[\frac{\{^{15}N/(^{15}N + {}^{14}N)\}_{sample}}{\{^{15}N/(^{15}N + {}^{14}N)\}_{standard}} - 1\right] \cdot 1000$$

It can be easily shown that the two modes of calculation differ by only 0.25‰ for delta values of 65‰ which are rarely met within natural settings. The two definitions can therefore be considered as equivalent. In order to be comparable from one laboratory to another, these measurements need to be done relative to a common standard. Such a standard should present several qualities; namely, perfect homogeneity and chemical and isotopic stability. Moreover, the isotopic composition of the standard must not be very different from the measured natural samples, in order to minimize systematic errors in the mass spectrometry, which increase with the 'isotopic distance' between standard and sample.

For these reasons, a standard which is abundant in nature and of simple chemical form (and coming from a medium where kinetic processes are rapid in order for the material to be well-mixed) is prefered.

We have carried out a series of measurements on atmospheric nitrogen from various origins in France and have assessed the constancy of its isotopic composition within mass spectrometer analytical precision: the standard deviation on 27 measurements is ± 0.026‰ (δ units). Therefore, the isotopic composition of atmospheric N_2 may be considered as practically constant and thus a reliable standard.

Expression of kinetic isotope fractionation. Isotope enrichment

Partition of isotopes between two compounds containing the same element (or between two phases) with different isotopic ratio is called isotope fractionation. This is due to different energy characteristics of the light and heavy molecule, from which comes a difference in the probability of reaction[2].

Consider a single step unidirectional reaction (*i.e.* the so-called 'kinetic reaction'):

substrate → product: S → P

Without any hypothesis concerning the intimate mechanism and reaction order, we represent the kinetic fractionation factor $\alpha_{p/s}$, as $\alpha_{p/s} = R_{pi}/R_s$ where R_{pi} is the isotope ratio of the increment of product which appears in an infinitely short time at time t, and R_s the isotope ratio of the substrate at the same time.

(Some authors[4,9,19] use the inverse ratio, *i.e.* $\alpha_{s/p}$).

We have:

$$\alpha_{p/s} = \frac{d^{15}N_p/d^{14}N_p}{^{15}N_s/^{14}N_s} \tag{1}$$

Since $d^{15}N_p = -d^{15}N_s$ and $d^{14}N_p = -d^{14}N_s$, one gets

$$\alpha_{p/s} = \frac{d^{15}N_s}{^{15}N_s} \bigg/ \frac{d^{14}N_s}{^{14}N_s} \tag{2}$$

If $\alpha_{p/s}$ is considered *a priori* as constant during the progress of the reaction, integration of (2) gives:

$$(\alpha_{p/s}) \ln\left\{\frac{^{14}N_s}{^{14}N_{s,0}}\right\} = \ln\left\{\frac{^{15}N_s}{^{15}N_{s,0}}\right\}, \tag{3}$$

where $^{15}N_{s,0}$ and $^{14}N_{s,0}$ stand for the quantity of the two isotopes at time zero.

Let f be the unreacted fraction of substrate at time t, and N_s the total quantity of nitrogen at time t ($N_s = {}^{14}N_s + {}^{15}N_s$).

$$f = \frac{N_s}{N_{s,0}} = \frac{^{14}N_s + {}^{15}N_s}{^{14}N_{s,0} + {}^{15}N_{s,0}}$$

In natural systems, ^{15}N abundance varies little around the value 0.3663 atom per cent (0.3663 ± 0.0183) and the approximation $^{14}N \simeq {}^{15}N + {}^{14}N$ is valid, so: $f \simeq {}^{14}N_s/{}^{14}N_{s,0}$.

Equation (3) becomes:

$$(\alpha_{p/s} - 1) \ln f = \ln \frac{R_s}{R_{s,0}}$$

or

$$\frac{R_s}{R_{s,0}} = f^{(\alpha_{p/s} - 1)} \tag{4}$$

In isotope geochemistry, this equation is classically termed the 'Rayleigh' equation. It was derived by Lord Rayleigh for the case of fractional distillation of mixed liquid.

With the δ‰ notation,

$$(\alpha_{p/s} - 1) \ln f = \ln \frac{10^{-3}\,\delta_s + 1}{10^{-3}\,\delta_{s,0} + 1} \tag{5}$$

As the fractionation factor generally differs by less than 5% from unity, it can be expressed, as an analogy with the delta notation, as:

$$\alpha_{p/s} = 1 + 10^{-3}\,\varepsilon_{p/s} \tag{6}$$

$\varepsilon_{p/s}$ is called the 'per mil enrichment factor' (either positive or negative) of the product relative to the substrate. This notation allows easy comparisons.

From (5) and (6), one gets:

$$\varepsilon_{p/s} = \frac{10^3 \cdot \ln \dfrac{10^{-3}\,\delta_s + 1}{10^{-3}\,\delta_{s,0} + 1}}{\ln f} \tag{7}$$

As $\ln(1 + u) \simeq u$ when u is small relative to 1 and $\ln \dfrac{1+u}{1+v} \simeq u - v$ when u and v are small relative to 1, for most δ values, the approximation:

$$\varepsilon_{p/s} = \frac{\delta_s - \delta_{s,0}}{\ln f} \quad \text{holds} \tag{8}$$

So in the hypothesis where α (and therefore ε) is constant during the reaction ($f \rightarrow 0$), the isotope composition of the substrate is a linear function of ln f and the enrichment factor ε is the slope of the straight line which passes through the origin ($\delta_s - \delta_{s,0} = 0$ for $\ln f = 0$).

A priori, the simplified relation (8) holds for $|\varepsilon|$ values $< 20‰$ and for $\delta_{s,0}$ values not too different from zero. Equation (7) is to be preferred in more extreme cases.

In a closed system ($P + S = S_0$), the isotopic abundance of the *accumulated* or *instantaneous* product is easily described:

Accumulated product

One has

$$f \cdot R_s + (1 - f)\overline{R_p} = R_{s,0} \tag{9}$$

which is the isotope balance equation, where $\overline{R_p}$ is the isotopic ratio of the accumulated product. Combining equations (4) and (9) gives:

$$\overline{R_p} = R_{s,0} \cdot \frac{1 - f^{\alpha_{p/s}}}{1 - f} \tag{10}$$

As equation (9) may be written

$$f \cdot \delta_s + (1 - f)\overline{\delta_p} = \delta_{s,0}$$

combination with (8) gives:

$$\overline{\delta_p} = \delta_{s,0} - \varepsilon_{p/s} \frac{f \ln f}{1 - f} \tag{11}$$

The expression $\frac{f \ln f}{1 - f}$ tends towards -1 when f tends towards 1, thus the isotopic composition of the first product formed differs from the isotopic composition of the initial substrate by ε:

$$\overline{\delta_{p(0+dt)}} = \delta_{s,0} + \varepsilon_{p/s} \tag{12}$$

Instantaneous product

Equation:

$$\alpha_{p/s} = R_{pi}/R_s$$

gives:

$$\alpha_{p/s} = \frac{\delta_{pi} + 1000}{\delta_s + 1000}$$

or

$$\delta_{pi} = \delta_s\left(1 + \frac{\varepsilon_{p/s}}{1000}\right) + \varepsilon_{p/s}$$

If $\varepsilon_{p/s}$ is small relative to 1000, one gets the approximation

$$\delta_{pi} \simeq \delta_s + \varepsilon_{p/s} \tag{13}$$

So, the curve for δ_{pi} as a function of f is derived from the curve for δ_s by a translation of $\varepsilon_{p/s}$ on the δ coordinate.

Fig. 1 shows these variations for the theoretical case $\varepsilon_{p/s} = -20‰$ and $\delta_{s,0} = 0‰$, with the approximation of Equation (8)[12,16].

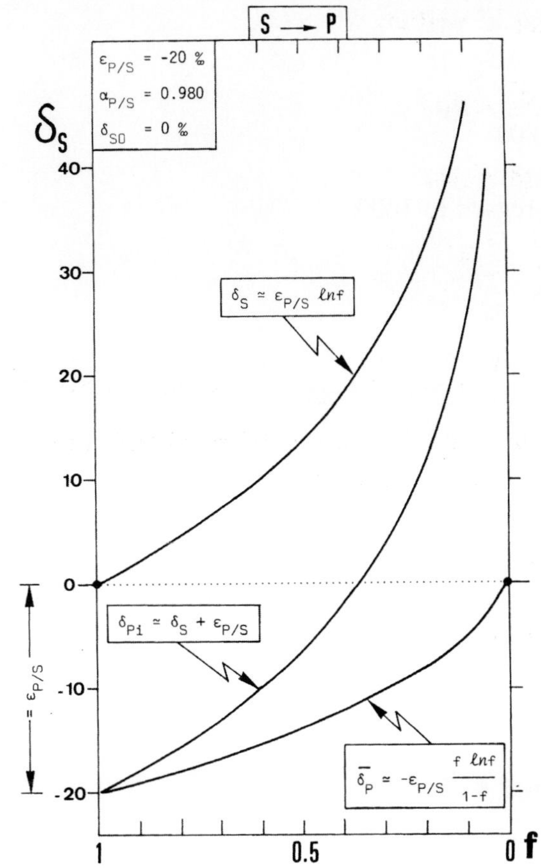

Fig. 1. Theoretical change in substrate and product isotopic composition in a single step unidirectional reaction. The approximation $\delta_s - \delta_{s,0} \simeq \varepsilon \ln f$ has been used with $\varepsilon_{p/s} = -20‰$ and $\delta_{s,0} = 0‰$. δ_s stands for the isotopic composition of the substrate, δ_p for the accumulated product and δ_{pi} for the instantaneous product.

In the case when the remaining substrate may be considered as an infinite reservoir with regards to the quantity of product (example: biological dinitrogen fixation by legumes, see ref. [15]), the ratio R_s of the substrate does not change, f is always $\simeq 1$ and one gets:

$$\delta_{pi} = \overline{\delta_p} = \delta_p \simeq \delta_{s,0} + \varepsilon_{p/s} = \delta_s + \varepsilon_{p/s}$$

More rigorously,

$$\alpha_{p/s} = R_p/R_s$$

whence,

$$\varepsilon_{p/s} = \frac{\delta_p - \delta_s}{\delta_s + 1000} \cdot 10^3$$

If δ_s is small with regard to 1000:

$$\varepsilon_{p/s} \simeq \delta_p - \delta_s$$

These relations are independent of the kinetics of the reaction. However, for a first order reaction, the following equations hold:

$$^{14}N_s = {}^{14}N_{s,0}\, e^{-kt}$$
$$^{15}N_s = {}^{15}N_{s,0}\, e^{-k't}$$

where k and k′ are respectively the rate constants for the reactions involving lighter and heavier isotopes, respectively.

Therefore, $f = e^{-kt}$, and

$$\frac{^{15}N_s}{^{14}N_s} = \frac{^{15}N_{s,0} \cdot e^{-k't}}{^{14}N_{s,0} \cdot e^{-kt}}$$

whence

$$R_s = R_{s,0}\, f^{(k'/k-1)}$$

or:

$$\ln \frac{10^{-3}\,\delta_s + 1}{10^{-3}\,\delta_{s,0} + 1} = \left(\frac{k'}{k} - 1\right) \ln f \qquad (14)$$

Comparison of equations (5) and (14) shows that in the case of first order reactions, $\alpha_{p/s}$ is the ratio of the rate constants of the heavy to the light isotope. The same formulation has been given by Tong and Yankwich[22] which call $r = k/k'$, a factor named β by various authors[5,8,21]:

$$\beta = 1/\alpha_{p/s} = \alpha_{s/p}.$$

Equations (5) and (14) are identical with those of Tong and Yankwich[22] and are easier to use.

EXPERIMENTAL RESULTS

They are related to the experimental determination of the isotopic enrichment factor during denitrification in soils samples and nitritation by pure *Nitrosomonas europaea* cultures.

1 *Experimental determination of the isotopic fractionation during denitrification in soils*

Selected soil samples were placed in experimental conditions favouring denitrification.

Materials and methods

100 ml of distilled water containing 0.1 ml N-serve (nitrification inhibitor) was added to 100 g of dried soil. After a few days of preincubation in order to eliminate the nitrates and nitrites (often present in the soil through nitrification), 10 ml of solution containing about 20 mg N–NO_3^- was added to the soil. The flasks (600 ml) were placed under a helium atmosphere, of which 50 ml was then replaced by the same volume of acetylene in order to stop denitrification at the N_2O step[7]. N_2O is easily measured through gas chromatography and will give the extent of reaction (f). The unreacted nitrate was extracted with 1 *N* KCl and measured on a Technicon analyser. Isotope analyses follow the protocol described by Mariotti and Létolle[14].

At each step of NO_3^- reduction, nitrite was also measured. It represents an intermediary step between NO_3^- and N_2O:

$$NO_3^- \underset{1}{\rightarrow} NO_2^- \underset{2}{\rightarrow} N_2O \underset{3}{\rightarrow} N_2$$

NO_2^- was never detected during the experiment. It may be concluded that step 1 was rate controlling for the overall reaction. Any step in the further reduction of NO_2^- cannot lead to an isotope effect since all the NO_2^- was reduced as rapidly as it was formed.

Under these conditions, the measured isotope fractionation will correspond to the nitrate-nitrite step (step 1). Here, the two-step unidirectional reaction may be reduced to a single step unidirectional one, which would not be the case if nitrites were present[3].

Results

Changes in the concentration of NO_3^- and N_2O, as well as the isotope composition of remaining substrate (expressed as $10^3 \ln\{10^{-3}\delta_s + 1\}/\{10^{-3}\delta_{s,0} + 1\}$) are reported on Figures 2 and 3 respectively (two different temperatures: 20°C and 30°C).

The sum, N–NO_3^- + N–N_2O, is constant within analytical precision, which is the expected case for a closed system ($P + S = S_0$).

Correlation coefficients for the line $Y_1 = \delta_s - \delta_{s,0}$ and $Y_2 = 10^3 \ln\{10^{-3}\delta_s + 1/10^{-3}\delta_{s,0} + 1\}$ as a function of ln f have been calculated (Table 1).

Correlation factors are significant at the probability level 0.01. This indicates

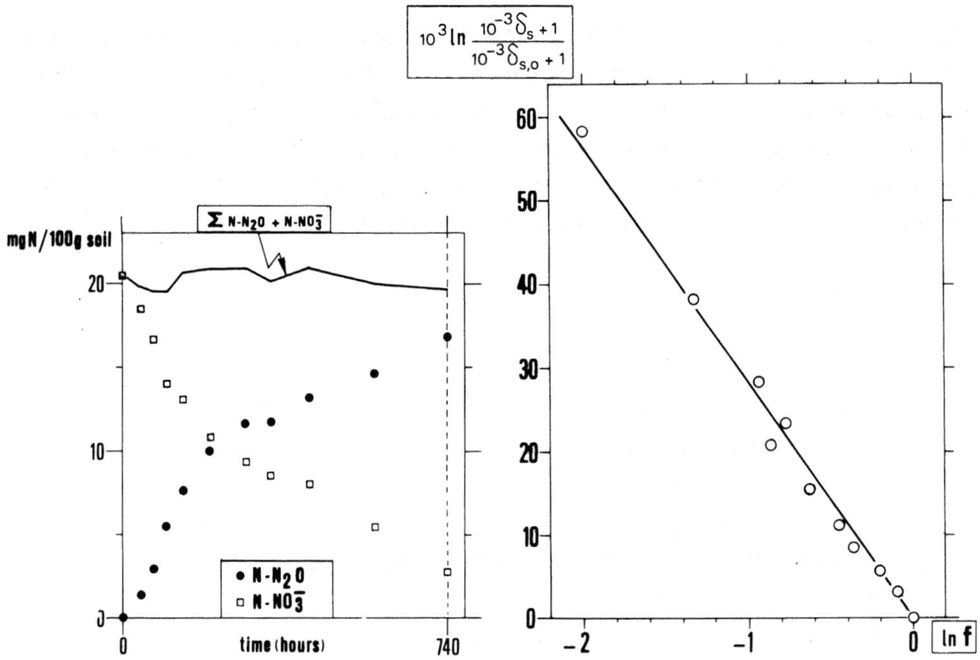

Fig. 2. Chemical and isotopic change in NO_3^- during denitrification; change in the concentration of the product N_2O and of the sum: $\Sigma(N\text{–}NO_3^- + N\text{–}N_2O)$. Incubation temperature was 20°C.

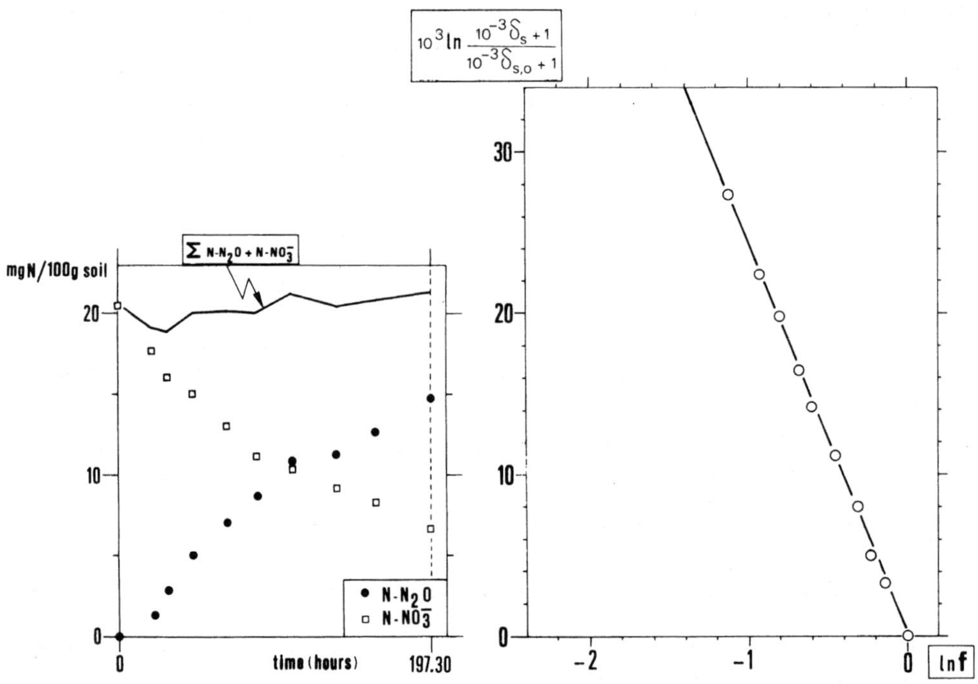

Fig. 3. Chemical and isotopic change in NO_3^- during denitrification; change in N_2O concentration and of the sum: $\Sigma(N\text{–}NO_3^- + N\text{–}N_2O)$. Incubation temperature is 30°C.

clearly that the assumption of the constancy of the isotope fractionation factor is valid, at least as far as the extent of reaction that has been studied, and therefore relations (7) and (8) are valid. The confidence limit for the slopes, ts_b, ($p = 0.05$) has been calculated. This gives the confidence limit for $\varepsilon_{p/s}$ experimentally measured. As these lines must pass at the origin, the slope was also calculated for the best fit of a line passing at this point. The slope is given by

$$m = \frac{\Sigma XY}{\Sigma X^2}$$

We have also tested the hypothesis that the intercept of the straight line calculated with experimental data is zero. Data are given in Table 1.

The difference of $\varepsilon_{p/s}$ values obtained either with equation (7) or approximation (8) is small and distinctly less than the confidence limit on $\varepsilon_{p/s}$. This confidence limit has been estimated for a series of 12 measurements of the denitrification fractionation factor, carried out under the same experimental conditions as described above (Mariotti *et al.*, to be published). Its mean value is 2.7‰ and may

Table 1. Regression parameters for experimental data of NO_3^- denitrification. (Step $NO_3^- \rightarrow N_2O$), reported on Figs. 2 and 3

Temperature	$Y = (\varepsilon \pm ts_b)\ln f + \beta$	Slope of the straight line passing through origin $\beta = 0$; $\varepsilon = \frac{\Sigma XY}{\Sigma X^2}$	Number of data	Correlation coefficient
20°C	$\delta_s - \delta_{s,o} = (-30.32 \pm 2.53)\ln f - 1.37$*	−29.12	11	−0.994**
	$10^3 \ln \frac{10^{-3}\delta_s + 1}{10^{-3}\delta_{s,o} + 1} = (-29.37 \pm 2.38)\ln f - 1.11$*	−28.39	11	−0.994**
30°C	$\delta_s - \delta_{s,o} = (-25.07 \pm 0.95)\ln f - 0.21$*	−24.79	10	−0.999**
	$10^3 \ln \frac{10^{-3}\delta_s + 1}{10^{-3}\delta_{s,o} + 1} = (-24.63 \pm 0.93)\ln f - 0.15$*	−24.43	10	−0.999**

* Intercept not statistically different from zero.
** Significant at $p = 0.01$.

reach 4.5‰. Due to analytical imprecisions, a relative precision of 10% of the value of $\varepsilon_{p/s}$ may be obtained at best.

One may conclude from the above experiments that the isotopic enrichment factor during denitrification, in the described conditions, is -29.4 ± 2.4‰ at 20°C and -24.6 ± 0.9‰ at 30°C.

In these experiments, the isotope composition of the gaseous product (N_2O) was not measured. However, in nitrification experiments it was possible to use the isotopic composition of the product (NO_2^-) in order to get two independant determinations of the fractionation factor: one based on the isotopic composition of the substrate, NH_4^+, and a second based on the isotopic composition of the product, NO_2^-.

2 *Determination of the isotope fractionation factor during nitrification with* Nitrosomonas europaea

Material and methods

The study has been carried on a pure culture of *N. europaea* (Rothamsted collection). The growing medium has been described by Lewis and Pramer[13], the only difference being the ammonium sulphate concentration. A 1.25 fold concentration of the parent medium was added to four series of six flasks each, as follows: 100 ml in 500 ml Erlenmeyers flasks for series A, 200 ml in 1000 ml Erlenmeyers for series B and C, 400 ml in 2000 ml Fernbach flasks for series D. Before autoclave sterilization, water was added such that the four series would attain a final volume of: 125 ml (A), 250 ml (B and C) and 500 ml (D) respectively, when ammonium sulphate was added later.

The ammonium sulphate solution was prepared and sterilized separately and aseptically added to series A to D in order to get a decreasing concentration scale as shown on Table 2.

In each series, one control flasks was not inoculated. The other flasks were inoculated at 1% volume with a culture grown for 72 hours on the Lewis and Pramer[13] medium with a 10^{-2} *M* concentration of $(NH_4)_2SO_4$. All flasks, including the control flasks, were incubated and stirred at 28°C.

Staggered samples were taken out of each series in order to follow the nitrification process, nitrites being measured through the modified Griess-Islovay method[1]. The incubation was concluded at different times for each series, according to the extent of the reaction: from 39 h to 208 h for series C and D, from 39 h to 264 h for series B and from 48 h to 471 h for series A. After incubation the cultures were cooled, and $HgCl_2$ added. The samples were then frozen until the time of analysis. The chemical analyses were performed after distillation with MgO (NH_4^+), then with MgO + Devarda alloy (for NO_2^-), the NH_4^+ nitrogen from the distillate being measured with an Orion model 95-10 specific electrode.

Results

The determination of the isotopic enrichment factors was carried out separately on the four experiments with a different ammonium substrate concentration. The four values are not statistically different one from another. Therefore a single value can be proposed based on the 19 pairs of experimental values.

In these experiments, the bacterial biomass was growing. However, the quan-

Table 2. Data for nitrification experiment with *Nitrosomonas europaea*. Four experiments with different initial substrate concentrations are shown. The values $N\text{–}NH_4^+$ and $N\text{–}NO_2^-$ mg represents the quantity of nitrogen for these two ions present in the volume chosen for incubation

Experiment	$N\text{–}NH_4^+$ concentration mg/l	Volume of medium ml	Time hours	$N\text{–}NH_4^+$ mg	$\delta^{15}N$ NH_4^+	f	$N\text{–}NO_2^-$ mg	$\delta^{15}N$ NO_2^-	$\Sigma(N\text{–}NH_4^+ + N\text{–}NO_2^-)$ mg	$10^3 \ln \frac{10^{-3}\delta_s+1}{10^{-3}\delta_{s,0}+1}$	$\delta_s - \delta_{s,0}$	$\delta_p - \delta_{s,0}$	$-\frac{f \ln f}{1-f}$
A	350.0	125	0	43.8	5.1	1.000	0.4	—	44.2	0.00	0.0	—	—
			48	41.7	5.5	0.952	2.7	−24.7	44.4	0.40	0.4	−29.8	0.976
			87	30.8	15.4	0.703	7.9	−18.8	38.7	10.20	10.3	−23.9	0.834
			162	20.8	32.1	0.475	19.2	−17.3	40.0	26.51	27.0	−22.4	0.673
			246	20.8	33.1	0.475	19.2	−15.1	40.0	27.48	28.0	−20.2	0.673
			471	24.0	35.1	0.548	20.7	−22.5	44.7	29.41	30.0	−27.6	0.729
B	226.4	250	0	56.6	4.0	1.000	0.0	—	56.6	0.00	0.0	—	—
			39	55.7	5.4	0.984	4.0	−27.8	60.6	1.39	1.4	−31.8	0.992
			70	36.4	21.9	0.643	7.2	−10.3	43.6	17.67	17.9	−14.3	0.795
			117	13.4	56.1	0.237	26.4	−10.7	39.8	50.59	52.1	−14.7	0.447
			162	11.0	66.2	0.194	32.2	−10.8	43.2	60.11	62.2	−14.8	0.395
			264	7.0	79.1	0.124	37.5	− 7.1	44.5	72.14	75.1	−11.1	0.295
C	100	250	0	25.0	4.7	1.000	0.0	—	25.0	0.00	0.0	—	—
			39	22.9	10.6	0.916	3.1	−23.3	26.0	5.86	5.9	−28.0	0.957
			48	18.4	17.0	0.736	3.9	−17.2	22.3	12.17	12.3	−21.9	0.855
			70	6.8	49.2	0.272	9.2	−12.8	16.0	43.34	44.5	−17.5	0.486
			96	4.4	61.1	0.176	15.2	−13.5	19.6	54.62	56.4	−18.2	0.371
			208	1.3	—	0.052	20.7	4.7	22.0	—	—	0.0	0.162
D	66.2	500	0	33.1	1.3	1.000	0.0	—	33.1	0.00	0.0	—	—
			39	22.0	19.7	0.665	8.4	−23.2	30.4	18.21	18.4	−24.5	0.810
			48	13.8	41.9	0.417	16.1	−18.8	29.9	39.75	40.6	−20.1	0.626
			63	4.9	77.3	0.148	21.3	− 8.4	26.2	73.16	76.0	− 9.7	0.332
			72	3.4	87.0	0.103	21.6	− 5.6	25.0	82.12	85.7	− 6.9	0.261
			208	2.9	—	0.088	23.8	2.1	26.7	—	—	0.8	0.234

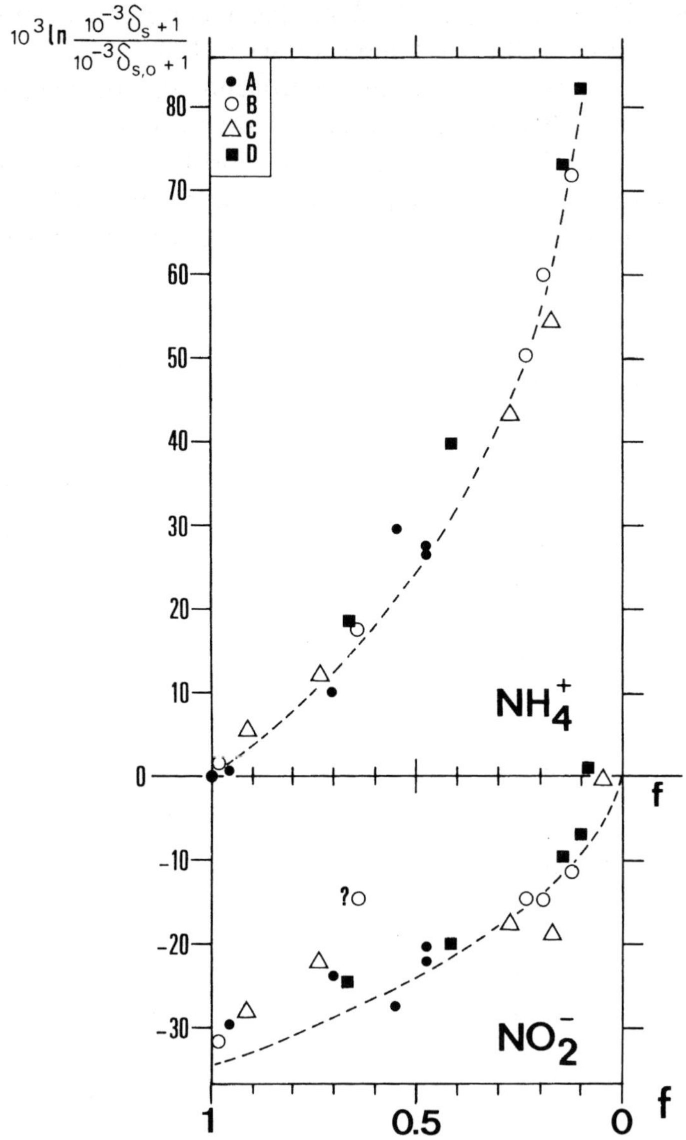

Fig. 4. Change in isotopic composition of substrate NH_4^+ and product NO_2^- as a function of unreacted fraction of substrate (f), during nitritation with *Nitrosomonas europaea.*

tity of nitrogen assimilated by the biomass was always negligible in comparison with the mineral species (NH_4^+ or NO_2^-). Data are presented on Table 2 and Figs. 4 and 5. Parameters of the regression line, using equation (7) are $\varepsilon_{p/s} = -34.7 \pm 2.5‰$ with the intercept value 1.8 which is not statistically different from zero. The correlation coefficient is 0.99. The same parameters calculated with equation (8) are $\varepsilon_{p/s} = -36.2 \pm 2.6‰$; the intercept is 1.5 (not statistically different from zero) and the correlation coefficient, 0.99.

The isotopic enrichment factor and consequently the fractionation factor is therefore constant with the extent of reaction in the nitrogen concentration range from $2.5 \cdot 10^{-2}\,M$ to $5 \cdot 10^{-4}\,M$. The isotope enrichment factor may also be calculated using the isotopic composition of the product: NO_2^-. From the equation (11), the isotope enrichment factor is deduced from the slope of the

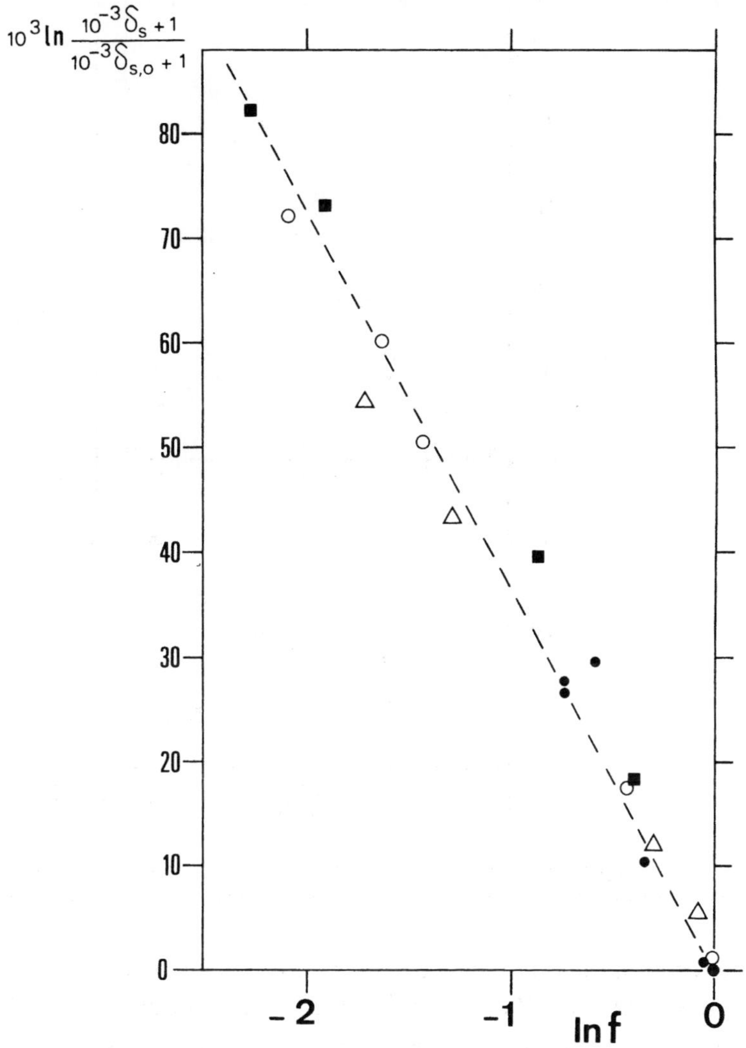

Fig. 5. Change in substrate NH_4^+ isotopic composition as a function of ln f during nitrification. The slope of the straight line is the isotopic enrichment factor $\varepsilon_{p/s}$.

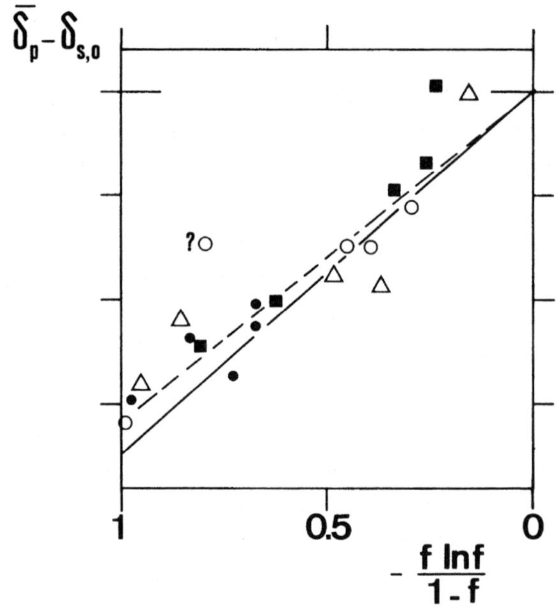

Fig. 6. Nitrification experiment. Isotopic composition of product (NO_2^-) as a function of $-f \ln f/1 - f$) (cf. equation 11). The slope of the solid straight line represents the isotope enrichment factor calculated from the change in the substrate (−34.7‰, see Fig. 5); the dotted line gives the isotopic enrichment factor calculated from the change in the product (−31.9, see text).

straight line $Y = \overline{\delta_p} - \delta_{s,0}$, $X = -f \ln f/(1 - f)$. f will be taken as $[NH_4^+]_t/[NH_4^+]_0$ (Fig. 6). One of the experimental values obtained for the isotopic composition of NO_2^- (experiment B, 70 h, Table 2) seems incorrect (see Fig. 4 and 6) and has not been taken into account in the calculation.

One gets:

$$\overline{\delta_p} - \delta_{s,0} = 0.64 - (31.91 \pm 6.39) \cdot \left(- \frac{f \ln f}{1 - f} \right); \quad r = -0.931$$

The intercept is not statistically different from zero.

It may be concluded that the estimation of the isotopic enrichment factors, calculated from changes in isotopic composition of either the substrate or the product, are not significantly different. Nevertheless, the first estimation is more precise. A second conclusion is that the system under study could be considered, within analytical errors, as a closed system. The isotope enrichment factor during nitrification in a pure culture of *N. europaea* is therefore about −35‰ in the given conditions, very close to the value published by Domenach and Chalamet[6].

CONCLUSION

Isotopic kinetic reaction studies, such as the one presented here, are needed in order to ascertain the magnitude of isotope fractionation associated with important N transformations. This knowledge is essential for balance estimations of the nitrogen budget.

The present study shows that under the experimental conditions we used, the isotope effects associated with nitrification and denitrification are constant with extent of reaction and with substrate concentration and that the substrate-product relation fits the model of a simple and unidirectional reaction.

The isotope fractionation factor for a reaction, the mechanism of which is not well known, cannot be estimated through a single determination of the isotopic enrichment of the product *versus* substrate, at any step in the progress of the reaction, since there are no *a priori* reasons for this fractionation factor to be constant.

The present data also show that the kinetic isotope fractionation factor cannot be estimated presently with a precision better than 10% of the measured value.

ACKNOWLEDGEMENTS

We thank Kohl, D. H. and Shearer, G. for their critical reading of the manuscript and Guillemot, J. for technical assistance.

Received 25 February 1981

REFERENCES

1 Barnes, H. and Folkard, A. R. 1951 The determination of nitrite. Analyst **76,** 599–603.
2 Bigeleisen, J. 1952 The effect of isotopic substitution on the rates of chemical reactions. J. Chem. Phys. **20,** 823–828.
3 Blackmer, A. M. and Bremner, J. M. 1977 Nitrogen isotope discrimination in denitrification of nitrate in soils. Soil Biol. Biochem. **9,** 73–77.
4 Chien, S. H., Shearer, G. and Kohl, D. H. 1977 The nitrogen isotope effect associated with nitrate and nitrite loss from waterlogged soils. Soil Sci. Soc. Am. J. **41,** 63–69.
5 Delwiche, C. C. and Steyn, P. L. 1970 Nitrogen isotope fractionation in soils and microbial reactions. Environ. Sci. Technol. **4,** 929–935.
6 Domenach, A. M. and Chalamet, A. 1977 Rapports isotopiques naturels de l'azote. I.-Premiers résultats: sols de Dombes. Rev. Ecol. Biol. Sol **14,** 279–287.
7 Federova, R. I., Milekhina, E. I. and Il'yuklina, N. I. 1973 Possibility of using the 'gas-exchange' method to detect extraterrestrial life: identification of nitrogen fixing organisms. Akad. Nauk. SSR Izvestia Ser. Biol. **6,** 797–806.
8 Freyer, H. D. and Aly, A. I. M. 1975 ^{15}N studies on identifying fertilizer excess in environmental systems. IAEA Isotope ratios as pollutant source and behaviour indicators. SM **191/9,** 21–33.

9 Friedman, I. and O'Neil, J. 1977 Data of Geochemistry (Sixth edition). Chapter KK. Compilation of stable isotope fractionation factors of geochemical interest. Geological Survey professional paper 440 KK. US Government Printing Office, Washington.

10 Junc, G. A. and Svec, H. J. 1958 Nitrogen isotope abundance measurements. US atomic energy Commission. Office of technical information ISC **1138.**

11 Karamanos, R. E. and Rennie, D. A. 1980 Changes in natural ^{15}N abundance associated with pedogenetic processes in soil. I Changes associated with saline seeps. Can. J. Soil Sci. **60,** 337–344.

12 Krouse, H. R. 1980 Sulphur isotopes in our environment. *In* Handbook of Environmental Isotope Geochemistry. Eds P. Fritz and J. Ch. Fontes. pp 435–471, Chapter 11, Elsevier.

13 Lewis, R. F. and Pramer, D. 1958 Isolation of Nitrosomonas in pure culture. J. Bacteriol. **76,** 524–528.

14 Mariotti, A. and Letolle, R. 1978 Analyse isotopique de l'azote au niveau des abondances naturelles. Analysis **6,** 421–425.

15 Mariotti, A., Mariotti, F., Amarger, N., Pizelle, G., Ngambi, J. M., Champigny, M. L. et Moyse, A. 1980 Fractionnements isotopiques de l'azote lors des processus d'absorption des nitrates et de fixation de l'azote atmosphérique par les plantes. Physiol. Vég. **18,** 163–181.

16 McCready, R. G. L., Laishley, E. J. and Krouse, H. R. 1976 Biogeochemical implications of inverse sulfur isotope effects during reduction of sulfur compounds by *Clostridium pasteurianum.* Geoch. Cosmoch. Acta **40,** 979–981.

17 McKinney, C. R., McCrea, J. M., Epstein, S., Allen, H. A. and Urey, H. C. 1950 Improvements in mass spectrometers for the measurements of small differences in isotope abundance ratio. Rev. Sci. Instr. **21,** 724–730.

18 Nier, A. O. 1947 A mass spectrometer for isotope and gas analysis. Rev. Sci. Instr. **18,** 398–411.

19 Rees, C. E. 1973 A steady state model for sulphur isotope fractionation in bacterial reduction processes. Geoch. Cosmoch. Acta **37,** 1141–1162.

20 Rennie, D. A., Paul, E. A. and Johns, L. E. 1976 Natural nitrogen-15 abundance of soil and plant samples. Can. J. Soil Sci. **56,** 43–50.

21 Shearer, G., Duffy, J., Kohl, D. H. and Commoner, B. 1974 A steady state model of isotopic fractionation accompanying nitrogen transformations in soils. Soil Sci. Soc. Am. Proc. **38,** 315–322.

22 Tong, J. Y. and Yankwich, P. E. 1957 Calculation of experimental isotope effects for pseudo first-order irreversible reactions. J. Phys. Chem. **61,** 540–543.

E Riparian Buffering of Surface and Subsurface Flow

Commentary on Paper E1

Lowrance, R. R., Todd, R. L. & Asmussen, L. E. (1984) Nutrient cycling in an agricultural watershed: I. Phreatic movement.

This article is one of three foundation papers, with Peterjohn & Correll (1984, Paper E2) and Jacobs & Gilliam (1985, Paper E3), reporting field evidence of the nitrogen buffering capacity of the riparian zone. These three teams were testing the same hypothesis at the same time, but in different climatic and hydrogeomorphological conditions. These parallel and contemporary studies on three different sites have strengthened the individual findings and provided the basis for further integrated hydrological and biogeochemical research on the role of riparian zones in buffering nitrogen fluxes from the drainage basin. Indeed, these three papers were and are still often cited together.

This study, like the other two, reports the monitoring of nutrient concentrations in well transects from upslope to stream and assumes that that subsurface flow paths have been sampled. The difference from the other studies, apart from its location in a 1568-ha agricultural catchment in Georgia, USA, is that they monitor subsurface water quality along nine field/forest or field/pasture interfaces to the stream. Interestingly, or sadly, they then reported their results as "field", "forest" and "streamside", hampering the evaluation of the role of land cover in the riparian zone on nitrogen buffering capacity. Nevertheless, this study design set the pattern for further investigations which tried to decipher the respective roles of local geomorphology and land cover on nitrogen buffering capacity (e.g. Sabater *et al.*, 2003).

They found that nitrate concentration was always significantly higher in field wells than in forest or streamside wells. Comparing these nitrate changes with natural chloride concentration, they showed that nitrate decrease was not due to dilution but to biological processes. They also found that organic N concentration was significantly higher under forest and streamside sites. They concluded that nitrate uptake and incorporation in organic matter was certainly one of the key mechanisms of nitrate removal. This seemed to be confirmed by the similar percentage of organic N in the total N fractions under streamside sites and in-stream. Yet, the organic N concentration increase reported in Fig. 3 under forest and streamside sites did not account for all the nitrate decrease. Therefore, they also mentioned the possible role of denitrification to remove nitrate. Of course, it is hazardous to draw a definite conclusion on concentration change with no information on flow paths and discharge.

In a companion paper (Lowrance *et al.*, 1984b), the authors calculated that subsurface flow represented less than 1% of the streamflow within the catchment. However, they calculated that annual N load in stream flow was about 4 kg N ha^{-1} $year^{-1}$, corresponding to less than one third of the precipitation input, let alone counting the fertilizer load. They concluded that in-stream and riparian zones converted inorganic N to organic N and removed it through denitrification. This is quite a paradox given their estimation that subsurface flow is marginal.

This paradox still holds today and requires further consideration, given that these three papers, i.e. Lowrance *et al.* (1984, Paper E1), Peterjohn & Correll (1984, Paper E2), Jacobs & Gilliam (1985, Paper E3), have for decades significantly influenced the focus on hydrological and biogeochemical processes in riparian zones. These papers inherently possessed a rationale for considering other N buffering landscape structures; unfortunately, we have followed the Chinese proverb for quite a while: "The Master shows the moon; the students look at the finger".

GP

References

Jacobs, T. C. & Gilliam, J. W. (1985) Riparian losses of nitrate from agricultural drainage waters. *J. Environ. Quality* **14**(4), 472–478.

Lowrance, R. R., Todd, R. L. & Asmussen, L. E. (1984b) Nutrient cycling in agricultural watershed: II. Streamflow and artificial drainage. *J. Environ. Quality* **13**(1), 27–32.

Peterjohn, W. T. & Correll, D. L. (1984) Nutrient dynamics in an agricultural watershed: observations on the role of a riparian forest. *Ecology* **65**(5), 1466–1475.

Sabater, S., Butturini, A., Clément, J. C., Burt, T., Dowrick, D., Hefting, M., Maître, V., Pinay, G., Postolache, C., Rzepecki, M. & Sabater, F. (2003) Nitrogen removal by riparian buffers under various N loads along a European climatic gradient: patterns and factors of variation. *Ecosystems* **6**, 20–30.

Nutrient Cycling in an Agricultural Watershed: I. Phreatic Movement[1]

R. RICHARD LOWRANCE, ROBERT L. TODD, AND LORIS E. ASMUSSEN[2]

ABSTRACT

Much of the runoff from agricultural fields in the southeastern Coastal Plain is carried to a stream channel system in a shallow phreatic aquifer. This subsurface runoff often passes through a band of riparian forest before becoming streamflow. It is hypothesized that the riparian ecosystem acts as a nutrient sink and reduces the concentrations and loads of nutrients in the shallow aquifer before the nutrients reach the stream channel. Concentrations and loads (kg/m^2) of NO_3-N, NH_4-N, organic N, dissolved molybdate–reactive P, total P, Ca, Mg, K, Cl, and SO_4-S were measured in shallow phreatic wells at 37 locations on an agricultural watershed near Tifton, Ga. Total water volume moving off the watershed in subsurface flow was < 1% of streamflow with corresponding small amounts of nutrients. Nitrate-N, Ca, and Mg had significantly higher concentrations in field wells than in forest or streamside wells. Concentrations of Cl were not reduced as water moved from field to forest. Processes within the riparian zone apparently converted primarily inorganic N from fields (76% NO_3-N, 6% NH_4-N, 18% organic N) into primarily organic N in streamside wells (10% NO_3-N, 14% NH_4-N, and 76% organic N). Concentration differences between field and forest wells indicated the riparian forest's ability to act as a sink for NO_3-N, Ca, Mg, K, and SO_4-S. Due to their role as nutrient sinks, riparian forests are important in maintaining stream water quality on agricultural watersheds.

Additional Index Words: subsurface runoff, nitrogen cycling, riparian ecosystem, nitrate, phosphorus, calcium, magnesium, potassium, chloride, sulfate.

Lowrance, R. R., R. L. Todd, and L. E. Asmussen. 1984. Nutrient cycling in an agricultural watershed: I. Phreatic movement. J. Environ. Qual. 13:22–27.

Nutrient movement from agricultural areas can take place either through surface or subsurface routes. The segregation of runoff between above- and belowground depends on soil physical properties, climate, and management practices. Previous studies have shown that readily soluble ions such as NO_3-N are likely to move in subsurface flow, while less soluble nutrients such as P will move primarily with surface sediment transport (Jackson et al., 1973; Benoit, 1973; Schuman et al., 1973).

Soils that have a horizon (or horizons) of higher permeability above a horizon of lower permeability generally show appreciable phreatic flow (Whipkey,

[1] Research supported by Natl. Sci. Foundation grants DEB 78-10841 and DEB 82-07210 to the Univ. of Georgia. Additional support from USDA-ARS, Southeast Watershed Res. Lab., Tifton, GA 31793; and Inst. of Ecology and Dep. of Agron., Univ. of Georgia, Athens, GA 30602. Received 14 Feb. 1983.

[2] Former Research Associate, Inst. of Ecology, presently Ecologist, USDA-ARS, Southeast Watershed Res. Lab., former Associate Professor of Agronomy and Associate Ecologist, Univ. of Georgia, presently Professor of Microbiology, South Dakota State Univ., Brookings, SD 57007, and Laboratory Director, USDA-ARS, Southeast Watershed Res. Lab., respectively.

1966; Weyman, 1970). The soils of the Atlantic and Gulf Coastal Plain are generally sandy and have high infiltration rates (Rawls et al., 1976). On the Little River Watershed (LRW) in the Tifton Upland, the more permeable surface horizons are underlain by a plinthite aquiclude at depths of 0.9 to 1.5 m (Rawls & Asmussen, 1973), and much of the incident precipitation infiltrates and moves laterally as phreatic flow above the plinthite. This phreatic flow enters the alluvial zones near the stream and eventually moves to the stream channel and off the watershed in surface flow. Stream channels of the Tifton Upland are more or less continuously bordered by a layer of mixed hardwood riparian forest; thus, water and associated nutrients from upland fields must flow through the forest as either subsurface or surface flow. This study reports on the concentrations and loads of nutrients moving in the phreatic zone from field to forest to stream channel. These concentrations and loads are the products of nutrient cycling processes in an agricultural watershed. Changes in these concentrations and loads are evidence for the ability of the riparian forest to remove nutrients from phreatic water. Land use effects on nutrients in the phreatic aquifer are also discussed.

STUDY AREA

This study was conducted during 1979 on watershed N, a subwatershed of the Little River near Tifton, Ga. (Fig. 1). Rainfall in 1979 was 1238 mm, near the 50-y average of 1203 mm (Batten, 1980). Watershed N is 1568 ha with 45% of the watershed in row crops [corn (*Zea mays* L.), peanuts (*Arachis hypogaea* L.), soybeans (*Glycine max* (L.) Merr.), milo (*Sorghum bicolor* L.), winter vegetables, and tobacco (*Nicotiana tabacum* L.)], 30% occupied by the riparian forest ecosystem, 14% in pasture, and the remainder devoted to pine forests, roads, and residences. The well-drained upland soils are primarily Plinthic Paleudults and the poorly drained bottomland soils are primarily Plinthic Paleaquults. Major soil series on the watershed are Tifton (51% of area), Alapaha (18%), and Dothan (8%). Previous studies have shown the importance of subsurface flow as a nutrient transport mechanism on this watershed. Asmussen and Ritchie (1969) found that for the period 1 Jan. to 31 Mar. 1968, 32% of the total precipitation was yielded as interflow. Working on a 0.34-ha plot in Tifton, Jackson et al. (1973) found from 17 to 41% of the total annual rainfall moved in subsurface flow and that subsurface movement accounted for 81% of the total runoff from the small field-sized watershed. This water carried 99% of the NO_3-N that was lost from the plot in runoff.

MATERIALS AND METHODS

Thirty-seven wells to sample the shallow perched water table on watershed N were located along nine transects extending from the field/forest or field/pasture interface to the stream. Wells were located at the edge of each field to avoid damage by farm equipment. Each well was hand augered to the apparent depth of the plinthite in August and September 1978. Field wells were 1.52 m deep; wells in the floodplain were as shallow as 0.8 m since the saturated zone was encountered at higher levels. Well casings were formed by 10-cm polyvinyl chloride (PVC) pipe perforated with 0.64-cm holes on the bottom 0.5-m and capped on the aboveground end. The perforated end of the pipe was wrapped with nylon windowscreen in order to avoid caving in through the holes, and the soil profile outside the casing was restored by replacing soil in reverse order of its removal. Plastic collars were fitted around each well and buried under a few centimeters of soil to prevent seepage of rainwater around the well casing.

Subsurface water movement off watershed N was measured by six wells located along the lower boundary of the catchment formed by a county road. The wells were installed at a depth of approximately 2.5 m and cased with perforated Al pipe except the sampling well, which was PVC. A seventh well was fitted with a strip chart recorder that provided a continuous record of water table heights. A water quality sample was taken along the roadside on each sampling date. Wells were sampled on 33 dates in 1979 and well logs were taken on six other dates. Samples (500-mL) were taken after rainfall events or at least every 2 weeks between events using a hand pump.

Concentrations of NO_3-N, NH_4-N, Cl, dissolved molybdate-reactive P (DMRP), and SO_4-S were determined by standard spectrophotometric techniques (APHA, 1976). Total Kjeldahl N and total P were determined by analysis of a digestate (Technicon, 1977). Calcium, magnesium, and potassium were determined using a Perkin-Elmer model 5000 atomic absorption spectrophotometer.

Darcy's equation was used to estimate water movement in the phreatic zone. This method is generally thought to be adequate for representing homogeneous flow (Davis & DeWiest, 1966). This equation uses values of water table slope or hydraulic head (I), hydraulic conductivity (K), and saturated area (A) to calculate water movement (Q_u) through a porous medium:

$$Q_u = KIA \qquad [1]$$

Records from the recording well were translated into hourly gage heights and a simple linear regression was used to predict the daily height of each observation well from the mean daily height of the recording well. Correlations of water table elevation in the sampling wells with the recording well were done to estimate the response of groundwater to precipitation and produce a more accurate record of saturated area at each well. Regressions were calculated for four well logs during one storm event on 22–25 March. These equations had r^2 values > 0.90 in most cases. The intercept of each regression line was adjusted seasonally by determining the average difference between the actual and the predicted gage height for each season. The timing of well hydrographs on different parts of the watershed varied considerably depending on both the position of a transect on the watershed and the position of the well on the transect. Correlation with the recording well assumed that the general shapes of the responses were similar with differences in hydrograph timing unimportant for monthly and annual calculations. Saturated area was estimated by multiplying the daily saturated depth by a width of 1 m so that unit loads could be compared among wells. Saturated depth was estimated by assuming that each soil profile conducted water through the top 1.5 m. Descriptions of each soil on the watershed (McCreery, 1967) indicate significant percentages of plinthite at this depth. Hydraulic conductivity of wells was determined by slug tests (Bouwer & Rice, 1976). Slope was calculated as the change in water table elevation ($h_1 - h_2$) over the distance between adjacent wells. When no upslope wells were avail-

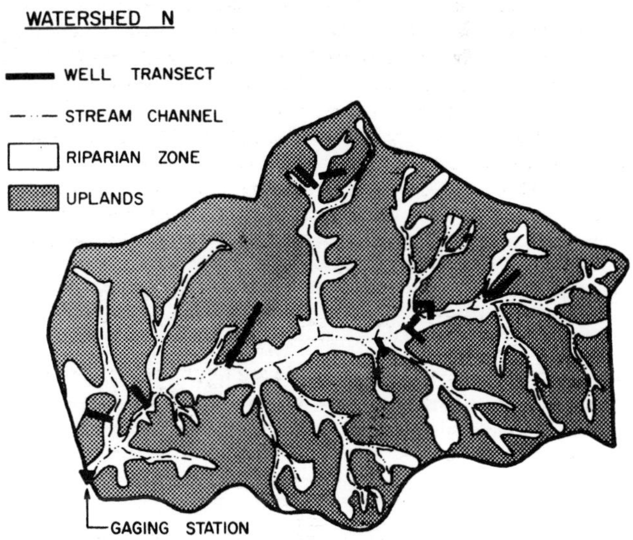

Fig. 1—Watershed N, a subwatershed of the Little River near Tifton, Ga., showing the location of well transects.

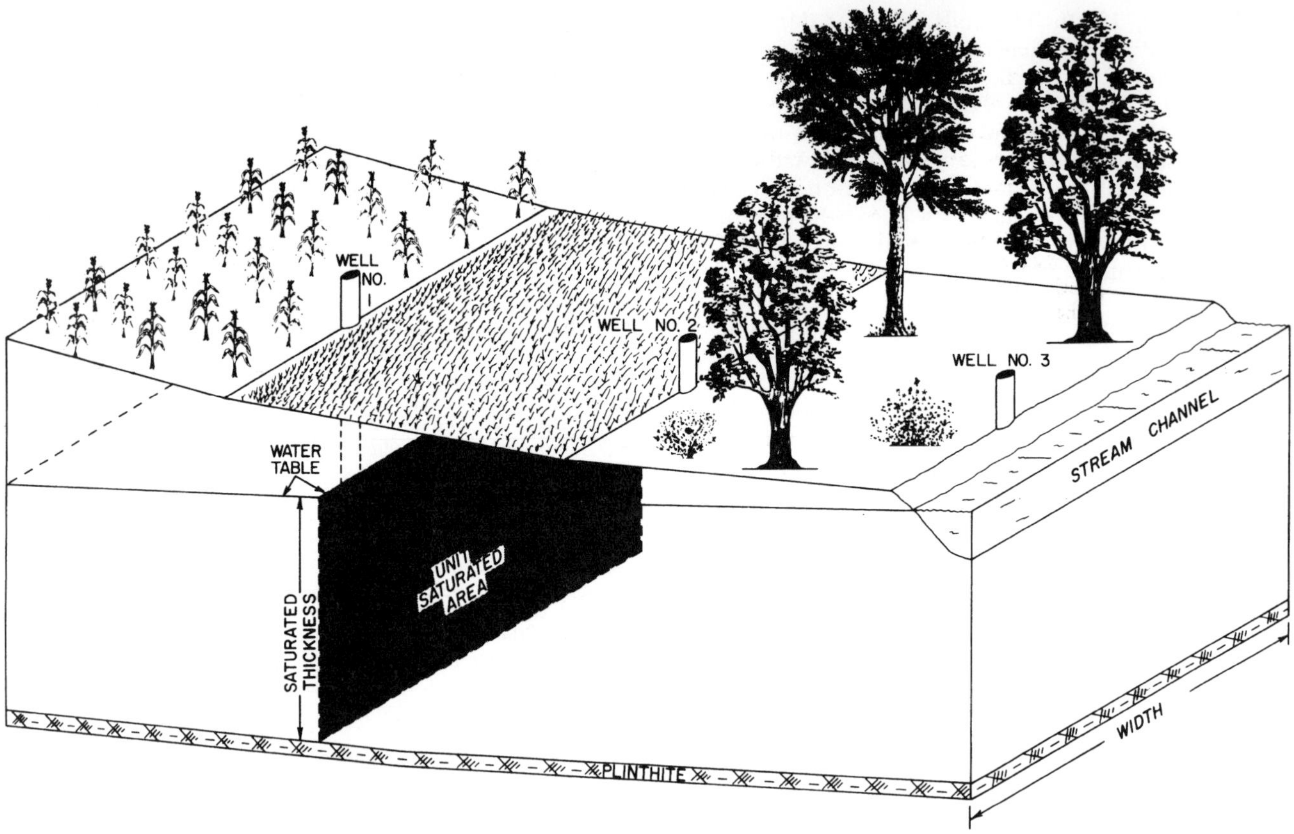

Fig. 2—Schematic representation of the calculation of unit discharge (Q_u) at a well. Well no. 1 is at the field/pasture interface; well no. 2 is at the pasture/forest interface; well no. 3 is in a streamside position. Slope (I) is the difference in water table elevation between two wells divided by the distance between the wells. Unit saturated area (A) is the product of saturated thickness (in meters) times a width of 1 m. The vertical scale is exaggerated.

Table 1—Summary of ranks of wells on seven transects based on Duncan's Multiple Range Test. Figures represent number of times a particular type of well had highest concentration for the nutrient species.

Nutrient	Field	Forest	Streamside
NO_3-N	7	0	0
NH_4-N	3	1	3
Organic-N	3	1	3
Cl	4	1	2
P	1	0	0
DMRP	3	1	3
Ca	7	0	0
Mg	5	0	1
K	4	1	2
SO_4-S	4	0	1

able, as for wells at the field/forest interface, the ground surface slope was used in calculating discharge. Figure 2 represents schematically the calculation of water volume at a well site. Subsurface nutrient loads were calculated by multiplying mean monthly concentrations by total monthly subsurface discharge through the unit cross section.

Saturated area for water transport off the watershed was graphically calculated for each sampling date from logs of the wells located along the watershed boundary. These saturated areas (y, in m^2) were related to water table elevation (x, in m) in the recording well by the linear regression:

$$y = 1\,908x - 539\,051, r^2 = 0.81 \quad [2]$$

Total subsurface discharge off the watershed was estimated by Darcy's equation using the mean hydraulic conductivity of all streamside wells and the slope of watershed N along the stream channel.

RESULTS AND DISCUSSION

Phreatic Zone Nutrient Concentrations

Each well was in either a field, pasture, forest, or streamside position. Duncan's multiple range test at the 0.05 level was used to group wells of each category according to their mean nutrient concentrations in 1979. Table 1 summarizes the results of this testing for seven transects. Calcium and nitrate-N represent one extreme, where field well concentrations were always significantly higher than forest or streamside wells. The other extreme was represented by total P, for which only one transect had significantly different concentrations. Organic N, NH_4-N, and DMRP showed no significant trend with higher concentrations in each type of well on various transects. Potassium had peak concentrations at four field and two streamside sites. Potassium is applied in large quantities to row crops (e.g., 196 kg ha^{-1} y^{-1} for tobacco), but is also recycled in canopy leachate by throughfall within the riparian zone (Brinson et al., 1980). Concentrations of Cl peaked in all three well types, indicating that different processes are controlling the amounts of Cl and NO_3-N found along the transects. The moderately mobile ions Ca and Mg were apparently stored in large quantities at depths in the soil.[3] Calcium and magnesium were applied as dolomitic lime and had consistently higher concentrations in field wells. Nutrients such as NH_4-N, DMRP, and organic N

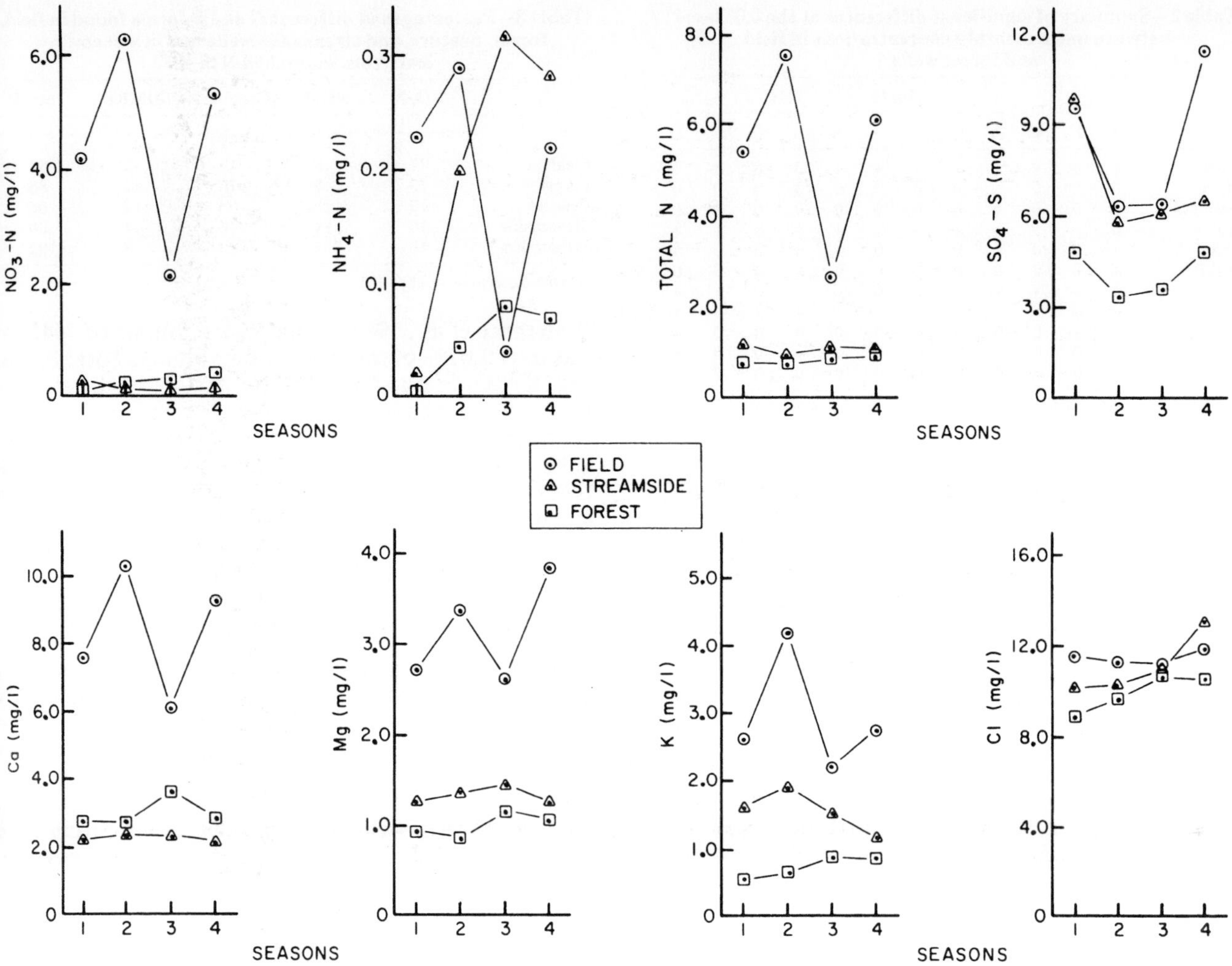

Fig. 3—Seasonal means of nutrients in field, forest, and streamside wells. Each season is 3 months long beginning with January in season 1.

were apparently not moving into the phreatic zone in response to management practices in the fields.

Mean seasonal concentrations of NO_3-N, NH_4-N, total N, Ca, Mg, and K in field wells (Fig. 3) showed the same general trend—increases (usually to a maximum) during season 2 (April, May, June) and decreasing concentrations during season 3 (July, August, September). Chloride and sulfate-S did not exhibit this trend, although lowest concentrations of SO_4-S were found in April through September. Total P and DMRP showed almost no change between seasons. Seasonal concentrations in streamside and forest wells exhibited no general trend, but forest wells had higher means of NH_4-N, Ca, Mg, and K during the summer months, possibly due to higher rates of mineralization of forest floor litter. The sharp increase of NH_4-N in streamside wells might be caused by lower rates of nitrification found close to streams due to increased wetness and a more reduced soil environment near the streams.[4]

[3]E. A. Carter. 1981. Effect of slope and landscape upon selected soil characteristics of the Little River Watershed. M.S. Thesis. Univ. of Georgia, Athens.

[4]O. Q. Hendrickson. 1981. Flux of nitrogen and carbon gases in bottomland forests of an agricultural watershed. Ph.D. Dissertation. Univ. of Georgia, Athens.

Streamflow concentrations of SO_4-S had similar seasonal behavior as streamside wells with peaks in winter and minima during the growing season (Lowrance et al., 1984). Streamflow NO_3-N was generally highest in season 1 and lowest in season 3, but streamside wells showed little change between seasons. Decreases in NO_3-N and SO_4-S during the summer months could have been due to increased microbial activity reducing more SO_4-S and NO_3-N to gaseous forms. Hendrickson[4] found peaks of denitrification occurring in June through September in forest and streamside locations.

Seasonal mean concentrations of NO_3-N, total N, Ca, Mg, and K in field wells always exceeded forest wells. Significant differences of monthly means (0.05 level for a Student's *t*-test) are summarized in Table 2. Magnesium was higher in fields every month. Organic N, NH_4-N, total P, and DMRP rarely had significantly different concentrations. Fields had higher Ca and NO_3-N during most months. Chloride and nitrate-N represent an interesting comparison since both ions are extremely mobile, moving 2 to 25% faster than water in some soils (Tullock et al., 1975; Smith & Davis, 1974). Field levels of Cl were only slightly higher than Cl in forest and streamside wells and the difference was only significant

Table 2—Summary of significant differences at the 0.05 level between mean monthly concentrations in field and forest wells.†

	Month											
Nutrient	J	F	M	A	M	J	J	A	S	O	N	D
NO_3-N	+	+	+	+	+	+	+	+	0	0	+	+
NH_4-N	0	0	0	0	0	0	0	0	0	0	0	0
Organic N	0	0	0	0	0	0	0	0	0	0	0	0
Total N	0	+	+	+	+	+	+	+	0	0	+	+
DMRP	+	0	0	0	0	0	0	−	0	0	0	0
Total P	+	0	0	0	0	0	0	0	0	0	0	0
Ca	+	+	+	+	+	0	0	+	+	0	+	+
Mg	+	+	+	+	+	+	+	+	+	+	+	+
K	+	+	0	0	+	0	0	0	0	0	0	+
SO_4-S	+	0	0	0	0	0	0	0	+	0	0	0
Cl	+	0	0	0	0	0	0	0	0	0	0	+

† + Indicates field wells higher; − indicates forest wells higher; 0 indicates no significant difference.

during 2 months (Table 2). Seasonal means of Cl concentrations were higher in fields, but showed a less distinct seasonal trend than NO_3-N (Fig. 3). The difference in behavior of NO_3-N and Cl indicates the effect of biological processes such as plant uptake and denitrification which removed NO_3-N in the riparian zone.

Phreatic Nutrient Loads

Hydraulic conductivities obtained by the slug test ranged from 0.09 m/d to 3.94 m/d. Due to coarser soils, wells in the uplands had higher conductivities (mean = 1.52 m/d) than forest wells (mean = 0.79 m/d) or streamside wells (mean = 1.15 m/d). Higher values at streamside than at forest locations are probably due to greater amounts of coarse sediments deposited during floods. Hydraulic conductivities integrated the values of a number of horizons through the perforated portion of the well casing. Conductivity was probably lower in deeper horizons approaching the plinthite, but the water table would have been found there more often than in the upper horizons.

Mean monthly water table slopes varied among wells by > 1 order of magnitude. The lowest value for the year was 0.007 m/m in August; the highest value of 0.087 m/m occurred in December. The ground surface slope used for upland wells was within the range of values derived from predicted water table elevations.

The percentages of total N found as NO_3-N, NH_4-N, and organic N varied considerably in field, forest, pasture, and streamside wells (Table 3). Streamflow data (Lowrance et al., 1984) showed 80% of streamflow N was organic—about the same as forest wells. Streamside wells had a higher percentage of NH_4-N and lower NO_3-N, but organic-N was about the same as streamflow. The ratio of DMRP to non-DMRP (total P minus DMRP) showed little difference among well types. The switch from inorganic N in field wells to organic N in the forest was due to uptake by forest vegetation and microbial populations and subsequent decomposition of the organic material. This transformation was an important factor in producing a predominance of organic N in streamflow.

Nutrient loads and water volumes transported off watershed N in subsurface flow were small compared with volumes and loads transported in streamflow

Table 3—Percentages of different N and P forms found in field, forest, pasture, and streamside wells and in streamflow load from watershed N in 1979.

	NO_3-N	NH_4-N	Organic N	DMRP	Total P
	%				
Field	77	6	17	12	88
Forest	20	5	75	12	88
Pasture	10	22	68	12	88
Streamside	10	14	76	14	86
Streamflow†	18	2	80	8	92

† From Lowrance et al., 1984.

(Lowrance et al., 1984). Total water volume of 3643 m^3 was only 0.07% of the streamflow volume. Total P, Ca, Mg, K, SO_4-S, and Cl were all < 0.07% of the streamflow load, indicating that they were relatively less concentrated in subsurface flow than in streamflow. Organic N, NO_3-N, NH_4-N, total N, and DMRP were all relatively more concentrated in subsurface flow than streamflow. In general, subsurface discharge through the watershed boundary carried negligible quantities of nutrients due to the small volume of water movement.

CONCLUSIONS

Nitrate-N, Ca, and Mg were always found in significantly higher concentrations in field wells, indicating that they were both more available for leaching from fields and that they were removed from the shallow aquifer in the riparian forest zones. Nitrate-N, NH_4-N, total N, Ca, Mg, and K generally increased during April through June in field wells and decreased during July through September. These seasonal dynamics were related to fertilizer application and subsequent crop uptake of nutrients. The higher concentrations of NO_3-N in field wells than in forest or streamside wells were expressed in the relative amounts of the total N load provided by different N species. Fields had 78% NO_3-N, 6% NH_4-N, and 16% organic N. Forest and streamside wells had 80 and 76%, respectively, of total N as organic N—about the same as streamflow. Apparently, the riparian forest zone acted as a filter for NO_3-N, Ca, Mg, K, and SO_4-S, while N was transformed from inorganic to organic forms in the bottomland forest. Almost all water and nutrient movement from the watershed was in streamflow with subsurface discharge off the watershed accounting for < 1% of streamflow.

LITERATURE CITED

1. American Public Health Association (APHA). 1976. Standard methods for the examination of water and waste water. APHA, Washington, D.C.
2. Asmussen, L. E., and J. C. Ritchie. 1969. Interflow or shallow phreatic flow in the Coastal Plain of Georgia. J. Hydrol. 9:182–193.
3. Batten, H. L. 1980. Little River research watersheds. USDA Misc. Publ. Southeast Watershed Res. Lab., Tifton, Ga.
4. Benoit, G. R. 1973. Effect of agricultural management of wet sloping soil on nitrate and phosphorus in surface and subsurface water. Water Resour. Res. 9:1296–1303.
5. Bouwer, H., and R. C. Rice. 1976. A slug test for determining hydraulic conductivity of unconfined aquifers with completely or partially penetrating wells. Water Resour. Res. 12:423–428.
6. Brinson, M. M., H. D. Bradshaw, R. N. Holmes, and J. B. Elkins, Jr. 1980. Litterfall, stemflow, and throughfall nutrient fluxes in an alluvial swamp forest. Ecology 61:827–835.

7. Davis, S. N., and R. J. M. DeWiest. 1966. Hydrogeology. John Wiley & Sons, Inc., New York.

8. Jackson, W. A., L. E. Asmussen, E. W. Hauser, and A. W. White. 1973. Nitrate in surface and subsurface flow from a small agricultural watershed. J. Environ. Qual. 2:480–482.

9. Lowrance, R. R., R. L. Todd, and L. E. Asmussen. 1984. Nutrient cycling in an agricultural watershed: II. Streamflow and artificial drainage. J. Environ. Qual. 13:27–32 (this issue).

10. McCreery, R. A. 1967. Notes on data for soils from Little River Watershed, Tift and Turner Counties, Georgia. Report to Southeast Watershed Res. Center, Athens, Ga. Univ. of Georgia, Athens.

11. Rawls, W. J., and L. E. Asmussen. 1973. Subsurface flow in the Georgia Coastal Plain. Proc. Am. Soc. Civ. Eng. 99:375–385.

12. Rawls, W. J., P. Yates, and L. E. Asmussen. 1976. Calibration of selected infiltration equations for the Georgia Coastal Plain. USDA Agric. Res. no. ARS-S-113, Southern Region, New Orleans, La.

13. Schuman, G. E., R. G. Spomer, and R. F. Piest. 1973. Phosphorus losses from four agricultural watersheds on Missouri Valley loess. Proc. Soil Sci. Soc. Am. 37:424–427.

14. Smith, S. J., and R. J. Davis. 1974. Relative movement of bromide and nitrate through soils. J. Environ. Qual. 3:152–155.

15. Technicon. 1977. Industrial Method no. 376-75W/B. Technicon Industrial Systems, Tarrytown, N.Y.

16. Tullock, R. J., N. T. Coleman, and P. F. Pratt. 1975. Rate of chloride and water movement in southern California soils. J. Environ. Qual. 4:127–131.

17. Weyman, D. R. 1970. Throughflow on hillslopes and its relation to the stream hydrograph. Bull. Int. Assoc. Sci. Hydrol. 15: 25–33.

18. Whipkey, R. Z. 1966. Subsurface stormflow from forested slopes. Bull. Int. Assoc. Sci. Hydrol. 10:74–85.

Commentary on Paper E2

Peterjohn, W. T. & Correll, D. L. (1984) Nutrient dynamics in an agricultural watershed: observations on the role of a riparian forest.

Dave Correll in this paper recalls that this study was conducted after an attempt to relate surface soil and subsoil composition to nutrient discharge for three watersheds of differing land use. No correlation was found. This implied that significant nutrient transformations were occurring as water moved through the catchment. This was really a breakthrough which led him to propose this research programme with Peterjohn. This article is probably one of the most cited field research works, and the most cited one on riparian studies, with over 600 citations to date.

This study consists of a mass balance budget of carbon, nitrogen and phosphorus between an agricultural field and its adjacent riparian zone in a very small drainage basin (16.3 ha). The originality of this study was to address this budget in a hillslope context in order to determine if significant nutrient transformation was occurring at this upland/wetland interface. A comparison of nutrient stocks in and out of the riparian zone allowed the authors to estimate a nutrient buffering capacity. However, the absence of coupled hydrological measurement, notably subsurface flow, weakened the estimation of the subsurface fluxes. For instance, they assumed that groundwater flow paths were perpendicular to the surface slope and that their piezometers were hydrologically connected.

The C, N and P budgets for standing stocks, i.e. surface and groundwater, soil and plants, were established on an annual basis and results for the three elements were equally discussed in the paper. However, the main result, referred to in numerous later studies on the subject, was that 75% of the groundwater nitrate had been removed within 19 m in the riparian zone. The authors proposed that plant uptake and denitrification were the main mechanisms responsible for this removal. They estimated that plant uptake would contribute to about one third of the removal, although they did not consider that mature forest would release most of this as organic N via litter fall and root exudates. They estimated that two thirds of this removal would be due to denitrification because they did not measure any significant increase in ammonia or dissolved organic N which could reveal assimilation or dissimilation processes.

They also emphasised that most of the removal was achieved within a short distance, despite differences between transects. A detailed groundwater flow analysis would probably have shown even more local variability within this 19-m width, but it did seem clear that groundwater nitrate removal could be complete within the riparian zone.

This paper was also very inspiring because it formulated the two key questions which still keep us busy 25 years later: (i) What caused the decrease in groundwater nitrate concentration in the riparian forest? (ii) How widely applicable are the effects they observed?

Indeed, even if nowadays denitrification is considered as the main removal process within riparian zones, the importance of anammox (Strous & Jetten, 2004), nitrifier denitrification (Wrage *et al.*, 2001) and oxidation by the ferrous wheel (Clément *et al.*, 2005) in removing nitrogen is still not clear. Moreover, the applicability of riparian zones as N buffers under different geomorphological conditions is still the subject of debate (Sabater *et al.*, 2003). Note, however, that the robustness of this temperate model (where a combination of plant uptake and denitrification act as the main removal mechanisms) under tropical climate for instance, has been rarely tested (McClain *et al.*, 1994).

GP

References

Clément, J. C., Shrestha, J., Ehrenfeld, J. G. & Jaffe, P. R. (2005) Ammonium oxidation coupled to dissimilatory reduction of iron under anaerobic conditions in wetland soils. *Soil Biology & Biochem.* **37**, 2323–2328.

McClain, M. E., Richey, J. E., & Pimentel, T. P. (1994) Groundwater nitrogen dynamics at the terrestrial-lotic interface of a small catchment in the Central Amazon Basin. *Biogeochemistry* **27**, 113–127.

Sabater, S., Butturini, A., Clement, J. C., Burt, T. P., Dowrick, D., Hefting, M., Maître, V., Pinay, G., Postolache, C., Rzepecki, M. & Sabater, F. (2003) Nitrogen removal by riparian buffers under various N loads along a European climatic gradient: patterns and factors of variation. *Ecosystems* **6**, 20–30.

Strous, M. & Jetten, M. S. M. (2004) Anaerobic oxidation of methane and ammonium. *Ann. Rev. Microbiol.* **58**, 99–117.

Wrage, N., Velthof, G. L., van Beusichem, M. L. & Oenema, O. (2001) Role of nitrifier denitrification in the production of nitrous oxide. *Soil Biol. & Biochem.* **33**, 1723–1732

Ecology, 65(5), 1984, pp. 1466–1475

NUTRIENT DYNAMICS IN AN AGRICULTURAL WATERSHED: OBSERVATIONS ON THE ROLE OF A RIPARIAN FOREST[1]

William T. Peterjohn and David L. Correll
Smithsonian Environmental Research Center, P.O. Box 28, Edgewater, Maryland 21037 USA

Abstract. Nutrient (C, N, and P) concentration changes were measured in surface runoff and shallow groundwater as they moved through a small agricultural (cropland) watershed located in Maryland. During the study period (March 1981 to March 1982), dramatic changes in water-borne nutrient loads occurred in the riparian forest of the watershed. From surface runoff waters that had transited ≈50 m of riparian forest, an estimated 4.1 Mg of particulates, 11 kg of particulate organic-N, 0.83 kg of ammonium-N, 2.7 kg of nitrate-N and 3.0 kg of total particulate-P per ha of riparian forest were removed during the study year. In addition, an estimated removal of 45 $kg \cdot ha^{-1} \cdot yr^{-1}$ of nitrate-N occurred in subsurface flow as it moved through the riparian zone.

Nutrient uptake rates for the cropland and riparian forest were estimated. These systems were then compared with respect to their pathways of nutrient flow and ability to retain nutrients. The cropland appeared to retain fewer nutrients than the riparian forest and is thought to incur the majority of its nutrient losses in harvested crop. The dominant pathway of total-N loss from the riparian forest seemed to be subsurface flux. Total phosphorus loss from the riparian forest appeared almost evenly divided between surface and subsurface losses.

Nutrient removals in the riparian forest are thought to be of ecological significance to receiving waters and indicate that coupling natural systems and managed habitats within a watershed may reduce diffuse-source pollution.

Key words: carbon; cropland; diffuse source pollution; groundwater; mass balance; nitrogen; nutrient cycling; phosphorus; riparian forest; surface runoff; watershed.

Introduction

Excessive nutrient loading can have significant ecological effects on the receiving waters of lakes (Edmondson 1972, Powers et al. 1972, Schelske and Stoermer 1972), streams (Hynes 1969), and estuaries (Fraser and Wilcox 1981, Myers and Iverson 1981). Nutrient loadings from managed watersheds have contributed an increasing amount of nutrients to receiving waters as agricultural practices have intensified and residential development has expanded to accommodate a growing human population. Furthermore, nutrient losses from agricultural lands represent both a monetary and energy loss to society.

In an effort to understand how nutrient loss is affected by internal watershed structure, Correll (1984) attempted to relate surface soil and subsoil composition to nutrient discharge for three watersheds of differing land use. Little or no correlation was found, and this was thought to indicate that significant nutrient transformations were occurring as water moved through the watersheds as surface runoff and groundwater. Thus, one objective of this study was to investigate nutrient (N, P, and C) transformations occurring as water moved through an agricultural (cropland and riparian forest) watershed as surface runoff and shallow groundwater. Another objective was to synthesize past data about this watershed to gain a better perspective of the overall nutrient dynamics of an agricultural ecosystem.

Site Description

The study site is a small subwatershed of the 3,332-ha Rhode River drainage basin. The Rhode River lies within the mid-Atlantic Coastal Plain along the western shore of Chesapeake Bay ≈20 km south of Annapolis, Maryland (38°53′N, 76°33′W). The subwatershed studied (Fig. 1) is a 16.3-ha basin of which 10.4 ha was planted with corn. Cropland was tilled prior to planting and preemergent herbicides were used to control subsequent weed growth. Riparian forest and hedgerows were composed of broadleaved, deciduous trees and accounted for the remaining 5.9 ha. The soils are a fine sandy loam and are extremely deep (>600 m). An underlying clay layer (the Marlboro Clay) is thought to create an effective aquiclude near sea level for the entire basin (Chirlin and Schaffner 1977). At the weir the clay layer is at a depth of ≈2.0 m. Therefore, the site contains a perched, shallow aquifer. Soils above the aquiclude are noncalcareous and contain no phosphate minerals (Pierce 1982). Surface soil organic matter content and pH in the cultivated fields were 1.9% and 5.6, respectively, when measured in 1977. The basin slope is 5.44% and the channel slope is 2.65%.

Material and Methods

Sampling

Bulk precipitation for chemical analysis was continuously sampled at an elevation of 13 m at the central

[1] Manuscript received 31 January 1983; revised 14 September 1983; accepted 16 September 1983.

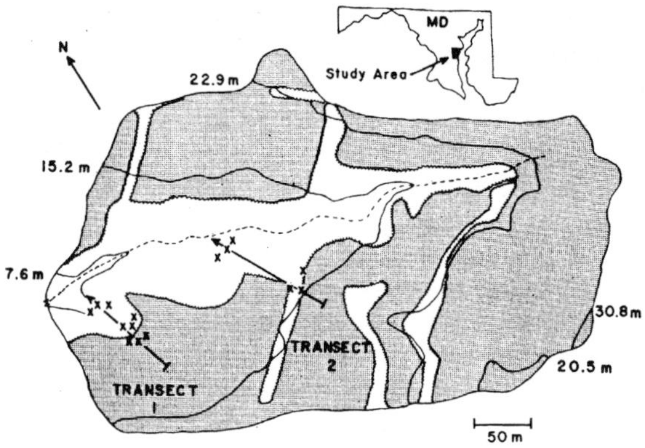

FIG. 1. General location of the study area (inset), and map of the watershed showing the positions of sampling transects (←) and goundwater wells (×). Shading indicates cultivated fields.

weather station for the Rhode River watershed (2.3 km from the study site). Collection was made in a clean polyethylene bottle fitted to a 28 cm diameter polyethylene funnel. Fiberglass window screen covered the funnel to keep insects out. Samples were collected following each rain event, and the sampler was cleaned and returned for use. The amount of rainfall was measured with a Belfort mass-recording rain gauge and a standard weather bureau manual gauge at the central weather station, and with a Steven's tipping-bucket gauge at the study site itself.

Stream discharge was monitored and water samples taken at a 120° sharp-crested V-notch weir, the foundation of which rested upon the Marlboro Clay layer. The associated instrument shed was equipped with a stilling well, depth monitor, and flow meter. The weir was built and instrumented in the spring of 1976 and both water and nutrient discharges have been measured since that time. Depth measurements were taken every 5 min and recorded as digital data on a punch tape. The flow meter was modified to close a sampling switch every 38 m^3 of discharge. Switch closure activated a sampling cycle in which a fixed volume of stream water was pumped from the base of the V-notch into sample containers. One sample bottle contained sulfuric acid preservative for biologically labile chemical species, the other contained no preservative. Samples were composited over weekly intervals for the 1-yr study period beginning in March 1981 and ending in March 1982. During a typical storm week over 2000 m^3 are discharged and over 50 samples taken. During a single, intense, summer storm that occurred during the study period 46 samples were taken in <12 h. During periods of very low discharge (<10 m^3/wk) weekly grab samples were taken. Quickflow and slowflow rates were calculated graphically from the hydrograph (Barnes 1940). In this paper we equate surface runoff with quickflow and subsurface or groundwater flows with slowflow. Subsurface (groundwater) flows in the system represent soil water percolation above the aquiclude. Seasons are defined in this paper as 3-mo periods (winter = December, January, and February).

Two transects of groundwater wells, ≈180 m apart, were established in the watershed (Fig. 1). Each transect consisted of several clusters of samplers at different elevations along the expected direction of flow. Each cluster consisted of a central groundwater well (piezometer) and two lateral wells located ≈10 m to either side and normal to the transect axis. Surface-water collectors located to collect only overland flow were associated with each groundwater well along transect 1. A total of 18 wells and nine surface water collectors were used. Surface-water collectors consisted of 4-L polyethylene bottles placed into holes in the soil in an inverted position. A rectangular slot was cut in the uphill side of the bottle at ground level and a short plastic apron was glued to the bottom of the slot and spread uphill to funnel surface flow into the bottle. Plastic tubing sealed into the cap of the inverted bottle allowed samples to be withdrawn without disturbing the sampler. Air was briefly bubbled through the sample to suspend particulates before the sample was withdrawn. Samplers were emptied prior to expected storm events and periodically cleaned. Groundwater wells consisted of 3.8 cm (inside diameter) polyvinyl chloride pipe perforated with 2.5-mm holes for ≈8 cm on the lower end. The bottom of the pipe was capped. Well holes were bored with a bucket auger either to the top of the Marlboro Clay or to a 4 m depth, whichever came first. Most wells were between 1.5 and 3.0 m deep. The pipes were inserted into the augered holes and clay packed around the pipe at the soil surface. The pipes extended ≈0.4 m above the ground and a loose cap was placed over the top. In all but one case the cluster of wells also had a fourth, shallower, groundwater well. On the day following a storm, water table heights were measured and wells pumped free of water. Samples were taken of the fresh groundwater the following day. When groundwater levels were low, equal volumes from each well within a cluster were composited to obtain enough sample for chemical analysis. Samples were returned to the laboratory and stored in the dark at 1.1°C. Aliquots for dissolved-nutrient determinations were filtered through prewashed Millipore HA membrane filters (0.45 μm nominal pore size) the day of collection and then returned to cold storage.

Nutrient analysis

Unfiltered rainfall samples were analyzed for nitrate, total Kjeldahl-N, ammonium-N, total-P, orthophosphate-P and organic matter concentrations. Surface runoff and groundwater samples were filtered and the filtrate analyzed for "dissolved" nitrite, nitrate, total Kjeldahl-N, ammonium-N, total-P, orthophosphate-P and organic matter concentrations. Unfiltered (whole) surface runoff samples were also analyzed for exchangeable ammonium-N, total Kjeldahl-N, exchangeable orthophosphate-P, total-P, total suspended particulate, and organic matter concentrations. Unfiltered, acidified, stream samples were analyzed for total Kjeldahl-N, ammonium-N, nitrate, total-P, orthophosphate-P, and organic matter.

Dissolved nitrite and nitrate were determined in nonacidified samples with a Dionex model 16 ion chromatograph. Nitrite and nitrate were determined in acidified samples by reduction on cadmium amalgam and colorimetry (APHA 1976). Since nitrite was present only in trace amounts, nitrite and nitrate are routinely summed and referred to as nitrate. Dissolved ammonium-N was determined by the oxidation of ammonia and labile amino compounds to nitrite and the subsequent colorimetric determination of nitrite (Richards and Kletsch 1964). Total Kjeldahl nitrogen, which includes ammonium nitrogen and organic amines, was determined by digestion with sulfuric acid and hydrogen peroxide (Martin 1972), distillation, and Nesslerization (APHA 1976). Organic nitrogen was calculated as the difference between total Kjeldahl nitrogen and ammonium nitrogen.

Dissolved total phosphorus was determined by a perchloric acid digestion (King 1932) and reaction with ammonium molybdate and stannous chloride (APHA 1976). Total phosphorus in unfiltered samples was determined by perchloric acid digestion, the development of a phosphomolybdic acid-blue complex and extraction of the blue complex with isobutyl alcohol (Correll and Miklas 1976). Determination of dissolved orthophosphate followed the same procedure used for dissolved total phosphorus except that samples were not digested.

Organic matter was measured as chemical oxygen demand (in milligrams per litre) (Maciolek 1962). An empirical conversion factor of 0.417 was used to obtain organic carbon concentration, in milligrams per litre. To determine extractable ammonium-N and orthophosphate-P, measured volumes of whole samples were filtered through polycarbonate nucleopore filters (0.4-μm pore size) and the particulate residue was washed with 1 mol/L KCl for ammonium-N or 0.5 mol/L HCl for orthophosphate-P. Following extraction and neutralization the same procedures used for the determination of dissolved ammonia and orthophosphate were followed. Suspended particulate concentration was calculated from the gain in mass of a prewashed, dried, and weighed millipore filter after filtration of a measured volume of sample. Filters both before and after filtration were dried for at least a week in a desiccator. All particulate parameters were calculated as the difference between concentrations found in unfiltered and filtered aliquots.

Vegetative sampling

In early April, 1982, prior to leafing out, ≈0.4 ha of the riparian forest near transect 1 was intensively sampled. The point-centered quarter technique (Cottam and Curtis 1956) was used to determine the relative dominance, density, frequency, size distribution, and importance value of the tree species present. Two increment cores were also taken from 130 of the 134 trees sampled. The average ring width was recorded for each tree and the dbh of the previous year calculated. The mean dbh for each of 13 size classes was then determined for the beginning and end of the study year. Weighted averages were calculated for both time periods using the number of trees in each size class as the weighting factor. These values and the allometric relations developed by Harris et al. (1973) were used to develop a rough estimate of the standing biomass of various tree components (branches, boles, large roots, and leaves). The difference in branch, bole, and root biomass prior and subsequent to the study year was considered the production of each component. Leaf production was considered to be the leaf biomass calculated by using the dbh at the end of the study period. Estimates of component biomass production were multiplied by the total density of trees and expressed in units of kilograms per hectare per year.

Cropland management, production, and nutrient uptake

Estimates of fertilizer applications and crop yields were obtained by direct inquiry from the farmer. These estimates were used in conjunction with data available from an intensive study in 1976–1978 that directly measured management inputs, crop nutrient uptake, crop production, and agricultural exports. A comparison of data obtained from the farmer with actual measurements of application rates showed an average difference of ≈2% for nitrogen and phosphorus.

Results

Mass balance

During the study period 10 040 m^3/ha of rainfall were measured. Of this 2540, 2970, 1890, and 2650 m^3/ha fell in the spring, summer, fall, and winter seasons, respectively (Table 1). The long-term (160-yr) average for seasonal precipitation in the Rhode River watershed is 2800, 3140, 2450, and 2460 m^3/ha in the spring, summer, fall, and winter, respectively (Higman and Correll 1982). Thus, seasonal precipitation was slightly above average in the winter, but below average for all other seasons. The net result was that annual rainfall was 800 m^3/ha below the long-term annual average.

Of the 10 040 m^3/ha of water that fell on the watershed during the study year, 23% was discharged in stream flow. Seasonally, 27, 13, 3.5, and 44% of precipitation inputs were discharged during the spring, summer, fall, and winter, respectively. Quickflow comprised 7.1% of the annual discharge and was greatest during the summer. Annual slow flow and quick flow measured during the study period were both below

TABLE 1. Watershed mass balances for the whole watershed in 1981–1982.

Season	Precipitation (100 m³/ha)*		Nutrient flux for watershed: Nitrate-N (kg/ha)	Ammonium-N (kg/ha)	Organic-N (kg/ha)	Total-P (kg/ha)	Orthophosphate-P (kg/ha)	Organic-C (kg/ha)
Bulk precipitation								
Spring	25.4		1.20	1.12	3.39	0.284	0.0584	13.9
Summer	29.7		1.26	0.785	1.43	0.0675	0.0252	11.2
Fall	18.9		0.95	0.540	0.111	0.0549	0.0349	5.67
Winter	26.5		1.36	0.385	1.67	0.0290	0.0190	5.58
Year	100.4		4.77	2.83	6.60	0.435	0.138	36.4
Long-term annual mean†	108.0 ± 21.8		4.79 ± 1.18	2.59 ± 0.56	6.01 ± 1.15	0.810 ± 0.278	···‡	43.1 ± 6.81
Management inputs								
Year	···		10.1	14.9	42.2	12.5	12.5	···
Watershed discharge	Slow flow (100 m³/ha)*	Quick flow (100 m³/ha)*						
Spring	6.26	0.687	0.532	0.197	0.577	0.251	0.116	153
Summer	3.20	0.804	0.428	0.0890	1.04	0.926	0.111	9.26
Fall	0.620	0.037	0.00675	0.00662	0.0264	0.0102	0.00656	0.424
Winter	11.5	0.119	1.53	0.0605	0.387	0.104	0.0548	3.56
Year	21.6	1.64	2.50	0.353	2.03	1.29	0.288	166
Long-term annual mean†	29.6 ± 13.9	4.97 ± 4.51	4.15 ± 3.06	0.434 ± 0.124	9.17 ± 10.2	1.72 ± 1.47	0.622 ± 0.407	58.2 ± 61.3

* Precipitation and stream flow expressed as 100 m³/ha is equivalent to the more traditional units of cm.
† 1973–1980 yearly means ± 1 SD for nitrogen bulk precipitation parameters (Correll and Ford 1982).
1973–1978 yearly means ± 1 SD for phosphorus and organic-C bulk precipitation parameters (D. L. Correll, *personal communication*).
1977–1981 yearly means ± 1 SD for nutrient and water discharge (D. L. Correll, *personal communication*).
Long-term (160-yr) yearly precipitation mean ± 1 SD for the vicinity of the Rhode River watershed (Higman and Correll 1982).
‡ ··· indicates no data are available.

their respective long-term means, but did fall within one standard deviation of their expected values (Table 1).

During the study year 14.2 kg/ha of total nitrogen (nitrate-N + ammonium-N + organic-N) was delivered to the watershed by bulk precipitation. Of this input of total nitrogen, 34% was nitrate-N, 20% ammonium-N, and 46% organic N. Input of total phosphorus by bulk precipitation was 0.435 kg/ha, of which 32% was orthophosphate-P. Organic-C dominated the measured inputs from bulk precipitation with an influx of 36.4 kg/ha. Nutrient inputs by bulk precipitation were all similar (±1 SD) to their respective long-term means, except for total phosphorus, which was almost half its expected value. When compared to the nutrient influx from bulk precipitation, management inputs were the most important source of nutrients to the system except for organic-C (Table 1). Management inputs occurred in three pulses of fertilizer application during the spring and summer of the study year.

The watershed as a whole discharged 17, 2, 4, and 10% of the total inputs of nitrate-N, ammonium-N, organic-N and total-P, respectively (Table 1). Output of organic-C was over four times the input in bulk precipitation. Seasonal area yield discharges ranged over an order of magnitude for all nutrients. Peak output occurred in the winter for nitrate-N; the spring for ammonium-N, orthophosphate-P, and organic-C; and the summer for organic-N and total-P.

Intrawatershed changes

Annual and seasonal average nutrient concentrations in surface runoff and groundwater are presented in Tables 2 and 3. In surface runoff flowing between the first and last clusters of samplers, reductions of 94, 78, 86, 84, 74 and 64% were observed in the mean annual total particulate, exchangeable ammonium-N, particulate organic-N, total particulate-P, exchangeable orthophosphate-P, and particulate organic carbon concentrations, respectively. Most of the total changes in concentration occurred within the first 19 m of riparian habitat. Dissolved nitrogen compounds in surface runoff also declined in concentration after traversing the riparian forest, with the greatest change occurring in the first 19 m. Total reductions of 79% for nitrate, 73% for ammonium-N, and 62% for organic-N were observed. The mean annual concentration of dissolved total phosphorus changed little in surface runoff. Dissolved organic carbon concentration increased 2.8-fold, but only 41% of this change occurred within the initial

TABLE 2. Seasonal and yearly mean nutrient concentrations in mg/L for surface runoff in watershed 109 (transect 1).

Position	Season	Total Sus. Part.	Nitrate-N	Ammonium-N		Organic-N		Total-P		Exch. Part. Ortho-phos-phate-P	Organic-C	
				Exch. Part.	Diss.	Part.	Diss.	Part.	Diss.		Part.	Diss.
Entering riparian forest	Spring	8 840	3.73	0.734	3.63	27.7	1.47	3.22	0.256	0.354	67.2	12.1
	Summer	11 500	10.5	0.524	1.17	32.1	2.72	11.9	0.127	0.740	148.1	10.0
	Fall	3 830	1.57	0.301	0.896	16.8	0.779	3.29	0.128	0.863	101.1	6.75
	Winter*	1 760	1.99	0.048	0.250	1.32	2.04	0.860	0.320	0.675	63.2	19.1
	Year	6 480	4.45	0.402	1.49	19.5	1.75	4.82	0.208	0.658	94.9	12.0
19 m into riparian forest	Spring	1 380	2.60	0.218	1.23	6.47	1.18	2.31	0.081	0.456	35.9	12.0
	Summer	966	1.93	0.120	0.409	5.06	1.44	2.09	0.093	0.406	72.4	9.90
	Fall	122	0.343	0.038	0.069	2.61	0.529	0.604	0.393	0.134	5.97	4.09
	Winter*	176	2.18	0.042	0.158	0.37	1.33	0.065	0.375	0.108	···†	56.6
	Year	661	1.76	0.104	0.466	3.63	1.12	1.27	0.236	0.276	38.1	20.6
Leaving riparian forest	Spring	372	0.742	0.076	0.404	2.54	1.18	0.449	0.251	0.163	27.9	23.8
	Summer	524	1.03	0.108	0.175	3.46	0.713	1.04	0.183	0.244	45.6	16.0
	Fall	···	···	···	···	···	···	···	···	···	···	···
	Winter*	360	1.05	0.078	0.651	2.02	0.081	···	···	0.109	29.9	59.2
	Year	419	0.941	0.087	0.410	2.67	0.658	0.744	0.217	0.172	34.5	33.0

* Data from winter 1981. No samples were taken in winter 1982.
† ··· indicates no data are available.

portion of riparian forest. Although mean annual particulate concentrations of phosphorus, carbon, and organic-N in surface runoff decreased after moving through the riparian zone, the concentrations of these nutrients per unit of sediment increased. The effect of the riparian forest on particulate composition in surface runoff was to increase the proportion that was organic-C from 1.5 to 8.2%, organic-P (particulate total-P − exchangeable orthophosphate-P) from 0.064 to 0.14%, and organic-N from 0.30 to 0.64%. Furthermore, mean annual concentrations of exchangeable ammonium-N and orthophosphate-P per unit of sediment increased approximately two-fold after transit through the riparian zone. These results indicate that the particulates leaving the forest were more organic in composition and had a greater exchange capacity.

A *t* test was used to compare annual mean concentrations between the clusters of surface runoff collectors along transect 1. Noncomposited samples were used in the statistical analyses so that both spatial and temporal variability were included. Mean annual concentrations were significantly different ($P < .05$) between the first cluster (on the edge of the cropland) and the second cluster (≈19 m into the riparian forest) for total suspended particulates, exchangeable ammonium-N, dissolved organic-N, total particulate-P, exchangeable orthophosphate-P, and particulate organic-C. Significant differences in mean annual concentrations between the second and third clusters were found for only nitrate-N and total particulate-P.

Another way of assessing the role of the riparian forest is to estimate the amount of nutrients or total particulates trapped or released per hectare. This was done by multiplying the change in mean concentration (concentration entering − concentration leaving) by the volume of surface runoff (quickflow) or groundwater flow (slow flow) measured at the weir for each season (see Tables 1–3). For surface runoff the result was that during the study year an estimated 4.13 Mg of particulates were trapped along with 0.219, 11.2, 2.98, 0.172 and 32.2 kg/ha of exchangeable ammonium-N, particulate organic-N, total particulate-P, exchangeable orthophosphate-P, and particulate organic-C, respectively. Calculated removals of dissolved nutrients in surface waters were 2.71, 0.827, and 0.568 kg/ha, respectively, for nitrate-N, ammonium-N, and organic-N, whereas 0.0160 kg/ha of dissolved total-P and 4.84 kg/ha of dissolved organic-C were released. These values are probably underestimates of the actual change in surface runoff loads because of reductions in surface flow due to evapotranspiration and/or infiltration in the riparian forest.

Surface runoff leaving the cropland had peak concentrations of total particulates, total particulate-P, and all nitrogen parameters in either the spring or summer. These concentration peaks thus corresponded with the time of fertilizer application and intense storm activity. Peak spring or summer concentrations were also apparent after runoff had traversed the 1st 19 m of riparian forest, but were less pronounced.

Changes in all mean annual groundwater concentrations while traversing the riparian zone were qualitatively similar for both transects, although differences in magnitude were evident (Table 3). In general, nitrate concentrations declined dramatically while ammonium-N increased in concentration. The remaining nutrient concentrations were relatively constant except for a four-fold increase in total-P along transect 1 and

TABLE 3. Seasonal and yearly mean nutrient concentrations in mg/L for groundwater in watershed 109.

		Nitrate-N		Ammonium-N		Organic-N		Total-P		Organic-C	
Position	Season	Transect 1	Transect 2	Transect 1	Transect 2	Transect 1	Transect 2	Transect 1	Transect 2	Transect 1	Transect 2
Entering riparian forest	Spring	5.43	5.78	0.097	0.058	0.312	0.226	0.024	0.120	1.41	1.36
	Summer	6.96	6.21	0.119	0.076	0.145	0.122	0.006	0.106	2.38	1.76
	Fall	6.89	7.22	0.045	0.132	0.169	0.121	0.016	0.152	1.97	1.90
	Winter	10.3	7.81	0.038	0.028	0.202	0.115	0.013	0.144	2.46	1.50
	Year	7.40	6.76	0.075	0.074	0.207	0.146	0.015	0.130	2.06	1.63
19 m into riparian forest	Spring	0.000	...*	0.098	...	0.213	...	0.012	...	2.43	...
	Summer	0.355	...	0.165	...	0.173	...	0.005	...	2.72	...
	Fall	0.282	...	0.114	...	0.214	...	0.018	...	1.82	...
	Winter	1.44	...	0.035	...	0.125	...	0.007	...	2.14	...
	Year	0.519	...	0.103	...	0.181	...	0.010	...	2.28	...
Leaving riparian forest	Spring	0.153	0.248	0.196	0.269	0.487	0.291	0.074	0.082	1.72	2.86
	Summer	0.453	0.000	0.498	0.478	0.223	0.419	0.065	0.378	2.81	4.58
	Fall	1.47	0.052	0.248	0.621	0.213	0.136	0.076	0.345	3.08	3.26
	Winter	0.978	0.105	0.156	0.396	0.144	0.125	0.032	0.183	2.77	3.08
	Year	0.764	0.101	0.274	0.441	0.267	0.243	0.062	0.247	2.60	3.44

* ... indicates no data are available.

a two-fold increase in organic-C along transect 2. A 90% total decrease in the mean annual nitrate concentration was observed along transect 1 and a 98% decrease was observed along transect 2. Essentially all nitrate loss occurred within the first portion of riparian forest. Mean annual concentrations of ammonium-N increased in transect 1, with the overall increase being greater than three-fold. A six-fold increase occurred in transect 2. Overall, the annual average nitrate concentration declined 6.6 mg/L while traversing the forest, while the average concentration of combined forms of reduced nitrogen only increased by 0.36 mg/L. Thus, nitrate loss was apparently not due to its conversion into reduced forms of nitrogen that were measured.

Using a t test, comparisons were made to detect significant differences in mean annual groundwater concentrations between clusters of wells along each transect. Only noncomposited samples were used in the statistical analyses. Along transect 1 significant differences ($P < .05$) in mean annual concentrations between the first and second clusters were found only for nitrate-N, and between the second and third clusters for ammonium-N and total P. Significant differences between the two clusters along transect 2 were found for nitrate-N, ammonium-N, total-P, and organic-C.

The changes in concentration along transect 1, when related to seasonal groundwater discharge (slow flow; Table 1) give an estimated annual removal of 45.5 $kg \cdot ha^{-1} \cdot yr^{-1}$ of nitrate-N and an estimated release of 0.917, 0.194, 0.209, and 2.09 $kg \cdot ha^{-1} \cdot yr^{-1}$ of ammonium-N, organic-N, total-P, and organic-C, respectively, from the riparian forest. These values are probably underestimates because of reductions in subsurface flow due to evapotranspiration in the riparian forest.

Riparian vegetative production and nutrient uptake

Seventy-two percent of the sampled trees in the riparian forest were <22 cm in dbh, and the forest was clearly dominated by sweetgum (43.2% Importance Value) and red maple (16.0% I.V.) (Peterjohn 1982). Net primary production (NPP) was calculated as 18 600 $kg \cdot ha^{-1} \cdot yr^{-1}$ and was the sum of 5650 leaf, 3770 branch, 8100 bole, and 1030 large root production per hectare per year.

Estimates for the net production of respective tree components were multiplied by the mean midsummer (June, July, and August) nutrient concentrations found in a nearby forest (Table 4) and summed to estimate total nitrogen and phosphorus requirements. Sapwood concentrations were used for both branch and bole components. Uptake is defined as the annual elemental increment associated with the bole, branches, and large roots, plus the annual loss in leaf litterfall and throughfall. Nitrogen uptake by forest trees was calculated as ≈77 $kg \cdot ha^{-1} \cdot yr^{-1}$. Nitrogen return in leaf litter was calculated as 62 $kg \cdot ha^{-1} \cdot yr^{-1}$. Throughfall was not measured at the study site, so a literature value of 0.04 for the ratio of annual net nitrogen removal to foliage nitrogen content was used as an estimator (Henderson et al. 1977). Using this ratio, net nitrogen flux from leaves as throughfall was found to be 3.7 $kg \cdot ha^{-1} \cdot yr^{-1}$.

Using the same procedure employed for nitrogen, total phosphorus uptake by forest trees was calculated to be ≈9.9 $kg \cdot ha^{-1} \cdot yr^{-1}$. Phosphorus return in leaf litter was calculated as 7.8 $kg \cdot ha^{-1} \cdot yr^{-1}$. The flux of phosphorus from leaves as throughfall was estimated as 0.53 $kg \cdot ha^{-1} \cdot yr^{-1}$, using 0.06 as the ratio of annual net phosphorus removal to foliage phosphorus content

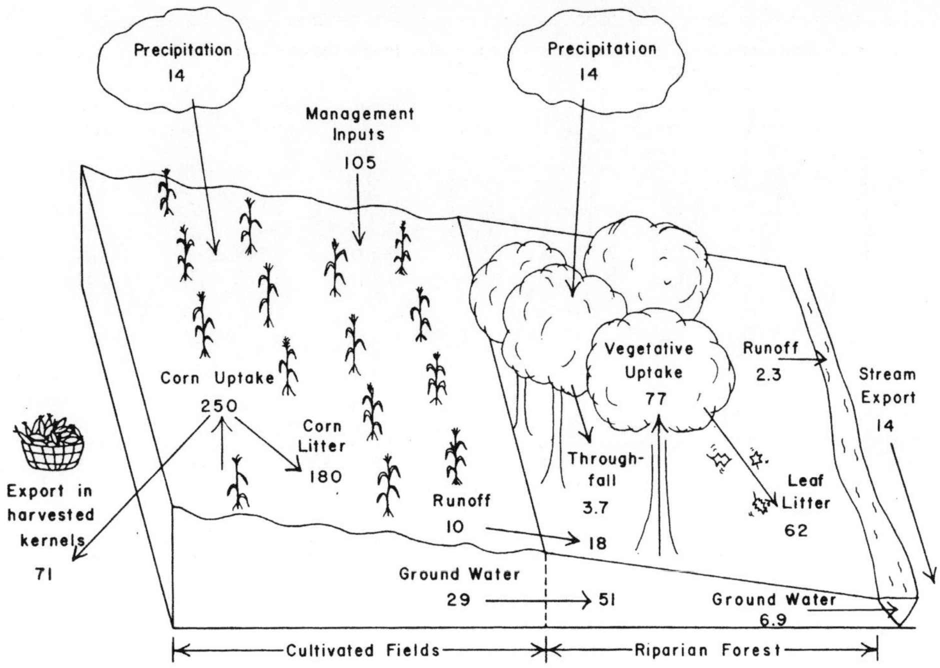

FIG. 2. Diagram of total-N flux and cycling in the study watershed from March 1981 to March 1982. All values are kilograms per hectare of the respective habitats (cropland and riparian forest/streams).

(Henderson et al. 1977). It should be emphasized that the uptake and production values presented are rough estimates. Nutrient uptake and turnover in the understory shrubs, herbaceous vegetation, and fine roots; nutrient return in branch, fruit, and flower litter; and whole-tree mortality were all factors not considered. Although rough, our production and uptake estimates are surprisingly reasonable when compared to literature values. Whittaker and Likens (1975) report 6000–25 000 $kg \cdot ha^{-1} \cdot yr^{-1}$ as the normal range of NPP in temperate deciduous forests, and the mean uptake for 14 temperate deciduous International Biological Program sites has been reported as 75.4 $kg \cdot ha^{-1} \cdot yr^{-1}$ of N and 5.6 $kg \cdot ha^{-1} \cdot yr^{-1}$ of P by Cole and Rapp (1981). A measurement of 6350 $kg \cdot ha^{-1} \cdot yr^{-1}$ for litter production in a nearby forest also compared well with our estimate of 5650 $kg \cdot ha^{-1} \cdot yr^{-1}$ (D. F. Whigham, *personal communication*).

TABLE 4. Mean Kjeldahl-N and total-P concentrations ($\mu g/g$) in trees in a mature deciduous forest in the Rhode River watershed, Maryland.*

Tissue	Kjeldahl-N	Total-P
Midsummer (June, July and August)		
sapwood	814	99.9
large roots	1 567	441
leaves	16 300	1 551
Winter (November and January)		
sapwood	1 080	159
large roots	2 070	435
leaf litter (November only)	10 996	1 386

* Samples are from 1977. Species represented are *Liriodendron tulipifera, Fagus grandifolia, Acer rubrum, Liquidambar styraciflua, Nyssa sylvaticus,* and *Quercus falcata.*

Cropland management, production, and nutrient uptake

Limestone (1680 kg/ha) was applied to the agricultural fields in early April, granular 2-10-13 fertilizer in mid-April, liquid N and herbicides in mid-May and urea in mid-July. Of the 105 kg/ha of total-N applied; 63.0% was organic-N, 22.3% was ammonium-N, and 15.0% was nitrate-N. Total-P loading was 19.6 kg/ha. In 1976 and 1977, the respective loadings were 142 and 189 kg/ha of total-N, and 39.7 and 33.7 kg/ha of total-P. Thus, nutrient loadings for this study year were low when compared to previous years.

Crop nitrogen and phosphorus uptake, losses of nitrogen and phosphorus as annual harvest, and crop NPP were estimated using data from a previous study of the same watershed. The same farmer has cultivated this basin since 1970 and little change in farming practices has been observed. Nutrient uptake by corn plants (above and belowground) was measured directly in 1976 and 1977; N uptake was found to be $\approx$250 kg and P uptake 45.0 $kg \cdot ha^{-1} \cdot yr^{-1}$ (Peterjohn 1982). The NPP of corn plants was also directly measured and found to be $\approx$24,000 $kg \cdot ha^{-1} \cdot yr^{-1}$. Nitrogen loss in harvested crop was estimated as 71.3 $kg \cdot ha^{-1} \cdot yr^{-1}$ by multiplying the yield of 7.8 m^3 (222 bushels)/ha in 1981 by a 1976

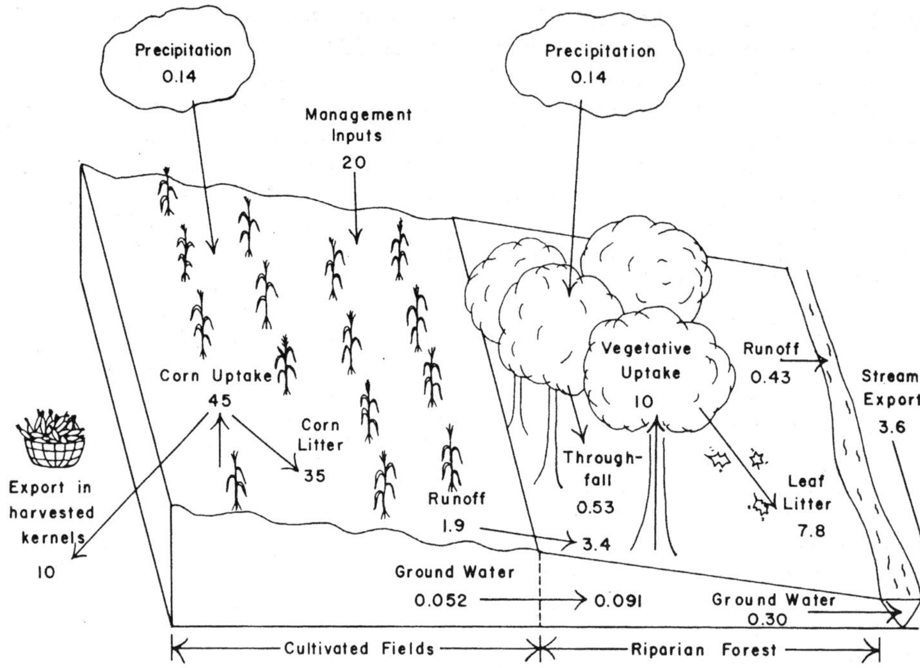

FIG. 3. Diagram of total-P flux and cycling in the study watershed from March 1981 to March 1982. All values are kilograms per hectare of the respective habitats (cropland and riparian forest/stream).

measurement of 33.3 kg of kernels per bushel, then multiplying the resultant product by a 1977 nitrogen measurement of 9.65 g/kg of kernels. Phosphorus loss in the harvested crop was estimated as 10.3 $kg \cdot ha^{-1} \cdot yr^{-1}$ using the same procedure. The value used for P content of kernels was 1.40 g/kg. Thus, only 28% of the total nitrogen and 22.9% of the total phosphorus uptake by corn plants was exported from the system in harvested kernels.

Discussion

To synthesize information about the N and P dynamics of this watershed, two schematic diagrams were drawn (Figs. 2 and 3). Although the values presented are estimates with unknown error terms, these diagrams are useful to compare pathways of nutrient flux between and within habitats, and to develop falsifiable hypotheses for future research. It is important to note that flux estimates are given per unit area of each component habitat (cultivated fields or riparian forest). Thus, because there is 1.76 × more cultivated land than riparian forest, N export from the agricultural habitat in groundwater was 29 kg/cultivated ha, whereas 51 kg/ha were gained by the riparian forest (Fig. 2).

The nitrogen dynamics for both the cultivated and riparian habitats are illustrated in Fig. 2. For the cultivated fields, input of total-N from bulk precipitation was small (12% of estimated total inputs) when compared to management applications. Of the estimated total nitrogen exports from the cropland, 64% (28% of corn uptake) was in harvested crop, 9.2% in surface runoff, and 26% in groundwater flow. Thus, the harvest seems to be the major pathway of total-nitrogen loss from the cropland. Groundwater appears to be the dominant pathway of total nitrogen flux between the cropland and riparian forest. Nitrogen retention (1 − outputs/inputs) for the cropland was calculated from our flux estimates and found to be low (8%), which is consistent with general ideas about disturbed ecosystems.

For the riparian forest, 17% of the estimated total-N inputs came in bulk precipitation, 61% in groundwater, and 22% in surface runoff. Of the estimated total-N losses from the riparian forest, 75% was lost in groundwater flow. Thus, it appears that the major pathway of nitrogen loss from the riparian forest was in subsurface flow. The calculated nitrogen retention by the riparian forest was 89%, and much higher than the retention of the cropland (8%).

The phosphorus dynamics for both the cultivated and riparian habitats are depicted in Fig. 3. For the cultivated fields, input of total-P from bulk precipitation was only 0.7% of the estimated total inputs. Of the estimated total phosphorus exports from the cropland, 84% (22% of corn uptake) was apparently lost in harvested crop, 16% in surface runoff and <1% in groundwater flow. Thus, the harvest seems to be the dominant pathway of phosphorus loss from the cropland, and surface runoff the dominant pathway of phosphorus flux between cropland and riparian forest. The calculated phosphorus retention by the cropland was 41% and higher than the same value for total nitrogen.

For the riparian forest, 3.8% of the estimated total phosphorus inputs came in bulk precipitation, 94% in surface runoff, and 2.5% in groundwater flow. Unlike the loss of phosphorus from the cropland, phosphorus export from the riparian forest was nearly evenly divided between the estimated losses in surface runoff (59%) and groundwater flow (41%). The calculated phosphorus retention by the riparian forest was 80%. This value is twice as high as the retention by the cropland and slightly lower than the calculated total-nitrogen retention for the riparian zone.

From the observations made in this study two important questions arise: (1) What caused the decrease in groundwater nitrate concentration in the riparian forest? (2) How widely applicable are the effects we observed? Data from this study are insufficient to determine conclusively the reason for nitrate loss from groundwater in the riparian forest. Only two theoretical possibilities, however, appear likely: uptake by the vegetation and denitrification. To assess the approximate effect of vegetative uptake we estimated the annual incremental storage of N in the boles, branches, and large roots of the riparian trees. This was done by multiplying our production estimates by the mean winter tissue concentrations presented in Table 4. Winter tissue concentrations were used to correct for the effects of any retranslocation of nutrients from the leaves. From this calculation we estimate that N at the rate of 15 $kg \cdot ha^{-1} \cdot yr^{-1}$ was removed due to incremental growth, which, if true, accounts for only 33% of the nitrate-N removed from the groundwater. Considering that groundwater was sampled below the fine root zone at a depth of at least 1 m, it would seem that the occurrence of significant denitrification is a viable hypothesis.

An insufficient number of studies have been conducted to assess adequately how common the effects we observed are. Reductions in sediment loads and their associated nutrients in surface runoff should be a fairly universal effect of riparian forests because of the physical nature of the processes involved; a few studies present evidence that riparian zones reduce sediment and phosphorus loads in adjacent streams (McColl 1978, Schlosser and Karr 1981*a, b*). Nitrate removal from shallow groundwater is probably a biological process requiring an appropriate environment. Therefore, the extent to which riparian forests serve as sites of subsurface nutrient change is difficult to predict. It is encouraging to note, however, that results similar to those presented here (particularly subsurface nitrate-N removal) were obtained by a closely analogous study conducted in Georgia (Lowrance et al. 1984).

Nutrient loss from diffuse sources (such as cornfields) can have significant ecological effects which are generally understood as a threat to most bodies of water. Therefore, the estimated removal in surface runoff of 4.1 Mg of particulates, 11 kg of particulate organic-N, 0.83 kg of dissolved ammonium-N, 2.7 kg of nitrate-N and 3.0 kg of total particulate-P per hectare of riparian forest is potentially an extremely important ecological function. In addition, the removal of nitrate-N at an estimated rate of 45 kg/ha in subsurface flow is especially important. Release of this amount would have doubled the amount of nitrate-N discharged during the study year.

Acknowledgments

We thank the staff at the Smithsonian Environmental Research Center for their assistance and technical support throughout this study. For financial support we are grateful to Miami University, the Smithsonian Institution's Environmental Sciences Program, and National Science Foundation grant DEB-79-11563.

Literature Cited

APHA (American Public Health Association). 1976. Standard methods for the examination of water and waste water. 14th edition. American Public Health Association, New York, New York, USA.

Barnes, B. S. 1940. Discussion on analysis of runoff characteristics by O. H. Meyers. Transactions of the American Society of Civil Engineers **105**:104–106.

Chirlin, G. R., and R. W. Schaffner. 1977. Observations on the water balance for seven sub-basins of Rhode River, Maryland. Pages 277–306 *in* D. L. Correll, editor. Watershed research in eastern North America. Volume 1. Smithsonian Press, Washington, D.C., USA.

Cole, D. W., and M. Rapp. 1981. Elemental cycling in forest ecosystems. Pages 341–409 *in* D. E. Reichle, editor. Dynamic properties of forest ecosystems. Cambridge University Press, Cambridge, England.

Correll, D. L. 1984, *in press.* N and P in soils and runoff of three coastal plain land uses. *In* R. L. Todd, R. Leonard, and L. Assmussen, editors. Nutrient cycling in agroecosystems. University of Georgia Press, Athens, Georgia, USA.

Correll, D. L., and D. Ford. 1982. Comparison of precipitation and land runoff as sources of estuarine nitrogen. Estuarine, Coastal and Shelf Science **15**:45–56.

Correll, D. L., and J. J. Miklas. 1976. Phosphorus cycling in a Maryland deciduous forest subjected to various levels of mineral nutrient loading. Pages 642–657 *in* F. G. Howell, J. B. Gentry, and M. H. Smith, editors. Mineral cycling in southeastern ecosystems. Symposium Series (CONF-740513), Energy Research and Development Administration, Washington, D.C., USA.

Cottam, G., and J. T. Curtis. 1956. The use of distance measures in phytosociological sampling. Ecology **37**:451–460.

Edmondson, W. T. 1972. Nutrients and phytoplankton in Lake Washington. American Society of Limnology and Oceanography Special Symposium **1**:172–188.

Fraser, T. H., and W. H. Wilcox. 1981. Enrichment of a subtropical estuary with nitrogen, phosphorus and silica. Pages 481–498 *in* B. J. Neilson and L. E. Cronin, editors. Estuaries and nutrients. Humana, Clifton, New Jersey, USA.

Harris, W. F., R. A. Goldstein, and G. S. Henderson. 1973. Analysis of forest biomass pools, annual primary production and turnover of biomass for a mixed deciduous forest watershed. Pages 41–46 *in* H. Young, editor. Proceedings of the working party on forest biomass of IUFRO. University of Maine Press, Orono, Maine, USA.

Henderson, G. S., W. F. Harris, D. E. Todd, and T. Grizzard. 1977. Quantity and chemistry of throughfall as influenced by forest-type and season. Journal of Ecology **65**:365–374.

Higman, D., and D. L. Correll. 1982. Seasonal and yearly variation in meterological parameters at the Chesapeake Bay Center for Environmental Studies. Pages 1–159 *in* D. L. Correll, editor. Environmental data summary for the Rhode River ecosystem. Volume A, Part I. Chesapeake Bay Center for Environmental Studies, Edgewater, Maryland, USA.

Hynes, H. B. N. 1969. The enrichment of streams. Pages 188–196 *in* Eutrophication: causes, consequences, correctives. National Academy of Sciences, Washington, D.C., USA.

King, E. J. 1932. The colorimetric determination of phosphorus. Biochemical Journal **26**:292–297.

Lowrance, R. R., R. L. Todd, and L. E. Asmussen. 1984. Nutrient cycling in an agricultural watershed: I. Phreatic movement. Journal of Environmental Quality **13**:22–27.

Maciolek, J. A. 1962. Limnological organic analyses by quantitative dichromate oxidation. Fish and Wildlife Service Publication, Washington, D.C., USA.

Martin, D. F. 1972. Marine chemistry. Volume 1. Marcel Dekker, New York, New York, USA.

McColl, R. H. S. 1978. Chemical runoff from pastures: the influence of fertilizer and riparian zones. New Zealand Journal of Marine and Freshwater Research **12**:371–380.

Myers, V. B., and R. I. Iverson. 1981. Phosphorus and nitrogen limited phytoplankton productivity in northeastern Gulf of Mexico coastal estuaries. Pages 569–582 *in* B. J. Neilson and L. E. Cronin, editors. Estuaries and nutrients. Humana, Clifton, New Jersey, USA.

Peterjohn, W. T. 1982. Nutrient transformations in three single-land-use watersheds. Thesis. Miami University of Ohio, Oxford, Ohio, USA.

Pierce, J. W. 1982. Geology and soils of the Rhode River Watershed. Pages 181–216 *in* D. L. Correll, editor. Environmental data summary for the Rhode River ecosystem. Volume A, Part I. Chesapeake Bay Center for Environmental Studies, Edgewater, Maryland, USA.

Powers, C. F., D. W. Schults, K. W. Malueg, R. M. Brice, and M. D. Schuldt. 1972. Algal responses to nutrient additions in natural waters. American Society of Limnology and Oceanography Special Symposium **1**:141–154.

Richards, F. A., and R. A. Kletsch. 1964. The spectrophotometric determination of ammonia and labile amino compounds in fresh- and seawater by oxidation to nitrite. Pages 65–81 *in* Sugawara festival volume. Maruza, Tokyo, Japan.

Schelske, C. L., and E. F. Stoermer. 1972. Phosphorus, silica, and eutrophication in Lake Michigan. American Society of Limnology and Oceanography Special Symposium **1**:157–170.

Schlosser, I. J., and J. R. Karr. 1981*a*. Water quality in agricultural watersheds: impact of riparian vegetation during base flow. Water Resources Bulletin **17**:233–240.

Schlosser, I. J., and J. R. Karr. 1981*b*. Riparian vegetation and channel morphology impact on spatial patterns of water quality in agricultural watersheds. Environmental Management **5**:233–243.

Whittaker, R. H., and G. E. Likens. 1975. The biosphere and man. Page 306 *in* H. Lieth and R. H. Whittaker, editors. Primary productivity of the biosphere. Springer-Verlag, New York, New York, USA.

Commentary on Paper E3

Jacobs, T. C. & Gilliam, J. W. (1985) Riparian losses of nitrate from agricultural drainage waters.

The aim of this study was to evaluate the fate of nitrate leaving cultivated fields in a poorly-drained Lower Coastal Plain watershed and a moderately well-drained Middle Coastal Plain watershed of North Carolina, USA. The context of the study was to quantify the role of different drainage systems in regulating the nitrate transfer from agricultural fields to the streams. Interestingly, they followed the water quality along different agricultural hillslope transects perpendicular to different man-made or natural landscape structures from high to low drainage capacity, i.e. tube drain, V-shaped ditch and riparian forest,. They did not use any tracing experiment to confirm groundwater flow paths between wells but they used the DRAINMOD model (Skaggs, 1978) to estimate quantities of subsurface drainage and surface runoff from each field.

Low soil permeability in the Lower Coastal Watershed meant that there was almost no subsurface flow into the drainage systems and the low surface water nitrate concentration resulted from denitrification within the poorly drained soils. More interestingly, in the moderately well-drained Middle Coastal Plain watershed, subsurface flow occurred and revealed that nitrate could be entirely removed in less than 16 m flow through the riparian zone. The authors considered that plant uptake would only account for approx. 20%; most of the removal was probably due to denitrification in reduced soil as confirmed by the measured redox and the presence of reduced iron. This result stresses the importance of soil permeability and the resulting subsurface flow within riparian zones, which have unfortunately been overlooked for too long. This has misled several studies measuring no nitrate in the middle of riparian zone and then extrapolating an average denitrification activity as a distance from the upland edge without checking if these zones were hydrologically connected. Therefore, this approach only provides potential removal activity since nitrate is often the factor limiting denitrification in the riparian zone due to the small amount or even complete absence of subsurface flow. Measuring the difference between real and potential activity is the main challenge in evaluating the riparian zone efficiency to remove nitrate.

Jacobs and Gilliam proposed that, in moderately well-drained agricultural soils, the best form of drainage was the natural one with poorly drained soil and dense riparian vegetation. However, the least retentive system as far as nitrate is concerned is the tube drain. The drainage ditch could nevertheless lower N concentration in drainage waters to some extent because of denitrification within the ditch bank and sediment. This emphasises that hot spots of denitrification can occur within different landscape structures, as long as nitrate encounters a reduced environment (McClain *et al.*, 2003).

In their conclusion Jacobs and Gilliam wrote: "*Finding restrictive subsurface movement of N in this important agricultural region should indicate possible expansion areas for farming as needs arise*". So far this has been interpreted mostly as the fact that using riparian zones can allow current intensive agricultural practices to be maintained or even increased upslope. Unfortunately, relying solely on riparian zones to buffer intensive agricultural N input has proved to be mostly a failure because of the lack of consideration of the trade-off between N removal efficiency and surface and subsurface water transit time.

GP

References

McClain, M. E., Boyer, E. W., Dent, C. L., Gergel, S. E., Grimm, N. B., Groffman, P. M., Hart, S. C., Harvey, J. W., Johnston, C. A., Mayorga, E., McDowell, W. H. & Pinay, G. (2003) Biogeochemical hot spots and hot moments at the interface of terrestrial and aquatic ecosystems. *Ecosystems* **6**, 301–312.

Skaggs, R. W. (1978) A water management model for shallow water table soils. Rep. 134. Water Resources Research Institute, University of North Carolina, Raleigh, North Carolina, USA.

Riparian Losses of Nitrate from Agricultural Drainage Waters[1]

T. C. JACOBS AND J. W. GILLIAM[2]

ABSTRACT

Increased nutrient levels in surface streams and eutrophication of some Coastal Plain waters has led to inquiries about both the amount and control of nitrate losses from agricultural fields. Nitrate concentrations in shallow groundwaters beneath cultivated fields and in the drainage waters from those fields were examined to determine the fate of nitrogen lost to drainage waters. From a Middle Coastal Plain watershed where well- and moderately well-drained soils dominate agricultural fields, 10 to 55 kg ha^{-1} yr^{-1} NO_3-N moved from the fields in subsurface drainage water. However, most fields are bordered by forested buffers between the cultivated areas and streams which consist of poorly and very poorly-drained soils covered by dense vegetation. The evidence strongly indicated that a substantial part of the nitrate in the drainage water was denitrified in the buffer strip and that assimilation by vegetation was insignificant. Buffer strips of < 16 m were effective for inducing significant losses of nitrate before drainage water reached the stream. A field containing subsurface drainage tubing which emptied into open ditches moved more nitrogen into surface water than those fields without subsurface drainage improvements. From a Lower Coastal Plain watershed, a dense clay layer below the surface horizon reduced subsurface drainage resulting in total losses from the field of only 6 to 12 kg ha^{-1} yr^{-1} NO_3-N. These losses were mostly in surface runoff. The extensive floodplain of the natural stream had a high capacity to reduce large quantities of N but the low total loss from the watershed is largely a result of low input to the drainage water from nonpoint sources. Soils included in this study were Typic Paleudults, Arenic Paleudults, Aquic Hapludults, and Aeric Paleaquults.

Additional Index Words: **denitrification, subsurface drainage, DRAINMOD, buffer strips, tile drainage.**

Jacobs, T. C., and J. W. Gilliam. 1985. Riparian losses of nitrate from agricultural drainage waters. J. Environ. Qual. 14:472–478.

The importance of nitrogen to the potential eutrophication of North Carolina estuaries and Coastal Plain streams (Harrison and Hobbie, 1974; Hobbie, 1974) and other estuaries (Carpenter et al., 1969) are well documented. Previous work in North Carolina (Gambrell et al., 1975b; Gilliam et al., 1978) has shown that a significant amount of fertilizer N leaves agricultural fields in drainage waters. Kaushik et al. (1981) has an excellent review of other locations with similar losses.

In recent times, increases in fixed nitrogen have come from two sources. Soybean production has increased by 760% since 1945 and yields are much higher (Anonymous, 1945–1982) so significantly more N is being fixed by soybeans. Fertilizer N usage in North Carolina has increased about 400% between 1945 and 1983 (Anonymous, 1945–1983); however, there has not been a proportional increase in the N content of most Coastal Plain streams. Similar observations have been made elsewhere in the USA, United Kingdom, and Canada (Kaushik et al., 1981). Aldrich (1980) reviewed data from some watersheds where NO_3-N concentrations have increased. Apparently some Coastal Plain streams have a buffer system that inhibits an increase in N where others do not.

In an era where the importance of interactions between agricultural operations and resource contribu-

[1] Paper no. 9574 of the Journal Series of the North Carolina Agricultural Research Service, Raleigh, NC 27695-7619. The work upon which this publication is based was supported in part by the Water Resources Research Inst. of the Univ. of North Carolina as authorized under the Water Resources Act of 1964. Received 19 Nov. 1984.

[2] Graduate research assistant and professor, Soil Science Dep., North Carolina State Univ., Raleigh, NC 27695.

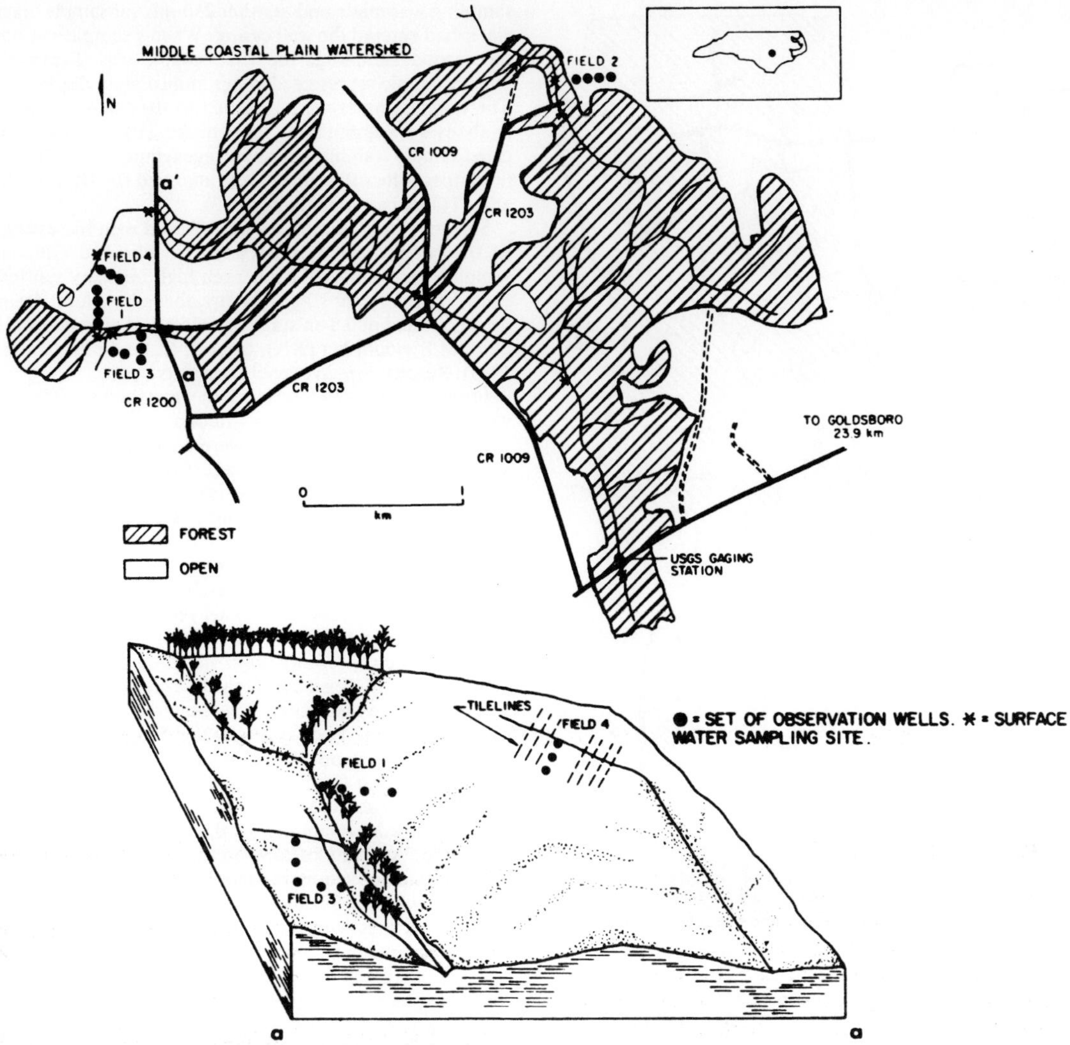

Fig. 1. Map of Middle Coastal Plain watershed with landscapes of three fields inset.

tions are recognized, the mechanism of control or influence over stream nutrient concentrations must be understood. Therefore, this study was undertaken to determine the fate of N leaving cultivated fields in two watersheds in the North Carolina Coastal Plain.

METHODS AND MATERIALS

Two watersheds in the North Carolina Coastal Plain were selected for a 3-yr study. A Middle Coastal Plain watershed was selected as typical of a large area of well-drained to moderately well-drained agricultural soils. This site was Beaverdam Creek watershed near Goldsboro, NC (Fig. 1). A Lower Coastal Plain watershed typical of a large agricultural production area of somewhat poorly-drained soils was also selected—Creeping Swamp watershed near Vanceboro, NC (Fig. 2). Field and soil characteristics of both watersheds are described (Table 1).

Three different agricultural drainage systems were identified in the Middle Coastal Plain watershed. These were surface and subsurface drainage to: a natural stream without improved drainage, crudely improved drainage with ditches cut across the fields, and V-ditches with drain tubing emptying into them. To measure movement of N in subsurface drainage water in and from each of the four fields at the Middle Coastal Plain watershed, a transect of shallow groundwater wells was placed on the slope between the highest elevation in the field and the stream or ditch to which it drained (Fig. 1). Each of these transects consisted of three or four locations with three to six different depths of wells in each location with the deepest well at 4.25 m. The soils were described and well casings of 38-mm polyvinyl chloride pipes installed. A well point was placed at the top of each restrictive soil layer or where a color change indicated potential differential water movement through the horizons. The bottom 0.25 m of the pipes were perforated with 4.7-mm holes every 50 mm. Soil was replaced around the pipes and then the surface around the pipes was packed with a 2:1 expanding clay to prevent channelization from the surface. Pipes were set with 0.25 m protruding above the surface and were stoppered between samplings.

Two predominant drainage systems were identified at the Lower Coastal Plain watershed. Two fields were chosen with drainage ditches cut both through and around them. These ditches carried both surface and subsurface runoff and had bottom grade set for maximum area drainage. The second type of drainage was grassed waterways used primarily to carry surface water.

Sampling frequency for the wells was approximately every 30 d, or after a rainfall event sufficiently large to cause water movement to the shallow water tables. The sampling was done less frequently during dry periods because of the small amount of water movement in the watershed. There were 24 sampling events from February 1980 to January 1983 at the Middle Coastal Plain watershed and 10 sampling events at the Lower Coastal Plain watershed between May 1980 and February 1981.

At each sampling date the depth to the water table was measured using an electronic resistance meter. The wells were emptied completely utilizing a portable vacuum pump and a 250-mL subsample was taken. After all wells in the set had been drained a second

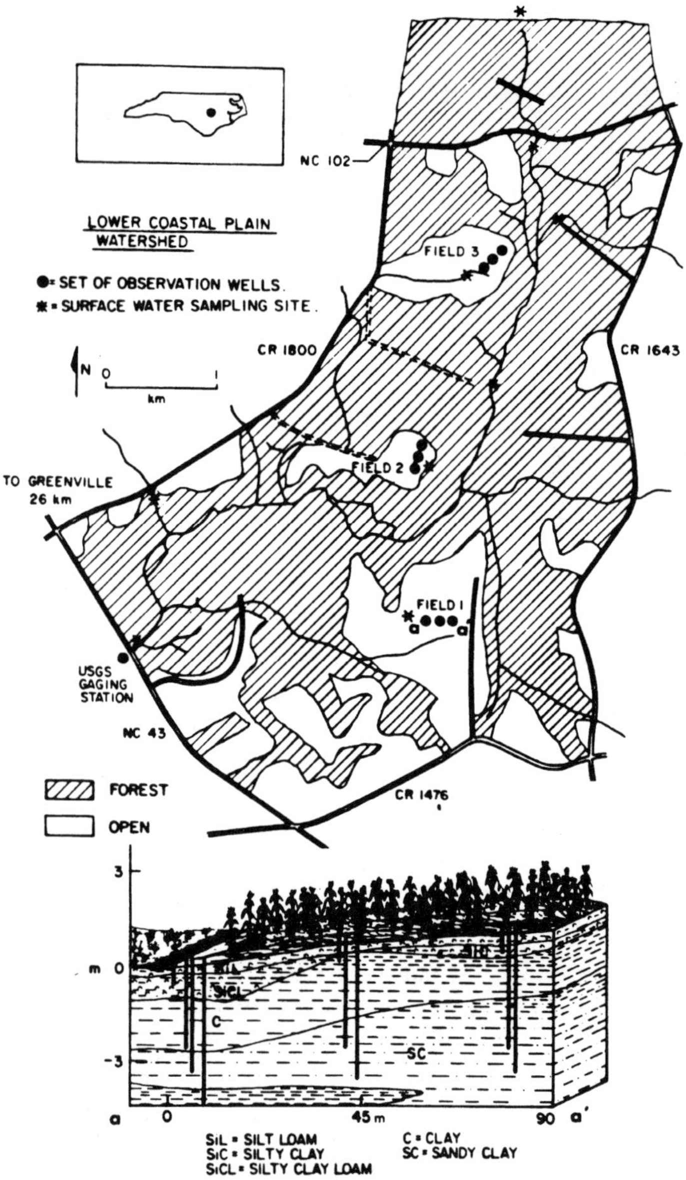

Fig. 2. Map of Lower Coastal Plain watershed with insert of stratigraphy of Field 1.

sampling was made and another 250-mL subsample taken if sufficient water had entered the well casing. When a sample was obtained during the second sampling, the first sample was discarded because the second sample represented water immediately displaced from the soil. The water samples were returned to the lab and stored at 4°C until analyses were completed. Spot checking of samples after storage indicated sample stability at this storage temperature without additions of inhibitors. The subsamples were analyzed for NO_3-N, NH_4-N, total P, and Cl (Jacobs and Gilliam, 1983).

Surface runoff samples were collected with increasing stage in each of the seven field drainageways (Jacobs and Gilliam, 1983). The samples were collected after each high intensity rainfall event which caused a significant stage increase above base level drainage. Samples were collected at 0.1-m stage increases and analyzed for NO_3-N, NH_4-N, total Kjeldahl N (TKN), total P, Cl, and pH (Jacobs and Gilliam, 1983). Results were averaged for events.

From the Middle Coastal Plain watershed surface seepage samples were taken at the edge of two floodplain areas downslope from cultivated fields. Water samples were collected from seepage pools considered characteristic of seepage found in many floodplain areas. When possible, samples were taken from standing pools at upper, middle, and lower elevations and analyzed for NO_3-N and Cl. Oxidation reduction potentials were measured in the soil and saturated surface litter when these water samples were collected. The redox potential was measured with a portable specific ion meter using a saturated potassium chloride calomel reference electrode and laboratory constructed bright platinum electrodes.

The Pt electrodes were constructed by placement of 20 mm of 20-gauge Pt wire in direct contact with a copper conductor and bound by heat shrinkable insulation. This required three coverings of the heat shrink insulation. Each electrode had 5 to 6 mm of Pt wire exposed for soil contact. Electrodes were tested in distilled water and ferrous-ferric chloride solution (Light, 1972) before field use. The electrodes were cleaned after each use with a small brush and a paste of commercial scouring powder (Gambrell et al., 1975a) and stored in distilled water until used. At five locations on each watershed, which included each type of drainage sequence identified, redox potentials were measured seven times from 1981 to 1983. These measurements were made at two or three subsites with platinum electrodes set at three depths: 6, 12, and 25 mm.

Riparian vegetation was harvested from 15 drainageway locations during the late growing season (14 October) so that maximum vegetative growth was present. These samples were taken from 1 to 3 m^2 areas, weighed, dried, and analyzed for N, P, and K.

A drainage model, DRAINMOD, (Skaggs, 1978; Skaggs and Gilliam, 1981) was used to estimate quantities of subsurface drainage and surface runoff from each field on both watersheds. The model was developed at North Carolina State University and tested extensively in the North Carolina Coastal Plain. It has been implemented nationally by the U.S. Department of Agriculture-Soil Conservation Service (USDA-SCS, 1980) as a criterion for designing drainage and sub-

Table 1. Field information for both Middle (M) and Lower (L) Coastal Plain watersheds.

Watershed	Field	Mapping unit	Subgroup classification of principle component	Crops	Drainage
M	1	Norfolk ls, 0–2% slopes†	Typic Paleudults	Grain crops, winter cover	No improved drainage to a natural stream
M	2	Norfolk ls, 0–6% slopes† Wagram ls, 0–6% slopes†	Typic Paleudults Arenic Paleudults	Vegetable crops, no winter cover	No improved drainage to a natural stream
M	3	Goldsboro ls, 0–2% slopes† Ruston sl, 2–6% slopes†	Aquic Paleudults Typic Paleudults	Grain crops, winter cover	Surface ditch and no improved drainage to a natural stream
M	4	Norfolk l, 0–2% slopes† Norfolk sl, 0–2% slopes†	Typic Paleudults Typic Paleudults	Grain and vegetable double crop, no winter cover	Surface shaped field, tile lines, and V-ditch
L	1	Craven sil, 1–4% slopes‡ Lenoir sil, 1–4% slopes‡	Aquic Hapludults Aeric Paleaquults	Grain crops, no winter cover	Grassed waterway
L	2	Lenoir l, 0–1% slopes§	Aeric Paleaquults	Grain crops, no winter cover	Ditched surface drains
L	3	Lenoir l, 0–1% slopes§ Aycock fsl, 0–1% slopes§	Aeric Paleaquults Typic Paleudults	Grain crops, no winter cover	Ditched surface drains

† As mapped, Soil Survey of Wayne County, North Carolina (1974).
‡ As mapped, Soil Survey of Craven County, North Carolina (1985).
§ As mapped, Soil Survey of Pitt County, North Carolina (1974).

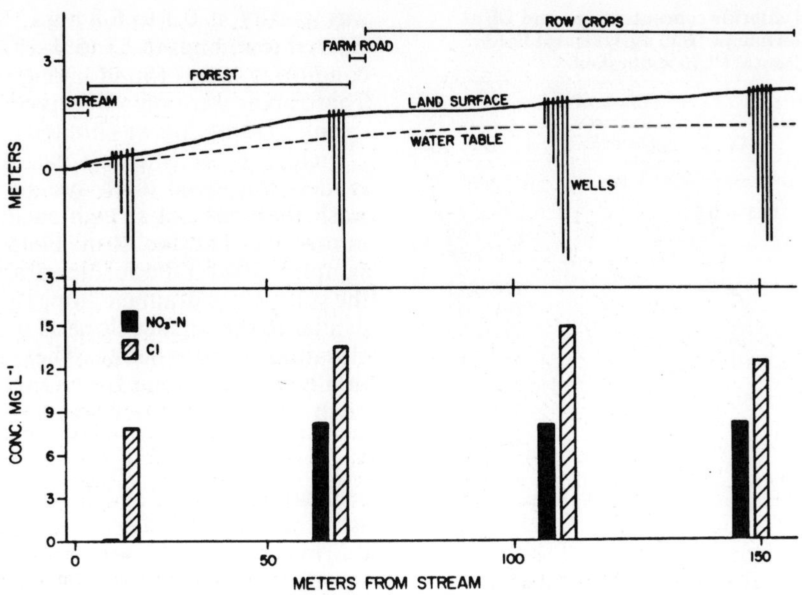

Fig. 3. Land use, water table, well positions, and mean NO_3-N and Cl concentrations for Field 1 in the Middle Coastal Plain watershed.

surface irrigation systems. The computer simulations of surface and subsurface runoff using DRAINMOD utilized site specific data such as soil physical characteristics, climatological data, surface elevations, identified impermeable layers, and crops specific to each field.

RESULTS

Middle Coastal Plain Watershed

By on-site examination, two fields were identified as having drainage by diffuse entry into a natural stream. Fields 1 and 2 exhibited similar results with respect to water movement, nutrient concentrations, and N losses. These results were observed even with different physical properties and management levels (Table 1). Field 1 had a finer loam texture, grain crops, and winter cover crop while Field 2 had a coarser loamy sand texture, was planted in vegetable crops and was plowed and left fallow each winter.

The 3-yr mean NO_3-N concentrations of well samples beneath the cropped areas of Fields 1 and 2 were 7.9 ± 0.3 and 7.3 ± 0.2 mg L^{-1}, respectively, with little variation between sampling dates and well depths. Combining the mean NO_3-N concentration for all shallow ground water wells with the subsurface drainage volumes computed by DRAINMOD resulted in estimates of subsurface losses of 17.1 and 11.9 kg ha^{-1} yr^{-1} of NO_3-N from Fields 1 and 2, respectively (Table 2). These values are consistent with those measured by Gambrell et al. (1975b) and Gilliam et al. (1978). Losses by surface runoff were 0.9 and 1.0 kg ha^{-1} yr^{-1} of NO_3-N from Fields 1 and 2, respectively (Table 2).

The mean 8.0 ± 0.5 and 6.9 ± 0.4 mg L^{-1} NO_3-N concentrations at the sampling sites near the field edge, but not in the cropped areas, approximated the mean 7.9 ± 0.3 and 7.3 ± 0.2 mg L^{-1} NO_3-N concentration measured from the two sets of field wells (Field 1 and 2, respectively). This indicated that NO_3-N was moving out of the field in the subsurface drainage water toward the streams as expected from the hydraulic gradient measured at the well sites (Fig. 3). These NO_3-N concentrations did not extend from the edge of the field to the well sites located 47 and 16 m (Fields 1 and 2, respectively) closer to the natural surface drainageway. The concentrations of NO_3-N measured in samples from these wells were usually < 0.1 mg L^{-1} NO_3-N. This short distance traveled through soils with a seasonal high water table and a heavy growth of riparian vegetation resulted in little N being measured at the wells near the stream. There was not a proportional loss of Cl between the edge of the fields and sites near the streams

Table 2. Estimates of surface and subsurface movement of nitrate from the monitored fields at the Middle (M) and Lower (L) Coastal Plain watersheds.

Watershed	Field	Estimated lateral conductivity	Subsurface† drainage	Mean subsurface conc of NO_3-N	No. well samples analyzed	Subsurface† NO_3-N loss	Surface† runoff	NO_3-N conc surface runoff	No. surface runoff samples analyzed	NO_3-N† loss by surface runoff
		mm h^{-1}	mm yr^{-1}	mg L^{-1}		kg ha^{-1} yr^{-1}	mm yr^{-1}	mg L^{-1}		kg ha^{-1} yr^{-1}
M	1	80	216	7.9 ± 0.3‡	221	17.1	100	0.9 ± 0.3‡	5	0.9
M	2	160	163	7.3 ± 0.2	242	11.9	78	1.3 ± 0.3	19	1.0
M	3	80	128	7.7 ± 0.3	193	9.9	224	1.7 ± 0.8	14	3.8
M	4	80	371	14.8 ± 0.4	249	54.9	19	1.2 ± 0.3	23	0.3
L	1	8	2	0.3 ± 0.1	87	<0.01	520	1.6 ± 0.7	28	8.4
		80	20	0.3 ± 0.1	87	<0.1	432	1.6 ± 0.7	28	6.9
L	2	8 @ 1% slope	1	0.3 ± <0.1	40	<0.01	493	1.2 ± 0.4	59	5.9
		8 @ 3% slope	5	0.3 ± <0.1	40	<0.1	490	1.2 ± 0.4	59	5.9
L	3	8	1	0.6 ± 0.1	80	<0.1	493	2.5 ± 0.5	52	12.5

†DRAINMOD was used to obtain these estimates. ‡ Standard error of the mean.

Table 3. Mean nitrate and chloride concentrations and Eh at selected seepage areas downslope from agricultural fields at the Middle Coastal Plain watershed.

Date	NO_3-N	Cl	Depth of electrode placement	Eh
	mg L^{-1}		mm	mV
22 Feb. 1982	4.4 ± 2.0	14.5 ± 0.5	6	229 ± 35†
			12	279 ± 45
			25	269 ± 6
3 Mar. 1982	2.2 ± 0.6	11.8 ± 0.4	6	276 ± 23
			12	294 ± 21
			25	257 ± 13
30 Apr. 1982	1.3 ± 0.3	9.9 ± 0.2	6	306 ± 23
			12	216 ± 30
			25	220 ± 30
4 Jan. 1983	0.9 ± 0.7	14.0 ± 1.8	6	326 ± 45
			12	327 ± 38
			25	305 ± 37

† Standard error of the mean.

(Fig. 3). This pattern was observed by Lowrance et al. (1984) for phreatic movement in the Georgia Coastal Plain.

A portion of the NO_3-N moving in the subsurface drainage was certainly utilized by the riparian vegetation on the flood plain. However, this utilization accounts for only a small percentage of the loss. Total biomass samples collected from 15 drainage locations ranged from 28.5 to 80.6 kg ha^{-1} of total nitrogen. Because most of this growth was annuals with few perennials, most of this N was in equilibrium. In addition, the total area encompassed by the drainageway and buffer areas was small compared to the fields which drained through them. Even with the fields on the drainage divides and the wooded areas next to the streams it was estimated < 20% of the woods could benefit from the field losses. The utilization of fertilizer N by riparian vegetation was also limited because of winter leaching when plants were dead or dormant. These conclusions are supported by Kaushik and Robinson (1976) and Kaushik et al. (1975) where no more than 1% of the N losses from a watershed were accounted for in the standing macrophyte.

Losses of NO_3-N to deep seepage were not indicated nor thought to be significant. The chloride concentrations beneath the cropped areas of the field and in the flow lines to the natural stream were consistently high indicating lateral flow from the field. This conclusion is supported by previous work in the Coastal Plain (Gilliam, 1975; Daniels, Gamble, and Gilliam, unpublished data) where NO_3-N moved laterally in better drained soils above an impermeable or semi-impermeable horizon to a drainage outlet at a lower elevation.

The loss of NO_3-N by denitrification between the edges of the fields and the streams was indicated by lower concentrations of nitrate in the streams and a change in the nitrate to chloride ratio. Reduction-oxidation potentials (Eh) were measured on both flood plain areas near the stream. The Eh values recorded near the surface on Fields 1, 2, and 3 indicated that denitrification was probable (Gambrell et al., 1975a). On the flood plains beside the fields there were large, depressional areas which remained wet long after surface runoff stopped. The water which seeped into the depressions was usually < 0.1 to 6.8 mg L^{-1} NO_3-N which was considered low, high in Cl (5.5–18.3 mg L^{-1}), and reduced conditions were found along the bottom (Table 3). These areas were only observed in late winter and early spring so sampling was limited to four events.

Field 3 at the Middle Coastal Plain watershed had a crudely improved ditch system which followed an old wash that was not straightened as it was mechanically improved. The wells in Field 3 were installed and sampled first on the surface drainage slope then later on the subsurface drainage slope (Fig. 1) which was perpendicular to the surface slope as indicated by ground water elevation measurements. There was a mean concentration of 2.4 ± 0.1 mg L^{-1} NO_3-N in the samples near the ditch because fertilizer was not applied adjacent to the wells which were set on the ditch bank and subsurface drainage was not in the direction of the wells. This concentration was ~33% of that measured in the field. After the relocation of wells, a concentration of 2.0 ± 0.5 mg L^{-1} NO_3-N was measured at the edge of the woods, 3 m downslope of the last rows in the field. This area was < 1 m higher in elevation than the stream and the water table near the edge of the field remained high long enough for denitrifying conditions during extended wet periods of the year. Soil mottling indicated that iron was reduced in this location and denitrification was likely. An even lower concentration of 0.1 ± < 0.05 mg L^{-1} NO_3-N was measured from wells on the boggy floodplain near the stream, but again the concentration of Cl remained almost constant at 10.7 mg L^{-1} as it moved through this same area. The lower elevations on the flood plain had organic matter contents of 2 to 46% in the top 2.5 cm. This, combined with a high water table, would cause reduction of NO_3-N much of the year. Redox measurements in this area indicated that at some periods there were areas where NO_3-N reduction would occur.

Applying DRAINMOD to Field 3, which drained surface water in one direction and subsurface in another, resulted in a calculated loss of only 128 mm of subsurface drainage each year. The slope and soil characteristics caused most runoff water to leave as surface drainage. The 7.7 mg L^{-1} NO_3-N concentration in groundwater below the field resulted in a 9.9 kg ha^{-1} yr^{-1} NO_3-N movement from the cultivated area (Table 2). Samples collected adjacent to the stream with very low concentrations of nitrate support the conclusion that the amount added to the stream through surface and subsurface flow was nearly zero.

Nitrate losses from Field 4 at the Middle Coastal Plain watershed were different from the other fields described. The depth to the water table was greater because of the tube drainage system. The average concentration of 14.8 ± 0.4 mg L^{-1} NO_3-N was higher than the other fields (Table 2) because of the intense double cropping system which required more fertilizer-N. The concentrations of NO_3-N in the wells immediately adjacent to the V-ditch were almost equal to those measured in the field. This was in contrast to all other sites where Duncan NMR Test indicated the NO_3-N concentrations were significantly lower ($p = 0.05$) just before entering the surface water. This indicated NO_3-N was moving into the ditch laterally as well as by the tube drainage system.

The subsurface drainage predicted for this field with improved subsurface drainage was greater than any other site. The drainage loss combined with an average concentration of 14.8 ± 0.4 mg L^{-1} NO_3-N resulted in a 54.9 kg ha^{-1} yr^{-1} subsurface NO_3-N loss from the field (Table 2).

One subsurface drainage line was sampled before it emptied into the drainage ditch. The effluent from this line consistently had a NO_3-N concentration lower than the water in the profile. The 3-yr mean was 10.3 ± 0.6 mg L^{-1} NO_3-N. The mean concentration of NO_3-N in the V-ditch (2.9 ± 0.1 mg L^{-1}) was always lower than in the profile (14.8 ± 0.4 mg L^{-1}) or in the drainage line effluent. Three examples showing differences in concentrations are shown in Table 4. This was observed many times when there was no surface runoff and the discharge from the drainage lines was sufficient for all surface runoff to be displaced from the ditch. The V-ditch concentration was also lower than the tube drainage effluent. Although the V-ditch sediments were credited with reducing some of the NO_3-N after the water reached the channel, much more was probably lost as the water carrying the NO_3-N moved through the bottom and side-wall materials into the ditch. This mechanism has been observed by Chen and Keeney (1974) with lake sediments.

Often flow was observed from the V-ditch when the tube lines were dry (Table 4). During these periods, a lower concentration of NO_3-N was usually observed in the ditch than in the field. The lower concentration indicated that the ditch sediments were reducing the NO_3-N upon entry into the surface water as previously described. Many measurements in the ditch sediments indicated that Eh was sufficiently low (200–280 mV) for denitrification.

Lower Coastal Plain Watershed

Three fields representing two types of drainage, V-ditches or grassed waterways, were identified in this watershed (Fig. 2). Both drainage systems produced almost identical results. The fields with V-ditches cut to 2 m and the field with the grassed waterway drained mostly surface waters. This was initially expected for the shallow grassed waterway but not for the deeper V-ditches. The bore hole descriptions at each location indicated a massive clay layer, 0.25 to 0.45 m, below the surface horizon which conducted a low volume of water (Fig. 2). It was noted that little or no flow came from these ditches shortly after the conclusion of a rainfall event. Typically ditch flow stopped while the fields were still too wet for traffic. This was not observed at the Middle Coastal Plain watershed.

The model DRAINMOD was used to compute drainage from a range of slopes, lateral conductivities, and soil series characteristics for the watershed. The amount of subsurface drainage was only 1.0 to 20.0 mm yr^{-1} even with full magnitude changes in assumed lateral conductivities (Table 2). Therefore, significant NO_3-N was not being lost from these fields by this route. Plotting of water levels and ground surfaces of the wells for 10 sampling periods indicated little water movement through the soil profile as predicted by the model. There was usually not a well defined hydraulic gradient in the

Table 4. Nitrate concentrations in the field wells, tube drain, and V-shaped collector ditch from Field 4 in the Middle Coastal Plain watershed both with and without tile flow.

Date	Conc of NO_3-N in wells	Conc of NO_3-N in tube drain	Conc of NO_3-N in V-shaped collector ditch
	mg L^{-1}		
25 Mar. 1980	16.3	12.2	7.2
9 Feb. 1982	16.4	10.5	8.2
11 Mar. 1982	13.9	10.8	8.7
9 May 1980	15.2	no flow	3.3
23 Dec. 1980	5.4	no flow	4.2
16 Feb. 1981	7.3	no flow	7.6

direction of the surface drainage. The low mean concentrations of NO_3-N (0.3 ± 0.1, 0.3 ± < 0.1, and 0.6 ± 0.1 mg L^{-1} for Fields 1, 2, and 3, respectively) and low chloride concentrations (5.5 to 10.5 mg L^{-1}) at the well points indicated little movement of NO_3-N to the shallow ground water. After sampling the field wells for 1 yr, it was concluded that significant NO_3-N was not being leached down through the soil profile in any of the fields, so sampling of these wells was discontinued. Although concentrations of NO_3-N in surface runoff was usually low, field ditches were monitored for 3 yr. Discharge from field drains were distributed across a wide flood plain. This is discussed in Jacobs and Gilliam (1985, this issue).

DISCUSSION

In the Middle Coastal Plain watershed, the type of field drainage system has little influence upon the amount of N leaving the field in the subsurface drainage water, but has a tremendous influence upon the amount entering the surface water. From an environmental view the best system by far is a natural drainageway bordered by poorly-drained soil and dense riparian vegetation. In the three fields with drainage into natural drainageways, the N was almost completely removed from the subsurface water before entering the stream. The high water table on the flood plain and the accumulation of organic matter provided by the vegetation resulted in reducing conditions in the sediments much of the time. Thus NO_3-N that moved into this zone with the drainage water was lost by denitrification before it could enter the stream. Removal through this process is largely responsible for the loss of most of the N from agricultural drainage water in the Coastal Plain. Previous observations in North Carolina have been that well- and moderately well-drained agricultural soils have 16 to 46 kg ha^{-1} yr^{-1} N moving from them in drainage water. It is these soils that most often do not require improved drainage and thus drain to natural outlets. In North Carolina, and much of the southeastern Coastal Plain, these natural outlets are nearly always bordered by poorly-drained soils and natural vegetation. Thus, unintentionally, but fortunately, the N removal mechanism has been largely left intact where it is needed the most.

The field drainage system that allows the most N to enter surface water is a tube drained moderately well- to somewhat poorly-drained soil. The tube drainage water monitored in this study averaged 10.2 mg L^{-1} of NO_3-N which is consistent with values measured in other studies on similar soils (Gambrell et al., 1975b, Gilliam et al., 1978).

The field soil water that moves into the drainage ditch without passing through the tube line has a lower N concentration than that moving through the tube line because of denitrification during entry into the ditch. However, this reduction is much less than when the water passes through a relatively narrow vegetated riparian area to reach a surface outlet. There is almost certainly some loss of NO_3-N from the water in the improved field drainage ditches through diffusion and denitrification in the bottom sediments because these sediments frequently were highly reduced with a very low concentration of NO_3-N in the interstitial water. This was thought to result from the accumulation of organic detritus in the ditch bottom which served as an energy source for microorganisms to reduce available O_2; however, this loss was not quantified.

Essentially all of the drainage from agricultural fields in the Lower Coastal Plain watershed was by surface drainage. This surface drainage water had a low N concentration, so little enters the surface waters from agricultural fields. Previous work (Kuenzler et al., 1980) had given the wide flood plain between the edge of the fields and the stream much of the credit for the relatively pure water draining from this watershed. Although the flood plain does have a large capacity for removing NO_3-N from water, the very low N content of the stream water was, for the most part, a result of a lack of entry of large quantities of N into the drainage water. Finding restrictive subsurface movement of N in this important agricultural region should indicate possible expansion areas for farming as needs arise.

ACKNOWLEDGMENT

The excellent help and cooperation of the U.S. Geological Survey, Water Resources Division, Raleigh, NC, throughout this project is gratefully acknowledged.

REFERENCES

Aldrich, S. R. 1980. Nitrogen in relation to food, environment, and energy. Special Pub. 61. College of Agriculture University of Illinois at Urbana-Champaign, IL.

Anonymous. 1945–1982. North Carolina crop and livestock reporting service. North Carolina Agricultural Statistics. North Carolina Department of Agriculture, Raleigh, NC.

Anonymous. 1945–1983. North Carolina fertilizer tonnage report. North Carolina Agricultural Statistics. North Carolina Department of Agriculture, Raleigh, NC.

Carpenter, J. H., D. W. Pritchard, and R. C. Whaley. 1969. Observations of eutrophications and nutrient cycles in some Coastal Plain estuaries. p. 210–221. *In* Eutrophication: Causes, Consequences, Correctives. Proceedings of a Symposium. University of Wisconsin. 1967. National Academy of Science, Washington, DC.

Chen, R. L., and D. R. Kenney. 1974. The fate of nitrate in lake sediment columns. Water Resour. Bull. 10:1162–1172.

Gambrell, R. P., J. W. Gilliam, and S. B. Weed. 1975a. Denitrification in subsoils of the North Carolina Coastal Plain as affected by soil drainage. J. Environ. Qual. 4:311–316.

Gambrell, R. P., J. W. Gilliam, and S. B. Weed. 1975b. Nitrogen losses from soils of the North Carolina Coastal Plain. J. Environ. Qual. 4:317–323.

Gilliam, J. W. 1975. Nitrogen fertilizers and ground water quality in North Carolina. p. 76–81. *In* A. G. Wollum (ed.) Proc. of the 18th Annual Meeting of the Soil Sci. Soc. of North Carolina. Raleigh, NC. February 1975. Soil Science Society of North Carolina, Raleigh, NC.

Gilliam, J. W., R. W. Skaggs, and S. B. Weed. 1978. An evaluation of the potential for using drainage control to reduce nitrate loss from agricultural fields to surface waters. Rep. 128. Water Resources Research Institute of the University of North Carolina, Raleigh, NC.

Harrison, W. G., and J. E. Hobbie. 1974. Nitrogen budget of a North Carolina estuary. Rep. 86. Water Resources Research Institute, University of North Carolina, Raleigh, NC.

Hobbie, J. E. 1974. Nutrients and eutrophication in the Pamlico River estuary, North Carolina, 1971–1973. Rep. 100. Water Resources Research Institute, University of North Carolina, Raleigh, NC.

Jacobs, T. C., and J. W. Gilliam. 1983. Nitrate loss from agricultural drainage waters: Implications for nonpoint source control. Rep. 209. Water Resources Research Institute, University of North Carolina, Raleigh, NC.

Jacobs, T. C., and J. W. Gilliam. 1985. Headwater stream losses of nitrogen from two coastal plain watersheds. J. Environ. Qual. 14: 467–472 (this issue).

Kaushik, N. K., J. B. Robinson, P. Spain, H. R. Whiteley, and W. Stammers. 1975. A quantitative study of nitrogen loss from water on a small spring-fed stream. p. 110–117. *In* Water Pollut. Res. in Canada. Proc. 10th Canadian Symp., February 1975. Toronto. Institute of Environmental Studies of the University of Toronto, Toronto.

Kaushik, N. K., and J. B. Robinson. 1976. Preliminary observations on nitrogen transport during summer in a small spring-fed Ontario stream. Hydrobiologia 49:59–63.

Kaushik, N. K., J. B. Robinson, W. N. Stammers, and H. R. Whiteley. 1981. Aspects of nitrogen transport and transformations in headwater streams. p. 113–139. *In* M. A. Lock and D. D. Williams (ed.) Perspectives in running water ecology. Plenum Press, New York.

Kuenzler, E. J., P. J. Mulholland, L. A. Yarbo, and L. A. Smock. 1980. Distributions and budgets of carbon, phosphorus, iron and manganese in a floodplain swamp ecosystem. Rep. 157. Water Resources Research Institute, University of North Carolina, Raleigh, NC.

Light, T. S. 1972. Standard solution for redox potential measurements. Anal. Chem. 44:1038–1039.

Lowrance, R. R., R. L. Todd, and L. E. Asmussen. 1984. Nutrient cycling in an agricultural watershed: I. Phreatic movement. J. Environ. Qual. 13:22–27.

Skaggs, R. W. 1978. A water management model for shallow water table soils. Rep. 134. Water Resources Research Institute, University of North Carolina, Raleigh, NC.

Skaggs, R. W., and J. W. Gilliam. 1981. Effect of drainage system design and operation of nitrate transport. Trans. Am. Soc. of Ag. Eng. 24:929–934, 940.

Soil Curvey of Craven County, North Carolina. 1985. United States Department of Agriculture Soil Conservation Service in cooperation with North Carolina Agricultural Experiment Station. U.S. Government Printing Office, Washington, DC.

Soil Survey of Pitt County, North Carolina. 1974. United States Department of Agriculture Soil Conservation Service in cooperation with North Carolina Agricultural Experiment Station. U.S. Government Printing Office, Washington, DC.

Soil Survey of Wayne County, North Carolina. 1974. United States Department of Agriculture Soil Conservation Service in cooperation with North Carolina Agricultural Experiment Station. U.S. Government Printing Office, Washington, DC.

U.S. Department of Agriculture-SCS. 1980. DRAINMOD reference report methods for design and evaluation of drainage-water management systems for soils with high water tables. USDA-SCS. South National Tech. Center, Fort Worth, TX.

Commentary on Paper E4

Warwick, J. & Hill, A. R. (1988) Nitrate depletion in the riparian zone of a small woodland stream.

In an earlier article, Alan Hill (1979) had reported that the removal of nitrate-N by denitrification in the downstream reaches of the Duffin Creek near Toronto (Ontario, Canada) represented 5–6% of the annual export of total nitrogen from this river basin. In this paper Hill and his colleague evaluate more specifically *in situ* and potential N removal capacity of a riparian zone on a groundwater spring source flowing through the upper soil horizon of a riparian forest. This is a rather peculiar situation compared to related papers published on a similar subject at the same period and presented in this volume (Lowrance *et al.*, 1984, Paper E1; Peterjohn & Correll 1984, Paper E2; Jacobs & Gilliam, 1985, Paper E3) because it deals with surface flow through a riparian zone rather than groundwater or subsurface flow. As such, it is more related to nutrient cycling in stream or ditch sediments. Yet, it is a very relevant paper because it addresses many of the key issues associated with nitrogen removal in riparian zones. Moreover, it is one of the first papers which combine nitrate addition both in the field and microbiological experiments in the laboratory.

The authors dripped an enriched nitrate solution with chloride as a conservative tracer to evaluate *in situ* capacity of riparian zone to remove nitrate. They did not find any significant decrease in nitrate concentration along 50- to 100-m flow paths. This result underlines that nitrate diffusion transfer to the biologically active sediment was too slow compared to the surface flow to entail a significant concentration decrease. The low rate of removal of nitrate added to soil cores taken from the riparian zone confirmed that diffusion was the factor limiting denitrification. This is a common feature which often limits the N buffering capacity of riparian zones. Indeed, there is a trade-off between water residence time and the rate of microbial processes recycling nitrate. Long water residence time within the riparian zone entails full N uptake or denitrification, but the overall contribution to the catchment water budget is marginal because of the limited water flow.

An adaptation of the potential denitrification measurement method proposed by Smith & Tiedje (1979, Paper D3) and Yoshinari & Knowles (1976) (see articles and related commentaries in this volume) allowed the authors to confirm that a denitrifying community was present in the riparian soils and that its activity was limited by nitrate availability. A 2-day soil slurry incubation demonstrated that the denitrifying community was somehow rapidly adapting (either by producing more denitrifying enzymes or by cell multiplication, the experiment could not conclude which) to reduce most of the added nitrate. They also mentioned that the discrepancy they measured between the nitrate removal and N_2O production under acetylene inhibition could be due to a dissimilation process, a hypothesis which is still under investigation today.

This study combining *in situ* and laboratory experiments was very inspiring and we can only agree with the authors' conclusion when they emphasise that "*reliable generalization about the role of riparian areas as nutrient filters will require the study of nutrient flux and transformation processes in relation to routes of groundwater movement and other site characteristics in a range of riparian environments*".

GP

References

Hill, A. R. (1979) Denitrification in the nitrogen budget of a river ecosystem. *Nature* **281**(5729), 291–292.

Jacobs, T. C. & Gilliam, J. W. (1985) Riparian losses of nitrate from agricultural drainage waters. *J. Environ. Qual.* **14**(4), 472–478.

Lowrance, R. R., Todd, R. L. & Asmussen, L. E. (1984) Nutrient cycling in an agricultural watershed. I. Phreatic movement. *J. Environ. Qual.* **13**, 27–32.

Peterjohn, W. T. & Correll, D. L. (1984) Nutrient dynamics in an agricultural watershed: observations on the role of a riparian forest. *Ecology* **65**(5), 1466–1475.

Smith, M. S. & Teidje, J. M. (1979) Phases of denitrification following oxygen depletion in soil. *Soil Biology & Biochem.* **11**, 261–267.

Yoshinari, T. & Knowles, R. (1976) Acetylene inhibition of nitrous oxide reduction by denitrifying bacteria. *Biochem. Biophys. Res. Communication* **69**, 705–710.

Hydrobiologia 157: 231–240 (1988)

Nitrate depletion in the riparian zone of a small woodland stream

June Warwick & Alan R. Hill
Department of Geography, York University, Toronto, Ontario, Canada M3J 1P3

Received 30 July 1986; in revised form 22 April 1987; accepted 23 April 1987

Key words: denitrification, ground water, nitrate, organic soils, riparian zone, stream

Abstract

Field enrichments with nitrate in two spring-fed drainage lines within the riparian zone of a small woodland stream near Toronto, Ontario showed an absence of nitrate depletion. Laboratory experiments with riparian substrates overlain with nitrate enriched solutions revealed a loss of only 5–8% of the nitrate during 48 h incubation at 12 °C. However, 22–24% of the initial nitrate was depleted between 24 and 48 h when a second set of substrate cores was incubated at 20 °C. Short-term (3 h) incubations of fresh substrates amended with acetylene were used to estimate *in situ* denitrification potentials which varied from 0.05–3.19 $\mu g\ N\ g^{-1}\ d^{-1}$ for organic and sandy sediments. Denitrification potentials were highly correlated with initial nitrate content of substrate samples implying that low nitrate levels in ground water and riparian substrates may be an important factor in controlling denitrification rates. The efficiency of nitrate removal in spring-fed drainage lines is also limited by short water residence times of < 1 h within the riparian zone. These data suggest that routes of ground water movement and substrate characteristics are important in determining nitrate depletion within stream riparian areas.

Introduction

The importance of the riparian zone in relation to land-water interactions has received increased attention in recent years (Cummins, 1980; Swanson *et al.*, 1982). The riparian zone in forested headwater catchments can exert a major control on the stream ecosystem in several ways. Inputs of allochthonous detritus form the main source of energy and also exert a considerable influence on community structure in small woodland streams (Fisher & Likens, 1973; Cummins, 1974; Molles, 1982; Conners & Naiman, 1984). Large woody debris derived from riparian vegetation can influence channel morphology and the movement of sediment and water in streams (Keller & Swanson, 1979; Swanson *et al.*, 1982).

Riparian zones also have a high potential to regulate nutrient fluxes between upland areas and the stream. Ground water is frequently present at a shallow depth beneath the riparian area and vegetation and soil processes may therefore modify the chemistry of ground water before it enters the stream (Swanson *et al.*, 1982; Lowrance, *et al.*, 1985). Ground water which may be influenced by passage through the riparian zone maintains base flow in headwater streams. Hydrological research also indicates that localized areas adjacent to the stream channel are of major importance as contributors of storm and snowmelt runoff to streams in humid landscapes (Dunne & Black, 1970; Dunne, 1983). Some studies have used environmental isotopes to show that 50–90% of the storm runoff in many streams is contributed by ground water displaced from the riparian zone (Sklash & Farvolden, 1979; Bottomley *et al.*, 1985).

Several recent studies have revealed that forested

riparian areas in lowland agricultural catchments in the Southeastern United States were effective in retaining several nutrients, particularly nitrates which entered the riparian zone in ground water from adjacent fertilized fields (Lowrance, *et al.*, 1984a; Peterjohn & Correll, 1984; Jacobs & Gilliam, 1985). Analysis of the processes of nitrate depletion in the riparian zone of these catchments suggested that a substantial part of the nitrate was denitrified.

The influence of the riparian zone on nutrient fluxes entering streams in small forested headwater catchments has not been investigated. It is likely that factors influencing nutrient transport in the riparian zone of such catchments may differ from those in lowland agricultural catchments. Nutrient concentrations in ground water entering the riparian zone from upland forest communities are likely to be low and in upland catchments, ground water often emerges as seeps and springs creating areas of diffuse saturated overland flow in the riparian areas (Dunne & Black, 1970; Rigler, 1979). Hynes (1983) has suggested that the movement of water by different routes in the riparian zone may have an important influence on the supply of nutrients to streams. This paper presents the results of a study designed;

1. to examine nitrate depletion in spring-fed zones of overland flow in the riparian zone of a small woodland stream;
2. to evaluate the importance of denitrification as a mechanism of nitrate removal in this environment.

Nitrate transformations were investigated by experimentally enriching riparian zone springs with nitrate. Laboratory experiments with riparian substrates were also used to provide quantitative evidence of denitrification activity.

The study area

The study area was located in a small forested headwater catchment of 2 km^2 on the south-facing slope of the Oak Ridges moraine about 50 km northeast of Toronto, Ontario. The catchment varies from 268 to 346 m in altitude and the topography consists of moderately to steeply sloping hummocky moraine dissected by dry valleys. The catchment is drained by a second order stream and three perennial tributaries. Stream discharge at the basin outlet varied between 40 – 50 l sec^{-1} and nitrate concentrations were < 100 $\mu g\ l^{-1}$ during the summer of 1984.

The moraine slopes are covered with well drained Grey-Brown Luvisols, whereas organic soils (Histosols) occur on the valley floors of the perennial streams. The forest cover on the upland slopes in the eastern portion of the catchment consists of mature stands of sugar maple-beech (*Acer saccharum-Fagus grandifolia*) and younger stands of white pine (*Pinus strobus*) and birch (*Betula papyrifera*). An old-field community of grasses and forbs with scattered patches of hawthorn (*Crataegus* spp.) and white pine occurs in the western section. The riparian zone adjacent to the perennial streams is forested with hemlock (*Tsuga canadensis*) and cedar (*Thuja occidentalis*).

A ground water discharge zone at the base of the moraine slopes produces a permanently saturated riparian area 20 – 100 m wide along the perennial stream channels. The water table intersects the ground surface at the upland margin of this riparian zone and then slopes toward the stream at a depth of 10 – 30 cm beneath the surface of the organic soils. Springs and seepage areas at the base of the moraine slopes are connected to the perennial streams by numerous drainage lines which cross the riparian area. The term drainage line has been defined as a line of flow having some depth but not necessarily contained within banks formed by water erosion (McCaig, 1983). The drainage lines in this study are often less than 40 cm in width and have water depths of 2 – 15 cm and discharges of < 1 l sec^{-1}. Surface flow frequently passes into shallow pipes and cavities formed by dead tree roots and partially uprooted trees. Flow from these drainage lines makes up approximately 50% of the ground water entering the second order stream from the riparian zone (Dr. N. Roulet, pers. comm.). Ground water also flows upward to the water table beneath the riparian zone and reaches the stream as bank and bed seepage.

Enrichment experiments were performed on two representative drainage lines referred to as DL1 and DL2 within the riparian zone of the second order stream. DL1 flows for about 50 m in a series of alternating surface and pipe segments before dividing into two separate flow lines which follow an erratic course mainly in pipes before entering the stream at outlets 25 m apart. DL2 which contains no pipe sections flows for 105 m through the riparian forest before entering the stream. Water in the drainage lines flows over a variety of substrates which include organic-rich sediments, accumulations of coarse woody debris and areas of sand and gravel sediment.

Flow in the initial 5–10 m of the drainage lines occurs through beds of watercress (*Nasturtium officinale*). Elsewhere the drainage lines are usually heavily shaded and aquatic plants are absent. Water temperature increased from 3–6 °C in April to 10–13 °C in August and pH varied between 7.6 and 8.0. During the enrichment experiments discharge in DL1 increased from 0.5–0.7 l sec^{-1} at 50 m downstream to 0.8–1.0 l sec^{-1} for the combined discharge of the two outlets entering the stream. Discharge in DL2 was usually 0.1–0.2 l sec^{-1} at 15 m below the source and 0.3–0.5 l sec^{-1} at 80 m. Background concentrations of NO_3-N in DL1 were 160–200 $\mu g\ l^{-1}$ at the spring source declining to 60–80 $\mu g\ l^{-1}$ at the two outlets, whereas concentrations in DL2 decreased from 20–40 $\mu g\ l^{-1}$ at the spring to $< 5\ \mu g\ l^{-1}$ at the stream.

Materials and methods

Field enrichment procedures

Nitrate enrichments were conducted on 7 occasions between April and August 1984 in order to detect any seasonal variations in response to enrichment. Enrichments were performed for periods of 4–4½ hours on April 16, June 19, July 11 and August 14, in DL1 and on April 18, June 21 and August 15 in DL2. The enrichment solution was prepared at the study sites by adding nitrate as $Ca\ (NO_3)_2$ to water from the drainage lines in a 22 litre carboy. Sodium chloride, mixed with the nitrate,was used as a conservative tracer. The solution was dripped through a flow regulated tube into drainage lines at a constant rate of 1 ml sec^{-1}. Solution flow rates were increased in DL2 after the initial two hours of enrichment on June 21 and August 15 in order to evaluate nitrate removal at two contrasting enrichment levels. Samples of the enrichment solution showed that concentration of nitrate and chloride were constant throughout the enrichment experiment. The enriched solution was added to each drainage line within a few metres of the spring source at a point where turbulent flow ensured thorough mixing.

Water samples were collected from the drainage lines before the experiment began and at 20 minute intervals until one hour after the enrichment ceased. Sample stations on DL1 were located at 50 m below the enrichment point and at 96 and 101 m downstream on the west and east branch respectively where these entered the second order stream. Samples were taken from DL2 at 15, 60 and 105 m downstream. An automated cadmium reduction method was used to measure $NO_3 + NO_2-N$ on a Technicon Autoanalyzer (APHA, 1976). Tests were also made separately for NO_2-N in the water samples. Chloride was analyzed using an automated thiocyanate procedure (Environment Canada, 1979). Nitrate transformations in the drainage lines were evaluated by comparing measured concentrations of NO_3-N with concentrations predicted from the Cl tracer using the formula of Richey, *et al.* (1985).

Laboratory experiments

Several experiments were conducted with 0–5 cm depth substrate samples collected from the two drainage lines in order to measure changes in nitrate concentration in water overlying substrates and to evaluate nitrate removal mechanisms. Attention was focused on the 0–5 cm depth because previous studies of nitrate loss from water overlying stream sediments and organic soils has shown that nitrate removal usually occurs in the upper few cms of the substrate (Sain *et al.*, 1977; Knowles, 1982).

Experiment 1

This experiment was designed to evaluate nitrate removal by drainage line substrates under conditions similar to the field enrichments. A plexiglass column (4.8 cm ID) was used to remove intact cores of 0–5 cm depth at 5 sites on each drainage line. These sites were selected to include sandy and organic substrates as well as sediments from a watercress bed near the spring source of each drainage line. The cores were returned to the laboratory and overlain with 500 ml of a solution containing 2 mg l^{-1} NO_3-N as KNO_3 to replicate the NO_3-N concentration observed during the enrichment experiments (Figs. 1 and 2). Air bubbles were introduced close to the water-substrate interface to simulate the turbulence and aeration characteristics of the drainage lines. The columns were incubated for 48 hrs at a temperature of 12 °C which was similar to the summer ground water temperature in the riparian zone. Solution samples were collected at 5,24 and 48 hrs and stored at 3 °C until analyzed for NO_3-N at the end of the experiment.

Experiment 2

The procedures of experiment 1 were repeated using a second set of 0–5 cm cores. However, in this experiment the incubation temperature was increased to 20 °C to assess the effect of a higher temperature on nitrate removal rates.

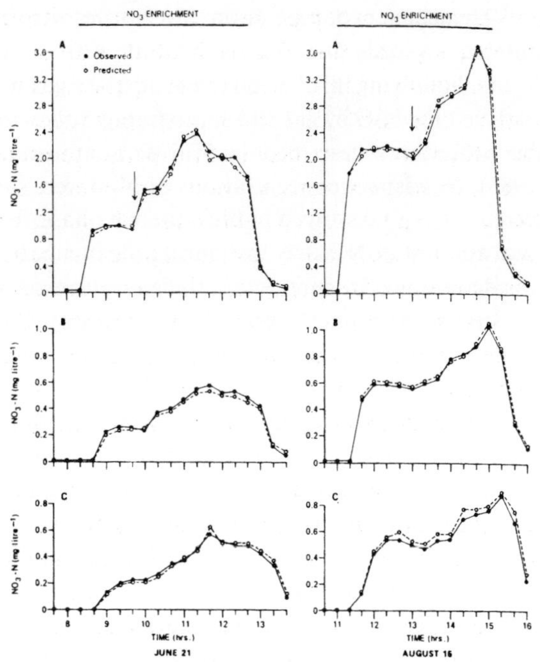

Fig. 2. Observed and predicted concentrations of NO_3-N in drainage line 2, A. 15 m downstream; B. 60 m downstream; C. 105 m downstream. At 2h (arrow) the enrichment solution flow rate was increased.

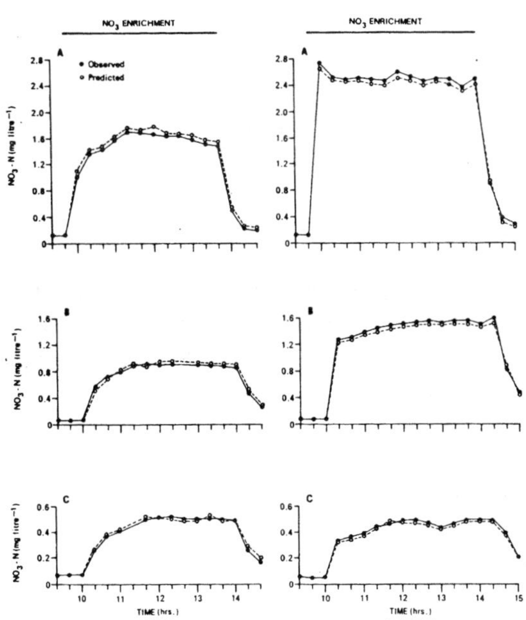

Fig. 1. Observed and predicted concentrations of NO_3-N in drainage line 1, A. 50 m downstream; B. west branch 96 m downstream; C. east branch 101 m downstream. Predicted concentrations were estimated from observed Cl concentrations.

Experiment 3

The disappearance of added nitrate in column incubations is usually attributed to denitrification. However, dissimilatory reduction of nitrate to ammonium and assimilatory reduction to ammonium that is incorporated into microbial biomass can also occur (Knowles, 1982). The importance of denitrification with respect to nitrate removal by the riparian substrates was evaluated using the acetylene blockage technique (Yoshinari & Knowles, 1976). Acetylene inhibits the reduction of N_2O to N_2 and the resulting accumulation of N_2O can be used as a measure of denitrifying activity. Five samples of 0–5 cm depth substrate were collected from each drainage line and returned to the laboratory. Approximately 30 g of fresh substrate from each sample was added to a 125 ml serum bottle and covered with 15 ml of a solution containing 5 mg l^{-1} NO_3-N as KNO_3. These additions produced 1 cm layer of solution overlying 1–2 cm depth of substrate. The bottles which were sealed with septum stoppers remained air-filled and acetylene was added to produce a 10%

(V/V) headspace concentration. The bottles were incubated statically at room temperature (20 °C) for 48 hrs. Following incubation the headspace gas was analyzed for N_2O by gas chromatography following the procedures described in Hill & Sanmugadas (1985). Headspace concentrations of N_2O were corrected for N_2O dissolved in the aqueous phase. The total amount of NO_3-N lost during the incubation was determined by subtracting the amount remaining after 48 hrs from the sum of the amounts initially present in the substrate and added in the nitrate solution.

Experiment 4

This experiment evaluated potential denitrification rates under conditions of ground water temperature and nitrate concentration typical of the drainage lines within the riparian zone. The procedures used were identical to those outlined in experiment 3 with the following exceptions. The substrate in each bottle was covered with 15 ml of water containing 0.1 mg l^{-1} NO_3-N which had been collected from DL1. The bottles were incubated at 11 °C for a three hour period following which activity was stopped by injecting 3 ml of 30% NaOH (Messer & Brezonik, 1983). Samples of headspace gas were analyzed for N_2O by gas chromatography. Smith & Tiedje (1979) have shown that denitrification rates measured during the first three hours of incubation of fresh soils or sediments are linear and provide an indication of the *in situ* denitrification potential.

Sediment Analysis

Subsamples of the substrates used in experiment 4 were dried at 80 °C for 48 hrs. Substrate pH was measured with a glass electrode on a 1:2 substrate and water mixture. Organic carbon was estimated by the Walkley-Black method and soluble carbon was determined by water extraction following the procedure of Waring & Gilliam (1983). The initial amount of nitrate in the substrates was measured from 2M KCL extracts using the Autoanalyzer.

Results

Field enrichments

Data collected during the nitrate enrichment experiments are shown for four dates in Figs. 1 and 2. Data obtained on three other dates were similar and are therefore not presented. During the enrichments NO_3-N and Cl concentrations below the injection point were 15–20 and 5–13 times greater than measured background levels in DL1. Higher initial NO_3-N and Cl concentrations which were 65–105 and 7–41 times greater than background levels occurred in DL2. However, the dilution effect of considerable ground water inputs to the drainage lines within the riparian zone produced a rapid downstream decrease in element concentrations (Figs. 1 and 2). The leading edge of the injected NO_3-N in the drainage lines travelled through the riparian zone in 30–40 minutes. Minor fluctuations in element concentration throughout the enrichment period result from an erratic solution supply from the carboy.

Predicted NO_3-N concentrations were up to 100 $\mu g\ l^{-1}$ greater than observed concentrations on some dates, for example DL1 at 50 m downstream

Table 1. Mean nitrate concentration ± SD of water overlying substrate cores as affected by temperature at an initial NO_3-N concentration of 2 mg l^{-1}.

Site	Number of samples	Temperature	NO_3-N concentration (mg l^{-1})		
			5 hours	24 hours	48 hours
DL1	5	12 °C	1.94 ± 0.06	1.90 ± 0.03	1.89 ± 0.04
	5	20 °C	1.91 ± 0.07	1.86 ± 0.08	1.53 ± 0.17
DL2	5	12 °C	1.92 ± 0.05	1.88 ± 0.05	1.84 ± 0.09
	5	20 °C	1.91 ± 0.05	1.83 ± 0.04	1.55 ± 0.09

Table 2. Percentage of added nitrate recovered as N_2O during aerobic incubation of substrate samples with C_2H_2 at 20 °C for 48 hours.

	NO_3-N added (μg bottle^{-1})	N_2O-N evolved (μg bottle^{-1})	NO_3-N lost (μg bottle^{-1})	% of lost NO_3-N recovered as N_2O-N
Drainage line 1				
Mean*	84.7	52.7	71.0	72.8
Range	82.2 – 87.3	28.8 – 70.2	55.1 – 85.4	52.3 – 83.0
Drainage line 2				
Mean	83.8	52.1	73.2	71.8
Range	80.9 – 85.6	39.2 – 68.6	56.0 – 84.8	47.6 – 87.0

*Mean of five substrate samples.

on July 11 and DL2 at 105 m on August 15 (Figs. 1 and 2). However, on other occasions observed NO_3-N concentrations were slightly greater than expected from dilution of Cl. The lack of any consistent pattern and the small magnitude of the differences between observed and predicted NO_3-N suggests that nitrate depletion is either absent or occurs at very low rates in spring-fed drainage lines within the riparian zone.

Laboratory Experiments

Data from the column experiments are summarized in Table 1. A very small decrease in nitrate was observed in solutions overlying most substrate cores after 5 hr of incubation at 12 °C. Nitrate concentrations remained constant in some columns or declined very slowly for the rest of the 48 hr incubation. The decline in nitrate concentration during the first day of incubation at 20 °C was similar to the pattern observed at 12 °C (Table 1). However, there was a considerable increase in the rate of nitrate loss between 24 and 48 hrs at 20 °C. The nitrate losses revealed in these experiments should be considered as net removal rates because concurrent nitrification may occur in the aerated water-sediment systems.

Considerable production of N_2O occurred in the riparian substrates which were incubated aerobically with a nitrate amended solution in the presence of

Table 3. In situ denitrification potentials and substrate characteristics.

Substrate type	N_2O evolved* μg N g^{-1} day^{-1}	Initial NO_3 μg N bottle^{-1}	Water soluble C μg g^{-1}	Organic C %	pH
Sandy sediment	0.11	5.35	93	3.2	7.4
	0.15	7.35	65	2.3	7.1
	0.53	5.30	120	3.0	7.5
	0.05	4.70	129	4.6	6.9
Organic sediment	0.10	2.50	331	12.0	7.0
	0.08	7.58	825	36.0	7.1
	1.20	12.70	421	21.7	6.9
	0.21	8.10	360	10.7	6.8
Watercress bed surface detritus and sediments	3.19	19.30	101	3.9	7.5
	1.05	16.00	75	1.8	7.6

*Extrapolated daily rates are based on 3 hr aerobic incubations with C_2H_2 + ambient NO_3 at 11 °C.

C_2H_2 (Table 2). The amount of N_2O production ranged from 28.8 to 70.2 μg N bottle^{-1} for the 48 hours with almost all the initial nitrate supply being depleted in most bottles. Denitrification measured as N_2O accounted for greater than 70% of the nitrate consumed in the majority of substrate samples. However, one sample from each drainage line exhibited an N_2O recovery equivalent to only about 50% of the nitrate loss.

Experiment 4 which was designed to measure *in situ* denitrification potentials revealed N_2O accumulation rates which constituted a loss of 2 to 41% of the initial nitrate present during the 3 hr incubation. Extrapolated daily denitrification rates were highly variable ranging from 0.05 to 3.19 μg N g^{-1} of substrate (Table 3). The highest denitrification potential was exhibited by sediments collected from watercress beds adjacent to springs. Organic and sandy sediments showed considerable variation in denitrification, although rates were low in most samples (Table 3). Denitrification rates had a highly significant positive correlation with initial nitrate levels in the substrate ($r = 0.87$, $P < 0.01$) but no significant relationship was observed with organic carbon, water soluble carbon or pH.

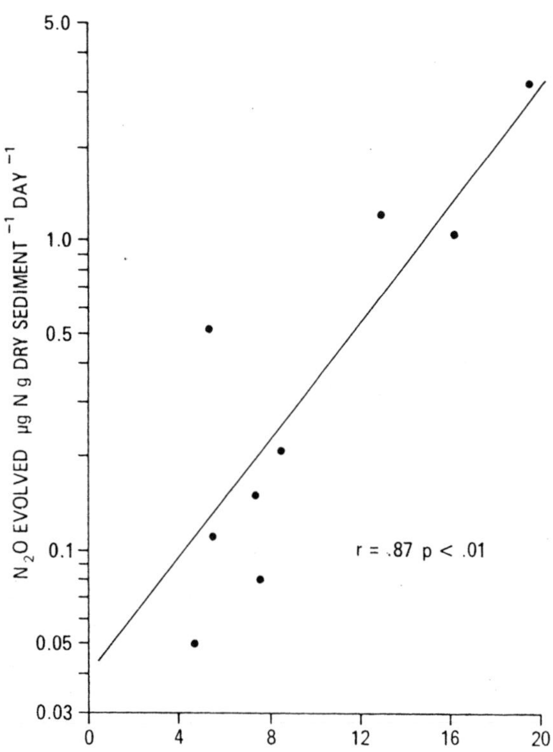

Fig. 3. Relationship between N_2O evolved during a 3 h aerobic incubation of drainage line substrates at 11 °C and initial NO_3-N.

Discussion

Although the field enrichment experiments produced nitrate concentrations which varied from 0.6 to more than 3.0 mg N l^{-1} at individual sampling stations on the riparian drainage lines, there was no evidence of nitrate depletion during transport through the riparian zone. This conclusion is supported by the column experiments which revealed a loss of approximately 3% of initial nitrate after 5 hrs. These rates of removal are too small to be detected by measurements of nitrate and chloride concentrations during field enrichments.

The considerable increase in nitrate loss from solutions during the second day of the column experiment at 20 °C indicates the presence of a lag phase in nitrate removal suggesting that nitrate reducing bacteria are physiologically inactive or are present at low initial population levels in the riparian substrates. The combination of high nitrate concentrations and warmer temperatures may have produced a population growth capable of accelerated nitrate removal.

Data from the 48 hr acetylene blockage experiment with nitrate substrates shows that denitrification was the main pathway of nitrate loss in the column experiments. The occurrence of nitrate losses which were considerably greater than N_2O evolution in some substrate samples may reflect dissimilatory or assimilatory reduction of NO_3 to NH_4. However, it is more likely that the importance of denitrification was underestimated in these samples because the acetylene block was incomplete. Considerable N_2O loss in the presence of acetylene has been observed in soils and sediments during long-term incubations (1–10) days particularly at low nitrate concentrations (Knowles, 1979; Oremland *et al.*, 1984).

The 3 hr acetylene blockage experiment which

was conducted under conditions similar to those found in the riparian zone provides the best estimate of the *in situ* denitrification potential of the drainage line substrates. However, it is possible that the mixing and aeration of the substrates which occurred during transfer to the serum bottles could have reduced denitrification rates below actual field levels. Potential denitrification rates based on the surface area of substrate in the bottles varied from 0.2 to 7.6 mg N m^{-2} day^{-1} for the organic and sandy sediments. Much higher rates of 17.3 and 29.5 mg N m^{-2} day^{-1} were recorded for the two substrate samples removed from patches of watercress. However, watercress beds occupy < 5% of the drainage line area.

Data from the field enrichments and laboratory experiments indicate that drainage lines within the riparian zone of this headwater stream have a low capacity to remove nitrate under field conditions. Daily NO_3-N fluxes downstream from the spring source were 7–9 g in DL1 and 0.3–0.5 g in DL2 at ambient nitrate concentrations during the summer. Calculations based on *in situ* denitrification potentials of substrates and measurements of drainage line area suggest that denitrification removes < 5% and 10–20% of the daily nitrate load in DL1 and 2 respectively. In contrast recent studies of riparian forests in lowland agricultural catchments have shown a large decrease in nitrate load in subsurface ground water traversing the riparian zone (Peterjohn & Correll, 1984; Lowrance *et al.*, 1984a). Laboratory incubation experiments have also revealed high potential rates of denitrification in flooded organic soils from a freshwater swamp and areas of drained peat and muck which contained high nitrate concentrations (Engler & Patrick, 1976; Bartlett *et al.*, 1979; Terry & Tate, 1980).

Nitrate removal in the riparian drainage lines examined in the present study is probably limited by two major factors. First, the drainage lines transport ground water rapidly through the riparian forest zone. The short residence time of this water reduces the efficiency of nitrate removal prior to entry into the stream channels. Previous studies have shown that ditches and tile drains connected to stream channels in agricultural catchments reduce the time and area of contact of water with riparian substrates allowing a greater amount of NO_3-N to enter streams (Lowrance *et al.*, 1984b; Cooper *et al.*, 1986). Second, *in situ* denitrification potentials measured in drainage line substrates are generally low. Moreover, the data from the field and laboratory experiments suggest that denitrifying activity does not increase rapidly in response to short-term enrichments.

The low denitrification rates in the riparian zone substrates are probably influenced mainly by the temperature and nitrate concentration of the ground water. Other factors which may control denitrification such as pH and soluble organic carbon are suitable for the growth of denitrifying bacteria in the substrates. Water temperature in the drainage lines varied from 3 to 13 °C in the April–August period. Previous research has shown that denitrification rates increase with increasing temperature (Terry & Nelson, 1975; Sain *et al.*, 1977). Rates of nitrate loss in laboratory incubations of stream sediments at 11 °C were only 45% of losses at 20 °C (Hill, 1983). Ground water nitrate concentrations were usually less than 200 μg N l^{-1} and riparian substrates contained very small amounts of nitrate. Nitrate amended sediments produced a significant stimulation of denitrification during a 48 hour incubation (Table 2). Extrapolated daily rates of denitrification were up to 2– to 7– fold higher than under *in situ* nitrate concentrations. The significant correlation between the short-term N_2O evolution rates and the initial nitrate content of substrate samples also implies that denitrification activity was limited by nitrate. Knowles (1982) has reported that at low nitrate concentrations, denitrification rates are dependent on nitrate concentration. Insignificant denitrifying activity in some headwater streams has also been linked to low nitrate concentrations (Meyer *et al.*, 1981; Duff *et al.*, 1984).

The absence of significant nitrate depletion in the riparian drainage lines still leaves open the question of whether the riparian forest zone is important in regulating the nitrate concentration of ground water as it approaches the streams. A substantial component of the ground water in the riparian zone reaches the stream by subsurface routes. The longer residence time of this water within the riparian zone could contribute to greater nitrate depletion by denitrification. Further research involving the analysis of a nitrogen mass balance for the riparian zone and the measurement of nitrate transformations in subsurface ground water is being initiated to answer this question. Our data suggest that reliable generalizations about the role of riparian areas as nutrient filters in forested upland catchments will require the study of nutrient flux and transformation processes in relation to routes of ground water movement and other site characteristics in a range of riparian environments.

Acknowledgements

We thank K. Sanmugadas for laboratory assistance, P. Lefaive for field assistance, M. Kellman and N. Roulet for their comments on the manuscript and the cartographic staff in the Department of Geography for the figures. Thanks are also due to the Metropolitan Toronto and Region Conservation Authority for access to the catchment. This research was supported by grants from the Natural Sciences and Engineering Research Council of Canada to Alan R. Hill.

References

APHA, 1976. Standard Methods for the Examination of Water and Wastewater, 14th edition. American Public Health Association, New York.

Bartlett, M. S., L. C. Brown, N. B. Hanes & N. H. Nickerson, 1979. Denitrification in freshwater wetland soil. J. Envir. Qual., 8: 460–464.

Bottomly, D. J., D. Craig & L. M. Johnston, 1985. Neutralization of acid runoff by groundwater discharge to streams in Canadian Precambrian shield watersheds. J. of Hydrol., 75: 1–26.

Conners, M. E. & R. J. Naiman, 1984. Particulate allochthonous inputs: Relationships with stream size in an undisturbed watershed. Can. J. Fish. Aquat. Sci., 41: 1473–1484.

Cooper, J. R., J. W. Gilliam & T. C. Jacobs, 1986. Riparian areas as a control of nonpoint pollutants. Watershed Research Perspectives (ed.) D. L. Correll, Smithsonian Institution Press, Washington D.C., 166–190.

Cummins, K. W., 1974. Structure and function of stream ecosystems. Bioscience, 24: 631–641.

Cummins, K. W., 1980. The multiple linkages of forests to streams. Forests: Fresh Perspectives from Ecosystem Analysis (ed.) R. H. Waring, Oregon State University Press, Corvallis, Oregon: 191–198.

Duff, J. H., F. J. Triska & R. S. Oremland, 1984. Denitrification associated with stream periphyton: Chamber estimates from undisturbed communities. J. Envir. Qual., 13: 514–518.

Dunne, T., 1983. The relation of field studies and modelling in the prediction of storm runoff. J. Hydrol., 65: 25–48.

Dunne, T. & R. D. Black, 1970. Partial area contributions to storm runoff in a small New England watershed. Water Resour. Res., 6: 1296–1311.

Engler, R. M. & W. H. Patrick, 1974. Nitrate removal from floodwater overlying flooded soils and sediments. J. Envir. Qual., 3: 409–413.

Environment Canada, 1979. Analytical Methods Manual. Inland Waters Directorate, Water Quality Branch, Ottawa, Canada.

Fisher, S. E. & G. E. Likens, 1973. Energy flow in Bear Brook, New Hampshire: an integrative appraoch to stream ecosystem metabolism. Ecol. Monogr., 43: 421–439.

Hill, A. R., 1983. Nitrate-nitrogen mass balances for two Ontario rivers. Dynamics of Lotic Ecosystems (ed.) T. D. Fontaine and S. M. Bartell, Ann Arbor Science Publishers Inc. Ann Arbor, Michigan: 457–477.

Hill, A. R. & K. Sanmugadas, 1985. Denitrification rates in relation to stream sediment characteristics. Water Res., 19: 1579–1586.

Hynes, H. B. N., 1983. Groundwater and stream ecology. Hydrobiologia 100: 93–99.

Jacobs, T. C. & J. W. Gilliam, 1985. Riparian losses of nitrate from agricultural drainage waters. J. Envir. Qual., 14: 472–478.

Keller, E. M. & F. J. Swanson, 1979. Effects of large organic material on channel form and fluvial processes. Earth Surface Processes, 4: 361–380.

Knowles, R., 1979. Denitrification, acetylene reduction, and methane metabolism in lake sediments exposed to acetylene. Appl. Envir. Microbiol., 38: 486–493.

Knowles, R., 1982. Denitrification. Microbiol. Rev., 46: 43–70.

Lowrance, R. R., R. L. Todd & L. E. Asmussen, 1984a. Nutrient cycling in an agricultural watershed: I. Phreatic movement. J. Envir. Qual., 13: 22–27.

Lowrance, R. R., R. L. Todd & L. E. Asmussen, 1984b. Nutrient cycling in an agricultural watershed: II Streamflow and artificial drainage. J. Envir. Qual., 13: 27–32.

Lowrance, R. R., R. Leonard & J. Sheridan, 1985. Managing riparian ecosystems to control nonpoint pollution. J. Soil Water Conserv., 40: 87–91.

Messer, J. & P. L. Brezonik, 1983. Comparison of denitrification rate estimation techniques in a large shallow lake. Water Res., 17: 631–640.

Meyer, J. L., G. E. Likens & J. Sloane 1981. Phosphorus, nitrogen and organic carbon flux in a headwater stream. Archiv. Hydrobiol., 91: 28–44.

McCaig, M., 1983. Contributions to storm quickflow in a small headwater catchment – the role of natural pipes and soil macropores. Earth Surface Processes and Landforms, 8: 239–252.

Molles, M. C., Jr. 1982. Trichopteran communities of streams associated with aspen and conifer forests: long-term structural change. Ecology, 63: 1–6.

Oremland, R. S., C. Umberger, C. W. Culbertson & R. L. Smith, 1984. Denitrification in San Fransisco Bay intertidal sediments. Appl. Envir. Microbiol., 47: 1106–1112.

Peterjohn, W. T. & D. L. Correll, 1984. Nutrient dynamics in an agricultural watershed: observations on the role of a riparian forest. Ecology, 65: 1466–1475.

Richey, J. S., W. H. McDowell & G. E. Likens, 1985. Nitrogen transformations in a small mountain stream. Hydrobiologia, 124: 129–139.

Rigler, F. H., 1979. The export of phosphorus from Dartmoor catchments: a model to explain variations of phosphorus concentrations in streamwater. J. Mar. biol. Ass. U.K., 59: 659–687.

Sain, P., J. B. Robinson, W. N. Stammers, N. K. Kaushik & H.R. Whiteley, 1977. A laboratory study on the role of stream sediment in nitrogen loss from water. J. Envir. Qual., 6: 274–278.

Sklash, M. G. & R. N. Farvolden, 1979. The role of groundwater in storm runoff. J. Hydrol., 43: 45–65.

Smith, M. S. & J. M. Tiedje, 1979. Phases of denitrification following oxygen depletion in soil. Soil Biol. Biochem., 11: 261–267.

Swanson, F. J., S. V. Gregory, J. R. Sedell & A. G. Campbell, 1982. Land-water interactions: the riparian zone. Analysis of Coniferous Forest Ecosystems in the Western United States (ed.) R. L. Edmonds US/IBP Synthesis Series 14. Hutchinson Ross Publishing Co. Stroudsburg, PA: 267–291.

Terry, R. E. & D. W. Nelson, 1975. Factors influencing nitrate transformations in sediments. J. Envir. Qual., 4: 549–554.

Terry, R. E. & R. L. Tate, 1980. Denitrification as a pathway for nitrate removal from organic soils. Soil Sci., 129: 162–166.

Waring, S. A. & J. W. Gilliam, 1983. The effect of acidity on nitrate reduction and denitrification in lower coastal plain soils. Soil Sci. Soc. Am. J., 47: 246–250.

Yoshinari, T. & R. Knowles, 1976. Acetylene inhibition of nitrous oxide reduction by denitrifying bacteria. Biochem. Biophys. Res. Commun., 69: 705–710.

Commentary on Paper E5

Dillaha, T. A., Reneau, R. B., Mostaghimi, S. & Lee, D. (1989) Vegetative filter strips for agricultural nonpoint source pollution control.

Agricultural scientists have for decades attempted to estimate soil loss from individual fields and slopes to determine land-use practices that will ensure sustainable soil productivity. By definition, the focus has been on soil erosion. In the United States, soil erosion research was consolidated in the 1950s, leading to what remains the most widely used method of soil loss prediction, the Universal Soil Loss Equation (USLE, see review in Mitchell & Bubenzer, 1980). Far less attention has, until recently, been paid to sediment transport beyond the field boundary or to on-slope deposition. Of course, for aquatic ecologists, the focus is in-stream and so it is net sediment transport to the stream channel that is important, i.e. net sediment yield to the channel, rather than the gross amount of erosion occurring on farmland. More recently, work by geomorphologists has altered this balance of interest somewhat (see below), with studies of the redistribution of soil being used to "unlock the black box of sediment delivery" (Walling, 1983).

In semi-arid areas, there is little to stop eroded material from entering stream channels. In complete contrast, in temperate zones, especially where the land is intensively farmed, there may be all sorts of barriers (e.g. hedgerows, ponds, riparian woodland) to prevent eroded soil moving far across the landscape. Because of this, only a proportion, perhaps a rather small proportion, of eroded soil makes it as far as the stream channel (let alone to the basin outlet). Nevertheless, accurate estimation of sediment delivery remains a challenge (Walling, 1990). The recognition that on-slope deposition could happen led some scientists to investigate the controls operating, in order to assess whether deposition could be significantly enhanced using appropriate buffer zone management techniques.

Theo Dillaha and colleagues were not the first to study the effectiveness of "vegetative filter strips" (VFS) for agricultural non-point source pollution control, but their paper is notable because it combines coverage of sediment and nutrient losses and because it covers both solid and soluble pollutants. Not surprisingly, the experimental approach reflected USLE soil-erosion plot studies, and involved three adjacent plots in three different situations. Each set of three comprised one plot with no VFS, one with a VFS of 4.6 m and the third with a VFS of 9.1 m. Quite reasonably, it was assumed that the input discharge to the two filter strips was the same as that from the plot with no VFS. The approach therefore allowed the influence of filter strip width to be assessed. It is acknowledged that the plots do not fully replicate field conditions because real cropland will have longer slope lengths and upslope areas. The authors felt that this would be partially offset in that the simulated rainfall was very high, both in terms of intensity (50 mm/hour) and total amount (200 mm in one week) over the study period.

It is clear (Tables 2 and 3) that VFS are effective for the removal of sediments being transported in surface runoff, with the longer strip being more effective. Given the very high rainfall intensities involved, even over small areas, the filter strips perform very well indeed. Nevertheless, it is clear that the shorter VFS were quickly inundated with sediment, showing that wider buffer strips might well be needed for some farmland, depending on soil, topography, climate and crop management. VFS effectiveness thus decreased with time, at least within the time frame of the field experiment; of course, (re)growth of vegetation on the deposited soil might restore trapping efficiency in the longer term, but some management of buffer zones to sustain effective trapping does seems likely. Total nitrogen and phosphorus were highly retained too, since most was in organic form and transported as part of the suspended sediment load (97% of total P and 78% of total N was sediment-bound). In some ways more interesting, the results for nitrate also show a decrease (Table 4), on average by 27% and 57% for the 4.6 and 9.1 m plots, respectively. Results for phosphate are less clear, with some suggestion of net release of P from the filter strip. Evidence of solute loss from surface runoff when flowing across a buffer zone remains rare.

A buffer strip can perform a multitude of functions, one of the most important of which is to trap soil eroded upslope. Studies aimed specifically at sediment trapping in riparian buffer zones remain relatively rare, but a body of research in hillslope geomorphology provides helpful supporting evidence. The use of Caesium-137 (^{137}Cs) in studying soil erosion is now well established. ^{137}Cs was produced as a bi-product of atmospheric testing of nuclear weapons from the 1950s to the 1970s. It was deposited uniformly and is strongly bound to clay minerals in the soil. When soil particles are eroded, the ^{137}Cs moves with the particles. Inventories of soil ^{137}Cs can be mapped to identify areas of soil erosion and deposition (Walling & Quine, 1990). Figure 1 provides an example and shows how soil has redistributed from interfluves into the hollow on the lower slope. There is therefore significant on-slope deposition in the riparian zone, and only a fraction of the soil eroded is delivered to the stream channel. This sort of observation provides the basis for the notion of the sediment delivery ratio (Walling, 1983).

Loss of nutrients and sediments from agricultural runoff causes eutrophication in surface water. Vegetated buffer zones adjacent to a stream can effectively remove and retain nutrients and sediments. It is, therefore, important to study design criteria which optimise the effect of buffer zones (see Dosskey *et al.*, 2008 for a recent example). The paper by Dillaha and colleagues widened coverage of sediment trapping by filter strips to include

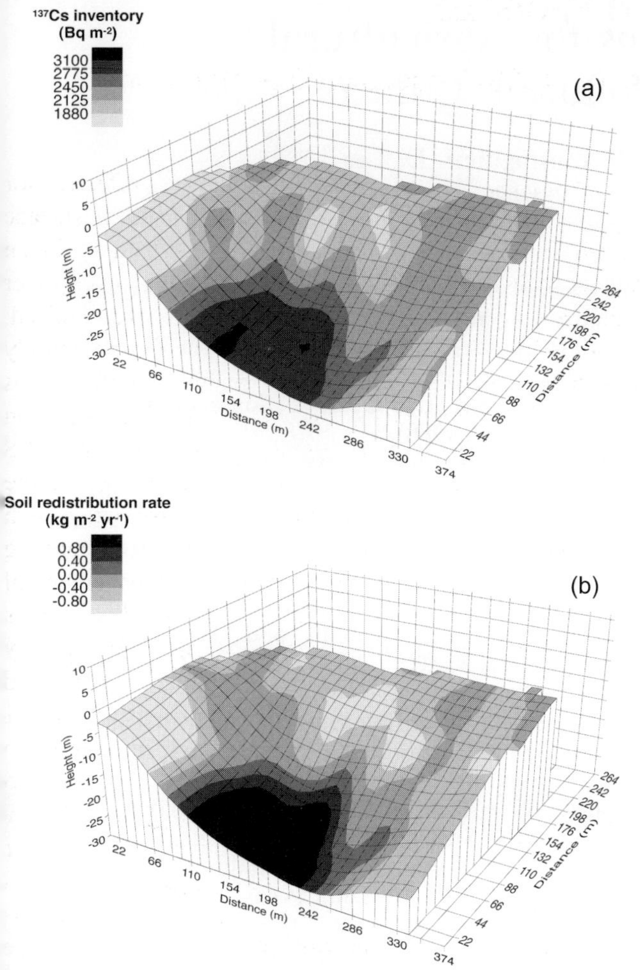

Fig. 1 *The pattern of soil redistribution within a cultivated field in Devon UK, documented using Cs-137 measurements. A network of >100 cores was used to define the spatial variability of Cs-137 inventories within the field (a), and these data were used to establish the pattern of soil redistribution (negative values represent erosion) averaged over the past approx. 40 years (b). Figure provided by Professor Des Walling, University of Exeter, UK.*

soluble nutrients. Relatively few studies have followed up this work, compared to the large number of studies of nutrient loss from subsurface runoff. This points to the continuing need for multifunctional studies of riparian buffer zones in which complete budgets for N and P are compiled. Bracken (2010) provides an up-to-date review of soil erosion by overland flow including buffer zones.

TPB

References

Bracken, L. J. (2010) Overland flow and soil erosion. In: *Sediment Cascades* (ed. by T. P. Burt & R. J. Allison), 181–216. Wiley, Chichester, UK.

Dosskey, M. G., Helmers, M. J. & Eisenhauer, D. E. (2008) A design aid for determining width of filter strips. *J. Soil & Water Conserv.* **63**(4), 232–241.

Mitchell, J. K. & Bubenzer, G. D. (1980) Soil loss estimation. In: *Soil Erosion* (ed. by M. J. Kirkby & R. P. C. Morgan), 17–62. Wiley, Chichester, UK.

Walling, D. E. (1983) The sediment delivery problem. *J. Hydrol.* **69**, 209–237.

Walling, D. E. (1990) Linking the field to the river: sediment delivery from agricultural land. In: *Soil Erosion on Agricultural Land* (ed. by J. Boardman, I. D. L. Foster & J. A. Dearing), 129–152, Wiley, Chichester, UK.

Walling, D. E. & Quine, T. A. (1990) Use of Caesium-137 to investigate patterns and rates of soil erosion on arable fields. In: *Soil Erosion on Agricultural Land* (ed. by J. Boardman, I. D. L. Foster & J. A. Dearing), 33–53. Wiley, Chichester, UK.

Vegetative Filter Strips for Agricultural Nonpoint Source Pollution Control

T.A. Dillaha, R.B. Reneau, S. Mostaghimi, D. Lee
ASSOC. MEMBER ASAE MEMBER ASAE

ABSTRACT

A rainfall simulator was used to evaluate the effectiveness of vegetative filter strips (VFS) for the removal of sediment, nitrogen (N), and phosphorus (P) from cropland runoff. Simulated rainfall was applied to nine experimental field plots on an eroded Groseclose silt loam soil (*clayey, mixed, mesic Typic Hapludalt*) with a 5.5 by 18.3 m bare cropland source area and either a 0, 4.6, or 9.1 m VFS located at the lower end of each plot. Fertilizers were applied to the plots at rates of 222 kg/ha of liquid N and 112 kg/ha of P_2O_5 and K_2O. Water samples were collected from the base of each plot and analyzed for sediment and nutrient content. One set of plots was constructed to encourage concentrated rather than shallow uniform flow.

The 9.1 and 4.6 m VFS with shallow uniform flow removed an average of 84 and 70% of the incoming suspended solids, 79 and 61% of the incoming P, and 73 and 54% of the incoming N, respectively. Soluble nutrients in the filter effluent were sometimes greater than the incoming soluble nutrient load, presumably due to lower removal efficiencies for soluble nutrients and the release of nutrients previously trapped in the filters. Concentrations of soluble inorganic N and P in filter strip effluent were sufficient to cause eutrophic plant growth in aquatic ecosystems. Observation of existing VFS showed that on-farm VFS were not likely to be as effective as experimental VFS because of problems with flow concentrations.

INTRODUCTION

Vegetative filter strips (VFS) are bands of planted or indigenous vegetation situated downslope of cropland or animal production facilities to provide locatlized erosion protection and filter nutrients, sediment, and other pollutants from agricultural runoff. Due to their low installation and maintenance costs and perceived effectiveness in removing pollutants, conservation and regulatory agencies are encouraging VFS use. Filter strips are approved cost-share practices in Virginia and Maryland's Chesapeake Bay Programs and were recently incorporated into the USDA's Conservation Reserve Program (Alder, 1988).

Vegetated filter strips have been shown to be an effective best management practice for the control of some nonpoint source pollutants, especially sediment and sediment-bound contaminants. Researchers have investigated the effectiveness of VFS for sediment removal from strip mine runoff, nutrient and solids removal from feedlot runoff, and the treatment of municipal wastewaters (Tollner et al., 1977; Young et al., 1980; Abernathy, 1984; USEPA, 1977). Their effectivensss for removing sediment and nutrients from cropland runoff, however, is much less certain, as limited research has been conducted in this area. Because of a lack of research and verified design procedures, VFS design is generally based on experience and there are many system failures. Inadequate knowledge of VFS dynamics has resulted in VFS installation in areas where they are inappropriate because of topographic limitations. Before VFS can be effectively used for reducing agricultural NPS pollution from cropland, the dynamics of sediment and nutrient transport processes in VFS must be better understood. Dillaha et al. (1987) and Magette et al. (1987) give a thorough review of previous research related to VFS.

The goal of this research was to investigate the effectiveness of VFS in reducing sediment and nutrient losses from cropland. To achieve this goal, plot studies were conducted using rainfall simulation techniques to investigate sediment, nitrogen (N), and phosphorus (P) transport in VFS receiving runoff from cropland. A survey of existing VFS in Virginia was also conducted to qualitatively evalute VFS performance in field situations.

EXPERIMENTAL PROCEDURES

Experimental studies were conducted on nine field plots during spring 1985 on an eroded Groseclose silt loam (*clayey, mixed, mesic Typic Hapudult*) at the Prices Fork Research Farm near Blacksburg, Virginia. Each plot had a cropland source area and an orchardgrass (*Dactylis glomerata*) VFS of known length. Figure 1 is a sketch of one set of experimental plots. Plots were prepared by installing metal borders to a depth of 15 cm along the boundaries and a metal gutter with a pipe outlet at the base of each plot. A 15 cm H-flume equipped with a FW-1 stage recorder was used for flow measurement.

The cropland portions of each plot were 5.5 m wide and 18.3 m long (0.01 ha). One plot in each set had no

---

Article was submitted for publication in June, 1988; reviewed and approved for publication by the Soil and Water Div. of ASAE in November, 1988.

The work upon which this paper is based was supported by the Virginia Agricultural Experiment Station through funds provided by the U.S. Environmental Protection Agency and the Virginia Water Resources Research Center.

The authors are: T.A. DILLAHA, Assistant Professor, Agricultural Engineering Dept.; R.B. RENEAU, Professor, Agronomy Dept; S. MOSTAGHIMI, Assistant Professor, Agricultural Engineering Dept., Virginia Polytechnic Institute and State University, Blacksburg; and D. LEE, Research Associate, Institute of Ecology, University of Georgia, Athens.

Acknowledgments: The authors also acknowledge the support by CSRS Regional Projects S-164 and S-211.

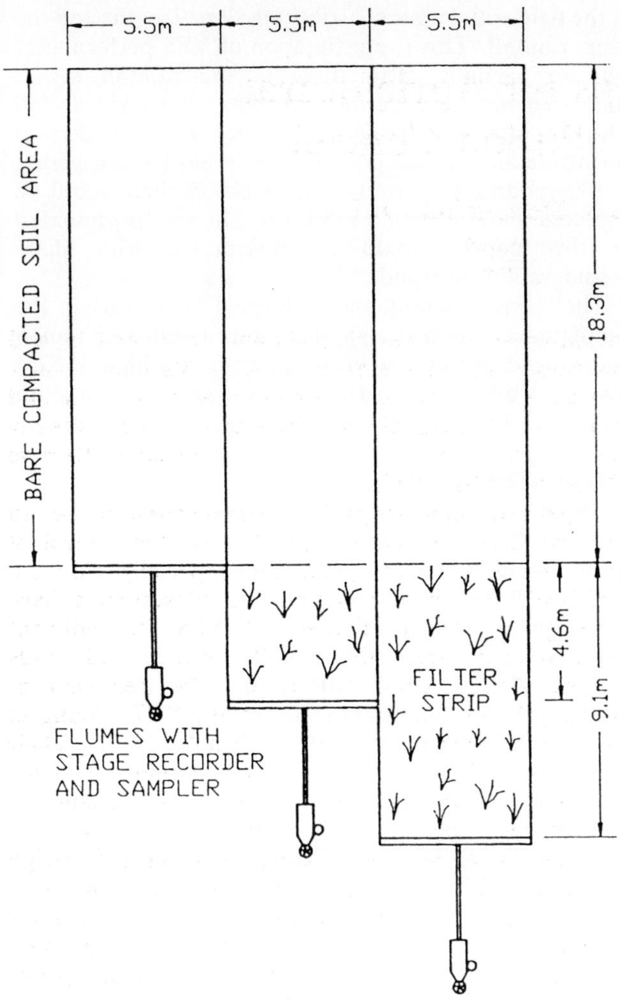

Fig. 1—Schematic diagram of experimental field plots.

VFS, another a 4.6 m VFS, and the third a 9.1 m VFS. To evaluate VFS performance, it was assumed that the discharge from the plot with no VFS was the input to the VFS of the adjacent two plots in the same set. This assumption is a potential source of error as soil erodibility and infiltration are spatially variable even within the same contiguous soil unit. The present study assumes that this potential error did not affect our results.

The plots were installed so that the cropland portions of the plots were located in an area that had previously been in no-till corn, while the VFS portions of the plots were located in previously established orchardgrass strips. The cropland areas were tilled in the fall of 1984 and left fallow over the winter. In April 1985, the bare portions of the plots were tilled conventionally and granular P_2O_4 and K_2O fertilizer were each applied at rates of 112 kg/ha. The plots were tilled again to incorporate the fertilizer into the soil. Two to three days before the simulations, 222 kg-N/ha of non-pressurized N solution (7% NH_4^+–N, 7% NO_3^-–N, 14% urea-N) was applied to the plots with precision liquid application system.

A summary of the physical characteristics of each plot, fertilizer application rates, and simulated rainfall intensities and duration is presented in Table 1. The first two sets of plots, QF1-QF3 and QF4-QF6, had negligible cross slopes. The third sets of plots (QF7-QF9) had a

TABLE 1. Plot Characteristics and Operating Conditions

	Plot								
	QF1	QF2	QF3	QF4	QF5	QF6	QF7	QF8	QF9
Filter length, m	9.1	4.6	0	9.1	4.6	0	0	9.1	4.6
Slope, %	11	11	11	16	16	16	5	5	5
Cross slope, %	<1	<1	<1	<1	<1	<1	4*	4*	4*

Filter strip vegetation	- Orchardgrass (trimmed to 10 cm)
Soil type	- Groseclose silt loam
Fertilization	- 222 kg-N/ha
	- 112 kg-K_2O/ha
	- 112 kg-P_2O_5/ha
Simulated rainfall intensity	- 50 mm/hr
Simulated rainfall duration	- Run 1 - 60 min
	- Run 2 - 30 min
	- Run 3 - 30 min
	- Run 4 - 60 min
	- Run 5 - 30 min
	- Run 6 - 30 min

*Cross slope used in an attempt to simulate concentrated flow in "real world" VFS

cross slope of 4% that caused runoff to concentrate and flow along one side of each plot. The resulting concentrated flow was intended to be an indicator of the effects of concentration flow on VFS performance. This was a concern in the present study because on-farm VFS are crossed by natural drainageways, which concentrate flow causing filter inundation and poor performance.

Simulated rainfall was applied to each set of plots for 1 h (run 1), followed 24 h later by two 30 min runs (runs 2 and 3), which were 30 min apart. These three runs were repeated 4 to 5 days later on each set of plots and constituted runs 4, 5 and 6. Approximately 200 mm of rainfall was applied to each plot over the one week simulation period. A rainfall intensity of approximately 50 mm/h was used during all simulations. The first and fourth simulated rainfall events (runs 1 and 4) closely approximate 2 to 5 yr recurrence interval, 1 h duration storms in Virginia and should approximate critical conditions for nutrient loss as fertilizer had just been applied to the plots.

Water samples were collected manually from the H-flumes at 3-min intervals throughout the runoff process for sediment and nutrient analysis. The water samples were frozen immediately after collection and stored up to 3 months before analysis. Samples were analyzed for: total suspended solids (TSS), ammonium (NH_4^++), nitrate-nitrite (NO_3^-3−), total Kjeldahl nitrogen (TKN), soluble TKN (TKN_s), orthophosphorus (PO_4^-−), total P (P_t), and soluble P_t (P_{ts}). Total N (N_t) was determined as the sum of TKN and NO_{33}−. All samples were analyzed in accordance with standard analytical techniques (USEPA, 1983). Detailed descriptions of analytical techniques, plot construction, the rainfall simulator, and monitoring procedures employed in this study are given by Dillaha et al. (1987).

The effects of VFS on sediment and nutrient yield reductions were tested using the SAS procedure GLM (SAS, 1985). The statistical models used in analyzing the data were constructed following procedures developed by Miller and Johnson (1984). Correlations were tested using the SAS procedure CORR. Differences were considered significant if at or beyond the 0.05 probability level.

After completion of the field experiments, existing VFS in Virginia were visited quarterly for one year to qualitatively evaluate VFS effectiveness for sediment and nutrient removal. Filter strips chosen for inspection were selected at random, although all were within the Chesapeake Bay and Chowan River basins. Over 24 km of VFS were inspected on 18 farms. The VFS inspected represented approximately 10% of the total of those in existence in Virginia.

RESULTS AND DISCUSSION

It is important to note that the experimental field plots used in this research do not exactly duplicate field conditions. The most significant difference is that real cropland will have larger areas upslope of the VFS contributing to runoff. The 4.6 and 9.1 m VFS used in the present research had contributing slope-lengths of 18.3 m. In actual field situations, slope-lengths contributing to shallow overland flow will be up to 100 m long. Surface runoff in fields with greater slope-lengths will tend to concentrate in natural drainageways within the field and cross the VFS in a few localized areas as concentrated flow. Filter strips are not expected to be effective under these concentrated flow conditions. The larger contributing slope-length of actual cropland will tend to increase surface runoff and sediment loading to VFS compared to the experimental plots. Consequently, the experimental VFS would be expected to be more effective for sediment and nutrient reduction than VFS in the field with larger contributing slope-lengths and the same rainfall. This overestimation of VFS performance will be partially offset because the rainfall events simulated on the experimental plots (200 mm rainfall with an intensity of 50 mm/h over a 7 day period) constitute an extreme precipitation event (return period > 100 yr) and will produce more runoff than would be expected under normal conditions. The results presented in this paper should be interpreted with these considerations in mind.

The rainfall simulator performed well during the simulations. The mean application rate was 47.9 mm/h and ranged from a low of 41.2 mm/h to a high of 52.4 mm/h. Uniformity coefficients averaged 93.3% with only 4 of 54 coefficients having values less than 90%.

Sediment Yield

As shown in Tables 2 and 3 and Fig. 2, all of the VFS were effective for sediment removal. Sediment losses from the plots without filters were 3.9, 8.9 and 2.1 Mg/ha for plots QF3, QF6 and QF7, respectively. Sediment yields for plots with VFS were significantly reduced ($\alpha<0.05$) and ranged from 0.10 (QF1) to 4.22 Mg/ha (QF5). Filter strip length also reduced sediment yield reductions significantly. The 4.6 m VFS (QF2, 5 and 9) reduced overall sediment yields by 86, 53 and 83% and the 9.1 m VFS, QF1, 4 and 8, reduced

TABLE 2. Sediment, Nutrient, and Water Yield

Plot	Runs	Filter length, m	TSS Mg/ha	NH_4^+-N	NO_3^--N	TKN	N_t	P_t	PO_4^--P	TKN_s	P_{ts}	Runoff mm
				kg/ha								
QF1	1-3	9.1	0.02	0.08	0.10	0.49	0.59	0.12	0.01	0.19	0.03	4.0
QF1	4-6	9.1	0.08	0.09	0.26	0.46	0.73	0.20	0.04	0.29	0.06	13.9
QF1	1-6	9.1	0.10	0.17	0.36	0.95	1.32	0.33	0.05	0.47	0.08	17.9
QF2	1-3	4.6	0.18	0.29	0.37	1.41	1.77	0.44	0.05	0.35	0.12	19.9
QF2	4-6	4.6	0.37	0.36	1.25	2.15	3.40	0.74	0.12	0.60	0.15	46.2
QF2	1-6	4.6	0.56	0.66	1.61	3.56	5.18	1.18	0.16	0.96	0.27	66.1
QF3	1-3	0	1.87	0.90	0.54	6.91	7.45	2.23	0.04	1.29	0.09	24.9
QF3	4-6	0	2.05	0.63	1.11	6.35	7.46	2.11	0.05	0.85	0.10	45.9
QF3	1-6	0	3.93	1.53	1.65	13.26	14.91	4.34	0.09	2.14	0.18	70.8
QF4	1-3	9.1	0.73	1.07	0.36	4.20	4.56	1.04	0.05	1.24	0.07	15.3
QF4	4-6	9.1	1.99	1.41	1.19	8.39	9.58	1.91	0.09	1.55	0.18	37.9
QF4	1-6	9.1	2.71	2.49	1.55	12.59	14.14	2.96	0.14	2.79	0.25	53.2
QF5	1-3	4.6	1.14	1.63	0.35	5.03	5.38	1.22	0.05	1.43	0.06	16.7
QF5	4-6	4.6	3.08	2.26	1.50	11.29	12.79	3.10	0.05	2.13	0.11	39.1
QF5	1-6	4.6	4.22	3.88	1.85	16.32	18.17	4.31	0.10	3.56	0.18	55.8
QF6	1-3	0	4.25	2.32	0.75	14.57	15.32	3.68	0.08	2.66	0.11	22.9
QF6	4-6	0	4.69	1.96	1.23	16.28	16.66	4.73	0.03	2.09	0.06	32.9
QF6	1-6	0	8.94	4.29	1.98	30.85	31.98	8.42	0.11	4.75	0.17	55.8
QF8	1-3	9.1	0.05	0.03	0.08	0.24	0.32	0.08	0.01	0.10	0.01	4.0
QF8	4-6	9.1	0.10	0.08	0.26	1.21	1.48	0.23	0.02	0.17	0.03	12.6
QF8	1-6	9.1	0.14	0.12	0.34	1.45	1.79	0.30	0.03	0.27	0.04	16.6
QF9	1-3	4.6	0.09	0.04	0.07	0.28	0.35	0.10	0.01	0.08	0.01	4.0
QF9	4-6	4.6	0.27	0.15	0.27	0.95	1.22	0.25	0.01	0.18	0.03	12.0
QF9	1-6	4.6	0.36	0.19	0.34	1.23	1.57	0.35	0.02	0.27	0.04	16.0
QF7	1-3	0	0.66	0.27	0.43	2.43	2.86	0.71	0.02	0.49	0.04	17.9
QF7	4-6	0	1.44	0.48	0.79	5.25	6.04	1.56	0.03	0.64	0.06	41.9
QF7	1-6	0	2.10	0.75	1.22	7.68	8.90	2.27	0.05	1.13	0.10	59.8

where: N_t = total N, P_t = total P, TKN_s = soluble TKN, P_{ts} = total soluble P, and Runs indicates which runs were summed to obtain the given yields.

TABLE 3. Percent Reduction in Sediment, Nutrient, and Water Yield

Plot	Runs	Filter length, m	TSS	NH_4^+-N	NO_3^--N	TKN	N_t	P_t	PO_4^--P	TKN_s	P_{ts}	Runoff
			%									
QF1	1-3	9.1	99	91	81	93	92	95	78	86	70	74
QF1	4-6	9.1	96	85	76	93	90	90	26	66	42	55
QF1	1-6	9.1	98	89	78	93	91	93	47	78	55	62
QF2	1-3	4.6	90	67	32	80	76	80	-31	73	-33	2
QF2	4-6	4.6	82	42	-13	66	54	65	-119	29	-59	-24
QF2	1-6	4.6	86	57	2	73	65	73	-83	55	-47	-15
QF4	1-3	9.1	83	54	52	71	70	72	43	53	38	0
QF4	4-6	9.1	58	28	3	48	42	60	-258	26	-192	-72
QF4	1-6	9.1	70	42	22	59	56	65	-31	41	-44	-42
QF5	1-3	4.6	73	30	54	65	65	67	43	46	41	9
QF5	4-6	4.6	34	-15	-22	31	23	35	-100	-2	-87	-48
QF5	1-6	4.6	53	9	7	47	43	49	8	25	-4	-24
QF8	1-3	9.1	93	88	81	90	89	89	71	80	82	65
QF8	4-6	9.1	93	83	67	77	76	86	30	74	47	55
QF8	1-6	9.1	93	84	72	81	80	87	48	76	60	58
QF9	1-3	4.6	86	84	84	88	88	86	79	83	79	72
QF9	4-6	4.6	81	69	66	82	80	84	60	71	55	63
QF9	1-6	4.6	83	74	72	84	82	85	69	76	64	66

where: N_t = total N, P_t = total P, TKN_s = soluble TKN, P_{ts}= total soluble P, and Runs indicates which runs were summed to obtain the corresponding percent reductions in yields.

sediment yields by 98, 70 and 93%, respectively. Doubling the filter lengths from 4.6 to 9.1 m reduced sediment yield by an additional 12, 23 and 10% for plots QF1, QF4 and QF8, respectively (Fig. 2).

Most of the sediment removed by the VFS was deposited just upslope or in the first few meters of the VFS. After vegetation in upper portion of the VFS was buried with sediment, sediment began to move into the uninundated area just below the previous zone of deposition. This deposition zone, typically 25 to 40 cm wide, continued to trap sediment until it too became inundated and the sediment would then spill over into the next region. This process continued until the simulations stopped or until the zone of deposition reached the end of the VFS. Similar observations were reported by Neibling and Alberts (1979) and Tollner et al. (1977).

The concentrated flow plots (QF8 and 9) were generally as effective as the uniform flow plots for sediment removal. They were more effective at sediment removal than the 16% slope uniform flow plots (QF4 and 5) and only slightly less effective than the 11% slope uniform flow plots (QF1 and 2). In the concentrated flow plots, overland flow was limited to a 1 to 2 m wide portion of the VFS. Apparently, the reduction in effective VFS area was offset by lower sediment loadings and runoff in the concentrated flow plots. Sediment delivery to the concentrated flow VFS was 2.10 versus 3.93 and 8.94 Mg/ha for the 11 and 16% uniform flow plots, respectively (Table 2). Foster and Meyer (1979) reported that sediment transport capacity was directly proportional to slope and flow rate. The concentrated flow plot slopes were 55 to 69% less than the uniform flow plots (5% versus 11 and 16%). These lower slopes would be expected to reduced sediment transport capacity in the simulated cropland portions of the plots and consequently sediment delivery to the VFS. The lower slopes would also reduce sediment transport capacity in the VFS resulting in increased sediment deposition. The concentrated flow VFS had the lowest runoff volumes of any plots (Table 3). These low runoff volumes also would be expected to reduce sediment transport capacity and consequently increase sediment deposition.

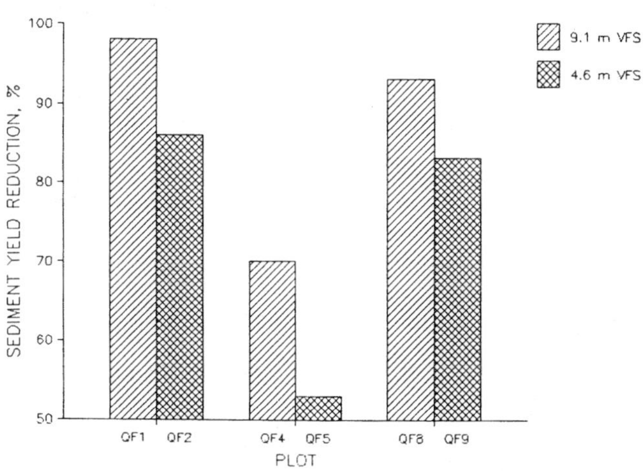

Fig. 2—Percent reduction in sediment yield.

The effectiveness of the VFS decreased with time as sediment accumulated in the VFS. As shown in Fig. 3, VFS effectiveness for sediment removal in the 4.6 m filter of plot QF5 decreased from 90% during run 1 to 5% during run six. Table 3 also demonstrates this effect as sediment reductions were always less during the second set of runs (4 to 6) than the first set (runs 1 to 3). This reduction was expected because of partial inundation of the VFS with sediment during the first set of runs and higher antecedent soil moisture and runoff rates during

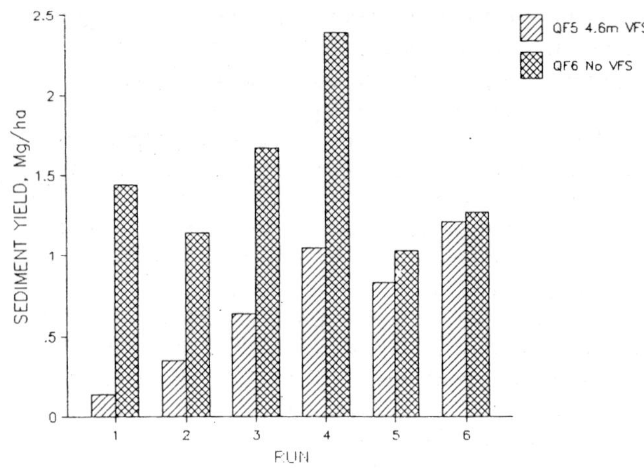

Fig. 3—Sediment yields for plots QF5 and QF6.

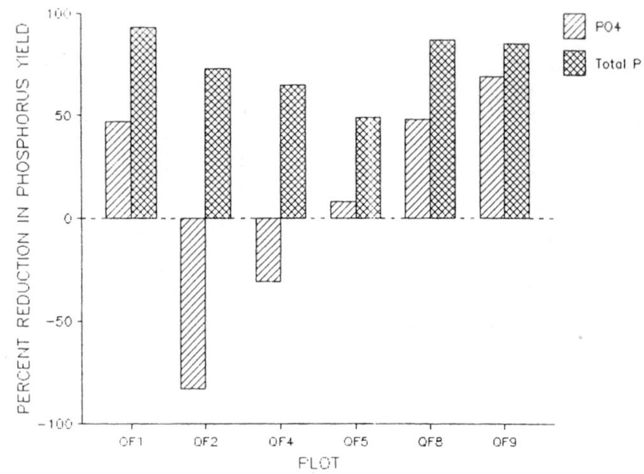

Fig. 4—Percent reduction in phosphorus yield.

the second set of runs.

Slope appeared to have an effect on VFS sediment yields and concentrations but this could not be tested statistically because of the experimental design. As shown in Fig. 2 and Tables 2 and 3, the steepest plots (QF4 and 5, 16% slope) had the lowest percent reductions and highest sediment yields. Plots QF1 and 2 (11% slope) had the highest percent reductions and lowest sediment yields. The concentrated flow plots (5% slope) were intermediate in effectiveness but are difficult to use in assessing slope effects because of the masking effects of concentrated flow.

Phosphorus Yield

Total phosphorus loss from the plots generally followed the same trends as sediment loss except that overall percent reductions in P_t were slightly less (0 to 17%) than sediment yield reductions and ranged from 49 to 85% and 65 to 95% for the 4.6 and 9.1 m VFS, respectively (Table 3 and Fig. 4). The good correlation between sediment and P_t yield (R=0.98 and α<0.0001) was expected since 97, 92 and 90% of the P leaving the 0, 4.6 and 9.1 m filter strips, respectively, was sediment-bound [$(P_t - P_{ts})/P_t$ from Table 2]. Since the filters were effective for sediment removal, they were also effective for P removal. Yields of P_t ranged from 2.27 to 8.42, 0.35 to 4.31, and 0.3 to 2.96 kg-P/ha for the 0, 4.6 and 9.1 m VFS, respectively (Table 2). The 4.6 and 9.1 m VFS reduced P_t yields by an average of 69 and 82%, respectively (Table 3). As with sediment loss, the steeper plots (QF4 and 5) were the least effective for P removal (Fig. 4). They, particularly plot QF5, quickly became inundated with sediment, which reduced their sediment and therefore sediment-bound P trapping efficiency.

Filter effectiveness in removing P_t was significant and decreased 2 to 32% between the first (1 to 3) and second (4 to 6) set of runs (Table 3). There appeared to be a tendency for previously trapped P to be released from the VFS vegetation and soil during latter runs as soluble P (P_{ts}). Consequently, yields of P_{ts} and PO_4^- were often higher than the inflows, especially during runs four to six shown in Tables 2 and 3 and Fig. 4.

Overall mean concentrations of PO_4^- in VFS runoff ranged from 0.11 to 0.17 and 0.08 to 0.20 mg-P/L for the 4.6 and 9.1 m VFS, respectively (Table 4). Soluble inorganic P concentrations of this magnitude are sufficient to cause eutrophic algae growth in lakes and reservoirs unless diluted by higher quality water. As shown in Table 4, mean P_t concentrations decreased as VFS length increased. The effects of VFS length on PO_4^- and P_{ts} concentrations, however, were much less certain.

Nitrogen Yield

Percent reductions in N_t were similar to those observed for P_t but generally 2 to 9% less. The 4.6 and 9.1 m VFS reduced N_t yields by an average of 63 and 76%, NO_3^- by

TABLE 4. Mean Concentration of Sediment and Nutrients in Runoff

Plot	Filter length, m	TSS	NH_4^+-N	NO_3^--N	TKN	N_t	P_t	PO_4^--P	TKN_s	P_{ts}
		mg/L								
QF1	9.1	354	0.62	1.34	3.51	4.85	1.20	0.17	1.75	0.30
QF2	4.6	676	0.80	1.97	4.34	6.31	1.44	0.20	1.16	0.33
QF3	0	5513	2.15	2.32	18.62	20.94	6.10	0.13	3.00	0.26
QF4	9.1	3404	3.12	1.95	15.80	17.75	3.71	0.17	3.50	0.31
QF5	4.6	6063	5.58	2.66	23.45	26.10	6.20	0.14	5.12	0.26
QF6	0	15929	7.64	3.53	55.00	57.00	15.00	0.19	8.47	0.31
QF8	9.1	582	0.47	1.38	5.83	7.21	1.22	0.11	1.07	0.16
QF9	4.6	1792	0.96	1.68	6.08	7.78	1.74	0.08	1.33	0.18
QF7	0	3538	1.27	2.06	12.95	15.01	3.83	0.09	1.90	0.17

where: N_t = total N, P_t = total P, TKN_s = soluble TKN, P_{ts} = total soluble P, and the mean concentrations are flow weighted over runs 1 to 6.

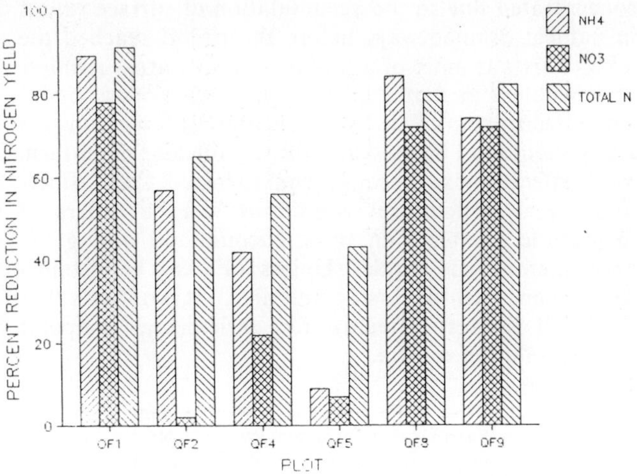

Fig. 5—Percent reduction in nitrogen yield.

27 and 57% and NH_4^+ by 50 and 72% respectively, as shown in Fig. 5 and Table 3. Yields of N_t ranged from 8.90 to 31.98, 1.57 to 18.17, and 1.32 to 14.44 kg-N/ha for the 0, 4.6 and 9.1 m VFS, respectively. Filter strip length had a significant effect on all forms of N yield. Total N yield, like P yield, was highly correlated with sediment yield (R=0.97 and α<0.0001) and 78, 65 and 66% of the N_t leaving the plots with 0, 4.6, and 9.1 m VFS, respectively, was sediment-bound [$(TKN - TKN_s)/N_t$ in Table 2]. As with P and sediment yield, the steepest plots (QF4-5) were least effective, the concentrated flow plots (QF8-9) were moderately effective, and the 11% slope plots (QF1-2) were the most effective for N removal (Fig. 5).

As shown in Table 2, 93% of the N_t leaving the bare portions of the plots and entering the filters was in the form of TKN (organic-N plus NH_4^+). This was expected because most of the N in the plots was residual organic N which had built up in the soils previously and because 75% of the N fertilizer applied to the plots was either urea or NH_4^+. Both NH_4^+ and urea are measured by the TKN test and have a tendency to bind to and be transported along with clay particles and organic matter in the soil. Also, urea is rapidly hydrolyzed to NH_4^+. By the end of the tests, most of the NH_4^+ and urea were probably mineralized to NO_3^- so residual organic N in the soil was presumably the primary source of N leaving the plots.

Concentrations of NH_4^+ and NO_3^- in VFS effluent were sufficient to cause eutrophic plant growth in aquatic ecosystems. Combined concentrations of NH_4^+ and NO_3^- ranged from 3.33 to 11.17, 2.74 to 8.24 and 1.85 to 5.07 mg-N/L, for the 0, 4.6, and 9.1 m VFS, respectively (Table 4).

Existing Vegetative Filter Strip Survey

The effectiveness of existing VFS in the Commonwealth of Virginia was qualitatively evaluated by visiting and observing VFS on 18 farms in Virginia. Filter strips were evaluated by interviewing landowners and soil conservationists and inspecting VFS to identify potential problems. All of the VFS surveyed were used in combination with cropland and were installed specifically for water quality improvement through state and federal cost-share programs. Filter strips were rarely used before 1983 on cropland as they were not recognized conservation practices eligible for cost-share money.

Filter strip performance was generally judged to fall into two categories, which depended on the topography of the site. In hilly areas, VFS were judged to be ineffective for removing sediment and nutrients from surface runoff because drainage usually concentrated in natural drainageways within the fields before reaching the VFS. Flow across these VFS during the larger runoff producing storms (the most significant in terms of water quality) was therefore primarily concentrated and the filters were locally inundated and ineffective. This assessment was confirmed by the fact that little sediment was observed to have accumulated in the majority of the filters observed. Filter strips in these areas, while not effective for trapping sediment and nutrients, were judged to be beneficial because they provide effective cover in areas immediately adjacent to streams which are often susceptible to severe localized channel and gully erosion. Filter strips also provide a narrow buffer between cropland and streams which may reduce the aerial drift of fertilizers and pesticides to streams during application.

In flatter areas, such as the coastal plain, VFS appeared to be more effective. Slopes were more uniform, and significant portions of stormwater runoff entered the VFS as shallow uniform flow. This observation was supported by the presence of significant sediment accumulations in many of the coastal plain VFS. Several one to three year old VFS were observed that had trapped so much sediment that they were higher than the fields they were protecting. In these cases, runoff tended to flow parallel to the VFS until a low point was reached where it flowed across as concentrated flow. In this situation, the VFS acted more like a terrace than a VFS.

Flow parallel to the VFS also was observed on several farms where moldboard plowing was practiced. When soil was turn plowed away from the VFS, a shallow ditch was formed parallel to the VFS. If this ditch was not removed by careful disking later, runoff once again concentrated and flowed parallel to the VFS until it reached a low point and crossed as channel flow.

SUMMARY AND CONCLUSIONS

Simulated rainfall was applied to a series of 5.5 by 18.3 m bare soil plots with 0, 4.6 or 9.1 m VFS located at the lower end of each plot as shown in Fig. 1. The plots were used to evaluate the effectiveness of VFS for controlling sediment and nutrient losses from cropland. For the cropland simulations, commercial fertilizer was applied to the bare tilled portions of the plots at rates of 112 kg/ha of granular P_2O_5 and K_2O and 222 kg-N/ha of non-pressurized N solution. Water samples were collected from H-flumes at the base of each plot to evaluate the effectiveness of the VFS in removing sediment, N, and P from cropland runoff. One set of plots was constructed with a cross slope so that flow through the filters would be deeper or concentrated rather than shallow and uniform. Observation of existing VFS in the Commonwealth of Virginia and analysis of the results of the plot studies led to the following conclusions:

1. Vegetative filter strips are effective for the removal

of sediment and other suspended solids from surface runoff if flow is shallow and uniform and the VFS have not been previously inundated with sediment. The 4.6 and 9.1 m VFS on the uniform flow plots removed 53 to 86 and 70 to 98%, respectively, of the incoming sediment. The 4.6 and 9.1 m VFS in the concentrated flow plots (which poorly simulated concentrated flow effects) reduced sediment yields by 83 and 93% respectively.

2. The effectiveness of VFS for sediment removal decreased with time as sediment accumulated in the VFS. Filter strip effectiveness with respect to sediment removal decreased from 0 to 39% between the first and second set of runs. One VFS (QF5) was almost totally inundated with sediment during the simulations and sediment trapping efficiency dropped from 90% during the first run to 5% during the last run. This may or may not be a problem in "real world" VFS because filter strip vegetation should be able to grow through most sediment accumulations. The success of VFS in surviving burial by sediment will be a function of random variables associated with rainfall, runoff, vegetation growth rate, depth of sediment accumulation, and other factors.

3. Concentrations of soluble inorganic N and P in the VFS effluents were sufficient to cause eutrophic growth of plants in aquatic systems unless diluted by higher quality water.

4. The VFS removed P_t and N_t nearly as effectively as sediment. This was expected because 97% of the P_t and 78% of the N_t entering the VFS was sediment-bound.

5. The VFS lengths used in this research were moderately effective in removing soluble N and P from cropland runoff. Soluble P and N in the outflow from the VFS was sometimes higher than the inflow, presumably due to the release of P and N which had been trapped in the filters previously.

6. Vegetative filter strips that are characterized by concentrated should be less effective for sediment, N, and P removal than filters with shallow uniform flow. Most on-farm VFS visited during this study were judged to be ineffective for sediment and nutrient removal because the majority of flow entering the VFS was concentrated due to the accumulation of surface runoff in natural drainageways before the runoff reached the VFS. This was more of a problem in hilly areas and less of a problem in flatter areas such as the coastal plain. The effectiveness of the experimental VFS used in this study should not be used as a direct indicator of on-farm VFS effectiveness because concentrated flow effects under real agricultural conditions will be orders of magnitude greater than those encountered during the experimental field studies. Unless VFS can be installed so that concentrated flow is minimized, it is unlikely that they will be very effective for agricultural nonpoint source pollution control.

References

1. Adler, K. 1988. *CRP water quality report: Signing-up for filter strips.* National Water Quality Evaluation Project Notes, No. 31. North Carolina Agricultural Extension Service, North Carolina State University, Raleigh.

2. Abernathy, A.R. 1984. Wastewater treatment by overland flow. In *Proc. 38th Ind. Waste Conf.* Purdue Univ. 1983. Boston: Butterworth Publishers.

3. Dillaha, T.A., R.B. Reneau, S. Mostahimi, V.O. Shanholtz and W.L. Magette. 1987. *Evaluating nutrient and sediment losses from agricultural lands: vegetative filter strips.* U.S. Environmental Protection Agency, Report No. CBP/TRS 4/87, Washington, D.C.

4. Foster, G.R. and L.D. Meyer, 1972. Transport of soil particles by shallow flow. *Transactions of the ASAE* 15(1):99-102.

5. Magette, W.L., R.B. Brinsfield, R.E. Palmer, J.D. Wood, T.A. Dillaha and R.B. Reneau. 1987. *Vegetative filter strips for agricultural runoff treatment.* U.S. Environmental Protection Agency, Report No. CBP/TRS 2/87, Washington, D.C.

6. Milliken, G.A. and D.E. Johnson. 1984. *Analysis of messy data, Vol. I: Designed experiments.* New York: Van Nostrand Reinhold Co.

7. Neibling, W.H. and E.E. Alberts. 1979. Composition and yield of soil particles transported through sod strips. ASAE Paper No. 79-2065. St. Joseph, MI: ASAE.

8. Tollner, E.W., B.J. Barfield, C.Vachirakornwatana and C.T. Haan. 1977. Sediment deposition patterns in simulated grass filters. *Transactions of the ASAE* 20(5):940-944.

9. USEPA. 1977. *Process design manual for land treatment of municipal wastewater.* Report No. EPA-625/1-77-008, U.S. Environmental Protection Agency, Washington, D.C.

10. USEPA. 1983. *Methods for the chemical analysis of water and wastes.* Report No. EPA-600/4-79-020, U.S. Environmental Protection Agency, Washington, D.C.

11. Young, R.A., T. Huntrods and W. Anderson. 1980. Effectiveness of vegetative buffer strips in controlling pollution from cropland runoff. *J. Environ. Qual.* 9:483-487.

Commentary on Paper E6

McDowell, W. H., Bowden, W. B. & Asbury, C. E. (1992) Riparian nitrogen dynamics in two geomorphologically distinct tropical rain forest watersheds: subsurface solute patterns.

This is one of the first papers dealing specifically with nitrogen cycling in tropical riparian zones. Yet, it is interesting to notice that the authors' main objective was not to challenge the buffer zone model developed from temperate zone studies, but to determine the effects of local geomorphology on nitrogen cycling in riparian zones. Smartly, they hit two birds with the same stone.

They made a very comprehensive hydrological study of two geomorphologically distinct riparian zones. One classified as an Inceptisol with strong textural zonation and redox status, i.e. oxidized clays at the surface graded to reduced clays and very reduced sands and gravel at depth. The other one classified as a Ultisol, with a mosaic of oxidized and reduced clays with similar textures in the surface and deeper soils. Measurements of groundwater table levels and saturated hydraulic conductivity allowed determination of groundwater flows. Groundwater nitrogen concentrations along transects between slope and riverbank showed much lower concentration than those generally measured under temperate climate. Dissolved organic nitrogen represented a significant part, approx. 20%, of the total dissolved nitrogen under both riparian zones while nitrate was always close zero; even when there was an upslope input, it disappeared within a few metres of flow under the riparian forest. Ammonia presented most variation between sites since it could represent between 5 to 90% of the total dissolved nitrogen in groundwater under the riparian zone.

The authors proposed a conceptual model to explain this significant difference in ammonia concentration between the two different sites. They suggested that the high ammonia concentration site resulted from a distinct spatial separation of oxidized and reduced zones, the latter being out of plant root reach, leading to ammonia accumulation. At the other site, they argued that oxidized and reduced zones could be found next to each other near the surface, allowing nitrification, denitrification and plant uptake to co-exist, limiting ammonia accumulation in the groundwater. In a companion paper, McDowell and his colleagues (Bowden *et al.*, 1992) measured a high potential for nitrous oxide emissions in soils, which they estimated to release between 8 to 25 kg N-N_2O ha^{-2} $year^{-1}$ via denitrification, an unknown and probably greater fraction being released as N_2. However, more recently Silver *et al.* (2001) have shown that dissimilatory nitrate reduction to ammonium (DNRA) is an important microbial process in tropical soils, three times greater than the combined N_2O and N_2 fluxes from nitrification and denitrification, avoiding gaseous loss and resulting in higher N retention via plant and microbe assimilation.

These results underpin the differences with the conceptual N model developed from temperate zones, especially when considering that naturally, nitrogen in tropical forests is relatively more available than in most temperate forests, leading to a more open N cycle (Martinelli *et al.*, 1999). Yet, there is still no clear evidence to date that riparian zones under tropical climate will buffer nitrogen input from fast developing intensive agricultural practices, and if so, by which mechanism.

GP

References

Bowden, W. B., McDowell, W. H., Sabury, C. E. & Finley, A. M. (1992) Riparian nitrogen dynamics in two geomorphologically distinct tropical rain forest watersheds: nitrous oxide fluxes. *Biogeochemistry* **18**, 77–99.

Martinelli, L. A., Piccolo, M. C., Townsend, A. R., Vitousek, P. M., Cuevas, E., McDowell, W. H., Robertson, G. P., Santos, O. C. & Treseder, K. (1999) Nitrogen stable isotopic composition of leaves and soil: Tropical versus temperate forests. *Biogeochemistry* **46**, 45–65.

Silver, W. L., Herman, D. J. & Firestone, M. K. (2001) Dissimilatory nitrate reduction to ammonium in upland tropical forest soils. *Ecology* **82**(9), 2410–2416.

Biogeochemistry **18**: 53—75, 1992

Riparian nitrogen dynamics in two geomorphologically distinct tropical rain forest watersheds: subsurface solute patterns

WILLIAM H. McDOWELL,[1] WILLIAM B. BOWDEN[1] & CLYDE E. ASBURY[2]

[1] *Department of Natural Resources, James Hall, University of New Hampshire, Durham, NH 03824;* [2] *Center for Energy and Environment Research, GPO Box 363682, San Juan, Puerto Rico 00936*

Accepted 30 November 1992

Key words: ammonium, dissolved organic nitrogen, groundwater chemistry, nitrate, riparian zone, tropical rain forest

Abstract. Nitrate, ammonium, dissolved organic N, and dissolved oxygen were measured in stream water and shallow groundwater in the riparian zones of two tropical watersheds with different soils and geomorphology. At both sites, concentrations of dissolved inorganic N (DIN; NH_4^+- and NO_3^--N) were low in stream water (< 110 ug/L). Markedly different patterns in DIN were observed in groundwater collected at the two sites. At the first site (Icacos watershed), DIN in upslope groundwater was dominated by NO_3^--N (550 ug/L) and oxygen concentrations were high (5.2 mg/L). As groundwater moved through the floodplain and to the stream, DIN shifted to dominance by NH_4^+-N (200—700 ug/L) and groundwater was often anoxic. At the second site (Bisley watershed), average concentrations of total dissolved nitrogen were considerably lower (300 ug/L) than at Icacos (600 ug/L), and the dominant form of nitrogen was DON rather than inorganic N. Concentrations of NH_4^+ and NO_3^- were similar throughout the riparian zone at Bisley, but concentrations of DON declined from upslope wells to stream water.

Differences in speciation and concentration of nitrogen in groundwater collected at the two sites appear to be controlled by differences in redox conditions and accessibility of dissolved N to plant roots, which are themselves the result of geomorphological differences between the two watersheds. At the Icacos site, a deep layer of coarse sand conducts subsurface water to the stream below the rooting zone of riparian vegetation and through zones of strong horizontal redox zonation. At the Bisley site, infiltration is impeded by dense clays and saturated flow passes through the variably oxidized rooting zone. At both sites, hydrologic export of nitrogen is controlled by intense biotic activity in the riparian zone. However, geomorphology appears to strongly modify the importance of specific biotic components.

Introduction

Work by Hynes (1983), Lowrance et al. (1984), Peterjohn & Correll (1984) and Grimm & Fisher (1984) demonstrates the potential importance of riparian and hyporheic zones in regulating terrestrial N fluxes to streams. More recently, several researchers have examined patterns of ammonium and nitrate in groundwater beneath and adjacent to streams and rivers (e.g. Stanford & Ward 1988; Triska et al. 1989; Ford & Naiman 1989; Cooper 1990; Duff & Triska 1990; Hill 1990; Triska et al. 1990; Valett et al. 1990; Hendricks & White 1991; Pringle & Triska 1991). Strong gradients in NO_3^--N concentration are often seen across the riparian zone, with nitrate loss as high as 95% for groundwater passing through organic mucks in an agricultural pasture (Cooper 1990). Similar losses are observed in other agricultural (Peterjohn & Correll 1984; Jacobs & Gilliam 1985) and forest soils (Davidson & Swank 1986; Hill 1990). Both denitrification by riparian microorganisms and plant uptake can reduce nitrate concentrations in the riparian zone (e.g. Lowrance et al. 1983). The magnitude of each process at a particular site has important implications for ecosystem nitrogen budgets, because denitrification represents a net loss of N from the terrestrial ecosystem to the atmosphere (in the form of N_2 or N_2O), while plant uptake conserves nitrogen.

Spatial patterns of ammonium and DON in the riparian zone are not as well documented as those of nitrate. Ammonium tends to show no strong spatial patterns across the riparian zone (Triska et al. 1989; Hill 1990), and concentrations in streamside groundwater generally do not exceed 100 ug/L. Very little is known about patterns of DON concentrations in riparian groundwater. In alder and old-growth redwood forests adjacent to Little Lost Man Creek, California, Triska et al. (1990) observed that concentrations of DON in groundwater were similar to those in adjacent stream water (50—100 ug/L), and showed no distinct spatial patterns.

Few studies have documented the effecs of geomorphology on nitrogen cycling in the riparian zone, although geomorphology is likely to be an important factor that regulates rates of production and loss of ammonium and nitrate in groundwater (Dahm et al. 1987). In this paper, we report patterns in NO_3^-, NH_4^+, dissolved organic nitrogen (DON), and dissolved oxygen (DO) in groundwater sampled across the riparian zone of two tropical rain forest catchments with different soils and geomorphology. We also examine the relationship of these patterns to differences in soil texture, hydrologic flow paths, and infiltration rates at the two sites.

Methods and materials

Site description

Luquillo Experimental Forest

Study sites were located in the Luquillo Experimental Forest, Puerto Rico (Fig. 1). Brown et al. (1983) provide a summary of research in the Luquillo Experimental Forest and describe the soils, vegetation, and climate of the area in detail. Scatena (1989) provides a detailed description of the Bisley watersheds of the Experimental Forest. Numerous ecological research projects have been conducted in the Forest, and an NSF-funded Long Term Ecological Research (LTER) site was established in the Luquillo Experimental Forest in 1988.

The Luquillo Experimental Forest is classified as subtropical moist, wet, lower montane wet, lower montane rain, and rain forest in the Holdridge Life Zone system (Ewel & Whitmore 1973). Elevation ranges from 200 to 1000 m. Plant communities vary with elevation; the tabonuco (*Dacroydes excelsa*) forest type dominates at elevations between 300 and 600 m and the colorado (*Cyrilla racemiflora*) forest type at elevations between 600 and 900 m (Wadsworth & Bonnet 1951). Palms (*Prestoea montana*) are common on steep slopes and in riparian zones. Vegetation

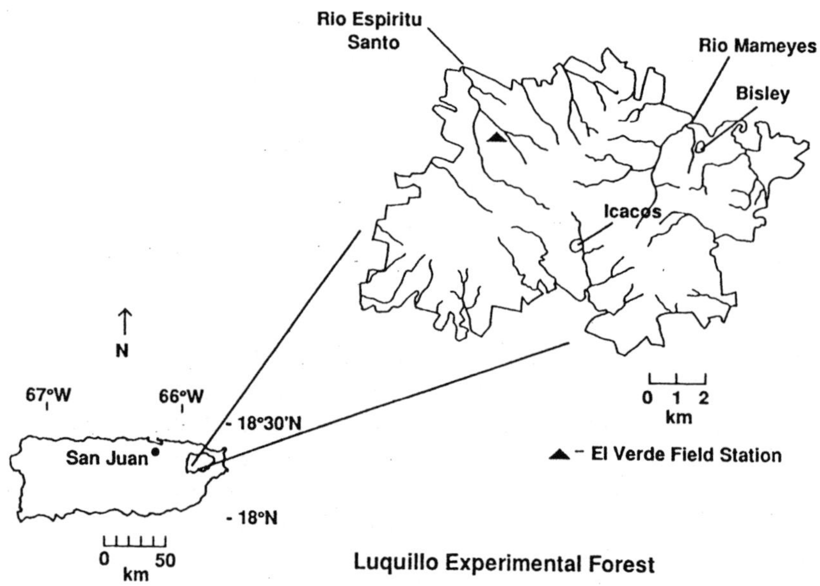

Fig. 1. Map of Puerto Rico showing locations of the study sites in the Luquillo Experimental Forest.

is shallow-rooted, and most roots (by number) are found within the uppermost 20 cm of the soil profile (Odum 1970; Frangi & Lugo 1985). Precipitation ranges from 250 to more than 450 cm/yr within the Experimental Forest and increases with elevation (Brown et al. 1983). Soils in the region are generally acidic clays (Ultisols) with low nutrient contents (Brown et al. 1983). Lower elevations in the tabonuco forest type were used for cultivation of coffee prior to the 1930's, but since that time, anthropogenic impacts on the Forest have been minimal.

Well fields

Study sites were established during August 1988 in the riparian zones of two watersheds in the Luquillo Experimental Forest: Icacos, and Bisley (site of the LTER project in Puerto Rico; Fig. 1). By riparian zone, we mean the ecotone between upslope forests and the open stream channel. We also refer to more specific geomorphological features (e.g. floodplain, stream bank) within each riparian zone as appropriate for the individual sites. The Icacos site is on a small floodplain of a tributary of the Rio Icacos (Fig. 2) at an elevation of 620 m above MSL; watershed area of the tributary adjacent to the well field is 12.4 ha. The Bisley site (elevation 240 m above MSL) is a variable source area (Fig. 2) draining a small catchment (<0.5 ha) downstream of the primary LTER study watersheds; its likely origin is a small landslide (F. Scatena, personal communication). Watershed area of the stream draining the well field at the point of sampling is approximately 50 ha. Stream channel substrate is sand with occasional boulders at the Icacos site, and clay, cobbles and boulders at the Bisley site. Rainfall averages 480 cm at Icacos (Bogart et al. 1964; Curtis et al. 1986) and averages 350—500 cm at Bisley (F. Scatena, personal communication). Bedrock of quartz diorite underliees the Rio Icacos drainage basin; andesitic to basaltic volcanic sandstone, mudstone, and breccia underlie the Bisley site and the rest of the Luquillo Forest (Seiders 1971). These mid- to late-Cretaceous formations are common in the Caribbean and much of the neotropics.

Vegetation in each site was tagged and identified to species as part of the wellfield installation. At the Icacos site, palms (*Prestoea montana*) were dominant in the floodplain, and upslope vegetation was dominated by several large *Cyrilla racemiflora.* At the Bisley site, vegetation was dominated by *Prestoea montana, Swietinia macrophylla* and *Guarea guidonia* in the well field, with tabonuco (*Dacroydes excelsa*) common on ridges.

At the Icacos site, soils are classified as Utuado clays (Inceptisol), while at the Bisley site, soils are classified as Los Guineos clays (Ultisol; Boccheciamp 1977). Chemistry of the Los Guineos series and a soil

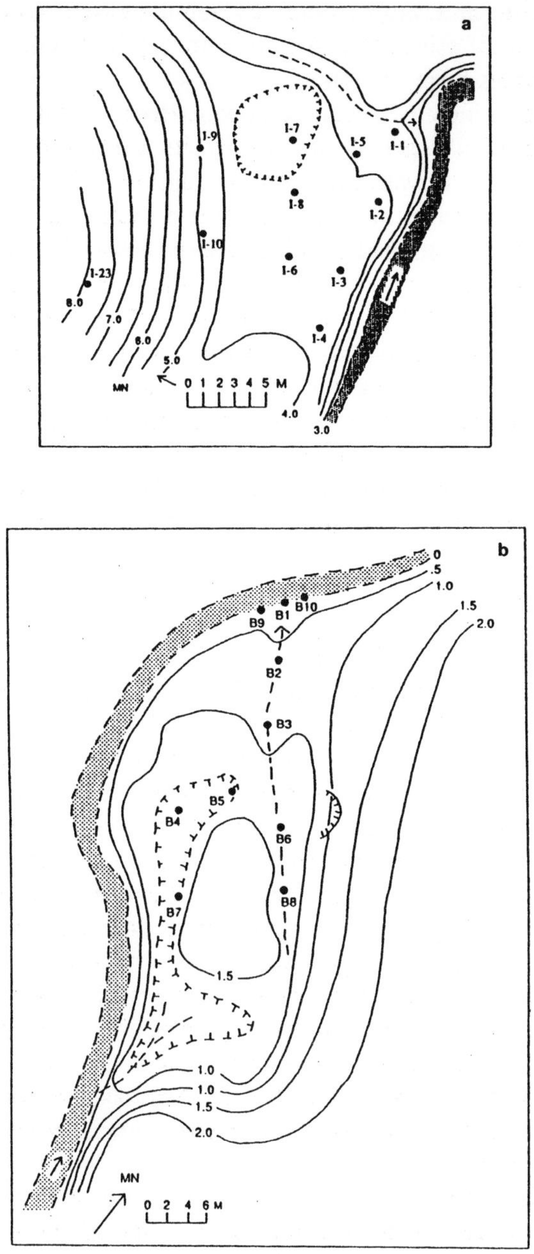

Fig. 2. Surface topography of the Icacos (a) and Bisley (b) well fields. Areas with frequent standing water are surrounded by hash marks. Dashed lines indicate intermittent drainage; shaded area is perennial stream channel. Elevations expressed relative to an arbitrarily chosen datum of approximately 620 m (Icacos) and 240 m (Bisley) above MSL.

similar to the Utuado clay (Picacho series) is described in detail by Fox (1982). Both C and N decrease rapidly with depth in both soils, and pH of soil-water slurries is 4.7—4.8.

Ten or eleven wells were installed at each site by hand augering with a 3.7 cm (1.5 in) bucket auger. At the Icacos site, wells were installed at the stream bank, in the floodplain, and up the adjacent ridge (Fig. 2a). At the Bisley site, wells were located at the stream bank and upslope into a variable source area (Fig. 2b). Depth of the wells was 180 and 210 cm in the Icacos watershed and 70 to 100 cm below the soil surface in the Bisley watershed. The well casings were 5.1 cm (2 in) diameter PVC with 61 cm (2 feet) of 0.25 mm (0.01 in) slotted screen and vented caps. Each screen was wrapped in a polyethylene geotextile fabric filter (Mirafi) to minimize clogging of the screens and turbidity in the samples. It was unnecessary to seal the wells due to the heavy clay texture of the surface soils at both sites. Deeper wells are impractical at the Bisley site because of numerous stones and dry, compact soil below about 40 cm.

Sampling regime

Wells were bailed 1 day (Icacos) or 2—3 days (Bisley) before sample collections. The longer recovery period was needed in Bisley to obtain sufficient water to sample. Prior to bailing, well elevations and dissolved oxygen concentrations were measured. Dissolved oxygen (DO) was measured in each well using a YSI Model 58 meter and a YSI probe and stirrer; readings were taken at the bottom of the well casing to minimize contamination due to atmospheric diffusion. At the Icacos site, tests showed differences in DO concentrations (average of 0.38 mg/L for initial values less than 1 mg/L; 0.95 mg/L for concentrations > 1 mg/L) between lower and upper depths in the wells, indicating either atmospheric contamination or vertical stratification of DO within the water table. Therefore, DO concentrations were probably maximum estimates, and may have been lower in situ. Wells in the Bisley watershed had too little standing water in the casing for similar comparisons. Due to the shallow water column, they were more likely to have been influenced by atmospheric oxygen.

Following the recovery period (1 day for Icacos or 3 days for Bisley) water samples were obtained from the bottom of each well using a portable peristaltic pump or Teflon bailer. Samples were pressure filtered for NH_4^+, because initial tests showed that during vacuum filtration there was a potential for loss of volatile NH_3 from some samples with high NH_4^+ concentration and pH 6—7. Vacuum filtration generally was used for all other analyses. An aliquot was acidified for NH_4^+ analysis; NO_3^- and total dissolved nitrogen were measured on an untreated aliquot. Data are reported for August 1988—1 September 1989, except for the Icacos wellfield. Hurricane Hugo passed over the Luquillo Experimental Forest on September 18—19, 1989, with major effects on vegetation and ecosystem-level processes (summarized by Lodge & McDowell 1991). Both the Bisley and Icacos watersheds suffered significant damage. The Icacos wellfield, however, was undamaged and data from August 1988 to June 1991 are reported here.

Hydrology

Saturated hydraulic conductivity in the vicinity of each well was determined by the Hvorslev method, in both bail and slug modes. Water levels in the Bisley wells were often below the top of the well screen, which introduces an error in the Hvorslev calculation for data from bail tests. In these cases an auger hole method was used to provide a second estimate of k_{sat} (Boersma 1965). We found that results from the Hvorslev method agreed well with the auger hole method when the Hvorslev calculation was corrected for the effective screen height when the water table fell below the top of the screen.

Infiltration rates of surface soils were determined using double-ring infiltrometers (12 cm diameter inner ring; 10 cm diameter outside ring). Infiltration rate was determined after allowing an initial 1 cm drop in water level in the inner ring (Bouwer 1986).

Analytical methods

Nitrate and ammonium were analyzed using automated Cd-Cu reduction (EPA 1983 method 353.2) and phenol-hypochlorite (EPA 1983 method 350.1), respectively. Dissolved organic N was estimated as the difference between dissolved inorganic N and total dissolved N measured by persulfate digestion of samples in sealed glass ampoules (Solorzano & Sharp 1980) followed by hydrazine reduction (EPA 1983 method 353.1) to measure NO_3^- in the digestate. Beginning in September 1989, single column ion chromatography with UV detection (214 nm) was used to measure nitrate and nitrite (EPA A-1000) using a mobile phase of buffered borate-gluconate (1.2 ml/min) and a Waters IC Pak A anion column. Extensive intercalibration showed excellent agreement between the two methods; detection levels improved by an order of magnitude with the ion chromatographic method. Duplicate analyses, internal standards, and blind analysis of EPA quality assurance samples were used to insure quality control.

Soil C and N content were analyzed on oven-dried (65 °C) samples using a Perkin Elmer 240B elemental analyzer with acetanilide as a standard. Soil pH was measured electrometrically in a 1:10 soil:water slurry in 0.01 N $CaCl_2$. Soil particle size distribution (pipet method; Day 1965) was determined on air-dried samples which were sonicated in Calgon (sodium metaphosphate) solution to disperse aggregates.

Results

Soils and hydrology in the well fields

At the Icacos site, soils showed strong vertical zonation in texture and redox status in the floodplain (Table 1). Oxidized clays at the surface graded to reduced clays and very reduced sands and gravel at depths of 1.8—2.1 m below the surface (Fig. 3a). In upslope sites, surface soils were oxidized (red and yellow) clays, grading to red sand and saprolite at depths greater than 2.5 m (Fig. 3a). In both floodplain and upslope soils, earthworm tubes were commonly observed to depths of 30 cm or more.

At the Bisley site, soils consisted of a mosaic of oxidized and reduced clays, with similar textures in surface and deeper soils (Table 1). An intermittent stratum with high gravel content was observed at 40—50 cm depth (Fig. 3b) at many locations.

Infiltration at the soil surface in non-ponded areas (Fig. 2) was more rapid at the Icacos (2—9 cm/min) than the Bisley site (0.07—1.5 cm/min). Although infiltration rates at the Icacos site were typically greater than at the Bisley site, infiltration rates at both sites exceed typical rainfall intensities (0.8 cm/hr or 0.013 cm/min, Brown et al. 1983). Soil C and N contents at the Icacos site were almost double those at the Bisley site, and declined sharply with depth at both sites (Table 1).

Results from bail tests indicated that saturated hydraulic conductivities near the wells were similar at both sites (about 1×10^{-5} to 1×10^{-4} cm/sec; Fig. 4). The conductivity of stream-side wells also tended to be higher at both sites than the conductivity of upslope wells. Results from slug tests were similar to those obtained with the bail tests at the Icacos site, but were higher by 10 to 100 times for wells at the Bisley site. In a homogeneous and saturated medium, there is no reason to expect that results from the two modes of measurement (bail and slug) should differ.

This difference may be related to differences in the structure of the Bisley soil compared to the Icacos soil. At the Icacos site, the well screens were always below the watertable and were embedded in a relatively homogeneous, conductive soil matrix. At the Bisley site, the watertable

Table 1. Characteristics of soils in the riparian zone in the Icacos and Bisley watersheds of the Luquillo Experimental Forest, Puerto Rico.

	Texture (% sand/silt/clay)		Carbon (%)		Nitrogen (%)		pH		Sulfide	
	Icacos	Bisley	Icacos	Bisley	Icacos	Bisley	Icacos	Bisley	Icacos	Bisley
Surface soils (0–10 cm)	20/43/37 (10/9/3)	11/37/51 (1/4/5)	8.1 (2.0)	4.1 (2.2)	0.36 (0.19)	0.27 (0.06)	4.5 (0.1)	4.6 (0.1)	–	–
Mid-level soils (45–55 cm)	22/30/49 (3/3/4)	8/38/54 (4/7/4)	6.1 (1.7)	0.90 (0.15)	0.24 (0.06)	0.08 (0.03)	4.2 (0.1)	4.4 (0.1)	+/–	–
Mid-level soils (90–100 cm)	36/27/37 (7/1/9)	10/39/51 (1/5/5)	3.2 (2.4)	0.53 (0.26)	0.12 (0.05)	0.04 (0.03)	4.3 (0.1)	4.4 (0.1)	+	–
Deep soils (1.8–2.2 m)	64/21/15 (25/13/13)	NS	2.1 (0.3)	NS	0.09 (0.02)	NS	4.9 (0.1)	NS	++	NS

NS indicates not sampled. C and N expressed as % dry weight; pH is measured in $CaCl_2$. Sulfide characterized as absent (–), present (+) and very strong (++) based on smelling the soil after smearing between two fingers. Mean (standard deviation) shown for three soil cores taken approximately 5 m from the stream channel in both watersheds.

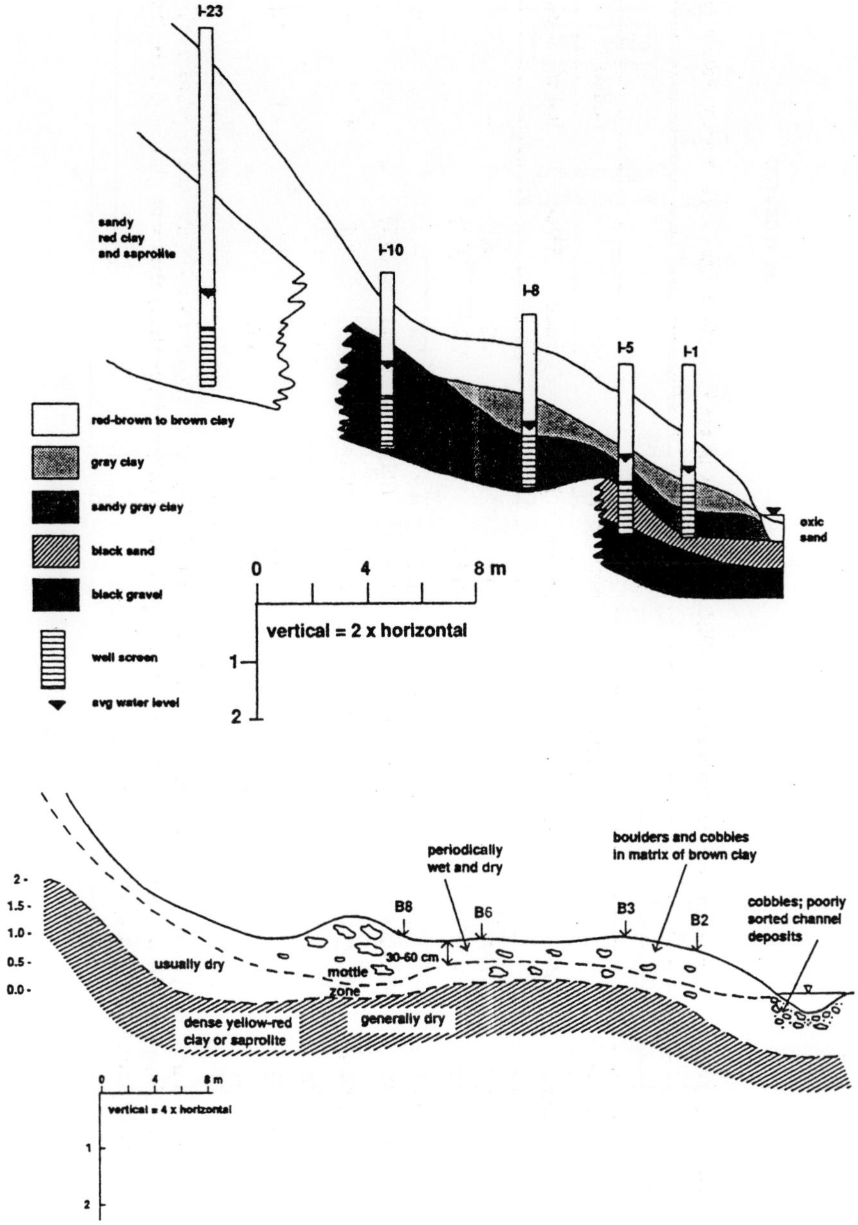

Fig. 3. Soil profiles in a transect across the Icacos (upper panel) and Bisley (lower panel) well fields, showing well locations and depths. For Icacos, average elevation of the water table is also included. Note differences in vertical exaggeration.

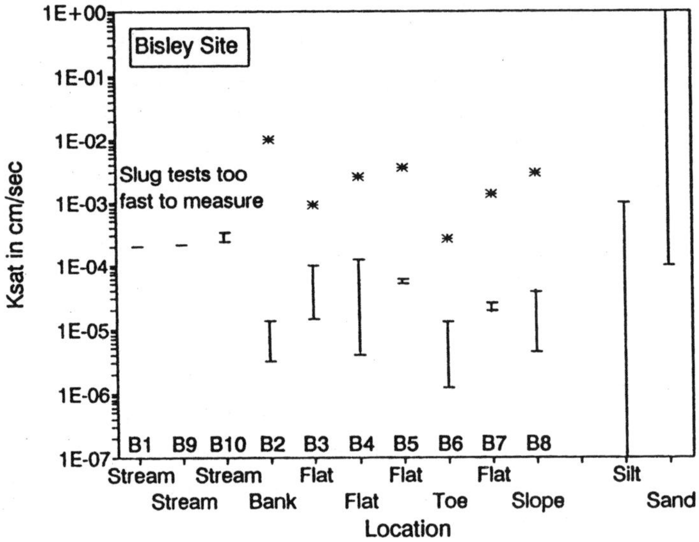

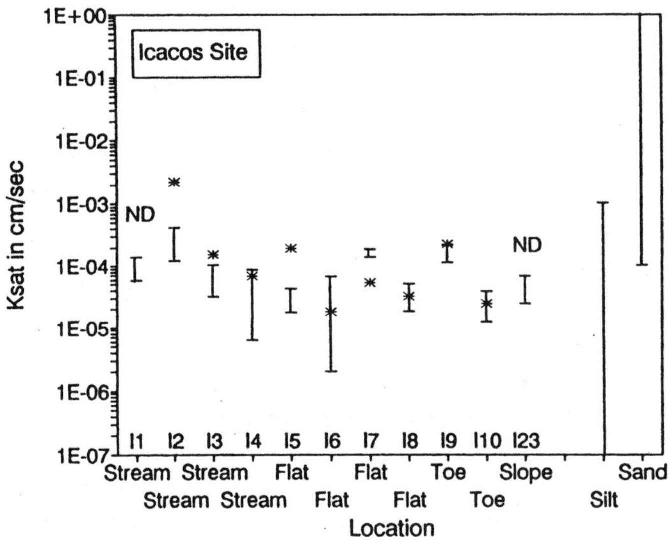

Fig. 4. Saturated hydraulic conductivity (k_{sat}) in the vicinity of individual wells in the two study sites. For reference, commonly reported values for silt and sand are also included (Boersma 1965). Vertical bars are +/− one standard error of the mean determined using bail tests. Asterisks represent values obtained from slug tests.

was often below the top of the well screen at the start of the bail-mode tests. This fact was taken into account in the Hvorslev calculation. However, we believe that the recovery of these wells following bailing was dependent on transport of water by relatively slow matrix flow through the structureless sub-soil near the bottom of the well screen. In contrast, soil nearer the surface (in the vicinity of the upper portion of the well screens) has a subangular, blocky structure that allows relatively rapid water flow through preferential flow paths around the soil blocks. Thus, in the slug mode, water inside the wells rapidly drained via this shallow, preferential flow network. We believe that this shallow flow network is responsible for the high infiltration rates at the Bisley site and promotes rapid, shallow interflow after storm events. Well levels in both watersheds responded very quickly to periods of intense rainfall (e.g. Fig. 5). Although some overland flow was observed in the well fields during such intense events, most incoming precipitation quickly entered subsurface flow paths, with shallower water movement through the Bisley site and deeper water movement at the Icacos site.

Groundwater surfaces sloped sharply toward the stream in the Icacos watershed (Fig. 6a). At the Icacos site, groundwater elevations indicated movement of water diagonally across the wellfield, with an average head differential of 2.1 m over 22 m. At the Bisley site, subsurface water movement likely followed the topographic surface, following the swale from upslope to downslope wells (Fig. 2b), with an average difference in groundwater elevation of 1.7 m over 24 m.

It appears that the sampling sites in both watersheds are in close hydrologic contact with the adjacent streams. Large gradients in water table elevation (Fig. 6) and moderate to high saturated hydraulic conductivities (Fig. 4) result in relatively rapid specific discharges (8.2×10^{-4} to 1.5×10^{-1} m/day at the observed hydraulic gradient). This conclusion is further supported by the rapid response of the well elevations to rainfall (Fig. 5).

Spatial and temporal variability in well chemistry

Large differences were observed in nitrate and ammonium concentrations of groundwater collected at the two sites. At all points in the riparian transect, concentrations of dissolved inorganic N were approximately five-fold greater at the Icacos site than at Bisley (Fig. 7), despite the fact that stream water concentrations at the two sites were similar. At Bisley, most dissolved nitrogen in groundwater was found as DON, while at Icacos, most was found as ammonium. Concentrations of dissolved organic N were comparable in both watersheds (100—300 ug/L; Fig. 7).

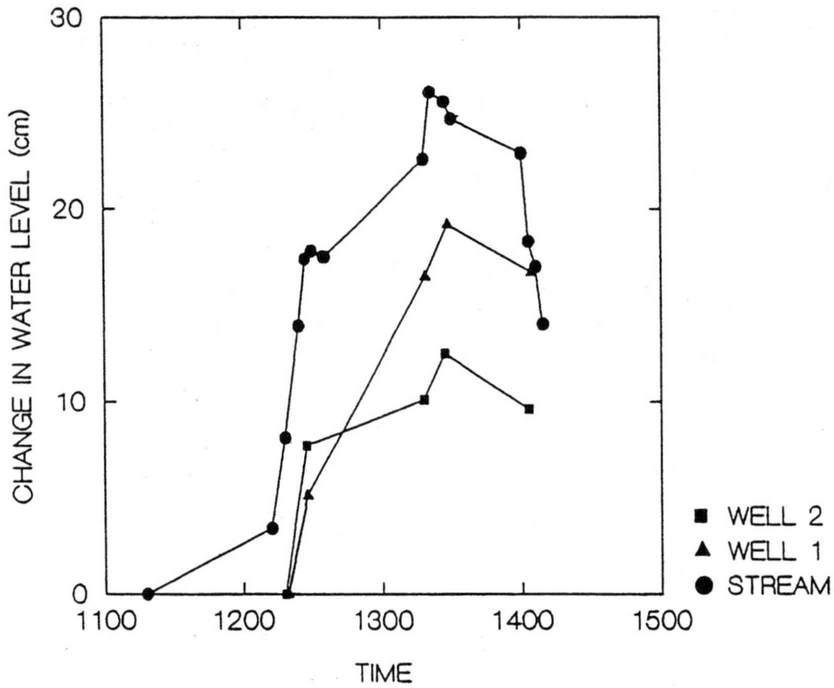

Fig. 5. Changes in elevation of stream and groundwater surface in the Icacos well field in response to a rainstorm on June 2, 1989.

Strong spatial patterns were observed across the riparian zone at the Icacos site for all constituents sampled. In upslope wells, 90—95% of total inorganic N was found as nitrate, and DO concentrations were high (averaging 5 mg/L in well 23). However, within the floodplain, NO_3^- and DO fell to concentrations of 9 ug/L and 1.4 mg/L, respectively. In contrast, NH_4^+ levels in upslope wells were typically very low (less than 30 ug/L) but concentrations increased significantly within the floodplain, to levels which often exceeded 500 ug/L. In the adjacent stream, nitrate concentrations were never as high as those observed in upslope wells, nor were ammonium concentrations ever as high as those observed in floodplain wells, even those 1—2 m from the stream bank (wells 1—4). There was a very consistent inverse relationship between nitrate and ammonium concentrations in samples from the Icacos wells, with very few samples containing both in appreciable quantities (Fig. 8a).

At the Bisley site, the relationship between NH_4^+ and NO_3^- was weak, with frequent co-occurrence of the two ions and generally low concentrations (Fig. 8b) compared to Icacos. At Bisley, there was a consistent decline in TDN and DON across the riprian catena, but no consistent

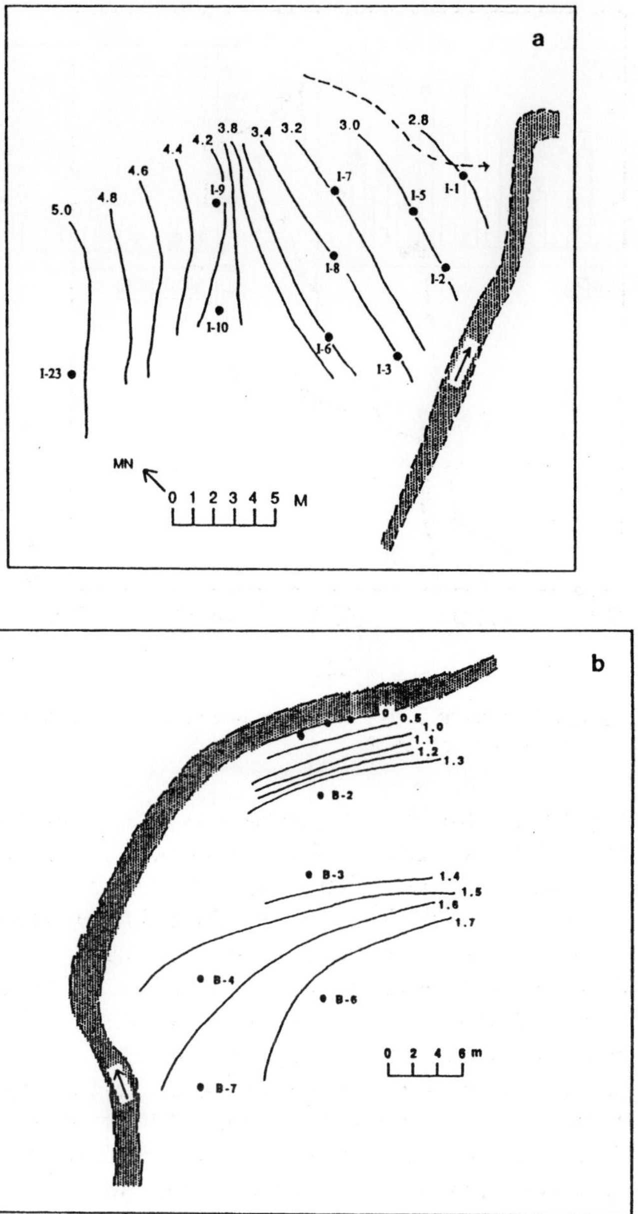

Fig. 6. Topography of the water table in the Icacos (a) and Bisley (b) well fields. Elevations are relative to the same datum used in Fig. 2. For Icacos, well 4 is anomalous (mean elevation 623.0 m) and is not shown. For Bisley, all wells which contained water on 75% or more of the sampling dates are shown.

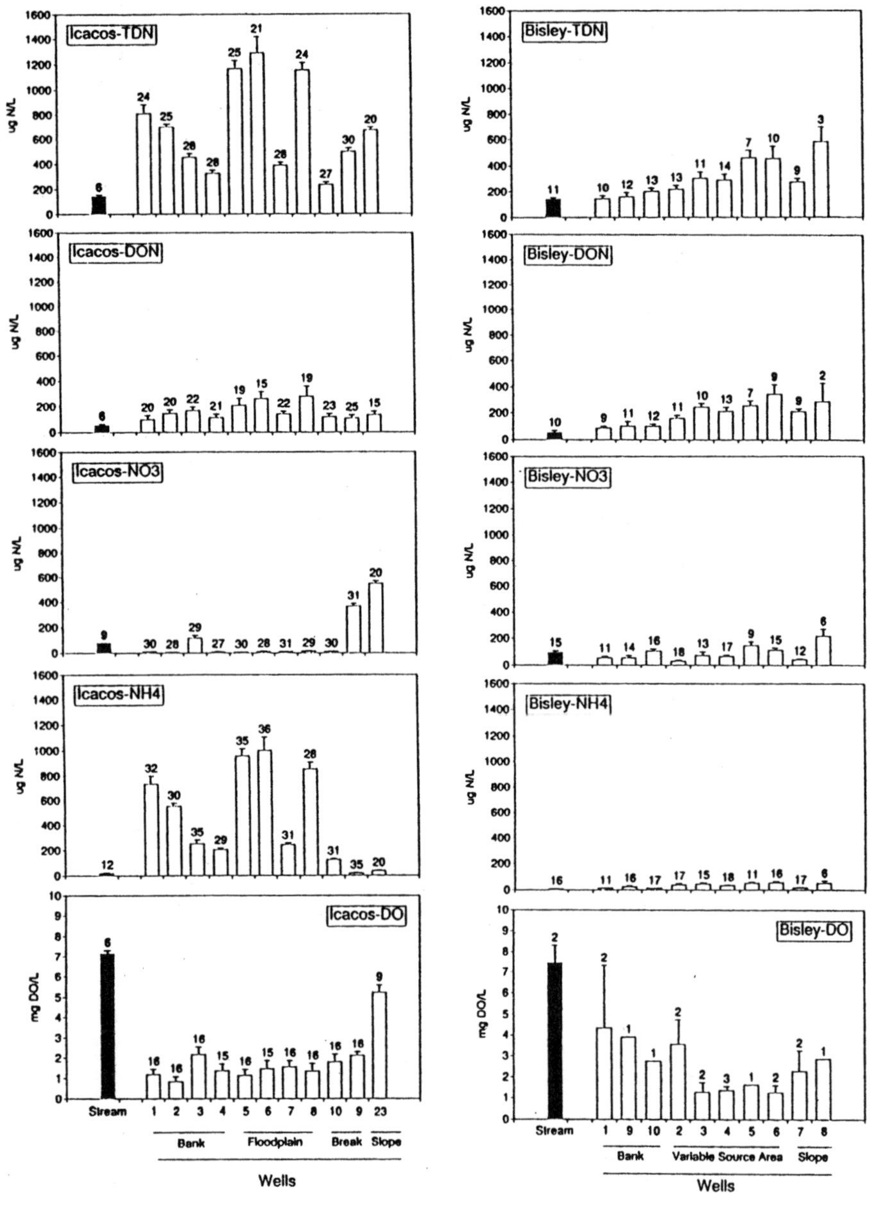

Fig. 7. Spatial variability in groundwater chemistry at the Icacos and Bisley sites. Well locations are shown in Fig. 2. Means are shown; error bars represent one standard error of the mean, and the number of observations is placed above each bar.

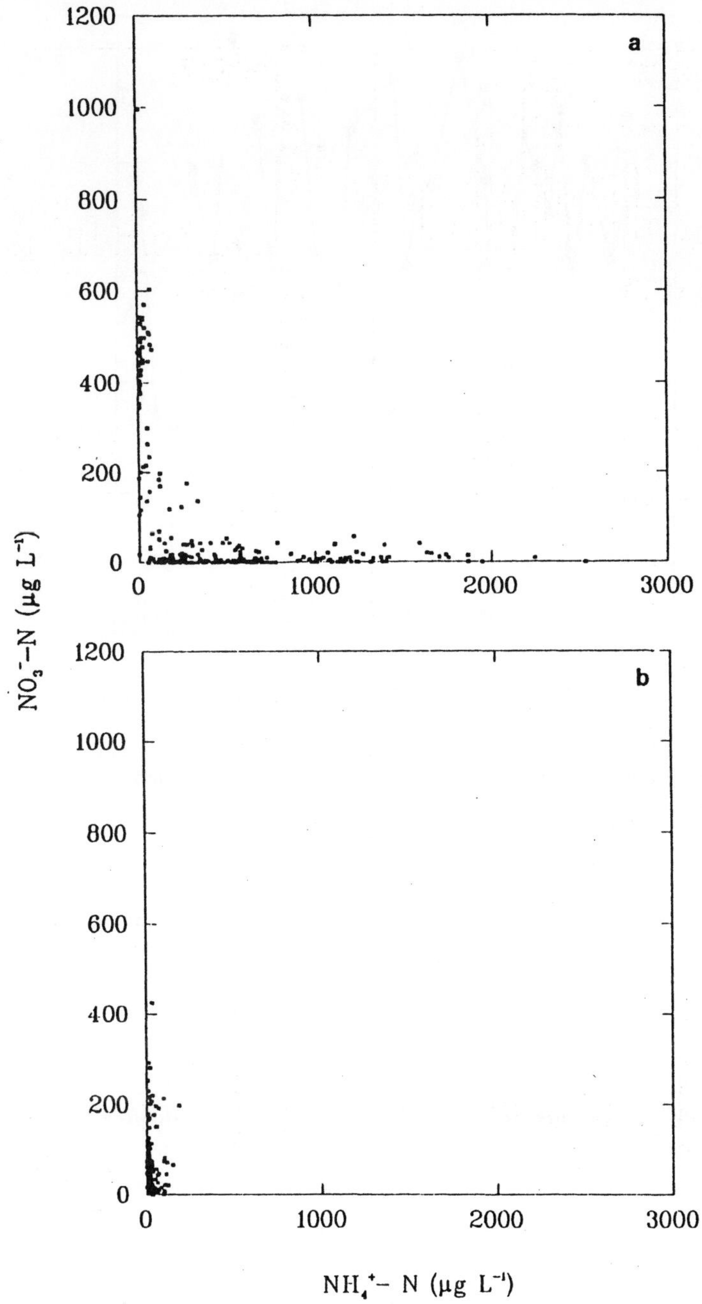

Fig. 8. Relationship between NO_3^- and NH_4^+ concentrations in groundwater collected in the Icacos (a) and Bisley (b) watersheds.

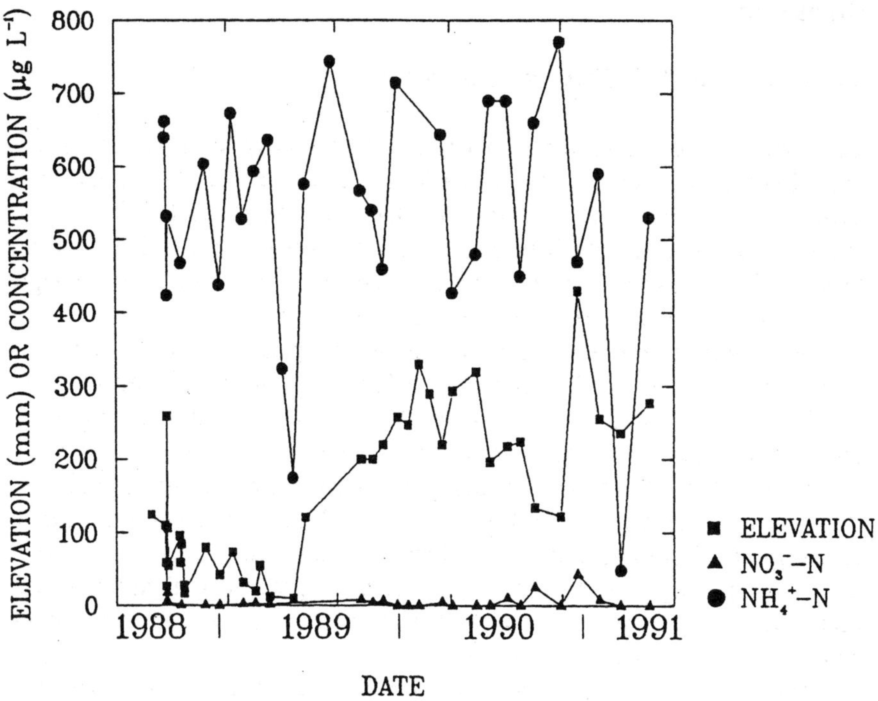

Fig. 9. Variation in elevation (mm above minimum recorded value) and inorganic nitrogen chemistry in well 2 of the Icacos site during the three-year study period.

trends were observed for NH_4^+ or NO_3^- concentrations (Fig. 7b). There was a strong gradient in dissolved oxygen from low levels in the variable source area to higher levels in the stream itself (Fig. 7b), as also observed at the Icacos site (Fig. 7a). However, unlike the Icacos site, there was no gradient in nitrogen species across this interface (Fig. 7b).

Concentrations of dissolved inorganic N showed no consistent seasonal variation or variation with water table elevation at either site (e.g. Fig. 9). Although several individual wells did show statistically significant variation in ammonium and nitrate concentrations as a function of piezometic head at the Icacos site, no consistent pattern emerged. Increases in DON concentrations were observed in several of the Icacos wells in early 1991, but similar patterns were not observed for NO_3^- or NH_4^+.

Discussion

Spatial patterns of N within the riparian zone

The riparian zone links terrestrial and aquatic ecosystems, and thus can alter hydrologic export from a basin (Gregory et al. 1991). Because hydrologic loss of nutrients from terrestrial ecosystems is widely used to measure the impact of disturbance or otherwise characterize nutrient losses, nitrogen transformation within the riparian zone may strongly affect perceptions of watershed-level nutrient dynamics.

We hypothesize that differences in geomorphology and solute flow paths strongly influence N flux and transformation in the riparian zones of the Icacos and Bisley watersheds. At the Icacos site, water quickly infiltrates through the rooting zone and into shallow groundwater, where it moves rapidly to the stream through highly conductive subsurface soils. At the Bisley site, surface infiltration is also rapid, but dense clays appear to retard infiltration to deeper soils. Subsurface flow at the Icacos site follows a deep trajectory past the rooting zone and through a highly reduced, conducting layer, while water at the Bisley site appears to follow a shallow trajectory through the rooting zone.

This difference in geomorphology appears to control both the form and concentration of N in shallow groundwater at the two sites, and suggests the following conceptual model. At the Icacos site, after infiltrating through oxic surface soils, subsurface runoff must flow through distinct oxic (upslope) and reduced (floodplain) zones with spatial scales of centimeters to meters. This subsurface flow occurs below most plant roots. Under these conditions mineralization, nitrification, plant uptake, and denitrification are segregated in space, so that ammonium and nitrate accumulate in distinct zones. In contrast, at the Bisley site, subsurface runoff remains near the surface in a variably oxidized zone in which redox conditions apparently vary over a scale of only fractions of a centimeter. Under these conditions, mineralization, nitrification, plant uptake, and denitrification might coexist so that ammonium and nitrate concentrations are persistently low and variable. This conceptual model is consistent with our hydrologic and geochemical data from the wells, but remains to be rigorously tested.

Spatial patterns of N at the stream-riparian interface

One of the most striking patterns observed in this study was the large decrease in concentrations of dissolved nitrogen between groundwater in the floodplain and adjacent stream water at the Icacos well field (Fig. 7a).

Declines in NH_4^+ concentrations were especially large. Several alternative hypotheses could explain these observations, including 1) our site is unrepresentative of the basin; 2) uptake by riparian vegetation removes NH_4^+, the dominant form of N in the floodplain; 3) in-stream processes remove NH_4^+; or 4) coupled nitrification and denitrification in the near-stream zone result in gaseous losses of N.

The first hypothesis that might explain the rapid loss of NH_4^+ at the Icacos site is that groundwater sampled at our study site was not representative of groundwater entering throughout the catchment. If this were the case, then the low DO, high-NH_4^+ groundwater we sampled would simply have been diluted by a much greater flow of high-DO, low-NH_4^+ stream water. We have observed, however, that large amounts of orange, flocculated iron accumulate throughout the length of the stream channel during low flow conditions. This indicates the presence throughout the stream basin of anoxic groundwater, which is likely to contain elevated NH_4^+ levels such as those observed in the floodplain.

The second hypothesis that might explain NH_4^+ loss at the Icacos site is vegetative uptake by palms and palm seedlings in the narrow strip between streamside wells and the stream channel. A simple mass balance calculation, however, suggests that N accumulation in riparian biomass cannot explain the entire NH_4^+ loss. Net annual biomass increment of palms in floodplain forest is about 2000 g/m^2 (Frangi & Lugo 1985), and riparian palms contain 0.7% N (Scatena et al. in press), for a total N accumulation of 14 g/m^2 in the riparian zone. Average annual runoff from the Icacos basin (Curtis et al. 1986) is 413 cm, or 5.1×10^8 L/yr from the 12.4 ha basin upstream of the well field. At an average change in NH_4^+-N concentration of 400 ug/L (difference between wells 1—4 and stream concentrations), a total loss of 204 kg NH_4^+-N must be accounted for. The area in which the observed N loss occurs is a 1.5-m wide strip at our study site (Fig. 2a). If similar losses of NH_4^+ occur along the entire stream channel above our site (600 m, or about 1800 m^2 of riparian stream bank), the rate of NH_4^+-N loss is equivalent to 113 g/m^2 in the riparian zone. Although these are only crude estimates, the difference between estimated plant uptake (14 g/m^2) and apparent NH_4^+ loss (113 g/m^2) is so large that mechanisms other than plant uptake must be responsible for losses of N from solution in the riparian zone.

The third hypothesis that might explain the large difference in NH_4^+ concentrations in streamside groundwater and open channel flow is that N removal (through microbial immobilization, nitrification-denitrification, or algal uptake) occurs within the open stream channel. Whole-stream enrichments (McDowell et al. in prep.), however, suggest that both NH_4^+ and NO_3^- loss in the open channel are small.

The fourth hypothesis that could explain the large apparent loss of NH_4^+ at the Icacos site is coupled nitrification and denitrification in the stream bank or beneath the channel in the hyporheic zone. This hypothesis requires the coexistence of both oxic and anoxic conditions in microzones (on the scale of millimeters), and implies that sharp gradients in NH_4^+ and NO_3^-, as well as large gaseous efflux of N (as N_2 or N_2O), occur in the stream bank and hyporheic zone. Ongoing work on groundwater chemistry suggests that the stream bank as well as the hyporheic zone may be sites of unusually active nitrogen transformation. The flux of N_2O measured by Bowden et al. (*this issue*) is equivalent to an annual flux of only 32 g/m^2 at the stream bank, far less than the 113 g/m of N which appears to be lost there annually (as calculated above). However, if the molar ratio for N_2O production is low (high proportion of N_2 production), denitrification might quantitatively account for this loss of inorganic nitrogen. We do not yet know what the molar ratio might be and so can not assess the importance of N_2 production. Thus, without further experimentation, none of the four alternate hypotheses outlined above can fully explain the losses of inorganic N we observed in shallow goundwater.

Conclusions

Patterns in nitrogen concentration and speciation in the Bisley watershed are similar to those observed at other forested sites, with moderate declines in some fractions and modest gradients in dissolved oxygen across the riparian zone (e.g. Triska et al. 1989; Hill 1990). In the Icacos basin, in contrast, gradients in dissolved oxygen and N species across the riparian zone are among the largest reported in the literature for undisturbed forested catchments (Triska et al. 1989; Hill 1990; Triska et al. 1990; Pringle & Triska 1991). Even in agricultural catchments, with large anthropogenic inputs of N, changes in NH_4^+ across the riparian zone are not as large as those observed at the Icacos site (Lowrance et al. 1983; Peterjohn & Correll 1984; Schnabel 1986), and NO_3^--N in groundwater is rarely lowered to levels observed in the Icacos floodplain (often less than 1 ug/L).

The different spatial patterns of N we measured in shallow riparian groundwater in the Luquillo Forest coincide with different dominant subsurface flow paths. It is likely that different flow paths dominate due to geomorphological differences within this heterogeneous, tropical landscape.

Acknowledgments

Funding was provided by grants BSR 8718396, BSR 8718395, and BSR 9007498 from the National Science Foundation. We thank A. Finley, S. Benjamin, M. Converse, M. McDowell, C. McSwiney, J. Nash, F. Scatena, and N. Sullivan for assistance in the field, and A. Finley, C. McSwiney, J. Pagano, and M. Soulia for laboratory analyses. C. T. Smith and F. Scatena provided valuable comments on the manuscript. Special thanks go to F. Scatena and A. Lugo for providing Forest Service personnel to conduct much of the sampling in Bisley. Use of laboratory facilities for some analyses was provided by the Center for Energy and Environment Research of the University of Puerto Rico, supported under grants LTER BSR 8811902 from NSF and Contract DE-AC05-760R 01833 from DOE.

References

Boccheciamp RA (1977) Soil survey of Humacao area of eastern Puerto Rico. 103 pp. U.S.D.A. Soil Conservation Service

Boersma L (1965) Field measurement of hydraulic conductivity below a water table, pp 222—233. In: Black CA (Ed) Methods of Soil Analysis Part 1 — Physical and Mineralogical Methods. 1st ed. American Society of Agronomy — Soil Science Society of America, Madison, WI

Bogart DB, Arnow T & Crooks JW (1964) Water Resources of Puerto Rico — A Progress Report. Water Resources Bulletin No. 4. 102 pp. U.S. Geological Survey, San Juan, PR

Bouwer H (1986) Intake rate: cylinder infiltrometer, pp 825—844. In: Klute A (Ed) Methods of Soil Analysis Part 1 — Physical and Mineralogical Methods. 2nd ed. American Society of Agronomy — Soil Science Society of America, Madison, WI

Bowden WB, McDowell WH, Asbury CE & Finley AM (1993) Riparian nitrogen dynamics in two geomorphologically distinct tropical rain forest watersheds: nitrous oxide fluxes. Biogeochemistry 18: 77—99

Brown S, Lugo AE, Silander S & Liegel L (1983) Research history and opportunities in the Luquillo Experimental Forest. General Technical Report SO-44. 128 pp. USDA Forest Service, New Orleans, Louisiana

Cooper AB (1990) Nitrate depletion in the riparian zone and stream channel of a small headwater catchment. Hydrobiologia 202: 13—26

Curtis RE Jr., Aquino Z, Diaz PL & Garcia R (1986) Water Resources Data — Puerto Rico and the U.S. Virgin Islands. Water Year 1986. U.S.G.S. Water-Data Report PR-86-1

Dahm CN, Trotter EH & Sedell JR (1987) Role of anaerobic zones and processess in stream ecosystem productivity, pp 157—178. In: Averett RC & McKnight DM (Eds) Chemical quality of water and the hydrologic cycle. Lewis Publishers, Chelsea, Michigan

Davidson EA & Swank WT (1986) Environmental parameters regulating gaseous nitrogen losses from two forested ecosystems via nitrification and denitrification. Applied and Environmental Microbiology 52: 1287—1292

Day PR (1965) Particle fractionation and particle-size analysis, pp 545—577. In: Black CA (Ed) Methods of Soil Analysis Part 1 — Physical and Mineralogical Methods. 1st ed. American Society of Agronomy — Soil Science Society of American, Madison, WI

Duff JH & Triska FJ (1990) Denitrification in sediments from the hyporheic zone adjacent to a small forested stream. Canadian Journal of Fisheries and Aquatic Sciences 47: 1140—1147

E.P.A. (1983) Methods for chemical analysis of water and wastes. E.P.A.-600/4-79-020. U.S. E.P.A. Environmental Monitoring and Support Laboratory, Cincinnati, OH

Ewel JJ & Whitmore JL (1973) The ecological life zones of Puerto Rico and the U.S. Virgin Islands. U.S.D.A. Forest Service Research Paper ITF-18. 72 pp. Institute of Tropical Forestry, Rio Piedras, P.R.

Fail JL, Hamzah MN, Haines BL & Todd RL (1986) Above and below-ground biomass, production, and element accumulation in riparian forests of an agricultural watershed, pp 193—223. In: Correll DL (Ed) Watershed Research Perspectives. Smithsonian Institution Press, Washington, DC

Ford TE & Naiman RJ (1989) Groundwater — surface water relationships in boreal forest watersheds: dissolved organic carbon and inorganic nutrient dynamics. Can. J. Fish. & Aqua. Sci. 46: 41—49

Frangi JL & Lugo AE (1985) Ecosystem dynamics of a subtropical floodplain forest. Ecological Monographs 55: 351—369

Gregory SV, Swanson FJ, McKee WA & Cummins KW (1991) An ecosystem perspective of riparian zones. BioScience 41: 540—551

Grimm NB & Fisher SG (1984) Exchange between interstitial and surface water: Implications for stream metabolism and nutrient cycling. Hydrobiologia 111: 219—228

Hendrickson OQ Jr. (1981) Flux of nitrogen and carbon gases in bottomland soils of an agricultural watershed. Ph.D. dissertation, University of Georgia

Hendricks SP & White DS (1991) Physicochemical patterns within the hyporheic zone of a northern Michigan river, with comments on surface water patterns. Canadian Journal of Fisheries and Aquatic Sciences 48: 1645—1654

Hill AR (1990) Groundwater flow paths in relation to nitrogen chemistry in the near-stream zone. Hydrobiologia 206: 39—52

Hynes HBN (1983) Groundwater and stream ecology. Hydrobiologia 100: 93—99

Jacobs TC & Gilliam JW (1985) Headwater stream losses of nitrogen from two coastal plain watersheds. Journal of Environmental Quality 14: 467—472

Lodge DJ & McDowell WH (1991) Summary of ecosystem-level effects of Caribbean hurricanes. Biotropica 23: 373—378

Lowrance RR, Todd RL & Asmussen LE (1983) Waterborne nutrient budgets for the riparian zone of an agricultural watershed. Agricultural Ecosystems and Environment 10: 371—384

Lowrance R, Todd R, Fail JJ, Henderickson OJ, Leonard R & Asmussen L (1984) Riparian forests as nutrient filters in agricultural watersheds. BioScience 34: 374—377

Odum HT (1970) Rain forest structure and mineral-cycling homeostasis, pp H3—H52. In: Odum HT & Pigeon RF (Eds) A tropical rain forest — a study of irradiation and ecology at El Verde, Puerto Rico. Division of Technical Information, U.S. Atomic Energy Commission, Washington, DC

Peterjohn WT & Correll DL (1984) Nutrient dynamics in an agricultural watershed: observations on the role of a riparian forest. Ecology 65: 1466—1475

Pringle CM & Triska FJ (1991) Effects of geothermal groundwater on nutrient dynamics of a lowland Costa Rican stream. Ecology 72: 951—965

Scatena FN (1979) An introduction to the physiography and history of the Bisley Experimental Watersheds in the Luquillo Mountains of Puerto Rico. General Technical

Report SO-72. 22 pp. USDA Southern Forest Experiment Station, New Orleans, Louisiana

Scatena FN, Silver W, Siccama T, Johnson AH & Sanchez MJ (1993) Biomass and nutrient content of Bisley Experimental Watersheds, Luquillo Experimental Forest. Biotropica (In press)

Schnabel RR (1986) Nitrate concentrations in a small stream as affected by chemical and hydrologic interactions in the riparian zone, pp 263—282. In: Correll DL Watershed Research Perspectives. Smithsonian Institution Press, Washington, DC

Seiders VM (1971) Geologic map of the El Yunque quadrangle, Puerto Rico. Miscellaneous Geological Investigation I-658. U.S. Department of the Interior, Geological Survey, Washington, DC

Solorzano L & Sharp JH (1980) Determination of total dissolved nitrogen in natural waters. Limnology and Oceanography 25: 751—754

Stanford JA & Ward JV (1988) The hyporheic habitat of river ecosystems. Nature 335: 64—66

Stark NM & Jordan CF (1978) Nutrient retention by the root mat of an Amazonian rain forest. Ecology 59: 434—437

Triska FJ, Duff JH & Avanzino RJ (1990) Influence of exchange flow between the channel and hyporheic zone on nitrate production in a small mountain stream. Canadian Journal of Fisheries and Aquatic Sciences 47: 2099—2111

Triska FJ, Kennedy VC, Avanzino RJ, Zellweger GW & Bencala KE (1989) Retention and transport of nutrients in a third-order stream in northwestern California: Hyporheic processes. Ecology 70: 1893—1905

Valett HM, Fisher SG & Stanley EH (1990) Physical and chemical characteristics of the hyporheic zone of a Sonoran Desert stream. Journal of the North American Benthological Society 9: 201—215

Wadsworth F & Bonnet JA (1951) Soils as a factor in the occurrence of two types of montane forest in Puerto Rico. Caribbean Forester 12: 67—70

Weaver PL & Murphy PG (1990) Forest structure and productivity in Puerto Rico's Luquillo Mountains. Biotropica 22: 69—82

Commentary on Paper E7

Haycock, N. E. & Burt, T. P. (1993) Role of floodplain sediments in reducing the nitrate concentration of subsurface run-off: a case study in the Cotswolds, UK.

This study presents one of the very first comprehensive hydrological analyses of a hillslope–flood plain interface, using classical hydrological methods over an 18-month period. Indeed, Haycock and Burt did not take the in-vogue perpendicular subsurface flow theory for granted and set up a grid of boreholes from the upslope edge to the stream bank to determine the flow paths within the flood plain, thanks to the iconoclast thinking of Haycock and the strong hydrology background of Burt. Combining results of long-term monitoring of water table contours and the use of chloride as a conservative tracer, they determined that hillslope groundwater entered the flood plain section and flowed into buried channel sediments, from where it drained in a down-valley direction. Therefore, groundwater flow within the flood plain was not perpendicular to the stream, as the dominant topographic gradient within the flood plain was down-valley. They found that the groundwater in the near-stream flood plain zone was not hydraulically connected to the hillslope, or the stream.

These findings were at odds with the currently published results which suggested that groundwater flowed perpendicularly from the hillslope to the stream via the flood plain. This particular hydrological set up led to a different interpretation of the nitrate concentration decrease to that previously published. In fact, the active removal zone was at the hillslope–flood plain interface; the near-stream flood plain being hydrologically disconnected, it did not contribute to nitrate reduction (cf. Clément *et al.*, 2003). What might have appeared at that time as a peculiar case study turned out to be the rule rather than the exception. Several further careful hydrological assessments of riparian zones have confirmed that in many situations subsurface flowpaths are very complex and strongly influenced by past hydro-geomorphological history, including meander pattern, buried channels and sediment deposits (Sabater *et al.*, 2003; Gurwick *et al.*, 2008).

Haycock and Burt found that nitrate reduction observed during the winter period was mainly due to biological processes (80–90%); the remaining decrease was due to dilution. They attributed this winter removal to denitrification, although they did not measure it directly. During the winter period, the rapid increase in hydraulic gradient and subsurface nitrate fluxes from hillslope to the flood plain could reduce significantly the removal capacity of the buried channel sediment. Yet, most of the time the N removal was above 80%. Indeed, these rapid decreases in removal capacity did not last very long. They hypothesized that the denitrifying community in the flood plain soils was very resilient and could rapidly adjust to nitrate increase as long as there was enough easily available organic carbon.

This paper offers three important take-home messages with crucial management implications:

(i) The hydrology of the riparian zone is generally very complex at the small scale. Therefore the correct evaluation of the nitrate buffering capacity of a riparian zone requires a careful hydraulic survey. Sole measurement of nitrate concentration decrease can often be misleading and overestimates the real capacity of the riparian zone.
(ii) The removal capacity of the riparian zone is mostly located at the hillslope–flood plain interface, a location where a trade-off exists between nitrate input, subsurface water residence time and anaerobic conditions.
(iii) The flood plain nitrate buffering capacity is independent of vegetation cover, as long as it is perennial and there is a plentiful supply of easily available organic carbon and anaerobic conditions.

GP

References

Clément, J. C., Aquilina, L., Bour, O., Plaine, K., Burt, T. P. & Pinay, G. (2003) Hydrological flowpaths and nitrate removal rates within a riparian floodplain along a fourth-order stream in Brittany (France). *Hydrol. Processes* **17**, 1177–1195.

Gurwick, N. P., McCorkle, D. M., Groffman, P. M., Gold, A. J., Kellogg, D. Q. & Seitz-Rundlett, P. (2008) Mineralization of ancient carbon in the subsurface of riparian forests. *J. Geophys. Res.* 113 (G2): Art. no. G02021.

Sabater, S., Butturini, A., Clément, J. C., Burt, T., Dowrick, D., Hefting, M., Maître, V., Pinay, G., Postolache, C., Rzepecki, M. & Sabater, F. (2003) Nitrogen removal by riparian buffers under various N loads along a European climatic gradient: patterns and factors of variation. *Ecosystems* **6**, 20–30.

HYDROLOGICAL PROCESSES, VOL. 7, 287–295 (1993)

ROLE OF FLOODPLAIN SEDIMENTS IN REDUCING THE NITRATE CONCENTRATION OF SUBSURFACE RUN-OFF: A CASE STUDY IN THE COTSWOLDS, UK

N. E. HAYCOCK

Department of Agricultural Water Management, Silsoe College, Cranfield Institute of Technology, Silsoe, Bedfordshire MK45 4DT, UK

AND

T. P. BURT

School of Geography, University of Oxford, Mansfield Road, Oxford OX1 3TB, UK

ABSTRACT

Discharge of groundwater from a limestone aquifer through floodplain sediments is associated with a large decrease in the nitrate concentration of the water. Results are presented to show that only a small amount of this reduction is caused by dilution of groundwater by water already present within the floodplain sediments; most of the effect is an active reduction process, most probably biological denitrification. The nitrate reduction process appears to operate independently of surface vegetation type and tends to be focused in specific regions of the floodplain where sediments are anaerobic and carbon-rich. The results suggest that active denitrification can operate throughout the winter, when nitrate concentrations in groundwater are at their highest and that the process remains effective even during periods of maximum run-off. The results show that undrained floodplains can be used as buffer zones to protect surface waters from groundwater polluted with agriculturally derived nitrate.

KEY WORDS Nitrate Floodplain Denitrification Buffer zones

INTRODUCTION

Within any catchment, nutrients are lost from one area only to be absorbed at another place. This applies not only for exchanges from one ecosystem to another—an exchange that is usually mediated by an ecotone (Risser, 1990)—but also within ecosystems (e.g. the river continuum concept; Vannote *et al.*, 1980). The management of nitrate within a catchment depends on managing catchment absorption processes and monitoring that they are working effectively. Inadequate absorption of nitrate by agricultural crops often leads to the loss of nitrate to subsurface water. The contamination of subsurface water represents a threat to groundwater and surface water quality. Assuming that the loss of nitrate from agricultural ecosystems is inevitable and assuming also that river water quality has to be preserved, it appears to be necessary to absorb nitrate in a zone between the agricultural and the surface water ecosystem—this area has been called a 'buffer zone' or riparian ecotone (Waikato Valley Authority, 1972). This paper presents evidence to support the notion that such zones are effective nitrate absorption zones.

The underlying hypothesis is that floodplain sediments have the potential to support a large population of denitrifying bacteria if the floodplain receives nitrate-contaminated groundwater. This potential prevails primarily because of the preponderance of anaerobic micro-sites within floodplain sediments, which are an ideal environment for bacterial denitrification. It follows that floodplain sediments have the potential to interact with nitrate-contaminated water before that same water enters the stream system (Pinay and Décamps, 1988; Pinay *et al.*, 1990). The aim of this investigation was to monitor the nitrate concentration of

0885–6087/93/030287–09$09.50

Received 24 October 1991
Accepted 10 July 1992

groundwater as it enters and leaves a floodplain section, and to evaluate the floodplain's nitrate absorption capacity. A thorough understanding of the flow of water within and through the floodplain section will be shown to be a prerequisite to a meaningful understanding of the pattern of nitrate concentration within the floodplain water. The data presented focus on the winter months and the nitrate reduction potential of the floodplain within this period, a time when nitrate leaching is most active (Addiscott *et al.*, 1991). If the floodplain is to offer protection for the stream ecosystem, it needs to be actively reducing nitrate in the winter months, and able to respond to the changes in nitrate loading stress in 'flush' periods (periods of high nitrate leaching; Foster and Walling, 1978). This final point is emphasized as previous investigations have suggested that biomass uptake is the predominate nitrate absorption process, and that denitrification is a minor process by comparison (Ambus and Lowrance, 1991; Peterjohn and Correll, 1984). Previous investigations have focused on the summer months, with little or no attention being paid to the dynamics of nitrate fluxes through floodplains in the winter months. If investigations have taken place in the winter, then the temporal resolution of field observations has been such that only a coarse comparison of seasons has been achievable (Jacobs and Gilliam, 1984; Lowrance and Leonard, 1988). In some instances the number of samples taken from the field has been so small (e.g. Ambus and Lowrance, 1991), that some of the conclusions drawn must be considered with great caution.

SITE DESCRIPTION AND METHODS

Floodplain sections were monitored on the River Leach, a headwater tributary of the Thames (OS SP 134132). Slopes above the floodplain are underlain by Great Oolite limestone (Figure 1); soils are of freely drained Sherborne Series with intensive arable farming (average N-fertiliser load was 120 kg ha^{-1} a^{-1}; average leaching loss was equivalent to 40 kg ha^{-1} a^{-1}; Haycock, 1991). The floodplain lies on impermeable Fullers' earth clay, with soils characteristic of the Evesham series (Avery, 1990).

In the grass vegetated floodplain site, a grid (24 × 16 m) of boreholes within the floodplain and adjoining slope allowed the elevation of the water-table to be measured and water samples from the wells to be collected. The grassland site consisted of a grid of 24 boreholes (1–4·5 m deep, depending on position), six holes up-slope by four holes up-valley (Figure 1). The boreholes were hand-augured unlined shafts, 0·15 m in diameter. The boreholes were capped with heavy duty PVC to prevent surface run-off entering the hole, but, more importantly, to minimize the oxidation of the soil at depth within the borehole (which may have affected the nitrate content of the borehole water). The boreholes remained unlined to allow the soil water to be in equilibrium with the well water and to allow depth-integrated water sampling, whereby 100 ml of water was removed from each well. The removal of larger amounts of water would have lowered the water-table locally around the borehole and may have induced an artifical wetting–drying cycle which may have increased the nitrate content of the well water. All 24 boreholes were sampled on average every five days (the average residence time of water within the site over the winter period was approximately five days). At the same time an automatic water sampler (Rock and Taylor Ltd, 48 bottle automatic water sampler) took samples at 24 hour intervals from a borehole at the down-valley outlet of the floodplain section (down-valley borehole of row 4, Figure 1).

Several springs and the main river were also monitored for discharge and water quality (weekly hand sampling, plus 24 hour automatic sampling at the catchment outlet) in an attempt to provide a wider context for the results collected at the floodplain, and to quantify in-stream nitrate reduction processes (Haycock, 1991). Another floodplain was monitored within this investigation (poplar vegetated), but is not discussed in this paper (see Haycock and Pinay, submitted).

Water samples were preserved with a biocide (mercuric chloride; 1 ml of 40 mg l^{-1} $HgCl_2$ to 100 ml of sample) and stored in cool boxes for a period of seven to 14 days until analysis. Nitrogen analysis of the water sample was divided into nitrate (NO_3-N), nitrite (NO_2-N), total oxidized nitrogen (NO_3-N + NO_2-N = TON-N) and ammonium (NH_4-N) species. This was undertaken on a continuous flow Chemlab series 4 autoanalyser (Chemlab, 1985).

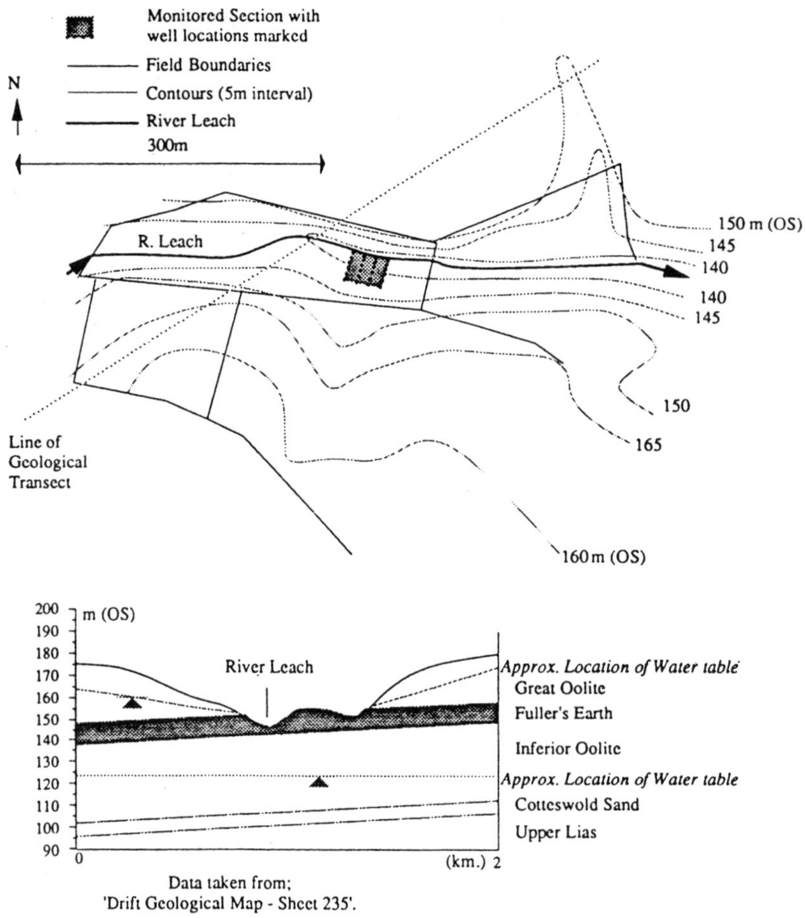

Figure 1. (a) Location and layout of the grass floodplain site relative to major topographical features, and (b) sketch of the geological section through the site with water-tables of the major aquifers marked

RESULTS

The data presented focus on the winter months of November 1989 to March 1990 (week 45 1989 to week 9 1990) and relate to the grass floodplain site.

Floodplain hydrology

The grass floodplain under investigation received water from three possible sources. The unconfined Great Oolite aquifer contributed groundwater along the hillslope edge of the floodplain section (Figure 1). The stream itself could contribute water to the floodplain by influent seepage from the channel; there was no evidence in the winter of flow from the stream into the floodplain within the section being monitored; however, this may have occurred at other points along the river. The last input to the floodplain section was along the upstream edge of the floodplain; there was ample evidence of floodplain groundwater moving down-valley, outflow from the floodplain section under investigation also draining down-valley rather than flowing directly into the stream. There was also the theoretical possibility of groundwater moving vertically up into the floodplain section through the Fuller's earth clays from the underlying Inferior Oolite aquifer. However, the water-table of the unconfined Inferior Oolite aquifer was about 20 m below the stream at the grass floodplain site and it was concluded that as there was 5 m of unsaturated Inferior Oolite above the water-table, there was no connectivity between the floodplain groundwater and the unconfined Inferior Oolite aquifer beneath (Figure 1).

The elevation of the water-table across the floodplain section varied considerably through the study period (Figure 2a). The lack of precipitation during the summer of 1989 in the Cotswolds led to a situation whereby water from the stream was flowing into the floodplain and then into the hillslope to become groundwater (flow from row 1 to row 6, Figure 2a). This situation was not reversed until mid-November when percolation into the Great Oolite raised the water-table to cause flow from the hillslope to the floodplain, a situation which persisted beyond March 1990.

For a transect across the floodplain, the mean water-table elevation over the study period shows that groundwater drains from the hillslope aquifer into the edge of the floodplain but does not flow across the floodplain to the stream directly (Figure 2a). Field evidence suggests that a permeable sediment exists at a depth of 1 m within the floodplain at a point corresponding to row 4 (Figure 2a). These sediments allow preferential drainage of the groundwater down-valley. These permeable sediments (K_{sat} approximately 2-3 m d^{-1}) also appear to control the direction of floodplain drainage between row 4 and row 2. The significance of the permeable sediment as a focus for drainage was emphasized by the level of the water-table within the floodplain which was lowest above the permeable sediments, especially at times of high flow. Field tracing experiments over the winter of 1990–1 showed the connectivity of the hillslope aquifer to these permeable sediments; the grid of boreholes allowed the tracer to be monitored as it flowed down-valley within the permeable sediment structure. These sediments appear to be associated with a buried channel, buried between 1820 and 1860 as part of a channelization scheme of the River Leach (Sherborne Estate Maps, 1820–60). The presence of buried sediments of varying permeability is a common feature of floodplain zones, and such flow complications are unavoidable (Schaffer, 1973).

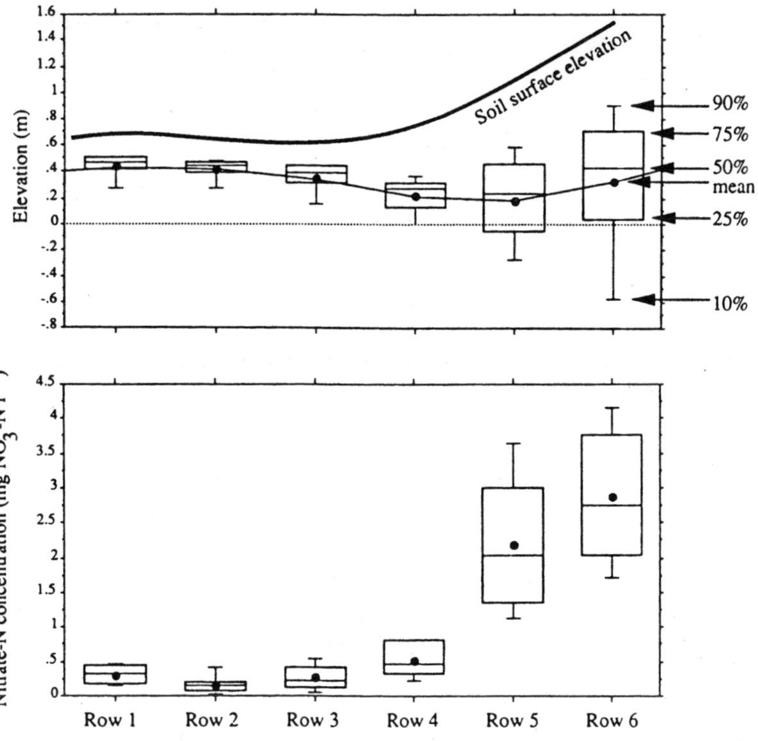

Figure 2. (a) Transect of water-table elevation (above an arbitrary datum) across a grass floodplain. The box-whisker plots depict the mean and 10, 25, 50, 75 and 90th percentiles of data from November 1989 to March 1990. (b) Transect of groundwater nitrate concentration through a grass floodplain. The box-whisker plots depict the mean and 10, 25, 50, 75 and 90th percentiles of data from November 1989 to March 1990

From the observations of water-table elevation within the floodplain, it was apparent that the buried channel played a major role in controlling the hydrology of the floodplain and acted as a drain to it. The buried channel receives hillslope groundwater, transporting it down-valley parallel to the present stream. It also controls the water elevation of rows 2 and 3, which usually drained preferentially into the buried channel rather than to the present stream. The high elevation of the water-table at row 2 represents the drainage divide between parts of the floodplain draining directly to the stream and those areas which drain down-valley via the buried channel. Only row 1 consistently drains directly towards the stream in the winter. Individual profiles of water-table elevation support this interpretation (Haycock, 1991).

Nitrate concentration of the floodplain groundwater

The input of nitrate to the floodplain section is principally controlled by subsurface water flow. Therefore nitrate could be transported into the floodplain from the hillslope, from up-valley or from the stream. Another local source of nitrate could be from the leaching of nitrate from the floodplain soils themselves.

From the data collected on the nitrate concentration of floodplain water, it appeared that hillslope inputs of nitrate were the most significant source. Indeed, the nitrate concentrations across the floodplain section showed a distinctive and consistent pattern throughout the winter; high winter mean levels of 2·0 and 2·3 mg NO_3-N l^{-1} in hillslope rows 5 and 6, respectively, and significantly lower concentrations in the remaining floodplain section (Figure 2b). Nitrate concentrations in rows 5 and 6 varied considerably (Figure 2b). The reverse flow of water from the stream to the hillslope in the first few weeks of November 1989 accounts for some of this variability (Figure 3b). Most of the fluctuation was caused by the episodic nature of nitrate leaching, a rising hillslope water-table coinciding with an increase in nitrate concentration and a falling water-table with a decrease in nitrate concentration. From the transect data it was evident that between rows 6 and 4 the mean winter concentration of nitrate in groundwater was lowered by 82% [mean concentration of nitrate at row 6 was 3·16 mg NO_3-N l^{-1} (SD 1·23) whereas the concentration over the same period at row 4 was 0·57 mg NO_3-N l^{-1}(SD 0·52)]. An explanation of the change in nitrate concentration in groundwater as it passed through the site is as follows.

Hillslope groundwater entered the floodplain section and flowed into the buried channel sediments, from where it drained in a down-valley direction. Groundwater flow within the floodplain was not perpendicular to the stream as the dominant topographical gradient within the floodplain is down-valley, as discussed by Anderson and Kneale (1982). The grid structure of boreholes within the floodplain allowed this down-valley water flow to be monitored; from this it was observed that the concentration of nitrate in hillslope groundwater seemed to decrease dramatically as the water flowed into the floodplain. As the hillslope water was clearly entering the floodplain (the connectivity of the flow was determined by the tracer experiment), the next question was to determine the dilution potential of the floodplain groundwater on the hillslope groundwater. The mean flow-rate into the floodplain section from up-valley (through the floodplain soils) was determined by saturated flow calculations (Anderson and Burt, 1978) and field determination of K_{sat} to be 22 l h^{-1} ($6{\cdot}02 \times 10^{-6}$ m^3 s^{-1}) for February 1990, whereas the mean flow from the hillslope into the floodplain was determined to be 1000 l h^{-1} ($2{\cdot}80 \times 10^{-4}$ m^3 s^{-1}) for the same period. No stream water flowed directly into the floodplain within the study section in February 1990; even if it had, its nitrate concentration would have had a mean value of 17·3 mg NO_3-N l^{-1} and would have offered no dilution potential Using chloride as a conservative tracer, it was concluded that the dilution potential of the floodplain groundwater derived from up-valley on hillslope groundwater increased to a maximum of 20% in storm periods, and had a mean value of 12%. Therefore, on average an additional 100 l h^{-1} must have been flowing through the permeable buried channel from up-valley to attain the dilution potential of approximately 12% of hillslope flow based on February 1990 figures. Although dilution accounts for an element of the observed change in groundwater nitrate concentration as it passes through the floodplain, it by no means accounts for most of the effect.

The conclusion drawn was that the nitrate reduction observed was due to dilution (12–20%) and biological processes (70–60%). The experiment was monitoring the same body of water as it moved through the site. The loss of nitrate from groundwater could have been due to uptake by vegetation, but considering the time of year this was considered to be unlikely. A further possibility was that the nitrate loss was due to it being

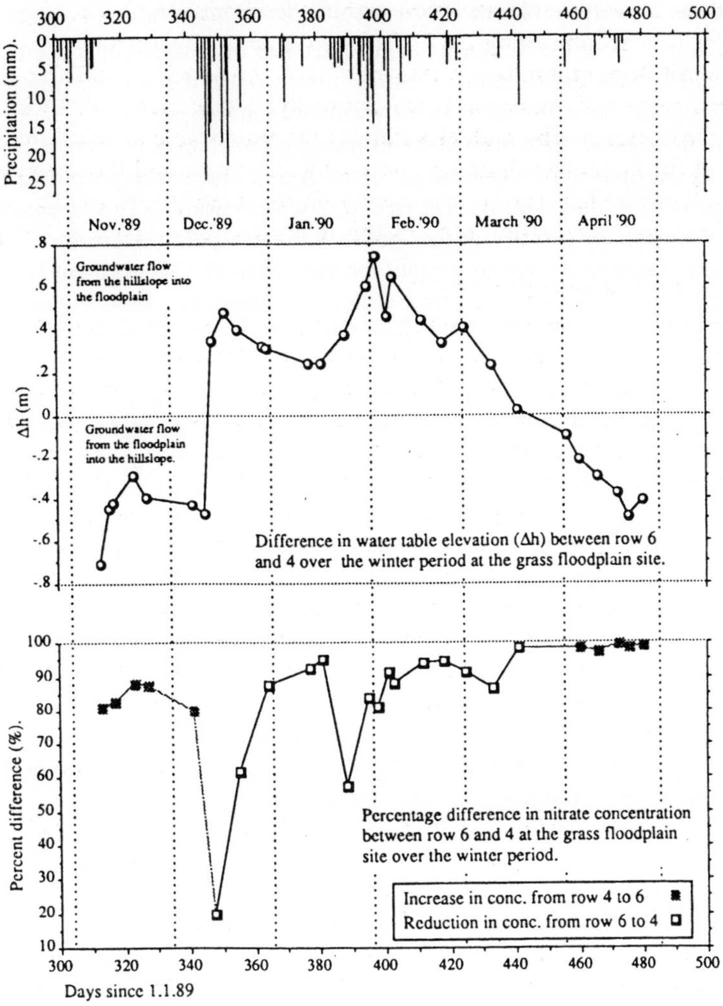

Figure 3. (a) Daily precipitation (Radcliffe Meteorological Station, Oxford, (b) water-table elevation difference and (c) percentage difference in nitrate concentration between rows 6 and 4 at the grass floodplain site over the winter period (November 1989 to April 1990 inclusive). Positive percentage values (c) imply loss of nitrate between rows 6 and 4

converted into ammonium (NH_4-N) or nitrite (NO_2-N), but there was no evidence of large increases in either of these. It was therefore concluded that the nitrate loss must be due either to its conversion into organic-N (i.e. uptake by soil microbial biomass), or to denitrification, or both.

Irrespective of the relative merits of either process, nitrate was clearly being absorbed within the floodplain. Over the monitoring period (18 months), no significant nitrate flush from the floodplain section was observed. Therefore it could be concluded that the most probable mechanism of nitrate removal was denitrification. If this conclusion is correct, then denitrification occurred within a very short distance after the hillslope groundwater flowed into the floodplain, and probably occurred within the buried channel sediments (carbon content 0·07–0·09 g g^{-1}). The highest monthly mean rate of loss of nitrate from groundwater within the floodplain section (February 1990), if assumed to take place within the plan area of the buried channel, was equivalent to 740 mg NO_3-N m^{-2} day^{-1}, which represented a 97% decrease in the nitrate load of the groundwater.

Effect of nitrate flushes on the nitrate absorption rate of the floodplain zone

During storms and shortly after, the discharge of groundwater from the aquifer increased as did its nitrate concentration. The ability of the floodplain to absorb nitrate when the concentration as well as the total loading increased was a point of particular interest within this investigation.

The efficiency of the floodplain was assessed by examining the change in nitrate concentration between rows 6 and 4 (Figure 2b), expressed as a percentage term. A decrease in the percentage difference of nitrate concentration between rows 6 and 4 only occurred on two occasions. This decrease in 'absorption' is characterized in Figure 3 when the percentage loss between the two points decreased to below 80%. On both of these occasions, this appears to be due to a rapid increase in hydraulic gradient between the two points, causing a decrease in the groundwater contact time with the sediment at row 4 (Figure 3). From this transect based data it appears that the grass floodplain under investigation did not maintain absorption rates at the mean level (82%) when the hydraulic gradient and nitrate concentrations increased. On closer investigation there was still a 60% (day 390; Figure 3) reduction in nitrate concentration, although 20% of this could be attributed to dilution. Even so, the nitrate absorption rate within the floodplain during storm periods was not significantly affected. Indeed, the concentration of groundwater flowing down-valley from row 4 only increased from 0·2 mg NO_3-N l^{-1} to 0·9 mg NO_3-N l^{-1} during the rising limb of the hillslope hydrograph, but soon recovered to a mean absorption rate of 70–80% at peak flow (Figure 4). The recovery of the absorption capacity of the floodplain site may be a function of the denitrifying bacterial population dynamics in the sediment and the fact that a change in water concentration and discharge will stimulate an increase in bacterial numbers to absorb incoming nitrate.

The low concentrations of nitrate in the floodplain (rows 1, 2 and 3; Figure 2a) can be generally explained by the fact that these rows are not hydraulically connected to the hillslope groundwater source. Nitrate that is present could be from the leaching of nitrate from the floodplain soil in the immediate vicinity of the borehole. Any leached nitrate from the soil would probably be denitrified rapidly. The nitrate within this zone was not thought to be derived from the stream as the mean concentration of the stream over the winter period was 14·7 mg NO_3-N l^{-1} and also water flow from the stream into the floodplain probably did not occur in this section of floodplain.

DISCUSSION AND CONCLUSIONS

The anaerobic, carbon-rich sediments which are characteristic of floodplains have the potential to support large populations of denitrifying bacteria. The grassed floodplain section discussed here has these characteristics and the observations made suggest that the winter loss of groundwater nitrate is a

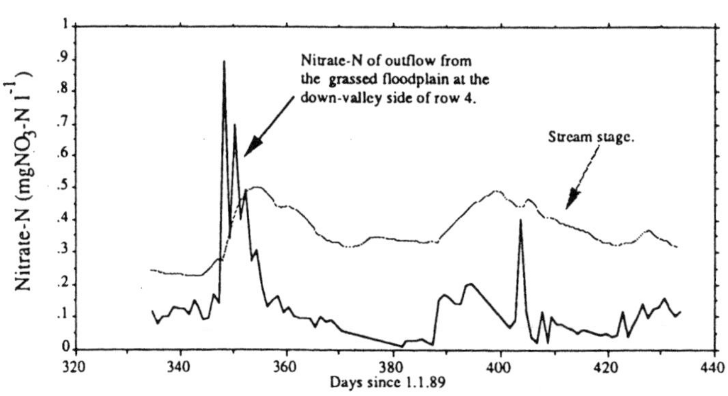

Figure 4. The nitrate concentration of subsurface outflow from the grass floodplain site over the winter period December 1989 to March 1990

microbiological process. The mechanism responsible for the loss of groundwater nitrate has proved to be dynamic, and it rapidly responded to increases in hydraulic loading stress. The absorption processes have also worked in such a way that ammonium has not accumulated in the groundwater leaving the floodplain section.

Work in a poplar plantation on a nearby section of floodplain (Haycock, 1991) supports the conclusion that the buried channel within the floodplain is the significant nitrate reduction zone, and not the floodplain soils in general. Comparison of two sites (Haycock and Pinay, submitted) suggests that nitrate transformation within floodplains is a sediment-related process and suggests that the effect may be partially dependent on the surface vegetation type, which is envisaged as providing dissolved organic carbon to sustain the denitrifying bacteria (Haycock and Pinay, submitted). This finding in turn may imply that the floodplain as a whole is not equally active in absorbing nitrate (Cooper, 1990; Pinay and Décamps, 1988), a conclusion at odds with other findings (Jacobs and Gilliam, 1985; Lowrance and Leonard, 1988; Schlosser and Karr, 1981), but that distinct sediment structures within it are responsible (Haycock, 1991).

Alternatively, what we have observed could be explained as follows: denitrifying bacteria operate best at the junction of anaerobic–aerobic zones, as seen for denitrification at the sediment–water interface within streams and lakes (Keeney, 1973; Van Kessel, 1977), where both carbon and nitrate are in abundance. It may be that the rapid loss of nitrate within the grassed site only coincides with the buried channel, as this is the zone where nitrate-rich water meets anaerobic carbon-rich sediments. Denitrification across the broader floodplain is excluded due to the lack of nitrate within the groundwater, which has been consumed at the junction of the floodplain and the hillslope. However, evidence from the poplar floodplain (Haycock, 1991) showed that when nitrate-rich water from the stream percolated into the floodplain, nitrate reduction only took place when the water passed through the buried sediments, suggesting that the type of sediment microsites, the degree of saturation and the presence of dissolved organic carbon and nitrate are the key factors facilitating the optimization of denitrification in floodplain environments.

If our interpretation is correct, then either the construction or management, or both, of suitable sedimentary zones within the floodplain sediments may increase the nitrate absorption capacity of these zones (Haycock *et al.*, submitted). The significance of carbon within the floodplain system is fundamental and, with that in mind, it can be suggested that floodplain vegetation is not directly involved in nitrate absorption but supplies carbon-rich material, both in solid and in dissolved forms, to the deeper denitrifying environment. Therefore, the type of vegetation present may at least have an indirect influence of the sustainability of the sedimentary bacterial processes (Haycock *et al.*, submitted). The denitrification processes within floodplain zones appears to be sediment dependent, not vegetation dependent. We conclude that the width of a 'nitrate buffer zone' need not be exactly defined, provided that the zone contains sediment of a suitable nature to encourage denitrification to proceed. Buried, permeable sediments are ubiquitous to most floodplain environments, and therefore the capacity to reduce diffuse agricultural pollution is implicit. What reduces this capacity is possibly the lack of a stable ground cover and drainage of this landscape, which assures that nitrate-rich water does not interact within floodplain sediments (Haycock *et al.*, submitted).

The nitrate concentrations of the groundwater reported here are low by UK standards, although the stream concentrations were high (Betton *et al.*, 1991). It is worth commenting, however, that these groundwater concentrations are mean values for the winter period; the maximum recorded nitrate concentration of water that entered the grass floodplain section over the 1989–90 winter was in excess of 15·0 mg NO_3-N l^{-1}. Both of these high nitrate surges in January 1990 corresponded to down-valley nitrate concentrations of less than 1·2 mg NO_3-N l^{-1}, implying strongly that the floodplain sediments are able to buffer even major leaching episodes.

The floodplain sections monitored were not underdrained. This is of considerable importance when it is considered that most of the floodplains in the UK are underdrained (Green, 1975; Newson, 1987). The results of this research suggests, as noted above, that if floodplains were not underdrained they may well have the capacity to reduce significantly the nitrate concentration of water draining nearby hillslope and aquifers. Our results suggest that if hillslope water is allowed to interact with floodplain sediments then denitrification could considerably reduce the nitrate content of the subsurface water, and so prove an effective water protection process. The implications of this conclusion, if supported by further studies, could be important for hydrologists and land managers alike. Reactivating the biological role of floodplains could be a potential remedy for one of the most pressing problems of surface water contamination in the UK and elsewhere.

ACKNOWLEDGEMENTS

Research funded under an NERC studentship (GT4/AAPS/64), which was a CASE award with the WRc, Medmenham. Additional support was provided by Keble College, Oxford, John Martin's Charity, Evesham and Rock and Taylor Instruments, Birmingham.

REFERENCES

Addiscott, T. M., Whitmore, A. P., and Powlson, D. S. 1991. *Farming, Fertilizers and the Nitrate Problem.* CAB International, Wallingford.

Ambus, P. and Lowrance, R. 1991. 'Comparison of denitrification in two riparian soils', *Soil Sci. Soc. Am. J.*, **55**, 994-997.

Anderson, M. G. and Burt, T. P. 1978. 'Towards more detailed field monitoring of variable source areas', *Wat. Resour. Res.*, **14**, 1123-1131.

Anderson, M. G. and Kneale, P. E. 1982. 'The influence of low angled topography on hillslope soil water convergence and stream discharge', *J. Hydrol.*, **571**, 65-80.

Avery, B. W. 1990. *Soils of the British Isles.* CAB International, Wallingford.

Betton, C., Webb, B. W., and Walling, D. E. 1991. 'Recent trends in NO_3-N concentration and loads in British rivers', *IAHS Pub. No. 203*, pp. 169-180.

Chemlab Laboratory Instrument 1985. *System 4 Sampler, User manual.* Chemlab Instruments, Hornchurch.

Cooper, A. B. 1990. 'Nitrate depletion in the riparian zone and stream channel of a small headwater catchment', *Hydrobiologia*, **202**, 13-26.

Foster, I. D. L. and Walling, D. E. 1978. 'The effects of the 1976 drought and autumn rainfall on stream solute levels', *Earth Surf. Process.*, **3**, 393-406.

Green, F. H. W. 1975. 'The effect of climatic and other environmental changes in water quality in rural areas' in Hey, R. D. and Davies, T. D. (Eds), *Science, Technology and Environmental Management.* Saxon House Lexington, pp. 123-136.

Haycock, N. E. 1991. 'Riparian land as buffer zones in agricultural catchments', *Unpublished DPhil Thesis*, University of Oxford.

Haycock, N. E. and Pinay, G., 'Nitrate retention in grass and poplar vegetated riparian buffer strips during the winter', *J. Environ. Qual.*, submitted.

Haycock, N. E., Pinay, G., and Walker, C. 'Nitrogen retention in river corridors: European perspective, *Ambio*, submitted.

Jacobs, T. C. and Gilliam, J. W. 1985. 'Riparian losses of nitrate from agricultural drainage waters', *J. Environ. Qual.*, **14**, 472-478.

Keeney, D. R. 1973. 'The nitrogen cycle in sediment-water systems', *J. Environ. Qual.*, **2**, 15-29.

Lowrance, R. R. and Leonard, R. A. 1988. 'Stream nutrient dynamics on coastal plain watershed', *J. Environ. Qual.*, **17**, 734-740.

Newson, M. D. 1987. 'Land and water: the "river look" on the face of geography', *Res. Pap. No. 52*, University of Newcastle upon Tyne, Department of Geography.

Peterjohn, W. T. and Correll, D. L. 1984. 'Nutrient dynamics in agricultural watershed: observations on the role of a riparian forest', *Ecology*, **65**, 1466-1475.

Pinay, G. and Décamps, H. 1988. 'The role of riparian woods in regulating nitrogen fluxes between the alluvial aquifer and surface waters: a conceptual model', *Regul. Riv.*, **2**, 507-516.

Pinay, G., Décamps, H., Chauvet, E., and Fustec, E. 1990. 'Functions of ecotones in fluvial systems' in Naiman, R. J. and Décamps, H. (Eds), *The Ecology and Management of Aquatic–Terrestrial Ecotones.* Parthenon Press, pp. 141-169.

Risser, P. G. 1990. 'The ecological importance of land-water ecotones' in Naiman, R. J. and Décamps, H. (Eds), *The Ecology and Management of Aquatic–Terrestrial Ecotones.* Parthenon Press, pp. 7-21.

Schäffer, W. 1973. 'Altrhein verbund am nördlichen Oberrhein', *Courier Forschungsinstitut Senckenberg*, **7**, 1-63.

Schlosser, I. and Karr, J. 1981. 'Water quality in agricultural watersheds, impact of riparian vegetation during baseflow', *Wat. Resour. Bull.*, **17**, 233-240.

Waikato Valley Authority 1973. *Lake Taupo Catchment Control Scheme, Appendix VI: the History, Principles and Status of the Lakeshore Reserves Proposals—the Taupo County Report.* Waikato Valley Authority, Hamilton.

Van Kessel, J. F. 1977. 'Factors affecting the denitrification rates in two water sediment systems', *Wat. Resour.*, **11**, 259-267.

Vannote, R. L., Minshall, G. W., Cummins, K. W., Sedall, J. R., and Cushing, C. E. 1980. 'The river continuum concept', *Can. J. Fish. Aq. Sci.*, **37**, 130-137.

F In-stream Processes

Commentary on Paper F1

Webster, J. R. & Patten, B. C. (1979) Effects of watershed perturbation on stream potassium and calcium dynamics.

The dynamics of stream solutes in the river has relevance beyond pure biogeochemical process. Solutes dynamics are a key component of the biological processes occurring within the river, but also in its associated compartments (the riparian and hyporheic zones). Solutes in river systems are affected by reactions and transformations, but also by the downstream transport of the water. This creates a remarkable difference between the nutrient cycling occurring in lakes and ponds (where the longitudinal transport is negligible, and the cycling is *in situ* and vertical) and that occurring in flowing water.

The paper by Webster and Patten was the first to define the conceptual basis for element spiralling as an expression of element dynamics in rivers. The concept was also applied to the processing of organic matter (Wallace *et al.*, 1977). Webster and Patten suggested that, if the stream was considered as a single site, cycling cannot occur since it would be prevented by unidirectional water movement. Only if the stream reach is considered as the appropriate unit for the definition of the dynamics, does a significant recycling become apparent. This recycling would therefore occur in a "rolling motion" manner, which if applied spatially, would show the shape of a spiral. Essentially, this concept stated that all elements (including nutrients) are transported, removed or returned from or to the water by abiotic or biotic processes. This movement, which integrates with the longitudinal water transport, configures a spiral. The whole stream nutrient dynamics becomes a balance between retention and transport.

The paper mostly tackled the dynamics of calcium and potassium in small headwater streams. In order to estimate the spatial dynamics of these elements, they used radioactive-labelled material as traceable surrogates of these two cations. With this approach, the dynamics of the two elements in the different compartments where they cycled (litter fall, Coarse Particulate Organic Matter, Fine Particulate Organic Matter, ingestion and egestion from invertebrates) were used as the key elements to determine their assimilation and recycling.

After this concept was formulated, the spiralling length was defined (Newbold *et al.* 1981; Elwood *et al.*, 1981) as an operational unit potentially characteristic of the nutrient dynamics in the stream reach. The spiralling length was therefore the average distance that an atom travels during its complete cycle between the water and the streambed. It has two main components: the uptake length and the turnover length. The spiralling length has been used since to describe the utilization efficiency of the elements in the water, in particular related to the nutrients. The form of the spiral depends on the turnover length and on the velocity of the water in transport. The formal outline of this concept was harmonized by means of a workshop report (Stream Solute Workshop, 1990). Its applications were later expanded to organic carbon (Newbold *et al.*, 1982), the different forms of nitrogen (Valett *et al.*, 1996, Paper F7) and phosphorus (Newbold *et al.*, 1983).

Finally, it is worthy of note that this paper was another product of research at the Coweeta Hydrologic Laboratory in North Carolina (Swank & Crossley, 1987), where Jack Webster did the fieldwork for his PhD at the University of Georgia: See also Hursh & Brater (1941, Paper B2) and Dunford & Fletcher (1947, Paper B3), both in this volume.

SS

References

Dunford, E. G. & Fletcher, P. W. (1947) Effect of removal of stream-bank vegetation upon water yield. *Trans. Am. Geophys. U.* **28**, 105–110.

Elwood, J. W., Newbold, J. D., Trimble, A. F. & Stark, R. W. (1981) The limiting role of phosphorus in a woodland stream ecosystem: effects of P enrichment on leaf decomposition and primary producers. *Ecology* **62**, 146–158.

Hursh C. R. & Brater E. F. (1941) Separating storm-hydrographs from small drainage-areas into surface- and subsurface-flow. *Trans. Am. Geophys. U.* **22**, 863–871.

Newbold, J. D., Elwood, J. W., O'Neill, R. V. & van Winkle, W. (1981) Measuring nutrient spiralling in streams. *Canadian J. Fisheries & Aquatic Sciences* **38**, 860–863.

Newbold, J. D., Mulholland, P. J., Elwood, J. W. & O'Neill, R. V. (1982) Organic carbon spiralling in stream ecosystems. *Oikos* **38**, 266–272.

Newbold, J. D., Elwood, J. W., O'Neill, R. V. & Sheldon, A. L. (1983) Phosphorus dynamics in a woodland stream ecosystem: a study of nutrient spiralling. *Ecology* **64**, 1249–1265.

Stream Solute Workshop (1990) Concepts and methods for assessing solute dynamics in stream ecosystems. *J. North Am. Benthological Soc.* **9**, 95–119.

Swank, W. T. & Crosslet, D. A. (1987) *Forest Hydrology and Ecology at Coweeta.* Ecological Studies volume 66. Springer-Verlag,

Valett, H. M., Morrice, J. A., Dahm, C. N. & Campana, M. E. (1996) Parent lithology, surface-groundwater exchange, and nitrate retention in headwater streams. *Limnology & Oceanography* **41**, 333–345.

Wallace, J. B., Webster, J. W. & Woodall, W. R. J. (1977) The role of filter feeders in stream ecosystems. *Archiv fur Hydrobiologie* **79**, 506–532.

Ecological Monographs, 1979, pp. 51–72

EFFECTS OF WATERSHED PERTURBATION ON STREAM POTASSIUM AND CALCIUM DYNAMICS[1,2]

JACKSON R. WEBSTER[3] AND BERNARD C. PATTEN
Department of Zoology and Institute of Ecology, University of Georgia, Athens, Georgia 30602 USA

Abstract. Three small streams located at Coweeta Hydrologic Laboratory, North Carolina, USA, on an old field watershed, a pine plantation watershed, and a hardwood forest watershed were investigated to determine effects of watershed perturbation on K and Ca dynamics in the stream ecosystems. Data collected included measurements of litterfall inputs, large particulate organic matter and benthic organism standing crops, large particulate organic matter and organism drift, and insect emergence. We used ^{85}Sr and ^{134}Cs to estimate detritivore ingestion and elimination rates of Ca and K, respectively.

We found that watershed perturbations had altered stream inputs and caused accompanying changes in the stream fauna. Our results indicated that the perturbed streams had less efficient physical processing of allochthonous inputs, but greater biological utilization of inputs. The streams exhibited high resilience to perturbation with complete recovery limited by the recovery rate of allochthonous inputs.

Key words: calcium; detritus; North Carolina; potassium; stream; stability.

INTRODUCTION

The defining feature of streams, unidirectional flow of water, has presented a challenge in formulating a theoretical perspective for stream studies (Margalef 1960, Hutchinson 1963). Some recent stream studies have, however, focused on functional aspects of running water ecosystems (Teal 1957; Nelson and Scott 1962; Minshall 1967; Tilly 1968; Hall 1972; Fisher and Likens 1972, 1973; Boling et al. 1974; Fisher 1977). Out of these studies has emerged a consistent picture of headwater streams (1st to 3rd order, Strahler 1957) as processors of organic matter synthesized in riparian ecosystems (Cummins 1974). Lindeman's (1942) study led to a traditional view of ecosystem energy flow encompassing primary producers, herbivores, carnivores, and decomposers. This view is evident in many early stream studies but the accumulation of evidence now indicates that the primary energy source to many headwater streams is leaves from the surrounding terrestrial environment (e.g., Cummins 1974, Hynes 1975). Considerable stream research over the past 15 or 20 yr has concerned processing and utilization of this allochthonous energy.

Control of the rate at which organic energy is incorporated into ecosystems is different in headwater streams than in most other ecosystems. In autotrophic ecosystems where organic energy is internally generated, production is often controlled by rates of internal nutrient regeneration. Such ecosystems have been referred to as "circular causal systems" (Hutchinson 1948, Patten 1973). In a typically heterotrophic, headwater stream, no such feedback control over organic energy input exists. There is no feedback control over litterfall inputs, and stream ecosystems do not have the same homeostatic capabilities possessed by circular causal systems. Control of biological function in headwater streams is external to the stream ecosystem, in the surrounding watershed. The influence of the terrestrial ecosystem on a stream can be seen in a variety of other ways including plant growth nutrient concentrations, the hydrologic regime, available light energy, and water temperature. Hynes (1975) concluded, "in every respect the valley rules the stream."

Considering a stream at a single point, it is evident that material cycling is negligible. Cycling is prevented by unidirectional water movement. Turbulence and upstream movements of adult insects (e.g., Muller 1954, Roos 1957), benthic invertebrates (e.g., Ball et al. 1963, Minckley 1964, Elliott 1971), and fish (Hall 1972) contribute little to upstream material movement. Biotic and abiotic materials must be continually resupplied from upstream or from the watershed. Food received by an organism from upstream is utilized and passed on with very low probability it will again pass that point in the stream in contemporary time.

If, however, one considers the dynamics within a stream reach, significant recycling becomes apparent. Much of the energy ingested by detritivores is egested and again becomes part of the detritus pool. This detritus is available to other detritivores further downstream. Thus, the downstream movement of organic matter occurs in a "rolling motion" as has been described for nutrients on a larger scale (Leopold 1941). Because of the spatial aspect involved in this cycling,

[1] Manuscript received 8 July 1976; accepted 6 October 1978.
[2] University of Georgia, *Contributions in Systems Ecology*, Number 42.
[3] Current address: Department of Biology, Virginia Polytechnic Institute and State University, Blacksburg, Virginia 24061 USA.

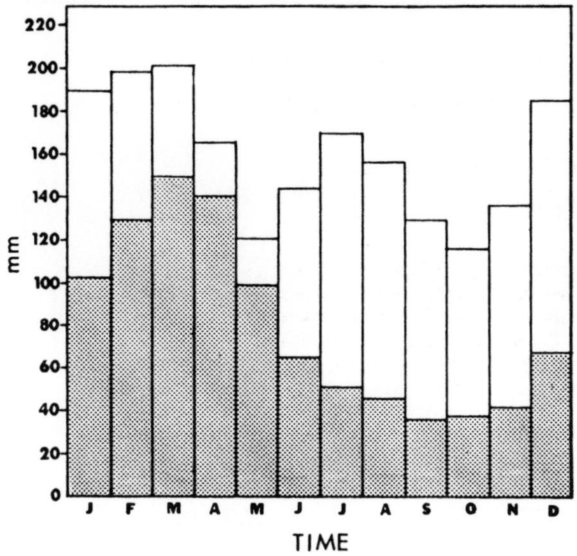

FIG. 1. Average precipitation (clear area) and discharge (shaded area) for the hardwood forest watershed (WS 18), based on 38 yr of precipitation data and 33 yr of stream flow records. Data from Coweeta Hydrologic Laboratory, North Carolina, USA, and Johnson and Swank (1973).

it might better be termed "spiralling" (Webster 1975, Wallace et al. 1977). Internal regulation of the stream ecosystem can be mediated by characteristics of the stream which shorten or tighten spiralling such as stream bed retention of particulate organic matter and insect filtering of fine particulate organic matter (Wallace et al. 1977).

Ecosystem stability is the ability of an ecosystem to (1) withstand and (2) recover from perturbation. These 2 properties have been designated as resistance and resilience, respectively (Webster et al. 1975). Streams are usually regarded as highly resilient ecosystems, a view based largely on the observation that if material such as a pollutant, is discharged into a stream, it is immediately carried away from the area of discharge (Cairns and Dickson 1977). Webster (1975) related ecosystem resilience to rapid turnover of ecosystem components. In streams, rapid turnover is produced by the continual flushing action of flowing water. Additionally, organisms found in headwater streams typically exhibit rapid turnover (Patrick 1970). The annual input of allochthonous detritus also contributes to stream resilience by replacing perturbation-removed biomass (Fisher and Likens 1973).

Webster et al. (1975) suggested that resistance and resilience are inversely related, that is, ecosystems with high resistance to perturbation typically exhibit low resilience and *vice versa*. In contrast to the high resilience exhibited by stream ecosystems, low resistance to perturbation is apparent. Margalef (1963) suggested that maintenance of ecosystem organization was dependent on biomass. This idea formed the basis for suggesting that the ability of an ecosystem to resist perturbation is related to standing crop (O'Neill et al. 1975, Webster 1975). Streams have a low standing crop biomass relative to other ecosystem types (cf., Whittaker 1975:224) though recent studies have shown that some headwater streams may have relatively high standing crops in the form of dead logs (J. Sedell, *personal communication*). Perturbations involving additions to or removals from streams, such as organic pollutant discharge or flood scouring, have dramatic influences on the streams. However, these effects are rapidly ameliorated by resiliency mechanisms.

The stability of stream ecosystems can be attributed to their resilience. In any specific stream, resilience is regulated by external and internal control mechanisms as described above. Following perturbation, biomass and nutrients are replaced by watershed inputs, and pollutants are removed by waterflow. Internal function is regained by retention mechanisms and rapid recolonization and regrowth by stream fauna.

Potassium and calcium dynamics in stream ecosystems.—Potassium is essential to all organisms except possibly some blue-green algae. It is the major cation of cytoplasm and is important in a number of cellular functions including enzyme activation and control of protein conformation (Bowen 1966, Phipps 1976). In animals, overall control of K levels is maintained by a balance of intake and elimination. Potassium is rapidly assimilated from the gastrointestinal tract (Davis 1963). In aquatic insects, regulation of K in the haemolymph is highly efficient, being primarily controlled by gut uptake and adsorption by the malpighian tubules and rectum (Shaw and Stobbart 1963). Potassium is relatively abundant and has never been suggested as a limiting nutrient for plant growth in natural waters (Hynes 1970), though in highly productive lakes and ponds, epilimnetic K concentrations may be reduced by algal uptake (Barrett 1957, Wetzel 1975).

Calcium is essential to all organisms (Bowen 1966). The many insoluble Ca compounds provide structure in both plants and animals. In addition to forming structural material of both endoskeleton and exoskeleton, Ca is important in a number of physiological functions (Clark 1958, Hsiao 1963, Wasserman and Taylor 1969). In contrast to K, the gastrointestinal epithelium is relatively impermeable to Ca and there is evidence of active transport of Ca based on *in vivo* and *in vitro* studies of mammalian intestinal tissue (Wasserman and Taylor 1969). In insects, excess Ca is excreted as a salt, most commonly suspended calcium carbonate granules (Clark 1958, Wigglesworth 1965). Calcium concentrations in lakes and streams are highly variable and dependent primarily on the substrate rock type. Early limnological studies of Wisconsin lakes demonstrated that Ca can affect plant and animal distributions and limit lake productivity (Hsiao 1963). Subsequently, the distribution of most freshwater organisms has been related to Ca concentrations (Wetzel 1975).

The objective of this study was to examine stream ecosystem response to perturbation of the watershed terrestrial vegetation. On 1 watershed, the hardwood vegetation was removed, replaced with grass, and subsequently allowed to undergo natural succession. On another watershed, the hardwood vegetation was replaced with white pine. We have compared the streams on these 2 watersheds with the stream on an unperturbed, hardwood forest watershed. These comparisons are based on the dynamics of K and Ca in the streams.

Description of Study Site

Coweeta Hydrologic Laboratory, North Carolina, is located in the Nantahala Mountains in the Blue Ridge Province of the Southern Appalachians. Average annual precipitation at Coweeta ranges from 170 cm at lower elevations to 250 cm on upper slopes. Precipitation is fairly evenly distributed through the year with an average of at least 75 mm falling each month (Fig. 1) and <2% of the total occurring as snow (Dils 1957). As a result, stream flow does not show great fluctuations typical of streams in other areas.

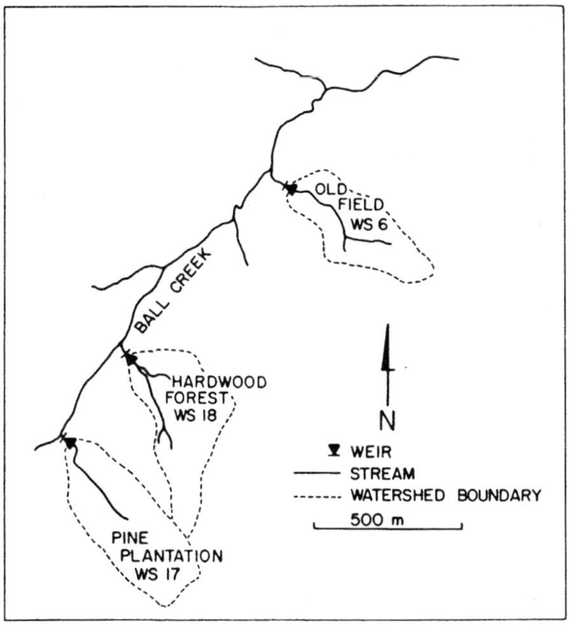

Fig. 2. Location of study streams, Coweeta Hydrologic Laboratory, Franklin, North Carolina, USA.

Table 1. Summary of physical characteristics of 3 watersheds and streams at Coweeta Hydrologic Laboratory, North Carolina, USA

	Watershed 6	Watershed 17	Watershed 18
Vegetation type	Old field	White pine	Hardwood
Watershed area* (ha)	8.86	13.48	12.46
Elevation, minimum* (m)	699	742	721
Elevation, maximum* (m)	905	1021	1006
Slope*	54	57	52
Total channel length (m)	450	210	564
Average channel width† (cm)	61.5	122.6	124.0
Average middle channel depth† (mm)	20.8	27.3	24.3
Total channel area (m^2)	276.8	379.9	699.2
Channel area/watershed area (%)	.31	.28	.56
Drainage density (m ha^{-1})	50.8	23.0	45.3
Channel gradient (m m^{-1})	.243	.249	.200
Average annual discharge (1 s^{-1})‡			
Water year 1973	3.69	5.03	6.19
Water year 1974	3.38	4.22	5.62
30-yr average	...	...	3.77
Maximum discharge (1 s^{-1})‡			
Water year 1973	79.9	84.7	105.6
Water year 1974	55.7	55.3	71.6
Maximum discharge (l s^{-1})‡			
Water year 1973	0.8	0.7	0.8
Water year 1974	1.0	0.6	0.8
Average annual elemental concentrations (mg l^{-1})§			
NO_3-N	0.672	0.127	0.003
NH_4-N	0.005	0.004	0.004
PO_4-P	0.002	0.002	0.002
Cl	1.296	0.503	0.538
SO_4	0.359	0.336	0.298
K	0.602	0.458	0.461
Na	1.102	0.784	0.883
Ca	1.074	0.518	0.626
Mg	0.649	0.227	0.296

* Data from Johnson and Swank (1973).
† Based on measurements made at 5-m intervals.
‡ Data from Coweeta Hydrologic Laboratory.
§ Data from Swank and Douglass (1975).

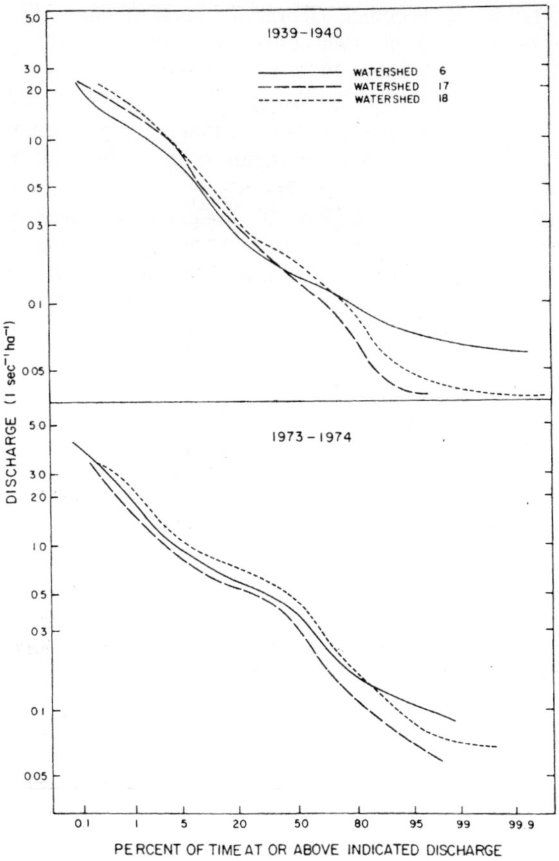

FIG. 3. Flow frequency diagrams for the 3 streams before (water years 1939 and 1940) and after (water years 1973 and 1974) perturbations. Data from Coweeta Hydrologic Laboratory, North Carolina, USA.

Our study was conducted on 3 watersheds with differing vegetation: a hardwood forest (Watershed 18), a white pine plantation (Watershed 17), and an old field (Watershed 6). The 3 watersheds are located within 2.59 km^2 and are all on NW facing slopes (Fig. 2). Physical characteristics of the watersheds and their streams are summarized in Table 1.

With exception of the chestnut blight, the hardwood forest watershed has remained undisturbed since at least 1924 (Johnson and Swank 1973). Prior to 1924, there was some selective logging. Dominant overstory vegetation includes chestnut oak (*Quercus prinus*), red maple (*Acer rubrum*), scarlet oak (*Quercus coccinea*), northern red oak (*Quercus rubra*), tulip poplar (*Liriodendron tulipifera*), pignut hickory (*Carya glabra*), and black oak (*Quercus velutina*), listed in order of contribution to basal area. Prominent understory species are rhododendron (*Rhododendron maximum*), mountain laurel (*Kalmia latifolia*), and dogwood (*Cornus florida*) (Day and Monk 1974). Rhododendron is especially dense in many places along the stream.

The pine plantation watershed was cleared of all shrub and forest vegetation in 1942. Annual sprout growth was cut back almost every year between 1942 and 1955. White pine was planted in 1956. Hardwood sprouts were cut or treated with herbicide as necessary to prevent competition with the pine (Johnson and Swank 1973).

The old field watershed has a long history of manipulations, beginning in 1942 when the riparian vegetation was cut. This was allowed to regrow until 1958 when all marketable timber was removed and the cover converted to Kentucky-31 fescue grass. From 1959 to 1965, various fertilizers were applied to the watershed (detailed in Johnson and Swank 1973) including dolomitic limestone. In May 1966, and for 2 subsequent yr, the grass cover was killed with herbicides (details described in Douglass et al. 1969). Since spring 1968, the watershed has been allowed to revert to successional vegetation. In 1968 and 1969 the vegetation consisted primarily of horseweed (*Erigeron canadensis*), fireweed (*Erechtites hierachifolia*), and grass species. By 1970, woody shrubs contributed up to 50% of the aboveground vegetation (Johnson and Swank 1973). During the period 1972–1974 blackberry (*Rubus* sp.) and black locust (*Robinia pseudoacacia*) were the predominant woody species.

Though the 3 streams and watersheds have some morphological differences (Table 1), their hydrologic properties are very similar (Fig. 3). The flow frequency diagrams (Fig. 3) indicate the quantity and variability of flow (Leopold et al. 1964). The relative shallowness of the slopes indicates the flow stability of these streams. The flow distribution is not log normal as would be indicated by a straight line, because of high frequencies of relatively high and low flows. This is apparently caused by high winter base flows and low summer base flows.

The flow frequency diagrams (Fig. 3) further show that the hydrologic properties of the streams on watersheds 6 and 17 have not been substantially changed by perturbations to these watersheds. Though water years 1939 and 1940 differed considerably from water years 1973 and 1974, the relationships of the flow frequency diagrams are similar. During both periods, there was somewhat less discharge per unit area from WS 17 than from WS 18. Careful analysis has shown the conversion to pine on WS 17 reduced flow (Swank and Douglass 1974), though it is not evident from these diagrams. The WS 18 and WS 17 graphs are nearly parallel, indicating similar flow variability. Discharge from WS 6 was very similar to the other watersheds, however the lesser slope of the graph indicates less flow variability, that is, WS 6 appears to exhibit greater flow stability, especially at low flows. Because this was true both before and after perturbation, it must be due to morphometric differences between WS 6 and WS 17 and 18. There is evidence of a greater groundwater reservoir on WS 6 than on the other watersheds (W. T. Swank, Coweeta Hydrologic Laboratory, *personal communication*).

Streams on the 3 watersheds are very small, 1st- and 2nd-order streams (Fig. 2). The substrate is sand and rocks ≤25 cm diameter. The gradient is steep in places and gentle in others with infrequent small pools. Exposed rocks are often covered with moss. In summer, the streams are heavily shaded, except for stretches in the old field watershed, and during fall and winter they are choked with leaves. Primary production in these streams is very low. In studies made on 2 similar Coweeta streams, no measurable chlorophyll was found during any season on artificial substrates incubated for 8 wk. Moss primary production is unknown but presumed low. Limitation of primary production appears to result from a combination of low light and nutrient levels (Table 1). To demonstrate the intense shading over the streams, on a clear day in August we made a series of solar radiation measurements with an Isco Model SR Spectroradiometer. At the point of measurement in the hardwood forest, only 0.17% of visible insolation reached the stream surface. Light penetration through the pine canopy was 1.51%. Three measurements made on the old field were 0.38, 2.69, and 26.26%, reflecting the patchiness of the old field vegetation.

Methods and Materials

Abiotic parameters.—Forest Service personnel at Coweeta Hydrologic Laboratory continuously measured stream flow with a sharp-crested, V-notch blade set in a concrete weir at the base of each watershed. Weekly samples for chemical analysis were collected in the stream directly above each weir pond. Cation concentrations from both precipitation and stream water were analyzed with a Bausch and Lomb Model AC2-20 atomic absorption spectrophotometer. More data can be found in Johnson and Swank (1973) and Swank and Douglass (1975).

Water temperatures in the old field and the hardwood forest streams were measured by Forest Service personnel from July 1971 through April 1973 with Bendix Model 1443 soil thermographs. The sensing elements were located above the weir ponds in free flowing water in the shade. Point measurements showed that water temperatures in the pine plantation and hardwood forest streams were very similar (Woodall 1971) and continuous measurements were not made in the pine plantation stream.

Litterfall inputs.—We collected litterfall (not including branches >5 mm diameter) in 15 × 300 cm V-shaped aluminum troughs. Four troughs were located above each of the 3 streams. Troughs were cleaned at least biweekly from August 1972 through January 1974. The biomass of collected litter was determined after drying at 60°C for 24 h. Subsamples were ground in a Wiley mill and analyzed for K and Ca by spark emission spectrophotometry.

Standing crops of large particulate organic matter and benthic fauna.—Standing crop samples were collected by R. Woodall, Georgia Power Company, Atlanta, Georgia, in 1968–1969 (Woodall 1971, 1972; Woodall and Wallace 1972, 1975). In addition to using Woodall's (1971, 1972; Woodall and Wallace 1972, 1975) data, we collected benthic fauna samples from the old field stream and detritus samples from all 3 streams.

During 1968–1969, Woodall collected 4 stratified random samples monthly from each stream using a 900 cm² Surber sampler with a mesh opening of 0.3 mm. The sampler was placed in the center of the stream channel and the substrate within disturbed by hand to a depth of ≈10 cm. Samples were transferred to a plastic bag, stored in crushed ice, and processed in the laboratory within 24 h. Organisms and large particulate organic matter (LPOM, > 0.3 mm) were sorted by hand from a white enamel pan. Dry weights of LPOM were obtained after drying at 105°C for 24 h. Nonquantitative samples were made for gut analysis and mineral analysis of organisms. Samples for mineral analysis were dried at 75°C overnight, weighed, acid digested and analyzed on a Bausch and Lomb Model AC2-20 atomic absorption spectrophotometer (Woodall 1972, Woodall and Wallace 1975).

During 1972–73, we collected 3 old field stream benthos samples per month. These samples were preserved with 10% formalin before sorting. After organisms had been removed, LPOM was elutriated from the samples and oven dried. We also collected 3 LPOM samples per month from the hardwood forest and pine plantation streams using the Surber sampler. These samples were oven dried at 60°C for 24 h immediately on return to the laboratory. All LPOM samples were subsequently washed to remove sand and gravel, and redried before weighing. Chemical analyses were made by spark emission spectrophotometry.

Drift of organisms and large particulate organic matter.—We measured drift of organisms and large particulate detritus by placing a 0.33-m wide drift net with a mesh opening of 0.7 mm in the stream at a point just above the weir pond. Sandbags directed total stream flow through the net. At the end of 24-h measurement periods, material collected in the net was emptied into a plastic bag and formalin added. In the laboratory, samples were treated in the same manner as materials collected with the Surber sampler. Organisms were sorted, identified, and counted and LPOM was dried, weighed, and analyzed. Sixteen 24-h samples were taken from the old-field stream, 17 from the pine plantation stream and 18 from the hardwood forest stream. An additional 8, 2, and 3 samples, respectively, were taken from each stream and analyzed for organism drift only.

Emergence of benthic insects.—We measured insect emergence with emergence traps modified from the design of Mundie (1971). Each pyramid-shaped trap had a 0.25-m² base with screen-covered sides (1.2 mm mesh opening). Insects were collected at the top

TABLE 2. Dissolved and sediment losses of potassium and calcium. Data from Coweeta Hydrologic Laboratory, North Carolina, USA and Monk (1975). Dissolved outputs are for water year 1973. Sediment data are from 1971 and 1972. Measurements are expressed per unit area of watershed

	Dissolved losses ($kg\ ha^{-1}$)		Sediment losses ($g\ ha^{-1}$)	
Watershed	K	Ca	K	Ca
Old field	7.4	13.9	8.7	161.1
Pine plantation	4.9	6.3	2.7	70.9
Hardwood forest	6.9	10.0	1.8	57.1

in an inverted 0.47-litre jar with a reservoir containing a solution of 10% formalin, glycerine, and detergent.

We placed 3 emergence traps in each stream, 1 near the weir, 1 near the head of the stream, and 1 in between. Traps were placed on exposed stones or stream bottoms so they were as close to the water as possible without blocking stream flow. Traps were operated continuously from July 1972 through August 1973, during which time we made collections at approximately biweekly intervals. In the laboratory, insects were counted and identified. To convert numbers data to biomass, K, and Ca, we collected adult insects from a black light. These insects were dried, weighed, wet ashed, and chemically analyzed by atomic absorption spectrophotometry.

Tracer determination of invertebrate elimination rates.—Behavior of radioisotopes in organisms and ecosystems is unique to specific isotopes. However, it is the assumption of radiotracer studies that chemically similar isotopes behave similarly and that measurements of radioisotopes reflect analogous behavior of nonradioactive isotopes.

In this study we tagged insects with ^{134}Cs and ^{85}Sr and used observed assimilation and elimination rates as estimates of K and Ca rates. Because of physical and chemical similarities between Cs and K, their physiological functions are grossly analogous (Davis 1963). Their similarity in metabolic processes has been recognized since studies by Ringer (1882), though there is no evidence that Cs and K follow the same metabolic pathways (Kornberg 1961). Discrimination between Cs and K has been shown in both plants and animals (McNeill and Trojan 1960, Bryan and Ward 1962, Davis 1963). However, because the radioisotopes of K are not very useful tracers (Phipps 1976), radiocesium has been widely used as a K analog.

Strontium is the closest chemical relative of Ca and its levels in organisms seem to be related to Ca levels (Comar and Wasserman 1964). Studies of Sr/Ca ratios have shown that differences are largely related to Sr/Ca ratios in the external medium or food supply (Nelson 1963, Ophel and Judd 1969) though discrimination has been demonstrated in a variety of organisms (e.g., Ophel and Fraser 1970). In general, Sr appears to be a quite acceptable tracer of Ca dynamics. Kornberg (1961) indicated that Sr and Ca follow very similar metabolic pathways. Schurr and Stamper (1962) related fluctuation of ^{85}Sr body burden of crayfish to normal accumulation, fixation, redistribution, and loss of Ca during the molting cycle.

Both radiostrontium and radiocesium dynamics in insects and insect food chains have been extensively studied (Crossley and Pryor 1960; Crossley and Howden 1961; Crossley and Schnell 1961; Crossley 1963, 1966, 1967; Reichle and Crossley 1965, 1967; Reichle 1967; Crossley and Reichle 1969; Van Hook and Crossley 1969; Wilhm 1970). These radioisotopes have been previously used as a basis for K and Ca models (Gist 1972, Crossley and Gist 1973).

Our methods were essentially modifications of those involving terrestrial insects. Work was concentrated on 4 detritivore species (food habits discussed below): nymphs of *Peltoperla maria* (Plecoptera) and *Stenonema* sp. (Ephemeroptera), larvae of *Tipula* sp. (Diptera), and crayfish (*Cambarus bartoni*). Birch (*Betula lutea*) leaves which had been collected at Coweeta and air dried were soaked in isotope solution ($\approx 25\ \mu Ci\ ml^{-1}$) for 24 h. They were then rinsed with tap water and air dried. Based on counts of small portions of these leaves at the time of feeding, the activity was $\approx 0.01\ \mu Ci\ mg^{-1}$ for ^{134}Cs and $1\ \mu Ci\ mg^{-1}$ for ^{85}Sr. Before feeding, leaves were soaked overnight in stream water. Animals used in the experiments were collected from Coweeta streams at least 1 day prior to beginning an experiment and placed in a plastic pan of stream water with aeration. At the beginning of an experiment, small pieces of tagged leaves were added to the pan. After 2 h, labeled animals were removed, counted, and isolated in plastic cups with running water. To test the possibility that animals could adsorb radioactive material which leached from the tagged leaves, we placed a screen down the middle of the feeding pan. Animals were placed on both sides of the screen and tagged leaves on 1 side. Over a 2 h period, none of the animals on the no-food side accumulated radioactivity significantly above background. A constant flow of fresh stream water was maintained through the plastic cups in which experimental animals were kept to assure proper oxygenation of the water and to prevent reingestion of egested materials. Temperatures in the cups were maintained as close as possible to 20°C; however, because water was taken directly from a nearby stream there was a 2°C to 4°C daily fluctuation.

Prior to counting, animals were placed in a small minnow net and rinsed with stream water. They were then placed in a 15 × 125 mm test tube containing 5 ml of stream water and counted in a Packard Auto Gamma Spectrometer Series 410A with a 5.08 × 5.08 cm sodium-iodide thallium-activated crystal enclosed in a lead shield (counting efficiencies: 12–13% for ^{134}Cs, 10% for ^{85}Sr; counting time = 1 min). Animals

TABLE 3. Litterfall into the 3 study streams

	Biomass (g m^{-2})			K concentrations (mg g^{-1})			Ca concentrations (mg g^{-1})		
	Old field	Pine planta-tion	Hard-wood forest	Old field	Pine planta-tion	Hard-wood forest	Old field	Pine planta-tion	Hard-wood forest
Winter	46.4	21.9	13.9	1.79	2.32	1.29	15.05	12.35	10.24
Spring	17.8	47.9	33.8	2.04	1.76	1.83	10.76	6.34	9.78
Summer	60.1	52.2	34.7	5.82	4.05	4.63	13.90	12.25	9.15
Fall	161.5	196.8	270.8	6.43	1.65	4.38	17.22	8.86	13.26
Total	285.8	318.8	353.2	5.28	2.10	4.48	15.77	9.28	12.42

were counted approximately every 2 h during the 1st day and 3× per day thereafter. Counting continued until the animal no longer contained sufficient radiotracer for a significant count or until sufficient data were obtained to estimate elimination rates.

RESULTS

Abiotic parameters.—Stream flows for the 3 streams are summarized in Table 1. From stream discharge records and stream water K and Ca concentrations, total export of dissolved K and Ca during water year 1973 was calculated (Table 2). Comparison with precipitation inputs indicated that all 3 watersheds had net losses of both elements, with little differences from analyses made by Johnson and Swank (1973) for water years 1970 and 1971. The major difference among the 3 watersheds was the greater Ca loss from the old field, likely due to limestone fertilization in 1959. (Hydrologic and chemistry data and analysis from W. T. Swank and J. E. Douglass, Coweeta Hydrologic Laboratory, *personal communication.*)

Sediment losses from the 3 watersheds were calculated from quarterly measurements and analysis of weir pond sediments (Table 2). Measurements were taken by Forest Service and University of Georgia personnel and reported by Monk (1975). Sediment losses were greatest from the old field, reflecting the least vegetational protection of the soil. When compared with dissolved nutrient outputs, losses of K and Ca as sediments were very small. Calcium sediment loss was ≈1% of dissolved loss and sediment loss of K was ≈0.1% of dissolved loss.

Mean annual temperatures for 1972 were 12.1°C and 10.8°C in the old field and hardwood forest streams, respectively. Highest recorded temperatures over 21 mo of record were 20°C in the old field stream in July 1972, and 17.2°C in the hardwood forest stream during August 1971 and July 1972. Lowest temperatures, 1.1°C and 2.8°C, occurred in the old field and hardwood forest streams, respectively, during January 1973. In general, temperature fluctuated more in the old field stream with maximum temperatures being higher and minimum temperatures lower than the hardwood forest stream. A maximum daily fluctuation of 6.7°C was measured in the old field stream. The greatest daily fluctuation recorded in the hardwood forest stream was 3.3°C.

Litterfall inputs.—Litterfall rate was greatest during autumn on all 3 watersheds (Fig. 4). Autumn inputs were greatest in the hardwood forest, where 77% of annual litterfall biomass entered the stream during the 3 fall mo (Table 3). The comparable figures for the pine plantation and old field were 62 and 56%, respectively. In both forest ecosystems, a small pulse also occurred during April and May, corresponding to leafing out and flower production. Total litterfall for the hardwood forest and pine plantation was very similar to measurements made by Cromack (1972, Cromack and Monk 1975) for the same watersheds (362 and 321 g m^{-2} respectively). Peak elemental concentrations of litter occurred in fall in the old field with minimums in spring. In the pine plantation peak concentrations were in summer and winter. We found the highest elemental concentrations of hardwood litter in fall. Annual average concentrations of both K and Ca were highest in the old field and lowest in the pine plantation. Cromack (1972) found K and Ca concentrations of 1.67 and 5.98 ml g^{-1} in pine needle litter and 4.7 and 12.3 mg g^{-1} in hardwood leaf litter. The higher elemental concentrations we found in the pine plantation litter (2.10 and 9.28 mg g^{-1}) were probably due to hardwood species invading the pine plantation along the stream where the pine canopy opened.

Standing crops of large particulate organic matter.—Large particulate organic matter (LPOM) standing crops during 1972–73 are summarized in Table 4. Corresponding to litter input, LPOM standing crop was highest in the hardwood forest stream and lowest in the old field stream, though differences between the old field stream and the pine plantation stream were slight. A paired t test of seasonal data showed no significant (α = .05) difference between LPOM standing crop in the old field and pine plantation streams nor between the hardwood forest and pine plantation streams. The hardwood forest and old field streams were significantly different.

In all 3 streams, fall biomass was greatest, followed by decrease during winter and an increase during spring. The spring increase was followed by another decrease during summer. The spring increase occurred

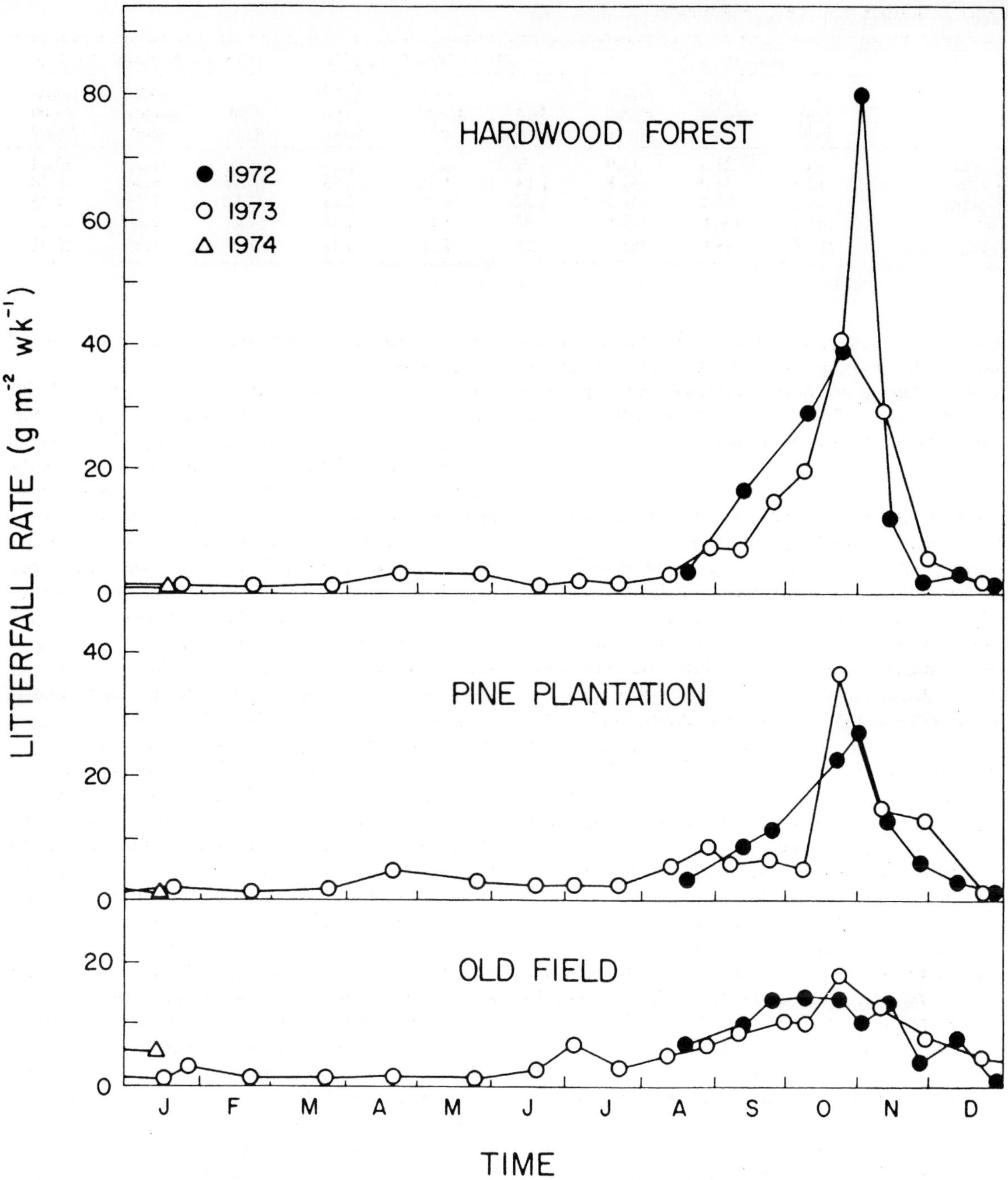

FIG. 4. Litterfall rates into the 3 study streams.

in all 3 streams, however, it was only significant (α = .05) in the old field. These spring increases may in part be due to lateral movement of leaf litter into the streams. Webster (1977), in a study of a 2nd-order stream on another hardwood forest watershed at Coweeta, found that blow-in of leaf litter was 1.6× as great as leaf fall and, though peaking in fall, continued through the winter and into spring.

Large particulate matter stored deep in the stream sediments was not measured in this study. Recent studies have shown that significant amounts of large particulate organic matter may be deposited and stored deep (>10 cm) in the streambed and on the floodplain and subsequently removed during major storms (recurrence interval >1 yr). This may lead to significant errors in annual organic matter budgets (J. Sedell and K. Cummins, *personal communication*).

Patterns of element concentrations in LPOM (Table

TABLE 4. Standing crops of large particulate organic matter

	Biomass (g m^{-2})			K concentrations (mg g^{-1})			Ca concentrations (mg g^{-1})		
	Old field	Pine plantation	Hardwood forest	Old field	Pine plantation	Hardwood forest	Old field	Pine plantation	Hardwood forest
Winter	44	70	125	0.27	0.75	<0.10[1]	9.64	9.03	13.85
Spring	93	98	130	0.33	0.16	<0.10[1]	6.97	7.94	10.25
Summer	71	107	111	0.38	0.73	0.34	11.06	12.36	12.31
Fall	181	148	294	0.71	0.37	<0.10[1]	12.06	8.64	14.34
Annual mean	97	106	165	0.52	0.48	0.14	10.41	9.46	13.10

[1] K concentrations for this period were below the detection limits (100 $\mu g \cdot g^{-1}$).

4) were almost identical to patterns of litterfall element concentrations (Table 3). Peak concentrations of both elements occurred in the old field stream in fall. In the pine plantation stream, the peaks were in summer and winter. In the hardwood forest stream, the peak K concentration occurred in summer and the peak Ca concentration in fall.

Potassium concentrations in LPOM were substantially <K concentrations in leaf litter, but Ca concentrations were in some cases > in leaf litter. In terrestrial litter bag studies Attiwill (1968) and Gosz et al. (1973) found significant K decreases in the 1st few mo and slower losses thereafter. In both studies there was little Ca loss even after a year. For K, Cromack (1972, Cromack and Monk 1975) reported exponential loss rates of 1.7 yr^{-1} and 1.8 yr^{-1} from litter bags of pine needles and mixed hardwood litter, respectively. He found no significant Ca loss from pine bags and Ca loss rate from mixed hardwood bags was .34 yr^{-1}. Two studies have compared element losses in terrestrial and aquatic sites. Witkamp and Frank (1969) found exponential loss rates of ^{134}Cs from oak leaves of 3.5 yr^{-1} in a forest, 8.3 yr^{-1} in a pond, and 16.6 yr^{-1} in a stream. In all 3 sites, ^{134}Cs losses were more rapid than weight losses. Thomas (1970) found that calcium loss rates were essentially identical to weight loss rates during leaf decomposition in a stream and the surrounding forest.

On an annual basis, LPOM standing crops in terms of biomass and Ca were higher in the hardwood forest stream than the other streams (Table 4). However, K standing crops were lowest in the hardwood forst stream. Standing crops of biomass, Ca and K in the pine plantation and old field stream were very similar. The only significant (α = .05) differences found by a paired t test were: LPOM biomass and Ca were greater in the hardwood forest than in the old field stream and K standing crop was higher in the pine plantation than in the hardwood forest stream.

Standing crops of benthic fauna.—The insect fauna of the 3 streams was dominated by *Peltoperla maria* (Plecoptera: Peltoperlidae), *Stenonema* spp. (Ephemeroptera: Heptageniidae), and tipulids (Diptera; Tipulidae), including *Tipula abdominalis, T. ignoblis, Eriocera longicornis, E. spinosa,* and *E. fultonensis* (Woodall and Wallace 1972). These 3 groups of benthic insects are all detritivores.

Peltoperla maria, like other stoneflies of the suborder Filipalpia, is a shredder (Cummins 1973). Woodall (1972) found nothing but detritus in the foreguts of 10 specimens. Wallace et al. (1970) found that *P. maria* fed readily on deciduous leaves, preferring some leaf species over others. From gut analysis, Elwood and Cushman (1975) also designated *P. maria* as a shredder. *Peltoperla maria* accounted for 11, 30, and 31% of the average annual insect biomass in the old field, pine plantation, and hardwood forest streams, respectively.

Various species of *Stenonema* apparently feed from the surfaces of submerged material, which depending on the stream, may be deposits of fine organic particles, diatoms or other algae growing on the surfaces, or possible heterotrophic microbes on the surfaces of large organic particles. Trama (1957) studied energy tranformation by *Stenonema pulchellum* fed cultured algal cells. Minckley (1963) found primarily detritus in the digestive tracts of *Stenonema*. Coffman et al. (1971) determined that 2 species of *Stenonema* fed primarily on algae in a small Pennsylvania stream. Cummins et al. (1973) showed that, with only whole leaves available as food, growth of *Stenonema fuscum* depended on presence of invertebrate shredders. Shapas and Hilsenhoff (1976) determined the foregut contents of 6 species of *Stenonema* to be principally detritus. Studying insects collected at Coweeta, Woodall (1972) found only detritus and fungal hyphae in the guts of *Stenonema* spp. In the heavily shaded streams at Coweeta, where instream primary production is very low, *Stenonema* apparently depends entirely on terrestrial leaf detritus as its food source. The biomass of *Stenonema* was 5, 9, and 9% of the average annual insect biomass in the 3 streams.

Minckley (1963) found primarily algae and higher plant material in the guts of *Tipula nobilis*?, however, tipulids are generally recognized as shredder detritivores (Cummins 1973). Vannote (1969) classified *T. abdominalis* as a shredder and studied the energy budget of *T. abdominalis* fed on deciduous leaves. Cummins et al. (1973) found that *T. abdominalis* and *T. caloptera* grew and survived well on a diet of decid-

TABLE 5. Average annual standing crops of insect detritivores and crayfish. All values are mg m^{-2}

	Insect detritivores	Crayfish
Old field	848	3909
Pine plantation	890	2915
Hardwood forest	1337	1607

uous leaves. Gut contents of *Tipula* collected at Coweeta were mostly detritus (Woodall 1972). The largest percentage of material in *Eriocera* gut contents was detritus, but Woodall (1972) also found insect fragments and algae. Tipulids accounted for a large percentage of the benthic insect biomass in the old field stream (56%) and an only somewhat smaller biomass in the pine plantation and hardwood forest streams (17 and 26%).

On an annual basis, these 3 groups of insects, *Peltoperla maria, Stenonema,* and tipulids, accounted for 73, 56 and 67% of the total insect biomass of the old field, pine plantation, and hardwood forest streams, respectively. Excluding predaceous insects (primarily odonates, setipalpian stoneflies, and rhyacophilid caddisflies), these dominant insects comprised 84, 71, and 76% of the primary consumer insect biomass.

Among the Trichoptera, net spinning caddisflies comprised 11, 13, and 9% of the average annual insect biomass in the old field, pine plantation, and hardwood forest streams, respectively. Species included *Diplectrona modesta, Parapsyche cardis, Dolophilodes* sp., *Chimarra* sp., and *Wormaldia* sp. Though the diets of these insects, especially *Parapsyche cardis* in the final instar, include other insects (e.g., Wallace et al. 1977), all except *P. cardis* are primarily detritivores when the complete life cycle is considered.

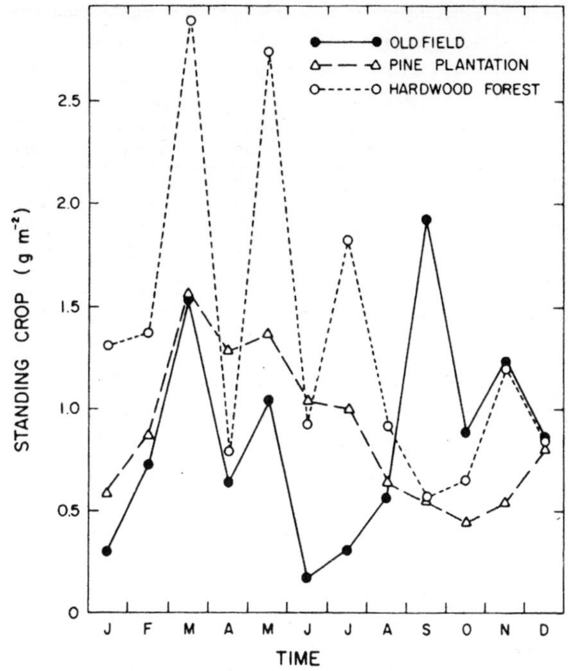

FIG. 5. Standing crops of benthic detritivore biomass in the 3 study streams.

There were a number of other regularly occurring insects found at Coweeta which Woodall (1972) or others have determined to be principally detritus feeders. These included the mayflies *Ephemerella* spp., *Baetis* sp., and *Paraleptophlebia* sp. Two other Filipalpia stoneflies were found, *Nemoura* sp. and *Leuctra* sp. Detritus feeding Trichoptera other than the net spinners included *Fattigia pele, Pycnopsyche* spp., and *Lepidostoma* sp. Diptera other than tipulids found in the 3 Coweeta streams included simuliids, dixids, cer-

TABLE 6. Dry weight (mg) and potassium and calcium concentrations ($\mu g\ g^{-1}$) of dominant stream detritivores. Data on crayfish and immature insects from Woodall (1972)

	Immatures			Adults		
Taxonomic group	Dry weight	K	Ca	Dry weight	K	Ca
Cambarus						
small (<2 cm)				57.7	1 900	100 300
medium (2–5 cm)				265.9	1 900	100 300
large (>5 cm)				2049.0	1 900	100 300
Tipulidae						
Eriocera	2.19	7100	3800			
Tipula	43.82	7300	4300	43.83	9 000	8 500
small tipulids				0.54	10 500	3 200
medium tipulids				5.18	9 200	1 200
Chironomidae	0.15	2100	7900	0.07	12 600	2 800
Stenonema	4.29	1900	3800	2.84	11 400	1 200
Peltoperla	2.88	1800	3500	4.10	12 400	4 100
Hydropsychidae				2.57	8 400	2 800
Diplectrona	1.33	2700	3500			
Parapsyche	4.33	1900	3200			

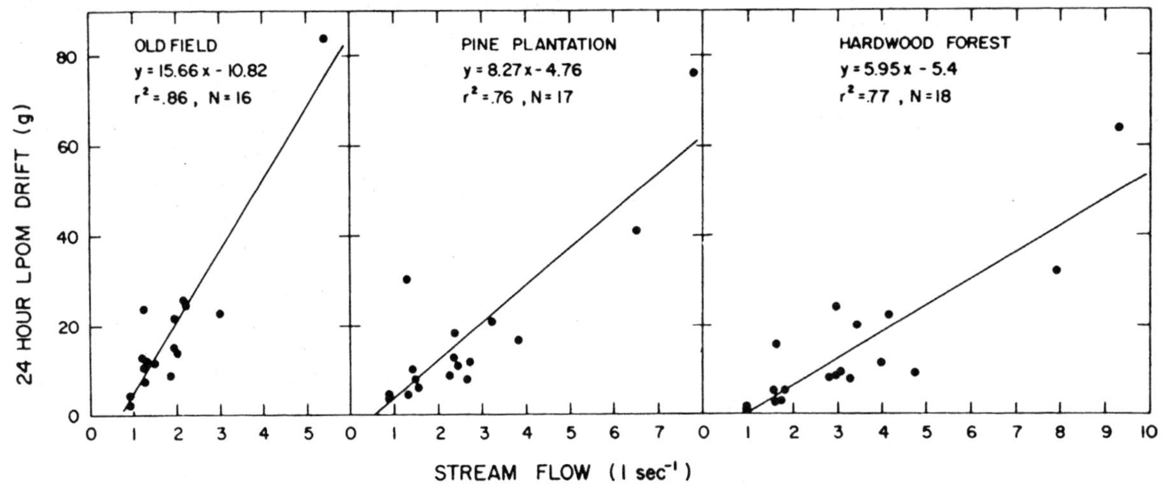

FIG. 6. The relationship of large particulate organic matter (LPOM) drift to stream flow in the 3 study streams.

atopogonids, and chironomids. Of the chironomids collected, only *Conchapelopia* sp. was in the Tanypodinae. Members of this subfamily are generally considered to be predators (Woodall and Wallace, 1972, Cummins 1973), but the other species of chironomids are probable detritivores. All these groups together along with other, less abundant, taxa, accounted for <15% of the detritivore biomass in any of the streams. However, because of rapid turnover, the chironomids especially may have a greater production than we have attributed to them.

Hynes (1970) described an idealized seasonal cycle of insect biomass in an unpolluted stream: minimum biomass occurs during summer when emergence and egg laying take place, followed by increase to an early spring peak. Detritivore biomass in the 3 study streams generally followed this trend (Fig. 5). In the hardwood forest and pine plantation streams, peak biomass occurred in early spring and minimum biomass was found in fall. In the old field stream, a biomass peak occurred in early spring, then a minimum in summer, and a second peak in fall. The September peak was due to the large number of tipulids present at that time.

The average standing biomass of insect detritivores was significantly (α = .05) greater in the hardwood forest stream than in the other 2 streams (Table 5). The old field and pine plantation streams were not significantly different.

The crayfish (*Cambarus bartoni*) was also a significant detritivore. Average standing crop biomass was the reverse of insect biomass being greatest in the old field stream and least in the hardwood forest stream (Table 5). In all 3 streams, crayfish biomass was > detritivore insect biomass, however, because of much slower turnover, their large biomass does not reflect their impact on the stream ecosystems.

Dry weights and K and Ca concentrations determined for the more abundant detritivores are shown in Table 6.

Large particulate organic matter drift.—Fisher and Likens (1973) found that concentrations of coarse particulate organic matter were proportional to discharge. Similarly, Hall (1972) reported that the log of leaf mass discharge was linearly proportional to water stage. However, we found a significant (α = .05) linear relationship between LPOM drift and discharge on all 3 streams (Fig. 6). In comparing these relationships by analysis of covariance, the regression slopes were similar for the hardwood forest and pine plantation streams. However, the slope was significantly greater for the old field stream. That is, a unit increase in stream flow in the old field stream produced a greater increase in LPOM discharge than in the other 2 streams.

We used the regression equations of LPOM drift as a function of stream discharge to estimate total output of particulate detritus. Monthly values were found from mean monthly discharge. Annual drift of LPOM was greatest from the old field stream and least from the hardwood forest stream (Table 7). It is difficult to evaluate the significance of these differences, noting especially that the sampling periods represented <5% of the annual cycle. However, these data are sup-

TABLE 7. Drift of large particulate organic matter (LPOM), insect detritivores, and crayfish from the study streams. All values in mg m^{-2} yr^{-1}

	LPOM	Insect detritivores	Crayfish
Old field	62 500	243.8	405
Pine plantation	35 300	52.6	158
Hardwood forest	17 100	50.5	57

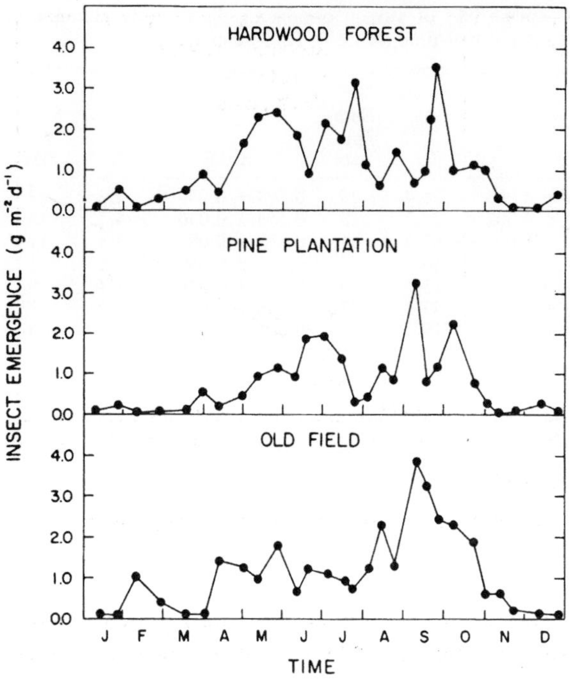

FIG. 7. Detritivore insect emergence from the 3 study streams.

ported by strong correlation with annual inorganic particulate discharge (Table 2).

Organism drift.—In a correlation analysis between insect detritivore drift and stream flow, we found no significant relationship for the pine plantation and hardwood forest streams (r^2 = .02 and .19, respectively). However, for the old field stream, there was a statistically significant relationship (r^2 = .63, N = 25). There was no significant relationship between drift and standing crop for any of the streams.

In calculating annual drift for the hardwood forest and pine plantation streams, a simple mean value was used because there was no apparent relationship to stream flow. For the old field stream, a linear regression equation was used to calculate drift from mean monthly stream flows. Annual insect detritivore drift was considerably greater from the old field stream than from the other 2 streams (Table 7). Tipulids, *Stenonema, Peltoperla,* and hydropsychids together comprised 55, 86, and 93% of the insect detritivore drift from the old field, pine plantation, and hardwood forest streams, respectively.

We found no significant correlations between crayfish drift and stream flow. Annual crayfish drift was calculated from a mean value of daily samples. Crayfish drift was greatest from the old field stream and least from the hardwood forest stream and, in all 3 streams, > insect drift (Table 7).

Insect and crayfish drift losses were small fractions of standing crops. Annual insect detritivore drift represented 29, 6, and 4% of the average standing crop in the old field, pine plantation, and hardwood forest streams, respectively. Crayfish drift was 10, 5, and 4% of standing crops.

TABLE 8. Emergence of adult detritivore insects from the 3 study streams

	Annual emergence (mg m^{-2})	Standing crop (%)
Old field	375.4	44
Pine plantation	254.7	29
Hardwood forest	389.7	29

Emergence of benthic insects.—Most insect emergence occurred during late spring, summer, and early fall, though some emergence, primarily philopotamids, winter stoneflies, and chironomids, took place during the winter (Fig. 7).

Annual emergence was greatest from the hardwood forest stream and least from the pine plantation stream (Table 8). Emergence from the old field and hardwood forest streams were not significantly different (α = .05), though emergence from each stream was significantly > emergence from the pine plantation stream. Annual emergence represented 29% of the average standing crop on the hardwood forest and pine plantation streams, but 44% on the old field stream. This difference can be attributed to the bias of emergence samplers. In particular, a large number of philopotamids were caught in the emergence traps in the old field stream, whereas this group was not abundant in the benthic samples. In general, the emergence traps selected against larger insects. For example, we seldom found the stonefly, *Peltoperla maria,* in emergence samples, though it was abundant in benthic samples, whereas the smaller and less abundant stoneflies, *Leuctra* and *Nemoura,* were common in emergence samples. Among Trichoptera, smaller psychomyiids and philopotamids were more commonly caught in emergence traps than larger and more abundant hydropsychids.

On an annual basis, *Peltoperta maria, Stenonema,* tipulids, and hydropsychids represented 55, 61, and 59% of total emergence biomass from the old field, pine plantation, and hardwood forest streams, respectively. From analysis of K and Ca concentrations in adult insects, we found that K concentrations were considerably greater in adults than in immatures (Table 6).

Tracer determination of invertebrate elimination rates.—The 2-compartment elimination model developed by Reichle and Crossley (1965) was used in analysis of radiotracer elimination data. In this model, the radioisotope is represented as being in 2 components, body and gut, with independent elimination from each component:

TABLE 9. Isotope elimination rates (h^{-1}) and assimilation efficiencies (%) of stream organisms. k_B is body elimination rate; T_B is body half-time ($ln\ 2/k_B$); k_G is gut elimination rate; T_G is gut half-time ($ln\ 2/k_G$). Half-times in hours

Species	Isotope	Assimilation efficiency ± SE	Sample size	k_B ± SE	T_B	Sample size	k_G ± SE	T_G	Sample size
Peltoperla maria	^{134}Cs	11.0 ± 4.2	9	0.026 ± 0.006	26.4	25	0.308 ± 0.050	2.2	27
	^{85}Sr	28.1 ± 10.0	5	0.020 ± 0.002	35.2	18	0.153 ± 0.030	4.5	18
Stenonema sp.	^{134}Cs	9.8 ± 1.5	4	0.011 ± 0.003	62.9	11	0.378 ± 0.089	1.8	11
	^{85}Sr	11.9 ± 9.8	3	0.041 ± 0.006	16.7	5	0.458 ± 0.090	1.5	5
Tipula sp.	^{85}Sr	27.3 ± 6.2	9	0.045 ± 0.006	15.4	9	0.491 ± 0.096	1.4	9
Cambarus bartoni	^{134}Cs	33.2 ± 7.8	5	0.0012 ± 0.004	587.4	7	0.046 ± 0.012	15.0	7
	^{85}Sr	83.3 ± 2.7	5	0.0010 ± 0.004	719.5	5	0.058 ± 0.034	12.0	5

$$\frac{dB(t)}{dt} = al(t) - k_B B(t), \quad (1)$$

and

$$\frac{dG(t)}{dt} = (1 - a)l(t) - k_G G(t), \quad (2)$$

where $B(t)$ is body content, $G(t)$ is gut content, a is the assimilation fraction, k_B is body elimination rate, k_G is the gut elimination rate, and $l(t)$ is the ingestion rate. Other models have been developed for invertebrate elimination, however, Webster and Crossley (1978) demonstrated that this model was satisfactory at the level of resolution of this study.

Time series data from each tracer experiment were fitted to a solution equation for Eqs. 1 and 2:

$$X(t) = al_0 e^{-k_B(t-t_0)} + (1 - a)l_0 e^{-k_G(t-t_0)}, \quad t \geq t_0, \quad (3)$$

where $X(t)$ is the sum of gut and body tracer content, l_0 is the activity of radioisotope ingested during the acute feeding period or the total activity of the animal at the end of a chronic feeding, t_0 is the time of the initial count, and other symbols are as above. After correction for background and physical decay, normalization, and logarithmic transformation, linear regression was used to find the slope (k_B) and intercept (a) of the long portion of the curve. This line was then "stripped" from the curve and linear regression used to find the slope (k_G) of the short component.

Results of the radioisotope elimination study are summarized in Table 9. For each species studied, we found 2-component elimination curves for both ^{134}Cs and ^{85}Sr. In some cases, assimilation efficiencies were not calculated because feeding occurred over a period >2 h (chronic feeding).

The crayfish, *Cambarus bartoni,* eliminated both isotopes much more slowly than any of the insects. Among the insects, *Tipula* sp., the largest insect studied, eliminated ^{85}Sr more rapidly from both body and gut than either *Peltoperla maria* or *Stenonema* sp. Elimination of ^{85}Sr by *Peltoperla* was considerably slower than by either *Tipula* or *Stenonema*. Gut elimination of ^{134}Cs was faster by *Stenonema* than by *Peltoperla* but body elimination was faster by *Peltoperla*.

Assimilation efficiencies were generally low with the exception of ^{85}Sr assimilation by crayfish which averaged 83.3% of ingested ^{85}Sr. In all cases assimilation of ^{85}Sr was > assimilation of ^{134}Cs.

Few measurements of radioisotope elimination by aquatic insects have been made. Kevern et al. (1964) found ^{134}Cs turnover rates for *Ephemera varia* and *Chironomus commatatus* which were somewhat slower (k_B = .08 d^{-1} and .06 d^{-1}, respectively) than insect turnover rates found in this study. Schurr and Stamper (1962) found a ^{85}Sr elimination rate (k_B) for *Cambarus longulus* of .13 d^{-1}, faster than found for *C. bartoni* in this study. Turnover rates of ^{134}Cs and ^{85}Sr by terrestrial insects are generally slower than the rates we measured (Crossley and Pryor 1960, Crossley and Schnell 1961, Hubbell et al. 1965, Reichle and Crossley 1965, Crossley 1966, Reichle 1967, Williams and Reichle 1968, Kowal 1969, Van Hook and Crossley 1969, Moulder et al. 1970, Crossley et al. 1971, Van Hook 1971, and Gist 1972). With the exception of *Cambarus bartoni,* the rates of ^{134}Cs and ^{85}Sr assimilation by the aquatic detritivores were in the same range of assimilation of elements, biomass, or energy found in other aquatic and terrestrial detritivores, 10–30%. (Aquatic: Kevern et al. 1964, Cummins 1969, Vannote 1969, and McDiffett 1970. Terrestrial: Gere 1956, Bocock 1963, Hartenstein 1964, Hubbell et al. 1965, Reichle 1967, Crossley and Van Hook 1970, Crossley et al. 1971, and Van Hook 1971.)

From Table 6, it is evident that crayfish concentrate Ca and higher assimilation and slower turnover of ^{85}Sr relative to ^{134}Cs were expected. However, among the insects, Ca and K concentrations were similar. Because Ca concentrations were generally an order of magnitude > K concentrations in their food source (Table 4), we expected that ^{85}Sr would be assimilated less or turned over more rapidly (Table 9). One possible explanation may be that Ca in plant tissue is not available to insects because it is mainly in cell walls, whereas the ^{85}Sr used in these experiments was just a surface tagging. An alternative explanation is that significant amounts of K were absorbed directly from the water.

By assuming steady state, that is, growth rates are

TABLE 10. Consumptive indices of benthic organisms estimated from radiotracer elimination rates

Organism	Consumptive index (d^{-1}) K	Ca	Biomass[2]	$B(ss)/[B(ss)+G(ss)]$[1]
Peltoperla maria	3.37		43.6	59
		1.28	0.34	75
Stenonema sp.	2.12		29.0	79
		4.97	1.43	60
Tipula sp.		3.18	1.04	80
Cambarus bartoni	0.081		1.10	95
		0.027	0.21	99

[1] Percent of element in body component at steady state.
[2] Consumptive index biomass = consumptive index element × concentration element in organism/concentration element in food.

negligible compared to elimination and egestion rates, ingestion rates necessary to support the turnover rates given in Table 9 can be calculated. The steady state solution to Eqs. 1 and 2 is:

$$X(ss) = \frac{a}{k_B} + \frac{1-a}{k_G} I(ss), \tag{4}$$

where (ss) indicates the steady state value. Rearranging Eq. 4 gives

$$\frac{I(ss)}{X(ss)} = \frac{k_B k_G}{a k_G + k_B - a k_B}, \tag{5}$$

where the ratio on the left, $I(ss)/X(ss)$, is Waldbauer's (1968) consumption index. Ingestion rates estimated by this method (Table 10) can be compared with rates found in other studies (Table 11). Estimates of biomass ingestion calculated from ^{85}Sr elimination are in the range of values in Table 11, though generally faster. However, biomass ingestion rates calculated from ^{134}Cs elimination are much higher. The explanation of these very high values is the low K concentration of leaf material (139 mg g^{-1}) used to convert from K ingestion to biomass ingestion. Again, the difficulty may be direct absorbtion of K from the water, though this was not indicated by the partitioned pan feeding experiments.

Ratios of body elemental content to total elemental content (Table 10) were calculated from Eqs. 1 and 2 as:

$$\frac{B(ss)}{B(ss)+G(ss)} = \frac{(k_G - k_B)a}{(k_G - k_B)a + k_B}. \tag{6}$$

Values calculated from this equation are shown in Table 10. In a comparable study, Hubbel et al. (1965) estimated that 77% of ^{90}Sr was in the body component of isopods. Elwood et al. (1976) found that gut contents of *Tipula* spp. comprised 8.5% of K, 60.3% of Cs, and 78.6% of Ca body burden.

DISCUSSION

Results of the field studies and experimental radioisotope study were used to calculate K and Ca dynamics for the 3 streams (Fig. 8). Standing crop values were calculated from Tables 4 and 5, organism and LPOM drift from Table 7, insect emergence from Table 8, and litterfall from Table 3. Elimination rates were calculated from rates in Table 9 as:

$$\text{Egestion} = \left(1 - \frac{B_i(ss)}{B_i(ss)+G_i(ss)}\right) k_B 2^{\frac{T-20}{10}} x_i(ss) \tag{7}$$

and

$$\text{Excretion} = \frac{B_i(ss)}{B_i(ss)+G_i(ss)} k_B 2^{\frac{T-20}{10}} x_i(ss), \tag{8}$$

TABLE 11. Consumptive indices of aquatic organisms

Organism	Food	Temperature (°C)	Consumptive index (d^{-1})	Reference
Banksiola crotchi	algae	18	1.41	Winterbourn 1971
Neophylax concinnus	algae and detritus	4–5	1.20[1]	Sedell 1971
Stenonema pulchellum	algae	20	0.19[1]	Trama 1957
Peltoperla maria	detritus	15	0.08[2]	Wallace et al. 1970
Pteronarcys scotti	detritus	10–15	0.062[1]	McDiffett 1970
Pycnopsyche guttifera	detritus	5	0.332	Cummins et al. 1973
Pycnopsyche gentilis	detritus	...	0.02–1.00	Mackay and Kalff 1973
Pycnopsyche luculenta	detritus	...	0.02–0.81	Mackay and Kalff 1973
Stenonema spp.	detritus	5	0.04–0.23	Cummins et al. 1973
Tipula spp.	detritus	5	0.16–0.19	Cummins et al. 1973
Tipula abdominalis	detritus	...	0.041[1]	Vannote 1969
Acroneuria californica	predator	18	0.03[3]	Heiman and Knight 1975
Chironomidae	predator	20	1.52–2.71	Luferov 1961[4]
Odonata	predator	...	0.3	Benke 1972
Banksiola crotchi	predator	18	1.66	Winterbourn 1971

[1] Index calculated by Cummins et al. (1973).
[2] Mean value.
[3] Assuming equivalent caloric values for *Acroneuria* and prey species.
[4] As cited in Monakov (1972).

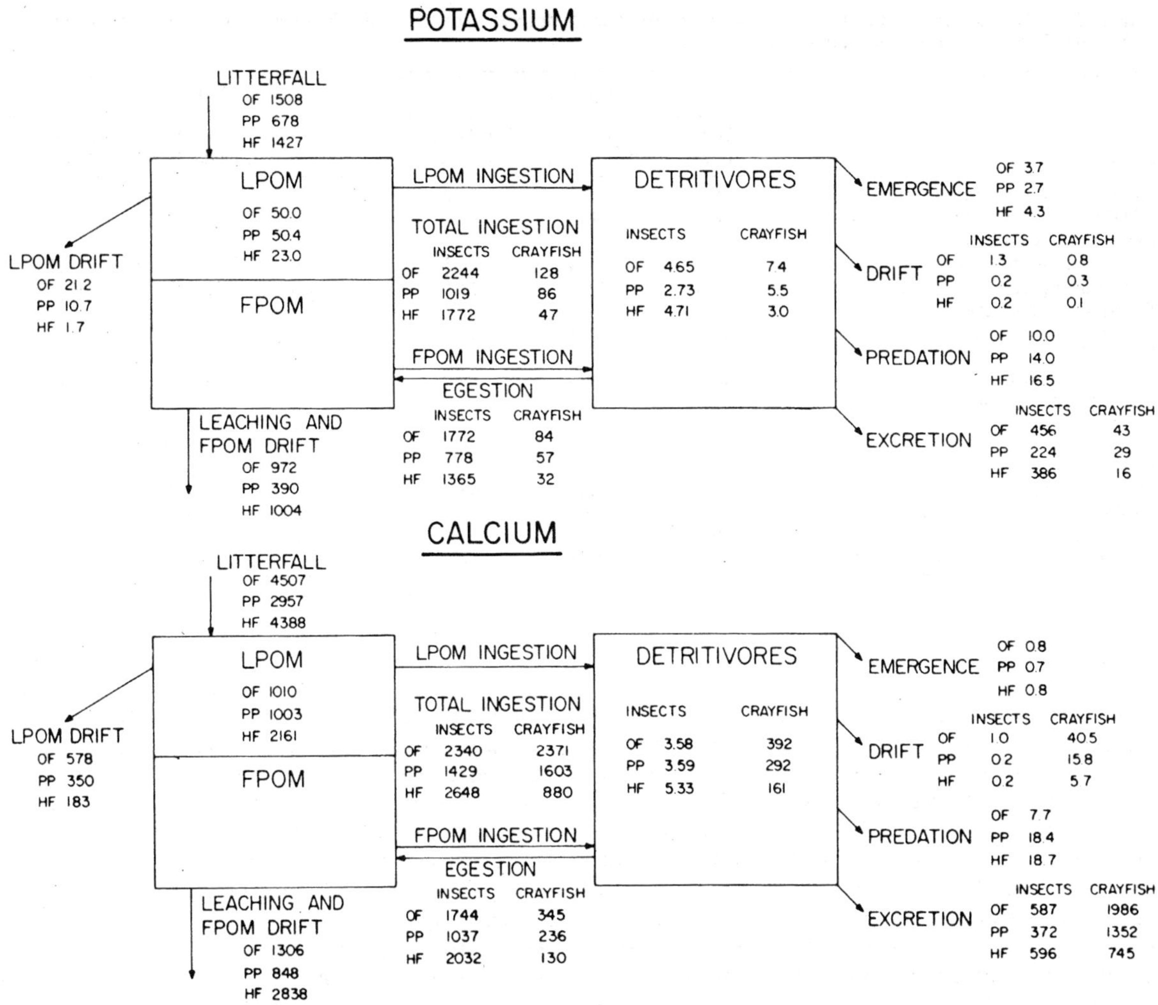

FIG. 8. Potassium and calcium dynamics in the old field (OF), pine plantation (PP), and hardwood forest (HF) watershed streams. Standing crops are mg m^{-2}; flows are mg m^{-2} yr^{-1}.

where T is the mean annual stream temperature and other symbols are as above. In Eqs. 7 and 8, the temperature term converts rates from experimental temperatures (20°C) to stream temperatures according to a Q_{10} of 2 (Reichle and Crossley 1965, Crossley 1966, Hasanen et al. 1967). For insect detritivores, elimination rates were calculated as means of *Peltoperla maria, Stenonema,* and *Tipula* rates, weighted according to their abundance.

Insect predation was calculated from measured insect predator standing crops (Woodall 1971, 1972; Webster 1975) and published ingestion rates (Lawton 1971, Winterbourn 1971, Heiman and Knight 1975). Salamander predation was similarly calculated from standing crops and published ingestion rates (Fitzpatrick 1973*a*, 1973*b*). Additionally, Woodall's (1972) analyses of salamander gut contents were used to derive the benthic insect fraction of salamander ingestion.

Detritivore ingestion was calculated to balance detritivore losses: emergence, drift, predation, excretion, and egestion. Leaching and FPOM losses, which were not measured, were estimated to balance litterfall inputs. For each element in each stream, we assumed that inputs (litterfall) and losses (emergence, organism drift, predation, excretion, particulate drift, and leaching) were equal for an annual cycle. This assumption is certainly only an approximation because bed storage of particulates and subsequent loss during extreme floods takes place over longer than annual periods (J. Sedell and K. Cummins, *personal communication*).

No allowance was made for detritivore uptake of dissolved ions. We assumed excretion of K and Ca occurs as dissolved ions, subsequently unavailable to the detritivores. Egested material, however, is recycled back into the particulate organic matter pool, available for subsequent uptake by filter and deposit feeding detritivores. The K data discussed below

TABLE 12. Parameters of potassium and calcium dynamics derived from Fig. 8. Parameters are explained further in the text. LPOM = large particulate organic matter

	Potassium			Calcium		
Parameter	Old field	Pine plantation	Hardwood forest	Old field	Pine plantation	Hardwood forest
LPOM turnover rate (yr^{-1})	36.5	13.3	62.0	4.5	2.9	2.0
LPOM processing efficiency (%)	98.6	98.4	99.9	87.2	88.2	95.8
Ingestion of crayfish (% of total ingestion)	5	8	3	50	53	25
Litterfall ingestion efficiency (%)	157.3	164.7	127.5	104.6	102.5	80.4
Detritivore production (mg element m^{-2})	15.8	17.1	21.0	49.9	35.1	25.4
Insect detritivore production (mg element m^{-2})	15.0	16.8	20.9	9.5	19.3	19.7
Crayfish production (mg element m^{-2})	0.8	0.3	0.1	40.5	15.8	5.7
Production efficiency (%)	1.0	2.6	1.5	1.1	1.2	0.6
Insect turnover (yr^{-1})	3.2	6.2	4.4	2.6	5.4	3.7
Crayfish turnover (yr^{-1})	0.11	0.05	0.03	0.10	0.05	0.04
Assimilation/litterfall (%)	34.2	40.3	29.7	58.2	59.6	31.1
Recycled (%)	78.3	75.5	76.7	44.3	42.0	61.3
FPOM and solution/litterfall (%)	64.4	58.1	70.3	29.0	28.7	64.7

should be evaluated in light of possible uptake of dissolved ions.

Input of K and Ca as litterfall were greatest in the old field stream and least in the pine plantation stream (Fig. 8). This contrasts with biomass inputs, which were greatest in the hardwood forest and least in the old field stream (Table 3), because of the high elemental concentrations of old field litter and low elemental concentrations of pine litter. Standing crops of large particulate organic matter K were very low because of the high degree of leaching of this element, especially from the hardwood forest litter. Calcium standing crops in LPOM were much higher, suggesting that this element was not leached. Calcium concentrations in LPOM were similar to litterfall concentrations throughout the year (Tables 3 and 4).

Turnover of LPOM, litterfall over LPOM standing crop, was much faster for K than Ca because of rapid K leaching, especially from hardwood litter (Table 12). Turnover of LPOM calculated from the Ca budgets should, as discussed above, approximate weight loss. From the Ca data, LPOM turnover was fastest in the old field stream and slowest in the hardwood forest stream. The LPOM turnover rates in Table 12 would be somewhat greater if lateral movement inputs were included and less if corrections were made for deep storage of detritus. In litterbag studies in a second-order stream on another forested Coweeta watershed, turnover rates of 0.9 to 7.2 yr^{-1} were found for various species (J. B. Waide, *personal communication*). Incorporating these rates into a model which predicted LPOM standing crop from species specific leaffall inputs, an average turnover rate of ≈ 1.5 yr^{-1} was found.

Drift of K and Ca as LPOM was greatest from the old field stream and least from the hardwood forest stream. Expressed as a percentage of litterfall input and subtracted from 100% to reflect percent processed, the LPOM processing efficiency (Webster 1977) was very high for both elements in all 3 streams (Table 12). That is, a large fraction of litterfall input was processed by combined chemical, mechanical, invertebrate, and microbial action to FPOM and dissolved material by the time it left the streams. Processing efficiencies of LPOM were greater for K than Ca, reflecting the solubility of K. For both elements, LPOM processing efficiencies were greatest in the hardwood forest stream.

The ingestion data further point to dissolved K uptake by the insect detritivores. The percent of total ingestion attributable to crayfish is much greater for Ca than for K (Table 12). If there were no dissolved uptake, these percentages would be identical for the 2 elements in each stream. The lower percent K ingestion by crayfish can probably be attributed to dissolved K uptake by the insects, as suggested by the high estimates of consumptive indices from ^{134}Cs elimination (Table 10).

Litterfall ingestion efficiencies were calculated as total detritivore ingestion over litterfall input (Table 12). As shown by these figures, total detritivore ingestion exceeded litterfall in all cases except Ca in the hardwood forest stream. Litterfall ingestion efficiencies > 100% indicated by K probably resulted from dissolved ion uptake. However, high Ca litterfall ingestion efficiencies appear to indicate reingestion of egested materials. For comparisons of ingestion at 1 trophic level to input at the previous level (I_n/I_{n-1}), efficiencies of 20% or less are usual (Kozlovsky 1968). However, in saprovore food chains, where consumer activities do not alter the rate of resource supply, higher food base utilization is possible (Wiegert and

Owen 1971). In the Coweeta stream ecosystems, high ingestion efficiencies probably result from filter feeder and deposit feeder ingestion of primarily shredder feces. Short and Maslin (1977) showed clearly that collectors utilize shredder feces. Because food that a stream detritivore ingests and subsequently egests is normally transported downstream prior to reingestion by another detritivore, the process is not cyclic but is better described as spiralling (Webster 1975, Wallace et al. 1977).

In terms of K, insect detritivore standing crops were similar in the old field and hardwood forest streams and lower in the pine plantation stream (Fig. 8). Crayfish standing crops were highest in the old field stream and lowest in the hardwood forest stream. Crayfish standing crop as K exceeded insect detritivores in the old field and pine plantation streams but not in the hardwood forest streams. Because of the slow turnover of K by crayfish (Table 9), uptake of K, whether ingestion or dissolved uptake, represented only a small portion of total K uptake by detritivores (Table 12). Because of a fairly low detritivore assimilation efficiency, most K was egested. Most of the remainder was lost as excretion with a small fraction being lost to emergent adults, drift, and predation.

Insect detritivore standing crops of Ca were greatest in the hardwood forest stream and similar in the other 2 streams (Fig. 8), reflecting biomass standing crops. Crayfish standing crops as Ca were highest in the old field stream and lowest in the hardwood forest stream. Due to the very high Ca concentrations of crayfish, Ca standing crops as crayfish greatly exceeded insect standing crops. Even with very slow Ca turnover (Table 9), estimated crayfish ingestion of Ca was about as great as insect ingestion because of the high crayfish standing crops. Crayfish ingestion represented 50% of total Ca ingestion in the old field stream, 53% in the pine plantation stream, but only 25% in the hardwood forest stream (Table 12). Because of their highly efficient assimilation of Ca, much of the detritivore Ca excretion was due to crayfish. Egestion of particulate Ca back to the particulate organic matter pool was mainly by insects.

Detritivore production was calculated as the sum of insect emergence, organism drift, and predation (Table 12). Insect production was greatest in the hardwood forest stream and least in the old field stream in terms of both elements. Crayfish contributed very little to production as K, however, because of high Ca concentration in drifting crayfish, their contribution to production as Ca was substantial. Because of the contribution from crayfish drift, total detritivore production as Ca was greatest in the old field stream and least in the hardwood forest stream. The production efficiency, the ratio of detritivore production to litterfall input, was very low for all 3 streams and for both elements (Table 12). For both K and Ca, the production efficiency was highest in the pine plantation stream, reflecting primarily the large number of predators and high predation in this stream, and low litterfall input.

Insect turnover ratios, annual production over mean annual standing crop, ranged from 2.6 to 6.2 (Table 12). These turnover ratios are very similar to estimates made for other streams (Waters 1966, 1969; Hynes and Coleman 1968; Hamilton 1969; Fisher and Likens 1973; Waters and Crawford 1973; Cushman et al. 1975; Resh 1977). Turnover ratios estimated from the K budgets were somewhat higher than the Ca budget estimates because of the relatively higher K concentrations of emerging adult insects (Table 6). Turnover ratios were greatest in the pine plantation stream and least in the old field stream. The main contributing factor to high insect turnover in the pine plantation stream was predation.

Crayfish turnover ratios were much $<$ insect turnover ratios, as might be predicted from the element turnover rates (Table 9). Turnover ratios for the 2 elements were identical (within rounding error) because they were calculated from the same biomass data. Crayfish turnover ratios were greatest in the old field stream and least in the hardwood forest stream. This is entirely attributable to crayfish drift because this was the only crayfish production parameter measured. Inclusion of other factors of crayfish production, such as predation by raccoons and loss of shed exoskeletons, would lead to higher crayfish production and turnover estimates.

More than either ingestion or production, the actual utilization of nutrients by the detritivores is best expressed by assimilation. Assimilated material is what is actually utilized, either in anabolic (production) or catabolic (maintenance) processes. Percent litterfall assimilation was calculated as detritivore production plus excretion as a percent of litterfall input (Table 12). For both elements, assimilation was greatest in the pine plantation stream and least in the hardwood forest stream. Thus, it appears that the detritivores were making less efficient use of available resources in the unperturbed stream in the hardwood forest than in the 2 perturbed streams. This conclusion becomes clearer as other aspects of Fig. 8 are considered.

The percent recycled was calculated as egestion over ingestion (Table 12). The percent of ingested K recycled was similar for all 3 streams. The recycling of Ca was lower because of the very high Ca assimilation of crayfish. The greater recycling of Ca in the hardwood forest stream can be attributed to the lower abundance of crayfish in this stream. This can also be interpreted as lesser utilization by the hardwood forest stream detritivores.

Input of litterfall not lost from the streams as LPOM, production, or excretion we assumed was lost as leachate and FPOM (Table 12). This loss was relatively high for K, and may have been even higher because much of the excretion loss may have come

TABLE 13. Comparison of various potassium and calcium losses. All values in g/yr

	Potassium			Calcium		
	Old field	Pine plantation	Hardwood forest	Old field	Pine plantation	Hardwood forest
Dissolved	65 260	66 680	86 410	122 830	84 950	121 820
Sediment	77.4	36.9	22.7	1 427	956	712
Large particulate organic matter	5.87	4.06	1.19	160	133	128
Fine particulate organic matter	...	...	...	361[1]	235[1]	1 984[1]
Organism drift[2]	0.63	0.30	0.35	13.5	7.3	5.7
Insect emergence[2]	1.05	1.22	3.26	0.23	0.40	0.77

[1] These figures are based on the assumption of no calcium leaching from organic material.
[2] These figures include drift and emergence of predators.

from dissolved K uptake, as previously discussed. Based on the similar Ca concentrations of LPOM and litterfall (Tables 3 and 4), there appears to be little leaching of Ca. Therefore, the estimates of Ca leaching and FPOM losses are mainly FPOM. Assuming no Ca leaching, FPOM losses relative to inputs were very similar for the old field and pine plantation streams but much larger for the hardwood forest stream (Table 12). This further exemplifies the lower detritivore utilization of available resources in the hardwood forest stream. However, this conclusion and the other conclusions summarized below must be evaluated according to the accuracy of measurements made in this study and with realization of the factors such as lateral movement and deep storage, which were not measured.

CONCLUSIONS

Perturbations to Watersheds 6 and 17, the old field and pine plantation watersheds, caused changes in the stream ecosystems of these watersheds. These changes were reflected in the K and Ca dynamics of the stream. Because of the indicated probability of dissolved K uptake, we have based our conclusions primarily on our determinations of Ca dynamics. The major changes correlated with watershed perturbation were:

1. Higher litterfall inputs in the old field stream despite a lower litterfall rate because of high elemental concentrations of the old field vegetation.
2. Lower litterfall inputs in the pine plantation stream because of lower elemental concentrations and a lower litterfall rate.
3. More rapid turnover of LPOM if leaching is not considered.
4. Greater inorganic sediment loss.
5. Greater LPOM drift, indicating lower LPOM processing efficiency.
6. Increased production and standing crop of crayfish.
7. Decreased benthic insect production and standing crop.
8. Increased utilization of litterfall inputs, as indicated by detritivore ingestion, assimilation, and production.
9. Less spiralling of Ca.

Taken together, these observed changes suggest that there was less efficient physical processing of inputs in the perturbed streams. Organic and inorganic particulate materials were passed through the streams more rapidly with less processing of the organic matter.

Biologically, there were shifts in the detritivore communities to greater importance of crayfish with lesser importance of the benthic insects. We do not know what caused this shift, though it is correlated with higher nutrient levels (Table 1). This shift illustrates the tendency for generalists to dominate perturbed ecosystems, whereas specialists dominate unperturbed ecosystems (Odum 1969). Another important consideration involves timing of inputs. As indicated in Fig. 4, watershed perturbations caused a less pronounced autumnal litterfall peak. Therefore, the litterfall available to benthic insects with life cycles adapted to autumnal inputs was a smaller fraction of total annual litterfall. Crayfish, because they are not univoltine, may be better adapted to the more uniform litterfall input.

Associated with the community change was an increase in biological utilization of litterfall inputs. A factor not considered in this study was the biological decomposability of the litterfall inputs. Leaf species such as black locust and blackberry decompose rapidly, whereas oaks and rhododendron are very slow to decompose (Petersen and Cummins 1974, J. B. Waide and R. L. Todd, *personal communication*). This factor certainly contributed to greater biological processing in the old field stream and lesser biological processing in the hardwood forest stream.

One justification for including stream studies in watershed nutrient studies has been the effect of instream processes on watershed budgets. We conclude from this study that effects of the stream ecosystems at Coweeta on whole watershed K and Ca budgets are very small. If there were no instream biological, chemical, or mechanical processing of litterfall inputs to dissolved K and Ca, the change in the net dissolved element losses from the watersheds would be $<4\%$. Losses of K and Ca from the streams and watersheds due to insect emergence and invertebrate drift were small in comparison to abiotic losses (Table 13). The major element losses for all 3 watersheds were dissolved ions. Sediment losses were next largest, except for the hardwood forest where organic particulate loss of Ca was > sediment loss.

Finally, we observed that the streams in the old field and pine plantation watersheds are functionally little different from the unperturbed stream. Though we do not know the magnitude of the stream perturbation, nor the resistance of the streams to the perturbations as reflected by functional changes during and immediately following perturbation, it is evident that any changes were followed by rapid recovery. This high resilience of the stream ecosystems can be attributed to the rapid turnover of the components. Any current differences between the perturbed and unperturbed streams reflect primarily differences in quantity, quality, and timing of allochthonous inputs. In general, the great resilience of stream ecosystems based on their internal characteristics is limited by the external regulation.

Acknowledgments

We are grateful to the many people who provided assistance and advice during the course of this research, including G. R. Best, S. Best, D. A. Crossley, Jr., C. S. Gist, F. B. Golley, C. D. Monk, J. B. Waide, J. B. Wallace, J. D. Webster, S. B. Webster, and R. G. Wiegert. We thank W. T. Swank, J. E. Douglass, and all the Coweeta staff for their assistance and for making their data available. W. R. Woodall, Jr., kindly provided his data and helped in initial phases of experimental design. K. W. Cummins, J. R. Sedell, and E. F. Benfield read versions of the manuscript and provided many helpful comments.

This work was supported by the Coweeta Site of the Eastern Deciduous Forest Biome, US-IBP, funded by the National Science Foundation under Interagency Agreement AG-199, BMS69-01147 A09 with the Energy Research and Development Administration, Oak Ridge National Laboratory. Support for the preparation of this version of the manuscript was provided by the Department of Biology and Center for Environmental Studies, Virginia Polytechnic Institute and State University. This is contribution No. 326 from the Eastern Deciduous Forest Biome, US-IBP.

Literature Cited

Attiwill, P. M. 1968. The loss of elements from decomposing litter. Ecology **49**:142–144.

Ball, R. C., T. A. Wojtalik, and F. F. Hooper. 1963. Upstream dispersion of radiophosphorus in a Michigan trout stream. Michigan Academician **48**:57–64.

Barrett, P. H. 1957. Potassium concentrations in fertilized trout lakes. Limnology and Oceanography **2**:287–294.

Benke, A. C. 1972. An experimental field study on the ecology of coexisting larval odonates. Doctoral dissertation. University of Georgia, Athens, Georgia, USA.

Bocock, K. L. 1963. The digestion and assimilation of food by *Glomeris*. Pages 85–91 *in* J. Doeksen and J. van der Drift, editors. Soil organisms. North Holland Publishing Company, Amsterdam, Netherlands.

Boling, R. H., Jr., E. D. Goodman, J. A. Van Sickle, J. O. Zimmer, K. W. Cummins, R. C. Petersen, and S. R. Reice. 1974. Toward a model of detritus processing in a woodland stream. Ecology **56**:141–151.

Bowen, H. J. M. 1966. Trace elements in biochemistry. Academic Press, New York, New York, USA.

Bryan, G. W., and E. Ward. 1962. Potassium metabolism and the accumulation of cesium-137 by decapod crustacea. Journal, Marine Biological Association of the United Kingdom **42**:199–241.

Cairns, J., Jr., and K. L. Dickson. 1977. Recovery of streams from spills of hazardous materials. Pages 24–42 *in* J. Cairns, Jr., K. L. Dickson, and E. E. Herricks, editors. Recovery and restoration of damaged ecosystems. University Press of Virginia, Charlottesville, Virginia, USA.

Clark, E. W. 1958. A review of literature on calcium and magnesium in insects. Annals, Entomological Society of America **51**:142–154.

Coffman, W. P., K. W. Cummins, and J. C. Wuycheck. 1971. Energy flow in a woodland stream ecosystem: I. tissue support trophic structure of the autumnal community. Archiv für Hydrobiologie **68**:232–276.

Comar, C. L., and R. H. Wasserman. 1964. Strontium. Pages 523–572 *in* C. L. Comar and F. Bronner, editors. Mineral metabolism. Volume II. Academic Press, New York, New York, USA.

Cromack, K., Jr. 1972. Litter production and litter decomposition in a mixed hardwood watershed and in a white pine watershed at Coweeta Hydrologic Station, North Carolina. Doctoral dissertation. University of Georgia, Athens, Georgia, USA.

———, and C. D. Monk. 1975. Litter production, decomposition, and nutrient cycling in a mixed hardwood watershed and a white pine forest. Pages 609–624 *in* F. G. Howell, J. B. Gentry, and M. H. Smith, editors. Mineral cycling in southeastern ecosystems. Energy Research and Development Administration Symposium Series (CONF-740513).

Crossley, D. A., Jr. 1963. Consumption of vegetation by insects. Pages 427–430 *in* V. Shultz and A. W. Klement, Jr., editors. Radioecology. Reinhold, New York, New York, USA.

———. 1967. Radioisotope measurement of food consumption by leaf beetle species, *Chrysomela knabi* Brown. Ecology **47**:1–8.

———. 1967. Comparative movement of ^{106}Ra, ^{60}Co, and ^{137}Cs in arthropod food chains. Pages 687–695 *in* D. J. Nelson and F. C. Evans, editors. Symposium on radioecology. CONF-670503, United States Atomic Energy Commission.

Crossley, D. A., Jr., and C. S. Gist. 1973. Use of radioisotopes in modeling soil microcommunities. Pages 258–278 *in* D. L. Dindal, editor. Proceedings, first soil microcommunities conference. CONF-711076, United States Atomic Energy Commission.

Crossley, D. A., Jr., and H. F. Howden. 1961. Insect-vegetation relationships in an area contaminated by radioactive wastes. Ecology **42**:302–317.

Crossley, D. A., Jr., and M. E. Pryor. 1960. The uptake and elimination of Cs-137 by a grasshopper *Romalea microptera*. Health Physics **4**:16–20.

Crossley, D. A., Jr., and D. E. Reichle. 1969. Analysis of transient behavior in insect food chains. BioScience **19**:341–343.

Crossley, D. A., Jr., D. E. Reichle, and C. A. Edwards. 1971. Intake and turnover of radioactive cesium by earthworms. Sonderdruck aus Pedobiologia **11**:71–76.

Crossley, D. A., Jr., and J. H. Schnell. 1961. Comparative elimination of radiocesium and radiostrontium by grasshoppers. Annals, Entomological Society of America **54**:459–461.

Crossley, D. A., Jr., and R. I. Van Hook, Jr. 1970. Energy assimilation by the house cricket, *Acheta domesticus*, measured with radioactive chromium-51. Annals, Entomological Society of America **63**:512–525.

Cummins, K. W. 1969. Energy budgets. Pages 30–37 *in* The stream ecosystem. Technical Report Number 7, Institute of Water Research. Michigan State University, East Lansing, Michigan, USA.

———. 1973. Trophic relations of aquatic insects. Annual Review of Entomology **18**:183–206.

———. 1974. Structure and function of stream ecosystems. BioScience **24**:631–641.

———, R. C. Petersen, F. O. Howard, J. C. Wuycheck, and V. I. Holt. 1973. The utilization of leaf litter by stream detritivores. Ecology **54**:336–345.

Cushman, R. M., J. W. Elwood, and S. G. Hildebrand. 1975. Production dynamics of *Alloperla mediana* Banks (Plecoptera: Chloroperlidae) and *Diplectrona modesta* Banks (Trichoptera: Hydropsychidae) in Walker Branch, Tennessee. Oak Ridge National Laboratory TM-4957, Oak Ridge, Tennessee, USA.

Davis, J. J. 1963. Cesium and its relationships to potassium in ecology. Pages 539–556 *in* V. Schultz and A. W. Klement, Jr., editors. Radioecology. Reinhold, New York, New York, USA.

Day, F. P., Jr., and C. D. Monk. 1974. Vegetation patterns on a southern Appalachian watershed. Ecology **55**:1064–1074.

Dils, R. E. 1957. A guide to the Coweeta Hydrologic Laboratory. Miscellaneous Publications, Southeast Forest Experimental Station, Asheville, North Carolina, USA.

Douglass, J. E., D. R. Cochrane, and D. W. Hill. 1969. Low herbicide concentration found in streamflow after a grass cover is killed. United State Department of Agriculture Forest. Service, Research Note SE-108.

Elliott, J. M. 1971. Upstream movement of benthic invertebrates in a lake district stream. Journal of Animal Ecology **40**:235–252.

Elwood, J. W., and R. M. Cushman. 1975. The life history and ecology of *Peltoperla maria* (Plecoptera: Peltoperlidae) in a small spring fed stream. Verhandlungen Internationale Vereinigung für Theoretische und Angewandte Limnologie **19**:3050–3056.

Elwood, J. W., S. G. Hildebrand, and J. J. Beauchamp. 1976. Contributions of gut contents to the concentration and body burden of elements in *Tipula* spp. from a spring-fed stream. Journal, Fisheries Research Board of Canada **33**:1930–1938.

Fisher, S. G. 1977. Organic matter processing by a stream-segment ecosystem: Fort River, Massachusetts, U.S.A. Internationale Revue der Gesamten Hydrobiologie **62**:701–727.

———, and G. E. Likens. 1972. Stream ecosystem: organic energy budget. BioScience **22**:33–35.

———. 1973. Energy flow in Bear Brook, New Hampshire: an integrative approach to stream ecosystem metabolism. Ecological Monographs **43**:421–439.

Fitzpatrick, L. C. 1973*a*. Energy allocation in the Allegheny mountain salamander *Desmognanthus ochrophaeus*. Ecological Monographs **43**:43–58.

———. 1973*b*. Influence of seasonal temperatures on the energy budget and metabolic rates of the northern two-lined salamander *Eurycea bislineata bislineata*. Comparative Biochemistry and Physiology: Part A, Comparative Physiology **43**:807–818.

Gere, G. 1956. The examination of the feeding biology and the humificative function of Diplopoda and Isopoda. Acta Biologica Academiae Scientiarum Hungaricae **6**:257–271.

Gist, C. S. 1972. Analysis of mineral pathways in a cryptozoan food web. Doctoral dissertation. University of Georgia, Athens, Georgia, USA.

Gosz, J. R., G. E. Likens, and F. H. Bormann. 1973. Nutrient release from decomposing leaf and branch litter in the Hubbard Brook Forest, New Hampshire. Ecological Monographs **43**:173–191.

Hall, C. A. S. 1972. Migration and metabolism in a temperate stream ecosystem. Ecology **53**:585–604.

Hamilton, A. L. 1969. On estimating annual production. Limnology and Oceanography **14**:771–782.

Hartenstein, R. 1964. Feeding, digestion, glycogen, and the environmental conditions of the digestive system in *Oniscus asellus*. Journal of Insect Physiology **10**:611–621.

Hasanen, E., S. Kolehmainen, and J. K. Miettinen. 1967. Biological half-times of ^{137}Cs and ^{22}Na in different fish species and their temperature dependence. Pages 912–920 *in* B. Aberg and F. P. Hungate, editors. Radioecological concentration processes. Pergamon Press, New York, New York, USA.

Heiman, D. R., and A. W. Knight. 1975. The influence of temperature on the bioenergetics of the carnivorous stonefly nymph, *Acroneura californica* Banks (Plecoptera, Perlidae). Ecology **56**:105–116.

Hsiao, S. C. 1963. The radioecology of calcium. Pages 525–532 *in* V. Schultz and A. W. Klement, Jr., editors. Radioecology. Reinhold, New York, New York, USA.

Hubbell, S. P., A. Sikora, and O. H. Paris. 1965. Radiotracer, gravimetric and calorimetric studies of ingestion and assimilation rates of an isopod. Health Physics **11**:1485–1501.

Hutchinson, G. E. 1948. Circular causal systems in ecology. Annals, New York Academy of Sciences **50**:221–246.

———. 1963. The prospect before us. Pages 683–690 *in* D. G. Frey, editor. Limnology in North America. University of Wisconsin Press, Madison, Wisconsin, USA.

Hynes, H. B. N. 1970. The ecology of running waters. University of Toronto Press, Toronto, Ontario, Canada.

———. 1975. The stream and its valley. Verhandlungen Internationale Vereinigung für Theoretische und Angewandte Limnologie **19**:1–15.

———, and M. J. Coleman. 1968. A simple method of assessing the annual production of stream benthos. Limnology and Oceanography **13**:569–573.

Johnson, P. L., and W. T. Swank. 1973. Studies of cation budgets in the southern Appalachians on four experimental watersheds with contrasting vegetation. Ecology **54**:70–80.

Kevern, N. R., N. A. Griffith, and T. Grizzard. 1964. Biological halflife of ^{134}Cs in carp and two aquatic insects. Pages 101–102 *in* Health Physics Division Annual Progress Report for period ending July 31, 1964. Oak Ridge National Laboratory ORNL-3697, Oak Ridge, Tennessee, USA.

Kornberg, H. A. 1961. The use of element-pairs in radiation hazard assessment. Health Physics **6**:46–62.

Kowal, N. E. 1969. Ingestion rate of a pine-mor oribatid mite. American Midland Naturalist **81**:595–597.

Kozlovsky, D. C. 1968. A critical evaluation of the trophic level concept. I. Ecological efficiencies. Ecology **49**:48–50.

Lawton, J. H. 1971. Ecological energetic studies on larvae of the damselfly *Pyrrhosoma nymphula* (Sulzer) (Odonata: zygoptera). Journal of Animal Ecology **40**:385–423.

Leopold, A. 1941. Lakes in relation to terrestrial life patterns. Pages 17–22 *in* A symposium on hydrobiology. University of Wisconsin Press, Madison, Wisconsin, USA.

Leopold, L. B., M. G. Wolman, and J. R. Miller. 1964. Fluvial processes in geomorphology. W. H. Freeman, San Francisco, California, USA.

Lindeman, R. L. 1942. The trophic-dynamic aspect of ecology. Ecology **23**:339–418.

Luferov, V. P. 1962. On feeding of Pelopiinae larvae (Diptera Tendipedidae). Transactions, Institute of Biology of Inland Waters, Academy of Science, USSR **4**:232–245.

Mackay, R. J., and J. Kalff. 1973. Ecology of two related species of caddisfly larvae in the organic substrates of a woodland stream. Ecology **54**:449–511.

Margalef, R. 1960. Ideas for a synthetic approach to the ecology of running waters. Internationale Revue der Gesamten Hydrobiologie **45**:133–153.

———. 1963. On certain unifying principles in ecology. American Naturalist **97**:357–374.

McDiffett, W. F. 1970. The transformation of energy by a stream detritivore *Pteronarcys scotti* (Plecoptera). Ecology **51**:995–998.

McNeill, K. G., and O. A. D. Trojan. 1960. The cesium-potassium discrimination ratio. Health Physics **4**:109–112.

Minckley, W. L. 1963. The ecology of a spring stream, Doe Run, Meade Co., Ky. Wildlife Monographs **11**:1–124.

———. 1964. Upstream movements of *Gammarus* (Amphipoda) in Doe Run, Meade County, Kentucky. Ecology **45**:196–197.

Minshall, G. W. 1967. Role of allochthonous detritus in the trophic structure of a woodland spring brook community. Ecology **48**:139–149.

Monakov, A. V. 1972. Review of studies on feeding of aquatic invertebrates conducted at the Institute of Biology of Inland Waters, Academy of Science, USSR. Journal, Fisheries Research Board of Canada **29**:363–383.

Monk, C. D. 1975. Nutrient losses in particulate form as weir pond sediments from four unit watersheds in the southern Appalachians. Pages 862–867 *in* F. G. Howell, J. B. Gentry, and M. H. Smith, editors. Mineral cycling in southeastern ecosystems. Energy Research and Development Administration Symposium Series (CONF-740513).

Moulder, B. C., D. E. Reichle, and S. I. Auerbach. 1970. Significance of spider predation in the energy dynamics of forest floor arthropod communities. Oak Ridge National Laboratory ORNL-4452, Oak Ridge, Tennessee, USA.

Muller, K. 1954. Investigations on the organic drift in North Swedish streams. Report, Institute of Freshwater Research Drottningholm **35**:133–148.

Mundie, J. H. 1971. Techniques for sampling emerging aquatic insects. Pages 80–93 *in* W. T. Edmondson and G. G. Winberg, editors. A manual for the assessment of secondary productivity in fresh waters. International Biological Project Handbook Number 17. Blackwell Scientific Publication, Oxford and Edinburgh, United Kingdom.

Nelson, D. J. 1963. The strontium and calcium relationships in Clinch and Tennessee River mollusks. Pages 203–211 *in* V. Schultz and A. W. Klement, Jr., editors. Radioecology. Reinhold, New York, New York, USA.

———, and D. C. Scott. 1962. Role of detritus in the productivity of a rock outcrop community in a piedmont stream. Limnology and Oceanography 7:396–413.

Odum, E. P. 1969. The strategy of ecosystem development. Science **164**:262–270.

O'Neill, R. V., W. F. Harris, B. S. Ausmus, and D. E. Reichle. 1975. A theoretical basis for ecosystem analysis with particular reference to element cycling. Pages 28–40 *in* F. G. Howell, J. B. Gentry, and M. H. Smith, editors. Mineral cycling in southeastern ecosystems. Energy Research and Development Administration Symposium Series (CONF-740513).

Ophel, I. L., and C. D. Fraser. 1970. Calcium and strontium discrimination by aquatic plants. Ecology **51**:324–327.

Ophel, I. L., and J. M. Judd. 1969. Strontium-calcium relationships in aquatic food chains. Pages 221–225 *in* D. J. Nelson and F. C. Evans, editors. Symposium on radioecology, CONF-670503. United States Atomic Energy Commission.

Patrick, R. 1970. Benthic stream communities. American Scientist **58**:546–549.

Patten, B. C. 1973. Need for an ecosystem perspective in eutrophication modeling. Pages 83–87 *in* E. J. Middlebrooks, D. H. Falkenborg, and T. E. Maloney, editors. Modeling the eutrophication process. Ann Arbor Science, Ann Arbor, Michigan, USA.

Petersen, R. C., and K. W. Cummins. 1974. Leaf processing in a woodland stream. Freshwater Biology **4**:343–368.

Phipps, D. A. 1976. Metals and metabolism. Clarendon Press, Oxford, England.

Reichle, D. E. 1967. Radioisotope turnover and energy flow in terrestrial isopod populations. Ecology **48**:351–366.

———, and D. A. Crossley, Jr. 1965. Radiocesium dispersion in a cryptozoan food web. Health Physics **11**:1375–1384.

———. 1967. Trophic level concentrations of cesium-137, sodium, and potassium in forest arthropods. Pages 678–686 *in* D. J. Nelson and F. C. Evans, editors. Symposium on radioecology. CONF-670503, United States Atomic Energy Commission.

Resh, V. H. 1977. Habitat and substrate influences on population and production dynamics by a stream caddisfly, *Ceraclea ancylus* (Leptoceridae). Freshwater Biology **7**:261–277.

Ringer, S. 1882. An investigation regarding the action of rubidium and cesium salts compared with the action of potassium salts on the ventricles of the frog's heart. Journal of Physiology (London) **4**:370–379.

Roos, T. 1957. Studies on upstream migration in adult stream dwelling insects. Report, Institute of Freshwater Research Drottningholm **38**:167–193.

Schurr, J. M., and M. N. Stamper. 1962. Model for the accumulation of strontium and calcium by recently molted crayfish (*Cambarus longulus longerostris* Ort.). Limnology and Oceanography **7**:474–477.

Sedell, J. R. 1971. Trophic ecology and natural history of *Neophylax concinnus* and *N. oligins* (Trichoptera: Limephilidae). Doctoral dissertation. University of Pittsburgh, Pittsburgh, Pennsylvania, USA.

Shapas, T. J., and W. L. Hilsenhoff. 1976. Feeding habits of Wisconsin's predominant lotic Plecoptera, Ephemeroptera, and Trichoptera. Great Lakes Entomology **9**:175–188.

Shaw, J., and R. H. Stobbart. 1963. Osmotic and ionic regulation in insects. Advances in Insect Physiology **1**:315–399.

Short, R. A., and P. E. Maslin. 1977. Processing of leaf litter by a stream detritivore: effect on nutrient availability to collectors. Ecology **58**:935–938.

Strahler, A. N. 1957. Quantitative analysis of watershed geomorphology. American Geophysical Union, Transactions **38**:913–920.

Swank, W. T., and J. E. Douglass. 1974. Streamflow greatly reduced by converting deciduous hardwood stands to pine. Science **185**:857–859.

———. 1975. Nutrient flux in undisturbed and manipulated forest ecosystems in the southern Appalachian Mountains. Publication de l'Association Internationale des Sciences Hydrologiques Symposium de Tokyo (Décêmbre 1975) **117**:445–456.

Teal, J. M. 1957. Community metabolism in a temperate cold spring. Ecological Monographs **27**:283–302.

Thomas, W. A. 1970. Weight and calcium losses from decomposing tree leaves on land and in water. Journal of Applied Ecology **7**:237–242.

Tilly, L. J. 1968. The structure and dynamics of Cone Spring. Ecological Monographs **38**:169–197.

Trama, F. B. 1957. The transformation of energy by an aquatic herbivore *Stenonema pulchellum* (Ephemeroptera). Doctoral dissertation. University of Michigan, Ann Arbor, Michigan, USA.

Van Hook, R. I., Jr. 1971. Energy and nutrient dynamics of spider and orthopteran populations in a grassland ecosystem. Ecological Monographs **41**:1–26.

———, and D. A. Crossley, Jr. 1969. Assimilation and biological turnover of Cesium-134, Iodine-131 and Chromium-51 in brown crickets, *Acheta domesticus* (L). Health Physics **16**:463–467.

Vannote, R. L. 1969. Detrital consumers in natural systems,

p. 20–23 *in* The stream ecosystem. Technical Report Number 7, Institute of Water Research, Michigan State University, East Lansing, Michigan, USA.

Waldbauer, G. P. 1968. The consumption and utilization of food by insects. Advances in Insect Physiology **5**:229–288.

Wallace, J. B., J. R. Webster, and W. R. Woodall, Jr. 1977. The role of filter feeders in stream ecosystems. Archiv für Hydrobiologie **79**:506–532.

Wallace, J. B., W. R. Woodall, Jr., and F. F. Sherberger. 1970. Breakdown of leaves by feeding of *Peltoperla maria* nymphs. Annals, Entomological Society of America **63**:562–567.

Wasserman, R. H., and A. N. Taylor. 1969. Some aspects of the intestinal absorption of calcium, with special reference to vitamin D. Pages 321–404 *in* C. L. Comar and F. Bronner, editors. Mineral metabolism. Volume III. Academic Press, New York, New York, USA.

Waters, T. F. 1966. Production rate, population density, and drift of a stream invertebrate. Ecology **47**:595–604.

———. 1969. The turnover ratio in production ecology of freshwater invertebrates. American Naturalist **103**:173–185.

———, and G. W. Crawford. 1973. Annual production of a stream mayfly population: a comparison of methods. Limnology and Oceanography **18**:286–296.

Webster, J. R. 1975. Analysis of potassium and calcium dynamics in stream ecosystems on three southern Appalachian watersheds of contrasting vegetation. Doctoral dissertation. University of Georgia, Athens, Georgia, USA.

———. 1977. Large particulate organic matter processing in stream ecosystems. Pages 505–526 *in* D. L. Correll, editor. Watershed research in eastern North America. Smithsonian Institution, Washington, District of Columbia, USA.

———, and D. A. Crossley, Jr. 1978. Evaluation of two models for predicting elemental accumulation by arthropods. Environmental Entomology **7**:411–417.

Webster, J. R., J. B. Waide, and B. C. Patten. 1975. Nutrient cycling and the stability of ecosystems. Pages 1–27 *in* F. G. Howell, J. B. Gentry, and M. H. Smith, editors. Mineral cycling in southeastern ecosystems. Energy Research and Development Administration Symposium Series (CONF-740513).

Wetzel, R. G. 1975. Limnology. Saunders, Philadelphia, Pennsylvania, USA.

Whittaker, R. H. 1975. Communities and ecosystems. MacMillan, New York, New York, USA.

Wiegert, R. G., and D. F. Owen. 1971. Trophic structure, available resources and population density in terrestrial vs. aquatic ecosystems. Journal of Theoretical Biology **30**:69–81.

Wigglesworth, V. B. 1965. The principles of insect physiology. Methuen, London, England.

Wilhm, J. L. 1970. Transfer of radioisotopes between detritus and benthic macroinvertebrates in laboratory microecosystems. Health Physics **18**:277–284.

Williams, E. C., Jr., and D. E. Reichle. 1968. Radioactive tracers in the study of energy turnover by a grazing insect (*Chrysochus auratus* Fab.; Coleoptera Chrysomelidae). Oikos **19**:10–18.

Winterbourn, M. J. 1971. An ecological study of *Banksiola crotchi* Banks (Trichoptera, Phryganeidae) in Marion Lake, British Columbia. Canadian Journal of Zoology **49**:637–645.

Witkamp, M., and M. L. Frank. 1969. Loss of weight, ^{60}Co and ^{137}Cs from tree litter in three subsystems of a watershed. Environmental Science and Technology **3**:1195–1198.

Woodall, W. R., Jr. 1971. A comparison of aquatic insects in four different watersheds. Master's thesis. University of Georgia, Athens, Georgia, USA.

———. 1972. Nutrient pathways in small mountain streams. Doctoral dissertation. University of Georgia, Athens, Georgia, USA.

———, and J. B. Wallace. 1972. The benthic fauna of four small southern Appalachian streams. American Midland Naturalist **88**:393–407.

———. 1975. Mineral pathways in small Appalachian streams. Pages 408–422 *in* F. G. Howell, J. B. Gentry, and M. H. Smith, editors. Mineral cycling in southeastern ecosystems. Energy Research and Development Administration Symposium Series (CONF-740513).

Commentary on Paper F2

Bencala, K. E. & Walters, R. A. (1983) Simulation of solute transport in a mountain pool-and-riffle stream: a transient storage model.

This paper makes up part of a series on the dynamics of solutes in natural stream channels. Bencala was the leading author of all of these papers (Bencala *et al.*, 1983, 1984; Bencala & Walters 1983; Bencala 1984). The ensemble is essential reading if one is to disentangle the contribution of the different compartments (water, sediments, hyporheic) to solute dynamics in streams.

The Bencala & Walters (1983) paper represents the first application of solute transport models in headwater streams. The main merit of the paper is to adapt some expressions derived from engineering, formerly applied to large rivers and estuaries, to low-order streams with a significant pool-riffle channel structure. Hence, the paper initiated the adaptation of the advection-dispersion model – already applied to large rivers – to systems with a remarkably transient storage area. This paper and its companions based their approach on the addition and monitoring of conservative tracers (mostly chloride) to describe and model the water flows through the channel and its associated compartments. This procedure made obvious evidence of the complex hydrology which is characteristic of small streams. The hydrology, as well as the interaction with sediments and channel structures, has its outcome in the solute dynamics of the stream.

The transient storage zones are defined as those compartments related to the main channel, which have particular physical and chemical properties. Amongst the elements included in this concept are the saturated zones from the riparian areas, the hyporheic zones, the pools and dams occurring naturally in the system, and in general all the compartments showing a differentiated hydrology but exhibiting a hydrological connection to the main stream flow. Considering the particular flow patterns of the transient storage areas, the paper described how the solutes are transported according to the first-order mass transfer exchange. The role of the storage zone was also defined by means of first-order equations, and therefore its definition was operational and not passively considered.

Modelling these processes was approached by means of two-dimensional transport and first-order mass-transfer equations. The expressions include the basic terms of convection, dispersion, and lateral inflow which govern the dynamics of solutes. In the description of the storage area, the equations considered both the relative areas as well as the rates of exchange between and within the compartments, and a sink term. The success of the approach was made clear by the close match between the predicted and observed patterns. The methodological approach and the formulas proposed in these papers were later incorporated by the Stream Solute Workshop (1990), and are now widely used in stream hydrology and ecology.

SS

References

Bencala, K. E. (1984) Interactions of solutes and streambed sediment. 2. A dynamic analysis of coupled hydrological and chemical processes that determine solute transport. *Water Resour. Res.* **20**, 1804–1814.

Bencala, K. E. & Walters, R. A. (1983) Simulation of solute transport in a mountain pool-and-riffle stream: a transient storage model. *Water Resour. Res.* **19**, 718–724.

Bencala, K. E., Jackman, A. P., Kennedy, V. C., Avanzino, R. J. & Zellweger, G. W. (1983) Kinetic analysis of strontium and potassium sorption onto sands and gravels in a natural channel. *Water Resour. Res.* **19**, 725–731.

Bencala, K. E., Kennedy, V. C., Zellweger, G. W., Jackman, A. P. & Avanzino, R. J. (1984) Interactions of solutes and streambed sediment. 1. An experimental analysis of cation and anion transport in a mountain stream. *Water Resour. Res.* **20**, 1797–1803.

Stream Solute Workshop (1990) Concepts and methods for assessing solute dynamics in stream ecosystems. *J. North Am. Benthological Soc.* **9**, 95–119.

WATER RESOURCES RESEARCH, VOL. 19, NO. 3, PAGES 718–724, JUNE 1983

Simulation of Solute Transport in a Mountain Pool-and-Riffle Stream: A Transient Storage Model

KENNETH E. BENCALA AND ROY A. WALTERS

U.S. Geological Survey, Menlo Park, California 94025

The physical characteristics of mountain streams differ from the uniform and conceptually well-defined open channels for which the analysis of solute transport has been oriented in the past and is now well understood. These physical conditions significantly influence solute transport behavior, as demonstrated by a transient storage model simulation of solute transport in a very small (0.0125 m^3 s^{-1}) mountain pool-and-riffle stream. The application is to a carefully controlled and intensively monitored chloride injection experiment. The data from the experiment are not explained by the standard convection-dispersion mechanisms alone. A transient storage model, which couples dead zones with the one-dimensional convection-dispersion equation, simulates the general characteristics of the solute transport behavior and a set of simulation parameters were determined that yield an adequate fit to the data. However, considerable uncertainty remains in determining physically realistic values of these parameters. The values of the simulation parameters used are compared to values used by other authors for other streams. The comparison supports, at least qualitatively, the determined parameter values.

INTRODUCTION

Extensive knowledge exists concerning the mechanisms and simulation of solute transport in streams [*Fischer et al.*, 1979]. From the applications perspective, interest has focused on larger streams located near population centers. As a generalization, these streams are physically low slope, deeper than the roughest bed feature, and relatively uniform (possibly due to flow regulation). Chemically, these streams contain and transport rather high concentrations of diverse constituents. Small mountain pool-and-riffle streams are described by opposing characteristics. Chemically, there are fewer significant constituents present and these are at lower concentrations. These streams cannot be characterized as uniform either in dimension or gradient, and the bed may significantly influence solute transport.

Of late there has been interest in studying the physical [*Keller and Tally*, 1979; *Keller*, 1975; *Heede*, 1972], chemical [*Hubbard and Striffler*, 1973; *Lewis and Grant*, 1979], and solute transport [*Day*, 1975, 1977] characteristics of small mountain pool-and-riffle streams. Certain larger mountain streams maintain the significant pool-and-riffle characteristic. Studies of the overall hydrochemical and transport characteristics of these larger streams are now appearing [*Janda*, 1977; *Kennedy and Malcolm*, 1977] and will presumably become more important as society continues to seek new locations to discharge its chemical and thermal pollution.

In this study, application of a transient storage model to a field experiment is presented. The model must reflect the large variability in the physical characteristics of the stream in order to simulate the field experiment. Concurrently, the model is useful in smoothing out this same variability in order to make the simulations tractable.

The importance of differing physical scales becomes evident, because the short distances and small volumes of water in the system suggest that similarly small-scale physical processes occurring in the stream may have significant consequences. Thus the application of a one-dimensional (or pseudo–two-dimensional) transport analysis remains empirical. The model equations have previously been applied by numerous workers, mostly to improve the shortcomings of the advection-dispersion model, in large streams. After presenting the model and the simulation of the experiment, we will discuss the apparent distinctions between previous applications and the one presented here. This comparison supports the conclusion that the transient storage mechanism is a viable first approximation to the complexity of the physical processes.

Paper number 3W0373.
0043-1397/83/003W-0373$05.00

MODEL

The typical starting point for the study of solute transport in streams centers on the one-dimensional convection-dispersion analysis. The 'tail' of a solute tracer pulse is often more pronounced than can be accounted for by this analysis. (See *Fischer et al.* [1979] for a complete discussion of solute transport in real streams, the one-dimensional analysis, and alternatives.) A common method for simulating these long tails has been to allow for storage zones or 'dead zones' along the stream channel. These storage zones are assumed to be stagnant relative to the longitudinal flow of the stream and to obey a first-order mass transfer type of exchange relationship. That is, the exchange of solute between the main stream channel and a storage zone is proportional to the difference in concentration between the stream and the storage zone. Implementation of this model requires an estimate of the storage zone cross-sectional area and an empirical exchange coefficient. Under a variety of nomenclatures, various theoretical, laboratory flume, and field aspects of dead zone models have been analyzed [*Hays et al.*, 1966; *Thackston and Krenkel*, 1967; *Thackston and Schnelle*, 1970; *Pedersen*, 1977; *Valentine and Wood*, 1977; *Sabol and Nordin*, 1978; *Valentine and Wood*, 1979*a*, *b*; *Tsai and Holley*, 1979; *Nordin and Troutman*, 1980].

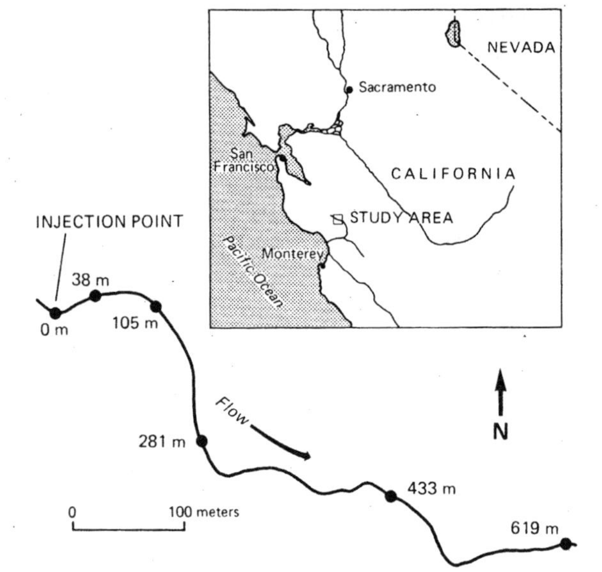

Fig. 1. Experimental reach of Uvas Creek (Santa Clara County, California). Injection point and five monitoring locations are indicated.

The model equations are

$$\frac{\partial C}{\partial t} + \frac{Q}{A}\frac{\partial C}{\partial x} = \frac{1}{A}\frac{\partial}{\partial x}\left(AD\frac{\partial C}{\partial x}\right) + \frac{q_L}{A}(C_L - C) + \alpha(C_S - C) \quad (1)$$

$$\frac{dC_S}{dt} = -\alpha\frac{A}{A_S}(C_S - C) \quad (2)$$

where

C solute concentration in the stream, mg l^{-1};
Q volumetric flow rate, $m^3 s^{-1}$;
A cross-sectional area of the channel, m^2;
D dispersion coefficient, $m^2 s^{-1}$;
q_L lateral volumetric inflow rate (per length), $m^3 s^{-1} m^{-1}$;
C_L solute concentration in lateral inflow, mg l^{-1};
C_S solute concentration in the storage zone, mg l^{-1};
A_S cross-sectional area of the storage zone, m^2;
α stream storage exchange coefficient, s^{-1};
t time, s;
x distance, m.

Equation (2) and the coupling term $\alpha(C_S - C)$ in (1) are deceptively simple, for they embody several physical principles and constraints. They ideally describe a system with the following characteristics.

1. There exists a storage zone that is not moving.
2. Within the storage zone, solute is uniformly and instantaneously distributed.
3. The transport of solute between the storage zone and the channel is determined simply by the difference in concentrations and an exchange coefficient.
4. It is possible to measure the cross-sectional area of the storage zone, the concentration of solute within the storage zone, and the exchange coefficient.

In a mountain pool-and-riffle stream, these characteristics are observed to varying degrees. It is easy to envision sizable storage zones of stagnant water that are not moving downstream. Such storage zones are located behind protruding logs, boulders, and vegetation in the shallows, along the edges of slowly moving pools, and in the thick gravel and cobble beds of swift riffles. However, it is not easy to envision a linear physical driving mechanism that simultaneously transfers mass between the stream and the storage zone, distributes it uniformly throughout the storage zone and yet prevents the storage zone from moving longitudinally.

In mountain streams, one does, nevertheless, observe significant tails from a solute pulse. Solute mass is removed during the rising phase of a solute pulse, temporarily retained in storage until the pulse passes, and then returned to the stream. From this observation we infer that there is in fact a mechanism that presents itself as transient storage of solute mass along the length of the stream. Hence we do not believe that a strict dead zone model is physically descriptive of the processes occurring in mountain streams, but rather that the observed 'transient storage' can be empirically simulated using the identical equations.

The distinction between describing physical dead zones and simulating observed transient storage is crucial to the determination and interpretation of the model parameters.

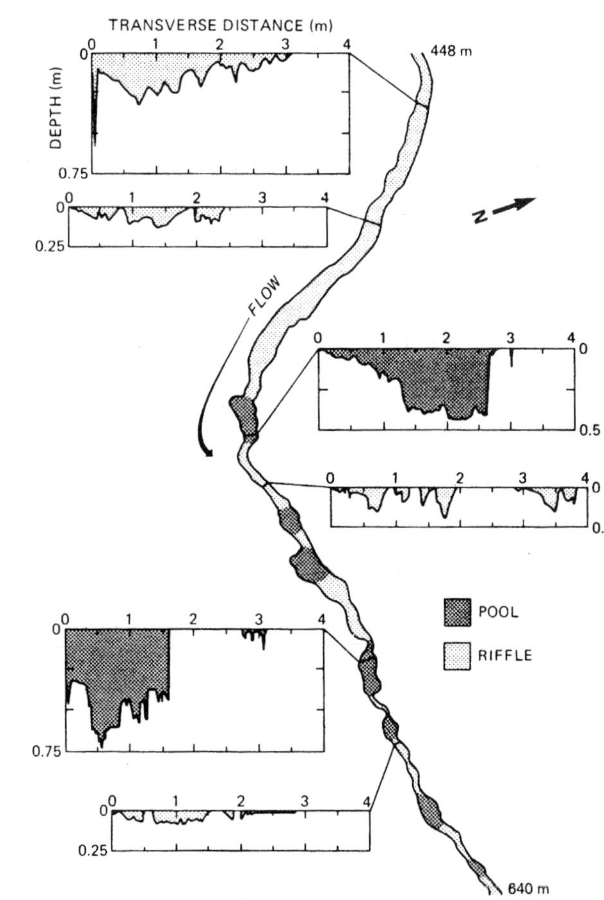

Fig. 2. Detailed mapping and transects (1973) of a reach in Uvas Creek. Note that the downstream distances measured in 1973 (Figures 2, 3, and 4) differ from those measured in 1972 by a few percent. These differences are due to the reworking of the stream channel during the intervening winter's storms and the imprecision of defining the longitudinal distance in a highly irregular channel.

A_S and α become operational parameters, with effective values determined by observing the response of the stream to the injection of a conservative solute.

APPLICATION

Zand et al. [1976] describe the steady injection of chloride tracer into Uvas Creek (see Figure 1). The experiment occurred in late summer during a period of low flow (0.0125 $m^3 s^{-1}$). The chloride was injected at a constant rate for three hours and reached a maximum concentration of 11.9 mg l^{-1} a short distance below the injection point. Background concentration was measured to be 3.7 mg l^{-1}.

An appreciation of the physical characteristics of Uvas Creek is crucial to the interpretation of both the experimental data and the simulations. Approximately one year after the 1972 field experiment a detailed mapping of the study reach was made in which stream elevations and over 100 channel transects were taken. Figures 2, 3, and 4 illustrate several important physical characteristics. The dominant feature is the spatial variability.

Figure 2 shows a detailed top view of a lower reach of Uvas Creek and six representative channel transects. The channel is highly irregular. It is composed of alternating pools and riffles. The transects show the pools to be relatively deep and uniform compared to the riffles. However, even in the pools, the bottom profile is irregular and small zones of water are separated from the main channel. In the riffles, the depth goes to 0.0 m at several locations across the transects in places where cobble and vegetation protrude through the flow. Figure 3 quantifies the image given by Figure 2 and shows the measured channel cross-sectional areas. (The actual values of A are not germane, as there was a higher flow rate in 1973.) The variability is so pronounced as to appear as scatter. Figure 4 shows the water surface elevation decrease over the study reach. The overall slope is steep (0.03 m m^{-1}) as expected for a mountain stream. On a small distance scale of tens of meters there are significant differences in slope ranging from virtually flat to several times the overall value.

These measurements support the picture of the complex water movement one observes in the stream. The water is alternately shooting through riffles with the cross section expanding as the water 'dumps' into the pools and then passing through the pools with the cross section contracting as the water falls into the next riffle. Particularly in riffle sections, the water is in contact with a gravel and cobble bed and can enter easily accessible void spaces.

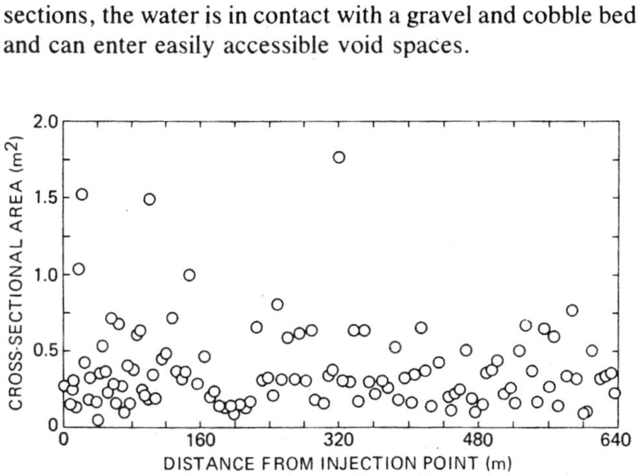

Fig. 3. Measured (1973) channel cross-sectional area in Uvas Creek.

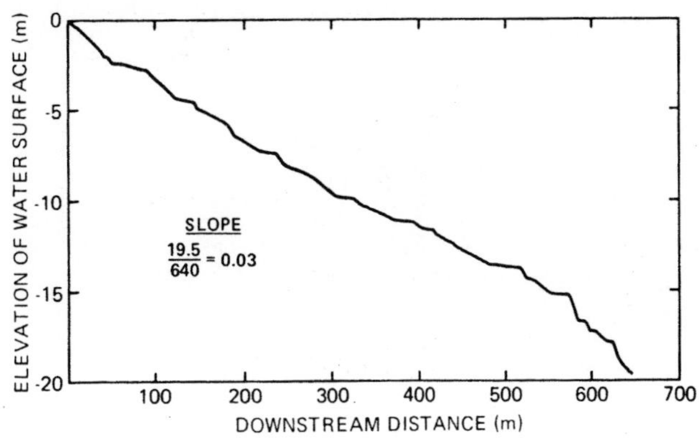

Fig. 4. Measured (1973) elevation of water surface in Uvas Creek.

The transient storage model will be used to simulate the transport of chloride in Uvas Creek resulting from the experimental steady injection. The model equations were solved by standard methods, using both finite difference and finite element (linear bases), spatial discretizations, and a Crank-Nicolson (time-centered) time integration. There were no essential differences between the finite difference and finite element simulations; results from the former are presented herein. The model parameters were selected by visually determining the set of parameters which yielded the 'best fit' to the concentration data. This approach is, of course, highly subjective. However, the spatial variability that exists in the stream makes it highly doubtful that a more complex technique would yield more meaningful results.

Several simulations were run with only the convection-dispersion mechanisms operating, i.e., no exchange with the storage zone (α = 0.0 s^{-1}). In these simulations, lateral inflow q_L was neglected because simulation experiments showed that physically plausible values of q_L did not alter the overall agreement of the simulations to the data. The model then has two parameters, A and D, with Q fixed by the determination of the chloride injection plateau at the monitoring station at 38 m. (Locations of monitoring stations are shown in Figure 1.) By using spatially uniform values of A and D, values for these parameters were determined by searching for the best fit to the leading edges of the pulses at two monitoring stations. The simulations based on the leading edges observed at 38 m and 619 m are shown in Figures 5*a* and 5*b*, respectively. The agreement with data is not good, but more importantly, the characteristics of the data and the simulation are different. The significant features of the concentration data are (1) the decrease in the maximum concentration at the downstream locations, (2) the clipping of the shoulders of the leading edges, (3) the extent of the tails, and (4) an apparent loss of mass at 619 m.

The application of the transient storage model is shown in Figures 6*a* and 6*b*. The best fit model parameters were determined in downstream sequence for each of the five reaches between successive monitoring locations (see Table 1). With parameters so determined, the simulation is excellent. Figure 6*a* compares the improvement of these results over identical parameters with no transient storage. Figure 6*b* shows the chloride concentration in the storage zones. The persistence of the tails in the storage zone is even more pronounced than in the channel. Elevated levels of chloride can remain in the storage zone long after the pulse has cleared the channel. These simulations exhibit all of the

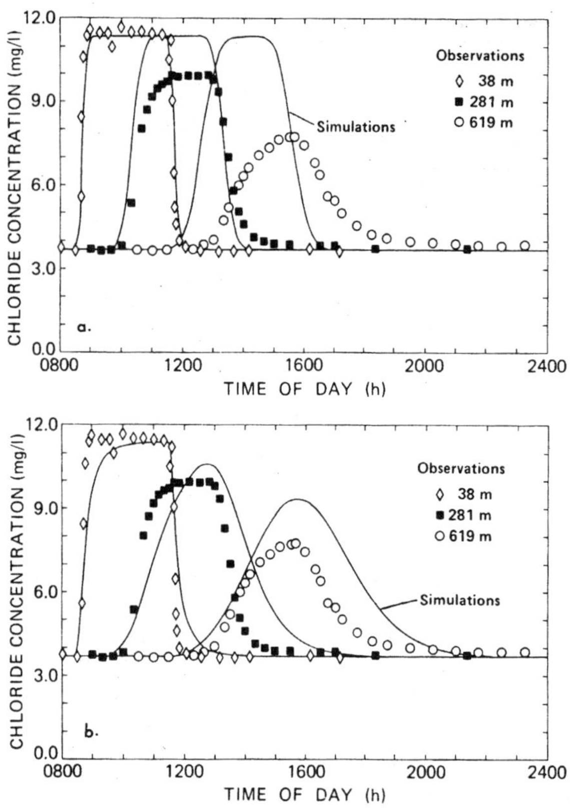

Fig. 5. Observed chloride concentrations and simulations at three locations for steady injection at Uvas Creek. Injection started at 08:30 and ended at 11:30. Data are presented from this period until 23:30 when rain began to fall. (See Table 1 for parameter values). (*a*) Simulation of convection-dispersion mechanisms; best fit to the leading edge of the pulse at 38 m. (*b*) Simulation of convection-dispersion mechanisms; best fit to the leading edge of the pulse at 619 m.

significant features of the stream concentration data listed above. In particular, Figure 6*b* illustrates that the storage zone acts as the needed temporary sink for the apparent loss of mass. In addition to the transient storage mechanism, the decreased maximum concentration at 281 m and 619 m could also be attributed to a chemical loss mechanism or increased discharge. The *p*H of the stream water was approximately 8; at this *p*H level the authors are unaware of a chemical loss mechanism for chloride. To simply invoke increased discharge does decrease the maximum concentration but will not account for the clipping of the shoulders of the leading edges and the extent of the tails observed in the data.

Reach-by-reach variation in the physical stream parameters was used in the simulation in Figure 6. The variation is substantial (on a relative basis). As intensive, concurrent monitoring at several locations is often physically or fiscally difficult, we tested the model's response to setting spatially uniform values for the parameters. Figure 7 shows the simulation yielding the best fit at 619 m. The simulation with reach-by-reach variation is, of course, better, but the uniform parameter simulation is also quite good. Uvas Creek, like other mountain streams, is characterized by a high degree of spatial variability and complex hydrology. Figure 7 indicates that although spatial variability in the parameters need not be a major consideration, its effects are, however, evident on scales of only hundreds of meters.

DISCUSSION

The simulations clearly indicate the importance of a transient storage mechanism for solute transport in Uvas Creek (and thus probably other mountain streams). The validity of this result rests heavily on the assertion that the model storage parameters are in fact physically realistic. We will discuss possibilities for the location of storage zones and indicate why the value of $A_S = 0.7$ m^2 is not unrealistic. Both A_S and the exchange coefficient α will then be discussed as they relate to parameter values used by other authors for applications to other streams.

That the effective A_S for Uvas Creek is greater than the A is a somewhat curious result of the simulation exercise. There are several possible mechanisms operating in Uvas Creek that allow this to be plausible. Five such mechanisms are listed below. The order is from the most general mechanisms that we would expect to exist in almost any stream to the most specific mechanisms likely to be most important in small mountain streams. Solute may be temporally stored by (1) turbulent eddies generated by large-scale bottom irregularities, (2) large but slowly moving recirculating zones along the sides of pools, particularly located immediately downstream of the entrance to a pool from a riffle section, (3) small but very rapidly mixing recirculating zones located behind flow obstructions, particularly located in riffle sections where cobble, small boulders, and vegetation commonly protrude through the flow, (4) side pockets of water effectively acting as dead ends for solute transport, and (5) flow into, out of, and through a coarse gravel and cobble

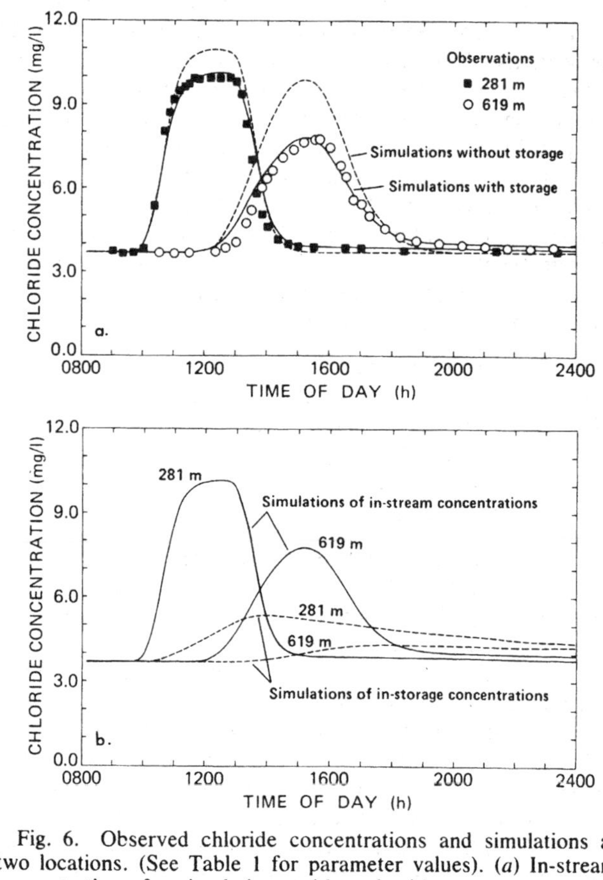

Fig. 6. Observed chloride concentrations and simulations at two locations. (See Table 1 for parameter values). (*a*) In-stream concentrations for simulations with and without storage mechanisms. (*b*) In-storage concentrations compared to in-stream concentration for simulation with storage mechanisms.

TABLE 1. Simulation Parameters

Simulation	Figure	Concentration at Injection Location C_{INJ}, mg l^{-1}	Reach, m	Flow Q, m^3 s^{-1}	Dispersion Coefficient D, m^2 s^{-1}	Cross-Sectional Areas: Stream A, m^2	Cross-Sectional Areas: Storage A, m^2	Exchange Coefficient α, s^{-1}
Convection-dispersion	5a	11.4	0–619	0.0125	0.12	0.30		0.00
	5b	11.4	0–619	0.0125	0.48	0.45		0.00
Storage	6a, 6b	11.4	0–38	0.0125	0.12	0.30	0.00	0.00
			38–105	0.0125	0.15	0.42	0.00	0.00
			105–281	0.0133	0.24	0.36	0.36	0.30×10^{-4}
			281–433	0.0136	0.31	0.41	0.41	0.10×10^{-4}
			433–619	0.0140	0.40	0.52	1.56	0.45×10^{-4}
Convection-dispersion	6a	*	*	*	*	*	*	0.00
Storage:	7							
variable parameters		*	*	*	*	*	*	*
uniform parameters		11.4	0–619	0.013	0.1	0.4	0.7	0.4×10^{-4}

*As above for storage simulation in Figure 6.

bed. All these mechanisms probably operate in Uvas Creek, and it is plausible that acting together they account for 0.7 m^2 of storage cross-sectional area even with a channel cross-sectional area of only 0.4 m^2.

We are unaware of any a priori physical arguments that would indicate whether or not the value for the exchange coefficient is plausible. The wide variety of spatial and temporal scales present in the above mechanisms require that an effective exchange coefficient will reflect a lumping of many values.

Several other authors have used storage models for a wide variety of stream conditions. Figures 8*a* and 8*b* show model parameters determined in three other papers for nearly 40 applications. These figures are revisions of figures appearing in the work of *Pedersen* [1977] and include additional data points. The individual stream parameters were measured by many different workers, reported in different formats, and attempt to represent variable parameters by only one value. Thus there is considerable uncertainty associated with any individual point on these plots. Trends, not absolute values, are significant.

Figure 8*a* is a plot of the ratio of storage area to channel area, A_S/A, against $f^{-1/2}$ where

- f friction factor, equal to $8(U_*/U)^2$;
- U stream velocity, equal to Q/A, m s^{-1};
- U shear velocity, equal to $(gSd)^{1/2}$, m s^{-1};
- g acceleration of gravity, m s^{-2};
- d channel depth, m;
- S water surface slope,

Taking U_* as the shear velocity, f is then a measure of friction [*Henderson*, 1966; *Fischer et al.*, 1979]. *Thackston and Schnelle* [1970], as well as *Pedersen* [1977], have found it instructive to relate storage area to various measures of friction. In Figure 8*a* there is a clear trend between decreased importance of storage area and diminishing 'friction.' The point for Uvas Creek follows this trend.

Henderson [1966] discussed many empirical relationships between measures of surface roughness and $f^{-1/2}$. In pipe flow, and to a lesser extent in open channel flow, there is empirical evidence supporting a linear relationship between $f^{-1/2}$ and the logarithm of the ratio of length scales for surface roughness to flow depth. Figure 8*a* shows considerable scatter in the data; however, the basic trend observed for surface roughness also appears with a measure of storage area.

Figure 8*b* is plot of dimensionless groupings relating a mass transfer grouping, the Nusselt number, to groupings of fluid, channel, and storage zone characteristics. The groupings are defined as follows.

- Nu Nusselt number, equal to $k4d/\varepsilon_V$;
- Re Reynolds number; equal to $U4d/\upsilon$;
- Sc Schmidt number; equal to υ/ε_V;
- d/B channel aspect ratio;
- d_S/B storage zone aspect ratio.

where

- k mass transfer coefficient, equal to αd, m s^{-1};
- B channel and storage zone breadth, m;
- d channel depth, equal to A/B, m;
- d_S storage zone depth, equal to A_S/B, m;
- ε_V vertical dispersion coefficient, equal to $0.0067\ dU_*$, m^2 s^{-1};
- υ kinematic viscosity, m^2 s^{-1}.

These groupings have been defined to be analogous to those used in chemical engineering operations for mass transfer from a flowing gas or liquid to a flat plat [*Bird et al.*, 1960; *McCabe and Smith*, 1967; *Welty et al.*, 1969]. In forming the

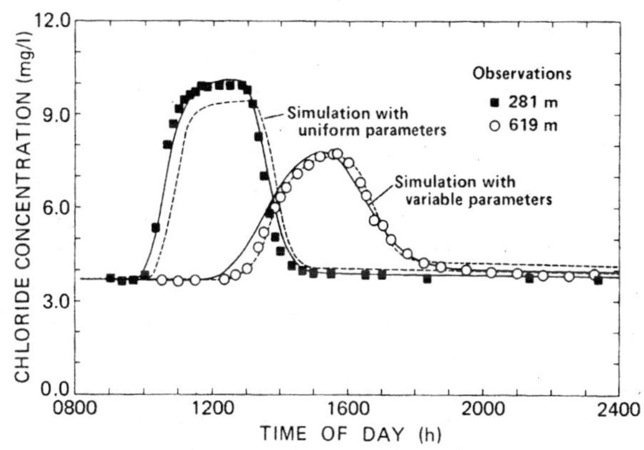

Fig. 7. Observed chloride concentrations and simulations at two locations. In-stream concentrations are compared for simulations with spatially varied and uniform parameter. (See Table 1 for parameter values).

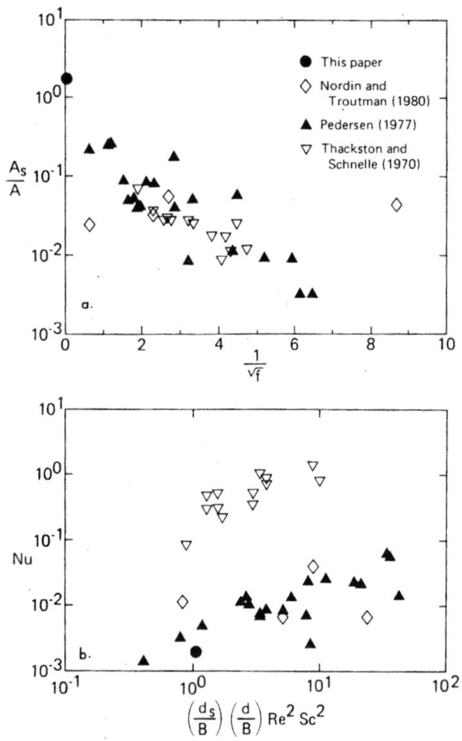

Fig. 8. Dead zone model parameters as determined in several applications. (Different authors define the dead zone model in terms of different parameters. For the cases shown here it was possible to convert to the groups A_S/A and Nu for comparison.) (*a*) Ratio of storage area to channel area plotted against a 'friction' grouping of channel parameters. (*b*) Mass transfer Nusselt number plotted against a grouping of fluid, channel, and storage zone parameters.

groupings, the channel and storage zone cross sections were assumed to be rectangular and $d \ll B$. The vertical dispersion coefficient is taken from an empirical relationship for natural channels [*Fischer et al.*, 1979] and a value of 10^{-6} m^2 s^{-1} was used for kinematic viscosity. (The Nusselt number is conventionally defined in terms of k, the mass transfer coefficient. Equations (1) and (2) can be written in terms of k instead of α. The use of the exchange coefficient in the model equations follows the water resources engineering convention for dissolved oxygen models using the reaeration coefficient.)

Figure 8*b* shows a positive relationship between Nu and the grouping $(d_S/B)(d/B)(Re)^2(Sc)^2$. Similar multiplicative power law relationships have been observed in many mass transfer operations [*Bird et al.*, 1960; *McCabe and Smith*, 1967; *Welty et al.*, 1969]. For mass transfer to a flat plate it has been shown by dimensional analysis [*Welty et al.*, 1969] that the Nusselt number is a function of the Reynolds and Schmidt numbers. In the storage zone problem two additional parameters, d_S and B, are present. Thus the appearance of the channel and storage zone aspect ratios is consistent with other mass transfer correlations.

In Figure 8*b* there is a separation of data points between points from *Thackston and Schnelle* [1970] and those from the papers of *Pedersen* [1977] and *Nordin and Troutman* [1980]. Each set follows the same trend. The point for Uvas Creek is closest to the grouping of the *Pedersen* [1977] and *Nordin and Troutman* [1980] points and it is located in the range of a very low Nusselt number. We cannot explain the separation of the data points; however, we can mention a few possibilities. First, there may in fact be fundamental limitations to the storage zone model that preclude any understanding in terms of measurable stream parameters. Second, the differences between authors in the methods by which the model parameters were determined may be so great as to bias or obscure certain physical trends. Finally, *Thackston and Schnelle* [1970] use data collected 'on relatively short, smooth, and uniform reaches.'

While we do not know the details of the physical characteristics of the other streams studied, we can assume from information given in the reports that the streams are not mountain pool-and-riffle streams and that they have considerably greater volumetric flows and velocities than Uvas Creek. Given the differences in physical conditions, we expect the points for Uvas Creek to be outliers.

Of course, none of this discussion 'proves' the model and, in fact, it shows the model to be a simplification of the physical processes of transport in a mountain stream. The simulation does reproduce the observed transient storage and it can be understood, at least qualitatively, in terms of the storage zone concept. As such, these simulations provide a good empirical transport framework for mountain streams, on which basis the analysis of other solute behavior, for example, solute-sediment interactions, can be pursued with confidence.

Acknowledgments. Most of the chemical data used in this report first appeared in a work by *Zand et al.* [1976]. Tabular summaries of this and other data on Uvas Creek were made available by G. W. Zellweger and R. J. Avanzino. All the information on Uvas Creek, including the previously unpublished detailed channel cross sections, were obtained under the direction of V. C. Kennedy in cooperation with S. M. Zand. The authors also benefitted from discussions with A. P. Jackman (University of California, Davis) regarding prior applications of storage models and results of his research (supported by the U. S. Geological Survey) on solute transport in Uvas Creek.

REFERENCES

Bird, R. B., W. E. Stewart, and E. N. Lightfoot, *Transport Phenomena*, John Wiley, New York, 1960.

Day, T. J., Longitudinal dispersion in natural channels, *Water Resour. Res.*, *11*(6), 909–918, 1975.

Day, T. J., Longitudinal dispersion of fluid particles in mountain streams, 1, Theory and field evidence, *N. Z. J. Hydrol.*, *16*(1), 7–25, 1977.

Fischer, H. B., E. J. List, R. C. Y. Koh, J. Imberger, and N. H. Brooks, *Mixing in Inland and Coastal Waters*, Academic, New York, 1979.

Hays, J. R., P. A. Krenkel, and K. B. Schnelle, Jr., Mass transport mechanisms in open-channel flow, *Sanitary and Water Resour. Eng. Tech. Rep. 8*, Vanderbilt Univ., Nashville, Tenn., 1966.

Heede, B. H., Flow and channel characteristics of two high mountain streams, *U.S. Dep. Agric. For. Serv. Res. Pap. RM-96*, 1972.

Henderson, F. M., *Open Channel Flow*, Macmillan, New York, 1966.

Hubbard, J. E., and W. D. Striffler, Cesium 137 in a mountain stream channel, *Water Resour. Res.*, *9*(5), 1440–1442, 1973.

Janda, R. L., Summary of watershed conditions in the vicinity of Redwood National Park, California, *U. S. Geol. Surv. Open File Rep. 78-25*, 1977.

Keller, E. A., Pools, riffles, and channelization, *Environ. Geol.*, *2*(2), 119–127, 1975.

Keller, E. A., and T. Talley, Effects of large organic debris on channel form and fluvial processes in the coastal redwood environment, in *Adjustments of the Fluvial System, 1979 Proceedings of the Tenth Annual Geomorphology Symposium*, edited by D. D. Rhodes and G. P. Williams, Kendall/Hunt, Dubuque, Iowa, 1979.

Kennedy, V. C., and R. L. Malcolm, Geochemistry of the Mattole

River of northern California, *U. S. Geol. Surv. Open File Rep. 78-205*, 1977.

Lewis, W. M., and M. C. Grant, Relationships between stream discharge and yield of dissolved substances from a Colorado mountain watershed, *Soil Sci.*, *128*(6), 353–363, 1979.

McCabe, W. L., and J. C. Smith, *Unit Operations of Chemical Engineering*, 2nd Ed., McGraw-Hill, New York, 1967.

Nordin, C. F., Jr., and B. M. Troutman, Longitudinal dispersion in rivers; the persistence of skewness in observed data, *Water Resour. Res.*, *16*(1), 123–128, 1980.

Pedersen, F. B., Prediction of longitudinal dispersion in natural streams, *Ser. Pap. 14*, Inst. of Hydrodyn. and Hydraul. Eng., Tech. Univ. of Denmark, Copenhagen, 1977.

Sabol, G. V., and C. F. Nordin, Dispersion in rivers as related to storage zones, *J. Hydraul. Div. Am. Soc. Civ. Eng.*, *104*(HY5), 695–708, 1978.

Thackston, E. L., and A. M. Krenkel, Longitudinal mixing in natural streams, *J. Sanit. Eng. Div. Am. Soc. Civ. Eng.*, *93*(SA5), 67–90, 1967.

Thackston, E. L., and K. B. Schnelle, Jr., Predicting effects of dead zones on stream mixing, *J. Sanit. Eng. Div. Am. Soc. Civ. Eng.*, *96*(SA2), 319–331, 1970.

Tsai, Y. H., and R. R. Holley, Temporal and spatial moments for longitudinal mixing in prismatic channels with storage in separation zones, *Hydraul. Eng. Ser. 35, UILU-ENG-79-2010*, Univ. of Ill. at Urbana-Champaign, 1979.

Valentine, E. M., and I. R. Wood, Longitudinal dispersion with dead zones, *J. Hydraul. Div. Am. Soc. Civ. Eng.*, *103*(HY9), 975–990, 1977.

Valentine, E. M., and I. R. Wood, Experiment in longitudinal dispersion with dead zones, *J. Hydraul. Div. Am. Soc. Civ. Eng.*, *105*(HY8), 999–1016, 1979*a*.

Valentine, E. M., and I. R. Wood, Dispersion in rough rectangular channels, *J. Hydraul. Div. Am. Soc. Div. Eng.*, *105*(HY12), 1537–1553, 1979*b*.

Welty, J. R., C. E. Wicks, and R. E. Wilson, *Fundamentals of Momentum, Heat and Mass Transfer*, John Wiley, New York, 1969.

Zand, S. M., V. C. Kennedy, G. W. Zellweger, and R. J. Avanzino, Solute transport and modeling of water quality in a small stream, *J. Res. U. S. Geol. Surv.*, *4*(2), 233–240, 1976.

(Received April 7, 1982;
revised November 18, 1982;
accepted December 6, 1982.)

Commentary on Paper F3

Hill, A. R. (1983) Nitrate-nitrogen mass balances for two Ontario rivers.

Nitrogen flows through the terrestrial ecosystem and is transported (and transformed) through the river network. The evaluation of this transport system has allowed us to conclude that nitrogen load contributions differ between biomes (Howarth *et al.*, 1996), and that headwater streams have a significant role in the catchment in spite of their moderate size (Alexander *et al.*, 2000). The river channels host relevant in-stream processes, which result in the reduction of the downstream nitrogen flux (Peterson *et al.*, 2001).

Alan Hill's (1983) paper was in fact a chapter in a benchmark book that collected novel views on multiple aspects of stream ecology (e.g. McIntire, 1983; Triska *et al.* 1983; Ward & Stanford 1983). The paper is a remarkable example of integration between field-based, large-scale nitrogen estimates, and small-scale, controlled laboratory measurements. The field approach determined that nitrate was lost during transport. This was inferred from the analysis of inputs and outputs estimated from the tributaries entering the river, as well as from those being contributed by the groundwater. The experimental approach carried out in the laboratory determined that the nitrate loss was mostly due to the activity of the biota inhabiting the sediments which transformed nitrate via denitrification. The combined approach (observational and experimental) showed that the differences occurring between two river systems differing in their groundwater contribution were related to the nitrate transported in baseflow by the river waters, but also to the sediment characteristics (in particular their granulometry and organic carbon content).

Nitrate loss was not steady throughout a hydrological cycle, but showed seasonal variations according to temperature (which either triggers or depletes the metabolic activity) and river discharge. Laboratory experiments were again able to determine that during high-discharge episodes, sediment disturbance increased as well as its associated aeration, and therefore its ability to denitrify decreased. The significance of these mechanisms was quantified at the catchment scale. Hence, the highest nitrogen exports occurred during the periods of the lowest denitrification activity. Denitrification accounted for only 1–5% of the annual export of total nitrogen in these Ontario Rivers. The moderate role of denitrification in the catchment context could not be separated from the role of climate. Hence, in other, milder areas, denitrification and nitrate assimilation could operate for longer periods and potentially have a more significant role during low-flow periods. This was later shown in several papers covering temperate environments (Howarth *et al.*, 1996).

SS

References

Alexander, B. N., Smith, R. A. & Schwarz, G. E. (2000) Effects of stream channel size on the delivery of nitrogen in the Gulf of Mexico. *Nature* **430**, 758–761.

Howarth, R. W., Billen, G., Swaney, D., Townsend, A., Jaworski, N., Lajtha, K., Downing, J. A., Elmgren, R., Caraco, N., Jordan, T., Berendse, F., Freney, J., Kudeyarov, V., Murdoch, P. & Zhu Zhao, L. (1996) Regional nitrogen budgets and riverine N and P fluxes for the drainages to the North Atlantic Ocean: Natural and human influences. *Biogeochemistry* **35**, 75–139.

McIntire, C. D. (1983) A conceptual framework for process studies in lotic ecosystems. In: *Dynamics of Lotic Ecosystems* (ed. by T. D. Fontaine & S. M. Bartell). Ann Arbor, Michigan, USA.

Peterson, B. J., Wollheim, W. M., Mulholland, P. J., Webster, J. R., Meyer, J. L., Tank, J. L., Marti, E., Bowden, W. B., Valett, H. M., Hershey, A. E., McDowell, W. H., Dodds, W. K., Hamilton, S. K., Gregory, S. & Morrall, D. D. (2001) Control of nitrogen export from watersheds by headwater streams. *Science* **292**, 86–90.

Triska, F. J., Kennedy, V. C., Avanzino, R. J. & Reilly, B. N. (1983) Effect of simulated canopy cover on regulation of nitrate uptake and primary production by natural periphyton assemblages. *Dynamics of Lotic Ecosystems* (ed. by T. D. Fontaine & S. M. Bartell), 129–159. Ann Arbor, Michigan, USA.

Ward, J. V. & Stanford, J. A. (1983) The serial discontinuity concept of lotic ecosystems. In: *Dynamics of Lotic Ecosystems* (ed. by T. D. Fontaine & S. M. Bartell), 29–42. Ann Arbor, Michigan, USA.

24. NITRATE-NITROGEN MASS BALANCES FOR TWO ONTARIO RIVERS

Alan R. Hill

Department of Geography
York University
Toronto, Ontario

ABSTRACT

Analysis of nitrogen transport revealed an average daily loss of 45 ± 9 kg NO_3–N (160 mg N/m^2) in Duffin Creek and 35 ± 21 kg NO_3–N (60 mg N/m^2) in the Nottawasaga River during low flows between May and October. Laboratory experiments suggested that most of the nitrate disappearance was caused by denitrification and nitrate reduction in stream sediments. Contrasts in nitrate removal rates in the two rivers probably resulted mainly from differences in sediment characteristics. The Nottawasaga River sediments had very low percentages of organic carbon and silt in comparison to sediments from pool reaches in Duffin Creek. Nitrate losses during stream transport rose rapidly to high levels in May and fell to lower levels in November as water temperature declined. Laboratory experiments suggested that denitrification and nitrate reduction may still occur, although at very low rates, during the winter months when water temperatures in the Nottawasaga River and Duffin Creek are 0 to 2°C. Denitrification was considered to be negligible during major flood events. The removal of nitrate represented <5% of the annual export of nitrogen from the two rivers. In low flow conditions, however, between May and October, nitrate losses corresponded to 50% of the average daily input of total nitrogen and 75% of the nitrate input in Duffin Creek. The Nottawasaga River had a much smaller assimilative capacity, removing only about 13% of the daily nitrate input.

457

INTRODUCTION

Nitrogen transformations, including such dissipative reactions as denitrification, have been investigated mainly in lakes (Keeney, 1973; Brezonik, 1977). There have been relatively few studies of nitrogen budgets and sink processes in rivers. This neglect has resulted from a tendency to regard rivers as inert pipelines rather than as functioning ecosystems with respect to nutrient transport (Likens, 1975).

Approximate nitrogen budgets based on a combination of measured and estimated fluxes revealed apparent losses of inorganic nitrogen in some rivers during low summer flows (Owens et al., 1972). The investigators attributed these losses primarily to bacterial denitrification in sediments. Several recent studies indicated that denitrification in anaerobic sediments may play a major role in removing nitrogen from streams. Van Kessel (1977b) measured a NO_3–N loss of 913 mg m^{-2} day^{-1} during a 20-day period for a small canal receiving treated sewage. Laboratory experiments with undisturbed water–sediment profiles indicated that the disappearance of nitrate was caused mainly by denitrification in the sediment (Van Kessel, 1977a). Field observations revealed considerable losses of nitrate during transport in two small headwater streams in Ontario, Canada (Kaushik et al., 1975; Robinson et al., 1978a). Laboratory studies showed rapid removal of nitrate from well-oxygenated water overlying sediments from these streams (Sain et al., 1977). Subsequently $^{15}NO_3$–N has been used to show that the losses were caused by denitrification (Chatarpaul and Robinson, 1979; Chatarpaul et al., 1980).

The studies cited do not provide the detailed field measurements of nitrogen transport in streams which are needed to assess the importance of denitrification and other sink processes in relation to seasonal and annual river nitrogen budgets. I have recently analyzed the magnitude of denitrification in relation to the annual export of nitrogen from Duffin Creek, Ontario (Hill, 1979). My data suggest that the removal of NO_3–N, primarily by denitrification, in the downstream reaches of the river represents 5 to 6% of the annual export of total nitrogen from this river basin. During low summer flows, however, nitrate sink processes removed approximately 75% of the daily NO_3–N input to the downstream reaches of the river (Hill, 1981).

A greater understanding of nitrogen flux and cycling in rivers requires field studies of nitrogen mass balances in a wide variety of rivers. In this paper I analyze nitrate sink processes in relation to nitrogen transport in the Nottawasaga River, Ontario. The data are compared with my measurements from Duffin Creek to examine the magnitude, seasonal variability, and relative importance of nitrate removal processes in these two river ecosystems.

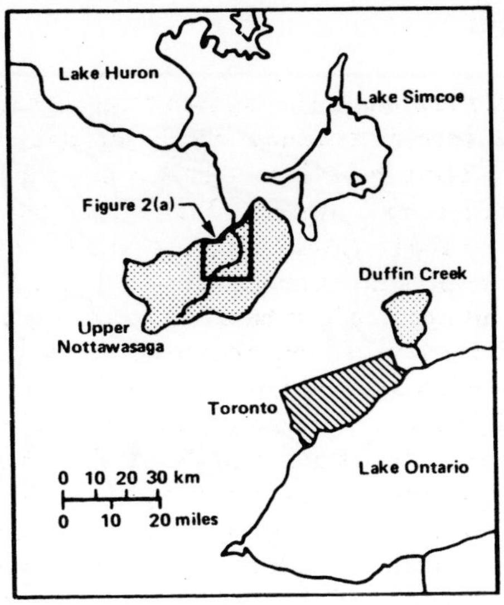

Figure 1. Location of study areas.

THE STUDY AREA

Nottawasaga River

The upper Nottawasaga River basin (1230 km^2) is located about 50 km north of Toronto (Figure 1). Research was restricted to the Alliston sand plain area of the basin, which is drained by the Nottawasaga River and its main tributaries, the Boyne River and Inisfil Creek (Figure 2a). The sand plain, which is underlain by a shallow groundwater table, is an area of intensive cash farming of potatoes, corn, and sod. Many of these crops are irrigated during the summer with water pumped from the major rivers. Alliston is served by a sewage treatment plant that discharges effluent into the Boyne River about 2 km upstream from station 33.

The Nottawasaga River is a hard-water sixth-order stream, ranging in width from 10 to 20 m and having a mean annual discharge of 7.9×10^5 m^3/day at station 34. The channels of the Nottawasaga River and its major tributaries are frequently incised to a depth of 10 to 20 m below the surface of the sand plain. The river between station 32 and the junction with the Inisfil Creek consists of alternating riffles and pools. Gravel riffles are generally absent in other reaches, and the river bed consists of unstable medium-to-coarse sandy sediments. Benthic algae occur on the

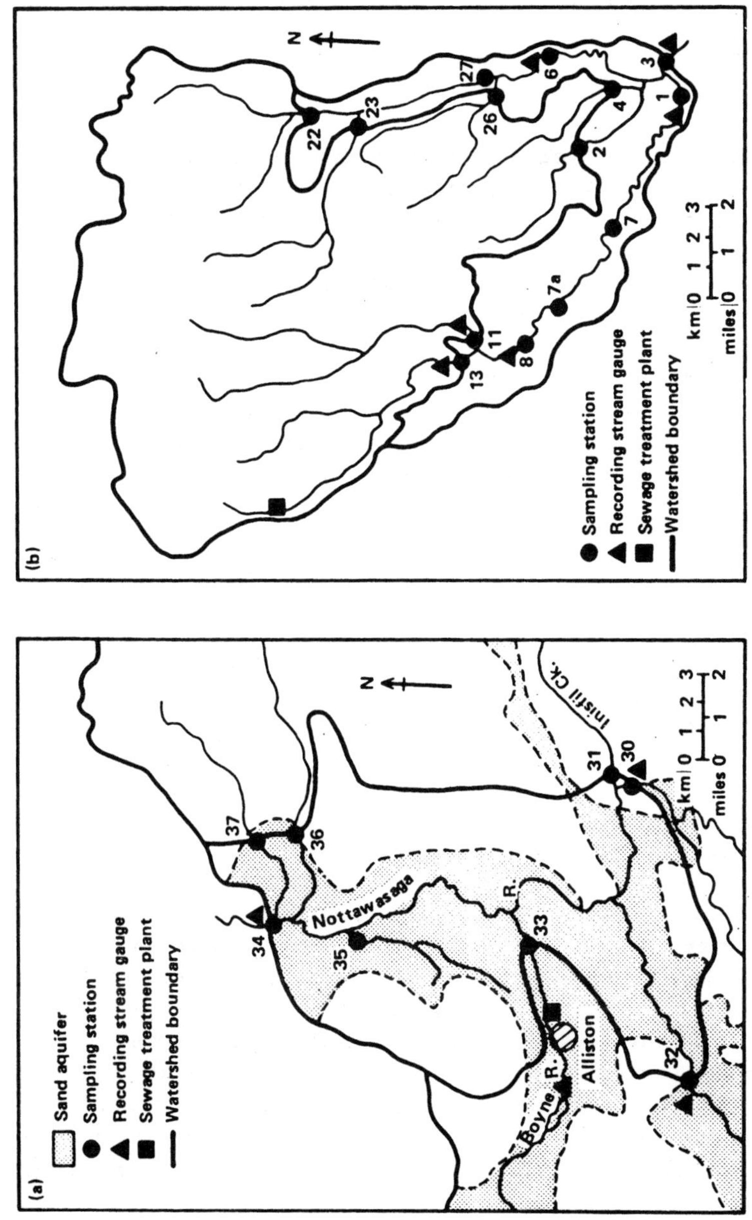

Figure 2. Location of sampling stations, recording stream gauges, and river reaches for which nitrogen budgets were calculated. (a) Nottawasaga River. (Sampling stations on four small sand plain streams are not shown). (b) Duffin Creek.

gravel riffles between station 32 and the junction with Inisfil Creek; elsewhere the river channels are largely devoid of benthic algae and macrophytes.

Duffin Creek

Duffin Creek is a shallow, hard-water, sixth-order stream that drains an area of 262 km^2 about 30 km east of Toronto (Figure 1). Land use for the drainage basin includes about 49% agricultural land, 21% forest, and 18% abandoned farm land in various stages of old-field succession. Stouffville, a small urban center is served by a tertiary sewage treatment plant that discharges effluent into the west branch of Duffin Creek (Figure 2b).

The mean annual discharge of the river is 2.2×10^5 m^3/day at the Pickering stream gauge (station 3). Both branches of the river are composed of riffles and pools, and the average channel width increases downstream from 4 m at stations 11, 13, and 22 to 30 m at station 3. Riffle substrates consist of gravel, ranging in size from a few millimeters up to 50 cm or more in diameter, and comprise approximately 50% of the total stream bed area. Calcareous pool sediments, varying between silt loam and sandy loam in texture, occupy areas between riffles. The river channel is generally unshaded and flows across a floodplain through unused pastures. Aquatic macrophytes are almost entirely absent from the river and the major component of the flora consists of benthic algae dominated by *Cladophora sp.*, which form a mat on the gravel riffles.

MATERIALS AND METHODS

The construction of a nutrient mass balance for river reaches involves measuring a number of input and output vectors. During low flows, the nutrient load of the river as it enters and leaves a particular reach constitutes the major input and output terms in the budget. In some river reaches additional inputs may be represented by small tributary streams and groundwater entering the main channel. The analysis of a mass balance for flood events was not attempted because it was impossible, in view of the number of sites involved, to measure adequately the rapid changes in chemistry and discharge which occur during storm runoff.

River Inputs and Exports

River discharge and nitrogen concentrations were measured at 13 stations in the Nottawasaga basin and at 14 stations on Duffin Creek (Figure 2). Government stream-flow recording gauges provided discharge records at a number of sampling stations. Current meter measurements coupled with cross-sectional areas were used to calculate discharge at times of stream water sampling for the remaining stations. The quantity of river water used for irrigation was estimated from water extraction permits submitted by individual landowners to the Ontario Ministry of Environment.

Groundwater Inputs

In the Nottawasaga study area the shallow Alliston sand plain aquifer represents an important nitrate input to the river. The concentration of NH_4–N and NO_3–N was measured at 60 seepage sites adjacent to the main river channels. The quantity of groundwater entering the river reaches was evaluated by using data from six small streams that drain portions of the sand plain. These streams are sustained by groundwater during summer low flows and their discharge, divided by drainage area, provides an estimate of groundwater flux per square kilometer. This estimate was then applied to areas of the sand plain which drain directly to the Nottawasaga River and its major tributaries, Inisfil Creek and the Boyne River. The input derived by this procedure agrees closely with the groundwater flux that can be calculated as the difference between daily inputs of water from major rivers and the sand plain tributaries and the export of water measured at station 34, together with estimates of withdrawals for irrigation.

Recent hydrogeologic studies indicate that there are no major inputs of groundwater in the downstream reaches of Duffin Creek (Sibul et al., 1977; Ostry, 1979). Small terraces composed of 1 to 3 m of gravel occur in the river valleys however, and contain perched groundwater tables that cause small increases in stream flow in some reaches during May and late October. No attempt was made to sample the concentration of nitrogen forms in seepages from these areas.

Water Chemistry

Water samples were collected in acid-washed polyethylene bottles and transported to the laboratory on ice. Nitrogen analysis was usually run within 2 days of sampling. An automated cadmium reduction method was used to measure NO_3 + NO_2–N (American Public Health Association, 1976). Tests were also made separately for NO_2–N in the water samples. Since concentrations of NO_2–N were very low, the NO_3 + NO_2–N value is henceforth referred to as NO_3–N. Water samples were analyzed for NH_4–N with an automated indophenol blue method (Technicon Industrial Systems, Method No. 98-70W). Measurements of all nitrogen forms were also available for the sampling stations on Duffin Creek from government water-quality monitoring studies (Ontario Ministry of Environment, 1973-1978).

Sediment Samples

A coring device was used to remove sediment samples from the river bed at ten sites in each study area. Composite samples for the 0 to 4 and 4 to 10-cm sediment depths were obtained by combining six samples collected at intervals across the river channel at each site. Ammonium-nitrogen and NO_3 + NO_2–N were determined by extracting moist sediment samples with 2*M* KCl followed by colorimetric analysis according to the procedures outlined for water samples. Organic nitrogen in sediments was determined with a block digestor, then the ammonia was determined on the Auto Analyzer (Issac and Johnson, 1976). Organic carbon was estimated by the Walkley–Black method (Allison, 1965). Sediment pH was measured with a glass electrode on a 1 : 2 soil and water mixture.

Laboratory Systems

The possible role of sediments in nitrate removal from stream water was investigated in a laboratory experiment. Intact cores were removed from the river bed at ten locations in each drainage basin. A stainless steel column of 5.8 cm OD and 0.1 cm in wall thickness was pushed into the sediment; its top end was closed with a rubber stopper; and the column was then removed, together with a core about 10 cm deep. The core was gently pushed out of the metal column into a tall glass cylinder with the same internal diameter as the metal coring device. Gravel samples were also collected in polypropylene pans from stream riffles at five sites on Duffin Creek. Additional details of riffle sampling procedures are given elsewhere (Hill, 1981).

To each glass cylinder, 500 ml of a solution containing 5 mg/l nitrate–N as KNO_3 was added to give a solution depth of about 15 cm over the sediment. To each pan, 4000 ml of the same solution was added to cover the gravel to a depth of 5 cm. Cylinders and pans containing only the solution were used as controls. Air bubbles were introduced near the water–sediment interface to reproduce the turbulence and aeration characteristic of river conditions. The cylinders and baths were incubated in the dark at 20° C, a temperature similar to the average summer water temperature of the rivers. Nitrate concentrations in the solutions were analyzed at intervals over a period of 6 days.

Material Mass Balances

The data used to examine mass balances were collected on days between late May and October. Discharge and nitrogen concentrations were measured only during periods of low stream flow, characterized by discharge fluctuations of less than 10% at stations 3 and 34 in the 24 to 36 h before sample collection.

Analysis of the mass balance for the Alliston sand plain area of the Nottawasaga River was based on a total of 21 days of observation in 1977 to 1979. The daily input for each day was derived from the summation of the individual fluxes at stations 30, 31, 32, 33, 36, and 37 on the major streams; the sand plain tributaries (e.g., station 35); and the groundwater contribution. The daily output from this study area consisted of the flux at station 34 and the estimated removal of water and nitrogen for irrigation (Figure 2a).

A mass balance was calculated for the downstream reach of Duffin Creek on the basis of data collected on a total of 18 days from 1973 to 1975 and 1978. The summation of the individual stream discharges and nitrogen fluxes at stations 2, 4, 11, 13, 22, 23, and 26 were compared with the output of water and nitrogen at station 3 for each day (Figure 2b).

RESULTS AND DISCUSSION

The water budget showed that average daily inputs and outputs were in balance for the Nottawasaga River, whereas Duffin Creek had a small but significant daily gain of about 8% (Table 1). Most of the daily water input to the downstream reaches of Duffin Creek was contributed by stations 22 (40%), 11 (27%), and 13 (14%). In the Alliston sand plain area stations 32 and 33 accounted for 48% and 24%, respectively, of the average daily input of water, and the other major streams contributed an additional 20%. Approximately 8% of the daily input was derived from groundwater, either in direct seepage to the main river channels (3%) or in sand plain tributary streams (5%).

Nitrate Budgets During Low Summer Flows

Low concentrations of NH_4–N (<0.08 mg/l) and trace levels of NO_2–N occurred at all sampling locations in both drainage basins. Mean concentrations of organic nitrogen varied between 0.25 and 0.61 mg/l for the Duffin Creek stations. Nitrate levels were generally in the range 0.10 to 1 mg N/l at most stream sampling stations. However, mean concentrations of NO_3–N at stations 13 and 33, downstream from sewage treatment plants, were 2.63 and 1.32 mg/l respectively. High NO_3–N concentrations (ranging from 3 to 40 mg/l) occurred in some small tributaries and groundwater seeps on the heavily fertilized Alliston sand plain.

The NO_3–N mass balance for the Nottawasaga River showed a loss on all 21 days of observation. Individual daily deficits ranged from less than 1 kg to a maximum of 70 kg, and the average daily NO_3–N mass balance indicated a loss of 35 ± 21 kg, which is equivalent to about 60 mg N/m^2 of stream bed (Table 1). In Duffin Creek significant NO_3–N losses were recorded on all 18 days of observation. The average daily loss for the 18 days was 45 ± 9 kg NO_3–N, which is equivalent to approximately 160 mg N/m^2 of stream bed (Table 1). Individual reaches on the east branch of

Table 1. Mean Daily Input and Output of Water and NO_3-N in the Nottawasaga River and Duffin Creek During Low Flows from Late May to October*

	Input	Output	Net gain or loss	Gain or loss as % of input
		Nottawasaga River		
Flow, l/s				
Total	3310 ± 561	3364 ± 595	+54 ± 141	2
Major streams	3058 ± 555			
Sand plain streams	147 ± 8			
Groundwater	105			
NO_3 – N, kg/day				
Total	267 ± 38	232 ± 45	-35 ± 21**	13
Major streams	172 ± 37			
Sand plain streams	16 ± 2			
Groundwater	79			
		Duffin Creek		
Flow, l/s	996 ± 170	1078 ± 249	+82 ± 127*	8
NO_3 – N, kg/day	60 ± 11	15 ± 9	-45 ± 9**	76

*Means are ± standard deviation. Mean daily gain or loss significantly different from zero at the 5% level is indicated by * and at the 1% level, by ** (a pair comparison t-test was used).

the river had average losses ranging from 40 to 110 mg N m^{-2} day^{-1}, whereas much higher losses (300 and 480 mg N m^{-2} day^{-1}) occurred downstream from the sewage treatment plant between stations 7 and 8 and between stations (11 + 13) and 8, respectively (Figure 2b). Analysis of the transport of organic nitrogen in Duffin Creek indicated that small daily gains and losses occurred with equal frequency and the average daily difference between inputs and outputs was less than 3%.

Mechanisms for Nitrate Removal in River Reaches

A variety of processes may be responsible for the occurrence of nitrate losses during river transport. Nitrate removal as a result of temporary immobilization during decomposition of organic matter is of minor significance during the summer in southern Ontario (Kaushik and Hynes, 1971; Kaushik et al., 1975). Nitrate assimilation by riparian tree roots in the streambed is also likely to be a negligible factor (Meyer, 1978). Aquatic macrophytes and benthic algae are unimportant in the Nottawasaga study area. Analysis of the role of benthic algae on gravel riffles in Duffin Creek suggests that nitrate uptake by the algae can account for only approximately 20% of the observed daily nitrate loss (Hill, 1981). The two-way exchange of water between a stream and a subsurface groundwater reservoir can cause element losses from river reaches (Rigler, 1979). Cation mass balances and hydrogeologic data indicate an absence of such water exchanges in the Nottawasaga River and Duffin Creek (Hill, 1982).

Laboratory experiments revealed disappearance of NO_3–N from solutions overlying sediment cores, with almost total loss occurring in some samples within 6 days despite the presence of high dissolved oxygen levels in the solutions. The control cylinders and pans containing no sediment did not exhibit any change in NO_3–N concentration during the experimental period. Less than 5% of the nitrate disappeared from solutions overlying sediment samples that had been sterilized with formaldehyde; this indicates that nitrate removal was caused by biological mechanisms.

In Duffin Creek average rates of NO_3–N loss over a 6-day period ranged from 80 to 250 mg/m^2 of sediment surface daily for twenty pool sediment cores and from 10 to 60 mg/m^2 for nine riffle gravel samples. The highest loss rates for the gravels were recorded in samples taken from small slackwater areas of the riffles where a thin veneer of silt, 5 to 10 mm thick, covered the surface of the gravel. The twelve sediment cores collected from the Nottawasaga River and its major tributaries exhibited a daily loss ranging from 10 to 90 mg/ m^2 of sediment surface over a 6-day period. The medium-to-coarse sands that occupy 80 to 90% of the stream bed area of the Nottawasaga study site have a mean nitrate loss rate (38 ± 27 mg N m^{-2} day^{-1}) which is only slightly higher than the average for the gravel riffles in Duffin Creek (26 ± 15 mg N m^{-2} day^{-1}). The pool sediments that cover approximately 50% of the bed area in Duffin Creek have a mean loss rate of 150 ± 53 mg N m^{-2} day^{-1}, which is almost four times greater than the Nottawasaga sediments.

The removal of nitrate from solutions overlying sediments has been attributed primarily to bacterial denitrification. (Sain et al, 1977; Van Kessel, 1977b). Laboratory nitrate losses may provide an overestimate of denitrification, however, since some assimilatory nitrate reduction and respiratory reduction to the level of ammonia may also occur (Keeney, 1973; Brezonik, 1977). Labeled $^{15}NO_3$–N experiments with sediments from Canagagigue Creek, about 70 km west of Toronto, indicated that 98% of the nitrate removed resulted from denitrification (Chatarpaul and Robinson, 1979). Van Kessel (1977c) also indicated that assimilatory reduction was responsible for less than 5% of NO_3–N losses in laboratory tests with sediments from a canal. A much greater proportion of immobilization and nitrate reduction to ammonia, ranging from 20 to 60% of the nitrate loss, has been recorded for lake sediments (Keeney et al., 1971; Chen et al., 1972b; 1979). Chatarpaul and Robinson (1979) suggested that these higher immobilization rates may result from the technique of adding the labelled nitrogen directly to the sediment rather than to the solution overlying the sediment. The latter technique was used in the studies that recorded negligible assimilatory reduction.

The rates of nitrate disappearance observed in the laboratory are of the same order of magnitude as the field losses recorded for the Nottawasaga River and Duffin Creek study reaches. Nevertheless, a precise comparison cannot be made between the laboratory and field data. The initial NO^3–N concentration of the solutions used in the laboratory experiment was considerably greater than the concentration in the rivers; consequently the laboratory experiment may overestimate denitrification rates in the rivers. (Van Kessel, 1977a). In other respects the laboratory experiment may seriously underestimate the real rate of denitrification in streams. Nitrate transport into sediments from the overlying water is a major determinant of denitrification rate. The turbulence and mixing caused by stream flow and the presence of benthic animals create much higher rates of nitrate transport into sediment under natural conditions (Van Kessel, 1977b).

Spatial Variations in Nitrate Removal Rates

Analysis of nitrate sink processes indicates that denitrification may be a major factor in nitrate removal in the Nottawasaga River and Duffin Creek. However, nitrate mass balances and my laboratory experiments suggest that the two rivers differ considerably with respect to denitrification and nitrate reduction rates in stream sediments during low summer flows.

Denitrification is affected by a number of environmental factors. The process requires an absence of oxygen, although not a highly reducing environment (National Academy of Sciences, 1978). It is influenced by temperature and pH, proceeding slowly at low temperatures and in acidic conditions (Keeney, 1973; Terry and Nelson, 1975; Van Kessel, 1977a). Carbon, although required as an energy source by denitrifers, does not appear to affect denitrification rates over a range from approximately 1 to 30% organic carbon content (Robinson, et al., 1978b). However, there must be some critical level of carbon below which denitrification rates will decrease. A positive correlation is evident between denitrification rates and nitrate concentrations in the overlying water (Van Kessel, 1977a). Stirring of sediment by benthic fauna aids rapid diffusion of nitrate into sediments and can have a positive effect on denitrification (Chatarpaul et al., 1979; 1980).

The wide variation in daily rates of nitrate removal (40 to 480 mg N/m^2) in the six downstream reaches of Duffin Creek results mainly from differences in stream nitrate concentration (Hill, 1981). Stream sediments are similar in all reaches; however, mean NO_3–N concentration varies from 0.5 to 1.0 mg/l upstream from station 7A, whereas NO_3–N levels are usually 0.2 to 0.4 mg/l in the other reaches of the river. The average NO_3–N concentration in the Nottawasaga River and its major tributaries is 0.5 to 1.0 mg/l, which is similar to or higher than levels in Duffin Creek. Despite the high nitrate concentration, the field and laboratory data indicate that nitrate removal rates in the Nottawasaga River are considerably lower than in Duffin Creek. A comparison of the two rivers suggests that they are similar in water retention times, temperature, and pH but differ in many sediment characteristics.

The Nottawasaga river sediments are medium-to-coarse sands with a much lower content of silt and organic carbon than the Duffin Creek pool sediments (Table 2). Differences in organic carbon and nitrogen are particularly evident in the 0- to 4-cm zone. Duffin Creek sediments in this zone contain about six times more organic carbon than those in the Nottawasaga River. Denitrification in stream sediments occurs mainly in the upper few centimeters of the river bed (Sain et al., 1977; Van Kessel,

Table 2. Characteristics of the Sediment at 0 to 4 and 4 to 10 cm Depth in Pool Reaches of Duffin Creek and the Nottawasaga River*

	Duffin Creek		Nottawasaga River	
	0–4 cm (N = 10)	4–10 cm (N = 10)	0–4 cm (N = 10)	4–10 cm (N = 10)
pH	7.9 ± 0.2	8.1 ± 0.2	8.1 ± 0.2	8.2 ± 0.1
NH_4–N, μg/g	9.0 ± 7.7	1.8 ± 2.5	1.8 ± 2.1	0.5 ± 1.1
NO_3 + NO_2–N, μg/g	0.07 ± 0.08	0.06 ± 0.09	0.16 ± 0.02	0.13 ± 0.13
Organic nitrogen, %	0.110 ± 0.07	0.015 ± 0.007	0.017 ± 0.016	0.011 ± 0.006
Organic carbon, %	1.40 ± 0.91	0.20 ± 0.10	0.22 ± 0.23	0.11 ± 0.08
C : N ratio	12.7 : 1	13.3 : 1	12.9 : 1	10.0 : 1
Gravel, %	16.1 ± 16.2	23.9 ± 25.6	4.6 ± 4.1	9.4 ± 16.8
Sand, %	53.2 ± 18.4	72.1 ± 24.4	91.0 ± 6.2	88.1 ± 16.3
Silt, %	25.1 ± 17.7	3.2 ± 1.9	3.8 ± 3.4	1.7 ± 1.4
Clay, %	5.6 ± 4.1	0.5 ± 0.3	0.6 ± 0.5	0.8 ± 0.4

*Means ± standard deviation (air-dry basis).

Table 3. Mean Daily Input and Output of Water and NO_3-N in the Nottawasaga River and Duffin Creek during Low Flows from Late November to March*

	Input	Output	Net gain or loss	Gain or loss as % of input
Nottawasaga River				
Flow, l/s	4078 ± 934	4319 ± 1124	+241 ± 249*	6
NO_3-N, kg/day	527 ± 97	535 ± 105	+ 8 ± 14	2
Duffin Creek				
Flow, l/s	1413 ± 255	1540 ± 306	+127 ± 93*	9
NO_3-N, kg/day	123 ± 16	137 ± 36	+ 14 ± 23	11

*Means ± standard deviation. Mean daily gain or loss significantly different from zero at the 5% level is indicated by * (a pair comparison t-test was used).

1977a); consequently the low rates of denitrification in the Nottawasaga River may result from insufficient levels of organic carbon in the surface sediment. In addition, the coarse texture of these sediments may reduce the number of anoxic microsites for denitrification near the surface.

Temporal Variations in Nitrate Removal Rates

Nitrate loss during stream transport may also occur during low flows in winter. Nitrate inputs and outputs were measured on 9 days in the Nottawasaga study area and on 5 days in Duffin Creek between late November and early March (Table 3). Stream discharge could not be measured accurately with a current meter at several stations because of a thick ice cover. In these cases daily discharge was estimated from regression analysis between nearby continuously gauged stations and the ungauged streams.

During winter low flows, small NO_3–N losses (1 to 12 kg) occurred on 3 days, and gains of 5 to 29 kg were observed on 6 days in the Nottawasaga River. A similar pattern of nitrate gains and losses was also evident in Duffin Creek. Mean daily nitrate outputs exceeded inputs by 2 and 11% in the Nottawasaga River and Duffin Creek, respectively (Table 3). These differences between inputs and outputs were not statistically significant, however; this indicates a balanced budget.

A more detailed analysis of the seasonal dynamics of nitrate sinks was undertaken in the late spring and early autumn of 1979 for the river reach between stations 7 and 8 on Duffin Creek (Figure 2b). Nitrate removal from the river reach became evident in early May as water temperatures

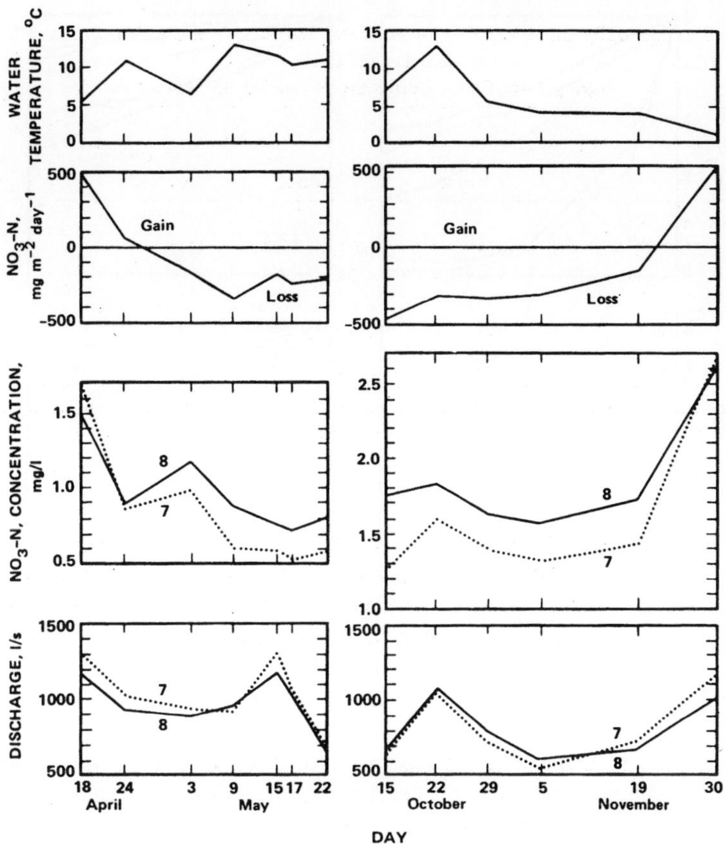

Figure 3. Variations in daily nitrate losses (mg N/m² of stream bed) for the reach between stations 7 and 8 on Duffin Creek during late spring and autumn 1979.

rose and stream discharge declined (Figure 3). A decrease in rates of nitrate removal also occurred in late October and early November.

These data suggest that denitrification and other sink processes decline during the autumn and increase in importance during May in Ontario rivers. The field data indicate an absence of nitrate removal between December and April. However, this evidence cannot be used to infer that nitrate sink processes cease during these months. The inaccuracy of discharge measurements on ice-covered streams and nitrate inputs from unsampled ephemeral tributaries create errors in the mass balance which preclude the detection of small losses of nitrate resulting from biological utilization under cold temperature conditions.

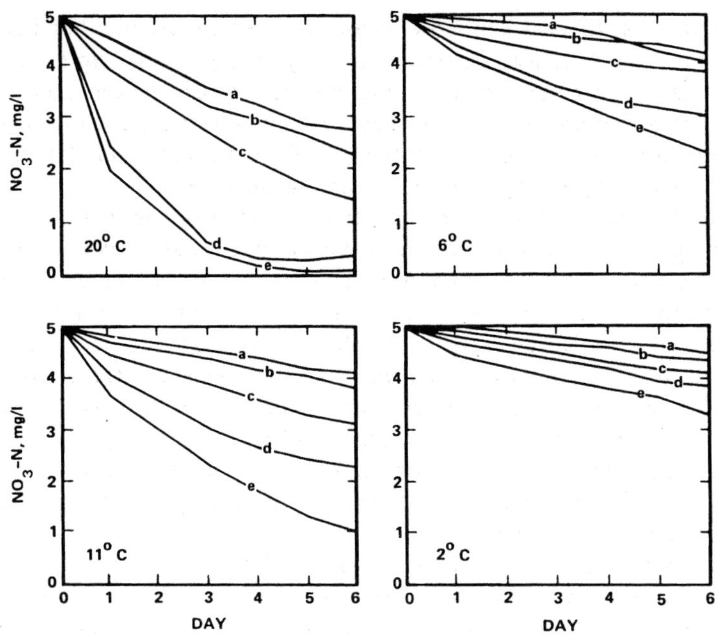

Figure 4. Changes in nitrate concentration in water overlying stream sediments incubated at temperatures between 2 and 20°C.

The possible occurrence of denitrification and nitrate reduction in sediments during the period from November to April in Ontario rivers was also examined by a laboratory experiment. The disappearance of nitrate was observed in twelve sediment–water profiles incubated at temperatures of 20, 11, 6, and 2°C. From December to March the water temperature in the Nottawasaga River and Duffin Creek fluctuates between 0 and 2°C. The laboratory data revealed that nitrate was lost from water in contact with sediments even at these low temperatures, although nitrate removal was only approximately 20% of the rate in summer water temperatures, which average 20°C (Figure 4). River water temperatures range from 4 to 11°C during November, April, and early May. Water–sediment profiles incubated at 6 and 11°C exhibited rates of nitrate loss that were 35 and 45%, respectively, of the nitrate removal at 20°C. These results suggest that river sediments may remove nitrate from stream water during the winter and spring, although at a much lower rate than is characteristic of the months from May to October.

Fluctuations in denitrification rates may also be influenced by river discharge. During major flood events the considerable disturbance and scouring of sediments probably creates aerated conditions that inhibit denitrification. Increases in stream velocity during smaller storm runoff events may still be sufficient to transport the upper few centimeters of sediment, although the stream bed at greater depths remains undisturbed. Since denitrification occurs mainly in the 0- to 5-cm sediment layer, this partial disturbance may reduce denitrification rates.

It is difficult to measure the effects of storm flows on denitrification and other sink processes in the field. However, the data collected for the reach between stations 7 and 8 on Duffin Creek provide some evidence of the effects of small discharge peaks in relation to nitrate removal rates (Figure 3). Higher discharges on May 15 and October 22 were not associated with increases in suspended sediment load, and there was only a small decrease in nitrate loss per square meter of stream bed in comparison to days with lower discharge. These two discharge peaks were approximately three times greater than average summer low flows in this reach; this suggests that nitrate sink processes may still be important during moderate stream flows.

Disturbance of stream sediments was simulated in the laboratory using beakers with 2 cm of sediment covered by 500 ml of solution containing 5 mg NO_3–N/l. The sediments were magnetically stirred at 20° C for 6 days. Analysis of stirred sediments from several sites on Duffin Creek revealed a 1- to 2-day lag phase, followed by an increase in nitrate; concentrations in the overlying solution rose to 10 mg N/l in some cases. Cessation of the disturbance resulted in a decline in nitrate concentration after a delay of 1 to 2 days. Subsamples of the sediment cores incubated in a nonstirred state showed rapid disappearance of nitrate over a 6-day period. The increase in nitrate levels in disturbed sediments is probably caused by rapid nitrificaiton. Similar results have been reported for stirred sediments from hardwater lakes (Chen et al., 1972a). These data suggest that denitrification is probably negligible in river channels during major flood events.

Importance of Nitrate Sink Processes in Relation to River Nitrogen Budgets

The mass balance for Duffin Creek during summer low flows indicates an average daily removal of 45 ± 9 kg NO_3–N (Table 1). In contrast, the Kjeldahl nitrogen load does not change significantly during stream transport. Bacterial denitrification and other minor nitrate sink processes remove approximately 50% of the average daily input of total nitrogen and 75% of the nitrate input received by the downstream river reaches (Hill, 1981). In the Alliston sand plain area of the Nottawasaga River the average daily removal of NO_3–N is 35 ± 21 kg, which represents approximately 13% of the nitrate input (Table 1). These nitrate losses, although substantial, probably underestimate the actual daily nitrate removal. The mass balance technique cannot measure the denitrification of nitrate generated within bottom sediments. Nitrate produced by nitrification in the thin aerobic surface layer of sediment may undergo denitrification in microsites with reducing conditions in this zone or after diffusion to deeper anaerobic sediments (Chatarpaul et al., 1979; 1980; Chen et al., 1972a).

Almost 80% of the annual nitrogen exports from Duffin Creek and the Nottawasaga River occur between November and April when denitrification and other nitrate sink processes are of minor importance because of cold temperatures and periods of high river discharge. In rivers that are not influenced by major point pollution sources, storm periods account for most of the annual nitrogen export. Approximately 65% of the nitrate is exported during 20% of the year when mean daily discharges exceeded 3 m^3/s in Duffin Creek and 14 m^3/s in the Nottawasaga River.

These element transport dynamics imply that nitrate removal by denitrification and other sink processes is relatively small in relation to annual nitrogen flux in most rivers. The average daily summer low flow loss of 35 kg NO_3–N in the Nottawasaga River would result in the removal of about 4600 kg NO_3–N between May and October. From the laboratory data on rates of denitrification in relation to water temperature, it is estimated that an additional 1000 kg is removed during low flows between November and April, for a total removal of 5600 kg. The average annual export of NO_3–N from the Nottawasaga basin at station 34 for the 1977–1979 period was about 410,000 kg (Hill, in preparation), and therefore nitrate removal during stream transport represents only about 1% of the annual nitrate export in this river. Similarly, nitrate losses in the downstream reaches of Duffin Creek represent only 5 to 6% of the annual export of total nitrogen (Hill, 1979).

This study focused on various aspects of nitrogen transport and transformations. The two rivers examined revealed considerable differences in the magnitude of nitrate sink processes during low flows. The high nitrate assimilative capacity of some rivers, as exemplified by Duffin Creek, is of considerable ecological significance because of the resulting reduction in nitrate concentration in stream water and the smaller export of nitrate to lakes and rivers downstream. Although a relatively small percentage of the annual nitrogen load of most rivers is transported during low discharges, the considerable time duration of these conditions enhances the ecological importance of nitrate sink processes. In southern Ontario low flows usually occur on 130 to 150 days between May and

October. In areas of the world that lack a cold winter, denitrification and other sink processes may be important in low flow conditions throughout the year.

Further research is required to evaluate nitrogen flux and cycling in rivers. The recent development of the acetylene blockage technique (Knowles, 1979; Yoshinari and Knowles, 1976) provides a relatively simple and inexpensive procedure for measuring denitrification rates, as well as evaluating the relative importance of NO_3–N immobilization and permanent loss by denitrification. The ecological effects of nitrate removal during low flows should be examined in relation to the functioning of downstream ecosystems. Increasing nitrogen inputs from agriculture, sewage treatment plants, and other sources are characteristic of many rivers; however, little is known about the capacity of streams to assimilate additional inputs. Studies are needed to evaluate the effects of increased nitrogen inputs on denitrification and other nitrate sink processes in rivers.*

ACKNOWLEDGMENTS

This work was supported by grants from the Natural Sciences and Engineering Research Council of Canada. I thank N. K. Kaushik and J. B. Robinson for their helpful advice during this study. I am also grateful to K. Sanmugadas, W. Scott, and T. Irvine for laboratory assistance; D. McLeish and L. Nordstrom for field assistance, and the Cartographic staff in the Department of Geography for the figures.

REFERENCES

Allison, L. E., 1965, Organic Carbon, in C. A. Black (Ed.), *Methods of Soil Analysis,* pp. 1367-1378, American Society of Agronomy, Madison, WI.

American Public Health Association, 1976, *Standard Methods for the Examination of Water and Wastewater,* 14th ed. New York.

Brezonik, P. L., 1977, Denitrification in Natural Waters. *Prog. Water Technol.,* 8: 373-392.

---

*Stream sediments from Duffin Creek and the Nottawasaga River have been incubated in the laboratory and monitored periodically for N_2O production using the acetylene blockage technique. Results indicate that >90% of the nitrate removal from solutions overlying the sediments can be attributed to denitrification.

Chatarpaul, L. and J. B. Robinson, 1979, Nitrogen Transformations in Stream Sediments: ^{15}N Studies, in C. D. Litchfield and P. L. Seyfried (Eds.), *Methodology for Biomass Determinations and Microbial Activities in Sediments,* pp. 119-127, Publication ASTM STP 673, Philadelphia.

______, J. B. Robinson, and N. K. Kaushik, 1979, Role of Tubificid Worms on Nitrogen Transformations in Stream Sediment, *J. Fish Res. Board Can.,* 36: 673-678.

______, J. B. Robinson, and N. K. Kaushik, 1980, Effects of Tubificid worms on Denitrification and Nitrification in Stream Sediment, *Can. J. Fish. Aquat. Sci.,* 37: 656-663.

Chen, R. L., D. R. Keeney, and J. G. Konrad, 1972a, Nitrification in Sediments of Selected Wisconsin Lakes, *J. Environ. Qual.,* 1: 151-154.

______, D. R. Keeney, D. A. Graetz, and A. J. Holding, 1972b, Denitrification and Nitrate Reduction in Wisconsin Lake Sediments, *J. Environ. Qual.,* 1: 158-162.

______, D. R. Keeney and L. J. Sikora, 1979, Effects of Hypolimnetic Aeration on Nitrogen Transformations in Simulated Lake Sediment-Water Systems, *J. Environ. Qual.,* 8: 429-433.

Hill, A. R. 1979, Denitrification in the Nitrogen Budget of a River Ecosystem, *Nature (London),* 281: 291-292.

______, 1981, Nitrate-Nitrogen Flux and Utilization in a Stream Ecosystem During Low Summer Flows, *Can. Geog.,* 25: 225-239.

______, 1982, Phosphorus and Major Cation Mass Balances for Two Rivers During Low Summer Flows, *Freshwater Biol.* 12: 293-304.

Issac, R. A., and W. C. Johnson, 1976, Determination of Total Nitrogen in Plant Tissue, Using a Block Digester, *Journal AOAC,* 59: 98-100.

Kaushik, N. K. and H. B. N. Hynes, 1971, The Fate of Dead Leaves that Fall Into Streams, *Arch. Hydrobiol.,* 68: 465-515.

______, J. B. Robinson, P. Sain, H. R. Whiteley and W. N. Stammers, 1975, A Quantitative Study of Nitrogen Loss from Water of a Small Spring-Fed Stream, *Proc. Can. Symp. Water Pollut. Res., Toronto,* 10: 110-117.

Keeney, D. R., 1973, The Nitrogen Cycle in Sediment-Water Systems, *J. Environ. Qual.,* 2: 15-29.

______, R. L. Chen, and D. A. Graetz, 1971, Denitrification and Nitrate Reduction in Sediments: Importance to the Nitrogen Budget of Lakes, *Nature (London),* 233: 66-67.

Knowles, R., 1979, Denitrification, Acetylene Reduction and Methane Metabolism in Lake Sediment Exposed to Acetylene, *Appl. Environ. Microbiol.,* 38: 486-493.

Likens, G. E., 1975, Nutrient Flux and Cycling in Freshwater Ecosystems, in F. G. Howell, J. B. Gentry, and M. H. Smith (Eds.), *Mineral Cycling in Southeastern Ecosystems,* ERDA Symposium Series, CONF-740513, pp. 314-348, NTIS, Springfield, VA.

Meyer, J. L., 1978, Transport and Transformation of Phosphorus in a Forest Stream Ecosystem, PhD. Thesis, Cornell University, Ithaca, NY.

National Academy of Sciences, 1978, *Nitrates: An Environmental Assessment,* Washington, DC.

Ontario Ministry of Environment, 1973-1978, *Water Quality Data for Ontario Lakes and Streams*, Vols. 8-14, Toronto.

Ostry, R. C. 1979, The Hydrogeology of the IFYGL Duffin Creek Study Area, Water Resources Report 5C, Ontario Ministry of Environment, Toronto.

Owens, M., J. H. N. Garland, I. C. Hart, and G. Wood, 1972, Nutrient Budgets in Rivers, *Symp. Zool. Soc. London*, 29: 21-40.

Rigler, F. H., 1979, The Export of Phosphorus from Dartmoor Catchments: A Model to Explain Variations of Phosphorus Concentrations in Streamwater, *J. Mar. Biol. Assoc. U.K.*, 59: 659-687.

Robinson, J. B., N. K. Kaushik, and L. Chatarpaul, 1978a, *Nitrogen Transport and Transformations in Canagagigue Creek*, International Joint Commission. Windsor, Ontario.

______, H. R. Whiteley, W. Stammers, N. K. Kaushik, and P. Sain, 1978b, The Fate of Nitrate in Small Streams and Its Management Implications, in R. C. Lohr et. al. (Eds.), *Best Management Practices for Agriculture and Silviculture*, pp. 247-259, Ann Arbor Science Publications, Ann Arbor, MI.

Sain, P., J. B. Robinson, W. N. Stammers, N. K. Kaushik, and H. R. Whiteley, 1977, A Laboratory Study of the Role of Stream Sediment in Nitrogen Loss from Water, *J. Environ. Qual.*, 6: 274-278.

Sibul, U., K. T. Wang, and D. Vallery, 1977, Ground-Water Resources of the Duffin-Rouge River Drainage Basins, Water Resources Report 8, *Ontario Ministry of Environment*, Toronto.

Terry, R. E., and D. W. Nelson, 1975, Factors Influencing Nitrate Transformations in Sediments, *J. Environ. Qual.*, 4: 549-554.

Van Kessel, J. F., 1977a, Factors Affecting the Denitrification Rate in Two Water-Sediment Systems, *Water Res.*, 11: 259-267.

______, 1977b, Removal of Nitrate from Effluent Following Discharge on Surface Water, *Water Res.*, 11: 533-537.

______, 1977c, the Immobilization of Nitrogen in a Water-Sediment System by Denitrifying Bacteria as a Result of Nitrate Respiration, *Prog. Water Technol.*, 8: 155-160.

Yoshinari, T., and R. Knowles, 1976, Acetylene Inhibition of Nitrous Oxide Reduction by Denitrifying Bacteria, *Biochem. Biophys. Res. Comm.*, 69: 705-710.

Commentary on Paper F4

Duff, J. H. & Triska, F. J. (1990) Denitrification in sediments from the hyporheic zone adjacent to a small forested stream.

The ability to retain particulate and dissolved material within a river system is a result of its overall complexity; this includes structures at different spatial scales, ranging from meanders or debris dams at the reach scale, to the processes occurring at the sediment grain scale. In this regard, the role of the subsurface compartment was not immediately recognised, probably because of the technical difficulties involved. Some papers using conservative tracers (e.g. Bencala, 1984; Triska *et al.*, 1989) had already outlined that the hyporheic area was a remarkable transient storage zone, potentially influencing some processes occurring in the stream. Flowing water beneath the river (the hyporheic zone) could be considered as a real interface for groundwater–stream water exchange. Its role was suspected in the early observations of the subsurface transformations of nitrogen (Grimm & Fisher, 1984; Lowrance *et al.*, 1984, Paper E1; Peterjohn & Correll, 1984, Paper E2). Triska *et al.* (1989) outlined a conceptual approach to the processes involved in the channel section, and suggested that the hyporheic area could be relevant in terms of nutrient cycling and biotic metabolism in the stream ecosystem.

The hyporheos is an area of advective exchange with the surface water. Its extent is a product of streambed roughness, as well as of its permeability and morphology. While the hyporheic zone is the area where the "transport is dominated by channel advection between sediment grains" (Triska *et al.*, 1989), the properties of groundwater are not influenced by the channel. Instead, the groundwater is dependent on hydrological links with the upland catchment. Duff & Triska (1990) analysed in detail the role of the hyporheos in the nitrogen metabolism of a forested stream.

The paper showed that nitrate was the mobile nitrogen form in the sediments. As such, nitrate was potentially used by algae and bacteria. Bacteria could use nitrate as an electron acceptor during DOC mineralization, while algae could use it for their biological synthesis. Nitrification occurring in the sediment could also cause the oxygen to decrease or even deplete within the hyporheos. That nitrate reduction to nitrogen gas in the hyporheic zone was related to the low oxygen levels and high DOC concentrations was further checked by measurement of *in situ* denitrification. The methodology involved the use of a series of wells specially equipped to avoid gas exchange between the hyporheic samples and the atmosphere. The use of this technique confirmed the link between the abundance of organic matter, oxygen depletion and denitrification. Further, the denitrification occurring in the stream channel was lower than that in the saturated upland area, this being related to their respective oxygen availability.

This paper confirmed that the hyporheos could function as an area of active nitrogen and carbon transformation. The fate of the chemical forms was related to the current redox state of the sediment. Therefore the hyporheos was not only an area of remarkable storage capacity but a place for high potential biological activity.

SS

References

Bencala, K. E. (1984) Interactions of solutes and streambed sediment. 2. A dynamic analysis of coupled hydrological and chemical processes that determine solute transport. *Water Resour. Res.* **20**, 1804–1814.

Grimm, N. B. & Fischer, S. G. (1984) Exchange between interstitial and surface water: implications for stream metabolism and nutrient cycling. *Hydrobiologia* **111**, 219–228.

Lowrance, R. R, Rodd, R. L., & Asmussen, L. E. (1984) Nutrient cycling in an agricultural watershed. I. Phreatic movement. *J. Environ. Qual.* **13**, 22–27.

Peterjohn, W. T. & Correll, D. L. (1984) Nutrient dynamics in an agricultural watershed: observations on the role of a riparian forest. *Ecology* **65**, 1466–1475.

Triska, F. J., Kennedy, V. C., Avanzino, R. J., Zwellweger, G. W. & Bencala, K. E. (1989) Retention and transport of nutrients in a third-order stream: hyporheic processes. *Ecology* **70**, 1892–1905.

Denitrification in Sediments from the Hyporheic Zone Adjacent to a Small Forested Stream

John H. Duff and Frank J. Triska

U.S. Geological Survey, Water Resources Division, 345 Middlefield Road, MS 496, Menlo Park, CA 94025 USA

Duff, J. H., and F. J. Triska. 1990. Denitrification in sediments from the hyporheic zone adjacent to a small forested stream. Can. J. Fish. Aquat. Sci. 47: 1140–1147.

Denitrification was assayed by the acetylene blockage technique in hyporheic sediments. Samples were obtained along transects perpendicular to the stream at two sites: (1) the base of a slope dominated by old-growth redwood and (2) the base of a slope dominated by alder regenerating from a clearcut in 1965. Denitrification was evident at in situ nitrate concentrations at all locations tested. Activity was stimulated by nitrate but nitrate plus glucose had no additional effect. Denitrifying potentials increased with increasing distance from the stream channel. Dissolved oxygen was 100% of the concentration expected in equilibrium with the atmosphere in water obtained from monitoring wells immediately adjacent to the stream but was as low as 7% of the expected value in water 11.4 m inland. Both nitrate and dissolved organic carbon decreased over summer in wells at the base of the alder-forested slope. A 48-h injection of nitrate-amended stream water into hyporheic water 8.4 m inland stimulated nitrous oxide production in the presence of acetylene. Nitrous oxide was generated as nitrate and acetylene were co-transported to a well 13 m down-gradient. The acetylene-block experiments coupled with the chemistry data suggest that denitrification can modify the chemistry of water during passage through the hyporheic zone.

Des essais de dénitrification ont été effectués par la technique de blocage à l'acétylène dans des sédiments hyporhéiques. Des échantillons ont été prélevés à deux endroits dans des transects perpendiculaires au cours d'eau : (1) à la base d'une pente dominée par des séquoias mûrs et (2) à la base d'une pente dominée par de jeunes aulnes repeuplant un parterre de coupe de 1965. La dénitrification était évidente pour des concentrations de nitrate du milieu naturel à tous les endroits analysés. L'activité était stimulée par le nitrate, mais l'addition de glucose au nitrate ne provoquait aucun autre effet. Les potentiels de dénitrification augmentaient avec l'éloignement du cours d'eau. L'oxygène dissous en équilibre avec l'atmosphère correspondait à 100% de la concentration dans l'eau prévue et mesurée dans des puits immédiatement adjacents au cours d'eau, mais atteignant à peine 7% de la valeur escomptée dans des eaux situées à 11,4 m à l'intérieur des terres. Le nitrate et le carbone organique dissous diminuaient pendant l'été dans les puits situés à la base de la pente boisée d'aulnes. Une injection d'eau additionnée de nitrate dans des eaux hyporhéiques, situées à 8,4 m dans les terres, pendant une période de 48 heures, a stimulé la production d'oxyde nitreux en présence d'acétylène. L'oxyde nitreux était produit sous forme de nitrate et l'acétylène était transporté conjointement à un puits situé à 13 m en bas du gradient. D'après les expèriences de blocage à l'acétylène combinées aux données chimiques, la dénitrification peut modifier les propriétés chimiques de l'eau lorsqu'elle circule dans la zone hyporhéique.

Received July 19, 1989
Accepted January 23, 1990
(JA235)

Reçu le 19 juillet 1989
Accepté le 23 janvier 1990

In upland watersheds, a variable fraction of the total stream discharge is exchanged between open-channel stream water and interstitial flows beneath and parallel to the stream channel (Bencala 1984; Kennedy et al. 1984; White et al. 1987; Triska et al. 1989). These subsurface zones, characterized by the direct and frequent contact of both stream and groundwater, comprise a dynamic and viable stream habitat — the hyporheic zone (Hynes 1983). At the base of the hyporheic zone, an irregular boundary is formed at the groundwater–stream water interface that constitutes the boundary of the stream ecosystem (Triska et al. 1989). Within the stream ecosystem, bidirectional exchange of nutrients can occur between stream water and the hyporheic zone.

The supply of dissolved nutrients to stream biota can be affected by the exchange between stream water and groundwater. For example, well-oxygenated hyporheic sediments foster nitrifying bacteria that can generate nitrate in excess of the biotic demand in the channel (Triska et al. 1990). Oxygen dissolved in stream water may be gradually depleted by physical, chemical, or biological processes as stream water is transported through the hyporheic zone. Exchange of oxygen between stream and subsurface habitats is important because oxygen availability commonly controls the oxidation and reduction of nitrogen compounds by bacteria (Knowles 1982). In areas of the hyporheic zone where dissolved oxygen is low, zones of denitrifying activity may be established.

In this study, we directly examine sediments from the hyporheic zone adjacent to a small forested stream for denitrification. We hypothesize that these sediments are potential sites of microbial denitrification and may act as a nitrate sink prior to reentry of subsurface water into the channel.

Materials and Methods

Study Site

Little Lost Man Creek is a pristine drainage near the town of Orick in northwestern California (Fig. 1). The stream flows

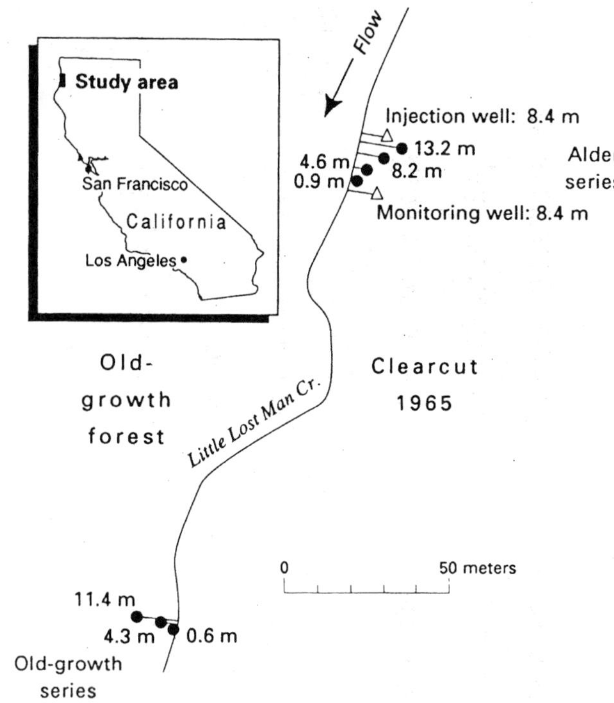

FIG. 1. The study site at Little Lost Man Creek, California.

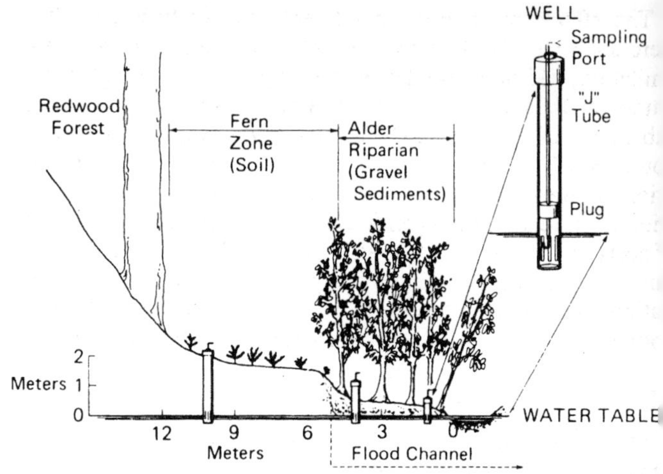

FIG. 2. A schematic cross section of the old-growth wells.

through Redwood National Park and is a third-order pool-and-riffle channel draining a long narrow coastal valley with steep slopes. The watershed area is 9.4 km^2 and ranges in altitude from 24 to 695 m. The overall channel gradient is 0.066%. The high permeability of the stream bed results in significant interflow during transport. Discharge ranges between 0.01 and 0.04 m^3·s^{-1} during base flow. Approximately 90% of the vegetation is old-growth coastal redwood forest (*Sequoia sempervirens*) that includes associated Douglas-fir (*Pseudotsuga menziesii*) and western hemlock (*Tsuga heterophylla*). Dominant riparian vegetation is red alder (*Alnus rubra*). About 10% of the watershed was clearcut in 1965. This study was done along a 220-m reach of Little Lost Man Creek that spans the clearcut and old-growth forest.

Sampling

In 1983, seven monitoring wells were placed in two distinct stream-side habitats across the riparian zone. One well series was placed at the base of a forested slope clearcut in 1965. Red alder has invaded the clearcut (Fig. 1). A second well series was located at the base of an old-growth coastal redwood forest typical of the watershed. The wells were located along transects perpendicular to the stream channel as far as 13.2 m inland.

Water samples were obtained from monitoring wells that were cased with 15-cm diameter polyvinyl chloride (PVC) pipe screened for a distance of approximately 20 cm below the water table at low flow (Fig. 2). The PVC casing was open at the base and capped above the ground. An aluminum plug (14 cm diameter) was positioned immediately above the water table to reduce gas exchange with the atmosphere. Water was sampled with peristaltic pumps through 0.64-cm diameter stainless steel tubing positioned in the center of the casing. For inorganic nutrients, water was pumped through a 142-mm diameter in-line filter (0.45-µm pore size Metricel filters, Gelman Sciences, Inc., Ann Arbor, MI[1]) into polyethylene bottles rinsed with sample three times. For dissolved organic carbon (DOC), water was pumped through a 142-mm diameter in-line glass fiber filter (Gelman A/E) housed in a stainless steel holder into glass bottles. Samples for DOC were subsequently filtered through 47-mm diameter silver filters into glass ampules. The glassware and glass fiber filters were combusted for 10 h at 450°C prior to sampling. Dissolved oxygen was measured with an oxygen probe in a polycarbonate cylinder. Well water was delivered continuously to the bottom of the cylinder until three to five times the capacity of the cylinder had overflowed and the oxygen reading had stabilized. Dissolved gases (acetylene and nitrous oxide) were extracted from water using the syringe shaking method of Rudd et al. (1974). Gas and nutrient samples were collected in triplicate.

Subsurface riparian sediment was collected from the cased wells by using a small plastic scoop at the end of a 1.8-m pipe. The cap, stainless steel tubing, and aluminum plug were removed to obtain a sample. Fresh sediment material was transferred to a nearby laboratory in Zip-lock bags. The upper 6 to 10 cm of stream sediment was collected in the same fashion after larger stones were removed from the surface. Fresh sediments were collected for each incubation.

Incubations

Denitrification was assayed by the acetylene blockage technique (Balderston et al. 1976; Yoshinari et al. 1976). Approximately 5 mL of sediment (9.6 g wet weight) was added to 20-mL glass bottles by using a wide-bore (1 cm) plastic syringe that had the end cut off. The sediment was slurried with 5 mL of water obtained from the same well. The bottles were flushed for 15 min with oxygen-free nitrogen and then sealed with recessed butyl rubber stoppers. Acetylene was generated from calcium carbide-saturated water and was added (2 mL) by syringe to the headspace of each bottle. The bottles were incubated on a shaker table at 18°C for 15 min after which 100-µL subsamples of the headspace were removed for nitrous oxide analysis. Headspace concentrations were corrected for nitrous oxide dissolved in the aqueous phase (Flett et al. 1976) using a Bunsen coefficient of 0.69.

[1]Use of brand names in this report is for identification purposes only and does not constitute endorsement by the U.S. Geological Survey.

The effects of various amendments on denitrification also ere analyzed. To differentiate carbon limitation from nitrate nitation, additional bottles were set up with acetate, glucose, trate, and nitrate plus acetate or nitrate plus glucose. For each bstrate, 100 μL of 60-mM stock solution was added to each ottle. Stock solutions were flushed with oxygen-free nitrogen ior to amendment. Some sediment slurries were incubated ith air in place of oxygen-free nitrogen and helium in place acetylene. Denitrification also was tested in unfiltered water mples enriched with nitrate and glucose (1 mM final concenation each). Fifteen millilitres of water were added to 20-mL ottles which were incubated anaerobically as described above.

itrate Injection Experiment

Stream water amended with nitrate was injected directly into bsurface water 8.4 m inland from the stream channel to assess e potential for denitrification by the indigenous bacterial population. A field modification of the acetylene blockage techique was used to inhibit bacterial reduction of nitrous oxide o-transported with acetylene. The injectate was prepared in a olyethylene carboy at the study site by adding 150 g of $NaNO_3$ 24 L of stream water. The solution was injected through a etering pump (Fluid Metering Inc., Oyster Bay, NY) directly to a cased injection well (Fig. 1) at a rate of 8 $mL \cdot min^{-1}$. cetylene (99.6 minimum purity, Matheson Gas Products, ewark, CA) was co-injected directly into the water in the cased jection well at a rate of approximately 10 $mL \cdot min^{-1}$. The olume of water in the injection well was approximately 3 L nd the subsurface velocity was 1–2 $m \cdot h^{-1}$. The subsurface elocity and flow path were determined in a 1984 injection xperiment with sodium bromide at the alder-forested site (G. V. Zellweger, U.S. Geological Survey, 345 Middlefield Rd., AS 496, Menlo Park, CA 94025, unpubl. data). Water samples or nitrate, dissolved nitrous oxide, and dissolved acetylene ere collected every 4 to 6 h in the monitoring well approxiately 13 m down-gradient. The injection ran for 48 h.

nalytical Procedures

Oxygen concentration and temperature were measured irectly in water samples collected in the field using an oxygen eter (Yellow Springs Model 57, Yellow Springs Instrument Company, Yellow Springs, OH) with a probe mounted on a tirring platform. Nitrate and ammonium were measured on an Autoanalyzer II (Technicon Corp. Inc., Tarrytown, NY) using cadmium-reduction method and a phenol-hypochlorite ethod, respectively. Nitrite was subtracted from itrate + nitrite. DOC (as CO_2) was analyzed on a Beckman odel 864 infrared analyzer (Beckman Instruments, Inc., Fulerton, CA) following sample digestion similar to the method of Menzel and Vaccaro (1964). Nitrous oxide and acetylene were determined on a gas chromatograph (Perkin-Elmer Sigma 2000, Perkin-Elmer, Norwalk, CT) equipped with a flame-ionzation detector and a ^{63}nickel electron capture detector. The electron capture detector was equipped with two Poropak® R stainless steel columns (244 cm long, 0.22 cm inside diameter) split by an eight-port backflush valve to vent the acetylene gas. The carrier was ultra high purity (UHP) oxygen-free nitrogen at a flow rate of 30 $mL \cdot min^{-1}$. The flame ionization detector was equipped with one Poropak® column (20% Q, 80% N; 305 cm long, 0.22 cm inside diameter) with UHP helium as the carrier gas (30 $mL \cdot min^{-1}$).

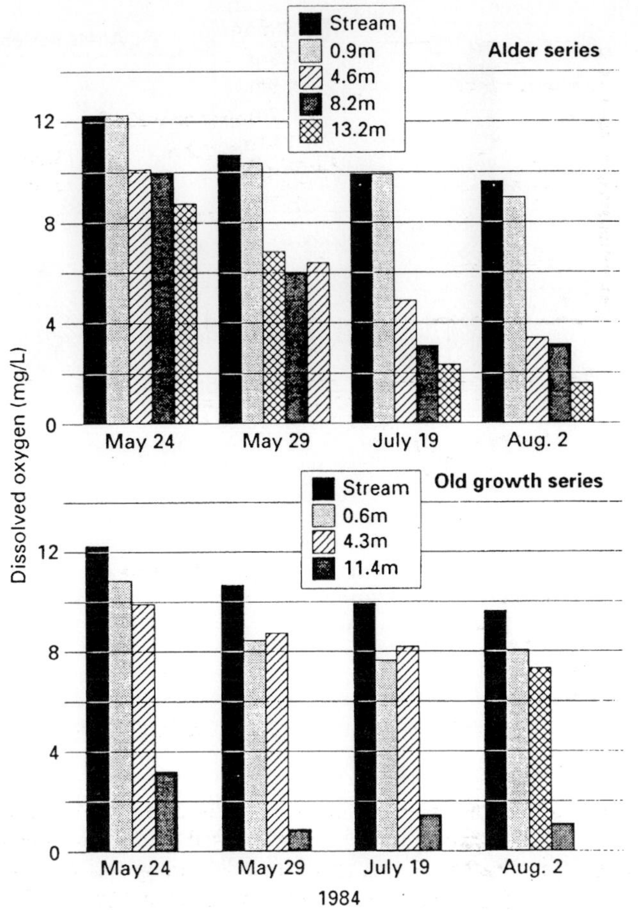

FIG. 3. Dissolved oxygen obtained from the alder and old-growth wells during base flow.

TABLE 1. Temperature (°C) and percentage saturation (%) of dissolved oxygen.

Date	5/24/84		5/29/84		7/19/84		8/14/84	
Distance (m)	(°C)	(%)	(°C)	(%)	(°C)	(%)	(°C)	(%)
Alder well series								
Stream	9.5	107	10.0	94	12.0	91	13.3	91
0.9	9.2	106	10.0	91	12.0	91	14.8	88
4.6	9.8	88	11.2	62	12.8	45	13.8	32
8.2	10.0	87	11.5	54	11.5	28	12.0	28
13.2	10.0	78	10.9	57	11.0	21	11.5	14
Old-growth well series								
Stream	10.3	107	12.2	94	15.5	91	15.0	91
0.6	9.4	94	10.0	74	14.2	75	13.8	78
4.3	9.4	85	10.1	77	13.0	77	13.4	69
11.4	10.0	28	10.1	7	11.2	13	12.0	9

Results

Chemical Gradients

Dissolved oxygen decreased laterally from the stream channel and temporally throughout summer (Fig. 3). Dissolved oxygen in the alder wells ranged from 12.2 $mg \cdot L^{-1}$ in stream and bankside water (0.9 m) in May to less than 2.0 $mg \cdot L^{-1}$ in the farthest well inland (13.2 m) in August. Dissolved oxygen in the stream and most bankside wells was 88 to 107% of the concentration expected in equilibrium with the atmosphere on

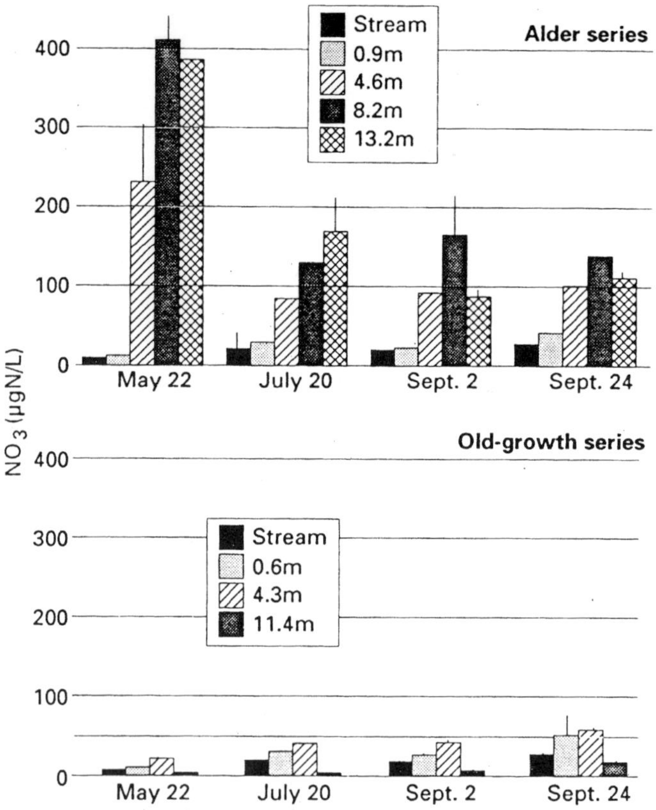

FIG. 4. Nitrate concentration obtained from the alder and old-growth wells during base flow. Error bars are ± 1 SD.

all dates (Table 1). The expected concentration is based on the measured temperature of the water. Dissolved oxygen in water 4.6 to 13.2 m inland was up to 88% of the concentration expected in equilibrium with the atmosphere in May, but was reduced to as low as 14% of the expected concentration by August. Similar results were obtained in the old-growth wells; one exception was in water 11.4 m inland, where dissolved oxygen was significantly below the expected concentration on all dates.

Nitrate concentration in the alder wells ranged from 8 µg $N{\cdot}L^{-1}$ in the stream to 411 µg $N{\cdot}L^{-1}$ in water 8.2 m inland in May (Fig. 4). The concentration of nitrate in wells 4.6 to 13.2 m inland was always greater than in the stream or bankside well (0.9 m). While the nitrate concentration of stream water and bankside water roughly tripled by August (11 to 41 µg $N{\cdot}L^{-1}$), there was a dramatic decrease in the nitrate concentration of water from wells 4.6 to 13.2 m inland over the same time. Both the alder and old-growth series had similar nitrate concentration patterns in water up to 4.6 m from the stream. However, beyond 4.6 m, the old-growth series differed from the alder series in three ways: (1) the range of nitrate concentrations from the stream to the farthest well inland was relatively small compared with the alder transect, (2) nitrate tended to increase rather than decrease throughout the summer as far inland as 11.4 m and (3) the concentration of nitrate in the farthest well inland was always significantly lower than that of stream water.

Dissolved organic carbon generally decreased during summer in the stream and in wells at both transects (Fig. 5). DOC ranged between 0.5 and 2.5 mg $C{\cdot}L^{-1}$ except in the farthest wells inland at each location (13.2 and 11.4 m), where DOC was approximately 4.5 mg $C{\cdot}L^{-1}$. In May and early August, DOC

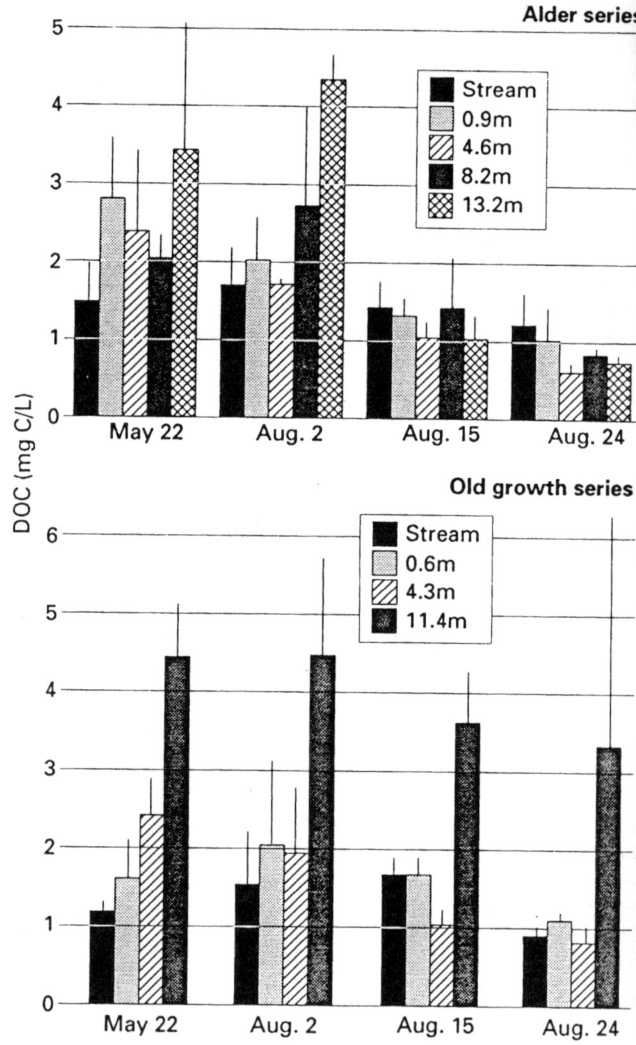

FIG. 5. Dissolved organic carbon obtained from the alder and old-growth wells during base flow. Error bars are ± 1 SD.

was higher in subsurface water compared with the stream. By late August, DOC in well water was higher than that in the stream at only one location, 11.4 m inland at the old-growth series.

Denitrification Assays

Slurries of subsurface sediment and well water obtained 8.2 m inland at the alder well series produced nitrous oxide when incubated in the presence of acetylene at in situ nitrate concentrations (Fig. 6a). Nitrous oxide production was linear up to 3.3 h, in most cases without a time lag. Production of nitrous oxide was stimulated by nitrate addition (1 mM final concentration). Further, addition of a readily utilizable carbon substrate to the incubation flasks (1 mM glucose final concentration) had negligible effect over nitrate alone. Nitrous oxide production by sediment slurries was inhibited by the presence of air in the headspace of the flasks (Fig. 6b); when the headspace was oxygen-free nitrogen, significant nitrous oxide accumulated in the absence of acetylene. Well water incubated without sediment did not produce detectable nitrous oxide at any site (data not shown).

The addition of 1 mM nitrate also stimulated denitrification in subsurface sediment obtained 11.4 m inland at the old-growth

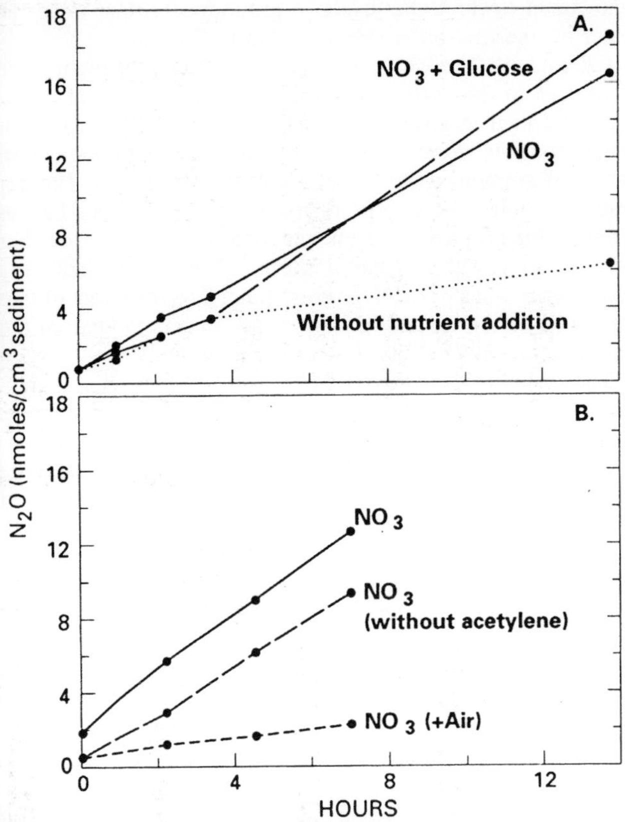

FIG. 6. Nitrous oxide formation in sediment slurries. Sediment was obtained from a cased well 8.2 m inland at the alder well series on two dates in August, 1984. Data points represent individual flasks. Flasks were incubated anaerobically (18°C) with acetylene except where noted. Final concentration of nitrate and glucose 1 mM each.

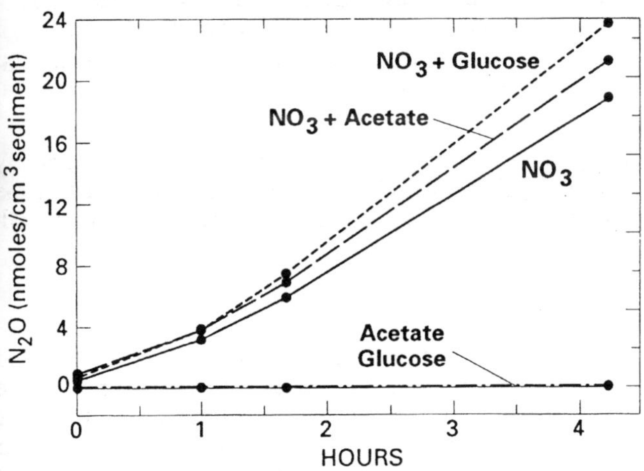

FIG. 7. Nitrous oxide formation in sediment slurries. Sediment was obtained from a cased well 11.4 m inland at the old-growth well series in August, 1984. Flasks were incubated anaerobically (18°C) in the presence of acetylene. Final concentration of nitrate, glucose and acetate 1 mM each.

series (Fig. 7). Neither nitrate plus glucose nor nitrate plus acetate had an effect on nitrous oxide formation over nitrate alone. Glucose- or acetate-amended slurries alone had no effect on denitrification rates.

Denitrifying activity under in situ nitrate and nitrate-amended conditions increased with distance from the stream channel at the alder well series (Fig. 8). Denitrifying activity was about

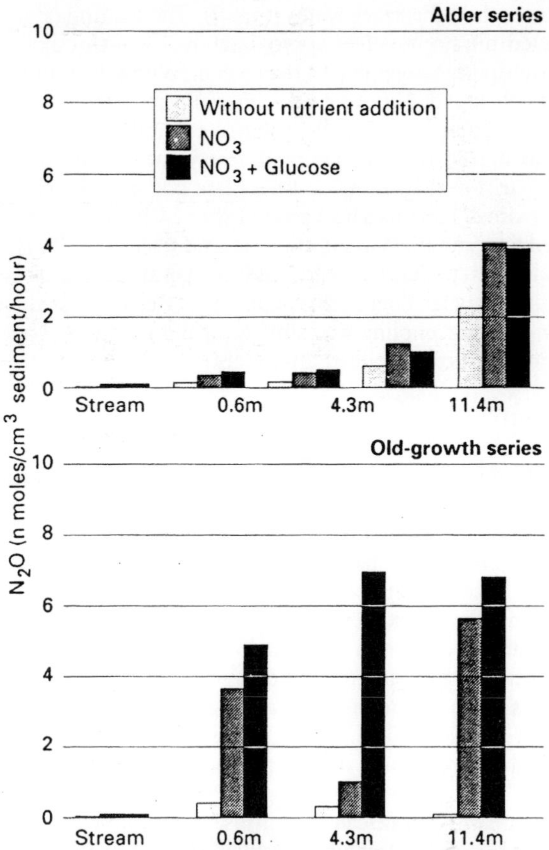

FIG. 8. Spatial gradients of nitrous oxide formation rates under various nutrient conditions in sediments slurries from the alder and old-growth wells. Rates were extrapolated from best fit linear regression of three to four time points from individual flasks.

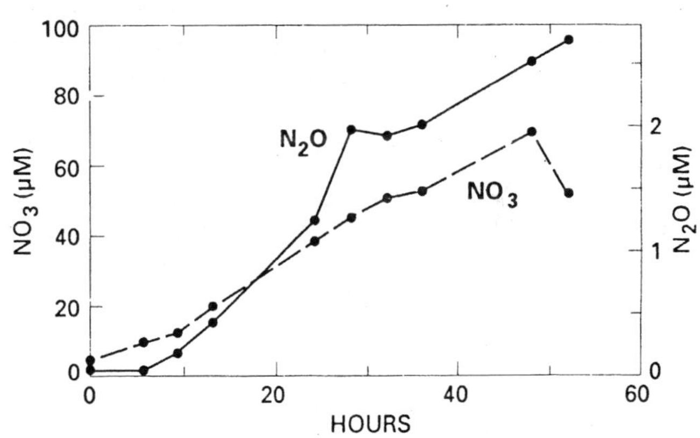

FIG. 9. Nitrate and nitrous oxide concentration 13 meters downgradient of a 48-h co-nitrate plus acetylene injection into a well located 8.4 m inland of the stream. The field experiment was performed in August, 1985.

three-fold higher 13.2 m inland than at the next closest site (8.2 m). Denitrifying activity at in situ nitrate concentrations was low in subsurface sediment from the old-growth series but was enhanced by nitrate addition. At one well, 4.3 m inland, glucose plus nitrate enhanced nitrous oxide formation over nitrate alone.

Field Injection

Data collected during the nitrate injection experiment indicated a close coupling between nitrate arrival and nitrous oxide formation in the subsurface water (Fig. 9). The leading edge of the injected nitrate travelled approximately 13 m through the subsurface riparian zone in less than 5.5 h. Within 9 h, nitrous oxide was observed. Nitrous oxide accumulated in the presence of acetylene at an increasingly faster rate as nitrate increased from 10 to 40 μM over 24 h (Fig. 9). Acetylene increased from 3.3 to 4.7 μM during the same time (data not shown). Nitrate was approximately 8 times background after 24 h. Nitrous oxide continued to accumulate as nitrate increased from 40 to 70 μM, but at a slower rate. Nitrous oxide dissolved in subsurface water was 55 times higher than background 48 h after its arrival and was rising when sampling was stopped. Both nitrate and acetylene concentrations declined after cutoff.

Discussion

Of all the dissolved inorganic nitrogen species found in headwater streams, nitrate is the most mobile. Its production begins with the decomposition of organic matter, which results in formation of ammonium (Blackburn 1983). In anoxic stream environments, the ammonium can be reassimilated by bacteria. In environments where oxygen is present, the ammonium is oxidized by nitrifying bacteria to nitrate (Schmidt 1982; Blackburn 1983). Because ammonium is readily sorbed onto sediments, the conversion to nitrate provides an easily transported source of nitrogen to stream biota. Plants, algae, and aerobic bacteria can assimilatively reduce nitrate for incorporation into cells (Payne 1973), a process that temporarily retards nitrate transport downstream. Nitrate can also diffuse into sediments. Here, if it encounters heterotrophic microbial activity at the appropriate redox potential, the nitrate is denitrified to dinitrogen by bacteria that use it as an electron acceptor (Firestone 1982). Because dinitrogen can be fixed only by certain bacteria, denitrification is a permanent sink for nitrate transported in streams.

The capacity for soils to denitrify is dependent on the in situ denitrifying enzyme content and the derepression of denitrifying enzyme synthesis (Smith and Tiedje 1979; Tiedje et al. 1982). Tiedje et al. (1982) suggest that carbon is more important than oxygen in determining the denitrifying enzyme content of habitats although oxygen inhibits enzyme activity and represses synthesis of new denitrifying enzymes (Payne 1973). The switch from aerobic to anaerobic or partially anaerobic conditions first eliminates oxygen inhibition then releases the synthesis of denitrifying enzymes (Smith and Tiedje 1979).

In sediments from the stream and hyporheic zone of Little Lost Man Creek, the content of DOC is apparently controlled by the influx of groundwater. Because of the long contact time with watershed soils, groundwater is high in reduced compounds such as DOC, DON, ammonium, and methane, as well as other major ions (Triska et al. 1990). An old-growth well, 11.4 m inland, was probably the only true groundwater well in either transect. Levels of DOC in groundwater are gradually depleted as groundwater seeps into the hyporheic zone. Decreases in concentration result from physical dilution as well as chemical and biological oxidations. The concentration gradients we commonly observe are consistent with reports by Wallis et al. (1981) and Dahm et al. (1987). Wallis et al. (1981) measured higher dissolved organic matter (DOM) concentrations in the saturated groundwater zone than in the surface waters of the Marmot Basin in Canada. Dahm et al. (1987) measured higher DOC concentrations in anaerobic and aerobic zones beneath the channel than in stream water at Sulfur Creek, Oregon. Our observations contrast with Meyer and Tate (1983), who found lower DOC levels in groundwater than in surface water in a second-order stream in North Carolina.

Lateral advection of stream water is probably the main source of oxygen to the hyporheic zone. The combination of ammonium supplied in groundwater and oxygen supplied in stream water produces an environment conducive to the biological oxidation of ammonium to nitrate (Triska et al. 1990). The generation of nitrate is important to biotic communities because nitrate is the primary nitrogen source to periphyton in the channel and an electron acceptor during DOC mineralization in anoxic microsites of the hyporheic zone. The net effect of these metabolic oxidation and reduction reactions is the partial control of nitrate availability downstream.

The content of dissolved oxygen in the hyporheic zone is controlled by a variety of physical, biological, and chemical factors. Physically, the content of dissolved oxygen in water is controlled by temperature. The solubility of dissolved oxygen in water decreases as temperatures rise. This physical control is evident in stream water where the oxygen concentration of the water is close to the value predicted on the basis of temperature alone, falling as temperatures rise over the summer. In contrast, oxygen concentrations of subsurface water up to 13.2 m inland fall well under the concentration expected from water temperature alone, indicating a strong demand for oxygen. One biological sink is microbial mineralization of DOC (including methane), whose concentration decreases as temperatures increase during summer. Secondarily, elevated levels of nitrate in the hyporheic zone compared with the stream indicate that nitrification can also constitute an oxygen sink. Chemically, iron may undergo oxidation in the hyporheic zone as groundwater mixes with stream water (C. N. Dahm, University of New Mexico, Albuquerque, NM 87131; A. S. Maest, U.S. Geological Survey, 345 Middlefield Rd., Menlo Park, MS 496 CA 94025, USA, pers. comm.). The impact of this reaction on oxygen concentration in the hyporheic zone is unknown.

In the absence of oxygen, mineralization proceeds through the reduction of nitrate (Knowles 1982). In order that bacteria use nitrate as a terminal electron acceptor, it is necessary that oxygen be near zero. Although oxygen levels below 1 $mg \cdot L^{-1}$ were detected in the hyporheic zone (7% of the concentration expected in equilibrium with the atmosphere), water was not without oxygen. Measurement of the oxygen status of water provides little direct information as to the oxygen status at microsites. Denitrification may still occur if the supply of oxygen to metabolically-active microsites is reduced both by the amount of water through which oxygen must diffuse (Firestone 1982) and by the intense respiratory activity at metabolically-active microsites (Greenwood 1961; Craswell and Martin 1974). In Little Lost Man Creek, the high oxygen demand as well as the demand for nitrate and DOC over time suggest that a nitrate reducing zone could exist within the hyporheic zone.

When assayed by the acetylene blockage technique, stream and subsurface sediments from the alder riparian series displayed a significant potential for denitrification. The activity was linear for 3 to 4 h, did not display an initial time lag, and was inhibited by the presence of air in the incubation flasks. Smith and Tiedje (1979) attributed the initial constant rate (1–3 h) for denitrification in soils to the activity of pre-existing denitrifying enzymes in the soil microflora. Denitrification potentials in unamended incubation flasks increased significantly with increasing distance from the stream channel, confirming the presence of nitrate-reducing activity in zones of low oxygen tension as far away as 13.2 m from the stream channel.

Denitrifying activity was highest at the most inland site in amended incubation flasks in spite of sufficient carbon to luce available nitrate at sites closer to the stream. This pattern as evident even in sediment slurries to which nitrate and glu- se were added. The latter conditions have been used to assay nitrification capacity or denitrifying enzyme quantity (Tiedje al. 1982). Smith and Tiedje (1979) found that the full suite denitrifying enzymes could be synthesized in aerobic soils der conditions in which they were not fully active. When ygen was removed from aerobic soils, active denitrification mmenced almost immediately. These results suggest that mbined hydrodynamic and biotic processes occurring along land flow paths result in the production and maintenance of gh levels of denitrifying enzymes in the bacterial population hich are available when the conditions become favorable for nitrification.

In the alder riparian series, denitrification potentials were trate-limited but not carbon-limited. Sufficient nitrate was ailable to support denitrification without the addition of trate to incubation flasks. However, enough oxidizable-car- on was present that a further stimulation in denitrification curred when additional nitrate was added to these flasks. otentially, the availability of nitrate could control mineraliza- on of DOC during summer base flow should oxygen become pleted.

Denitrifying activity was associated with stream and subsur- ce solids. The population of denitrifying organisms are, there- re, attached to sediments rather than existing as free-living ganisms in the water. An analogous situation was observed a predominant nitrate-reducing zone of a sand and gravel quifer receiving treated sewage (Smith and Duff 1988) and is situation is probably not different from other subsurface d surface water systems. Denitrification in the hyporheic zone Little Lost Man Creek is controlled by water flux from the roundwater table and lateral advection of stream water, hereas in estuarine and coastal sediments it is largely limited y the downward diffusion of nitrate into surface sediments Oremland et al. 1984; Anderson et al. 1984) or by the rate of ater movement past the particulate surfaces of an aquifer (Tru- ell et al. 1986; Smith and Duff 1988).

There is little direct evidence for denitrification in subsurface ater systems. Slater and Capone (1987) measured denitrifi- ation in aquifer soil and nearshore marine sediments influ- nced by groundwater, Smith and Duff (1988) measured deni- ification in slurried core material from a fresh water sand and ravel aquifer, and Ward (1985) reported denitrification in itrate-amended core samples obtained adjacent to a septic tile eld. In streams, there is direct evidence for denitrification in urface sediments (Van Kessel 1977; Chatarpaul and Robinson 979; Swank and Caskey 1982; Cooper and Cooke 1984; Chris- ensen and Sorensen 1988). This study is believed to be the first irect evidence reported for denitrification in the hyporheic zone f a small headwater stream. Prior to this, the only direct meas- rements of deep sediment metabolism were reported by Grimm nd Fisher (1984) for the hyporheic zone below the stream hannel. We are currently assessing the denitrifying capacity f stream, hyporheic, and groundwater sediments with regard o seasonal and spatial trends of nutrient chemistry and there- ore have not emphasized actual rate measurements.

In this study, the level of oxygen demand, the presence of educed compounds, and the availability of degradable organic ubstrates indicated that denitrifying conditions could be estab- ished as close as 4.6 m to the stream. Denitrifying potential in these zones was confirmed by incubating subsurface sediment slurries in the presence of nitrate and glucose. Manipulation of dissolved oxygen in slurry experiments can lead to uncertainties as to whether the activity exhibited in flasks is indicative of in situ conditions under ambient nitrate and carbon levels. Some potential limitations were overcome by directly assaying deni- trifying activity within the subsurface habitat. Field enrichment with nitrate provided evidence of in situ denitrifying activity as nitrate concentrations were elevated from 5 μM (background) to 70 μM, a level that we have observed in subsurface water subsequent to this experiment. Six hours after nitrate arrived, dissolved nitrous oxide in subsurface water increased 210-fold over the concentration expected in equilibrium with the atmos- phere. This was surprising because of the high bed permeability and rapid flow of water ($>1\ m \cdot h^{-1}$) during base flow. Because nitrous oxide dissolved in soil water rapidly degases from solu- tion (Bowden and Bormann 1986), the levels we measured rep- resent only nitrous oxide transported and thus underestimate total in situ production. Nitrate concentration and temperature clearly affect the rate of nitrous oxide formation in denitrifi- cation assays (Knowles 1982). While temperature was not measured continuously during this experiment, hyporheic water exhibits minimal temperature flux within short periods. In the four weeks preceeding and following this experiment, hypor- heic water ranged from 11.5 to 12°C at this site (11.4 m inland). While preliminary, the data suggest that denitrifying bacteria respond rapidly to short-term nitrate pulses and that hyporheic microbial processes can modify the nitrate concentration of sub- surface water.

In summary, denitrification occurred in shallow riparian sed- iments connected to the stream by the hyporheic zone. Increases of denitrifying potentials with increasing distance from the stream channel also correlated with decreasing oxygen concen- trations and increasing concentrations of reduced compounds. Denitrifying activity was associated with stream-channel sol- ids. Therefore, maintenance of a denitrifying capacity is dependent on the flux between groundwater and stream water which, in turn, controls the availability of DOC and generation of nitrate. Geochemical environments favorable to denitrifica- tion appear to be established through a combination of hydro- dynamic, chemical, and biotic processes. Microbial processes in the lateral hyporheic zone may, thereby, modify the chem- istry of water before reemergence from the subsurface.

Acknowledgements

We thank K. E. Bencala, G. W. Zellweger, and R. J. Avanzino for help with well installation, stream experiments, and nutrient analyses. We also thank R. J. Avanzino, J. S. Kuwabara, and R. S. Smith for reviewing the manuscript.

References

ANDERSEN, T. K., M. H. JENSEN, AND J. SORENSEN. 1984. Diurnal variation of nitrogen cycling in coastal, marine sediments. I. Denitrification. Mar. Biol. 83: 171–176.

BALDERSTON, W. L., B. SHERR, AND W. J. PAYNE. 1976. Blockage by acetylene of nitrous oxide reduction in *Pseudomonas perfectomarinus*. Appl. Environ. Microbiol. 31: 504–508.

BENCALA, K. E. 1984. Interactions of solutes and streambed sediment. 2. A dynamic analysis of coupled hydrologic and chemical processes that determine solute transport. Water Resour. Res. 20: 1804–1803.

BLACKBURN, T. H. 1983. The microbial nitrogen cycle, p. 63–90. *In* W. E. Krumbein [ed.] Microbial geochemistry. Blackwell Scientific Publications, St. Louis, MO.

BOWDEN, W. B., AND F. H. BORMANN. 1986. Transport and loss of nitrous oxide in soil water after forest clear-cutting. Science (Wash., DC) 233: 867–869.

CHATARPAUL, L., AND J. B. ROBINSON. 1979. Nitrogen transformations in stream sediments: ^{15}N studies, p. 119–127. *In* C. D. Litchfield, and P. C. Seyfried [ed.] Methodology for biomass determinations and microbial activities in sediments. American Society for Testing and Materials. Publication ASTM STP, Vol. 673.

CHRISTENSEN, P. B., AND J. SORENSEN. 1988. Denitrification in sediment of lowland streams: regional and seasonal variation in Gelbaek and Rabis Baek, Denmark. FEMS Microbiol. Ecol. 53: 335–344.

COOPER, A. B., AND J. G. COOKE. 1984. Nitrate loss and transformation in 2 vegetated headwater streams. N. Z. J. Mar. Freshwater Res. 18: 441–450.

CRASWELL, E. T., AND A. E. MARTIN. 1974. Effect of moisture content on denitrification in a clay soil. Soil Biol. Biochem. 6: 127–129.

DAHM, C. N., E. H. TROTTER, AND J. R. SEDELL. 1987. Role of anaerobic zones and processes in stream ecosystem productivity, p. 157–178. *In* R. C. Averett and D. M. McKnight [ed.] Chemical quality of water and the hydrologic cycle. Lewis Publishers, Inc., Chelsea, MI.

FIRESTONE, M. K. 1982. Biological denitrification, p. 289–326. *In* F. J. Stevenson [ed.] Nitrogen in agricultural soils. American Society of Agronomy, Madison, WI.

FLETT, R. J., R. D. HAMILTON, AND N. E. R. CAMPBELL. 1976. Aquatic acetylene reduction techniques: solutions to several problems. Can. J. Microbiol. 22: 43–51.

GREENWOOD, D. J. 1961. The effect of oxygen concentration on the decomposition of organic materials in soil. Plant Soil 14: 360–376.

GRIMM, N. B., AND S. G. FISHER. 1984. Exchange between interstitial and surface water: implications for stream metabolism and nutrient cycling. Hydrobiologia 111: 219–228.

HYNES, H. B. N. 1983. Groundwater and stream ecology. Hydrobiologia 100: 93–99.

KENNEDY, V. C., A. P. JACKMAN, S. M. ZAND, G. W. ZELLWEGER, AND R. J. AVANZINO. 1984. Transport and concentration controls for chloride, strontium, potassium, and lead in Uvas Creek, a small cobble-bed stream in Santa Clara County, California, U.S.A. 1. Conceptual model. J. Hydrol. 75: 67–110.

KNOWLES, R. 1982. Denitrification. Microbiol. Rev. 46: 43–70.

MENZEL, D. W., AND R. F. VACCARO. 1964. The measurement of dissolved organic and particulate carbon in seawater. Limnol. Oceanogr. 9: 138–142.

MEYER, J. L., AND C. M. TATE. 1983. The effects of watershed disturbance on dissolved organic carbon dynamics of a stream. Ecology 64: 33–44.

OREMLAND, R. S., C. UMBERGER, C. W. CULBERTSON, AND R. L. SMITH. 1984 Denitrification in San Francisco Bay intertidal sediments. Appl. Environ Microbiol. 47: 1106–1112.

PAYNE, W. J. 1973. Reduction of nitrogenous oxides by microorganisms. Bac teriol. Rev. 37: 409–452.

RUDD, J. W. M., R. D. HAMILTON, AND N. E. R. CAMPBELL. 1974. Measure ment of microbial oxidation of methane in lake water. Limnol. Oceanogr 19: 519–524.

SCHMIDT, E. L. 1982. Nitrification in soils, p. 253–288. *In* F. J. Stevenso [ed.] Nitrogen in agricultural soils. American Society of Agronomy, Mad ison, WI.

SLATER, J. M., AND D. G. CAPONE. 1987. Denitrification in aquifer soil an nearshore marine sediments influenced by groundwater nitrate. Appl Environ. Microbiol. 53: 1292–1297.

SMITH, R. L., AND J. H. DUFF. 1988. Denitrification in a sand and grave aquifer. Appl. Environ. Microbiol. 54: 1071–1078.

SMITH, M. S., AND J. M. TIEDJE. 1979. Phases of denitrification followin oxygen depletion in soil. Soil Biol. Biochem. 11: 261–267.

SWANK, W. T., AND W. H. CASKEY. 1982. Nitrate depletion in a second-orde mountain stream. J. Environ. Qual. 11: 581–584.

TIEDJE, J. M., A. J. SEXSTONE, D. D. MYROLD, AND J. A. ROBINSON. 1982 Denitrification: ecological niches, competition, and survival. Antoni Leeuwenhoek J. Microbiol. Serol. 48: 569–583.

TRISKA, F. J., V. C. KENNEDY, R. J. AVANZINO, G. W. ZELLWEGER, AND K. E BENCALA. 1989. Retention and transport of nutrients in a third order stream hyporheic processes. Ecology 70: 1893–1905.

TRUDELL, M. R., R. W. GILLHAM, AND J. A. CHERRY. 1986. An in-situ stud of the occurrence and rate of denitrification in a shallow unconfined san aquifer. J. Hydrol. 83: 251–268.

VAN KESSEL, J. F. 1977. Removal of nitrate from effluent following discharg on surface water. Water Res. 11: 533–537.

WALLIS, P. M., H. B. N. HYNES, AND S. A. TELANG. 1981. The importanc of groundwater in the transportation of allochthonous dissolved organi matter to the streams draining a small mountain basin. Hydrobiologia 79 77–90.

WARD, T. 1985. Characterizing the aerobic and anaerobic microbial activitie in surface and subsurface soils. Environ. Toxicol. Chem. 4: 727–737.

WHITE, D. S., C. H. ELZINGA, AND S. P. HENDRICKS. 1987. Temperature pat terns within the hyporheic zone of a northern Michigan river. J. N. Am Benthol. Soc. 6: 85–91.

YOSHINARI, T., R. HYNES, AND R. KNOWLES. 1976. Acetylene inhibition o nitrous oxide reduction and measurement of denitrification and nitroger fixation in soil. Soil Biol. Biochem. 9: 177–183.

Commentary on Paper F5

Mulholland, P. J. (1992) Regulation of nutrient concentrations in a temperate forest stream: roles of upland, riparian, and instream processes.

The nutrient dynamics occurring in a catchment are the result of different processes operating in interlinked spatial and temporal scales. Terrestrial processes that occur in the soil and in the vadose zone are a result of geological and biogeochemical factors. The nutrient forms and availability in the groundwater are mostly determined by biogeochemical processes. Finally, the processes occurring within the stream are relevant to the nutrient dynamics, since they include recycling and uptake. All of these compartments determine mass balance variations which had been evidenced by catchment studies (e.g. Triska *et al.*, 1984, Paper A5), but had not been analysed to the detail of the separate components. To determine the relevance of the different components to the overall process is not a trivial issue, since it requires a variety of approaches and specific methodologies. This is the merit of Mulholland's 1992 paper, the first to consider the respective roles of upland, riparian and in-stream processes in the nutrient dynamics during baseflow conditions. Analysis during storm events (e.g. Mulholland *et al.*, 1990) is useful to account for hydrological and chemical analyses, but less suitable for the detailed analysis of the biogeochemical processes. The study was carried out in the Walker Branch experimental catchment, eastern Tennessee. The experimental setting allowed for the accurate control of inputs and outputs, as well as for the appropriate upscaling from the site to the ecosystem measurement.

The paper determined that the sharing of responsibilities in the outcome of the nitrogen, or phosphorus, or carbon at the end of the channel was rather clear: it was mostly biogeochemical in the riparian and stream compartments; mainly geochemical in the upland area. Though a common origin of the elements was deduced from the similar cation concentrations in riparian and stream waters, the different redox conditions (evidenced by oxygen dynamics) quickly caused large differences between the compartments. This determined that the different forms (reduced, oxidized) of nitrogen were distributed differentially in surface water compared to the groundwater. Further, in some periods of the year these variations were more noticeable than in others. These variations could be observed even at the daily scale. Hence, areas such as the riparian zone could function either as a source or sink of nutrients, depending on the redox conditions.

The study highlighted the constraints that may appear in the quantification of the respective contributions of the different compartments to the resulting stream nutrient dynamics. Amongst the most important were: the diffuse inputs contributed (mostly nitrogen and carbon) from the soil to the stream water, the spatial variability of the soil, and the varying water flow paths. While some of these observations could be taken as updated observations to the particular Walker Branch catchment, others such as the varying flow paths were novel intuitions independently confirmed later on (Haycock & Burt, 1993, Paper E7). These constraints were seen as a deterrent to quantification, and today are still a major challenge for ecosystem research in nutrient dynamics.

SS

References

Haycock, N. E. & Burt, T. P. (1993) Role of floodplain sediments in reducing the nitrate concentration of subsurface run-off: a case study in the Costwolds, UK. *Hydrol. Processes* **7**, 287–295.

Mulholland, P. J., Wilson, G. V. & Jardine, P. M. (1990) Hydrogeochemical response of a forested watershed to storms: effects of preferential flow along shallow and deep pathways. *Water Resour. Res.* **26**, 3021–3036.

Triska, F. J., Sedell, J. R., Cromack K., Jr, Gregory, S. V. & McCorison, F. M. (1984) Nitrogenous budget for a small coniferous forest stream. *Ecol. Monographs* **54**, 119–140.

Limnol. Oceanogr., 37(7), 1992, 1512–1526

Regulation of nutrient concentrations in a temperate forest stream: Roles of upland, riparian, and instream processes

Patrick J. Mulholland

Environmental Sciences Division, Oak Ridge National Laboratory, P.O. Box 2008, Oak Ridge, Tennessee 37831-6036

Abstract

The roles of terrestrial and instream processes in controlling stream-water N and P concentrations were studied over a 2-yr period in a deciduous forest stream in eastern Tennessee. Upper soil horizons were highly effective sinks for inorganic N and P in throughfall, and weathering of the parent dolomite was the dominant source of inorganic P to the stream. The riparian zone was a potential source of NH_4^+ and P to the stream when dissolved oxygen concentrations in riparian groundwater were low, but a sink for P when dissolved oxygen concentrations were high. High rates of instream immobilization of inorganic N and P were observed from late autumn to spring, primarily as a result of uptake by microbes on decomposing leaves and secondarily by algae. Immobilization of inorganic N and P resulted in longitudinal declines in concentrations with distance downstream from groundwater inputs (springs), thereby increasing the importance of organic forms of these nutrients in stream water in downstream reaches. The seasonal pattern of winter minima and summer maxima in stream-water N and P concentrations observed here is opposite to the pattern observed in many northern streams, suggesting the importance of winter nutrient cycling processes in soils and streams in warmer climates.

The question of how nutrient concentrations in headwater streams are controlled is important to stream productivity (Elwood et al. 1981*a*; Grimm and Fisher 1986) as well as to the size and timing of nutrient inputs to larger aquatic ecosystems downstream (Jasper et al. 1983). Studies of small watersheds that have been experimentally perturbed demonstrate the importance of terrestrial processes on stream-water nutrient concentrations (Likens et al. 1977; Swank 1988). In most forested watersheds, biological and geochemical processes in upper soil horizons effectively retain N and P and thus reduce inputs to streams (Wood et al. 1984; Qualls et al. 1991), although in old-aged forests N retention is generally low (Vitousek 1977). Watershed geology also is an important regulator of P concentrations in streams (Dillon and Kirchner 1975).

Observations that nutrient concentrations in stream water are lower than in groundwater inputs to streams (Grimm and Fisher 1984; Ford and Naiman 1989) suggest that instream processes are also important in regulating stream-water nutrient concentrations. Mass balance studies of headwater streams or stream segments have indicated prolonged periods of net instream retention of nutrients punctuated by large net losses during storms (Meyer and Likens 1979; Grimm 1987). Experimental radiotracer injections (Newbold et al. 1983; Mulholland et al. 1985) and stable element additions (Triska et al. 1989; Munn and Meyer 1990) have demonstrated rapid instream uptake of the limiting nutrient by stream algae and microbes. Instream processes can also result in net transformation of nutrients transported downstream from inorganic to dissolved organic or particulate forms (Meyer and Likens 1979).

Most studies of stream nutrient dynamics have considered either terrestrial (inputs) or instream processes (retention and transformation), but not both, and few have addressed the role of the riparian zone in forested watersheds. Although riparian zones are effective sinks for nutrients transported

Acknowledgments

I thank Bonnie Lu for assistance with sampling and laboratory analyses. Lisa Au and Mark Gallyoun also helped with field sampling. Jerry Elwood, Alan Steinman, Robert Stauffer, and two anonymous reviewers provided comments on the manuscript.

This work was sponsored by the Walker Branch Watershed Project and the Oak Ridge National Environmental Research Park, Environmental Sciences Division, under contract DE-AC05-84OR21400 with Martin Marietta Energy Systems, Inc.

Publication 3907, Environmental Sciences Division, ORNL.

from upland sources in agricultural landscapes (Lowrance et al. 1984; Peterjohn and Correll 1984), their role in modifying the transport of nutrients to streams in forested watersheds is unclear.

Previous studies of nutrient concentrations in Walker Branch, a stream draining a forested watershed in eastern Tennessee, have focused on annual fluxes and variation in concentrations during storms (Henderson et al. 1977; Elwood and Turner 1989; Mulholland et al. 1990). However, baseflow periods dominate the annual hydrograph of Walker Branch, and nutrient concentrations during baseflow are probably more important to instream biological processes than concentrations during storms. In this study I investigated the roles of upland, riparian, and instream processes in regulating nutrient concentrations during baseflow periods in Walker Branch.

Study site

The study was conducted in the West Fork of Walker Branch watershed. The West Fork is a 38.4-ha forested watershed located on the Oak Ridge Reservation in the Ridge and Valley Province of eastern Tennessee (35°58′N, 84°17′W). The climate is typical of the humid southern Appalachian region, with mean annual temperature of 14.5°C and mean annual precipitation of 140 cm.

Watershed topography is characterized by broad ridges with relatively steep slopes (some exceeding 40%) and narrow valleys. The watershed is underlain by the Knox Group, a 610-m-thick sequence of Cambrian and Ordovician siliceous dolomite that weathers rapidly to form a deep mantle with abundant chert on the surface.

Soils are primarily Ultisols, with small areas of Alfisols (suborder Hapludalf) along the east side of the lower 150 m of the perennial stream (Johnson and Henderson 1989). Surface infiltration rates are high, as is the hydraulic conductivity of surface soils; however, hydraulic conductivity decreases rapidly with depth because of increasing clay content. The soils are acidic (pH 4–5) and low in exchangeable bases, N, and P (Johnson and Todd 1990). The predominant clay mineral is kaolinite, with lesser amounts of chloritized vermiculite and mica. Depth of the weathered zone ranges from a meter or less near the stream up to 30 m on the ridges.

Watershed vegetation is second-growth deciduous forest dominated by oak and hickory, with scattered pine on some of the ridges and mesophytic hardwoods such as tulip poplar and beech in the valleys. The watershed has been largely undisturbed since 1940. Land use before that was a mixture of row-crop agriculture, pasture, and woodlots.

The watershed is drained by a first-order stream about 360 m long (Fig. 1). Four perennial springs discharge to the stream; two form the headwaters (S4, S5) and two others (S3, S3A) discharge about 160 m downstream from the headwaters. Flumes have been installed on each of the springs to allow calculation of discharge. A flume has also been installed at a site on the upper portion of the perennial stream (WB300) to measure discharge. The stream flows on exposed bedrock over its lower 200 m (downstream from station WB200). Streamflow is continuously monitored via a v-notch weir at the base of the watershed (WB0) and varies from summer baseflows of ~3–5 liters s^{-1} to occasional stormflows exceeding 50 liters s^{-1} for short periods (usually <1 d). Streamflow averages ~70% of annual precipitation, with baseflow accounting for 77% of the annual streamflow (Luxmoore and Huff 1989). The relatively high streamflow (as a proportion of precipitation) is the result of interbasin transfer of groundwater, estimated to account for ~30% of the streamflow in the West Fork. During baseflow periods average stream depth is ~5–10 cm and average width ~1.5–3 m.

Previous studies of stream-water chemistry have shown that the dominant ions are Ca^{2+}, Mg^{2+}, and HCO_3^-, derived primarily from dolomite weathering (Elwood and Turner 1989). Concentrations of SO_4^{2-} and dissolved organic C (DOC) are relatively low during baseflow, with values of 1–3 and 0.2–0.6 mg $liter^{-1}$, respectively. Baseflow pH is generally 7.8–8.2 and alkalinity is ~2–3 meq $liter^{-1}$. During storms, base cation and HCO_3^- concentrations in stream water are diluted by shallow macropore flow through the acidic, base-poor soils, and SO_4^{2-} and DOC concentrations increase as a result of

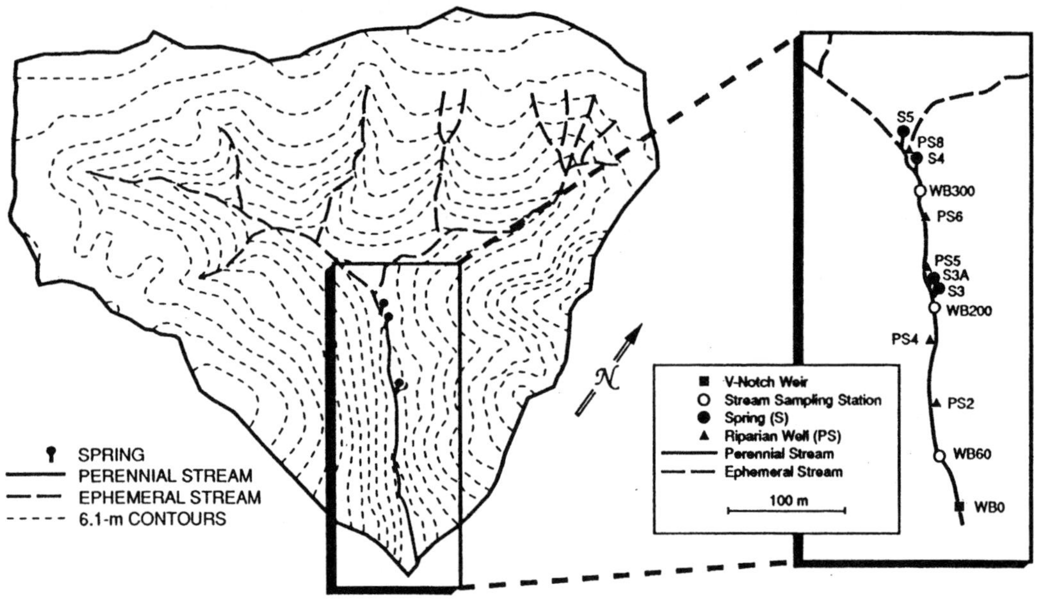

Fig. 1. Map of the West Fork of Walker Branch watershed showing the sampling sites.

canopy washout and leaching of organic soil horizons and stream channel debris (Elwood and Turner 1989; Mulholland et al. 1990). N and P concentrations are low (100–300 and 2–8 μg liter^{-1} for total soluble N and P) with organic forms making up 40–75% of the total (Henderson et al. 1977; Segars et al. 1986). Watershed mass balances of N and P have demonstrated strong net retention of atmospheric inputs (Johnson and Henderson 1989). Results of a 3-month P addition to stream water indicated that periphyton production and litter decomposition in the stream were P limited (Elwood et al. 1981*a*). Radiotracer injections to stream water have indicated that instream intake of PO_4 is primarily by microbes associated with decomposing leaves and is highly seasonal (Newbold et al. 1983; Mulholland et al. 1985).

Materials and methods

For a 2-yr period beginning in November 1987, water samples were collected monthly during baseflow (<10 liters s^{-1} at WB0) from three stream sites (WB60, WB200, WB300), four perennial springs (S3, S3A, S4, S5), and five shallow wells penetrating the groundwater in the riparian zone within 2 m of the stream bank (PS2, PS4, PS5, PS6, PS8) (Fig. 1). Unfiltered samples were collected for alkalinity. Water temperature, specific conductance (conductivity bridge), and pH (potentiometrically with a glass electrode) were measured in the field at the time of sample collection. For analysis of other solutes, samples were filtered in the field (Gelman type A/E glass-fiber filters, pore size, 1 μm), placed on ice, returned to the laboratory, and refrigerated until analysis. Samples were analyzed for N and P within 4 d of collection. Samples for cation analyses were preserved with 0.5% high-purity HNO_3, and cation and anion samples were analyzed within 4 months of collection.

Alkalinity was measured on unfiltered samples by titration with 0.1 N HCl to a pH of 4.5. Concentrations of Ca, Mg, Na, Fe, Mn, Si, and Al were measured by inductively coupled plasma emission spectroscopy and K by atomic absorption spectrometry. Concentrations of SO_4^{2-} and Cl^- were measured by ion chromatography. Concentrations of soluble reactive P (SRP) were measured by the ascorbic acid method and concentrations of total soluble P (TSP) were measured by the same method after persulfate oxidation (Am. Public Health Assoc. 1989). Samples were analyzed manually with a 10-cm spectrophotometer cuvette to achieve low analytical detection limits (0.5 μg P liter^{-1}) and high precision (SD of 10 replicates: 0.4 μg P liter^{-1}). Soluble unreactive P (SUP) was calculated as the difference between TSP and SRP and was assumed to be the organic P fraction. Concentrations of NH_4^+ were measured by automated phenate colorimetry and (NO_2^- + NO_3^-) were measured by automated Cu-Cd reduction followed by azo dye colorimetry (Am. Public Health Assoc. 1989). Concentrations of total Kjeldahl N (TKN) were measured by Hg-catalyzed acid digestion followed by automated phenate colorimetry to measure the evolved NH_4^+ (Am. Public Health Assoc. 1989). Detection limits and precision were as follows: NH_4^+—2 μg N liter^{-1}, 1 μg N liter^{-1}; (NO_2^- + NO_3^-)—1 μg N liter^{-1}, 0.5 μg N liter^{-1}; TKN—20 μg N liter^{-1}, 20 μg N liter^{-1}. Dissolved organic N (DON) was calculated as the difference between TKN and NH_4^+. Total inorganic C (TIC) was measured on unfiltered samples and dissolved organic C (DOC) on filtered samples using an OI model 700 total carbon analyzer (detection limit, 0.1 mg liter^{-1}; precision, 0.04 mg liter^{-1}).

Net instream retention of N and P was estimated over the stream reach WB200–WB60. Experimental injections of conservative tracers (^3H_2O, Cl^-) to the stream indicated that streamflow increases by a small and relatively constant percentage (mean = 12.1%, SD = 1.1%, n = 6) in this reach over a flow range of 6–20 liters s^{-1} measured at WB0 (Genereux et al. in press; P. J. Mulholland unpubl. data). The conservative tracer injections also indicated a flow increase of 10–25% in the reach from WB60 to WB0 over this flow range. Flow rates at WB200 and WB60 were calculated for each sampling date using relationships between flow rates at WB0 and WB60, determined from the conservative tracer injections, and assuming a constant inflow of 12.1% between WB200 and WB60. Instantaneous flux rates of N and P were computed for WB200, WB60, and the inflow between WB200 and WB60 by multiplying concentrations by flow rates. Concentrations of N and P in the inflow between WB200 and WB60 were taken as the average concentrations in PS2 and PS4, the riparian groundwater wells in this stream reach. Net retention (R) of N and P between WB200 and WB60 was calculated as

$$R = (F_{WB200} + F_{inflow}) - F_{WB60} \quad (1)$$

where F_{WB200} is the flux (μg liter^{-1} s^{-1}) at WB200, F_{inflow} the total lateral influx over the reach from WB200 to WB60, and F_{WB60} the flux at WB60.

Nutrient content of bedrock was determined on samples collected from outcrops at 11 locations along the stream. After removing the exposed surface material, the rock samples were ground to a fine powder with a mortar and pestle and digested in a 12% HNO_3/HF solution. Total Ca, Mg, Na, K, Fe, Mn, and P in the digested samples were analyzed by inductively coupled plasma emission spectroscopy. Samples of ground bedrock for N analysis were extracted in 0.02 N HCl and analyzed for NO_3^- by ion chromatography and for TKN as described above.

The effect of diel cycles in insolation on stream-water nutrient concentrations was determined by measuring NH_4^+, (NO_2^- + NO_3^-), SRP, and DOC at hourly intervals over a 40-h period on a clear day in mid-April, just before leaf emergence. Samples were collected at WB60 with an automatic sampler, filtered within 12 h of collection (Gelman type A/E glass-fiber filters), refrigerated, and analyzed within 2 d according to methods described above.

Statistical comparisons between mean water chemistry values were made with modified t-tests assuming unequal variances for comparison between two populations or Tukey's multiple range test for comparisons among three or more populations (SAS 1985).

Results

Precipitation and canopy leaching (via throughfall) were potentially large sources of N and P to the stream, whereas soils were a major sink, particularly for the inorganic forms (Fig. 2). Higher concentrations of SRP and inorganic N were found in riparian groundwater and springs than in upslope soil solution, suggesting within-watershed sources of these forms. The higher concentrations in riparian groundwater and springs

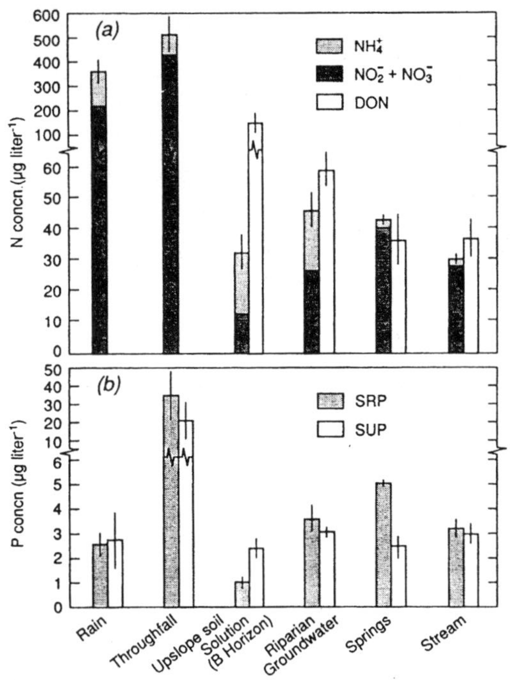

Fig. 2. Spatial patterns in soluble N and P concentrations across the Walker Branch watershed. Average N concentrations in rain and throughfall are for 1981–1983 (Lindberg et al. 1986). DON concentrations in rain and throughfall were not measured. Average N concentrations in upslope soil solutions are from 1983 to 1986 (Johnson and Todd 1990). Average P concentrations in rain, throughfall, and upslope soil solutions are for 1984–1985 (Segars et al. 1986). Average concentrations of N and P in riparian groundwater (2-yr mean of five sites), springs (2-yr mean of monthly discharge-weighted means of four sites), and the stream (2-yr mean at WB60) are from this study. Vertical bars denote 1 SE of the mean.

were probably not a result of evaporative concentration, because the soil solution samples were collected from the B horizon below the primary rooting zone (~1 m). Concentrations of SRP and inorganic N in the stream were lower than those in riparian groundwater and springs, suggesting additional removal of these forms by instream processes. Organic N concentrations were lower in springs and the stream compared with upslope soil solutions, and SUP concentrations were similar in soil solutions, riparian groundwater, springs, and the stream.

Discharge varied seasonally in springs and the stream, with higher flows during the winter–spring period and lower flows in late summer and autumn (Fig. 3). The seasonal pattern in streamflow was much more pronounced in 1989, a wetter than average year (precip., 168 cm), than in 1988, a relatively dry year (precip., 112 cm). On sampling dates from November 1987 to December 1988 the combined discharge of the four identifiable springs on the watershed accounted for ≥50% of the flow at WB0.

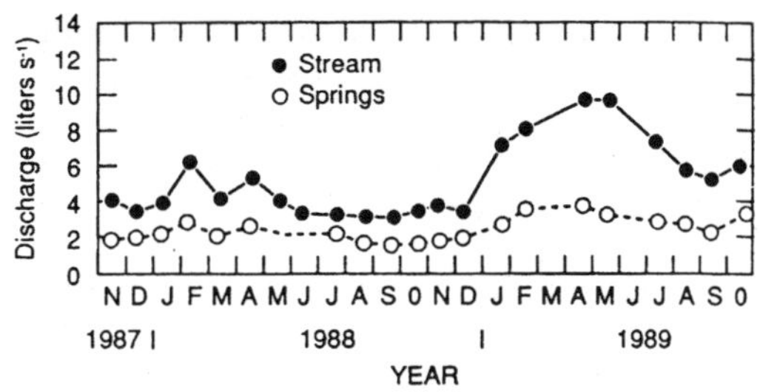

Fig. 3. Instantaneous discharge in the stream at the weir (WB0) and the sum of discharge in the four major springs during the period of study.

Concentrations of inorganic N and P in springs, riparian groundwater, and stream water were seasonally variable, with lower concentrations in late autumn, winter, and early spring and higher concentrations in summer (Fig. 4a,b,c). NH_4^+ exhibited the least variability, remaining low at all sites except during summer in riparian groundwater (Fig. 4a). The seasonal variations in (NO_2^- + NO_3^-) and SRP concentrations in stream water were considerably greater than variations in springs (Fig. 4b,c). Also, except in summer, concentrations of (NO_2^- + NO_3^-) and SRP were higher in springs than in riparian groundwater or stream water. Seasonal patterns in organic N and P were less distinct than for the inorganic forms, with SUP concentrations generally higher in summer than winter (Fig. 4d) and DON (not shown) having no consistent seasonal pattern.

Concentrations of SRP in springs were strongly correlated with Ca^{2+} (Fig. 5a), suggesting that bedrock weathering was an important source of SRP in groundwater. However, the average SRP:Ca ratio (by mass) in springs (1.64×10^{-4}, SD = 0.53×10^{-4}, $n = 132$) was about twofold lower than that measured in dolomite outcrops

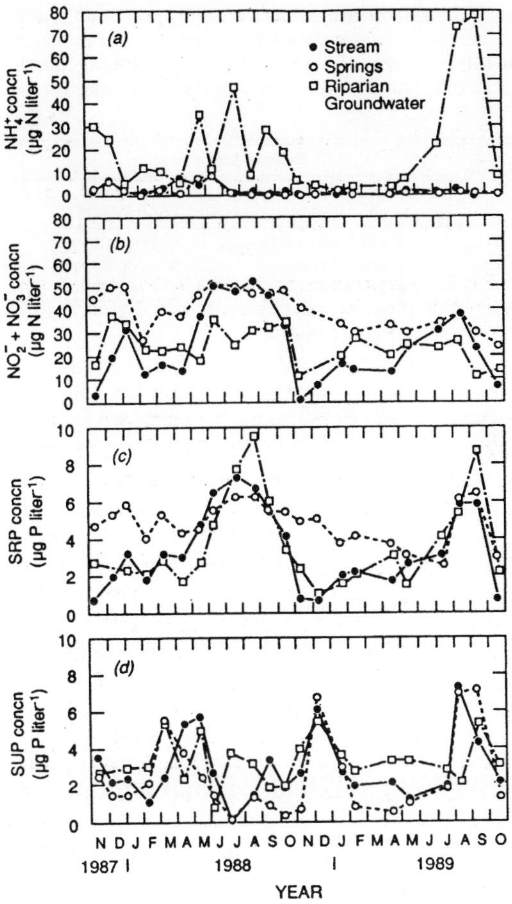

Fig. 4. Seasonal patterns in concentrations of NH_4^+, (NO_2^- + NO_3^-), SRP, and SUP in the stream (WB60), springs (discharge-weighted mean of four sites), and riparian groundwater (mean of five sites).

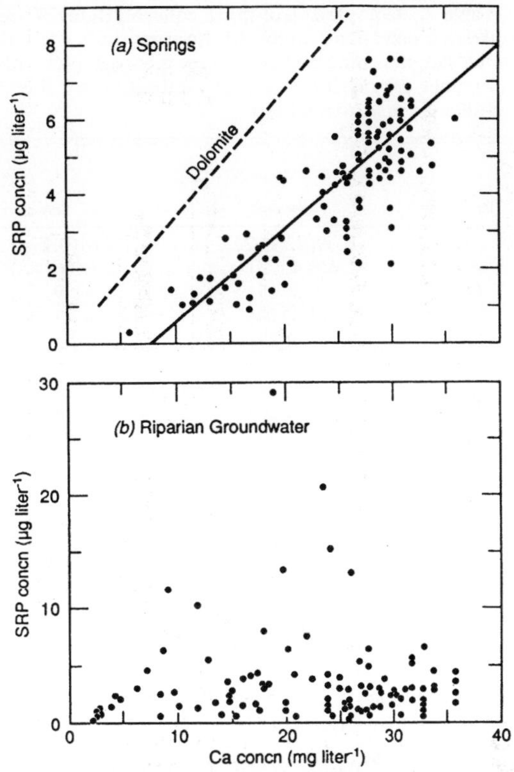

Fig. 5. Relationships between SRP and Ca^{2+} concentrations in springs (slope = 0.22±0.01, y-intercept = −1.3±0.4, R^2 = 0.67, F = 236, P < 0.001) and riparian groundwater (R^2 = 0, F = 0.002, P = 0.96). Dashed line shows the relationship between Ca and SRP measured in the dolomite.

near the stream (Table 1). In contrast to the springs, concentrations of SRP and Ca^{2+} in riparian groundwater were uncorrelated (Fig. 5b).

Differences in the chemistry of stream water, springs, and riparian groundwater suggest that redox conditions were important in regulating riparian groundwater chemistry (Table 2). Concentrations of alkalinity, Ca^{2+}, and Si were similar in stream water, springs, and riparian groundwater, indicating similar contact with soils and bedrock. However, dissolved oxygen concentrations in riparian groundwater were considerably lower and much more variable (range, 0.4–10.1 mg liter^{-1}) than in stream water and springs (range, 5.8–10.7 mg liter^{-1}). The higher concentrations of DOC and soluble Fe and Mn in riparian groundwater, particularly in samples with low dissolved oxygen concentration, suggest that

Table 1. Chemical composition of the dolomitic bedrock based on samples (n = 11) from outcrops along the stream (all values are mg g^{-1} unless otherwise given).

Element	Mean	SD
Ca	210	19
Mg	110	11
Na	0.5	0.1
K	4.0	2.2
Fe	1.9	1.2
Mn	0.1	0.04
P (µg g^{-1})	0.064	0.016
NO_3^--N (µg g^{-1})	<0.01	
P:Ca ratio (10^{-4} µg mg^{-1})	3.21	0.83

Table 2. Comparison of mean concentrations of various solutes in stream water (WB60), spring water, and riparian groundwater sampled from streamside wells. Riparian groundwater chemistry is presented as averages for all samples combined and averages for samples with dissolved oxygen (DO) concentrations <2.0 mg $liter^{-1}$. All values are mg $liter^{-1}$ except alkalinity (meq $liter^{-1}$) and pH. Sample standard deviation and number of samples given in parentheses.

Parameter	Stream	Springs	Riparian groundwater	
			All samples	DO (<2.0 mg $liter^{-1}$)
pH	8.09(0.29, 22)	7.69(0.27, 88)	7.18(0.40, 103)	6.92(0.23, 22)
Alk.	2.6(0.4, 22)	2.5(0.6, 88)	2.2(0.9, 103)	2.3(0.8, 22)
DO	9.2(0.6, 17)	8.3(0.9, 68)	4.4(2.8, 84)	1.2(0.5, 22)
Ca	26.3(3.4, 22)	24.5(5.8, 88)	22.6(9.0, 88)	21.7(8.8, 23)
Si	3.9(0.3, 22)	3.9(0.3, 88)	3.7(0.5, 102)	3.9(0.6, 21)
TIC	31.8(4.7, 22)	31.3(7.4, 87)	32.1(11.7, 98)	37.8(10.5, 18)
DOC	0.7(0.2, 22)	0.6(0.2, 88)	1.2(0.7, 103)	1.8(1.0, 22)
SO_4	1.9(0.4, 20)	1.8(0.6, 78)	3.3(1.8, 97)	2.8(2.1, 21)
Fe	0.001(0.003, 20)	0.003(0.009, 80)	0.30(0.89, 95)	0.99(1.54, 21)
Mn	0.004(0.004, 22)	0.003(0.006, 88)	0.13(0.27, 102)	0.40(0.45, 21)

anaerobic processes are important in riparian soils. Although SO_4^{2-} concentrations were higher in riparian groundwater than in springs, they were reduced somewhat at low dissolved oxygen concentrations as would be expected with increased extent of anoxic conditions in saturated riparian soils.

Differences in the concentrations of N and P between springs and riparian groundwater were at least partly related to differences in dissolved oxygen concentrations. Concentrations of NH_4^+ and DON were significantly higher and concentrations of (NO_2^- + NO_3^-) were significantly lower in riparian groundwater compared with springs, although for NH_4^+ only when riparian dissolved oxygen concentrations were low (Fig. 6a,b,c). Concentrations of SRP in riparian groundwater were significantly lower than those in springs when riparian dissolved oxygen concentrations were ≥2 mg $liter^{-1}$, but were significantly greater than those in springs when riparian dissolved oxygen was <2 mg $liter^{-1}$ (Fig. 6d).

Concentrations of stream-water (NO_2^- + NO_3^-) and SRP were generally lower than concentrations in spring water and declined between WB200 and WB60 during much of the year (Fig. 7a,b). Instream concentration declines were greatest immediately after autumn leaf fall, which occurs primarily from mid-October to early November, and absent in summer and early autumn. Concentrations of DON and SUP, as well as other major solutes (e.g. Ca^{2+}, Mg^{2+}, Na^+, K^+, alkalinity, SO_4^{2-}, Cl^-, Si, DOC), were similar at WB200 and WB60 (data not shown), and pH consistently increased from WB200 (mean = 7.94, SD = 0.16, n = 22) to WB60 (mean = 8.09, SD = 0.29, n = 22), primarily as a result of degassing of supersaturated levels of dissolved CO_2 in spring inputs (determined by substantial increases in pH when samples were bubbled with air).

There was considerable instream net retention of inorganic N and P from November to April (Fig. 8). For dates on which there was net uptake of both inorganic N and P, the mean N : P uptake ratio (molar) was 32.7 (SD = 20.9, n = 11). Ratios of the organic : total concentrations of N and P were significantly higher at WB60 compared with WB200, with the difference being greater in the November–April period than for all dates combined (Fig. 9).

Diel patterns in (NO_2^- + NO_3^-) and DOC

→

Fig. 6. Comparisons between springs (mean of all sites and dates), riparian groundwater samples with dissolved oxygen concentrations ≥2 mg $liter^{-1}$, and riparian groundwater samples with dissolved oxygen concentrations <2 mg $liter^{-1}$ for NH_4^+, (NO_2^- + NO_3^-), DON, SRP, and SUP concentrations. Histograms with different letters are significantly different (P < 0.05). Vertical bars denote 1 SE of the mean.

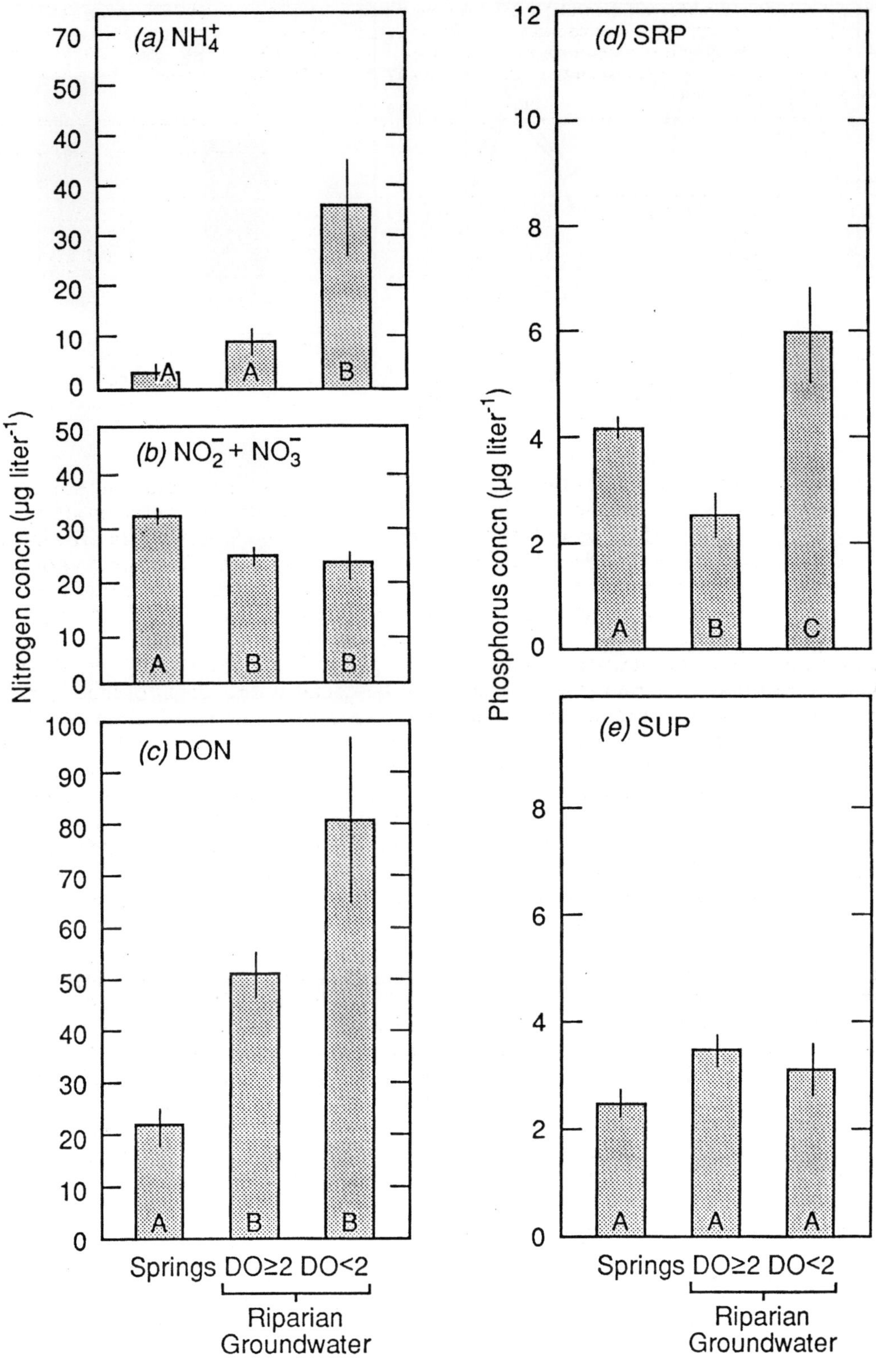
(a) NH$_4^+$
(b) NO$_2^-$ + NO$_3^-$
(c) DON
(d) SRP
(e) SUP
Nitrogen concn (µg liter^{-1})
Phosphorus concn (µg liter^{-1})
Springs DO≥2 DO<2
Riparian Groundwater
Springs DO≥2 DO<2
Riparian Groundwater

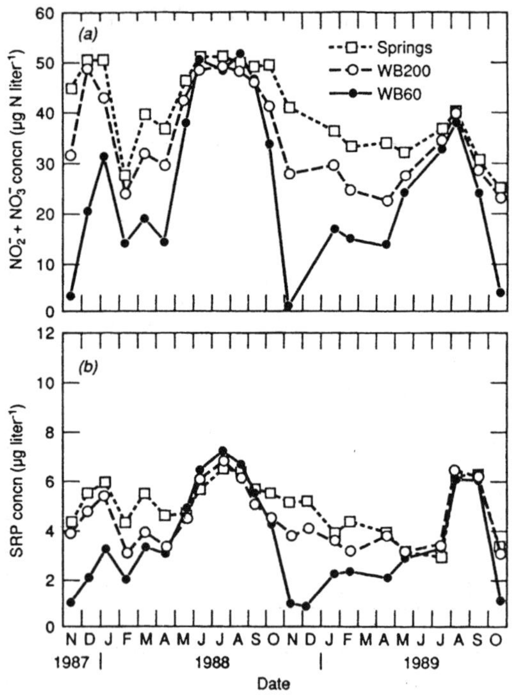

Fig. 7. Seasonal patterns in (NO_2^- + NO_3^-) and SRP concentrations in springs (discharge-weighted means) and at two stream sites (WB200 and WB60) downstream from springs.

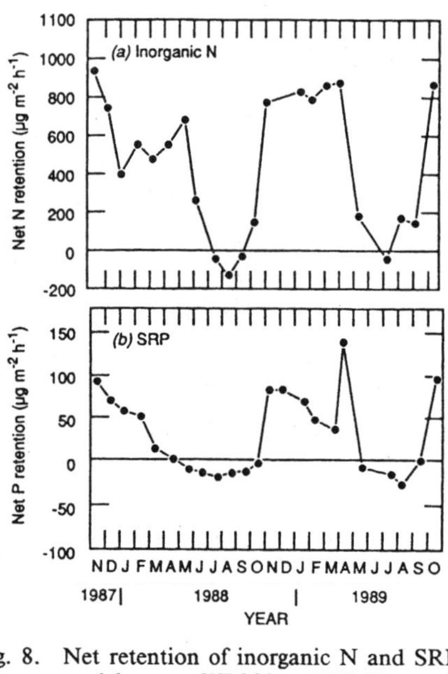

Fig. 8. Net retention of inorganic N and SRP in the stream reach between WB200 and WB60 over time.

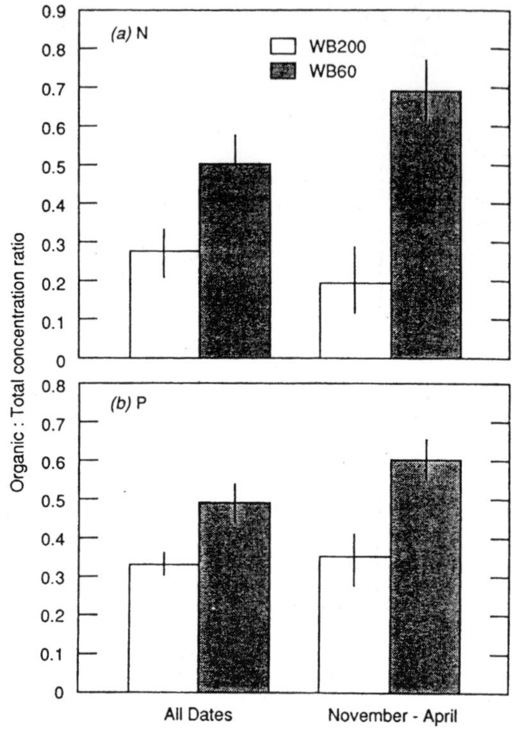

Fig. 9. Comparison between mean ratios of organic to total forms of N and P dissolved in stream water at WB200 and WB60. Comparisons were made for all dates and for November–April when net retention of inorganic forms of N and P was greatest. Differences in organic : total N and P ratios between sites were statistically significant at the $P < 0.05$ level for all dates and at the $P < 0.01$ level for November–April. Vertical bars denote 1 SE of the mean.

concentrations were observed in April at WB60 (Fig. 10). Concentrations of (NO_2^- + NO_3^-) were lowest during the day and highest at night. The opposite pattern was observed for DOC. Concentrations of NH_4^+ remained below the detection limit and SRP concentrations varied between 1.0 and 1.4 µg liter^{-1}, but showed no evidence of a diel pattern.

Discussion

Upslope soil and bedrock processes—The low concentrations of N and P in Walker Branch soil solutions indicate strong retention of throughfall inputs by soil processes—a finding consistent with reports from other forested watersheds (Wood et al. 1984; Qualls et al. 1991). Wood et al. (1984) sug-

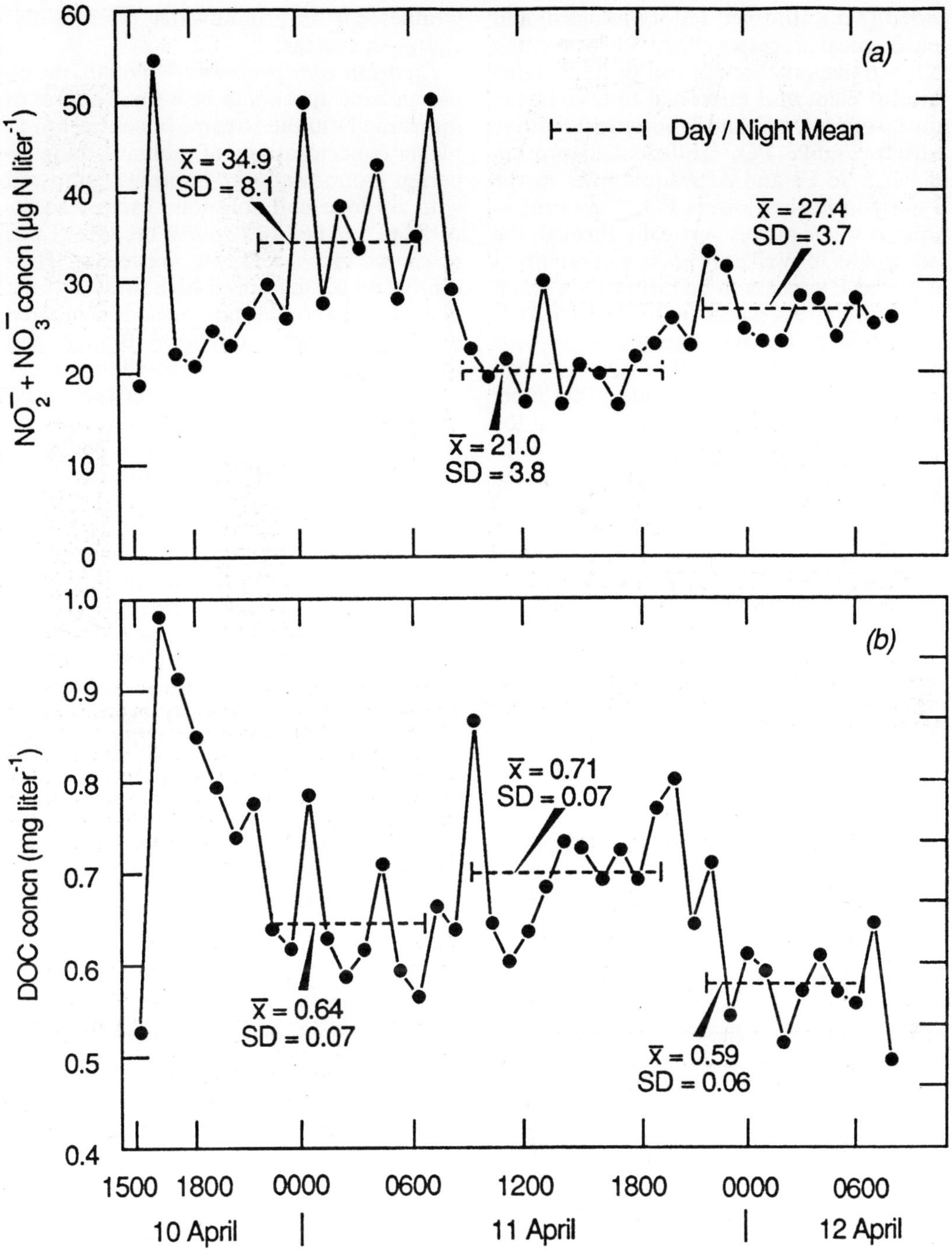

Fig. 10. Diel variation in concentrations of (NO_2^- + NO_3^-) and DOC at WB60 measured on 10–12 April 1991.

gested that a stratified set of biological and geochemical processes effectively prevented PO_4^{3-} transport through soil profiles. High rates of plant and microbial uptake in the upper soil layers (O and A horizons) remove most throughfall PO_4^{3-}. Efficient adsorption of PO_4^{3-} to Fe and Al sesquioxides in the B horizon further lowers PO_4^{3-} concentrations as water moves vertically through the soil profile. In Walker Branch, soil retention of N and P appears to be primarily a result of microbial uptake in upper horizons, as has also been shown in forested watersheds at Coweeta, North Carolina (Qualls et al. 1991). Segars et al. (1986) found that most of the throughfall SRP was removed in the litter layer and that there was little seasonality in soil solution P concentrations in Walker Branch. Soil solution NH_4^+ and NO_3^- concentrations in Walker Branch are low throughout the year and also show little seasonality (Mulholland et al. 1990; Van Miegroet et al. 1990). Although soil temperatures are lower in winter (generally 5°–10°C) compared with summer (20°–22°C, M. A. Huston pers. comm.), microbial nutrient immobilization rates may be quite high in winter due to the autumn input of fresh leaf litter with relatively high C:N and C:P ratios.

There appears to be an internal source of P in the watershed as evidenced by considerably higher concentrations of SRP in springs and riparian groundwater than in upslope soil solutions. The relatively high concentrations of P in the dolomitic bedrock and the strong correlation between SRP and Ca^{2+} in springs indicate that the source of P is probably bedrock weathering. Bedrock is the dominant source of Ca^{2+} in the watershed. Soils are acidic with very low base exchange capacity, and soil solution Ca^{2+} concentrations are 1–2 mg liter^{-1} (Johnson and Todd 1990; Mulholland et al. 1990), ~10–30 times lower than Ca^{2+} concentrations in springs, riparian groundwater, and the stream. However, the mean P:Ca ratio in bedrock was twice that in spring water, suggesting that either weathering is incongruent with respect to Ca and P, the outcrops are not representative of the bedrock undergoing weathering in the watershed, or there is some immobilization of weathered P in groundwater prior to discharge in springs.

Riparian zone processes—Overall, the riparian zone appears to be a small source of inorganic N to the stream. In contrast to P, higher concentrations of inorganic N in riparian groundwater and springs compared with upslope soil solutions cannot be attributed to a bedrock source because N was not found in the dolomite. Garten and Huston (in prep.) measured higher rates of net N mineralization in riparian soils compared with upslope soils in Walker Branch, and this may account for the higher inorganic N in riparian groundwater. Lower concentrations of DON in riparian groundwater and springs compared with upslope soil solutions suggest that DON mineralization could also account for higher inorganic N.

Redox conditions appeared to be important in regulating N and P in riparian groundwater. During summer dissolved oxygen concentrations in riparian groundwater were generally lowest, indicating more extensive areas of anoxia, and the riparian zone appeared to be a source of NH_4^+ and SRP to the stream (Fig. 4). Lower nitrification rates and mobilization of P as oxides of Fe and Mn were reduced probably account for the relatively high concentrations of NH_4^+ and SRP in riparian groundwater when dissolved oxygen concentrations were low (Fig. 6). However, when dissolved oxygen concentrations were high, the riparian zone appeared to be a sink for groundwater SRP, as indicated by the similarity in Ca^{2+} concentrations (and thus, source of P) but lower SRP concentrations in riparian groundwater compared with springs. Lower concentrations of (NO_2^- + NO_3^-) in riparian groundwater compared with springs suggests that denitrification may play a role in limiting groundwater inputs of inorganic N to the stream, although higher rates of microbial assimilation or lower rates of nitrification could also explain the lower riparian concentrations.

Previous studies have emphasized the riparian zone as a sink for nutrients in drainage water, but these have largely involved drainage water with high nutrient concentrations. For example, studies of Peterjohn and Correll (1984) and Lowrance et al.

(1984) demonstrated high rates of immobilization of N and P in runoff from upland agricultural areas along surface and subsurface flowpaths through riparian forests. In forested catchments, high immobilization rates of NH_4^+ (Hill and Warwick 1987) and NO_3^- (Duff and Triska 1990), injected at relatively high concentrations to surface water and groundwater in the riparian zone, have been reported. Results from my study broaden the view of riparian zones as regulators of nutrient flux to streams, suggesting that in watersheds with relatively low concentrations, the riparian zone can be a nutrient source or sink depending on redox conditions.

Despite differences in water chemistry between riparian groundwater and stream water, the importance of the riparian zone as a source or sink for N and P in Walker Branch cannot be quantified because of uncertainties in the contribution to total streamflow of diffuse seepage through riparian soils. Discharge of perennial springs contributed an average of 55% of streamflow on the sampling dates. However, the chemistry of spring water may be controlled primarily by upslope soil processes and bedrock weathering and minimally influenced by riparian soil processes. It is not known whether the remainder of inflow to the stream has a chemical composition similar to the four identified springs or to riparian groundwater. Water levels measured in each of the riparian wells indicated hydraulic gradients toward the stream at all times, suggesting some flow of riparian groundwater to the stream. However, hydraulic conductivity was highly variable across our riparian well network, ranging from 19 to >110 cm d^{-1} as determined by pump tests conducted on one date (Mulholland unpubl. data). The high spatial variability in hydraulic conductivity and chemistry in the riparian zone precludes quantitative determination of the contribution of riparian groundwater to nutrient inputs to the stream.

Instream processes—Instream processes strongly influenced the concentrations and forms of N and P in Walker Branch stream water. Stream-water concentrations of inorganic N and P were usually lower than those in springs and riparian groundwater, and mass balances over a 140-m stream reach indicated considerable retention of N and P during much of the year. Lower stream N and P concentrations cannot be attributed to dilution by shallow drainage from upslope soils having low N and P concentrations. Stream-water Ca^{2+} concentrations on all dates were $\geq 89\%$ ($\geq 95\%$ on all but two dates) of the discharge-weighted Ca^{2+} concentration in springs, indicating that stream water at baseflow was generated almost exclusively by discharge of relatively deep groundwater in contact with bedrock (with high Ca^{2+} concentrations) rather than by drainage directly from upper soil layers (with very low Ca^{2+} concentrations). However, there is some dilution of the deeper groundwater during winter and early spring as indicated by lower concentrations of Ca^{2+}, (NO_2^- + NO_3^-), and SRP and higher discharge in springs during this period.

Instream retention of inorganic N and P was strongly seasonal, with highest retention rates in late autumn and winter and lowest rates in summer. These results are consistent with previous studies in Walker Branch showing shortest PO_4^{3-} uptake lengths (most efficient uptake of PO_4^{3-} from stream water) in late autumn just after peak leaf input and longest uptake lengths in summer (Mulholland et al. 1985). Together, these studies highlight the importance of the seasonally pulsed input of terrestrial detritus and its subsequent decomposition by stream microbes in regulating concentrations and transport of stream-water N and P. The importance of detrital inputs to streams in regulating nutrient transport via instream immobilization and release has been demonstrated for other streams in forested watersheds as well (Munn and Meyer 1990).

The seasonal variation in (NO_2^- + NO_3^-) concentration in Walker Branch is opposite to the more commonly observed pattern in forested watersheds of dormant season maxima and growing season minima (Vitousek 1977; Foster et al. 1989). High stream NO_3^- concentrations during the dormant season are thought to result from reduced uptake by terrestrial vegetation and soil microbes at this time. However, seasonal NO_3^- patterns in streams may be a function of latitude. The winter maximum-summer minimum pattern has been documented primarily in streams in the northern U.S. and Canada. Seasonal patterns similar to that in Walker Branch (winter minimum and summer maximum) have been reported for a reference stream at Coweeta (Swank 1988) and for a stream on the Cumberland Plateau of Tennessee (J. M. Kelly pers. comm.). Although an earlier report on NO_3^- concentrations in Walker Branch indicated high concentrations in winter and low concentrations in summer (Elwood and Turner 1989), reanalysis has shown an error in interpreting sampling dates and the revised seasonal concentration pattern is similar to that reported here with winter minima and summer maxima (J. W. Elwood pers. comm.). In regions where soils remain unfrozen and stream-water temperatures remain well above 0°C during the dormant season, rates of N immobilization associated with microbial decomposition of autumn-shed leaves are high in soils and streams.

Uptake by stream autotrophs may also be an important mechanism for nutrient retention in Walker Branch during periods when light levels reaching the stream are relatively high. The diel pattern in (NO_2^- + NO_3^-) and DOC concentrations in April suggests that uptake by stream autotrophs may help account for the relatively high rates of instream nutrient retention observed in late winter and early spring when the abundance of leaf detritus in Walker Branch is low (Mulholland et al. 1985). Leaf emergence is usually complete by early May and the sharp decline observed in net retention of N and P in late spring may reflect reduced uptake by stream autotrophs after light levels decline with leaf emergence. Recent studies indicate that light is an important limiting factor for periphyton growth in Walker Branch in summer (Steinman 1992; A. D. Rosemond in prep.).

The role of abiotic exchange of P between water and sediments appears to be less important in regulating stream-water P concentrations in Walker Branch than in some other streams (Meyer 1979). Past studies of PO_4^{3-} uptake by Walker Branch sediments indicated that uptake is primarily a biological process (Elwood et al. 1981*b*; J. L. Meyer unpubl. data). Removal of PO_4^{3-} from stream water by coprecipitation with $CaCO_3$ is probably of minor significance because concentrations of Ca^{2+} and alkalinity were unchanged over the stream reach studied and stream water at both sites was undersaturated with respect to $CaCO_3$, except for a few dates in summer (when no instream net P retention was observed). Klotz (1991) suggested that sediment organic matter content may determine P uptake in streams, with physicochemical buffering dominating when organic content is low and biological processes dominating in streams with more organic-rich sediments. The results of my study suggest that organic matter quality (C : P ratio and the rate of microbial production associated with detritus) is also important in the biological regulation of stream-water P concentrations.

The importance of instream biological processes in regulating stream-water nutrient concentrations may be due in part to relatively stable streamflow in Walker Branch. Baseflow is maintained at $\geq$3 liters s^{-1}, even during the driest periods, partially due to groundwater transfer from adjacent watersheds (Luxmoore and Huff 1989)—a feature common in karst landscapes. Very high stormflows are also relatively uncommon, probably as a result of a buffering effect from the deep soils throughout much of the watershed (Mulholland et al. 1990). Such flow stabilization may result in greater detritus retention and more stable biological communities compared to more hydrologically flashy streams, thereby increasing the influence of instream biological processes on water chemistry.

Net retention of inorganic N and P in Walker Branch resulted in increasing importance of the organic forms of these nutrients with distance downstream from large groundwater inputs (springs). Although spatial gradients in nutrient concentration and form are probably more prominent in the spring-fed streams of karst regions, even in nonkarst regions groundwater inputs to streams are likely non-uniform. In stream segments with little groundwater input, upstream communities can reduce the concentrations of available nutrients, thereby increasing the dependence of downstream communities on remineralization and enzymatic hydrolysis of dissolved organic nutrients. Mulholland and Rosemond (in press) have shown a reduction in P content and an increase in phosphatase activity of periphyton with distance downstream from spring S3 in Walker Branch, thus demonstrating longitudinal linkage of stream biological communities via stream-water nutrient dynamics.

Instream retention of N and P is not necessarily long term. In Walker Branch, some of the N and P retained at baseflow during the autumn–winter–spring period was released back to the water in summer (Fig. 8). High discharge events have also been shown to remove attached biomass and sediments and the associated nutrients that accumulated during lower flow periods (Meyer and Likens 1979; Grimm 1987). Increases in concentrations of particulate N and P during storms have been observed in past studies of Walker Branch (Henderson et al. 1977), and transport of particulate N and P during storms likely accounts for at least a portion of the instream retention of dissolved N and P observed during periods of baseflow. Deposition and eventual remineralization of particulate organic nutrients in downstream reaches may provide important subsidies to these systems during subsequent baseflow periods when inorganic nutrient concentrations in water are low. For N, denitrification may also account for a portion of the instream retention, thus resulting in permanent removal.

In conclusion, although processes in upslope soils strongly limit the concentrations of inorganic N and P in drainage through the soil profile, bedrock weathering provides an additional source of P to the stream. The riparian zone appears to be a net source of NH_4^+ and inorganic P when dissolved oxygen concentrations in riparian groundwater are low, but a net sink for inorganic P when dissolved oxygen is high. Distinct spatial and seasonal variations in streamwater N and P concentrations, however, are primarily the result of instream biological processes, particularly uptake by microbes decomposing inputs of leaf litter from terrestrial vegetation.

References

AMERICAN PUBLIC HEALTH ASSOCIATION. 1989. Standard methods for the examination of water and waste water, 17th ed. APHA.

DILLON, P. J., AND W. B. KIRCHNER. 1975. The effects of geology and land use on the export of phosphorus from watersheds. Water Res. **9:** 135–148.

DUFF, J. H., AND F. J. TRISKA. 1990. Denitrification in sediments from the hyporheic zone adjacent to a small forested stream. Can. J. Fish. Aquat. Sci. **47:** 1140–1147.

ELWOOD, J. W., J. D. NEWBOLD, R. V. O'NEILL, R. W. STARK, AND P. T. SINGLEY. 1981*a*. The role of microbes associated with organic and inorganic substrates in phosphorus spiralling in a woodland stream. Int. Ver. Theor. Angew. Limnol. Verh. **21:** 850–856.

———, ———, A. F. TRIMBLE, AND R. W. STARK. 1981*b*. The limiting role of phosphorus in a woodland stream ecosystem: Effects of P enrichment on leaf decomposition and primary producers. Ecology **62:** 146–158.

———, AND R. R. TURNER. 1989. Streams: Water chemistry and ecology, p. 301–350. *In* D. W. Johnson and R. I. Van Hook [eds.], Analysis of biogeochemical cycling processes in Walker Branch watershed. Springer.

FORD, T. E., AND R. J. NAIMAN. 1989. Groundwater-surface water relationships in boreal forest watersheds: Dissolved organic carbon and inorganic nutrient dynamics. Can. J. Fish. Aquat. Sci. **46:** 41–49.

FOSTER, N. W., J. A. NICOLSON, AND P. W. HAZLETT. 1989. Temporal variation in nitrate and nutrient cations in drainage waters from a deciduous forest. J. Environ. Qual. **18:** 238–244.

GENEREUX, D. P., H. F. HEMOND, AND P. J. MULHOLLAND. In press. Spatial and temporal variability in streamflow generation on the West Fork of Walker Branch watershed. J. Hydrol.

GRIMM, N. B. 1987. Nitrogen dynamics during succession in a desert stream. Ecology **68:** 1157–1170.

———, AND S. G. FISHER. 1984. Exchange between interstitial and surface water: Implications for stream metabolism and nutrient cycling. Hydrobiologia **111:** 219–228.

———, AND ———. 1986. Nitrogen limitation in a Sonoran desert stream. J. N. Am. Benthol. Soc. **5:** 2–15.

HENDERSON, G. S., A. HUNLEY, AND W. SELVIDGE. 1977. Nutrient discharge from Walker Branch watershed, p. 307–320. *In* D. L. Correll [ed.], Watershed research in eastern North America: A workshop to compare results. Chesapeake Bay Center Environ. Stud., Smithson. Inst.

HILL, A. R., AND J. WARWICK. 1987. Ammonium transformations in springwater within the riparian zone of a small woodland stream. Can. J. Fish. Aquat. Sci. **44:** 1948–1956.

JASPER, S., AND OTHERS. 1983. Primary productivity in a large, temperate lake with river interflow: Kootenay Lake, British Columbia. Can. J. Fish. Aquat. Sci. **40:** 319–327.

JOHNSON, D. W., AND G. S. HENDERSON. 1989. Terrestrial nutrient cycling, p. 233–300. *In* D. W. Johnson and R. I. Van Hook [eds.], Analysis of biogeochemical cycling processes in Walker Branch watershed. Springer.

———, AND D. E. TODD. 1990. Nutrient cycling in forests of Walker Branch watershed, Tennessee: Roles of uptake and leaching in causing soil changes. J. Environ. Qual. **19:** 97–104.

KLOTZ, R. L. 1991. Temporal relation between soluble reactive phosphorus and factors in stream water and sediments in Hoxie Gorge Creek, New York. Can. J. Fish. Aquat. Sci. **48:** 84–90.

LIKENS, G. E., F. H. BORMANN, R. S. PIERCE, J. S. EATON, AND N. M. JOHNSON. 1977. Biogeochemistry of a forested ecosystem. Springer.

LINDBERG, S. E., G. M. LOVETT, D. D. RICHTER, AND D. W. JOHNSON. 1986. Atmospheric deposition and canopy interactions of major ions in a forest. Science **231:** 141–145.

LOWRANCE, R., AND OTHERS. 1984. Riparian forests as nutrient filters in agricultural watersheds. BioScience **34:** 374–377.

LUXMOORE, R. J., AND D. D. HUFF. 1989. Water, p. 164–196. *In* D. W. Johnson and R. I. Van Hook [eds.], Analysis of biogeochemical cycling processes in Walker Branch watershed. Springer.

MEYER, J. L. 1979. The role of sediments and bryophytes in phosphorus dynamics in a headwater stream ecosystem. Limnol. Oceanogr. **24:** 365–375.

———, AND G. E. LIKENS. 1979. Transport and transformation of phosphorus in a forest stream ecosystem. Ecology **60:** 1255–1269.

MULHOLLAND, P. J., J. D. NEWBOLD, J. W. ELWOOD, AND J. W. WEBSTER. 1985. Phosphorus spiralling in a woodland stream: Seasonal variations. Ecology **66:** 1012–1023.

———, AND A. D. ROSEMOND. In press. Periphyton response to longitudinal nutrient depletion in a woodland stream: Evidence of upstream-downstream linkage. J. N. Am. Benthol. Soc.

———, G. V. WILSON, AND P. M. JARDINE. 1990. Hydrogeochemical response of a forested watershed to storms: Effects of preferential flow along shallow and deep pathways. Water Resour. Res. **26:** 3021–3036.

MUNN, N. L., AND J. L. MEYER. 1990. Habitat-specific solute retention in two small streams: An intersite comparison. Ecology **71:** 2069–2082.

NEWBOLD, J. D., J. W. ELWOOD, R. V. O'NEILL, AND A. L. SHELDON. 1983. Phosphorus dynamics in a woodland stream ecosystem: A study of nutrient spiralling. Ecology **64:** 1249–1265.

PETERJOHN, W. T., AND D. L. CORRELL. 1984. Nutrient dynamics in an agricultural watershed: Observations on the role of a riparian forest. Ecology **65:** 1466–1475.

QUALLS, R. G., B. L. HAINES, AND W. T. SWANK. 1991. Fluxes of dissolved organic nutrients and humic substances in a deciduous forest. Ecology **72:** 254–266.

SAS. 1985. SAS user's guide: Statistics, version 5. SAS Inst.

SEGARS, J. E., R. A. MINEAR, J. W. ELWOOD, AND P. J. MULHOLLAND. 1986. Chemical characterization of soluble phosphorus forms along a hydrologic flowpath of a forested stream ecosystem. Oak Ridge Natl. Lab. ORNL/TM-9737.

STEINMAN, A. D. 1992. Does an increase in irradiance influence the periphyton in a heavily-grazed woodland stream? Oecologia **91:** 163–170.

SWANK, W. T. 1988. Stream chemistry responses to disturbance, p. 339–358. *In* W. T. Swank and D. A. Crossley, Jr. [eds.], Forest hydrology and ecology at Coweeta. Springer.

TRISKA, F. J., V. C. KENNEDY, R. J. AVANZINO, G. W. ZELLWEGER, AND K. E. BENCALA. 1989. Retention and transport of nutrients in a third-order stream: Channel processes. Ecology **70:** 1877–1892.

VAN MIEGROET, H., D. W. JOHNSON, AND D. W. COLE. 1990. Soil nitrification as affected by N fertility and changes in forest floor C/N ratio in four forest soils. Can. J. For. Res. **20:** 1012–1019.

VITOUSEK, P. M. 1977. The regulation of element concentrations in mountain streams in the northeastern United States. Ecol. Monogr. **47:** 65–87.

WOOD, T., F. H. BORMANN, AND G. K. VOIGT. 1984. Phosphorus cycling in a northern hardwood forest: Biological and chemical control. Science **223:** 391–393.

Submitted: 12 August 1991
Accepted: 30 March 1992
Revised: 18 June 1992

Commentary on Paper F6

Holmes, R. M., Fisher, S. G. & Grimm, N. B. (1994) Parafluvial nitrogen dynamics in a desert stream ecosystem.

Most of the studies carried out for the analysis of nutrient dynamics and the corresponding relevance of hydrology were developed in temperate systems, where water flow is permanent. Desert streams mostly differ from temperate systems because of their irregular hydrology, which includes flash floods and extended droughts. Further, desert streams show a well developed parafluvial area between the active channel and the riparian zone. The parafluvial zone is defined as that part of the active channel not occupied by surface water. Its extent changes markedly over time, ranging from being extremely wide during low flow episodes or drought, to disappear when the whole channel is occupied by stream water. In the case of Sycamore Creek (Arizona, USA), where this study was carried out, the hydraulic conductivity was high because of its coarse substratum, and therefore the parafluvial zone was an area of high surface–subsurface exchange.

The study showed that the parafluvial zone had a distinct hydrology and biogeochemistry, rather different from those in the surface water. Most of the flow paths analysed in the parafluvial zone had higher nitrate than the surface water, and dissolved oxygen declined as water flowed through the parafluvial zone. The elevated nitrate concentration was related to the nitrification occurring in the parafluvial zone. This mostly occurred during summer, both because of the favourable higher temperatures, the large activity and biomass of nitrifiers, and the large algal biomass. As a result, the hyporheic waters from these desert streams can be seen as nitrate sources to the surface waters. There are areas in the stream where there is discharge from the interstitial to surface waters (upwelling areas), and other areas where infiltration to the hyporheic (downwelling areas) is predominant. The hyporheic waters from these desert streams can be seen as nitrate sources to the surface waters, a relevant contribution in upwelling areas of the stream (Valett *et al.*, 1994).

Other studies have shown that the parafluvial can also be an active compartment in large rivers, even located in the temperate zone, and is not exclusive to dry river systems. Pinay *et al.* (1984) in the Garonne and Claret *et al.* (1997) in the Rhone showed that in this zone the concentrations of DOC and nitrate differed from those in the main channel. Claret *et al.* (1997) showed the influence of geomorphological heterogeneity (grain size and spatial distribution of fine sediments) for nutrient transformations as well as for the nutrient exchanges between parafluvial interstitial systems and the adjacent river.

SS

References

Claret, C., Marmonier, P., Boissier, J. M., Fontvieille, D. & Blanc, P. (1997) Nutrient transfer between parafluvial interstitial water and river water: Influence of gravel bar heterogeneity. *Freshwater Biology* **37**, 657–670.

Pinay, G., Haycock, N. E., Ruffinoni, C. & Holmes, R. M. (1994) The role of denitrification in nitrogen removal in river corridors. In: *Global Wetlands: Old World and New* (ed. by W. J. Mitsch), 107–116. Elsevier Science, The Netherlands.

Valett, H. M., Fisher, S. G., Grimm, N. B. & Camill, P. (1994) Vertical hydrologic exchange and ecological stability of a desert stream ecosystem. *Ecology* **75**, 548–560.

J. N. Am. Benthol. Soc., 1994, 13(4):468-478

Parafluvial nitrogen dynamics in a desert stream ecosystem

ROBERT M. HOLMES, STUART G. FISHER, AND NANCY B. GRIMM

Department of Zoology, Arizona State University, Tempe, Arizona 85287-1501 USA

Abstract. We investigated nitrogen dynamics over a 15-mo period in the parafluvial zone (the part of the active channel without surface water) of Sycamore Creek, Arizona, a nitrogen-limited Sonoran Desert stream. The parafluvial zone and surface stream are linked hydrologically; thus, nitrogen dynamics in the parafluvial zone potentially influence whole-system functioning. We identified discrete parafluvial flowpaths by following the movement of fluorescent dye through gravel bars over time, sampled subsurface water along these flowpaths, and collected parafluvial sediments for measurement of nitrification rate. Water samples were analyzed for nitrate-N, ammonium-N, dissolved oxygen, temperature, and conductivity. Nitrate-N concentration increased along parafluvial flowpaths, with the largest increases occurring in summer. Although ammonium-N concentration was low and did not vary with season or location on flowpath, dissolved oxygen declined as water moved through parafluvial gravel bars. Net nitrification rate was highest in the summer and at the heads of flowpaths where surface water entered the parafluvial zone, suggesting that nitrification may be dependent upon ammonium, dissolved organic nitrogen, or particulate organic nitrogen imported from the surface stream. Overall, the parafluvial zone of Sycamore Creek was a source of nitrate to the nitrogen-limited surface stream, and may play an important role in the productivity of the stream ecosystem.

Key words: Nutrient dynamics, desert streams, nitrate, nitrification, parafluvial zone, hyporheic zone.

Nutrient dynamics in stream ecosystems are a function of processes occurring in the surface stream as well as interactions between surface waters and subsurface regions (Fisher and Likens 1973, Likens and Bormann 1974, Hynes 1975, Peterjohn and Correll 1984). In nutrient-limited streams, these interactions are of particular importance if subsurface zones are sources or sinks of the limiting nutrient. For example, nitrogen-fixing cyanobacteria are a source of fixed nitrogen (Howarth et al. 1988, Paerl 1990), whereas denitrification in riparian sediments is a nitrogen sink (Peterjohn and Correll 1984, Lowrance et al. 19865, Cooper 1990, Duff and Triska 1990, Pinay et al. 1993). Because ecosystem properties such as primary productivity or nutrient transport and retention are products of the interactions of ecosystem components, it is useful to investigate functioning of individual system components to better understand the entire system.

Desert stream ecosystems are composed of four interacting subsystems: the surface stream, and hyporheic, parafluvial, and riparian zones (Fig. 1). Most research in desert streams and in stream ecology generally has focused on the surface stream. For example, we know that nitrogen availability limits primary productivity in surface waters of Sycamore Creek, Arizona (Grimm and Fisher 1986a), and that nitrogen is potentially limiting in many streams of the southwestern United States (Grimm and Fisher 1986b).

There is, however, a growing awareness of the importance of subsurface waters to stream structure and functioning (Hynes 1983, Ward 1989, Triska et al. 1989). Surface–hyporheic interactions have been investigated in several streams (e.g., Grimm and Fisher 1984, Triska et al. 1989, Hendricks and White 1991, Valett et al. 1994). In Sycamore Creek, the hyporheic zone is a source of nitrate to surface waters (Valett et al. 1990, Valett et al. 1994). This input alleviates nitrogen limitation at sites of hyporheic upwelling, and consequently algal productivity in Sycamore Creek is higher in upwelling than in downwelling zones (Valett et al. 1994). Algal recovery following disturbance is also accelerated in upwelling zones; thus, surface–hyporheic exchange increases ecosystem stability through enhanced resilience (Valett et al. 1994).

Less is known about processes in interstitial waters of the parafluvial zone, the region of the active channel without surface water. The parafluvial zone of Little Lost Man Creek, California, is a source of nitrate to the surface stream (Triska et al. 1989, Triska et al. 1990), whereas in the Garonne River, France, it is a nitrate sink (Pinay et al. 1994). In desert streams such as

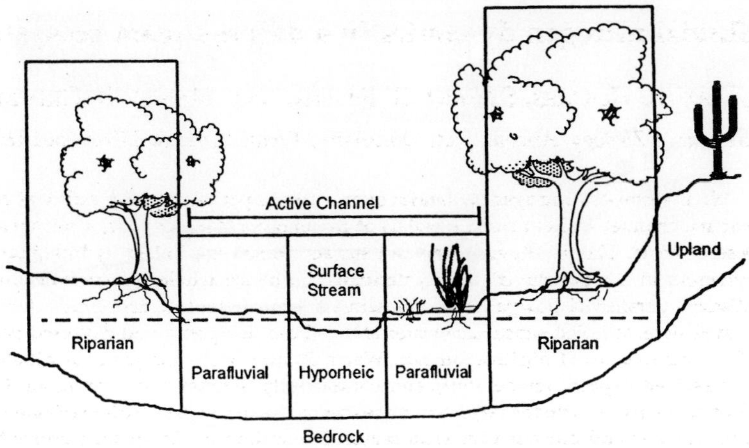

FIG. 1. Diagrammatic representation of surface stream, hyporheic, parafluvial, and riparian components of a desert stream ecosystem. The dashed line represents the water table, which may slope toward or away from the surface stream.

Sycamore Creek, the parafluvial zone is extensive and hydraulic conductivity is high, resulting in high potential for surface–subsurface exchange. The effect of this exchange depends in part upon the kinds and rates of parafluvial processes and their consequences for chemical change in interstitial water. Whether the parafluvial zone acts as a nitrogen source or sink to Sycamore Creek is the focus of this investigation.

The objectives of this study are 1) to determine hydrologic flowpaths within the parafluvial zone and demonstrate hydrologic linkage between parafluvial gravel bars and surface waters, 2) to investigate nitrogen dynamics along parafluvial flowpaths and determine whether the parafluvial zone is a source or sink of inorganic nitrogen to the surface stream, and 3) to evaluate the importance of parafluvial nitrogen dynamics to whole-system functioning.

Study Site

Sycamore Creek is about 35 km northeast of Phoenix, Arizona. Elevations reach 2164 m at the headwaters and drop to 427 m at its confluence with the Verde River. Catchment area is 505 km² and stream length is about 65 km. The geomorphology of Sycamore Creek is characteristic of Sonoran Desert streams and can be broadly divided into three sections (Wertz 1963). The channel in the upper section of the watershed is composed of boulders and exposed bedrock. Mid-reaches are characterized by coarse sand and gravel sediments averaging 80 cm deep. The surface stream in this region typically occupies only part of the active channel, creating extensive gravel bars that form the parafluvial zone. The downstream section of the stream channel has deep alluvial sediments that rarely support surface flow.

The study site is a 300-m reach in the middle section of Sycamore Creek. Elevation is ~650 m and alluvial sediments in the reach average >1.5 m deep. The active channel is ~25-m wide, with the water table generally being within 25 cm of the surface of parafluvial gravel bars (Fig. 1). Channel slope is 0.006 and typical baseflow discharge is ~0.05 m^3/s (range 0 to >400 m^3/s). Vegetation of parafluvial gravel bars has been sparse since 1991 when several large floods scoured the bars clear. However, at times the parafluvial zone is densely covered by seep-willow (*Baccharis salicifolia*) and/or Bermuda grass (*Cynodon dactylon*), as well as seedlings of common riparian species such as cottonwood (*Populus freemontii*), sycamore (*Platanus wrightii*), and mesquite (*Prosopis glandulosa*).

Methods

Parafluvial hydrology and linkage with the surface stream

The areal extent of the parafluvial zone was determined by measuring the percent of the

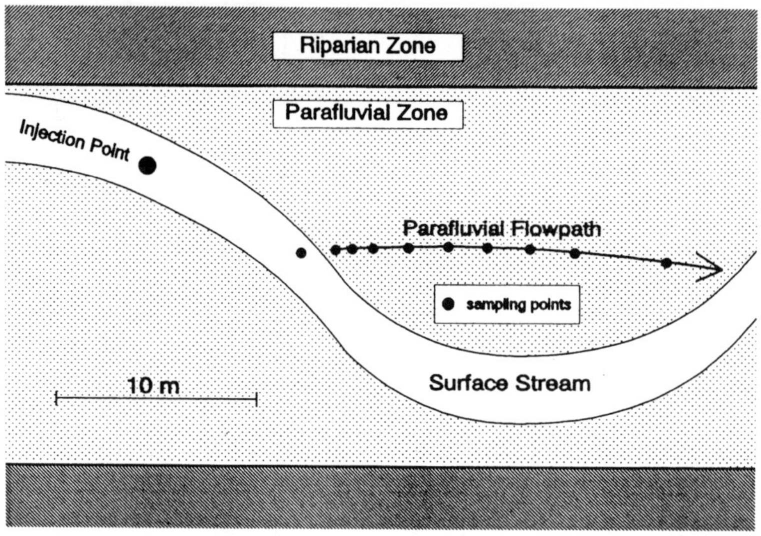

FIG. 2. Areal view of solute injection site. Solutes were injected into the surface stream and sampled in parafluvial wells (sampling points are indicated by solid circles).

active channel occupied by parafluvial gravel bars. Measurements were made using a series of 11 transects at 20-m intervals along the study reach, or estimates were made from photographs of the study site. These estimates were made 33 times during the study.

We sampled a total of 28 parafluvial flowpaths on 13 days from October 1992 to January 1994. Subsurface flow direction and rate in parafluvial gravel bars was determined by injecting fluorescent dye at the upstream margins of gravel bars and at several points within the bars and tracing its progress through the bar over time. We injected dye by adding it to shallow pits, and followed its movement by digging a series of trenches perpendicular to the predicted flowpath roughly 1 h after injection. Injection time, distance from injection point to the farthest trench where dye was seen, and time when dye was located were recorded to determine flow rate. Discrete flowpaths of at least 10 m in length were identified.

Hydrologic exchange from the surface stream to the parafluvial zone was investigated by injecting a conservative tracer (chloride as NaCl) into the surface channel and sampling parafluvial wells downstream from the injection point. Chloride injections were performed in July 1993, and parafluvial wells up to 14 m from the surface stream were sampled. Parafluvial to surface exchange was examined by searching for parafluvially injected dye in the surface stream.

Physico-chemical characteristics along parafluvial flowpaths

Physico-chemical characteristics of interstitial water along parafluvial flowpaths were determined by sampling with mini-piezometers (Lee and Cherry 1978) inserted to a depth of 25 cm below the water table. Piezometers, with tips covered by Nitex® mesh to minimize clogging, were inserted along discrete flowpaths at 0.1, 1.0, 2.0, 5.0, and 10.0 m (and occasionally up to 30.0 m) from the upstream parafluvial-stream interface. Samples were collected using a battery-powered peristaltic pump fitted with an in-line filter. At each sampling point we first measured dissolved oxygen (Leeds & Northrup Instruments D.O. probe), temperature, and conductivity, and then collected triplicate water samples in acid-washed 60-mL polyethylene bottles for nitrate analysis. Separate triplicate 25-mL water samples were collected and fixed in the field with phenol reagent before spectrophotometric ammonium analysis (Shimadzu UV160U) by the phenolhypochlorite method (Solorzano 1969). Before November 1993, samples were field-filtered through 0.8-μm cellulose-acetate filters (Geofilter®, 142 mm diameter), thereafter with Whatman® GFF filters (nominal pore size = 0.7 μm). Comparison of the two filtration methods showed their equivalence for nitrate and ammonium analyses. Nitrate samples were analyzed on a Bran and Luebbe TRAACS 800® autoanalyzer using a modified version of Technicon Industrial Method No. 818-87T.

Net nitrification

Net nitrification rate was determined using sediments from 15 parafluvial flowpaths. Sediments were taken from 5 to 15 cm below the water table, at 0.1 m and 5.0 m from the stream–parafluvial interface. We also collected sediments at 1.0 m, 2.0 m, and 10.0 m (in addition to 0.1 m and 5.0 m) from the stream–parafluvial interface along eight flowpaths during summer 1993. Net nitrification rate was measured by placing approximately 160 g (wet weight) sediments in 250-mL Erlenmeyer flasks, and incubating with 100 mL distilled water for 15–22 hours in the dark at 25°C. Sediments from each sampling point were run in triplicate. After the incubation period, a 5-mL water sample was collected from each flask and analyzed for nitrate-N. Net nitrification rate was expressed per unit dry sediment mass. Initial nitrate-N samples were not taken. We later determined that ~10% of the nitrate-N increase is due to leaching when distilled water is first added to sediments. Net nitrification rates for sediments collected at 0.1 m and 5.0 m were log-transformed to correct for non-normality and analyzed using two-factor analysis of variance, with season and location on flowpath as factors. Net nitrification along the eight intensively-sampled flowpaths was analyzed using one-way ANOVA. Season was not a factor in this analysis because all samples were collected in summer. All statistical analyses were done with the SYSTAT (1992) software package.

Co-injection of ammonium and chloride

We co-injected the conservative tracer chloride (as NaCl) and ammonium (as NH_4Cl) to the surface stream to demonstrate surface–parafluvial hydrologic linkage and to investigate potential ammonium limitation of nitrification. Our objective was to determine whether nitrate production in parafluvial sediments was limited

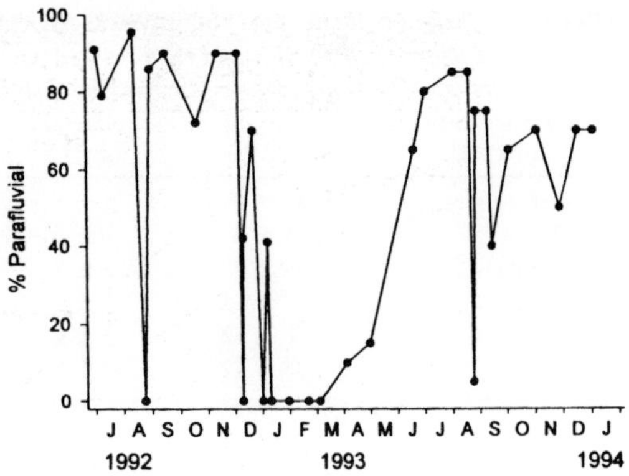

FIG. 3. Percent of the active channel composing the parafluvial zone. During spates when the entire active channel is underwater, the parafluvial zones does not exist. Conversely, when surface flow ceases completely, as it sometimes does, the parafluvial zone occupies 100% of the active channel.

by ammonium availability. On 19 July 1993, we sampled subsurface water at nine points along a parafluvial flowpath, as well as surface water adjacent to the head of the parafluvial flowpath (Fig. 2). All samples were analyzed for nitrate-N, ammonium-N, dissolved oxygen, temperature, and conductivity by previously described methods. Chloride was analyzed on a Bran and Luebbe TRAACS 800® using Technicon Industrial Method #783-86T.

Solute injection into the surface channel began at 9:30 am on 20 July 1993 and continued for 20 h. Surface water and parafluvial wells were sampled 1, 2, 4, 24, 48, and 72 h after the initiation of the injection. If nitrification was limited by ammonium availability, we expected to find increased nitrate-N concentrations in the parafluvial zone after initiation of the injection due to increased rates of nitrification. Conversely, lack of change in nitrate-N concentration would indicate that nitrification was independent of the rate of ammonium supply.

Results

Parafluvial hydrology and linkage with the surface stream

The parafluvial zone occupied from 0 to 95% of the active channel of the 300-m study reach over the period of the study (Fig. 3). Parafluvial coverage was generally greatest in summer, al-

TABLE 1. Variation of chemical and physical characteristics with distance along parafluvial flowpaths. Data are presented as means (±1 SE) over all sites and seasons ($n = 26$).

Distance along flowpath	NO_3-N (μg/L)	NH_4-N (μg/L)	D.O. (mg/L)	Temperature (°C)	Conductivity (μS/cm)
Surface stream	45 ± 9	2 ± 1	8.8 ± 0.3	21.5 ± 1.3	467 ± 10
0.1 m	51 ± 8	1 ± 1	8.7 ± 0.3	21.5 ± 1.4	469 ± 10
1.0 m	59 ± 9	3 ± 3	8.6 ± 0.3	20.8 ± 1.3	468 ± 10
2.0 m	67 ± 11	1 ± 0	8.5 ± 0.4	19.6 ± 1.2	470 ± 10
5.0 m	78 ± 12	5 ± 4	7.9 ± 0.4	18.9 ± 1.1	473 ± 11
10.0 m	98 ± 12	2 ± 1	7.2 ± 0.4	20.5 ± 1.3	476 ± 10

though the parafluvial zone often composed much of the active channel in winter as well. An unusually large flood (>400 m^3/s) beginning 8 January 1993 inundated the active channel for almost 3 months, and thus eliminated the parafluvial zone for that period (Fig. 3).

During the solute injection experiment, surface-injected chloride appeared in all parafluvial wells, verifying surface–parafluvial linkage. Likewise, parafluvially injected dye was often found in the surface stream, confirming parafluvial–surface linkage. Surface water typically entered gravel bars at meanders and re-entered the surface stream farther downstream. Parafluvial water velocity averaged 1.96 m/h and ranged from 0.5 to 5.0 m/h, although velocities at different locations along a flowpath on a given day were similar. No seasonal trends were apparent in flowpath velocity, although flow direction fluctuated as stream water level changed relative to the parafluvial water table. Visual inspection of dye plumes downstream of injection points suggested lateral dispersion of water flowing through gravel bars was minimal as dye plumes remained constricted even 5–10 m from the injection point.

Physico-chemical characteristics along parafluvial flowpaths

Physico-chemical parameters along parafluvial flowpaths are summarized in Table 1. Nitrate-N concentration increased as water flowed through parafluvial gravel bars (repeated measures ANOVA, test for repeated-factor, $p < 0.001$, Table 2). The mean nitrate-N increase along 10-m parafluvial flowpaths was >50 μg/L and the maximum exceeded 200 μg/L. There was a significant seasonal effect on parafluvial nitrate-N concentration (test for independent factor effect, $p < 0.05$, Table 2), with the greatest concentration increase in summer (Fig. 4A, B). Overall, 25 of the 26 flowpaths sampled exhibited higher nitrate-N concentration at 10 m than in surface-stream water. One winter-sampled flowpath showed a small nitrate-N decline (2 μg/L). In contrast to nitrate, ammonium concentrations were generally low, often below limits of detection (<5 μg/L, Table 1), and did not vary as a function of distance along flowpath or season (Table 2).

Dissolved oxygen declined as water flowed through the parafluvial zone ($p < 0.001$, Table

TABLE 2. Results of repeated measures ANOVA for chemical variables along parafluvial flowpaths. The independent variable is season and the repeated variable is distance along flowpath.

Dependent variable	Mean squared error		F statistic: Season	F statistic: Distance	F statistic: Season × distance
Nitrate	between	0.011	3.930*	—	—
	within	0.000	—	13.446***	3.352***
Ammonium	between	0.000	3.229	—	—
	within	0.000	—	0.821	0.698
Dissolved Oxygen	between	9.365	10.192***	—	—
	within	0.252	—	17.576***	3.941***

* $p < 0.05$; ** $p < 0.01$; *** $p < 0.001$.

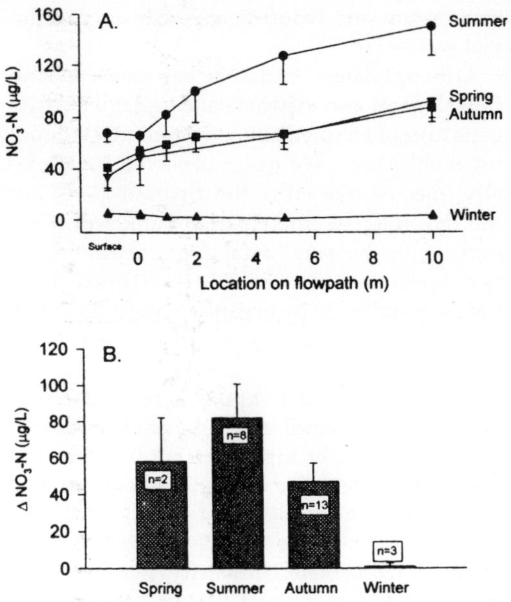

FIG. 4. A. Nitrate-N concentration along parafluvial flowpaths as a function of season. Each data point represents the mean (−1 SE) concentration of several flowpaths (see Fig. 4B for number of flowpaths sampled). B. Change in nitrate-N concentration between the surface stream and the 10.0-m sampling point along parafluvial flowpaths. Numbers of flowpaths sampled in each season are indicated. Error bars represent +1 SE.

2, Fig. 5A) and also varied as a function of season (repeated measures ANOVA, test for independent factor, $p < 0.001$, Table 2). The greatest decline occurred during summer, followed by spring, autumn, and winter (Fig. 5B). For flowpaths that were sampled over distances >10 m, the greatest dissolved oxygen declines were generally within the first 5–10 m of the flowpaths (Fig. 6). Conductivity and temperature did not vary significantly along flowpaths (Table 1, $p > 0.05$ in both cases).

Net nitrification

Net nitrification rate varied seasonally ($p < 0.001$), with summer rates being significantly higher than autumn or winter (Fig. 7). Mean summer net nitrification was much higher in sediments from the 0.1-m sampling point than from 5.0 m, but not in autumn and winter (Fig. 7). Thus, location was not a significant factor in explaining net nitrification in this analysis overall ($p > 0.05$). Location was significant, however, for the eight flowpaths that were intensively investigated in the summer of 1993 ($p < 0.05$). Tukey multiple comparison analysis showed sediments from 0.1 m exhibited higher net nitrification rates than sediments collected 5.0 m into the parafluvial flowpaths ($p < 0.05$, Fig. 8), which is consistent with the summer nitrification rates shown in Figure 7.

Ammonium and chloride co-injection

All parafluvial wells exhibited elevated solute concentrations in response to the surface stream injection. Pre-injection nitrate-N and ammonium-N in the surface stream were 36 µg/L and below detection level (<5 µg/L), respectively. During the injection, surface-water nitrate-N near the head of the parafluvial flowpath ranged from 49 µg/L to 82 µg/L, while ammonium-N reached 1554 µg/L. Surface-in-

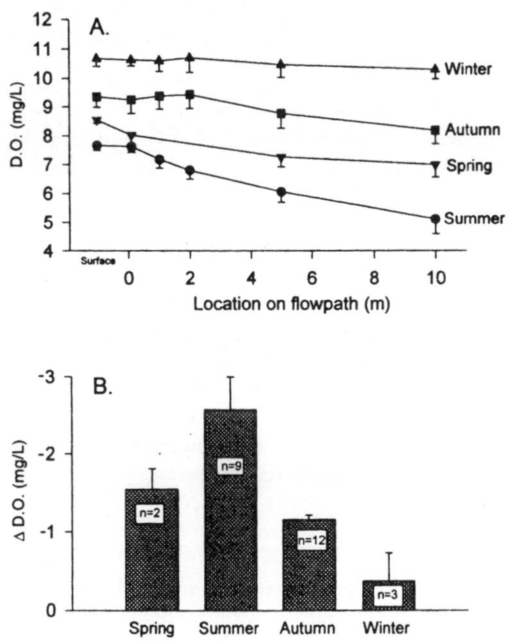

FIG. 5. A. Dissolved oxygen concentrations along parafluvial flowpaths as a function of season. Each data point represents the mean (−1 SE) concentration of several flowpaths (see Fig. 5B for number of flowpaths sampled). B. Change in dissolved oxygen concentration between the surface stream and the 10.0-m sampling point along parafluvial flowpaths. Numbers of flowpaths sampled in each season are indicated. Error bars represent +1 SE.

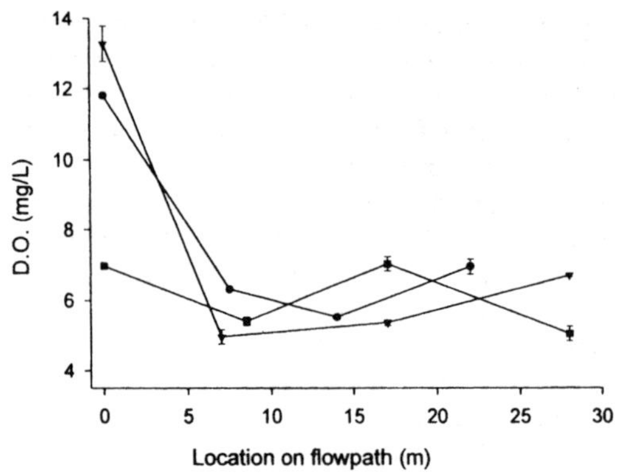

FIG. 6. Oxygen concentration along three parafluvial flowpaths that were sampled over distances greater than 10 m. Each line represents a distinct flowpath. Note the rapid initial dissolved oxygen decline followed by a levelling-off. Error bars represent ±1 SE, $n = 3$.

jected solutes arrived rapidly in parafluvial wells, with chloride appearing first, followed by increases in nitrate-N and ammonium-N. For example, at the sampling well 4 m along the flowpath, pre-injection nitrate-N was 67 μg/L and the highest recorded concentration (585 μg/L) was 24 h after the injection began (Fig. 9). Highest recorded ammonium concentration was also at 24 h (352 μg/L).

Discussion

Parafluvial hydrology and linkage with the surface stream

Hydrologic linkage is the primary interaction between parafluvial and surface-water environments. Aerial linkage may be important when parafluvial vegetation is dense and leaf litterfall is heavy; however, the absence of vegetation on parafluvial gravel bars made this vector insignificant.

Frequent hydrologic exchange between surface and subsurface environments increases the probability that subsurface processes will influence surface-stream functioning. Parafluvial to surface-stream linkage was demonstrated when parafluvially injected dye was detected in the surface stream. Surface to parafluvial exchange was suggested when dye injected at the stream–parafluvial interface was transported into parafluvial gravel bars. This linkage was verified by the tracer experiment when surface-injected ammonium and chloride appeared in parafluvial wells (Fig. 9).

Although lateral exchange between the parafluvial zone and riparian and upland ecosystems cannot be ruled out, we found no evidence for this linkage. We never observed parafluvially injected dye enter the riparian zone, nor did dye injected into riparian bank sediments move into the parafluvial zone. Sediments at the riparian–parafluvial interface are fine-grained and rich in organic matter, and may restrict the exchange of water between these zones.

Deep groundwater inputs were not investigated. However, indirect evidence suggests that deep groundwater inputs are not important in Sycamore Creek. For example, hyporheic and parafluvial concentrations of dissolved organic carbon are similar to surface-stream values (J. Jones, Arizona State University, personal communication), in contrast to systems with significant groundwater inputs (Ford and Naiman 1989, Fiebig and Lock 1991). In addition, conductivity and temperature did not vary significantly along flowpaths (Table 1, $p > 0.05$), which further suggests that a distinct groundwater source was not entering the stream ecosystem. Thus, it appears that most of the water in the

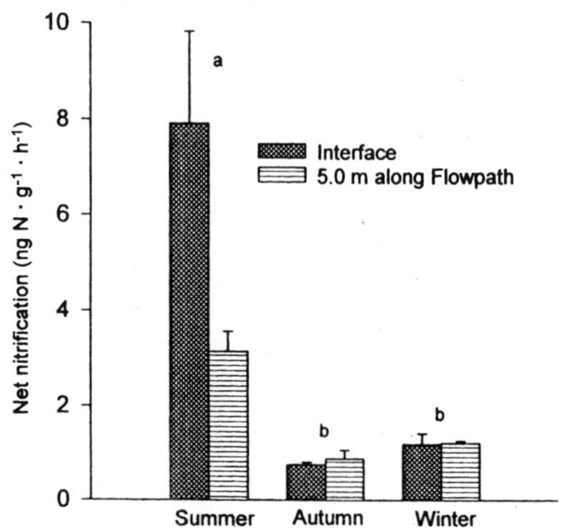

FIG. 7. Net nitrification rate (per unit sediment dry mass) in parafluvial sediments at the stream–parafluvial interface and at 5.0 m along parafluvial flowpaths. Error bars represent +1 SE. Statistically significant differences among seasons (Tukey's multiple comparison, $p < 0.05$) are denoted by different lower case letters.

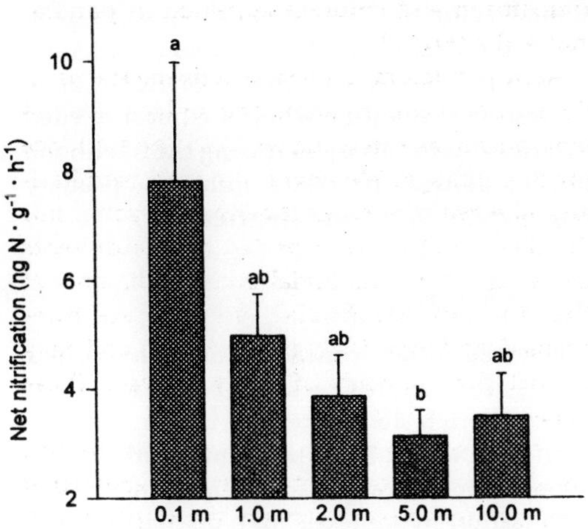

FIG. 8. Net nitrification rate (per unit sediment dry mass) for eight intensively sampled flowpaths. All of these flowpaths were sampled in the summer of 1993. Error bars represent +1 SE. Statistically significant differences (Tukey's multiple comparison, $p < 0.05$) are denoted by different lower case letters.

parafluvial zone had been within the active channel for an extended period, as opposed to recently entering the active channel from the riparian, upland, or groundwater systems.

Parafluvial water velocities in Sycamore Creek (mean = 1.96 m/h) were similar to those in the hyporheic zone (mean = 2.2 m/h; Valett et al. 1990). These rates are greater than typical groundwater velocities and probably are at the upper extreme for interstitial water velocity in stream sediments. The high parafluvial water velocity found in Sycamore Creek may influence nutrient dynamics. That is, rapid interstitial velocity decreases hydrologic residence time, and therefore may influence the relative contribution of aerobic and anaerobic processes, as shorter residence times should minimize dissolved oxygen declines.

Parafluvial flowpath length varied from less than 1 m to greater than 100 m. For logistical reasons, we sampled only flowpaths at least 10 m long and then generally restricted sampling to the first 10 m. This approach facilitates comparison between dates and locations, but it fails to describe the full range of variability present in the system. For example, the net effect of many small bars is probably different from the effect of a few large bars. At times the parafluvial zone occupies 95% or more of the active channel, whereas at other times the parafluvial zone is absent (Fig. 3). Moreover, the distribution of flowpath lengths is not constant but varies in response to flooding and drying as the configuration of the active channel changes, further influencing stream–parafluvial interactions.

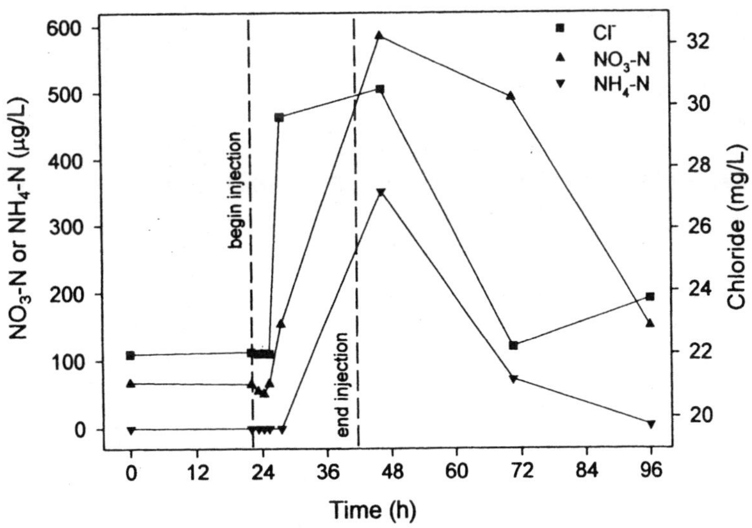

FIG. 9. Ammonium-N, nitrate-N, and chloride concentrations in parafluvial well 4 m from stream–parafluvial interface. Similar patterns were found at the other sampling locations.

Physico-chemical characteristics along parafluvial flowpaths

Dissolved oxygen concentration influences nitrogen dynamics, as some transformations are obligately aerobic whereas others are anaerobic. In Sycamore Creek, dissolved oxygen declined along parafluvial flowpaths, with the extent of the decline being seasonally variable (Fig. 5, Table 2). Lowest initial dissolved oxygen concentration, as well as the greatest decline along parafluvial flowpaths, occurred during summer. However, over the course of this study, dissolved oxygen seldom dropped below 5 mg/L. At these levels, aerobic processes (e.g., nitrification) would predominate over anaerobic processes (e.g., denitrification), except at active anoxic microsites.

For flowpaths that were sampled over distances greater than 10 m, dissolved oxygen generally decreased most rapidly within the first 6–8 m of the flowpaths (Fig. 6). Similarly, dissolved oxygen decline along 10-m flowpaths sampled in summer was greatest in the first few metres of the flowpaths (Fig. 5A). These observations are consistent with the hypothesis that subsurface respiration is dependent upon surface inputs of dissolved or particulate organic matter. Jones (Arizona State University, personal communication) found higher hyporheic respiration in downwelling zones (where surface water entered the hyporheic zone) than in upwelling zones (where hyporheic water entered the surface stream). He suggested that the higher respiration in downwelling zones is related to the import of labile, surface-derived organic matter. A similar dependence on stream–parafluvial linkage to support parafluvial respiration may be inferred from our oxygen data.

Although ammonium in surface and parafluvial waters of Sycamore Creek was always low (Table 1), and did not vary as a function of season or location along flowpath (Table 2), ammonium production through mineralization may be high, with rapid nitrification, adsorption, or uptake accounting for the low ambient concentration. At a minimum, ammonium-N supply must equal nitrate-N production, assuming that the entire nitrate increase is from nitrification.

In summer, algal productivity is generally highest and nitrogen is typically limiting in Sycamore Creek (Grimm and Fisher 1986a). Similarly, nitrate-N increases along parafluvial flowpaths were greatest in summer (Fig. 4A, B). Thus, nitrate production in the parafluvial zone is greatest at the time when it is most needed by surface autotrophs, increasing the likelihood that parafluvial processes influence primary productivity in the surface stream.

Net nitrification

Elevated nitrate concentration in hyporheic waters of Sycamore Creek may be due to storage of nitrate-rich floodwaters or in-situ nitrification (Grimm et al. 1991); substantial nitrification potential in hyporheic sediments supports the nitrification hypothesis. High discharge spates have 6.1 times higher nitrate-N concentration than baseflow (Grimm and Fisher 1986b), so spates also may recharge the hyporheic zone with nitrate. As pointed out by Grimm et al. (1991), these sources are not mutually exclusive. For example, organic matter could be imported into the stream ecosystem during floods and later mineralized and nitrified within interstitial sediments. However, storage of nitrate-rich floodwater alone would not explain the observed nitrate increases along parafluvial flowpaths. That is, if floodwater storage were responsible for elevated subsurface nitrate-N, we would expect nitrate-N in the hyporheic and parafluvial zones to be high and spatially constant immediately following a spate, and then to gradually decrease following flushing with lower nitrate-N surface water.

Results of the present study support the hypothesis that nitrification within parafluvial sediments in Sycamore Creek is responsible for elevated nitrate-N concentration in this zone. The period of highest net nitrification rate (summer) corresponded with the time of greatest nitrate increase along parafluvial flowpaths (Fig. 4). Although higher summer temperature might increase rates of microbially mediated reactions including nitrification (Focht and Verstraete 1977), temperature alone cannot explain the observed seasonal differences in net nitrification rate because all lab nitrification analyses were done at the same temperature. Therefore, nitrifier populations were either larger or more active in summer than in other seasons.

In summer, the highest net nitrification rates occurred at the stream–parafluvial interface, suggesting that nitrification in subsurface sediments is closely linked to surface-derived sources (ammonium, dissolved organic nitrogen [DON], or particulate organic nitrogen [PON]). Ammonium injection into surface water demonstrated that nitrate production (nitrification) was limited by ammonium supply (Fig. 9). Because surface water ammonium concentration is low (Table 1), it appears that surface-derived DON or PON mineralized in parafluvial sediments provides the ammonium substrate for nitrification. Algal production and standing crops are normally highest in the summer (Grimm and Fisher 1986a). Thus, algal-derived DOM or POM are potential sources of organic matter for fueling parafluvial metabolism. This creates the potential for a positive feedback loop (or spiral; sensu Newbold et al. 1981) between surface primary production, which is nitrogen-limited, and nitrification along parafluvial and hyporheic flowpaths. That is, algal leachates may fuel nitrification in subsurface sediments, leading to elevated nitrate-N in subsurface waters, which later re-enters the surface stream and stimulates primary production.

Ecosystem implications

Previous research has shown the importance of surface-hyporheic hydrologic exchange to the structure and functioning of stream ecosystems (Grimm and Fisher 1984, Valett et al. 1994). Our study demonstrates the significance of the parafluvial zone as an additional source of nitrate to a nitrogen-limited stream ecosystem. Casual observation suggested that algal biomass was higher in areas where parafluvial water entered the surface stream, as in upwelling zones where nitrate-rich hyporheic water enters the surface (Valett et al. 1994). As with hyporheic-surface exchange, parafluvial nitrate inputs might be expected to contribute to ecosystem resilience following disturbance (sensu Webster et al. 1983).

The relative importance of hyporheic and parafluvial subsystems to whole-stream ecosystem functioning depends upon channel configuration, rates of supply and distribution of sand and gravel substrates, hydraulic conductivity of bed materials, and channel geomorphology which generates the hydraulic heads that determine the direction and velocity of flowpaths. In wide, sandy streams of the desert Southwest, these attributes tend to accentuate the significance of hyporheic and parafluvial subsystems. Similar studies in streams in redwood forests of California (Triska et al. 1989, Triska et al. 1993), the Oak Ridges moraine of southern Ontario (Hill 1990), New Mexico highlands (C. Dahm, University of New Mexico, personal communication), and southwest France (Pinay et al. 1994) suggest that these processes are widespread. A comprehensive understanding of stream ecology will increasingly require a broadened view of what constitutes the stream ecosystem.

Acknowledgements

We thank Bryan Harper, Eugènia Martí, Tegan Blaine, John Schade, Janey Freeman, Sandra Clinton, Nicole Drake, and Amy Wiebel for assistance in the field and in the lab, and Jay Jones for comments throughout the project. We also thank John Alcock, James Elser, Rosemary Mackay, Alan Steinman, and two anonymous reviewers for constructive comments on the manuscript. This research was supported by National Science Foundation grants BSR-8818612, DEB-9108362, and DEB-9306909.

Literature Cited

COOPER, B. A. 1990. Nitrate depletion in the riparian zone and stream channel of a small headwater catchment. Hydrobiologia 202:13–26.

DUFF, J. H., AND F. J. TRISKA. 1990. Denitrification in sediments from the hyporheic zone adjacent to a small forested stream. Canadian Journal of Fisheries and Aquatic Sciences 47:1140–1147.

FIEBIG, D. M., AND M. A. LOCK. 1991. Immobilization of dissolved organic matter from groundwater discharging through the stream bed. Freshwater Biology 26:45–55.

FISHER, S. G., AND G. E. LIKENS. 1973. Energy flow in Bear Brook, New Hampshire: an integrative approach to stream ecosystem analysis. Ecological Monographs 43:421–439.

FOCHT, D. D., AND W. VERSTRAETE. 1977. Biochemical ecology of nitrification and denitrification. Advances in Microbial Ecology 1:135–214.

FORD, T. E., AND R. J. NAIMAN. 1989. Groundwater-surface water relationships in boreal forest watersheds: dissolved organic carbon and inorganic nutrient dynamics. Canadian Journal of Fisheries and Aquatic Sciences 46:41–49.

GRIMM, N. B., AND S. G. FISHER. 1984. Exchange between surface and interstitial water: implications for stream metabolism and nutrient cycling. Hydrobiologia 111:219–228.

GRIMM, N. B., AND S. G. FISHER. 1986a. Nitrogen limitation in a Sonoran Desert stream. Journal of the North American Benthological Society 5:2–15.

GRIMM, N. B., AND S. G. FISHER. 1986b. Nitrogen limitation potential of Arizona streams and rivers. Journal of the Arizona-Nevada Academy of Science 21:31–43.

GRIMM, N. B., H. M. VALETT, E. H. STANLEY, AND S. G. FISHER. 1991. Contribution of the hyporheic zone to stability of an arid-land stream. Verhandlungen der Internationalen Vereinigung für Theoretische und Angewandte Limnologie 24: 1595–1599.

HENDRICKS, S. P., AND D. S. WHITE. 1991. Physicochemical patterns within a hyporheic zone of a northern Michigan river, with comments on surface water patterns. Canadian Journal of Fisheries and Aquatic Sciences 48:1645–1654.

HILL, A. R. 1990. Ground water flow paths in relation to nitrogen chemistry in the near-stream zone. Hydrobiologia 206:39–52.

HOWARTH, R. W., R. MARINO, J. LANE, AND J. COLE. 1988. Nitrogen fixation in freshwater, estuarine, and marine ecosystems. I. Rates and importance. Limnology and Oceanography 33:688–701.

HYNES, H. B. N. 1975. The stream and its valley. Verhandlungen der Internationalen Vereinigung für Theoretische und Angewandte Limnologie 19:1–15.

HYNES, H. B. N. 1983. Groundwater and stream ecology. Hydrobiologia 100:93–99.

LEE, D. R., AND J. A. CHERRY. 1978. A field exercise on groundwater flow using seepage meters and mini-piezometers. Journal of Geological Education 27:6–10.

LIKENS, G. E., AND F. H. BORMANN. 1974. Linkages between terrestrial and aquatic ecosystems. BioScience 24:447–456.

LOWRANCE, R. R., R A. LEONARD, AND L. E. ASMUSSEN. 1985. Nutrient budgets for agricultural watersheds in the southeastern coastal plain. Ecology 66:287–296.

NEWBOLD, J. D., J. W. ELWOOD, R. V. O'NEILL, AND W. VAN WINKLE. 1981. Measuring nutrient spiralling in streams. Canadian Journal of Fisheries and Aquatic Sciences 38:860–863.

PAERL, H. W. 1990. Physiological ecology and regulation of N_2 fixation in natural waters. Advances in Microbial Ecology 11:305–344.

PETERJOHN, W. T., AND D. L. CORRELL. 1984. Nutrient dynamics in an agricultural watershed: observations on the role of a riparian forest. Ecology 65:1466–1475.

PINAY, G., N. E. HAYCOCK, C. RUFFINONI, AND R. M. HOLMES. 1994. The role of denitrification in nitrogen retention in river corridors. Pages 107–116 *in* W. J. Mitsch (editor). Global wetlands: Old World and New. Elsevier, Amsterdam.

PINAY, G., L. ROQUES, AND A. FABRE. 1993. Spatial and temporal patterns of denitrification in a riparian forest. Journal of Applied Ecology 30:581–591.

SOLORZANO, L. 1969. Determination of ammonia in natural waters by the phenolhypochlorite method. Limnology and Oceanography 14:799–801.

SYSTAT, INC. 1992. SYSTAT for windows: statistics. Version 5 Edition. Evanston, Illinois.

TRISKA, F. J., J. H. DUFF, AND R. J. AVANZINO. 1993. Patterns of hydrological exchange and nutrient transformation in the hyporheic zone of a gravel-bottom stream: examining terrestrial-aquatic linkages. Freshwater Biology 29:259–274.

TRISKA, F. J., J. H. DUFF, AND R. J. AVANZINO. 1990. Influence of exchange flow between the channel and hyporheic zone on nitrate production in a small mountain stream. Canadian Journal of Fisheries and Aquatic Sciences 47:2099–2111.

TRISKA, F. J., V. C. KENNEDY, R. J. AVANZINO, G. W. ZELLWEGER, AND K. E. BENCALA. 1989. Retention and transport of nutrients in a third-order stream in northwestern California: hyporheic processes. Ecology 70:1893–1905.

VALETT, H. M., S. G. FISHER, N. B. GRIMM, AND P. CAMILL. 1994. Vertical hydrologic exchange and ecological stability of a desert stream. Ecology 75: 548–560.

VALETT, H. M., S. G. FISHER, AND E. H. STANLEY. 1990. Physical and chemical characteristics of the hyporheic zone of a Sonoran Desert stream. Journal of the North American Benthological Society 9:201–215.

WARD, J. V. 1989. The four-dimensional nature of lotic ecosystems. Journal of the North American Benthological Society 8:2–8.

WEBSTER, J. R., M. E. GURTZ, J. J. HAINS, J. L. MEYER, W. T. SWANK, J. B. WAIDE, AND J. B. WALLACE. 1983. Stability of stream ecosystems. Pages 355–395 *in* J. R. Barnes and G. W. Minshall (editors). Stream ecology. Plenum Press, New York.

WERTZ, J. B. 1963. Mechanisms of erosion and deposition along channelways. Journal of the Arizona Academy of Science 2:146–163.

Received: 26 April 1994
Accepted: 22 August 1994

Commentary on Paper F7

Valett, H. M., Morrice, J. A., Dahm, C. N. & Campana, M. E. (1996) Parent lithology, surface–groundwater exchange, and nitrate retention in headwater streams.

To scale up results about nutrient dynamics which are obtained from a study site to the whole catchment requires careful consideration of hydrological, biological and chemical processes. The rationale for this study was that the variable conditions and processes occurring at the hyporheos interface had a hydrological basis which was related to the parent/bedrock geology of the basin.

The study departed from the earlier approaches on nutrient spiralling (Newbold *et al.*, 1981) and from the recognition of the relevance of the hyporheic zone as an active area of biological activity (Triska *et al.*, 1989). Processes which could be detectable at the watershed scale were to be related to the geology of the basin, in particular to characteristics such as the hydraulic conductivity and hydraulic gradient. These factors would define the extent and infiltration capacity of the hyporheic zone. Moreover, they would be essential in defining the extent of the transient storage zones in rivers, and as such influence the overall nutrient retention dynamics. Munn & Meyer (1990) and later Martí & Sabater (1996) identified that the parent geology was related to the nutrient uptake of the systems. In particular, Munn & Meyer (1990) found that the ratio of N to P largely determined the uptake of one or other nutrient: streams with a lower N:P ratio (volcanic parent geology) would have a higher uptake rate of N than P, while those of lower P availability (granitic parent geology) would show higher uptake rate of P than N. In other words, limitation of a given nutrient would increase its uptake rate.

Valett and colleagues determined that catchments of low hydraulic conductivity were prone to the depletion of dissolved oxygen in the hyporheos. This was related to the longer water residence time characteristic of aquitards. Hence, the nominal water travel time and the interstitial dissolved oxygen closely matched in the hyporheos. This finding indicated that the subsurface residence time was the primary factor responsible for oxygen availability and changes of redox state along the groundwater flow paths. This could explain the distribution of oxidized and reduced chemical forms of nitrogen; steeper redox gradients at the stream surface would exist in systems with lower hydraulic conductivity.

The paper clearly showed that nutrient retention in baseflow conditions was closely correlated to the width of the transient storage of these systems, and that nutrient ratios were only secondary factors in the nutrient retention occurring in the stream.

SS

References

Martí, E. & Sabater, F. (1996) High variability in temporal and spatial nutrient retention in Mediterranean streams. *Ecology* **77**, 854–869.

Munn, N. L. & Meyer, J. L. (1990) Habitat-specific solute retention in two small streams: an intersite comparison. *Ecology* **71**, 2069–2082.

Newbold, J. D., Elwood, J. W, O'Neill, R. V. & van Winkle, W. (1981) Measuring nutrient spiralling in streams. *Canadian J. Fisheries & Aquatic Sci.* **38**, 860–863.

Triska, F. J., Kennedy, V. C., Avanzino, R. J., Zwellweger, G. W. & Bencala, K. E. (1989) Retention and transport of nutrients in a third-order stream: hyporheic processes. *Ecology* **70**, 1892–1905.

Limnol. Oceanogr., 41(2), 1996, 333–345

Parent lithology, surface–groundwater exchange, and nitrate retention in headwater streams

H. Maurice Valett, John A. Morrice, and Clifford N. Dahm
Department of Biology, University of New Mexico, Albuquerque 87131

Michael E. Campana
Department of Earth and Planetary Sciences, University of New Mexico

Abstract

We address the ecological ramifications of variation in hydrologic interaction between streams and alluvial aquifers in catchments with alluvium derived from parent materials of contrasting geologic composition. We present a conceptual model in which solute retention in streams results from hydrologic retention (increased water residence time resulting from surface–groundwater exchange), biological nutrient cycling, and chemical processes. Solute injection experiments were done in study catchments comprised of sandstone-siltstone (site 1), volcanic tuff (site 2), and granite-gneiss (site 3). Distribution of an injected conservative tracer (Br) illustrated that rate and extent of surface-water penetration into the alluvial aquifer increased across study catchments as was predicted from increasing alluvial hydraulic conductivity. Concurrently, groundwater inputs at baseflow represented between 13 and 57% of aboveground discharge at upstream transects, indicating bidirectional hydrologic exchange along the study reaches. N : P ratios in surface water ranged from 4 to 16, suggesting strong biotic demand for inorganic N. Coinjection of NaBr and $NaNO_3$ revealed longest nitrate uptake length (S_W) at site 1, intermediate S_W at site 2, and shortest uptake length at site 3. Modeling of transient hydrologic solute storage revealed that S_W correlated with hydraulic storage, suggesting an important role for subsurface processes in total nitrate retention.

Retention of nutrients and organic matter is an insightful measure of lotic ecosystem functioning given the downstream transport characteristic of streams and rivers. The competing processes of hydrologic export and biological retention are expressed in nutrient spiraling concepts (Webster and Patten 1979; Elwood et al. 1983; Newbold et al. 1983) in which downstream transport adds longitudinal displacement to nutrient cycles so that they act as nutrient spirals. Studies of nutrient spiraling have used uptake length as a measure of nutrient retention; nutrient uptake length is defined as the average distance traveled by a nutrient molecule in dissolved form before it is incorporated into biomass (Elwood et al. 1983).

Benthic and hyporheic (i.e. groundwater zones that receive water from the aboveground channel, sensu Orghidan 1959; Triska et al. 1989*a*) regions of streams are linked by groundwater upwelling (discharge) and downwelling (recharge) that result from streambed geometry (Vaux 1968; White et al. 1987) and hillslope–channel water interaction (Harvey and Bencala 1993). Because subsurface sediments are metabolically active, hyporheic zones may be sinks or sources of nutrients, depending on the relative magnitude of processes generating or immobilizing biologically important solutes (Triska et al. 1989*b*; McDowell et al. 1992). The mixing of surface and hyporheic water integrates the biogeochemistry of these zones and generates nutrient fluxes that may influence processes in both subsystems (*see* Gibert et al. 1990; Valett et al. 1994). In this study, we combine the nutrient spiraling approach with hydrologic and biogeochemical analyses of the hyporheic zone to investigate the role of surface–subsurface interaction in nutrient retention.

The size of the hyporheic zone and the intensity of interaction between surface water and groundwater vary greatly among lotic systems. In a coastal piedmont stream, the hyporheic zone was restricted to 30–40-cm depth by essentially impermeable clay lenses at greater depths (Smock et al. 1992). In sandy-bottomed arid-land streams, hyporheic water volume was as much as 320 times that in the surface stream (Valett et al. 1990). In Little Lost Man Creek, California, 25% of the total discharge during baseflow was within the interstitial spaces of the hyporheic zone (Zellweger et al. 1989). Similarly, Castro and Hornberger (1991) calculated 47% of total catchment discharge was interstitial in a low-order mountain stream channel. These interactions occur farther down the drainage as well, and Stanford and Ward (1993) described large-scale (km) integration of rivers and their floodplains via hydrologic exchange through the hyporheic zone.

At a basin scale, the hydrologic nature of surface–groundwater exchange can be predicted by parent lithology of the stream catchments (Kelson and Wells 1989). Features such as discharge per unit area, stream power,

Acknowledgments

We thank the numerous individuals who contributed to the success of the field, laboratory, and analytical efforts that culminated in this manuscript, including M. Baker, G. Wroblicky, K. Henry, F. Lieurance, T. McDonald, J. Curro, M. Santistevan, J. N. Morrice, B. Horton, P. Suina, C. White, J. Craig, S. Hofstadt, G. Shore, and B. Kaehele. Special thanks to P. Valero and A. Bearce for access and care of sites in the Jemez and Sangre de Cristo mountains.

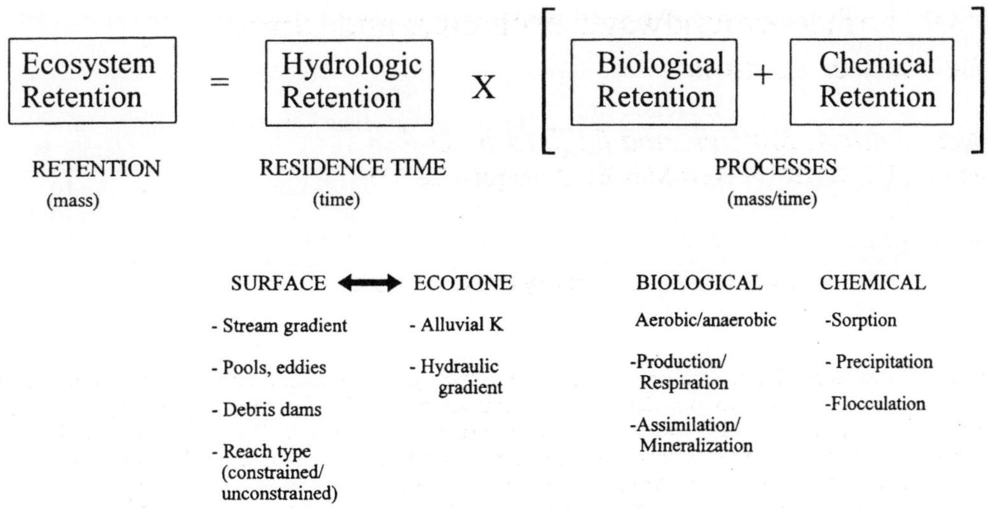

Fig. 1. A conceptual model for nutrient retention in stream ecosystems. Hydrologic retention results from structural features of the surface that retard flow and from the interaction of surface water and groundwater that increases water residence time. Nutrient cycling and metabolic processes promote biological retention, and chemical processes may be important retentive mechanisms for certain solutes and conditions. Total ecosystem retention (mass) is presented as the product of processing rate (mass/time) and water residence time (time).

and sediment transport vary predictably among catchments comprised of crystalline rocks (i.e. granite-gneiss) compared to sediments of weathered sedimentary materials (Kelson and Wells 1989). Just as catchment lithology influences the availability of biologically meaningful nutrients in surface water (Kruger and Waters 1983; Pringle and Triska 1991), the variable nature of conditions and processes in hyporheic zones may result from hydrologic fluxes that also differ according to catchment lithology.

Nutrient retention in lotic ecosystems results from the interaction of hydrologic, chemical, and biological properties (Fig. 1). In our model, hydrologic retention occurs when water enters flow paths moving slower than the advective velocity of the main channel, resulting in increased water residence times. In subsurface regions, residence times depend primarily on the hydraulic gradient and alluvial hydraulic conductivity (K). Steeper hydraulic gradients and higher K will promote stream–groundwater exchange that results in hydrologic retention and increased residence time for water. Biotic components of stream ecosystems influence retention by generating, immobilizing, transforming, or removing biologically active solutes. Comparably, the processes of sorption, flocculation, and precipitation result in chemical retention by retarding solute transport. All dissolved solutes are influenced by the hydraulic retention of water. Depending on the solute of interest, chemical and biological processes may be minor or major determinants of solute retention. Total ecosystem retention is the product of process rates (chemical and biological) and water residence time (Fig. 1).

We focus on hydrologic and biological influences on the transport and total retention of NO_3-N in three stream ecosystems in which interaction of surface and groundwater were expected to differ. Our central hypothesis is that surface–groundwater interaction is a strong determinant of nutrient retention in lotic ecosystems. Based on the conceptual model detailed above, we predict that stream-aquifer systems with higher alluvial K will have a greater extent of surface–hyporheic interaction and will retain more biologically important solutes. To test these predictions, we used long-term coinjections of NaBr and $NaNO_3$ in three catchments with differing parent lithology and alluvial K. Hydrologic interaction was assessed by quantifying tracer distributions and mathematically modeling solute transport. Biological retention was measured as nitrate uptake length (sensu Newbold et al. 1983) and compared to measures of stream–aquifer exchange.

Study sites

Surface water–hyporheic interaction and NO_3-N utilization were investigated in three perennial first-order streams located in western and north-central New Mexico (Table 1). Surface-water atomic N : P ratios indicated potential N limitation of primary productivity at each site (Gosz 1975; Coleman and Dahm 1990). Study streams were chosen to provide contrasts among fluvial systems in catchments with alluvial sediments derived from materials with different parent lithology. Replicate basin types were not established and statistical inferences among catchments do not attempt to assess the influence of parent lithology (*see* Hurlbert 1984). Spatial replication at the ecosystem level is often intractable (Carpenter 1989), but comparative efforts have revealed important aspects of ecosystem structure and functioning (Likens et al. 1970; Schindler 1974); thus we used a comparative approach for our research.

Table 1. Study catchment characteristics for three headwater streams and their associated alluvial aquifers.

Site		Elevation (m) Low	Elevation (m) High	Drainage area (ha)	Stream gradient (%)	Parent lithology	*K* (cm s^{-1})
1	Aspen Creek	2,377	2,718	322	2.0	Sandstone	1.3×10^{-4}
2	Rio Calaveras	2,475	2,926	3,760	1.3	Welded tuff	1.2×10^{-3}
3	Gallina Creek	2,524	3,586	618	11.1	Granite-gneiss	4.1×10^{-3}

Site 1, Aspen Creek, is in the Zuni Mountains where alluvial material is derived from Permian sandstone of the Yeso Formation and has the lowest alluvial hydraulic conductivity (Table 1). Alluvium at site 2, Rio Calaveras, is derived from Pleistocene volcanic tuff generated from cataclysmic eruption of the Valle Caldera in the Jemez Mountains; hydraulic conductivity of sediments at Rio Calaveras is intermediate (Table 1). Site 3, Gallina Creek in the Sangre de Cristo Mountains, consists of alluvium that is mainly poorly sorted granitic cobbles and boulders with high hydraulic conductivity (Table 1). More detailed information on alluvial properties is given by Wroblicky et al. (1992).

Methods

Wells and piezometers—In 1991 and 1992, study reaches were established in each catchment along segments of the streams lacking tributary inputs (Fig. 2). Lengths of study reaches were 175 m at site 1, 110 m at site 2, and 190 m at site 3. At sites 1 and 2, five transects were established that divided study reaches into four subreaches of roughly equal lengths (site 1, 37 m; site 2, 25 m). Four transects were established at site 3, creating three subreaches ~50 m long (Fig. 2). Discharge at the head of each study reach was measured with 5-cm cutthroat flumes installed 35 m upstream from the most-upstream transect at each site.

At sites 1 and 2, transects of piezometers and sampling wells were perpendicular to the stream at nearly equal intervals along the study reach. Each sampling transect consisted of a single surface-water sampling site and five wells. Wells were established 1 and 3 m to the left and right of the stream, and a single well was placed beneath the wetted perimeter (Fig. 2). At site 2, an additional sampling well was added to the upstream transect. For clarity, groundwater wells are identified sequentially from left to right by a two-digit number; the first digit indicates the transect number and the second digit refers to the well's position within a transect. At site 3, boulder-size alluvium prevented installation of wells at evenly spaced intervals or beneath flowing water, so 13 wells were placed along four transects at distances of 1–4 m from the active channel (Fig. 2).

At all sites, sampling wells consisted of 5-cm PVC casings with 50-cm slotted screens (slot size, 254 μm) inserted 50–75 cm beneath the water table by means of either a 10-cm diameter power auger or hand digging. At site 3, slotted screens were 20 cm long. Annular spaces were backfilled first with silica sand and then with native sediments. Bentonite seals were placed 0–20 cm beneath the ground surface to prevent infiltration along the well casing.

Solute injections—Conservative (NaBr) and nonconservative ($NaNO_3$) solutes were coinjected between June 1992 and August 1993. One long-term injection experiment was done under summer or fall baseflow conditions in each catchment. Solutes were introduced to surface water at a steady rate by a metering pump (Fluid Metering Inc.) that was calibrated frequently during injections. Background samples for Br and NO_3-N were collected from each surface transect and all monitoring wells before injection.

At site 1, solutes were added for 10 d in September 1992, and surface-water and groundwater chemistry was monitored throughout tracer addition and for 10 d after injection ended. At site 2, solutes were added for 10 d in August 1992, and concentrations were monitored for an additional 3 d thereafter. Solutes were added to site 3 in October 1992 for a total of 7 d, and water chemistry monitoring continued for 7 d after solute addition stopped.

During injections, Br concentrations were increased

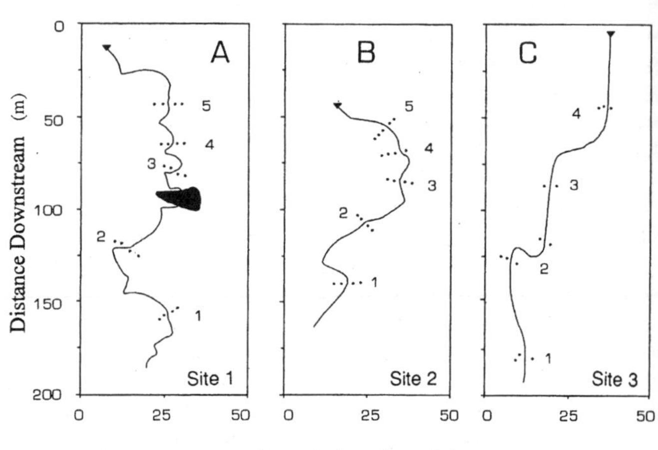

Fig. 2. Coinjection study sites with spatial distribution of sampling wells (•) along the numbered transects and the cutthroat flume (▼) established upstream of the final transect at each site. A. Aspen Creek, sandstone siltstone. B. Rio Calaveras, welded tuff. C. Gallina Creek, granite-gneiss.

from ~0.04 to 7 mg liter^{-1}. Background NO_3-N concentrations ranged from 10 to 120 ppb. Injected nitrate generated a 6.4-fold increase in surface NO_3-N at the upstream transect at site 1, a 6.8-fold increase at site 2, and a 24-fold increase at site 3. Surface-water samples were taken frequently during the initial phases of injection and during injection termination to characterize solute transport and quantify hydraulic storage (*see below*). Well and surface samples were taken daily during the remainder of the experiment.

Three additional injections were performed at site 3 to assess biological and hydrologic retention over the annual hydrologic regime. In addition to the baseflow injection in fall 1992, injections of 1, 1, and 7 d were done in winter (February 1993), spring (June 1993), and summer (August 1993). During these injections, plateau NO_3-N at the upstream transect was between 9 and 32 times background concentration.

Conservative solute analysis—Temporal and spatial patterns of the conservative tracer were analyzed to quantify the rate and extent of surface water infiltration and water movement in the alluvial aquifer. Data analyses followed those described by Triska et al. (1989*a,b*) and detailed by the Stream Solute Workshop (1990). Tracer data were analyzed graphically, and plateau concentrations (C_P) were determined as means for all values occurring during plateau periods. Plateau concentrations of the conservative tracer at sequential downstream surface transects were compared with concentrations at transects upstream to quantify the lateral inflow (Q_L) of unlabeled water (i.e. tracer dilution) along the subreach bounded by the designated transects. The C_P of Br in well water was compared to those at adjacent surface transects to determine percent surface water found in the well. Wells with <10% surface water were designated phreatic (sensu Danielopol 1980). Connected wells contained >10% surface water and were considered part of the hyporheic zone (sensu Triska et al. 1989*b*). Highly connected wells contained >50% surface water. Rate of tracer movement in surface and groundwater environments was represented by the nominal travel time (T_n) calculated as the time required for tracer to reach half of its eventual plateau concentration (i.e. $C_P/2$).

A hydrologic model of one-dimensional transport with terms for inflow and storage (OTIS; Runkel and Broshears 1991) was used to quantify properties of solute transport during these injections. The model is physically based and quantifies the bulk downstream flow (i.e. advection), dispersion, lateral inflow, and transient storage of water by simulating the temporal response of a conservative tracer at fixed points along the stream reach (Bencala et al. 1984; Runkel and Broshears 1991). Model parameters are manipulated iteratively until the model-generated concentration curve provides the best-fit to empirical tracer data as determined by visual inspection (sensu Bencala et al. 1990). Parameter values are then used to compare physical processes of solute transport. The transient storage term (A_S in m^2) in the model accounts for regions of the fluvial environment in which the tracer moves at rates much slower than the average advective velocity of the surface stream. A_S can be conceived of as the reach-integrated average cross-sectional area of a fluid storage zone (Stream Solute Workshop 1990) and is considered a good measure of hydrologic retention resulting from surface–groundwater interactions (Bencala et al. 1993). The magnitude of transient storage can be compared among streams or across changing hydrologic conditions by normalizing A_S to the average cross-sectional areas of the aboveground stream (A_S/A). In this paper, we examine measures of biological NO_3-N utilization in light of changing transient storage.

Dynamic solute analysis—Dissolved substances used by organisms or otherwise changed biotically or chemically are called nonconservative, reactive, or dynamic solutes (Stream Solute Workshop 1990). Biological influences on the retention of inorganic N were determined by calculating NO_3-N uptake lengths (S_W, length) for independent subreaches (during baseflow injections) and for the entire study reach. Uptake lengths represent the average distance traveled by a solute in the dissolved form (sensu Newbold et al. 1983) and were determined by analyzing Br and NO_3-N plateau concentrations at successive points along the stream. Once corrected for dilution (as described above for conservative solutes), ln-transformed plateau concentrations of NO_3-N can be fitted as a straight line against distance downstream. The slope of the line is the negative inverse of uptake length (Stream Solute Workshop 1990). Furthermore, the time spent in transport as a dissolved solute (i.e. turnover time, T_W) is obtained as uptake length normalized to water velocity (u, length/time) such that $T_W = S_W/u$ (Stream Solute Workshop 1990).

Interstitial NO_3-N utilization in connected wells was determined according to Triska et al. (1989*b*) where predicted values of NO_3-N in a well at time T were calculated as

$$NO_3\text{-}N_{w,T} = NO_3\text{-}N_{w,bkg} + \left[(Br_{w,T} - Br_{bkg}) \times \frac{NO_3\text{-}N_{s,T-T_n} - NO_3\text{-}N_{s,bkg}}{Br_{s,T-T_n} - Br_{s,bkg}}\right]. \quad (1)$$

Subscripts are: w—well; s—surface; bkg—background concentration; T—time; and T_n—nominal travel time. Equation 1 assumes that NO_3-N acts conservatively and that labeled water in a connected well at time T was present on the surface at $T - T_n$, reflecting the T_n required for surface water to reach the monitoring well. Deviations of observed NO_3-N from values predicted by Eq. 1 indicate production (observed > predicted) or consumption (observed < predicted) of the solute along flow paths linking surface and hyporheic environments. For all connected wells (i.e. >10% surface water), differences in observed and predicted NO_3-N were tested statistically by a nonparametric Mann-Whitney U-test for paired observations. Relative NO_3-N utilization ([observed/predict-

Table 2. Hydrologic connectivity between streams and alluvial aquifers of catchments of different parent lithologies and the surface-water content (%) and nominal travel time (T_n, h) for hyporheic wells.

Site 1			Site 2			Site 3		
Well	T_n	% surf	Well	T_n	% surf	Well	T_n	% surf
13	67.9	29.6	11	112.2	14.5	11	50	40.4
23	28.3	97.2	12	63.5	62.3	12	33.1	85.4
24	142.1	13.9	13	2.2	96.8	13	38.7	77.5
32	199.6	14.9	14	138.5	44.0	21	31.9	91.0
33	22.9	94.7	23	1.9	100.0	22	53	43.8
34	217.1	29.6	24	133.6	46.4	23	73.8	35.5
55	181.9	49.2	25	88.2	100.0	24	16.4	100.0
			45	176.0	12.5	25	64.1	59.4
						31	21.7	97.6
						32	19.3	18.1
						41	44.9	44.9
						42	28.5	85.4
						43	37.6	67.6

ed] × 100%) was compared to percent surface water, dissolved oxygen (DO) content, and T_n for all wells by calculating Pearson correlation coefficients between these variables after proper transformation (*see below*; Sokal and Rohlf 1981).

Concentrations of biogeochemically active solutes were compared among phreatic, hyporheic, and surface environments by Kruskal-Wallis nonparametric ANOVA followed by the Student-Newman-Keuls multiple comparison test (Day and Quinn 1989). Atomic N : P ratios were square-root transformed, and DO concentrations expressed as percent saturation were transformed (arcsin $x^{1/2}$) before analyses. Across sites, relationships between hydrologic properties and interstitial DO were assessed by calculating Pearson correlation coefficients for interstitial DO and T_n or interstitial DO and percent surface water when data from all sites were combined.

Spatial and temporal relationships between hydrologic and biological retention were analyzed by calculating Pearson correlation coefficients relating Q, u, S_W, A_S, A_S/A, and T_W for all pairs of variables that were not autocorrelated. Values for A_S, A_S/A, and S_W were ln-transformed for analyses.

Analytical techniques—Water temperature and DO were measured with a DO probe (YSI Inc.) or determined from modified Winkler titrations (Wetzel and Likens 1991). Chemical analyses of chloride (Cl), bromide (Br), NO_3-N, and SO_4-S were carried out on an ion chromatograph (Dionex Dx-100) equipped with a conductivity detector for anion analysis. Soluble reactive P (SRP) and ammonium N (NH_4-N) were analyzed on a Technicon AutoAnalyzer following the techniques of Murphy and Riley (1962) and Soloranzo (1969). Total inorganic N (TIN) was determined as the sum of NO_3-N and NH_4-N. Dissolved organic C (DOC) was measured by wet persulfate digestion (Menzel and Vaccaro 1964) on an Oceanographic International model 700 total carbon analyzer.

Results

Surface water–groundwater interaction—The number of connected wells increased from seven at site 1 to eight at site 2 and 13 at site 3 (Table 2) corresponding to 27, 31, and 100% of the number of monitoring wells at each site. Under baseflow conditions, there was decreased variation in and increased percentage of surface water in hyporheic wells across the three study basins progressing from site 1 and site 3 (Table 2). T_n indicates that the rate of exchange between surface water and groundwater increased across sites as well (Table 2).

Three of the connected wells at site 1 (sedimentary basin) were located beneath the wetted perimeter, and two of these wells were highly connected (i.e. contained >50% surface water; Table 2). Surface water content of hyporheic wells averaged 47.0±13.4% ($\bar{x}$ ± SE) and was highly variable (C.V. = 74%). Average T_n for wells at site 1 was 5.1±1.3 d ($\bar{x}$ ± SE).

Four of the connected wells at site 2 (welded tuff basin) contained >50% surface water, but only two of them were beneath the wetted perimeter (Table 2). Despite being more than 3 m from the wetted perimeter, well 25 contained 100% surface water, and an adjacent well closer to the stream (well 24) contained 46.4% surface water (Table 2). Average percent surface water in hyporheic wells was greater (59.6±12.9%, $\bar{x}$ ± SE) and less variable (C.V. = 61%) at site 2 than at site 1. Rate of interaction was increased at site 2, where average T_n for hyporheic wells was 3.7±0.9 d ($\bar{x}$ ± SE).

All monitoring wells at site 3 (granitic basin) were connected and the average percent surface water was highest (65.1±7.3%) and least variable (C.V. = 40%) at this site. Rates of surface water penetration and movement through the alluvial aquifer were also greatest at this site, where T_n was 1.6±0.2 d ($\bar{x}$ = SE).

Biogeochemistry—Groundwater temperature was significantly higher than surface-water temperature at sites

Table 3. Physical-chemical properties of surface (100% surface water), hyporheic (>10% surface water), and phreatic (<10% surface water) subsystems of three first-order alluvial aquifers. Values are means and *P*-values are for Kruskal-Wallis tests. Means in a row with identical superscripts are not significantly different (SNK, $\alpha = 0.05$).

	P	Surface	Hypo-rheic	Phreatic
Site 1				
n		5	7	18
NO_3-N (μg liter^{-1})	0.001	44^a	10^b	4^c
NH_4-N (μg liter^{-1})	0.018	34^a	55^{ab}	63^b
TIN (μg liter^{-1})	0.460	78^a	65^a	67^a
SRP (μg liter^{-1})	0.003	50^a	22^b	14^b
N:P	0.059	4^a	20^a	48^a
DOC (mg liter^{-1})	0.017	2.73^a	3.79^b	3.60^b
SO_4-S (mg liter^{-1})	0.011	8.50^a	7.25^a	6.06^a
DO (mg liter^{-1})	0.004	5.59^a	0.45^b	0.25^b
DO (% sat)	0.004	70.0^a	5.67^b	3.12^b
Temp. (°C)	0.019	8.97^a	10.7^b	10.5^b
Site 2				
n		5	8	18
NO_3-N (μg liter^{-1})	0.005	145^a	49^b	19^b
NH_4-N (μg liter^{-1})	0.121	41^a	116^a	118^a
TIN (μg liter^{-1})	0.201	186^a	167^a	137^a
SRP (μg liter$^{1-1}$)	0.004	20^a	28^a	20^a
N:P	0.593	16.6^a	12.1^a	13.8^a
DOC (mg liter^{-1})	0.045	2.06^a	1.58^a	1.27^a
SO_4-S (mg liter^{-1})	0.426	4.27^a	3.23^a	3.45^a
DO (mg liter^{-1})	0.002	6.53^a	0.95^b	0.32^b
DO (% sat)	0.002	87.6^a	12.5^b	4.2^b
Temp. (°C)	0.357	12.6^a	12.6^a	13.1^a
Site 3				
n		4	13	0
NO_3-N (μg liter^{-1})	0.165	5	15	—
NH_4-N (μg liter^{-1})	0.568	17	23	—
TIN (μg liter^{-1})	0.533	22	38	—
SRP (μg liter^{-1})	0.793	8	8	—
N:P	0.527	9.0	10.1	—
DOC (mg liter^{-1})	0.011	2.83	4.14	—
SO_4-S (mg liter^{-1})	0.112	121	180	—
DO (mg liter^{-1})	0.002	8.0	4.22	—
DO (% sat)	0.001	93.1	50.2	—
Temp. (°C)	0.001	6.8	7.7	—

1 and 3, where hyporheic water was 1.7 and 0.9°C higher than surface water (Table 3). At these sites, hyporheic and phreatic temperatures were not significantly different (Table 3). Furthermore, temperatures of surface and hyporheic water at site 2 were identical ($\bar{x}$ = 12.6°C), and there were no significant differences in temperature among subsystems.

Average surface-water concentrations of DO were 70.0% of saturation at site 1, 87.6% at site 2, and 93.1% at site 3 (Table 3). DO concentrations in the hyporheic zone were significantly lower than in the surface water for each stream, but there were no significant differences in DO between hyporheic and phreatic environments at sites 1 and 2. At site 1, DO averaged <0.5 ppm (<6% saturation) in the hyporheic zone and dropped to 0.25 ppm in phreatic wells. Well 13 at site 1 contained 29.6% surface water (Table 2), but was completely anoxic throughout the injection under baseflow conditions. At site 2, hyporheic DO was also low ($\bar{x}$ = 0.95 mg liter^{-1}) and percent saturation dropped to 4.2% in phreatic wells. By contrast, hyporheic DO at site 3 was >50% of saturation and averaged 4.22 mg liter^{-1}. Analyses on data combined across sites indicated that interstitial DO (% sat) was not significantly correlated with percent surface water ($r^2 = 0.10$, $P = 0.10$), but interstitial DO (% sat) was significantly correlated with T_n ($r^2 = 0.30$, $P = 0.002$).

There were no significant differences in TIN among surface, hyporheic, and phreatic environments of the three study sites, and TIN remained low for all locations (Table 3). Concentration of SRP was significantly higher in surface water at site 1 than in interstitial water at this site (Table 3). High SRP and low TIN concentrations resulted in an N : P of 4.0 in the surface water of site 1. Significantly lower SRP in groundwater raised the interstitial N : P at site 1, but differences in N : P among subsystems were not significant. At site 2, SRP concentrations were similar among surface, hyporheic, and phreatic water and were lower than at site 1. N : P ratios ranged from 12.1 to 16.6 and did not vary significantly among subsystems. Despite the lowest SRP concentrations in surface and hyporheic water at site 3 (Table 3), N : P ratios were 9.0 and 10.1 for surface and hyporheic water, illustrating the low TIN concentrations at this site (Table 3).

Despite the lack of differences in TIN among subsystems, there were distinct and significant differences in concentrations of inorganic N species in the subsystems at sites 1 and 2 (Table 3). At both sites, significantly higher NO_3-N was detected in aboveground stream water than in groundwater (Table 3). At sites 1 and 2 average NO_3-N decreased significantly across surface, hyporheic, and phreatic subsystems. NO_3-N concentrations in the surface and hyporheic zone at site 3 were very low and did not differ significantly. Concentrations of NH_4-N were always lowest in surface water and phreatic concentrations were significantly higher than surface concentrations at site 1 (Table 3). Interstitial NH_4-N concentrations were nearly three times surface concentrations at site 2, but mean values were not significantly different (Table 3). Surface and interstitial NH_4-N concentrations at site 3 were low and nearly identical (Table 3).

Although site 3 was the lowest in inorganic N and P, interstitial and surface waters at site 3 were enriched in DOC and SO_4-S compared to sites 1 and 2 (Table 3). SO_4-S concentrations were lowest in groundwater at site 2; surface concentrations were higher but not significantly greater than interstitial concentrations (Table 3). DOC concentrations were also lowest at site 2. The highest average DOC concentration at site 2 was in surface water (Table 3); DOC decreased in the hyporheic and phreatic zones, but differences among subsystems were not significant (Table 3). SO_4-S levels were moderately higher at site 1 than at site 2 (Table 3). Nonparametric ANOVA indicated a significant subsystem effect on SO_4-S at site 3 (Table 3), but multiple comparisons failed to show sig-

Table 4. NO_3-utilization in the hyporheic zone. Values are mean ± SE for well concentrations. *P*-values refer to results from Mann-Whitney *U*-tests of observed − predicted NO_3-N concentrations.

Well	*n*	Obs − pred (μg liter^{-1})	Obs/pred (%)	*P*
Site 1				
13	9	−48±10	19±6	0.004
23	11	−107±16	58±4	0.001
24	9	−17±4	39±7	0.004
32	6	−1±1	95±10	0.750
33	10	−262±44	8±1	0.002
34	7	−44±14	47±15	0.030
55	8	−11±22	63±26	0.540
Site 2				
11	5	−96±10	7±3	0.062
12	13	−388±46	4±1	0.002
13	21	−49±49	98±10	0.193
14	4	−297±59	4±2	0.125
23	17	−11±32	99±3	0.950
24	3	−345±75	4±3	0.250
25	7	−683±75	8±4	0.016
45	2	−103±18	16±8	0.500
Site 3				
11	9	−5±3	87±18	0.203
12	11	−6±4	99±21	0.193
13	9	−11±5	89±22	0.153
21	10	−7±4	99±14	0.110
22	7	−15±5	51±14	0.031
23	7	−11±4	57±15	0.047
24	13	−27±5	62±15	0.003
25	8	−20±6	51±16	0.031
31	12	−80±11	17±6	0.001
32	12	−7±1	72±4	0.001
41	11	−59±12	51±19	0.010
42	13	−94±19	80±39	0.004
43	9	−118±9	9±1	0.004

nificant differences among subsystems. Furthermore, DOC concentrations were higher at site 1 than at site 2. At site 1, DOC concentrations were significantly higher in groundwater than in surface water (Table 3). DOC concentrations at site 3 were similar to those at site 2, and groundwater DOC concentrations also were significantly higher than those in surface water at this site. Finally, we found no significant difference between hyporheic and surface water SO_4-S at site 3.

Hyporheic NO_3-N utilization—Comparing observed NO_3-N concentrations to those predicted by Eq. 1 showed that none of the hyporheic wells showed evidence of net nitrate production (Table 4). Five of the wells at site 1 had significant rates of NO_3-N consumption during the injection. The average NO_3-N consumed was 53% of the available amount across all seven wells at site 1 and was 65% for the five wells with significant differences between observed and predicted concentrations (Table 4). Only two of the connected wells at site 2 displayed significantly decreased NO_3-N levels, but these two wells consumed nearly 94% of injection-derived nitrate (Table 4). Additionally, average NO_3-N concentrations were only 44% of predicted totals in five wells where differences were not significant and sample sizes were small (e.g. n = 2–7; Table 4) due to long T_n (*see* Eq. 1, Table 2). Observed NO_3-N was significantly lower than predicted in nine of the hyporheic wells at site 3 (Table 4). Predicted and observed NO_3-N concentrations were very similar in the site 3 wells lacking significant consumption (Table 4). In these four wells, observed values were just 7 ppb less than predicted and only 6.5% of available nitrate was consumed. Conversely, the other nine wells consumed an average of 50% of injected NO_3-N.

Table 5. Lateral inflow (Q_L, liters s^{-1} and percent of discharge at head of subreach) and NO_3-N uptake lengths (S_W, m for study reaches and established subreaches under baseflow conditions. Surface discharge (liters s^{-1}) at the head of study reaches was 1.1 at site 1, 2.0 at site 2, and 0.75 at site 3.

Subreach	Transect	Q_L liters s^{-1}	Q_L %	S_W
Site 1				
4	5-4	0.20	18	1,408
3	4-3	0.00	0	1,727
2	3-2	0.02	1	530
1	2-1	0.17	13	−2,545
	5-1	0.39	36	1,178
Site 2				
4	5-4	0.02	1	247
3	4-3	0.04	2	227
2	3-2	0.04	2	−271
1	2-1	0.17	8	913
	5-1	0.27	13	782
Site 3				
3	4-3	0.21	27	110
2	3-2	0.16	17	130
1	2-1	0.06	5	126
	4-1	0.43	57	133

Consumption of NO_3-N, represented by average percent of predicted concentration for individual wells was significantly correlated to T_n ($r = 0.55$, $n = 27$, $P = 0.003$). Similarly, NO_3-N consumption was correlated to interstitial DO ($r = 0.56$, $n = 27$, $P = 0.002$) and moderately related to percent surface water ($r = 0.35$, $n = 27$, $P = 0.061$).

Baseflow groundwater discharge and NO_3-N uptake lengths—Uptake lengths calculated for entire study reaches under baseflow conditions decreased from a maximum of 1.178 m at site 1 to 782 m at site 2 to 133 m at site 3 (Table 5). Analysis of conservative tracer concentrations indicates that lateral inflow represented 36% of discharge at the head of the study reaches at site 1, 13% at site 2, and 57% at site 3 (Table 5).

Magnitude of lateral inflow varied greatly among subreaches within a site, as did calculated NO_3-N uptake lengths (Table 5). Groundwater inputs at site 1 occurred

Table 6. Surface discharge (Q, liters s^{-1}), velocity (u, cm s^{-1}), lateral inflow (Q_L, liters s^{-1} and percent of discharge at the head of the study reach), NO_3-N uptake length (S_W, m), and turnover time (T_W, h) during seasonal injections at site 3.

			Q_L			
Season	Q	u	liters s^{-1}	%	S_W	T_W
Fall	0.75	1.3	0.43	57	133	2.84
Winter	15.0	10.7	0.45	3	738	1.91
Spring	75.0	20.8	0.75	1	3,183	4.25
Summer	2.0	2.5	0.76	38	205	2.27

in the upstream and downstream subreaches (4 and 1); groundwater discharge in other subreaches was negligible. The NO_3-N uptake length for subreach 1 indicated net nitrate increase between transects 2 and 1 at site 1.

Total groundwater discharge at site 2 was the lowest of the three sites and occurred predominantly over the final subreach (Table 5). Uptake lengths calculated for site 2 varied among subreaches and identified a region of NO_3-N increase at subreach 2 as indicated by a negative uptake length (Table 5).

Surface-water dilution by groundwater inputs occurred throughout the study reach at site 3, but magnitude decreased with distance downstream. Lateral inflow at this site was comparable to that at site 1, but represented a 57% dilution due to lower baseflow discharge (Table 5). Uptake lengths among subreaches at site 3 were similar to each other and to the uptake length calculated for the entire study reach (Table 5).

Variation in hydrologic and biological retention—Hydrologic conditions at site 3 varied greatly across seasons (Table 6). Discharge during the spring snowmelt injection was two orders of magnitude higher than during fall baseflow conditions. Concurrently, average water velocity was 20 times greater in spring. During the winter injection, snow depth at site 3 was ~100 cm and stream discharge was 20 times baseflow (Table 6). Discharge and water velocity during the summer injection were about double what they were during baseflow.

Lateral inflow varied by a factor of 1.76 across seasons, and groundwater inputs were highest in spring and summer. Inflow represented minimal dilution in winter and spring, but was more substantial during lower discharge conditions (Table 6).

Uptake lengths for NO_3-N ranged from 133 m at baseflow to >3 km during spring runoff (Table 6). In contrast, T_W varied by a factor of only 2.2 across seasons. T_W was maximal in spring and the minimum T_W was in winter.

Results of solute transport modeling indicated that the magnitude of transient storage decreased exponentially with increasing discharge (Fig. 3). Following ln-transformation, absolute transient storage (A_S) and storage normalized to stream size (A_S/A) were significantly correlated with water velocity (A_S: $r = -0.95$, $n = 4$, $P = 0.041$; A_S/A: $r = -0.97$, $n = 4$, $P = 0.025$). Similarly, these storage measures were also closely related to stream discharge (A_S: $r = -0.93$, $n = 4$, $P = 0.066$; A_S/A: $r = -0.94$, $n = 4$, $P = 0.062$).

Uptake length was also heavily influenced by stream discharge (Fig. 3) and these variables were significantly correlated ($r = 0.999$, $n = 4$, $P = 0.0001$). In addition, uptake length was significantly correlated to stream velocity ($r = 0.96$, $n = 4$, $P = 0.043$). Although T_W was highest during maximal discharge (Table 6), correlations between T_W and hydrologic features such as discharge ($P = 0.135$) or transient storage ($P = 0.349$) were not significant. Following ln-transformation, uptake length for NO_3-N and storage zone size (i.e. A_S/A) were inversely related among basins under baseflow conditions, but correlation was not significant due to small sample size ($r = -0.987$, $n = 3$, $P = 0.101$). Uptake length for NO_3-N and storage zone size at site 3 were significantly correlated across hydrologic conditions ($r = -0.985$, $n = 4$, $P = 0.015$). Data combined from the six solute injections illustrate a strong relationship between transient storage and total nutrient retention (Fig. 4; $r = -0.88$, $n = 6$, $P = 0.022$).

Discussion

Variation in hydrologic exchange—Results of this research show that the size of the interactive zone and the amount of water exchanged between surface and groundwater environments increased with increasing alluvial hydraulic conductivity. Where gradients were similar, interaction was more extensive in the site with higher alluvial hydraulic conductivity (i.e. there was more exchange at site 2 than at site 1). At site 3, high alluvial K and stream gradient generated extensive infiltration and maximized the spatial extent of the hyporheic zone. At the same time, a 190-m reach of this stream received groundwater inputs equal to over half of its baseflow at the head of the reach. Modeling results emphasized this bidirectional exchange because transient solute storage was nearly two orders of magnitude greater in the highly connected alluvium at site 3 than in the sedimentary system of site 1. Surface–groundwater interactions at site 1 were spatially restricted and travel times along flow paths were lengthy, reflecting low interstitial water velocities. Solute transport modeling, travel time, and spatial distribution of tracer at site 2 indicated an intermediate degree of surface–groundwater interaction.

Biogeochemical structure of the stream-aquifer system—Triska et al. (1989*b*) provided a schematic illustration of the interface between groundwater and surface-water components of lotic ecosystems that emphasized the spatial distribution of biologically meaningful solutes in surface water, distant groundwater, and an interactive hyporheic zone characterized by strong gradients in nutrient concentrations. Later, they emphasized the influence of groundwater flow paths on N transformations taking place in interstitial environments (Triska et al. 1993). They found that travel times to groundwater wells and percent of surface water in wells were related to dis-

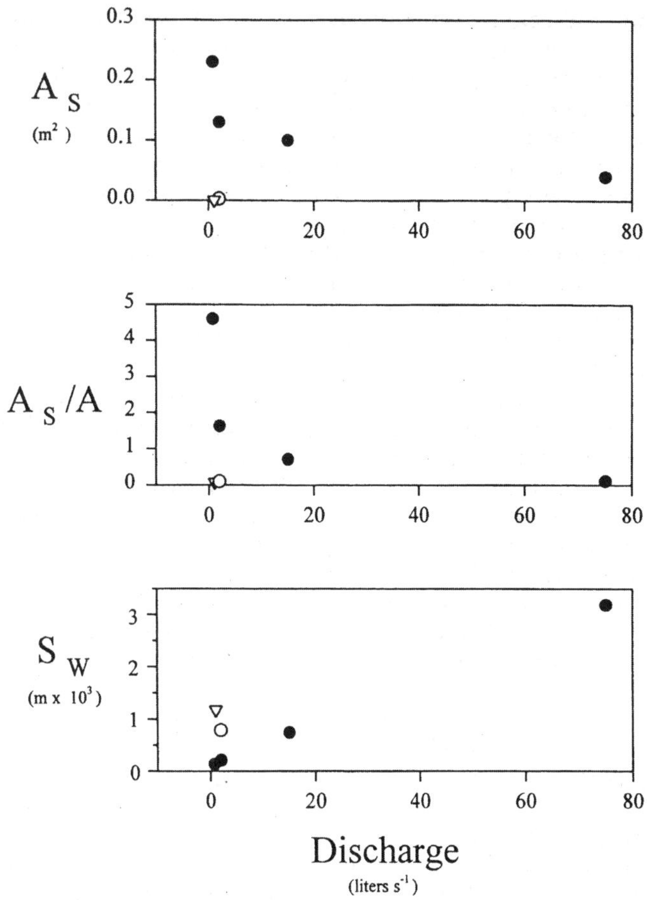

Fig. 3. Responses of biological and hydrologic retention to seasonal changes in stream discharge. Hydrologic retention is represented by absolute (A_S) and normalized (A_S/A) transient storage of conservative tracer. Biological retention is plotted as nitrate uptake length (S_W). Site 1, Aspen Creek—▽; site 2, Rio Calaveras—○; site 3, Gallina Creek—●.

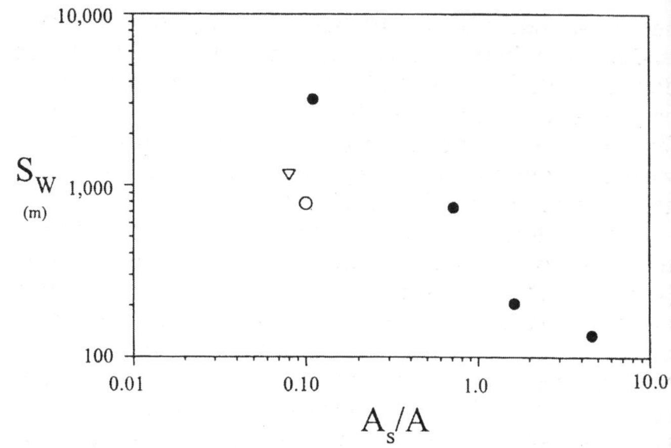

Fig. 4. Relationship between nitrate uptake length (S_W) and transient hydrologic storage normalized to stream size (A_S/A) for the six coinjections. Symbols as in Fig. 3.

tance inland from the stream channel. Furthermore, they reported a significant positive correlation between percent surface water and interstitial DO (Triska et al. 1993).

In contrast, the spatial distribution of connected wells in our study sites was not tied to distance from the wetted perimeter. Instead, patterns of hydrologic exchange were spatially complex within catchments, reflecting variation in horizontal and vertical hydraulic gradients (unpubl. data). At these sites, interstitial DO was not predictably related to percent surface water. A sampling well at site 1 that included nearly 30% surface water remained anoxic throughout this study, whereas a well located 3 m lateral to the stream at site 2 had 100% surface water with DO of only 9% saturation. These data suggest that in catchments of low *K*, DO is stripped from surface water as it enters the hyporheic zone. This loss of DO presumably reflects the influence of increased residence time in low-*K* aquifers. Significant correlation between interstitial DO and nominal travel time (T_n) calculated across sites suggests that subsurface residence time is primarily responsible for oxygen availability and thereby strongly influences redox potential along flow paths in alluvial aquifers.

Alternatively, variation in interstitial biological activity may account for differing patterns of interstitial DO among catchments. Recent studies assessing carbon dynamics in the hyporheic zone (e.g. Findlay et al. 1993; Vervier et al. 1993; Jones et al. 1995) and research reported here indicate that there is a clear need for increased characterization of material fluxes (i.e. DOC, nutrients) and interstitial metabolism at the stream-aquifer ecotone. Studies focusing on the sediment–water interface in lentic ecosystems have revealed that sediment biogeochemical processes are influenced by diffusive exchange and that these processes may have strong implications for the functioning of the aquatic system as a whole (e.g. Sweerts et al. 1991). In lotic systems, exchange at the sediment–water interface is driven by advective flow. Because this exchange occurs in a biogeochemically active interstitial environment, it can exert strong effects on the structure and functioning of the benthos and the surface stream (Grimm et al. 1991; Valett et al. 1994).

In our studies, variation in interstitial DO was reflected in the distribution of oxidized and reduced forms of inorganic N. Although TIN did not differ among surface, hyporheic, or phreatic environments within a single catchment, interstitial TIN was predominantly NH_4-N, concentrations of which did differ significantly among subsystems. Percentage of interstitial TIN that was NH_4-N decreased from 91% at site 1 to 77% at site 2, to 60% at site 3. This decrease corresponded to increased interstitial DO and decreased T_n. Analysis of a conservative tracer (Br) showed that water was actively exchanged bidirectionally along each of the study reaches. In contrast to interstitial water, inorganic N in surface water was predominantly NO_3-N. Change in the predominant form of inorganic N suggests that a dynamic zone of nutrient transformation and steep redox gradients exists in close proximity to the stream surface in low-*K* systems. We predict that the oxic–anoxic boundary may commonly be only centimeters below the benthic substratum at site 1 (sedimentary basin), tens of centimeters to a few meters distant at site 2 (welded tuff), and restricted to microzones only at site 3 (granite-gneiss). In this manner, alluvial features may dictate the occurrence of anaerobic zones with the potential to strongly influence ecosystem biogeochemistry (*see* Dahm et al. 1987).

Distributions of biogeochemically important solutes in the groundwater of the sedimentary (site 1) and volcanic (site 2) catchments support the notion that metabolic pathways in their respective alluvial aquifers are predominantly anaerobic. Significant variation in the concentrations of NO_3-N between stream water and groundwater suggests that production and consumption of NO_3-N may be a key aspect of system metabolism occurring at the surface–groundwater interface, as has been described for the stream-aquifer ecotone in other systems (e.g. Grimm et al. 1991; McDowell et al. 1992; Triska et al. 1993). In addition, subsurface sulfate concentrations were lower than surface concentrations in both low-*K* catchments, suggesting sulfate reduction as an active hyporheic process. Other research at these sites has shown elevated interstitial methane concentrations at the low-*K* sites, sometime >10,000 times atmospheric saturation (Baker et al. 1994). In contrast to the low-*K* systems, metabolism in the granitic systems seems to be predominantly aerobic. Microzones of anoxia undoubtedly exist, but the bulk of the water in the alluvial sediments is well oxygenated. In this well-oxygenated system, N concentrations may influence system metabolism by limiting decomposition, as has been illustrated for other stream systems (Kaushik and Hynes, 1971; Howarth and Fisher 1976; Triska et al. 1984). In all cases, hydrologic flow paths link the processes that affect nutrient cycling in subsurface and surface components of the ecosystem.

Hyporheic NO_3-N utilization—None of the wells in the hyporheic zone of the three study streams showed any sign of net nitrate production. Nearly half of the added nitrate was consumed along flow paths linking surface water and the hyporheic zone sampled at well locations. The intensity of hyporheic NO_3-N utilization was related to DO content and T_n to the sampling wells. These data suggest that regions of slow flow and anoxia are more active in NO_3-N consumption than in faster, more aerobic environments. Differences in NO_3-N utilization may reflect the fact that NO_3-N can be consumed by both dissimilatory and assimilatory reduction under proper conditions in anaerobic environments and that denitrification is generally suppressed in aerobic environments (Sprent 1987). We propose that assimilatory demand for inorganic N in the granitic system was so great that there was no measurable generation of NO_3-N in any of the monitoring wells despite well-oxygenated interstitial water. In this case, the hyporheic zone of high-*K* systems may also act as a sink for NO_3-N.

Spatial variation in NO_3-N retention—Hydrologic modeling and point-specific measures of tracer distribution indicate that hydrologic exchange between the aquifer and surface stream in montane catchments is bidirectional and spatially variable over short distances (i.e. 100–200 m). Uptake lengths calculated from upstream–downstream comparisons of NO_3-N concentrations are integrated measures of the hydrologic and biological conditions along the study reach (Stream Solute Workshop 1990). Therefore, nitrate uptake lengths in our study systems are influenced by hydrologic retention associated with exchange between surface and interstitial environments and by biological retention associated with metabolic activity of the benthic zone and within the alluvial aquifer.

We predicted that the intensity of hydrologic interaction and nutrient retention would decrease among basins with decreasing alluvia *K*. Results from solute transport modeling (i.e. storage zone size, A_S) and calculated NO_3-N uptake lengths support these predictions. Low T_n and spatially extensive hydrologic exchange in the granitic system resulted in the highest hydrologic retention. Less extensive interaction between surface and subsurface systems at sites 1 and 2 ultimately reduced hydrologic residence time, which contributed to decreased nutrient retention.

Munn and Meyer (1990) emphasized that uptake lengths for NO_3-N and SRP varied among streams in different geologic settings. They stated that parent lithology had the greatest influence on SRP uptake, but that N : P ratios were the primary determinant of NO_3-N uptake length. In their study, shortest NO_3-N uptake lengths occurred in streams with the lowest N : P ratio (i.e. those with highest potential for nitrogen limitation). Their study compared stream systems in which N : P ranged from 1.6 in a Cascade (Oregon) Mountain stream to >16 at Hubbard Brook, New Hampshire. Our study sites displayed a comparable range in N : P, but NO_3-N uptake lengths were longest in the stream with the lowest elemental ratio (i.e. site 1, where surface–groundwater interaction was least extensive). We contend that nitrate retention increased across sites as a result of increased surface–hyporheic interaction (hydrologic retention) and biotic utilization (biological retention) occurring in surface and interstitial environments.

Although measures of NO_3-N utilization in the hyporheic zone of the study streams failed to indicate NO_3-N production at depths of 50 cm below the water table, uptake lengths calculated for individual subreaches revealed specific regions of increased nitrate in surface water at sites 1 and 2; no increases in NO_3-N were observed at site 3. Surface NO_3-N increased at site 1 along subreach 1, where surface discharge increased by 16%. Groundwater concentrations of NO_3-N at site 1 were low, and lateral inflow of this water without tracer would have resulted in decreased NO_3-N stream water along subreach 1. Instead, surface NO_3-N concentrations at this location increased by 7% during the coinjection. Increased surface NO_3-N may have resulted from the import of NH_4-N from the largely anoxic alluvial aquifer and subsequent nitrification at the oxic–anoxic interface. Based upon the robust consumption of nitrate in wells 50 cm beneath the water table, we predicted this interface would occur within 50 cm of the wetted perimeter at site 1.

NO_3-N production was observed at site 2 as well. Despite NO_3-N consumption in hyporheic wells, we observed an 8% increase in surface nitrate for a single subreach. In contrast to the considerable lateral inflow at site 1, the increase in surface NO_3-N at Rio Calaveras was accompanied by only a 2% increase in surface discharge. Concentrations of interstitial NH_4-N were highest in the hyporheic zone of site 2, and hydrologic delivery to the surface–groundwater interface may have maintained rates of nitrification adequate to account for the 8% increase in stream water NO_3-N. This perspective is consistent with the observed spatial patterns of NO_3-N and NH_4-N in surface, hyporheic, and phreatic subsystems discussed earlier and suggests a potential mechanism to account for these patterns.

There was substantial lateral inflow in other subreaches of site 1 (18% of surface discharge at subreach 4) and site 2 (8% at subreach 1), but these reaches consumed between 4 and 5% of injected NO_3-N. These results suggest that spatial scales of hydrologic inputs and biogeochemical processes at the surface–groundwater interface may be highly heterogeneous and that subreaches ~30 m long may integrate across many different biogeochemical pathways. Clearly, uptake lengths calculated for the entire 100–150-m study reaches, in fact, do integrate considerable known hydrologic and biogeochemical variability.

Temporal variation in NO_3-N retention—We performed four injections at Gallina Creek (site 3) to investigate seasonal changes in stream discharge and its influence on hydraulic retention and NO_3-N utilization. The temporal variation in NO_3-N uptake length for this site was comparable to the variations in S_W observed among the three study sites at baseflow. Temporal variation in NO_3-N uptake length reflects the influence of changing stream discharge and water velocity. Other workers have emphasized that changing discharge influences nutrient uptake lengths (e.g. Newbold et al. 1983; D'Angelo and Webster 1991; D'Angelo et al. 1991). Some research has indicated that moderate increases from baseflow conditions may enhance uptake by increasing nutrient flux to sediments (Mulholland et al. 1985; Triska et al. 1990). More dramatic increases in discharge decrease the interaction between dissolved nutrients in the water column and metabolically active benthic sediments, resulting in decreased retention and longer uptake lengths (D'Angelo and Webster 1991; D'Angelo et al. 1991). Our results indicate that increasing discharge dramatically decreases the absolute and relative amount of hydraulic retention (i.e. decreases the size of the hyporheic zone) and subsequently minimized the contribution of interstitial processes to nitrogen retention in headwater streams. Turnover time for NO_3-N varied by a factor of only two over four injection experiments with markedly different flow regimes and temperature. Consistent turnover times suggest that biological demand for NO_3-N remained relatively constant and that discharge is a primary determinant of total NO_3-N retention.

Appropriately, downstream flow of water and associated transport of materials has been central in models of lotic ecosystem functioning (Vannote et al. 1980; Newbold et al. 1983). Recently, this emphasis has been expanded spatially and conceptually by recognizing the important interactions of surface water and groundwater and their implications for ecosystem structure and functioning (Triska et al. 1989*b*; Valett et al. 1994; Jones et al. 1995). Our results suggest that nutrient retention attributed to streams is strongly influenced by fluvial and biological interactions with the alluvial aquifers that are integral parts of the catchment.

Within basins, interactions between alluvial aquifers and streams are hydrologically variable over short distances with strong implications for biogeochemical cycling and ecosystem nutrient retention. Bencala (1993) presented contrasting views of streams as "pipes" vs. integrated systems that exchange water and solutes among surface, aquifer, and hillslope components. Our study suggests that streams are alternately similar to both models at different times of the year. A detailed understanding of nutrient cycling and retention in streams and aquifers of a single headwater catchment requires information on hydrologic exchange and the structure and functioning of the surface–groundwater interface. At larger scales, comparative studies among different catchments could be organized around properties of sediment alluvium, such as hydraulic conductivity.

References

BAKER, M. A., AND OTHERS. 1994. Spatial and temporal variation in methane distributions at the ground water/surface water interface in headwater streams, p. 29–38. *In* Ground water ecology. Proc. 2nd Int. Conf. Am. Water Resour. Assoc. Tech. Publ. Ser.

BENCALA, K. E. 1993. A perspective on stream-catchment connections. J. N. Am. Benthol. Soc. **12:** 44–47.

———, J. H. DUFF, J. W. HARVEY, A. P. JACKMAN, AND F. J. TRISKA. 1993. Modelling within the stream-catchment continuum, p. 163–187. *In* A. J. Jackman et al. [eds.], Modelling change in environmental systems. Wiley.

———, V. C. KENNEDY, G. W. ZELLWEGER, A. P. JACKMAN, AND R. J. AVANZINO. 1984. Interactions of solutes and streambed sediment 1. An experimental analysis of cation and anion transport in a mountain stream. Water Resour. Res. **20:** 1797–1803.

———, D. M. MCKNIGHT, AND G. W. ZELLWEGER. 1990. Characterization of transport in an acidic and metal-rich mountain stream based on lithium tracer injection and simulation of transient storage. Water Resour. Res. **26:** 989–1000.

CARPENTER, S. R. 1989. Replication and treatment strength in whole-lake experiments. Ecology **70:** 453–463.

CASTRO, N. M, AND G. M. HORNBERGER. 1991. Surface-subsurface water interactions in an alluviated mountain stream channel. Water Resour. Res. **27:** 1613–1621.

COLEMAN, R. L., AND C. N. DAHM. 1990. Stream geomorphology: Effects on periphyton standing crop and primary production. J. N. Am. Benthol. Soc. **9:** 293–302.

DAHM, C. N., E. H. TROTTER, AND J. R. SEDELL. 1987. Role of anaerobic zones and processes in stream ecosystem productivity, p. 157–177. *In* R. A. Averett and D. M. McKnight [eds.], Chemical quality of water and the hydrologic cycle. Lewis.

D'Angelo, D. J., and J. R. Webster. 1991. Phosphorus retention in streams draining pine and hardwood catchments in the southern Appalachian Mountains. Freshwater Biol. **26:** 335–345.

———, ———, and E. F. Benfield. 1991. Mechanisms of stream phosphorus retention: An experimental study. J. N. Am. Benthol. Soc. **10:** 225–237.

Danielopol, D. L. 1980. The role of the limnologist in groundwater studies. Int. Rev. Gesamten Hydrobiol. **65:** 777–791.

Day, R. W., and G. P. Quinn. 1989. Comparisons of treatments after an analysis of variance in ecology. Ecol. Monogr. **59:** 433–463.

Elwood, J. W., J. D. Newbold, R. V. O'Neill, and W. Van Winkle. 1983. Resource spiraling: An operational paradigm for analyzing lotic ecosystems, p. 3–27. *In* T. D. Fontaine III and S. M. Bartell [eds.], Dynamics of lotic ecosystems. Ann Arbor Sci.

Findlay, S., D. Strayer, C. Goumbala, and K. Gould. 1993. Metabolism of streamwater dissolved organic carbon in the shallow hyporheic zone. Limnol. Oceanogr. **38:** 1493–1499.

Gibert, J., M. J. Dole-Olivier, P. Marmonier, and P. Vervier. 1990. Surface water-groundwater ecotones, p. 199–225. *In* The ecology and management of aquatic-terrestrial ecotones. Man and the Biosphere. Ser. V. 4. UNESCO.

Gosz, J. R. 1975. Nutrient budgets for undisturbed ecosystems along an elevation gradient in New Mexico, p. 780–799. *In* Mineral cycling in southeastern ecosystems. Symp. Ser. CONF-74513. NTIS.

Grimm, N. B., H. M. Valett, E. H. Stanley, and S. G. Fisher. 1991. Contribution of the hyporheic zone to stability of an arid-land stream. Int. Ver. Theor. Angew. Limnol. Verh. **24:** 1595–1599.

Harvey, J. W., and K. E. Bencala. 1993. The effect of streambed topography on surface–subsurface water exchange in mountain catchments. Water Resour. Res. **29:** 89–98.

Howarth, R. W., and S. G. Fisher. 1976. Carbon, nitrogen and phosphorus dynamics during leaf decay in a nutrient-enriched stream microecosystem. Freshwater Biol. **6:** 221–228.

Hurlbert, S. H. 1984. Pseudoreplication and the design of ecological field experiments. Ecol. Monogr. **54:** 187–211.

Jones, J., S. G. Fisher, and N. B. Grimm. 1995. Vertical hydrologic exchange and hyporheic metabolism in a Sonoran desert stream. Ecology **76:** 942–952.

Kaushik, N. K., and H. B. N. Hynes. 1971. The fate of dead leaves that fall into streams. Arch. Hydrobiol. **68:** 465–515.

Kelson, K. I., and S. G. Wells. 1989. Geologic influences on fluvial hydrology and bedload transport in a small mountainous watershed, northern New Mexico, U.S.A. Earth Surf. Processes Landform **14:** 671–690.

Kruger, C. C., and T. F. Waters. 1983. Annual production of macroinvertebrates in three streams of differing water quality. Ecology **64:** 840–850.

Likens, G. E., F. H. Bormann, M. M. Johnson, D. W. Fisher, and R. S. Pierce. 1970. Effects of forest cutting and herbicide treatment on nutrient budgets in the Hubbard Brook watershed ecosystem. Ecol. Monogr. **40:** 23–47.

McDowell, W. H., W. B. Bowden, and C. E. Asbury. 1992. Riparian nitrogen dynamics in two geomorphologically distinct tropical rain forests watersheds: Subsurface solute patterns. Biogeochemistry **18:** 53–75.

Menzel, D. W., and R. F. Vaccaro. 1964. The measurement of dissolved organic and particulate carbon in seawater. Limnol. Oceanogr. **9:** 138–142.

Mulholland, P. J., J. D. Newbold, J. W. Elwood, L. A. Ferren, and J. R. Webster. 1985. Phosphorus spiraling in a woodland stream: Seasonal variations. Ecology **66:** 1012–1023.

Munn, N. L., and J. L. Meyer. 1990. Habitat-specific solute retention in two small streams: An intersite comparison. Ecology **71:** 2069–2082.

Murphy, J., and J. P. Riley. 1962. Determination of phosphate in natural waters. Anal. Chim. Acta **27:** 31–36.

Newbold, J. D., J. W. Elwood, R. V. O'Neill, and A. L. Sheldon. 1983. Phosphorus dynamics in a woodland stream ecosystem: A study of nutrient spiraling. Ecology **64:** 1249–1265.

Orghidan, T. 1959. Ein neuer Lebensraum des unterirdischen Wassers, de hyporheische Biotop. Arch. Hydrobiol. **55:** 392–414.

Pringle, C. M., and F. J. Triska. 1991. Effects of geothermal groundwater on nutrient dynamics of a lowland Costa Rican stream. Ecology **72:** 951–965.

Runkel, R. L., and R. E. Broshears. 1991. One-dimensional transport with inflow and storage (OTIS): A solute transport model for small streams. CADSWES Tech. Rep. 91-01.

Schindler, D. W. 1974. Eutrophication and recovery in experimental lakes: Implications for lake management. Science **184:** 897–898.

Smock, L. A., J. E. Galdden, J. L. Riekenberg, L. C. Smith, and C. R. Black. 1992. Lotic macroinvertebrate production in three dimensions: Channel surface, hyporheic, and floodplain environments. Ecology **73:** 876–886.

Sokal, R. R., and F. J. Rohlf. 1981. Biometry. Freeman.

Solóranzo, L. 1969. Determination of ammonium in natural waters by the phenolhypochlorite method. Limnol. Oceanogr. **14:** 799–801.

Sprent, J. I. 1987. The ecology of the nitrogen cycle. Cambridge.

Stanford, J., and J. V. Ward. 1993. An ecosystem perspective of alluvial rivers: Connectivity and the hyporheic corridor. J. N. Am. Benthol. Soc. **12:** 48–60.

Stream Solute Workshop. 1990. Concepts and methods for assessing solute dynamics in stream ecosystems. J. N. Am. Benthol. Soc. **9:** 95–119.

Sweerts, J. R., M. J. Bar-Gilissen, A. A. Cornelese, and T. E. Cappenberg. 1991. Oxygen-consuming processes at the profundal and littoral sediment–water interface of a small mesotrophic lake (Lake Vechten, The Netherlands). Limnol. Oceanogr. **36:** 1124–1133.

Triska, F. J., J. H. Duff, and R. J. Avanzino. 1993. Patterns of hydrological exchange and nutrient transformation in the hyporheic zone of a gravel-bottom stream: Examining terrestrial-aquatic linkages. Freshwater Biol. **29:** 259–274.

———, V. C. Kennedy, R. J. Avanzino, G. W. Zellweger, and K. E. Bencala. 1989*a*. Retention and transport of nutrients in a third-order stream: Channel processes. Ecology **70:** 1877–1892.

———, ———, ———, ———, and ———. 1989*b*. Retention and transport of nutrients in a third-order stream in northwestern California: Hyporheic processes. Ecology **70:** 1893–1905.

———, ———, ———, ———, and ———. 1990. In situ retention-transport responses to nitrate loading and storm discharge in a third-order stream. J. N. Am. Benthol. Soc. **9:** 229–239.

———, and others. 1984. Nitrogen budget for small coniferous forest stream. Ecol. Monogr. **54:** 119–140.

VALETT, H. M., S. G. FISHER, N. B. GRIMM, AND P. CAMILL. 1994. Vertical hydrologic exchange and ecological stability of a desert stream ecosystem. Ecology **75:** 548–560.

———, ———, AND E. H. STANLEY. 1990. Physical and chemical characteristics of the hyporheic zone of a Sonoran desert stream. J. N. Am. Benthol. Soc. **9:** 201–215.

VANNOTE, R. L., G. W. MINSHALL, K. W. CUMMINS, J. R. SEDELL, AND C. E. CUSHING. 1980. The river continuum concept. Can. J. Fish. Aquat. Sci. **37:** 130–137.

VAUX, W. G. 1968. Intergravel flow and interchange in water in a streambed. Fish. Bull. **66:** 479–489.

VERVIER, P., M. DOBSON, AND G. PINAY. 1993. Role of interaction zones between surface and ground waters in DOC transport and processing: Considerations for river restoration. Freshwater Biol. **29:** 275–284.

WEBSTER, J. R., AND B. C. PATTEN. 1979. Effects of watershed perturbation on stream potassium and calcium dynamics. Ecol. Monogr. **19:** 51–72.

WETZEL, R. G., AND G. E. LIKENS. 1991. Limnological analyses. Springer.

WHITE, D. S., C. H. ELZINGA, AND S. P. HENDRICKS. 1987. Temperature patterns within the hyporheic zone of a northern Michigan river. J. N. Am. Benthol. Soc. **6:** 85–91.

WROBLICKY, G. J., AND OTHERS. 1992. Remote monitoring of stream hyporheic zones with inexpensive pressure transducer-data acquisition systems, p. 267–277. *In* Ground water ecology. Proc. 1st Int. Conf. Am. Water Resour. Assoc. Tech. Publ. Ser.

ZELLWEGER, G. W., R. J. AVANZINO, AND K. E. BENCALA. 1989. Comparison of tracer-dilution and current-meter discharge methods in a small gravel-bed stream, Little Lost Man Creek, California. U.S. Geol. Surv. Tech. Rep. WRIR 89-4150.

Submitted: 29 July 1994
Accepted: 12 July 1995
Amended: 26 September 1994

IAHS, the International Association of Hydrological Sciences

IAHS is a nonprofit-making nongovernmental scientific organization committed to serving the science of hydrology and the worldwide community of hydrologists. IAHS, established in 1922, now has more than 5000 members from 124 countries.

The IAHS objective is to promote hydrology as part of the geosciences and as the basis for water management applications. The primary aims are to:

- initiate, facilitate, and coordinate research into hydrological problems requiring international cooperation
- provide for international discussion, comparison, and publication of hydrological research
- support hydrological sciences in developing countries
- bridge the gap between science and practice, including providing training opportunities and materials

International scientific commissions deal with surface water, groundwater, snow and ice hydrology, erosion and sedimentation, water quality, water resources, land–atmosphere systems and specific techniques, while working groups address particular issues. The Predictions in Ungauged Basins (PUB) Decade, 2003–2012, is a major initiative.

IAHS fulfils its mission of worldwide dissemination of research results and practice, and raises funds, by publishing and selling *Hydrological Sciences Journal* and books. Details of all publications are available at the web site, including abstracts of published papers. IAHS supports hydrological science in financially disadvantaged countries by: providing publications free of charge to libraries in those countries; and, offering discounts on publications purchased by members and contributing towards their costs when attending IAHS meetings.

Membership of IAHS is *free of charge* and open to anyone wishing to participate in IAHS activities, such as IAHS symposia and workshops.

IAHS Ltd is a UK registered charity (no. 1078635); all revenues are used to support the aims of the Association. Also, IAHS is one of the seven associations that comprise the International Union of Geodesy and Geophysics (IUGG, *www.iugg.org*).

IAHS AISH

www.iahs.info

Groundwater–Surface Water Interaction: *Process Understanding, Conceptualization and Modelling*

Edited by Corinna Abesser, Thorsten Wagener & Gunnar Nuetzmann

IAHS Publ. 321 (*2008*) ISBN 978-1-901502-59-6, 214 + x pp. Price £48.00

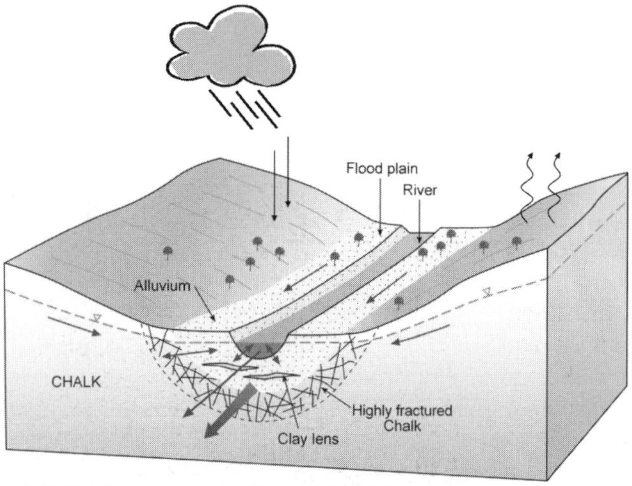

IPR/98-53C British Geological Survey © NERC. All rights reserved.

There is a pressing need to identify and develop methods that provide an appropriate framework for the integrated investigation, conceptualisation and modelling of surface–subsurface systems and their interfaces. The increasing focus on understanding the impacts of climate variability and change on water resources and ecosystems also requires a close connection of these systems to atmospheric variables. The goal is to improve the conceptual understanding of groundwater–surface water interactions in different landscapes and at different scales, ultimately leading to robust algorithms for simulating the effects of management strategies on surface water/groundwater systems.

To address these issues, a symposium: *A New Focus on Integrated Analysis of Groundwater/ Surface Water Systems* was held during the International Union of Geodesy and Geophysics (IUGG) General Assembly in Perugia, Italy, July 2007. Organised jointly by the IAHS International Commissions on Groundwater (ICGW) and Coupled Atmosphere–Land Systems (ICCLAS), the symposium provided a forum for scientists from many disciplines to advance the integrated analysis of groundwater/surface water systems.

The abstracts of the papers can be seen at: **www.iahs.info** *with information about other IAHS publications and IAHS activities.*

Please send book orders and enquiries to:

Mrs Jill Gash
IAHS Press, Centre for Ecology and Hydrology
Wallingford, Oxfordshire OX10 8BB, UK

jilly@iahs.demon.co.uk
tel.: + 44 1491 692442
fax: + 44 1491 692448

Status and Perspectives of Hydrology in Small Basins

Edited by *Andreas Herrmann & Sybille Schumann*

IAHS Publ. 336 (*May 2010*) ISBN 978-1-907161-08-7, 316 + xii pp. Price £65.00

Only in well-defined small basins with high-quality measurements can the complexities of combined physical, chemical and biological processes be adequately investigated. Small hydrological research basins provide interdisciplinary observatories, thus contributing to the study of the impact of environmental changes and to developing strategies for water and land-use management under such change.

This volume, the Proceedings of a Workshop held at Goslar-Hahnenklee, Germany, in 2009 focuses on:

- Presently operational small research basins
- Fundamental hydrological research results drawn from small basins
- Hydrological processes
- Importance of small basin data and results for hydrological modelling

and includes the *Braunschweig Declaration* on:

The need for a global network of long-term small hydrological research basins.

Abstracts of the contributions can be seen at: www.iahs.info with information about other IAHS publications and IAHS activities.

Please send book orders and enquiries to:

Mrs Jill Gash
IAHS Press, Centre for Ecology and Hydrology
Wallingford, Oxfordshire OX10 8BB, UK

jilly@iahs.demon.co.uk
tel.: + 44 1491 692442
fax: + 44 1491 692448

BENCHMARK PAPERS IN HYDROLOGY

The IAHS Series that collects together, by theme, the outstanding papers that provide the scientific foundations for hydrology today. Each volume reprints 30+ papers selected by an expert, with commentary on their significance.

Keith Beven's selection of 31 papers on the theme of *Streamflow Generation Processes*, spans the period from 1933 to 1984, commencing with Horton's early papers on infiltration and on maximum groundwater levels. With the aid of the Introduction and Commentaries, they provide a stimulating insight to how this part of the field of hydrology developed.

IAHS BM1 ISBN 978-901502-53-4 (2006) A4 format, softback, 432 + x pp., £40.00

The measurement techniques section commence with the Wagon Wheel Gap catchment water balance (1921) paper, and Dalton's seminal essay *On Evaporation* (1802) starts the selection of papers on evaporation estimation, which also reprints the original Penman and Thornthwaite papers.

IAHS BM2 ISBN 978-901502-98-5 (2007) A4 format, softback, 526 + x pp., £40.00

All aspects of groundwater are considered, from a translation of Darcy's experimental results that led to Darcy's law, and classic papers by Meinzer, Theis and Hubbert, to Slichter's (1905) contribution identifying dispersion in the field, and are linked with more recent attempts to represent dispersion and heterogeneity with models.

IAHS BM3 ISBN 978-901502-74-9 (2008) A4 format, hardback, 626 + x pp., £55.00

Loague notes that hundreds, if not thousands, of hydrologic-response models have been developed, but that not all were created equal. This volume reprints 30 papers that exemplify the best in rainfall–runoff modelling. It charts developments from Mulvany's (1851) rational method for estimating peak flow, probably the first rainfall–runoff model, up to 1989. Papers on other empirical approaches, such as Sherman (1932) and Mockus (1949), are reprinted. As are Richards (1931) and Smith & Parlange (1978), the innovative contributions of Alan Freeze, and later Keith Beven, and the seminal papers of Moore & Clarke (1981) and Abbott *et al.* (1986).

IAHS BM4 ISBN 978-1-907161-06-3 (2010) A4 format, hardback, 506 + vi pp, £65.00

Order from IAHS Press
CEH Wallingford
Oxfordshire OX10 8BB, UK

jilly@iahs.demon.co.uk
tel: +44 (0) 1491 692442
fax:+44 (0) 1491 692448

SAHRA

This IAHS series is sponsored by SAHRA